54th International Conference on Experimental Stress Analysis (EAN 2016)

AA004970

Srni, Czech Republic
30 May - 2 June 2016

ISBN: 978-1-5108-2670-0

Printed from e-media with permission by:

Curran Associates, Inc.
57 Morehouse Lane
Red Hook, NY 12571

Some format issues inherent in the e-media version may also appear in this print version.

Copyright© (2016) by the Czech Society for Mechanics
All rights reserved.

Printed by Curran Associates, Inc. (2016)

For permission requests, please contact the Czech Society for Mechanics
at the address below.

Czech Society for Mechanics
Dolejskova 5
Praha 8
182 00
Czech Republic

Phone: +420 266 053 045
Fax: +420 286 587 784

csm@csm.cz

Additional copies of this publication are available from:

Curran Associates, Inc.
57 Morehouse Lane
Red Hook, NY 12571 USA
Phone: 845-758-0400
Fax: 845-758-2634
Email: curran@proceedings.com
Web: www.proceedings.com

B

C

Č

L

M

N

O

P

Organization

Program Commitee

Chairman

- prof. Ing. František Plánička, CSc. (FAV ZČU v Plzni)

Members

- Ing. Karel Doubrava, Ph.D. (FS ČVUT v Praze)
- Prof. Dr. Josef Eberhardsteiner (TU Wien)
- Ing. František Fojtík, Ph.D. (FS VŠB-TU Ostrava)
- prof. Ing. Nikolaj Ganev, CSc. (FJFI ČVUT v Praze)
- prof. Dr. Ing. František Holešovský (FVTM UJEP v Ústí nad Labem)
- prof. Ing. Stanislav Holý, CSc. (FS ČVUT v Praze)
- Ing. Lubomír Houfek, Ph.D. (FSI VUT Brno)
- prof. RNDr. Miroslav Hrabovský, DrSc. (SLO UP a FZÚ AV ČR, Olomouc)
- prof. Ing. Josef Jíra, CSc. (FD ČVUT v Praze)
- prof. Ing. Vladislav Laš, CSc. (ZČU v Plzni)
- Ing. Aleš Lufinka, Ph.D. (FS TU v Liberci)
- Doc. Ing. Tomáš Návrat, Ph.D. (FSI VUT Brno)
- prof. Ing. Milan Růžička, CSc. (FS ČVUT v Praze)
- Dr.h.c. mult. prof. Ing. František Trebuňa, CSc. (SjF TU Košice)
- Ing. Jaroslav Václavík (VZÚ Plzeň)
- doc. Ing. Miloš Vlk, CSc. (FSI VUT Brno)

Organizing Committee

Chairman

- Ing. Vlastimil Vacek, CSc.

Members

- Ing. Miroslav Horák, Ph.D.
- Ing. Jan Krystek, Ph.D.
- Ing. Michal Novák.

Scientific Program

- Development of experimental methods in mechanics and biomechanics
- New methods and applications of deformation and stress analysis in mechanical, civil and other structures
- Experiment as a tool for the verification of analytical and numerical methods
- Experimental research and prediction of strength, life and operational reliability of structures and devices
- Monitoring of operational loadings and operational states of structures and devices
- Methods and means of teaching experimental methods.

UNIVERSITY OF WEST BOHEMIA
ČSM
HBP HBM měřicí technika
COMTES FHT
Zwick / Roell
InterCora
HOCHTIEF
1991–2016 25 LET
Plzeňský Prazdroj
PLZEŇSKÝ KRAJ

EAN 2016

54th international conference on Experimental Stress Analysis

Partners

UNIVERSITY OF WEST BOHEMIA

ČSM

HBP HBM měřicí technika

COMTES FHT
Complete Technological Service · Forming, Heat Treatment

Zwick / Roell

InterCora

HOCHTIEF

25 LET 1991–2016 ŠUMAVA

Plzeňský Prazdroj

PLZEŇSKÝ KRAJ

Isothermal LCF, Relaxation and Thermomechanical Fatigue Tests of Si-Mo Based Cast Iron

M. Bartošák[1,*], M. Španiel[1], K. Doubrava[1], C. Novotný[1]

[1] *Czech Technical University in Prague, Faculty of Mechanical Engineering, Department of Mechanics, Biomechanics and Mechatronics, Technická 4, Praha 6, Czech Republic*

[*] *michal.bartosak@fs.cvut.cz*

Abstract: High temperature behaviour of the tested material can be influenced by various processes, such as oxidation, cyclic creep and phase transformations, according to the selected temperature and strain rate. Isothermal low cycle fatigue (LCF), cyclic relaxation and thermomechanical fatigue (TMF) tests were conducted on Si-Mo based cast iron for the wide temperature range on the constructed thermomechanical fatigue test stand.

Keywords: cast iron; low cycle fatigue; thermomechanical fatigue; relaxation; viscoplasticity; TMF test stand.

1 Introduction

Ductile cast irons are often used for turbocharged diesel engine components, e.g. turbine housing of a turbocharger. Very good castability and relatively low price makes silicon-molybnedum cast iron a good choice for these components. Turbocharger applications are characterized by operating temperature usually below 650°C, but temperature may exceed up to 800°C.

Silicon-molybdenum cast iron components are often subjected to a complex mechanical and thermal loadings during the lifetime. Inhomogenous temperature distribution, mechanical loads, restricted thermal strains and boundary conditions of the components during variable operating conditions, containing start up, load, partial load and shutdown phases, often lead to a state, that is usually termed as a thermomechanical fatigue [1, 2]. High operating temperatures pottentially induces creep and time dependent effects. The durability assesment is increasingly demanded to ensure the field reliability targets [3]. This or a similar type of loading is usually simulated at uniaxially loaded specimens on a test stand, which enables to independently control temperature and mechanical loading, e.g. [3, 4, 5, 6].

First the paper presents construction and control of the TMF test stand. In a second part, the cyclic mechanical behaviour and durability are studied for the investigated material based on the performed tests.

2 Experiments

2.1 Test Stand

All tests were carried out on Instron PL 160 K servo-hydraulic actuator. The heating of the specimen is achieved by a direct electrical resistance heating, i.e. by passing a electric current through the specimen. This is attained by the additional heating device added to the Instron servo-hydraulic actuator, Fig. 1 and Fig. 2. The cooling of the specimen is passive, under the ambient air, meanwhile. The maximum achievable load for the machine is 160 kN and the maximum allowable temperature is up to 1000°C, depending upon the diameter of the specimen.

Temperature is measured by K-type spotwelded thermocouple [7] in the center of the specimen. Strain is controlled by the axial high temperature extensometer (up to 1200°C) with ceramic bars. Force is measured by the built-in load cell of the testing machine. PC based system is accompanied by NI-DAQ measurement card to ensure the collection of data from the sensors. The card is also used for the heating control as well as strain control. Closed-loop PID control is achieved by the combination of LabView and Instron controlling software (i.e. RS Console), outlined in Fig. 3. Among others, Instron software provides the safety limits for the tests.

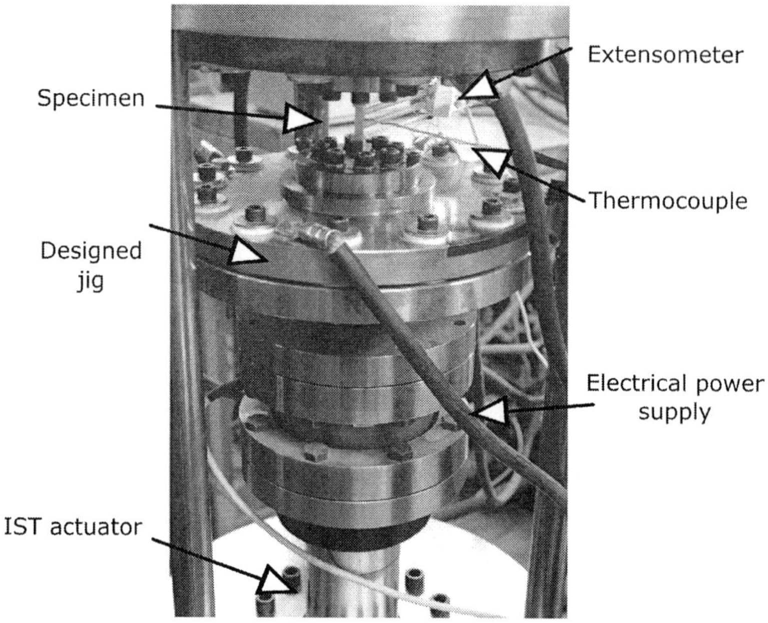

Figure 1: Test stand.

Several measurements by the infrared thermography and numerical simulations of the heated specimen temperature field preceded in order to verify the temperature uniformity in the gauge length of the specimen. The radial temperature differences are assumed to be negligible, also because of the relatively high thermal conductance of the investigated material. Further, value of the supply voltage for the heating was recorded for the selected geometry of the specimen and for the selected temperature range for a purpose of controlling the signal from the thermocouple. The thermocouple is removed and the new one is welded, if the value deviates more than approximately 5% from the known static characterictic. In the case of TMF test, the supply voltage is controlled for the maximum and minimum temperature of the cycle.

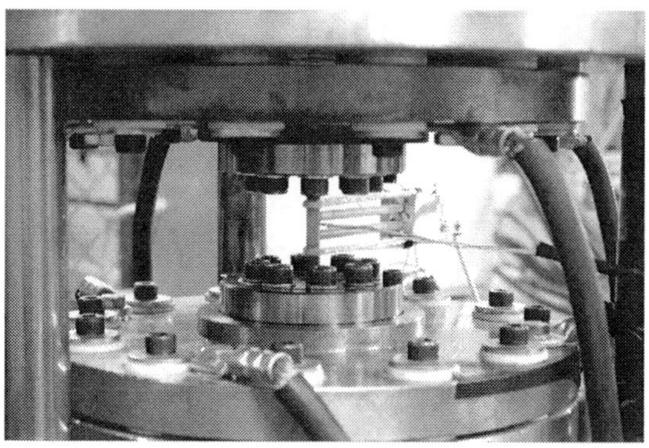
Figure 2: Heated specimen.

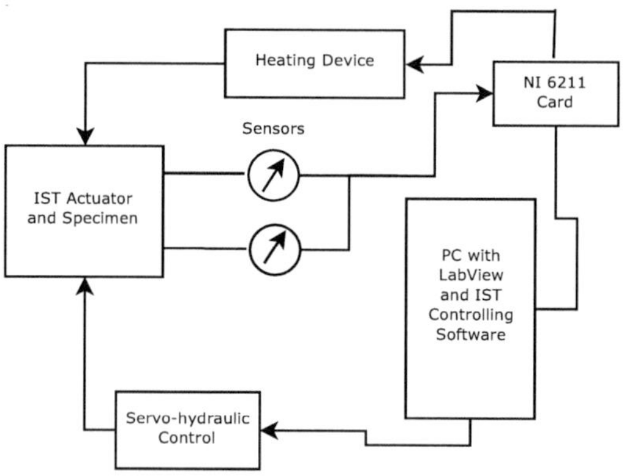

Figure 3: Outline of the proposed closed-loop control of the test rig.

2.2 Investigated Material

The material under investigation is the silicon molybdenum cast iron with a spherical graphite, annealed. The chemical composition of the material is given in Tab. 1. Size of the spherical graphite was approximately 15 μm. Tests were performed for the round solid specimens with the gauge diameter 6.5 mm and gauge length 8.9 mm. Specimens were machined for the testing.

Table 1: Chemical composition of Si-Mo cast iron.

Si [%]	C [%]	Mo [%]	Mn [%]	Cr [%]	Cu [%]	Mg [%]	P [%]	Ni [%]	Al [%]
4.10	3.21	0.555	0.394	0.085	0.066	0.048	0.038	0.024	0.018

2.3 Test Conditions

Following uniaxial tests were performed on unnotched specimens

- isothermal LCF,
- (cyclic) relaxation tests,
- out of phase TMF tests.

All tests were attained as strain controlled. Isothermal LCF tests were carried out under ambient condition, extending the temperature from 20°C up to 750°C, as fully reversed (R_ϵ =-1), for the variable mechanical strain ranges $\Delta\epsilon$ from 0.005 to 0.02. Relaxation tests were performed as isothermal at 550°C, 650°C and 750°C for the strain range $\Delta\epsilon$= 0.012 with 300 s hold time in a tension, during which a creep behaviour is observed in the form of a relaxation. LCF and cyclic relaxation tests were performed with the constant strain rate, $\dot{\epsilon} = 0.003\ s^{-1}$, as a triangular and trapezoidal shape, respectively. The waveformes are presented in Fig. 4. The reason for the high strain rates, in the case of isothermal LCF, is to minimize creep effects.

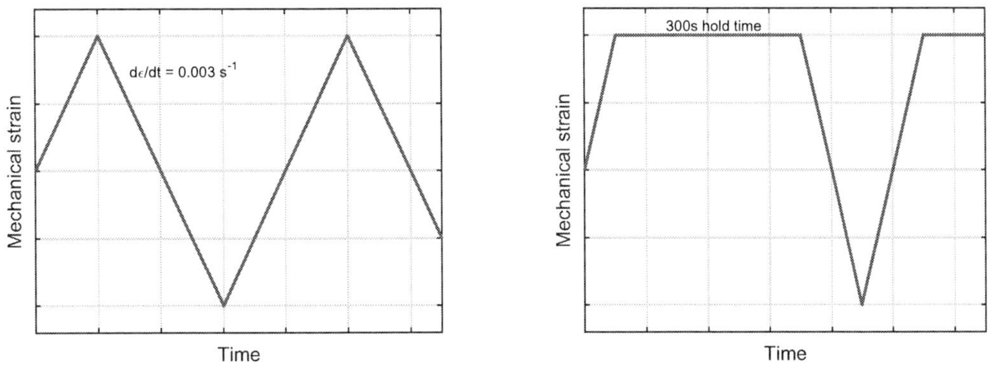

(a) (b)

Figure 4: Strain-time paths for isothermal LCF (a) and relaxation tests (b).

Thermomechanical tests were performed with the minimum test temperature 100°C , maximum temperature varied from 550°C to 800°C, with no dwell periods. For the test between 100° and 650°, the heating phase lasted approximately 104 s, followed by the cooling on air, 240 s total time. The thermal strain was totally suppresed for these tests, so the total strain ϵ_t was kept constant. Total mechanical strain ϵ_m, according to the condition $\epsilon_t = \epsilon_{th} + \epsilon_m$, was then shifted against the thermal strain ϵ_{th} by 180° with amplitude equal to the thermal strain amplitude, Fig. 5. This kind of test is often called as out of phase TMF (OP-TMF).

3 Experimental Results

For the isothermal LCF tests, cyclic softening was observed for the temperatures higher than approximately 500°C, softening was more significant with the increasing temperature. On the other hand cyclic hardening was observed for the temperatures below 500°C. Specimens showed monotonically increasing durability in the high strange ranges with the increasing temperature. Decrease in the lifetime was observed with the increasing temperatures for the low strain range tests. Strain-life curves for the selected temperatures are presented in Fig. 12, where N_f are cycles to a failure, determined classically when a 10% drop in stress occured compared to the stabilized stress.

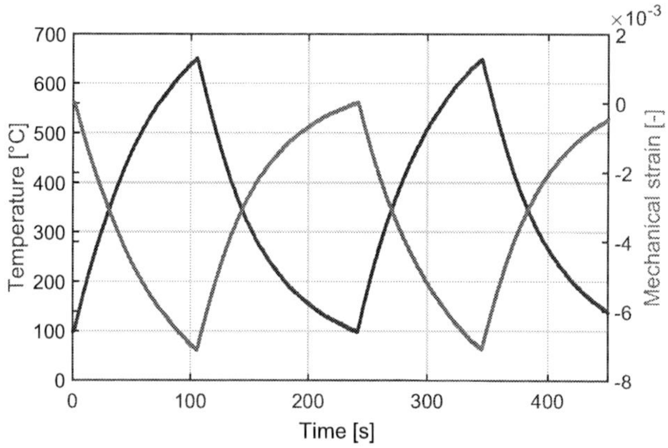

Figure 5: Temperature-, strain-time path for TMF test.

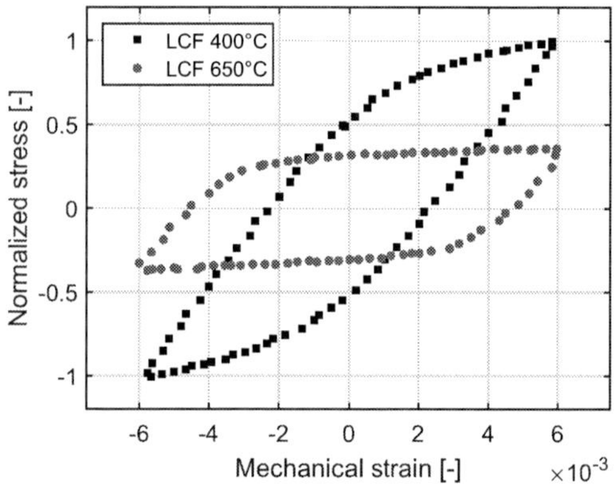

Figure 6: LCF at 400°C and 650° - stabilized hysteresis stress-strain loops.

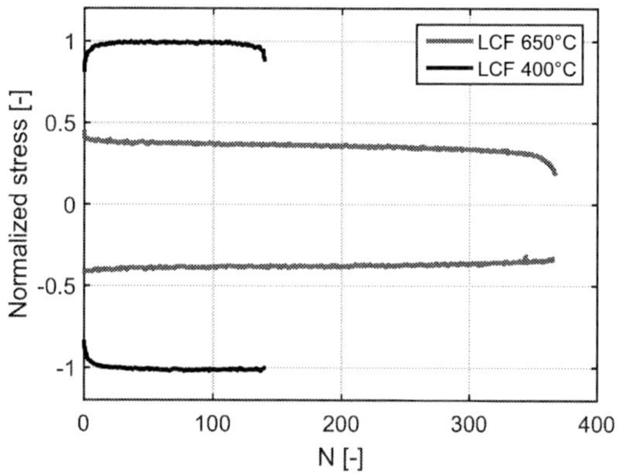

Figure 7: Maximum and minimum stress as a function of the number of cycles.

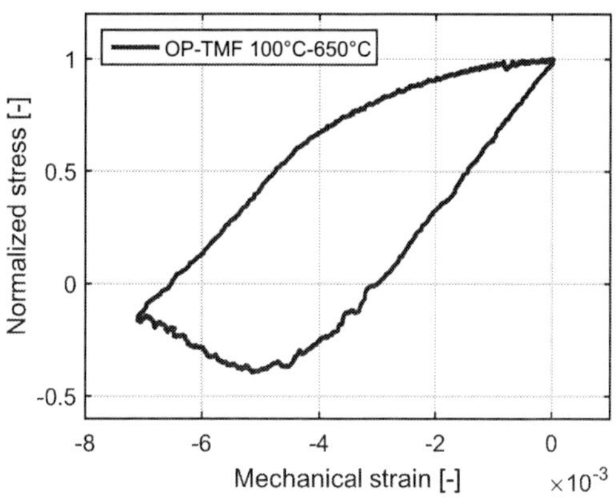

Figure 8: Out of phase TMF between 100°C and 650°C; hysteresis stress-strain loop.

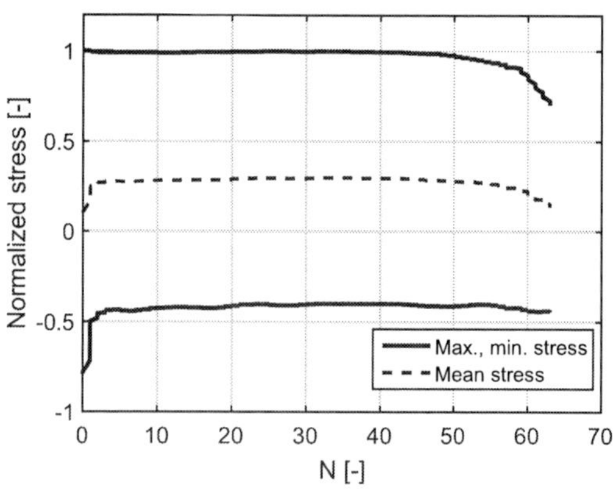

Figure 9: Maximum, minimum and mean stress as a function of the number of cycles.

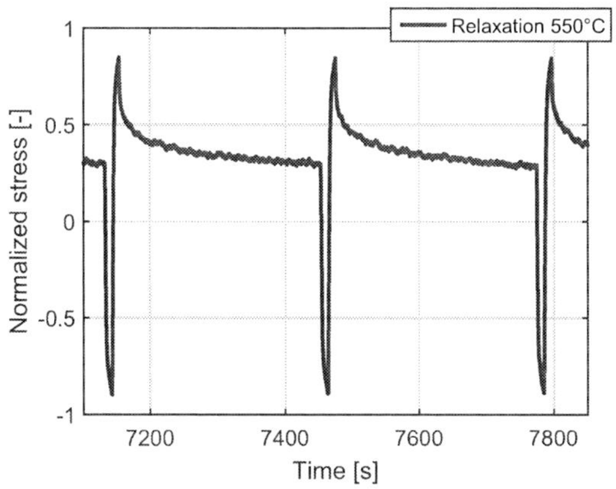

Figure 10: Cyclic relaxation test; 300s hold time in tension at 550°C.

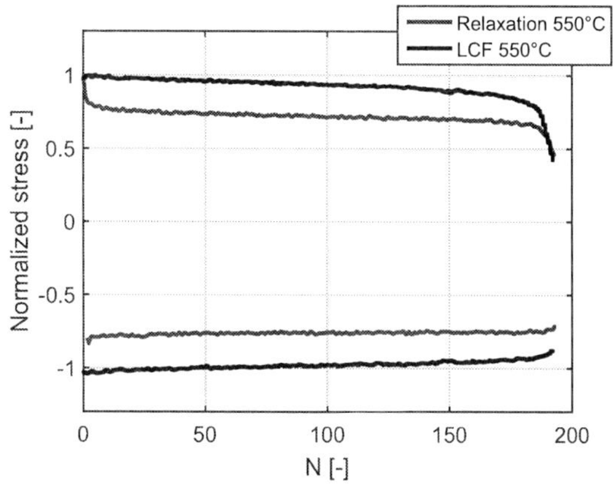

Figure 11: Maximum and minimum stress as a function of the number of cycles.

Fig. 6 presents the stabilized stress-strain hysteresis loops for the selected strain range $\Delta\epsilon$=0.012 at the temperature 400°C and 650°C, respectively. Maximum and minimum stress as a function of the number of cycles is for the selected LCF tests presented in Fig. 7. Mean stress is close to zero for the above mentioned tests.

Out of phase TMF test were conducted with the thermal strain equal to the negative total mechanical strain, this should represent extreme conditions, i.e. as a fully constrained material element during a heating and cooling down phases. Stabilized stress - strain hysteresis loop for the cycle between 100°C and 650°C is presented in Fig. 8. Maximum, mean and minimum stress as a function of cycles are shown in Fig. 9. TMF lifetime is drastically reduced compared to the isothermal LCF tests, one can note that this is due to the significant influence of the mean stress. Stabilization is achieved already after the first two cycles for the selected TMF test.

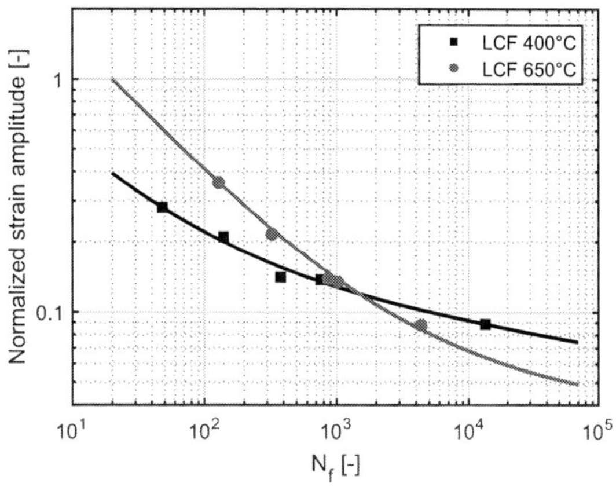

Figure 12: Manson-Coffin-Basquin curves at 400°C and 650°C.

Axial stress as a function of time for the cyclic relaxation test with 300 s hold time in a tension at 550°C is presented for the strain range $\Delta\epsilon$=0.012 in Fig. 10. One can see that more than 50% of the stress relaxes during the first tens of seconds during the hold time, this is even more significant for a higher temperature, at 750°C it is over the 70%. Fig. 11 presents the maximum and minimum stress as a function of the cycles, the test with the hold time is compared to the isothermal LCF without hold time for the same strain range and temperature. Interaction of the creep and plasticity [8, 9] is obvious, also resulting in the overall decrease in the stress amplitude for the test with the hold time compared with the LCF.

4 Conclusion

The obtained experimental data allow to calibrate and identify commonly used elastoplastic and/or viscoplastic material models [8, 9], mostly based on isothermal LCF data and verified on performed TMF tests.

The constructed TMF test stand compromises the capatibilities and price. However, it's planned to install a cooling air jet to ensure the faster cooling and strain rate for TMF tests. As a contrary to the passive cooling, heating can be much faster than was shown.

For the specimens, a various life reduction has been observed depending on the selected strain range and temperature. Effect of the mean stress on the lifetime is significant for the out of phase TMF tests. Further, significant stress relaxation was shown for the specimens with the increasing temperature for the tests with the hold time in a tension. The material is experiencing a relatively strong time dependent effects above approximately 500°C. Interaction of creep and plasticity could be also observed in the form of the overall decrease in the stress amplitude for the cyclic relaxation test compared with the LCF test without hold time. More TMF tests with different temperature/mechanical strain phase and LCF tests at different strain rates are planned for the silicon-molybdenum cast iron.

Acknowledgement

The authors would like to acknowledge the support from the Technology Agency of the Czech Republic, grant No. TE01020020, and support from the Grant Agency of the Czech Technical University in Prague, grant No. SGS15/187/OHK2/3T/12.

References

[1] J.-J. Thomas et al., Thermomechanical design in the automotive industry, Fatigue and Fracture of Engineering Materials and Structures 27 (2004) 887–895, doi: 10.1111/j.1460-2695.2004.00746.x.

[2] R.-W. Neu, H. Sehitoglu, Thermomechanical fatigue, oxidation, and creep: Part I. Damage mechanisms, Metallurgical transactions A 20.9 (1989) 1755–1767.

[3] S. Tabibian et al., TMF–LCF life assessment of a Lost Foam Casting A319 aluminum alloy. International Journal of Fatigue 53 (2013) 75–81, doi: 10.1016/j.ijfatigue.2012.01.012.

[4] T. Beck et al., Damage mechanisms of cast Al–Si–Mg alloys under superimposed thermal–mechanical fatigue and high-cycle fatigue loading, Materials Science and Engineering: A 468 (2007) 184–192, doi: 10.1016/j.msea.2006.05.177.

[5] X. Wu et al., Thermomechanical Fatigue of Ductile Cast Iron and Its Life Prediction, Metallurgical and Materials Transactions A 46.6 (2015) 2530–2543, doi: 10.1007/s11661-015-2873-9.

[6] S.-L.Mannan, M. Valsan, High-temperature low cycle fatigue, creep–fatigue and thermomechanical fatigue of steels and their welds, International Journal of Mechanical Sciences 48.2 (2006), 160–175, doi: 10.1016/j.ijmecsci.2005.08.004.

[7] T. Beck, K. Rau, Temperature measurement and control methods in TMF testing – a comparison and evaluation, International Journal of Fatigue 30 (2008) 226–233, doi: 10.1016/j.ijfatigue.2007.01.038.

[8] J. Lemaitre, J.-L. Chaboche, Mechanics of solid materials, Cambridge, Cambridge university press, 1994.

[9] J.-L. Chaboche, Constitutive equations for cyclic plasticity and cyclic viscoplasticity, International journal of plasticity 5.3 (1989), 247–302, doi: 10.1016/0749-6419(89)90015-6.

Statistical based approach to material identification of composite materials

L. Bek [1,a], R. Kottner [1], J. Krystek [1], V. Laš [1]

[1] *University of West Bohemia, Faculty of Applied Sciences, NTIS – New Technologies for the Information Society, Europian Center of Excellence, Univerzitní 22, 306 14 Pilsen, Czech Republic*

[a] *lukasbek@kme.zcu.cz*

Abstract: Experimentally investigated material properties of unidirectional carbon fibre composite samples manufactured in an autoclave were subjected to the statistical analysis. The Maximum normed residual method was used for the identification of outliers within each batch. Furthermore, the k-sample Andreson-Darling test was used to determine whether the properties from different batches are combinable. Combinable material properties were again tested for outliers. Resulting set of data was tested using the Anderson-Darling goodness-of-fit test in order to evaluate whether the approximation by the Normal distribution, the Lognormal distribution or the Weibull 2-parameter distribution would be appropriate. Material properties were compared within one batch as well as between batches.

Keywords: Carbon fibre composite; Anderson-Darling; distribution; identification; statistic.

1 Introduction

Due to the benefits in form of high stiffness and strength compared to low density, composite materials are widely used in many applications in sport, automotive, marine, and aerospace industries. A composite production process consists of the positioning of dry or pre-impregnated fibres, a potential resin infusion and a curing process. The positioning of fibres usually performed by human workers or robots introduces certain material variability factor into each composite part which may result in stiffness and strength variability of a final structure.

This material variability may result from number of sources as batch-to-batch variability of raw materials, batch-to-batch variability caused by operators, batch intrinsic variability and testing variability [1]. For the approximation of material property variability, the Normal distribution, the Lognormal distribution and the 2-parameter Weibull distribution are the most appropriate selections [1]. The Normal distribution was used in [2] for the approximation of stiffness and strength parameters of carbon fibre composite. Results of the approximation were later implemented into the finite element method software and used for a fatigue simulation. The Weibull distribution was used in [3], [4], and [5]. The paper [3] was focused on the approximation of strength of unidirectional carbon fibre samples and on the parameter estimation of Weibull distribution using the linear regression. The Maximum likelihood method was used for the Weibull distribution parameters estimation in [4]. The distribution was used for the approximation of bending stiffness of carbon fibre composite having layup $[0/90]_{2S}$. Also in [5], the Weibull distribution with the Maximum likelihood method was used for the approximation of bamboo fibre strength.

For the purposes of using own manufactured composite material for research of riveted joints, this work was focused on manufacturing of test samples and the investigation of their material properties. Furthermore, the detection of outliers, the test for combinability of results obtained from measurements from different batches, and the selection of appropriate statistical distribution was carried out.

2 Preparation of test samples

Two unidirectional (UD) carbon fibre plates were manufactured by author in the ASC Econoclave autoclave (see Fig. 1) from non-crimped fabric (NCF) Saertex with fibres TENAX-J IMS60 E13 24K and from epoxy resin MSG L285. For the manufacturing process, Airbus patented Vacuum Assisted Process (VAP®) was selected [6]. This process is characterized by using VAP® membrane which brings a benefit of better resin distribution during the infusion resulting in better fibre to resin volume ratio than in case of a standard infusion procedure. Furthermore, this process reduces the amount of air bubbles closed in composite and therefore decreases the material property variability.

Test samples were cut using a CNC milling machine from different plates. For the testing, dimensions following ASTM D3039 were determined as 200 mm × 25 mm × 2.6 mm. Placement and orientation of samples on the cut plates were selected in order to obtain samples having fibre orientation 0°, 45° and 90° with respect to the longer edge of a sample (see Fig. 2). Six samples per each fibre orientation and batch were created and numbered to guarantee full traceability.

Dimensions and weight of each sample were measured for the purposes of the fibre volume ratio calculation and the sample stiffness evaluation. Resulting mean fibre volume ratios with standard deviations for each sample batch are listed in Tab. 1.

Fig. 1: VAP® manufacturing process of UD carbon fibre plate.

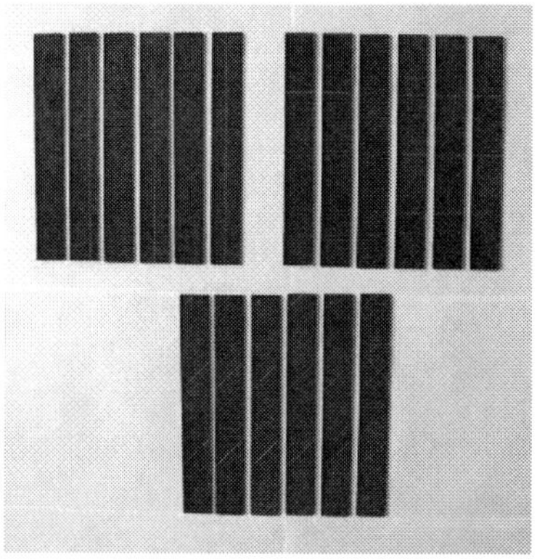

Fig. 2: Example of sample batch having fibre orientation 0°, 90° and 45°.

Tab. 1: Mean fibre volume ratios and standard deviations of fibre volume ratios of test samples.

Samples	Mean fibre volume ratio [%]	Standard deviation of fibre volume ratio [%]
0° fibre orientation - Batch 1	60.7	1.5
90° fibre orientation - Batch 1	60.9	0.4
45° fibre orientation - Batch 1	60.6	0.3
0° fibre orientation - Batch 2	63.7	1.2
90° fibre orientation - Batch 2	63.7	0.2
45° fibre orientation - Batch 2	64.7	0.2

3 Theoretical background

3.1 Maximum normed residual test

The Maximum normed residual test (MNR) is a test for the identification of outliers in a set of unstructured data $x_1, x_2, \ldots, x_n$. The MNR can be calculated as:

$$\mathrm{MNR} = \max_i \frac{|x_i - \bar{x}|}{s}, \qquad i=1, 2, \ldots, n \tag{1}$$

where $\bar{x}$ denotes to the sample mean and s to the standard deviation. This number is compared to the critical value (CV) calculated for the sample size n as:

$$\mathrm{CV} = \frac{n-1}{\sqrt{n}} \sqrt{\frac{t^2}{n-2+t^2}}, \tag{2}$$

where

$$t = [1 - \alpha/2n]. \tag{3}$$

α is the significance level and recommended $\alpha = 0.05$ for this test. If the MNR is lower than CV then it can be concluded that no outlier was detected. Otherwise, the sample associated with highest $|x_i - \bar{x}|$ is detected as an outlier. Once the outlier is detected, it has to be removed from calculations and the MNR test must be repeated [1].

3.2 The k-sample Anderson-Darling test

The k-sample Anderson-Darling test is used for the verification of the hypothesis whether the samples from different batches can be combined into one group of observations $z_1, z_2, \ldots, z_L$. This hypothesis is confirmed if the ADK number is lower than critical value ADC. ADK can be calculated as [1]:

$$\mathrm{ADK} = \frac{n-1}{n^2(k-1)} \sum_{i=1}^{k} \left\{ \frac{1}{n_i} \sum_{j=1}^{L} h_j \frac{\left(nH_{ij} - n_i H_j\right)^2}{H_j(n-H_j) - nh_j/4} \right\}, \tag{4}$$

where

n	is the number of combined samples,
k	is the number of batches,
L	is the number of observations,
h_j	is the number of values in the combined samples equal to z_j,
H_j	is the number of values in the combined samples less than z_j plus one half the number of values in the combined samples equal to z_j,
F_{ij}	is the number of values in the i-th group which are less than z_j plus one half the number of values in this group which are equal to z_j.

The critical value ADC can be evaluated as:

$$\mathrm{ADC} = 1 + \sigma_n \left[1.645 + \frac{0.678}{\sqrt{k-1}} - \frac{0.362}{k-1}\right], \tag{5}$$

where

$$\sigma_n^2 = \mathrm{Var(ADK)} = \frac{an^3 + bn^2 + cn + d}{(n-1)(n-2)(n-3)(k-1)^2}, \tag{6}$$

$$
\begin{aligned}
a &= (4g-6)(k-1) + (10-6g)S, \\
b &= (2g-4)k^2 + 8Tk + (2g-14T-4)S - 8T + 4g - 6, \\
c &= (6T+2g-2)k^2 + (4T-4g+6)k + (2T-6)S + 4T, \\
d &= (2T+6)k^2 - 4Tk,
\end{aligned} \tag{7}
$$

$$S = \sum_{i=1}^{k} \frac{1}{n_i}, \tag{8}$$

$$T = \sum_{i=1}^{n-1} \frac{1}{i}, \tag{9}$$

3.3 Anderson-Darling goodness-of-fit test for Normal and Lognormal distribution

For the hypothesis whether the observations can be approximated by the Normal or the Lognormal distribution, the Anderson-Darling goodness-of-fit test can be used. The hypothesis is confirmed with five percent of risk of an error if the Observation significance level (OSL) is lower than 0.05. The OSL can be calculated for values [1]:

$$z_i = \frac{x_i - \bar{x}}{s}, \quad \text{in case of the Normal distribution,} \tag{10}$$

$$z_i = \frac{\ln(x_i) - \bar{x}_L}{s_L}, \quad \text{in case of the Lognormal distribution,} \tag{11}$$

where $\bar{x}$ and s are the sample mean and the standard deviation, respectively. $\bar{x}_L$ and s_L denote to the mean and the standard deviation calculated from observations $\ln(x_i)$, respectively.

The OSL is then evaluated as follows:

$$\mathrm{AD} = \sum_{i=1}^{n} \left[\frac{1-2i}{n} \{ ln[F_0(z_i)] + \ln[1 - F_0(z_{n+1-i})] \} \right] - n, \tag{12}$$

$$\mathrm{AD}^* = \left[1 + \frac{4}{n} - \frac{25}{n^2} \right] \mathrm{AD}, \tag{13}$$

$$\mathrm{OSL} = 1/\{1 + \exp[-0.48 + 0.78\ln(\mathrm{AD}^*) + 4.58\mathrm{AD}^*]\}. \tag{14}$$

3.4 Anderson-Darling goodness-of-fit test for 2-parameter Weibull distribution

For the hypothesis whether the observations can be approximated by the 2-parameter Weibull distribution, the Anderson-Darling goodness-of-fit test can be used as well. The OSL can be calculated for values [1]:

$$z_i = \left(\frac{x_i}{\hat{\alpha}} \right)^{\hat{\beta}}, \tag{15}$$

where $\hat{\alpha}$ and $\hat{\beta}$ represent the scale and the shape parameters of the Weibull distribution, respectively [7]. The OSL is then evaluated as follows:

$$\mathrm{AD} = \sum_{i=1}^{n} \left[\frac{1-2i}{n} \{ ln[1 - \exp(-z_i)] - z_{n+1-i} \} \right] - n, \tag{16}$$

$$\mathrm{AD}^* = \left[1 + 0.2/\sqrt{n} \right] \mathrm{AD}, \tag{17}$$

$$\mathrm{OSL} = 1/\{1 + \exp[-0.10 + 1.24\ln(\mathrm{AD}^*) + 4.48\mathrm{AD}^*]\}. \tag{18}$$

4 Experiments and evaluation

Experiments were carried out in the testing machine Zwick/Roell Z050. All test samples were subjected to tensile loading with standard head speed rate of 2 mm/min. Except of samples having 0° fibre orientation (the ultimate load of these samples was beyond the capacity of the testing machine), samples were loaded up to the failure. The force was measured using load cell and the displacement using an extensometer. The clamp length was 100 mm and the extensometer measurement area was 60 mm long. In total, 36 experiments with samples having 3 different fibre orientations were performed (see fig. 3).

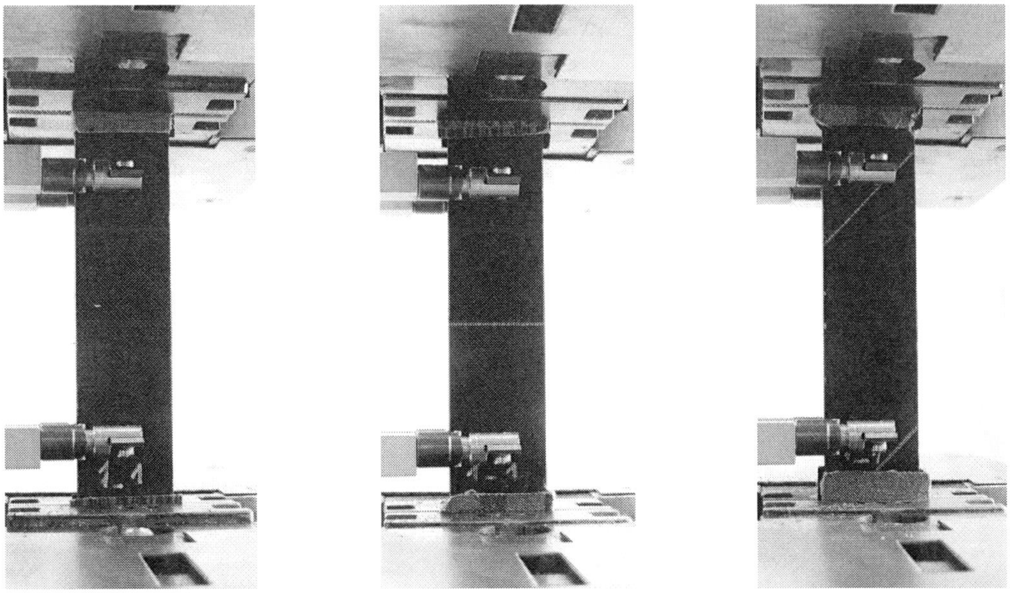

Fig. 3: Experiments with samples having 0° (left), 90° (middle) and 45° (right) fibre orientations.

Investigated force-displacement curves were transformed into the nominal stress-strain curves. In fig. 4, the stress-strain curves for samples having 0° fibre orientation are displayed. Samples were loaded up to 20 kN (approx. 320 MPa). Fig. 5 displays samples having 90° fibre orientation and fig. 6 displays samples having 45° fibre orientation. Here, the non-linear material behaviour of the composite during the shear loading can be observed.

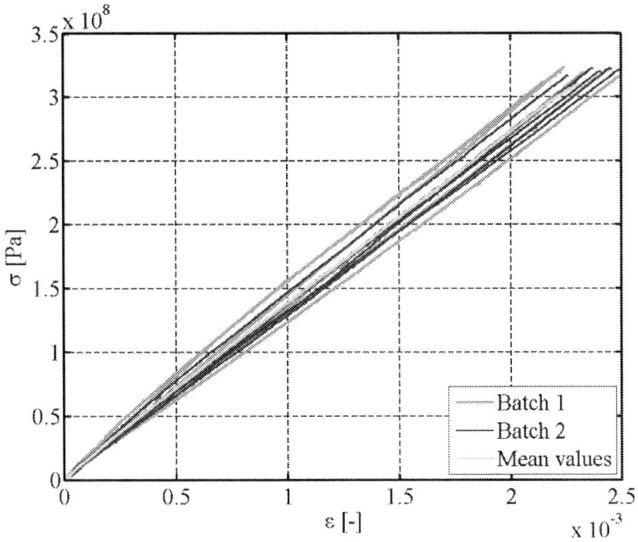

Fig. 4: Stress-strain curves for samples having 0° fibre orientation and their mean values.

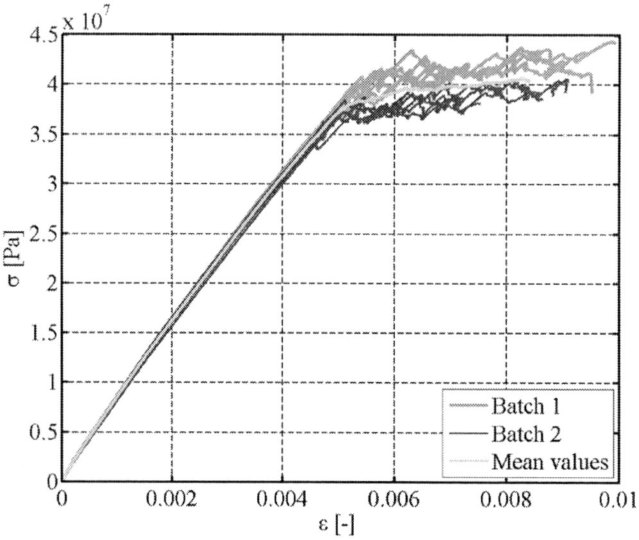

Fig. 5: Stress-strain curves for samples having 90° fibre orientation and their mean values.

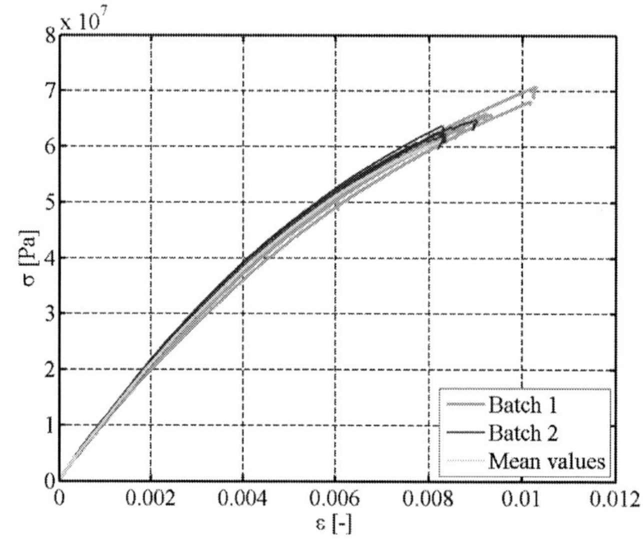

Fig. 6: Stress-strain curves for samples having 45° fibre orientation and their mean values.

From the stress-strain curves, stiffness and strength (except of samples having 0° fibre orientation) parameters of the composite material were determined. All values were tested by the MNR test for the detection of outliers. Only the strength of the sample no. 2 having 45° fibre orientation from the batch 1 was detected as an outlier. Therefore, it was not included into further calculations. The summary of the mean values and the standard deviations calculated from the rest of samples for each batch and fibre orientation are listed in Tab. 2.

Tab. 2: Mean values and standard deviations of stiffness and strength parameters.

Samples	Stiffness mean [MPa]	Stiffness standard deviation [MPa]	Strength mean [MPa]	Strength standard deviation [MPa]
0° fibre orientation - Batch 1	134724.3	5409.1	-	-
0° fibre orientation - Batch 2	132339.4	4562.4	-	-
90° fibre orientation - Batch 1	7671.6	96.8	43.1	1.0
90° fibre orientation - Batch 2	7568.9	118.6	39.5	0.9
45° fibre orientation - Batch 1	12612.6	198.0	67.0	2.5
45° fibre orientation - Batch 2	12916.8	279.4	62.5	1.9

Furthermore, the k-sample Anderson-Darling test was applied in order to determine whether the data obtained from measurements on the sample batches 1 and 2 can be combined and assumed as one set of data. Calculated ADK and ADC numbers are listed in Tab. 3. Due to the requirement to fulfil the ADK < ADC condition, it can be concluded that the stiffness parameters can be combined into one set of data and the strength cannot be combined into one set of data. Therefore, only the stiffness parameters were possible to subject to further tests because the strength parameters would have to be tested separately, which would lead to unprecise results due to the low number of tests within batches.

Tab. 3: Results of k-sample Anderson-Darling test.

Samples	ADK for stiffness param. [-]	ADC for stiffness param. [-]	ADK for strength param. [-]	ADC for strength param. [-]
0° fibre orientation	0.5109	2.2882	-	-
90° fibre orientation	1.4161	2.2882	4.7827	2.2882
45° fibre orientation	2.2154	2.2882	3.7754	2.2710

Corresponding stiffness parameters from the batches 1 and 2 were combined into sets of data and again tested for outliers using the MNR test. No outliers were detected in any of the data sets. Therefore, data were tested by the Anderson-Darling goodness-of-fit tests in order to evaluate the hypothesis whether they can be approximated by the Normal, the Lognormal or the 2-parameter Weibull distributions. For each data set, the OSL number for pointed distributions were calculated and listed in Tab 4.

Tab. 4: Results of Anderson-Darling goodness-of-fit test.

Data set	OSL Normal distribution [-]	OSL Lognormal distribution [-]	OSL Weibull distribution [-]
0° fibre orientation	0.5014	0.5423	0.1447
90° fibre orientation	0.6527	0.6510	0.6550
45° fibre orientation	0.4146	0.4391	0.1724

Based on results listed in Tab. 4, it can be concluded that any of the distributions are able to approximate prescribed set of data. Therefore, the B-values for each distribution were calculated. These values are listed in Tab. 5

Tab. 5: Calculated B-values.

Data set	B-value according to Normal dist. [MPa]	B-value according to Lognormal dist. [MPa]	B-value according to Weibull dist. [MPa]
0° fibre orientation	122630.0	123067.6	110168.7
90° fibre orientation	7363.0	7366.4	7130.6
45° fibre orientation	12145.0	12160.2	11443.1

Conclusion

Stiffness and strength parameters of the manufactured composite were determined and statistically evaluated. One sample was detected as an outlier from the strength point of view. Corresponding samples from two batches were combined into sets of data and then goodness-of-fit tests were performed. Unfortunately it was concluded that the data sets cannot be approximated with any of the listed distributions and therefore B-values were calculated. For future purposes, more sample batches will be manufactured and measured.

Acknowledgement

L. Bek was supported by the grant project SGS-2016-038. R. Kottner, J. Krystek and V.Laš were supported by the project LO1506 of the Czech Ministry of Education, Youth and Sports.

References

[1] U.S. Department of Defense. *Military Handbook - MIL-HDBK-17-1F: Composite Materials Handbook; Volume 1. Polymer matrix composites guidelines for characterization of structural materials*. U.S. Department of Defense, 2002.

[2] W. Lian and W. Yao. Fatigue life prediction of composite laminates by FEA simulation method. *Internal Journal of Fatigue*, 2010.

[3] M.H. Dirikolu, A. Aktas, and B. Birgören. Statistical Analysis of Fracture Strength of Composite Materials Using Weibull Distribution. *Turk. J. Eng. Environ. Sci, Vol. 26*, 2002.

[4] E. Barbero, J. Fernádez-Sáez, and C. Navarro. Statistical analysis of the mechanical properties of composite materials. *Composites Part B: Engineering, Vol. 31*, 2000.

[5] F. Wang and J. Shao. Modified Weibull Distribution for Analysing the Tensile Strength of Bamboo Fibers. *Polymers, Vol. 6*, 2014.

[6] Retrieved from: http://www.vap-info.com.

[7] M. Fusek, *Statistika extrémních hodnot*, Diploma thesis, Brno University of Technology, 2009.

Strength of thermal drilled threaded holes

R. Bělík [1,a], M. Kováčik [1]

[1] *VŠB–Technical University of Ostrava, Faculty of Mechanical Engineering, Department of Applied Mechanics, 17. listopadu 15/2172, 708 33 Ostrava, Czech Republic*

[a] *radomir.belik@vsb.cz*

Abstract: The paper presents experiments on strength of bolted connections with threaded holes. The results compare different types of technology used for thread manufacturing. Especially thermal drilling technology. The method is tested on various types of steels, including both mild grade and high strength steel types. The results are analysed compared to standardized values. As a conclusion, this paper presents a technique to make estimations about such connections strength.

Keywords: Flowdrill; thermal drilling; Thread Stripping; DOCOL; Bolted Joint

1 Introduction

In case of bolted joints, there are basically three types of static failures according to ISO standards [1]. It is desirable though, to design a joint so, that a failure occurs in the bolt shank and thus preventing a thread stripping. To achieve this, several procedures have been developed by Alexander [2] and verified and optimised by Hagiwara [3]. These works are however focused mainly on bolt-nut connections to optimise each configuration according to its property class and geometry. When nuts are replaced by threaded holes for convenience, the thread stripping is no longer avoided automatically and assessments of the thread stripping have to be made.

Acc. to Alexander's theory [2], if the shear area of a thread is insufficient, the stripping occurs before a tensile failure in the bolt shank. There are many factors contributing to the shear area, however, a so called effective length of engagement of a thread is one of essential variables. Achieving a proper length of engagement with holes drilled in thin plates was rather difficult until introduction of the thermal friction drilling technique [5].

The friction drilling technique is a very useful and quick way to create a threaded hole in thin plates without using any additional material or removing any. The basic principle is to use the material which would be removed otherwise to extend the depth of the hole by creating an extruded sleeve. This is achieved via plastic deformation of the metal that is being machined. The thread itself could be then cut or formed again using plastic deformation Fig. 1.

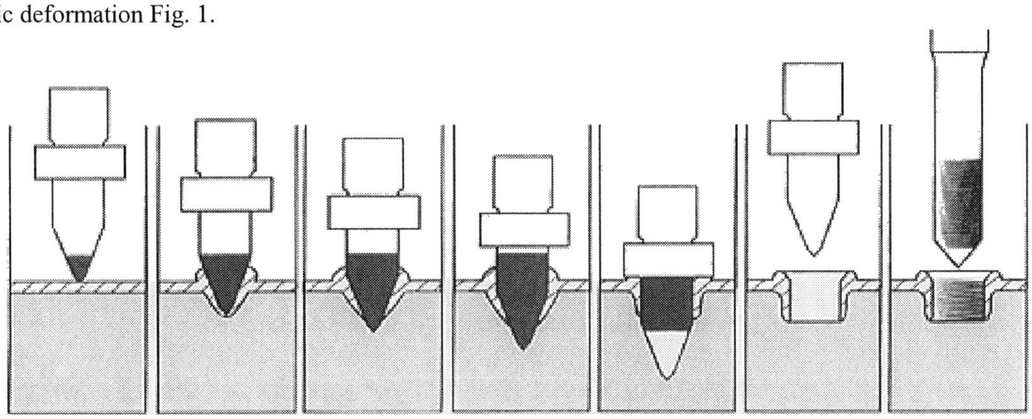

Fig. 1: Thermal drilling process illustration [5].

However, many calculations for the thread stripping in such as in [4,6] do not take into account main characteristics of this technology, such as sleeve geometry, thermal softening, etc. Thus, an experimental verification is very useful to determine if standard procedures could be extended to this case as well. Such study is described below on various sets of geometries and materials.

2 Method of Evaluation

Following many procedures and standards [1,6], joint is required to withstand certain amount of axial tensile force according to its size and property class. A similar approach is used to evaluate joints presented in this study on both mild and high strength steel types. There are three material types considered in this study. Two mild steel grades S235 and S355 according to EN 10025 [7] and a DOCOL high strength steel acc. to its datasheet [8]. In addition, each material specimen was subjected to a uniaxial test for obtaining reference values of tensile strength for verifying calculations.

It is obvious, that the reliability of this method depends on the thickness of the base material very much. It affects the sleeve length and has a direct relationship to the effective length of engagement. However, not all types are available for all sets of materials, 2mm and 3mm thick specimen (before drilling) were considered.

To study this problem more thoroughly another set of parameters was added to the evaluation. Three metric thread sizes were considered: M5x0.8, M6x1, M8x1.25. All these parameters listed above were included into the specimen fabrication.

3 Experimental setup

As many standards refer to axial force values as a measure of joint strength [1,2,6], the same approach was used and presented in this paper, e.g. an axial tensile force associated with any mode of failure. The joints were evaluated in a tensile testing device in order to obtain a force-displacement relationship.

It is clear, that the design of tested specimens could affect the results heavily. The design was focused on minimising negative influences such as eccentricity or stiffness effects. A radially symmetrical specimen constraint at its edges was proposed for such solution, see Fig 2. below. An interface between a specimen and the testing device was specially designed for this experiment to satisfy clamped constraints. High property class 12.9. bolts were used for all configurations. The load rate was constant for all measurements 5mm/min.

a) specimen b) test rig

Fig. 2: The tensile testing device with the interface for a clamped specimen.

4 Results

A relationship with its failure load is obtained from each measurement. Each measurement was done thrice. Since this is a destructive testing, there were three samples total for each measurement.

Since high strength bolts were used on purpose to avoid tension failures, the thread stripping is a major type of failure both on internal or external threads. None of tested configurations resulted in a breakage of the bolt shank. In addition to standard failures described in standards[1], threaded holes fabricated using the thermal drilling technology showed a new type of failure, where the sleeve formed by the technology separated from the base material resulting in a fatal loss of its functionality.

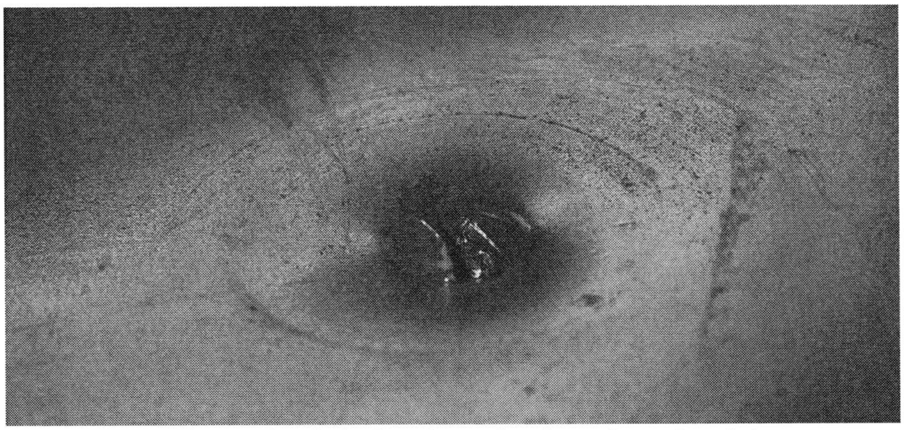

Fig. 3: Failure mode of a broken sleeve. Specimen shown from below.

All measurements were done thrice, where results are evaluated as an arithmetical average. Although force-displacement relationships show a distinct stiffness behaviour for mild and high strength steel types, it can be observed however, that there are no major discrepancies among each sample result sets.

Although all specimen were subjected during testing to relatively large plastic deformations around the thread (see Fig. 3), Graphs in Fig. 4 do not show any significant softening during the whole process of testing. All configurations with mild steel grades showed a relationship similar to Fig. 4a. The situation was similar in case of high strength steels with thickness 2 mm, where all configurations showed the same initial hardening as in Fig. 4b. The graphs below show results for samples of size M5 only, since the relationships are similar as explained above.

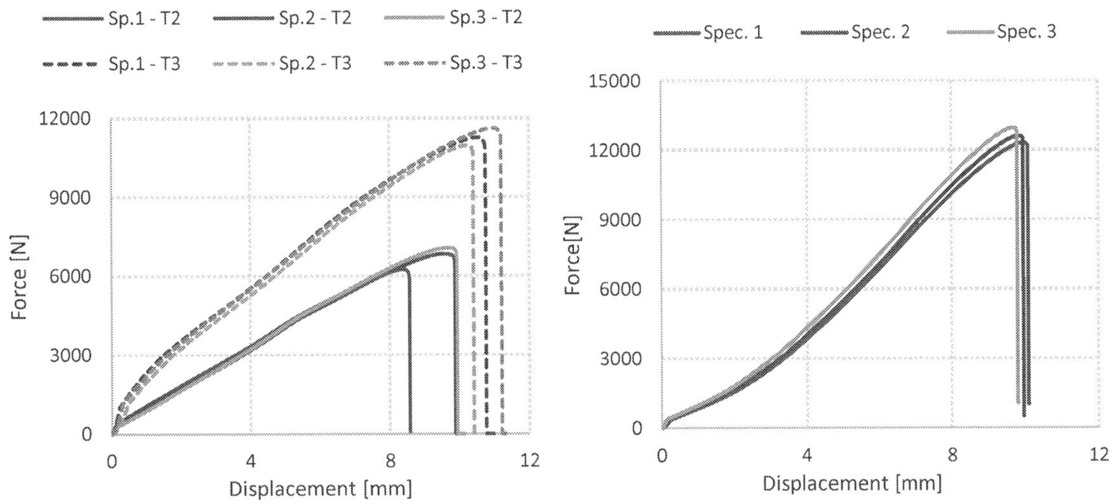

a) S235JRG1 mild steel – thickness 2 and 3 mm b) SSAB DOCOL 1200M high strength steel

Fig. 4: An example of force-displacement relationship for size M5

Overall results were summarized in the Tab 1. The table shows both averages of maximal values for each configuration. It should be mentioned, that in many cases, various types of failures occurred in the same configuration. Therefore, only the prevailing type of failure is displayed in the table below. Unmarked values refer to an external thread stripping failure.

Due to large deformations, it can be assumed that the effective length of engagement did not remain constant during the testing which puts some challenge into computational verification. It can be easily shown, that results in the Tab. 1 regarding same geometry sets are not exactly proportional to the shear strength of the material as expected, especially in cases, where the thread stripping occurred on bolt in one case but in hole in the other.

Tab. 1: Average of measured values of joint strength for several geometries

Thickness	2 mm			3 mm	
Material	S235JRG	S355J0	DOCOL 1200M	S235JRG1	S355J0
M5	6 700* N	8 110 N	12 630 N	11 270* N	12 720 N
M6	8 420* N	9 140 N	15 000 N	13 420* N	15 250 N
M8	9 580* N	11 600** N	16 720* N	14 200* N	18 510** N

* *failures with dominant internal thread stripping*

***failures with dominant sleeve breakage*

5 Conclusion

There are some facts, which are clearly apparent from the results. It is obvious, that thickness has a major influence on the joint strength. Since increase of thickness is associated with an increase of effective length of engagement, it can be assumed that the strength is a function of a shear area in this case as well, which is correspondence to Alexander's theory [1].

However, it can be observed from Tab. 1, that same geometry configurations (as in M5, 2mm thickness, S355J0 a DOCOL1200M) associated with an external stripping (taking the same reference material of the bolt into account) provide quite distinct values. Therefore Alexander's theory cannot by simply extended to this case and requires some additional variables.

In addition, some configurations did break in very a different manner. The failure, where the sleeve separated from the base material brings new uncertainties to the estimation of strength of joints produced by the thermal drilling. There are possibilities to follow this study with different types of constraints. Which would allow evaluation of the sleeve failure. Measurements could be carried out on a different sets of specimen thicknesses, however choice of configurations presented in this paper were believed to be most effective for the technology.

Acknowledgements

The authors would like kindly to show their gratitude to company Borcad CZ s.r.o. for fabrication of the specimen and fabrication of the interface needed for actual measurement. This work was also supported by a project funded by the Ministry of Education, Youth and Sports of the Czech Republic (No. SP2016/145).

References

[1] ISO 898-2: 1992, Mechanical properties of fasteners-Part 2: Nuts with specified proof load values - coarse thread. Switzerland: International Organisation for Standardization, 1992.

[2] Alexander ,E. M., Analysis and Design of Threaded Assemblies, SAE Technical Paper 770420, doi:10.4271/770420, 1977.

[3] Hagiwara, M., Sakai, H., Verification of the Design Concept on Nuts in Bolt/Nut Assembly for the Revision of ISO 898-2 and ISO 898-6, Journal of Advanced Mechanical Systems, and Manufacturing, Vol. 1, No. 5 (2007), pp. 755-762

[4] Oberg, E. et al., Machinery's Handbook, 27thEdition, New York, Industrial Press Inc., 2004

[5] Flowdrill technology from http://www.flowdrill.com/eu_en/, 14-04-2016

[6] VDI-2230, Systematische Berechnung Hochbeanspruchter Schraubenverbindungen Zylindrische, Einschrauben verbindungen, *VDI-Handbuch Konstruktion,* November 1998.

[7] EN 10025-2:2004 - Hot rolled products of structural steels

[8] SSAB Swedish Steel, Data sheet 8251 Docol 1200M

Determination of Transmission Mechanical Efficiency by Strain Gauges Measurement

O. Berka[1,*], M. Dub[1], F. Lopot[1], M. Janda[1], P. Burda[1], V. Dynybyl[1]

[1] *Czech Technical University in Prague, Faculty of Mechanical Engineering, Department of Designing and Machine Components, Technická 4, 166 07, Prague 6, Czech Republic*

* *ondrej.berka@fs.cvut.cz*

Abstract: This paper is focused on determination of the transmission mechanical efficiency during the test and as well as usual operation. Efficiency is one of the main indicators of the transmission quality. In general, that is determined either by analytical calculations with conditions or by value from the electric converter or torque sensor.

Keywords: Efficiency; Transmission; Measurement; Strain Gauge.

1 Introduction

The mechanical efficiency is usually monitored during transmission laboratory testing. For this purpose values from the torque sensor or electrical quantity of the electric motor from the converter are used most often. However, values from electric converter are influenced by mistake, because parameters are included in the electric motor efficiency and it is not only finding transmission mechanical efficiency. Also using torque sensor is not suitable for the universal measurement because in most cases torque sensors are used during laboratory testing. Potential customer wants this transmission efficiency in usual machine operation. There is not space to modify assembly of the complete device which is usually necessary in the process of torque sensors installation. For this purpose the new method was created. This method is created on the basis of measurement with strain gauges on input and output shafts.

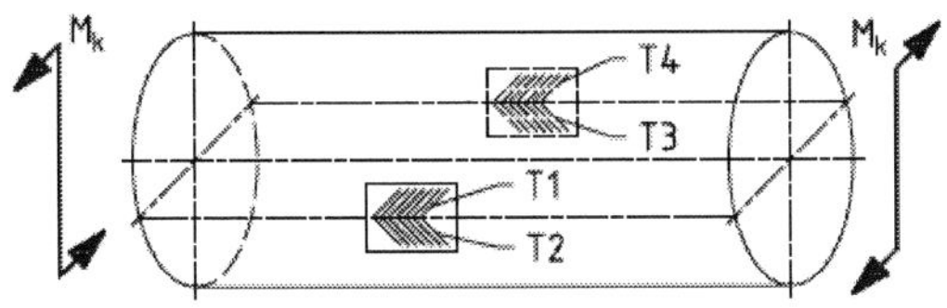

Fig. 1: Shaft under torsion with X rosettes strain gauge [6].

2 Measurement

2.1 Sensor

The sensor consists of several strain gauges. Special X rosettes strain gauges are used for measurements on a shaft under torsion (Fig. 1). Connection is carried out to the full Wheatstone bridge circuit (Fig. 2). The full bridge circuit configuration is the most sensitive bridge and the best in compensation signal from additional normal and bending loading.

Experiment design comes from loading values during measurement. The first step is calculation of maximum stress (deformation) of the measured shaft. For the shaft under torsion this equation is used (M_k is torque, τ is shear stress, ε is strain, G is shear modulus and W_k is torsional section modulus)

$$M_k = W_k \cdot \tau = W_k \cdot G \cdot 2 \cdot \varepsilon \tag{1}$$

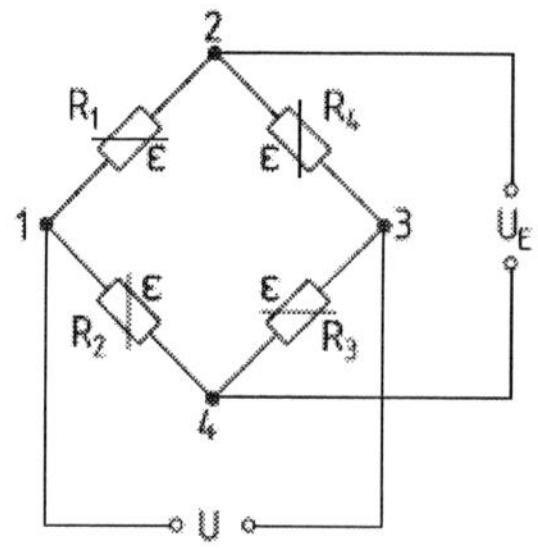

Fig. 2: Shaft under torsion with X rosettes strain gauge.

For full bridge circuit this equation is used (U_E is bridge excitation voltage, U is bridge output voltage, R_i is resistance, ΔR_i is resistance change)

$$U = \frac{U_E}{4}\left(\frac{\Delta R_1}{R_1} - \frac{\Delta R_2}{R_2} + \frac{\Delta R_3}{R_3} - \frac{\Delta R_4}{R_4}\right) \tag{2}$$

For X rosettes strain gauge in full bridge circuit is used (k is $k - factor$ of the strain gauge)

$$\frac{\Delta R}{R} = k \cdot \varepsilon \tag{3}$$

After that, this gives:

$$\varepsilon = \frac{1}{4}\frac{4}{k}\frac{U}{U_E} \tag{4}$$

And for the torque

$$U = \frac{U_E \cdot M_k \cdot k}{W_k \cdot G \cdot 2} \tag{5}$$

For example maximum torque is $3000Nm$, $U_E = 2.5V$, shaft diameter is $100mm$, $k = 1.4$, $G = 81000MPa$, for this parameters output voltage is $0.33mV$. These are real parameters for common loading. Measuring signal output voltage is very low and for this measurement the result will be with low sensitivity.

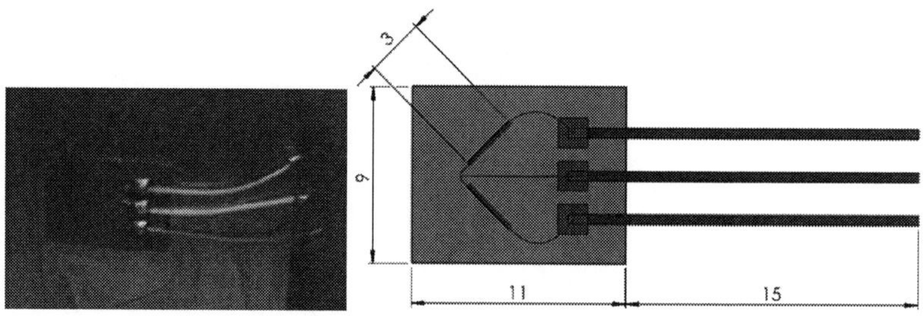

Fig. 3: Semiconductor strain gauge for torque measurement

There is one more option, to use more sensitivity strain gauge. Semiconductor strain gauges are used for this purpose (Fig. 3). In general, these gauges have 60x higher strain sensitivity. Equation for semiconductor strain gauge is different from common strain gauges. There are two gauge factors C_A and C_B. Basic equation is

$$\frac{\Delta R_i}{R_i} = 1 + C_{iA}\varepsilon_i + C_{iB}\varepsilon_i^2 \tag{6}$$

and in equation for full bridge circuit this gives (with $\varepsilon = -\varepsilon_1 = -\varepsilon_2 = -\varepsilon_3 = -\varepsilon_4$ from strain gauges installation on shaft)

$$U = \frac{U_E}{4}\left(\varepsilon(C_{1A} + C_{2A} + C_{3A} + C_{4A}) + \varepsilon^2(C_{1B} - C_{2B} + C_{3B} - C_{4B})\right) \tag{7}$$

and

$$U = \varepsilon \cdot V + \varepsilon^2 \cdot W \tag{8}$$

$$U = \left(\frac{M_k}{2 \cdot G \cdot W_k}\right) V + \left(\frac{M_k}{2 \cdot G \cdot W_k}\right)^2 W \tag{9}$$

Gauge factors have approximately values $C_A = 130$ and $C_B = 4000$. With semiconductor strain gauges gives output voltage $31mV$. Measuring signal from semiconductor strain gauges is about 100x higher and also therefore these gauges are better for measurement of efficiency. These develop to semiconductor strain gauges for measuring a shaft under torsion. These strain gauges are custom-made.

2.2 Measuring device

Unconventional measuring devices have to be used for this measurement because there is small space for installation around shafts and also because shafts rotate with high speed and there is not possible to use wire connection. Special autonomous measuring apparatus were created and device has own power source (batteries) and own recording part (memory card). At the same time device enables to transmit on-line signal to computer. Maximum sampling frequency is 2000Hz for 6 measuring channels. Device is resistant to centrifugal force caused by rotation.

2.3 Efficiency measuring

Sensors with strain gauges are installed on an input haft and on an output shaft. Efficiency is determined on the basis of compare input torque with output torque. Equation for this is

$$\eta = \frac{M_{kout}}{M_{kin} \cdot i} \tag{10}$$

where η is a mechanical efficiency and i is a gear ratio.

Measurement of a torque is based on the determination of a deformation caused only by torque without another loading. However shaft loading is combined by bending, torsion and tension/compressing. If full bridge circuit is used then bending and tension/compressing are compensated. This bridge gives on output only deformation from torsion loading. After measuring with this bridge there is no difficulty to calculate a torque from output signal.

Moreover one difficulty can appear. Input and output transmission shafts have several possibilities of geometrical change of their shape. Computation of torque is based on geometrically regular shape and it is circle for a shaft. On the shaft there are grooves for key, diameters changes a many others possibilities. It is a case of stress distribution surroundings of notch. In the beginning FEM computation should be made for to set a place for strain gauge installation, which are not affected by notch (Fig. 4).

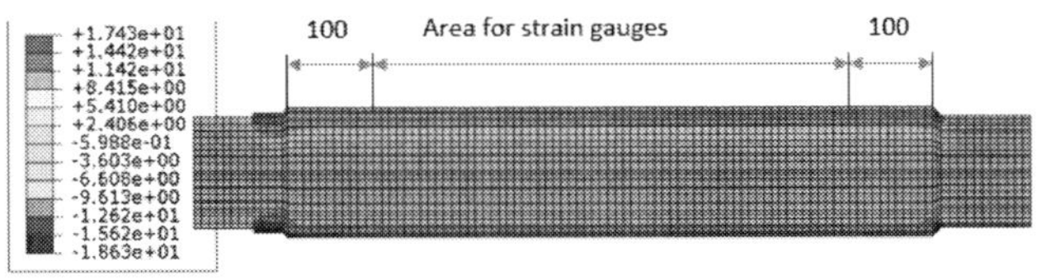

Fig. 4: Determination of area for strain gauges (Shear stress [MPa])

3 Results and Conclusion

As stated above several measurements were made according to this method. Measurements are made in more loading levels. This is very useful measurement tool for compare efficiency with vibration, noise, temperatures individual parts or oil and e.g. with mesh quality which is monitored in teeth transmission [1], [3], [5] or monitoring of gearbox external forces (hinge of tram gearbox) [2], [4]. Presentation of torque measurement is in Fig. 5. Range efficiency value was 84 %.

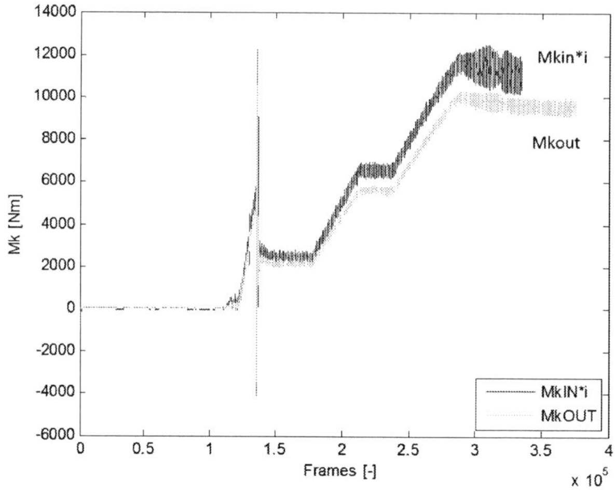

Fig. 5: Time behaviour of torque

The results of these measurements are review of the designed method. The adjust method gives relevant results. Used semiconductor strain gauges have sufficient sensitivity. The designed procedure (measuring method) can be repeated.

Next tasks will be mainly set of measurements that enables to determine the real measurement error.

Acknowledgement

This work was supported by the industrial cooperation.

References

[1] O. Berka, F. Lopot, M. Dub, Experimental Analysis of Gear Loading in Planetary Transmission, in proc.: EAN 2014 - 52nd International Conference on Experimental Stress Analysis (2014), P. Polach, VZU Plzen, Marianske Lazne.

[2] M. Dub, at al., Experimental analysis of stress state and motion of tram gearbox hinge, in proc.: EAN 2014 - 52nd International Conference on Experimental Stress Analysis (2014), P. Polach, VZU Plzen, Marianske Lazne.

[3] O. Berka, at al., Analysis of satellites loading in planetary transmission, in proc.: EAN 2015 - 53nd International Conference on Experimental Stress Analysis (2015), P. Padevet, CTU in Prague, Cesky Krumlov, 16-19.

[4] M. Dub, at al., Backlash and torsional stiffness of a gearbox, in proc.: EAN 2015 - 53nd International Conference on Experimental Stress Analysis (2015), P. Padevet, CTU in Prague, Cesky Krumlov, 74-77.

[5] O. Berka, at al., Mesh quality monitoring in evolventn teeth in gearboxes under operation -signal identification, in proc.: XXI IMEKO World Congress "Measurement in Research and Industry" (2015), J. Holub, CTU in Prague, Prague, 1582-1586.

[6] K. Hoffmann, An Introduction to Stress Analysis and Transducer Design using Strain Gauges, Pfungstadt, HBM, 1987.

Tensile Test of Motorcycle Garment Leather

T. Bońkowski[1,*], R. Kottner[1], J. Krystek[1], L. Hubík[2], T. Görner[3], L. Hynčík[1]

[1] *European Centre of Excellence, NTIS - New Technologies for Information Society, Faculty of Applied Sciences, University of West Bohemia, Univerzitní 8, 306 14 Plzeň, Czech Republic*

[2] *PSi Hubík, s.r.o., Daimlerova 3, 301 00 Plzeň, Czech Republic*

[3] *Regional Technological Institute, Faculty of Mechanical Engineering, University of West Bohemia, Univerzitní 8, 306 14 Plzeň, Czech Republic*

[*] *tomasz@ntis.zcu.cz*

Abstract: This work describes tensile tests of random selected leather samples. Tests were performed according to ISO 3376 standard. Dimensions of cross sections were measured (mean thickness = 1.4 mm, mean width = 19.8 mm) and then samples were grouped with a view on texture (with and without perforation). The aim of the tests was to obtain average strength and distribution of stiffness in the tested group of samples. Additionally cyclic (3 cycles) tests were performed. Three sets of Ogden model parameters were identified to obtain material models for the leather having minimum, average and maximum stiffness.

Keywords: leather; tensile test; motorcycle garment; ISO 3376; cyclic test; Ogden model.

1 Introduction

To create a realistic tool for motorcycle driver injury assessment, it is necessary to take into account influence of all types of personal protective equipment (PPE). Natural leather is the main material which is used to produce motorcycle protective garments. Therefore, the investigation of leather mechanical properties which allow the creation of the numerical model of the garments was the subject of this work. Since leather is a heterogeneous and anisotropic material with nonlinear behavior [2, 3, 4, 5, 6, 7], providing data for the numerical model is a complex problem. Moreover, application randomness of leather (treatment of leather, angle of cutting in the relation to the direction of majority of collagen fibers, the location of cutting in the relation to the part of an animals body, etc.) in the garment manufacturing process does the task more complicated.

2 Methodology

The study was carried out in order to obtain average strength and fit a constitutive material model of leather samples cut from a motorcycle garment. Moreover, differences of ultimate strength between samples having different texture were examined. Seven steps can be distinguished in the research methodology :

- samples preparation according to ISO [1] standard,
- determination of the sample texture and condition,
- measurement of real dimensions of the samples,
- uniaxial tensile tests (ultimate strength, cyclic response),
- experimental results evaluation and selection of 3 representative characteristics,
- fitting hyperelastic constitutive material model,
- validation of material model in finite element method (FEM) environments (Abaqus, Virtual Performance Solution ⟨ VPS ⟩ - former PamCrash).

3 Samples Preparation

Samples of the leather were randomly cut from a motorcycle garment using a press knife. Fig. 1 describes the geometry of tested samples. "Large" dimensions series of samples were selected according to ISO [1] standard.

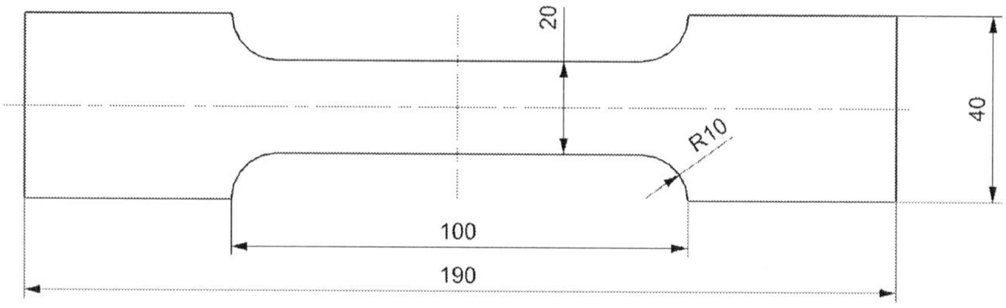

Fig. 1: Sample shape and dimensions [mm].

4 Sample Texture Determination

In addition to the protective function of motorcycle garments, different types of leather textures are used to provide thermal comfort. Specimens were grouped with a view on the texture (with and without perforation - Fig. 2).

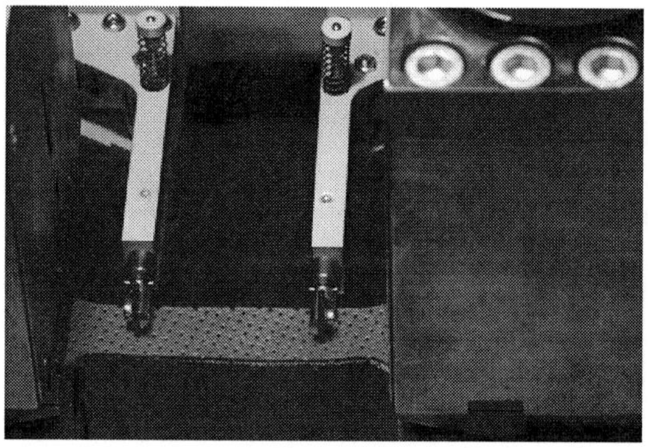

(a) Perforated sample.

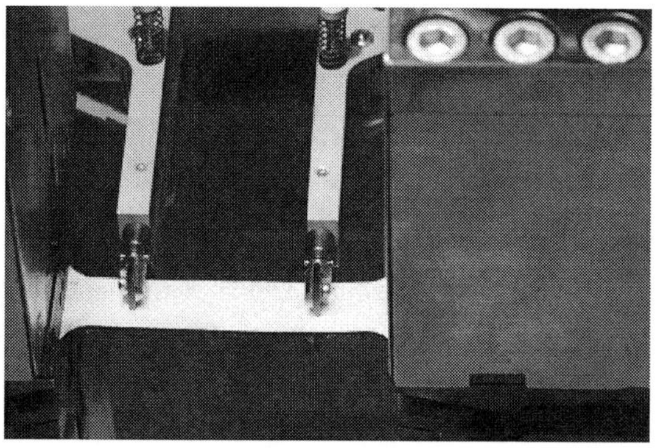

(b) Non perforated sample.

Fig. 2: Differences in samples texture.

5 Real Dimensions Measurement

The sample cross-section area is essential to obtain stress-strain relationship from force-displacement characteristics. Real dimensions (Tab. 1) of each sample (thickness, width) were measured using vernier calipers at three positions on the grain side and three on the flesh side.

Tab. 1: Average dimensions of samples cross-section.

	mean thickness[mm]	mean width[mm]	mean area[mm^2]
non perforated samples	1.42	19.76	28.07
perforated samples	1.36	19.82	26.91

6 Uniaxial Tensile Test

6.1 Ultimate Strength Test

18 samples were selected for the ultimate strength test (Fig. 3). The test was performed with speed of 100 mm/min, temperature was equal to 23 °C, relative humidity was 50%. Preload of 1N was applied to each sample before the test.

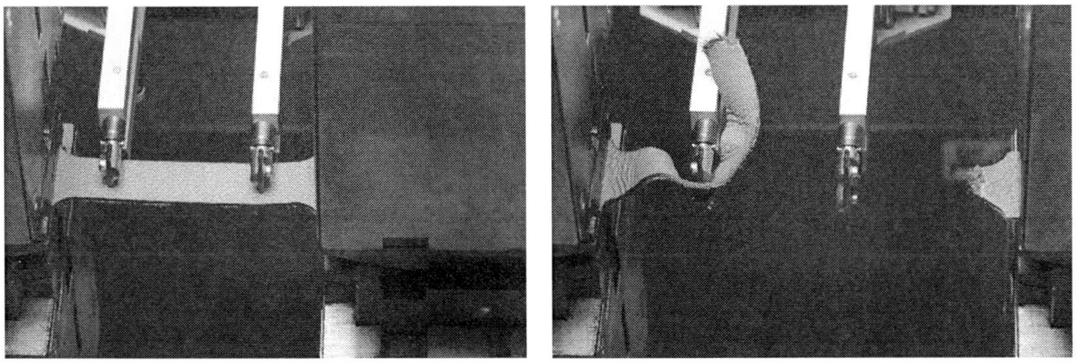

(a) Sample before elongation (b) Sample after failure

Fig. 3: Ultimate strength test.

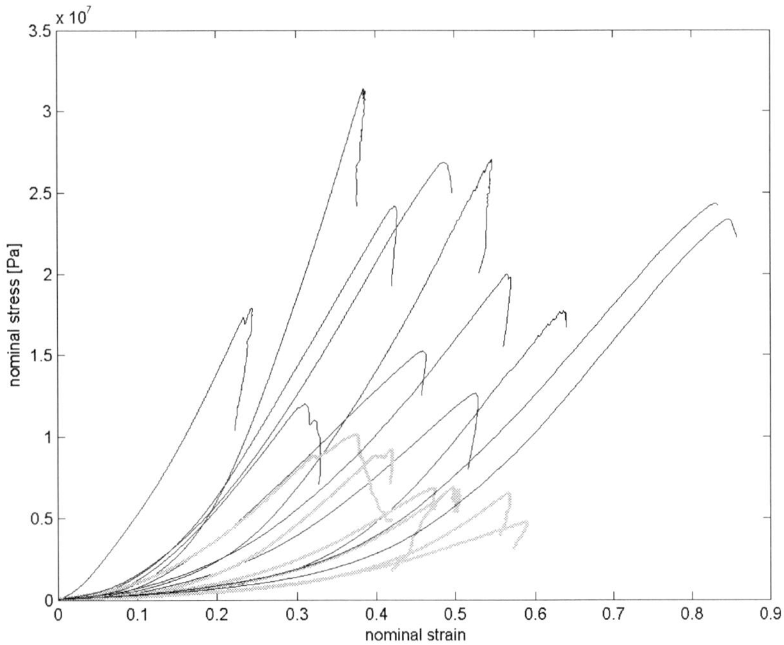

Fig. 4: Stress-strain curves (gray - perforated, black - non perforated).

Performed tests (Fig. 4) shown that leathers used to produce motorcycle protective garments have significantly inconsistent mechanical properties. Failure occurred in wide range of strains (0.23–0.86) and stresses (5–32 MPa). Non perforated samples had average ultimate strength approximately 21 MPa, average ultimate strength in case of perforated specimens was 3 times lower.

6.2 Cyclic Loading Test

Additional set of samples was tested using cyclic loading (3 cycles). Value of nominal strain 0.18 was prescribed as a upper limit for the cyclic loading. After reaching the upper limit of strain, testing machine was maintaining strain for 1 minute. Then a sample was unloaded with speed of 100 mm/min (Fig. 5).

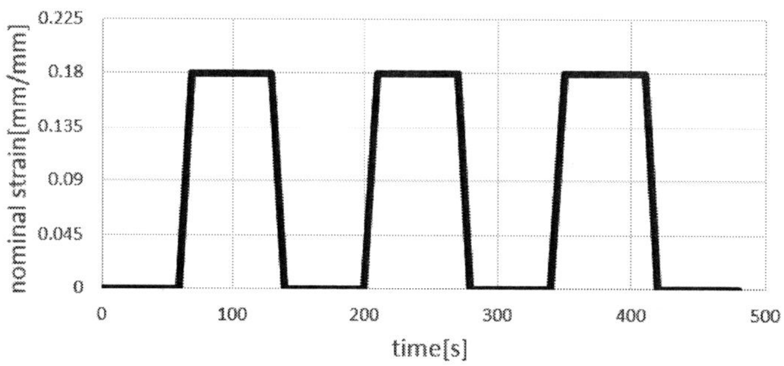

Fig. 5: Prescribed loading in cyclic test.

Temperature was equal to 21 °C and relative humidity was 46.5% during this test. Fig. 6 describes the behavior of specimens (representatives for each group) during the cyclic loading. It is obvious that plastic strain of 5.4% remained in the samples.

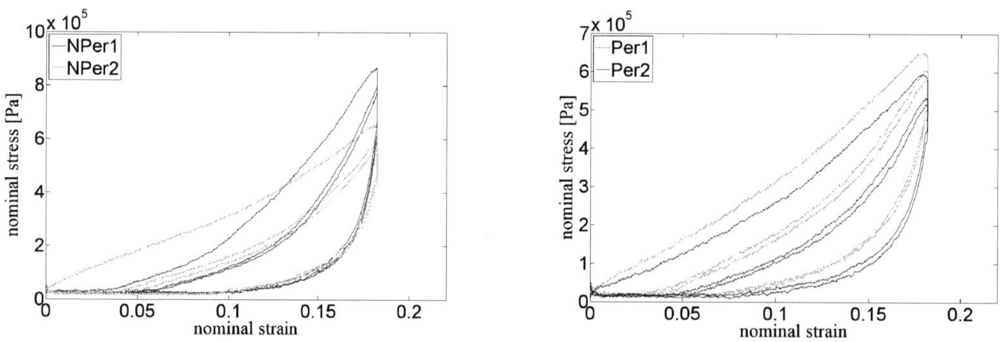

(a) Samples without perforation

(b) Samples with perforation

Fig. 6: Cyclic test response of representative samples.

7 Experimental Results Evaluation

In order to identify material model parameters, three (B2, Y3, Y4) experimental stress-strain characteristics were selected as representatives (Fig. 7). Each of the selected samples represented different stiffness:

- B2 sample - represented leather with highest stiffness,
- Y3 sample - represented leather with average stiffness,
- Y4 sample - represented leather with lowest stiffness.

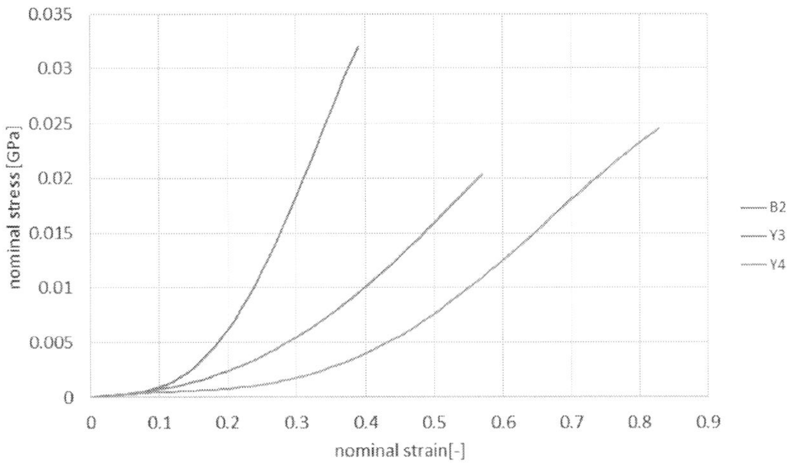

Fig. 7: Stress-strain curves of samples used to material model identification.

8 Hyperelastic Constitutive Model Fitting

The aim of these simulation was to fit a hyperelastic constitutive material model. With respect to the reference [2, 3, 5, 6], the Ogden hyperelastic material model was selected. The Hyperelastic constitutive material model using the Ogden formulation is based on the strain energy functional [8]:

$$W(\lambda_1, \lambda_2, \lambda_3) = \sum_{i=1}^{N} \frac{\mu_{i,o}}{\alpha_i} \left(\lambda_1^{\alpha_i} + \lambda_2^{\alpha_i} + \lambda_3^{\alpha_i} - 3\right) \tag{1}$$

where $\lambda_1, \lambda_2, \lambda_3$ are principal stretches, μ_i is Ogden parameter, α_i is Ogden exponent and N is model order. To determine the Ogden parameters from the tests, the material was assumed to be incompressible. The formulation with order N=1 was used.

In order to fit the model parameters to the experimental data, the linear least-square optimization method was used. The optimization was carried out under the build-in preprocessor of VPS (former PamCrash). The least square method minimizes the error in stress described by Eq. 2 [9]:

$$E = \frac{1}{2} \sum_{i=1}^{ND} \left(P_i^{test} - P_i^{derived}\right)^2 \tag{2}$$

where P_i^{test} is the measured nominal stress, $P_i^{derived}$ is the stress obtained from the strain energy functional W and ND is the number of experimental data points.

It should be noted that the implementation of the Ogden hyperelastic model is different in commercial software (VPS and Abaqus). The Basic description of strain energy functional (Eq. 1 [8]) assumes parameter μ_i in slightly different way than commercial implementations. The VPS solver describes the strain energy functional by the equation [9]:

$$W(\lambda_1, \lambda_2, \lambda_3) = \sum_{A=1}^{3} \sum_{i=1}^{N} 2\frac{\mu_{i,p}}{\alpha_i} \left(\bar{\lambda}_A^{\alpha_i} - 1\right) \tag{3}$$

where A are principal directions and $\mu_{i,p}$ is the Ogden parameter in the VPS implementation. On the other hand the Abaqus is using the Ogden model in the following formulation [10]:

$$W(\lambda_1, \lambda_2, \lambda_3) = \sum_{i=1}^{N} \frac{2\mu_{i,a}}{\alpha_i^2} \left(\bar{\lambda}_1^{\alpha_i} + \bar{\lambda}_2^{\alpha_i} + \bar{\lambda}_3^{\alpha_i} - 3\right) \tag{4}$$

where $\mu_{i,a}$ is the Ogden parameter in the Abaqus implementation. To harmonize results of the parameters optimization, a correlation table has been developed (Tab. 2).

Tab. 2: Correlation of Ogden model parameter between different implementation.

	Ogden	VPS(PamCrash)	Abaqus
Ogden		$\mu_o = 2\mu_p$	$\mu_o = 2\frac{\mu_a}{\alpha}$
VPS (PamCrash)	$\mu_p = 0.5\mu_o$		$\mu_p = \frac{\mu_a}{\alpha}$
Abaqus	$\mu_a = 0.5\alpha\mu_o$	$\mu_a = \alpha\mu_p$	

9 Model Validation - FEM Simulation

In order to validate the material model, the simulations were performed in FEM the commercial software. Fig. 8 presents comparison between the numerical simulations (solid lines) and experimental data (dashed lines) for the three selected representatives samples. Material models sufficiently reflects the behavior of samples.

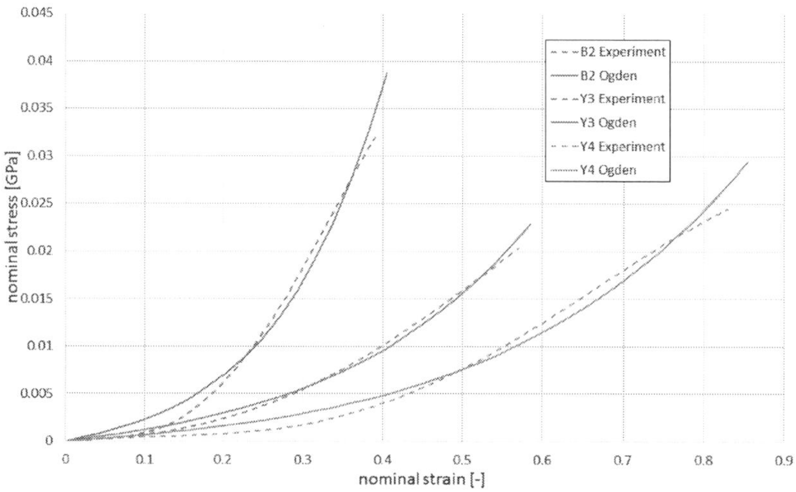

Fig. 8: Comparison between model and experiment.

Tab. 3: Identified Ogden parameters.

		B2	Y3	Y4
	VPS [GPa]	5.1360E-04	4.7720E-04	2.9530E-04
μ	Abaqus [GPa]	6.0040E-03	3.7747E-03	2.1640E-03
	Ogden [GPa]	1.0272E-03	9.5440E-04	5.9060E-04
	VPS [-]			
α	Abaqus [-]	1.1690E+01	7.9100E+00	7.3280E+00
	Ogden [-]			

10 Conclusion

The Average ultimate tensile strength of non perforated leather was approximately 21 MPa. Perforated samples had the average ultimate strength 3 times lower than non perforated. 5.4% of plastic strain remains in the leather samples during cycling loading with maximum strain on 0.18. The Cross-solver correlation table for the Ogden model has been developed. Ogden model parameters of the three samples were identified (Tab. 3). The model was validated using numerical simulations of performed tests.

Acknowledgement

The research leading to these results has received funding from the People Programme (Marie Curie Actions) of the European Unions Seventh Framework Programme FP7/2007-2013/ under REA grant agreement no 608092.

References

[1] Leather - Physical and mechanical tests - Determination of tensile strength and percentage extension, ISO 3376:2002(E).

[2] N. F. A. Manan, J. Mahmud, A. Jumahat, Biomechanical Behaviour Of Bovine Skin: An Experiment-theory Integration And Finite Element Simulation, Jurnal Teknologi 76:10 (2005) 103–111.

[3] Z. Li, D. Paudecerfb, J. Yangc, Mechanical behaviour of natural cow leather in tension, Acta Mechanica Solida Sinica 22:1 (2009) 37–44.

[4] A.G. Ward, The mechanical properties of leather, Rheologica Acta 13 (1974) 103-112.

[5] N. F. A. Manan, J. Mahmud, M. H. Ismail, Quantifying the Biomechanical Properties of Bovine Skin under Uniaxial Tension, Journal of Medical and Bioengineering 2:1 (2013) 45-48.

[6] R. B. Groves et al., An anisotropic,hyperelastic model for skin: Experimental measurements, finite element modelling and identification of parameters for human and murine skin, Journal of the Mechanical Behavior of Biomedical Materials 18 (2013) 167-180.

[7] Y. Wenge., "The mechanical properties of leather in relation to softness" (PhD diss., University of Leicester, 1999).

[8] R.W. Ogden, Non-Linear Elastic Deformations, New York, Dover Publications Inc., 1984.

[9] Virtual Performance Solution 2014 - Solver Reference Manual Volume III, 2014.

[10] Abaqus 6.11 - Theory Manual, 2011.

Fracture Aspects in Testing Welds

Petr Brož[1, a]

[1]University of West Bohemia, Faculty of Applied Sciences, Plzen, Czech Republic

[a]broz.petr@tiscali.cz

Keywords: Base metal, crack growth law, effective *R* – ratio, fracture resistance, post weld treatment, secondary stresses, weld defect

Abstract. For weldments, the paper is specialized in fracture testing. As well, the effect of specific weld characteristics on concept of test issues is given. Especially, consideration is paid to the effect on testing of some of the attributes such as experiments on definite microstructural zones, elimination of both residual stress influences on fracture toughness measurement and misalignment effects. It is also addressed determination of crack growth and weldment tensile qualities.

Introduction

Fracture toughness testing of weldments is focused on the effect of definite mictrostructural zones and elimination of both residual stress influences on fracture toughness measurements and misalignment effects (Fig.1). Characteristic specimens for discussed testing are stated (Figs 2,3). It is worthwhile to recall the advantages of testing specific microstructures by shallow notched specimens, when applying *K, J* integral and *CTOD* solutions [1]. An extra to conventional testing stand for the choice of the notch location in connection with test metallography and the possible necessity to modify the residual stress state before or in the course of pre – cracking. A knowledge of the various parameters influence is essential so as to achieve conservative limitations in structural assessment and testing.

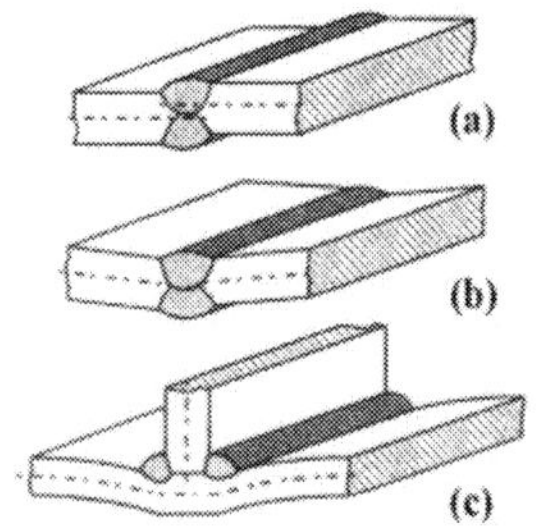

Fig. 1 Axial and angular misalignment illustrations

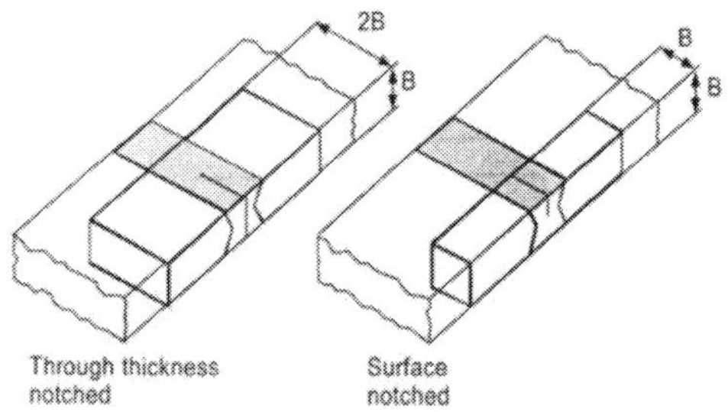

Fig. 2 Through – thickness and surface notched specimens [1]

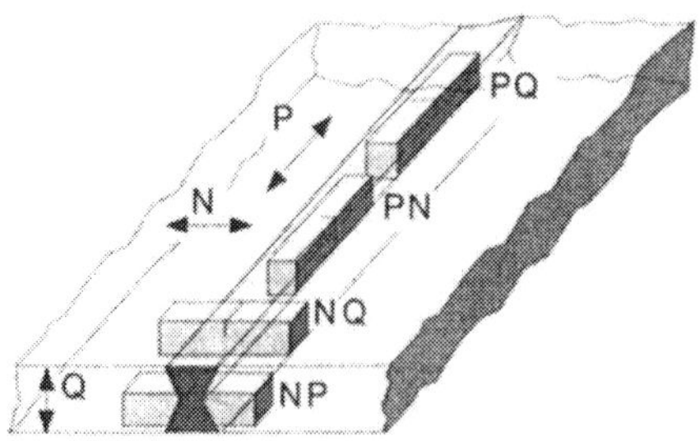

Fig. 3 Ranking for specimen outlook regarding the weld axis [3]

Testing definite microstructures

The purpose is to stipulate the fracture toughness as regards crack initiation or extension for a characteristic target region, for example the weld centre line, the heat affected zone, and characteristic microstructure, eg coarse grainy affected zone. Welded joints that depict some linear misalignment or angular torsion should be straightened before testing [2]. This is made by local bending of the specimen ends to local bending moments owing to misalignment and reinforcement in the construction must be taken into account at the applied side, that is, while stipulating the crack driving force in the component [3].

Elimination of weld misalignment

Before fatigue pre – cracking, it may be essential to alter the welding residual stresses in the specimen prior to testing. There are some stimulations. The welding residual stresses demonstrate an inhomogenous design over the section which may affect fatigue crack growth in the way that the pre – crack front will evolve an irregular form.
The residual stresses are in part relieved and redistributed owing to the extraction of the specimen and the insertion of the crack initial notch, the remaining residual stresses may even be large enough to considerably affect the shape of the pre – crack front and the outcome of the fracture toughness test. An evident technique for residual stress relief is post – weld heat interpretation.

Pre – cracking

In the course of local compress before pre – cracking, hardened steel plates are pressed on one or both side surfaces of the specimen over 86 – 92 % of the ligament ahead of the crack tip thus that a plastic strain of 1 % of the specimen thickness is put out [4]. The reverse bending method did not obtain its line into the standards as it is considered like relatively wasteful in both, the decrease of residual stresses and setting up of a straight crack front. The stepwise high R – ratio technique is consisted of two steps of pre – cracking for different R – ratios. Usually, no pre – treatment of the specimens is essential for welds that have been post – weld heat handled for the reason of stress relief. The ISO test standard advices the application of the shortest fatigue crack length allowed in the document ISO 12135 for the purpose to minimize crack front curving and also crack deviation from the precise target zone analyses.

Predicting crack initiation and propagation lifespans in butt welds

The crack actions in fatigue that initiate at the toe of a butt weld usually involves two main phases: (i) the crack initiation life (N_i) and (ii) the crack propagation life (N_p) [5]. The initiation period consists of the cycles indispensable to shake down the residual stresses in the weld. The crack propagation period consists of two trajectories (1) through thickness (N_{px}) and (2) transverse (width direction) (N_{py}). The notched – root stress method is employed to estimate the crack initiate life, whilst the fatigue crack propagation lives were estimated applying fracture mechanics interpretation [6], [7]. The experiments were performed for Grade AISI 304L Stainless Steel (analogous to the Grade CSN 17 249), with constant amplitude and stress ratio $R = 0$. The analysis confirms that the crack growth parameters are dependent on the crack propagation courses for the same material. Butt welds tested in fatigue with their reinforcement intact demonstrate less fatigue resistance than flat plate or butt welds with reinforcement removed owing to the notch associated with the toe of the weld.

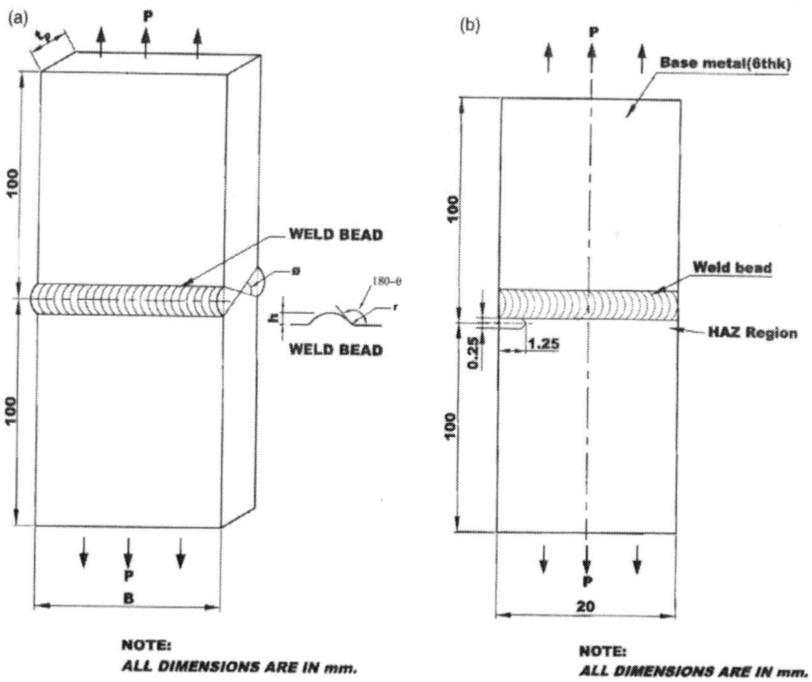

Fig. 4 Particulars of specimens: (a) without notch, (b) with notch [5]

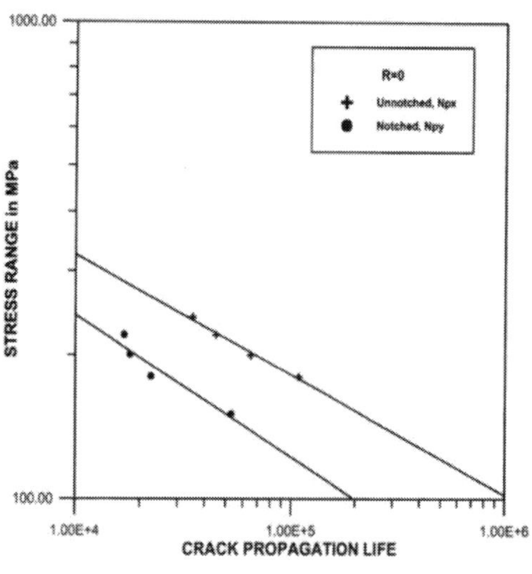

Fig. 6 Crack propagation lifespans of unnotched and notched specimens [5]

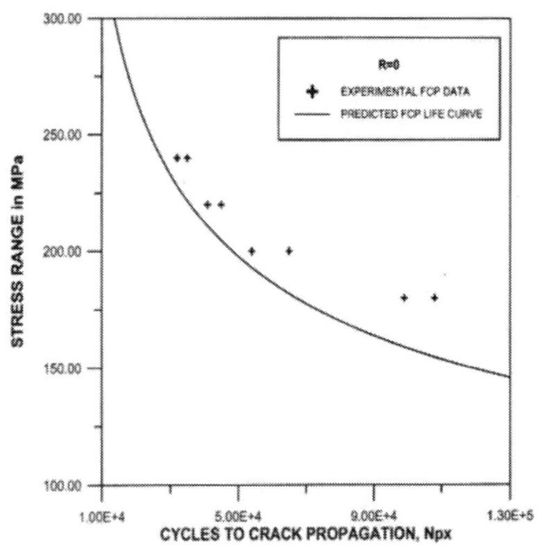

Fig. 7 Comparison between the predicted curve for fatigue crack propagation lifespan and test data for unnotched specimens

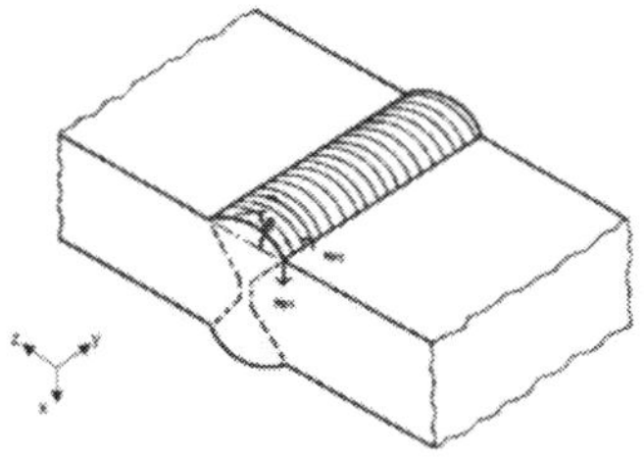

Fig. 5 Crack propagation trajectories in butt-welded joints

Conclusion

Both, the fracture toughness and determination of stress – strain dependences for heterogenous samples, namely for notched test bar weldmens, are outlined. To predict the fatigue crack propagation lifespan in the *y* – direction (N_{py}) of the butt welds, a new test routine employing fracture mechanics outlooks is summarized. The fatigue crack initiation and propagation lifespans of the stainless steel butt welds under constant amplitude load can be to a considerable degree predicted. Since the mictrostructure alongside the crack trajectory and the stress concentration factor are exceedingly different for unnotched and notched specimens, the fatigue crack propagation properties change in both cases for the same material.

Acknowledgment

The author acknowledges the support from the University of West Bohemia (grant no. SGS – 2016 – 038 : Modelling heterogenous materials and complex mechanical structures in mechanical and civil engineering).

References

[1] BS7910. Guide on methods for assessing the acceptability of flaws in metallic structures. British Standards Institution (BSI), London; 2005.

[2] HS Reemsnyder, HG Pisarski, MG Dawes. Residual stresses and fatigue precracking techniques for weldment fracture toughness specimens. J Test Eval 1991; 416-23.

[3] P. Brož, D. Dobiáš: Some sides of fatigue damage, Acta Technica, Vol. 60 (2015) p. 253.

[4] ISO 12135. Metallic materials – unified method of test for the determination of quasistatic fracture toughness. International Organization for Standardization (ISO); 2002.

[5] Zheng Xiulin, Lu Baotong, Cui Tianxie, Lin Chao. Fatigue tests and life prediction of 16 Mn steel butt welds without crack – like defect. International Journal of Fracture 1994; 68: 275-85.

[6] G Bhuyan, O Vosikovsky. Prediction of fatigue crack initiation lives for welded plate T – joints based on the local stress – strain approach. Int J Fatigue 1989; 11(3): 153-9

[7] Zheng Xiulin. A further study on fatigue crack initiation life – mechanical model for fatigue crack initiation. Int J Fatigue 1986; 8(1):17-21.

Identification of complex modulus using of the Hopkinson Split Pressure Bar.

J. Buchar [1,a], J. Trnka [2b], R. Ridky [1c]

[1] *SVSFEM s.r.o., Škrochova 42; 615 00, Brno; CZ*

[3]*ÚT AV ČR, Dolejškova 5; 182 00 Praha;CZ*

[a]*buchar@ svsfem.cz*

Abstract: A split Hopkinson pressure bar procedure was used for non-parametric identificon of complex modulus under conditions of non-equilibrium and axially non-uniform stress Tests were carried out with aluminium bars and with specimens of composite materials (Low viscosity epoxy resin and Baytube multi-walled carbon nanotubes) and having diameter 15 mm and length 7 mm. The influence of the nanotubes concentration on the dynamic response of given composites was evaluated.

Keywords: viscoelasticity; complex modulus; Hopkinson bar;

1 Introduction

Viscoelastic components employed for vibration isolation or shock absorption in automobiles, machines and buildings are often subject to a dynamic loading. From the dynamic design point of view, it is essential to be able to predict complex stiffness of the viscoelastic components accurately and efficiently, which requires, first of all, information about the complex modulus of the viscoelastic materials under operational environmental conditions.

The complex modulus represents the behaviour of viscoelastic materials in frequency domain [1] :

$$E^*(\omega) = E'(\omega) + iE''(\omega) \tag{1}$$

Where $E'(\omega)$ is the storage modulus and $E''(\omega)$ the loss modulus

There are many experimental techniques for the evaluation of this quantity. Most of these techniques are limited to relatively narrow band of frequencies [2]. In order to obtain the values of complex modulus in the broad extent of frequencies experimental methods based on the wave propagation in long bars were developed [3]. In many cases it is impossible to obtain specimens of the tested materials in form of long bars. This is a reason for the use of Hopkinson Split Pressure Bar which uses specimens of relatively small dimensions. This method is commonly used for study of the constitutive properties of materials under conditions of high strain rates and large strains [4]. For a SHPB test to be considered valid, it is required that several conditions concerning the specimen be approximately fulfilled, namely the stress equilibrium in the specimen. Because of the time and strain required to establish equilibrium, it is difficult to estimate the modulus of elasticity at high strain rates from a classical SHPB test . In specimens made of polymeric materials it is particularly difficult to achieve equilibrium because of the low wave speeds. For cases where the classical SHPB analyses do not give acceptable results, a parametric identification method [5] has been developed which does not require equilibrium or axial uniformity of stress and strain.
The aim of this study was to use this procedure, which does not require equilibrium, for non-parametric identification of the complex modulus of a viscoelastic composites.

Material and Experimental Technique

For the experiments composite material composed from the low viscosity epoxy resin BASF Mastertop P605 was used as the binder and the Baytube multi-walled carbon nanotubes C150P as the reinforcement were prepared. The main mechanical properties of the tubes are given in the Table 1.

Tab. 1: Properties of the Baytube MWCNT

MWCNT	Tensile strength [GPa]	E modulus [TPa]	No. of walls	Outer diameter [nm]	Density [kg.m^{-3}]	Length [μm]
C150P	>10	>1	3-15	5-20	150-350	1->10

Their distribution is shown in the Fig.1.

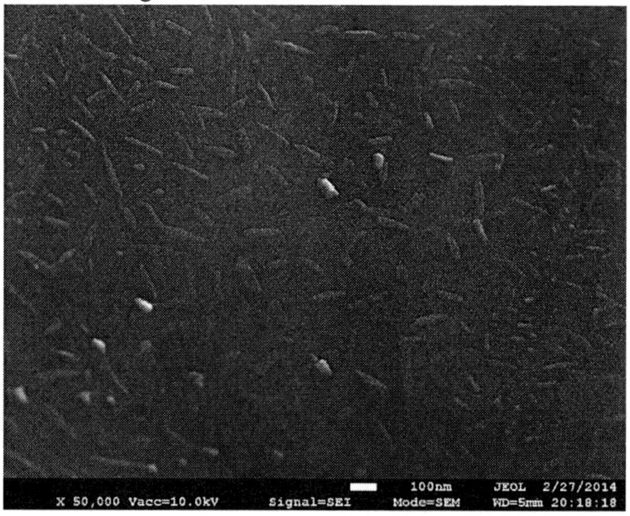

Fig.1: Distribution of carbon nanotubes.

The composites with different volume part of carbon nanotubes were prepared (1,2,3,4%). The different volume content was used to study the differences in mechanical parameters and behavior under blast load.

The main mechanical properties have been determined quasi static compression test. The results in the form of a dependence stress – strain are presented in the Fig.2.

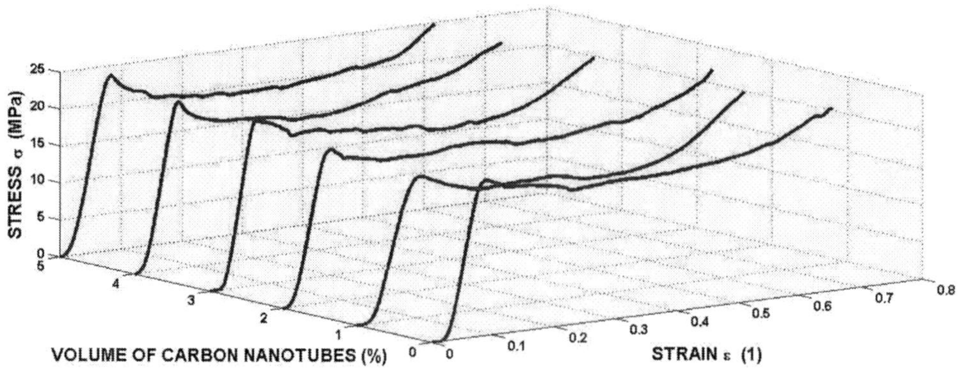

Fig.2: Stress strain curves at static compression.

It can be seen that the stress increases rapidly up to some value and than interval of nearly constant stress occurs. The content of the nanotubes does not influence the qualitative features of this dependence.The increase in the carbon nanotubes volume lead to some increase in the stress in the first region. In this region no permanent changes in the material structure occurs. The deformation is elastic and/or viscoelastic.

In the next step the dynamic response of the tested material was studied using of the split Hopkinson pressure bar (SHPB) technique. In this method a short test specimen is placed between two long bars, called pressure bars, and a wave is generated in one of the bars through axial impact – see Fig.3. with a striker bar. This incident wave propagates along the length of the incident bar until it encounters the specimen, whereupon the wave is reflected from and transmitted through the specimen. The wave reflected from the specimen is measured by means of a pair of strain gauges attached to the input bar, and the wave transmitted through the specimen is measured similarly by means of a pair of strain gauges attached to the output bar.

Experimental tests were carried out with bars made of Aluminium alloy. The behaviour of bars is linearly elastic. The penny shaped specimens ,15 mm in diameter and 7.5 mm in length were used.

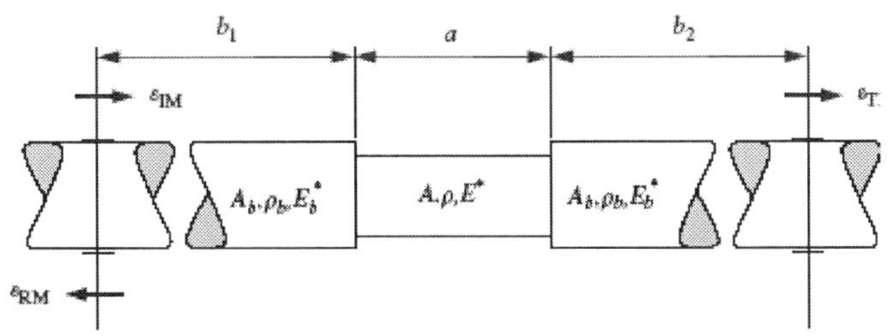

Fig. 3: Schematic of the central part of SHPB setup showing test specimen placed between two pressure bars.

Specimen with the different nanotubes volume were loaded by the stress pulses σ_I nearly of the same amplitude – see Fig.4.

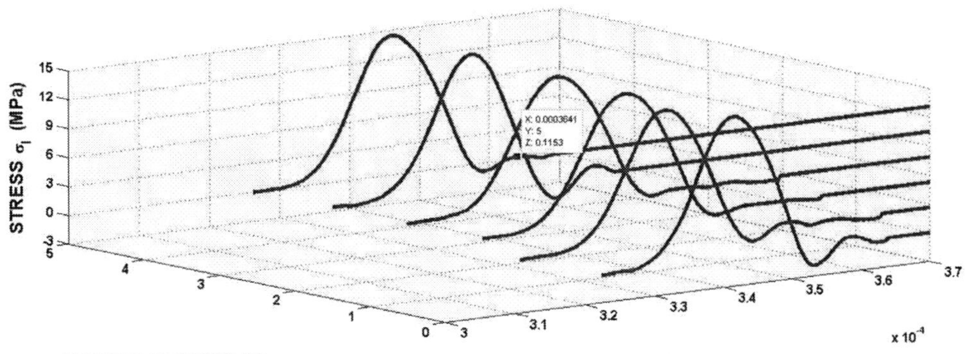

Fig.4: Loading stress pulses used in the given research.

The transmitted and reflected pulses were recorded.

Experimental Results.

The transmitted and reflected stress pulses are displayed in the Fig.5.

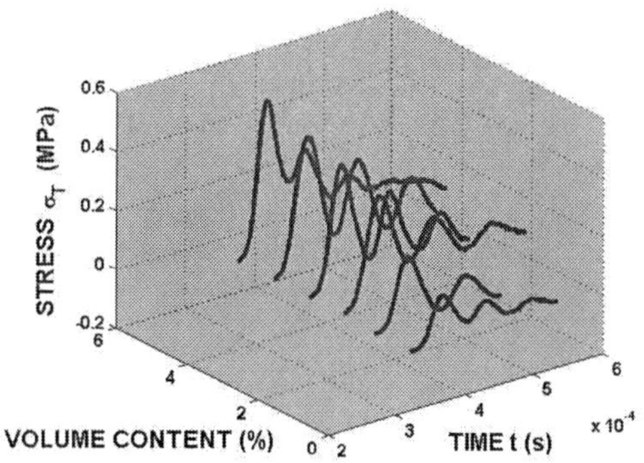

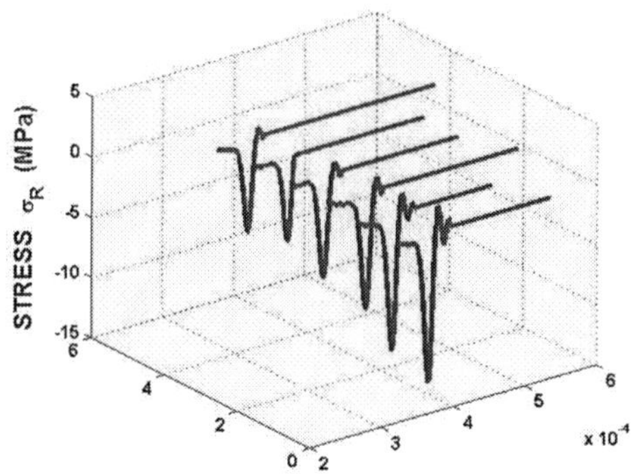

Fig.5: Experimentally recorded transmitted,σ_T, and reflected,σ_R,,stress pulses.

The classical evaluation of the given records follows in the determination stress,strain and strain rate in the specimen. This procedure is based on the assumption of the stress equilibrium in the specimen. This condition is very difficult to achieve namely in the specimen made from polymeric materials. In this case a new procedure was developed . By considering the propagation of waves in the bars and the specimen, and the continuity of velocity and force at the bar–specimen interfaces, one obtains the frequency–domain relation (Mousavi et al.,2005) :

$$\frac{\hat{\varepsilon}_{RM}}{\hat{\varepsilon}_{TM}} = \frac{1}{2}\left(\frac{Z}{Z_b} - \frac{Z_b}{Z}\right)\sinh\left(i\omega a\sqrt{\frac{\rho}{E^*}}\right)$$

where $\hat{\varepsilon}_{RM}$ and $\hat{\varepsilon}_{TM}$ are the Fourier transforms of the measured strains associated with the reflected and transmitted waves, respectively,ω is the angular frequency, and

$$Z = A\sqrt{\rho E^*} \qquad Z_b = A_b\sqrt{\rho_b E_b^*}$$

are the characteristic impedances of the specimen and the bars, respectively.

The use of this procedure is based on the frequency representation of the transmitted and reflected stress pulses. The representation of these pulses in the frequency domain is given by the Fourier transform. Owing to the fact that the bars behave linearly elastic we can use the identity :

$$\frac{\hat{\varepsilon}_{RM}}{\hat{\varepsilon}_{RM}} = \frac{\hat{\sigma}_R}{\hat{\sigma}_T}$$

Where $\hat{\sigma}_{R,T} = \int_{-\infty}^{+\infty} \sigma_{R,T}(t) e^{-i\omega t} dt$

The Fourier transform of the stress pulse is generally a complex function which can be expressed as :

$$\hat{\sigma} = |\hat{\sigma}| e^{i\varphi}$$

Where $|\hat{\sigma}|$ is the magnitude and φ is the phase.

Frequency dependence of the magnitude and phase is shown in the Figs.6 -8.

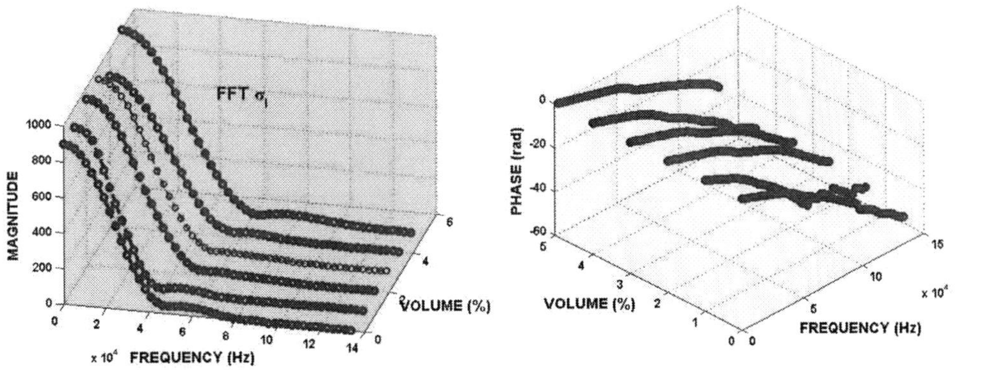

Fig.6: Spectral characteristics of the incident stress pulses.

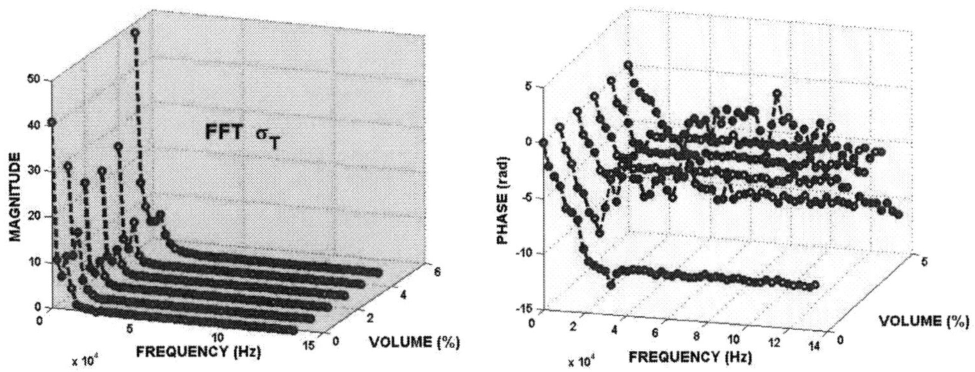

Fig.7: Spectral characteristics of the transmitted stress pulses.

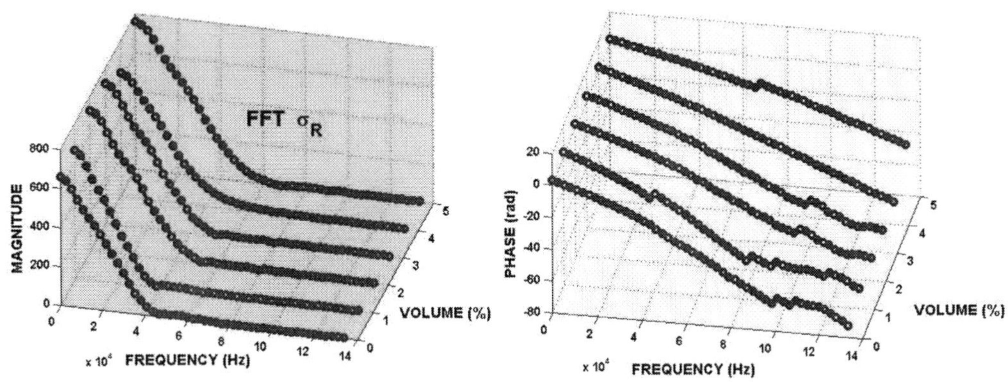

Fig.8: Spectral characteristics of the reflected stress pulses.

It is obvious that the most important are frequencies up tu about 10^4 Hz. Fig.8 shows the frequency dependence of the complex modulus for the specimens with the different volume content of the nano carbon tubes.

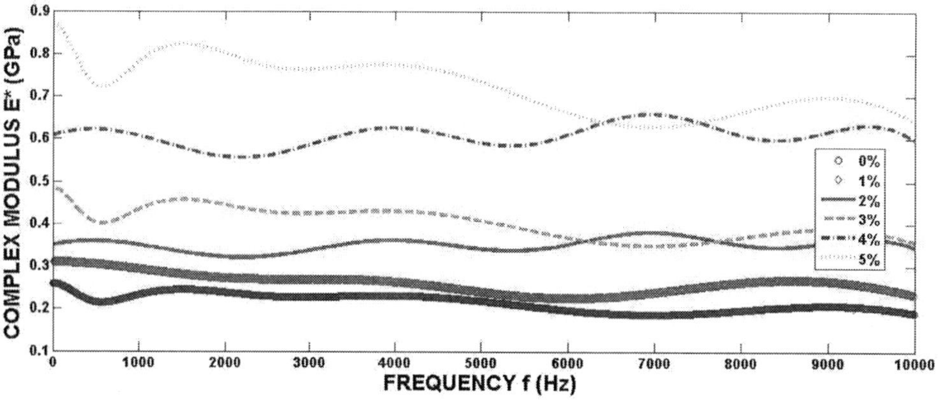

Fig.8. Complex modulus of tested specimens.

One can see that the complex modulus exhibits an oscillations. Its value increases with the value of volume content of nano carbon tubes for the most of frequencies. The detail analysis shows that the storage modulus E' is greater than the loss modulus E''.

Conclussion.

In the given paper a method method for identification of complex modulus using of the Hopkinson Split Pressure Bar technique was applied to the study of the dynamic response of composites filled bz nano carbon tubes. Spectral analysis of the waves reflected from and transmitted through the specimen was carried out to extract the complex modulus. Results show influence of the volume kontent of the carbon nanotubes on the value of the komplex modulus. At the same time the influence of the frequency is relatively less signifiant.

Acknowledgement.

The authors gratefully acknowledge the financial support from the Grant Agency of the Czech Republic through the project of the Grant Agency of the Czech Republic GA13-22945S. The authors should also to acknowledge the support of the Institute of Thermomechanics AS CR, v.v.i. of the Czech Academy of Sciences through the project No. RVO:61388998

References

[1] Nashif, A.D., Jones, D.I.G., Henderson, J.P., Vibration Damping. Wiley Interscience, New York (1985)

[2] Kulik, V.M., Semenov, B.N., Boiko, A.V., Seoudi, B.M., Chun, H.H., Lee, I., Measurement of dynamic properties of viscoelastic materials. Exp. Mech. 49, (2009) 417–425.

[3] Ahonsi B, Harrigan JJ, Aleyaasin M., On the propagation coefficient of longitudinalstress waves in viscoelastic bars. Int J Impact Eng .45(2012) 39-51.

[4] Gray III GT., Classic split Hopkinson pressure bar testing. ASM Handbook Vol 8, Mechanical testing and evaluation. ASM, 2000. p. 462–76.

[5] Mousavi,S.,Welch,K.,Valdek,U.,&Lundberg,B., Non-equilibrium split Hopkinson pressure bar procedure for non-parametric identification of complex modulus.Int.J.Impact Engng.31(2005)1133-1151.

On-line Monitoring of Resin Flow Using Capacitive Sensor Array During VaRTM Production Process

J. Cagáň [1,a], J. Rosa [1,b]

[1] *Výzkumný a zkušební letecký ústav, a. s. (Aerospace Research and Test Establishment), Beranových 130, 199 05 Praha – Letňany, Czech Republic*
[a] *cagan@vzlu.cz*, [b] *rosa@vzlu.cz*

Abstract: Advanced composite products with high fibre volume often suffer from dry spots formation during VaRTM production process. In order to facilitate mould design and process control, numerical simulations are usually used. Commonly used simulations of the resin transfer moulding process relies on a set of permeability coefficients. It is necessary to calibrate the permeability coefficients for each individual laminate stacking sequence and mould geometry separately. This paper describes the development of an on-line monitoring system for calibration of the permeability coefficients. Moreover, it can monitor a cure state of the composite. Monitoring system functionality is demonstrated on a transparent mould and is also compared with the simulation.

Keywords: vacuum-assisted resin transfer moulding; monitoring system; permeability coefficient; synchronous demodulation, fringing electric field; capacitive sensor.

1 Introduction

The vacuum-assisted resin transfer moulding (VaRTM) production process is widely used production method in the segment of advanced fiber-reinforced plastic composite materials. The VaRTM production method utilises a pressure gradient between inlets and outlets for resin matrix transportation in the mould. A fabric weaving style and its laminate stacking sequence strongly affect the fabric permeability. Increased attention is focused on the permeability because this parameter directly affects the occurrence of so-called dry spots (see Fig. 1). Those uneven or insufficiently saturated places may occur especially in the case of products featuring more complex forms. To avoid the dry spots, numerical simulations of filling process are widely used for process design or control. The permeability coefficients are one of the most critical entries into the simulation.

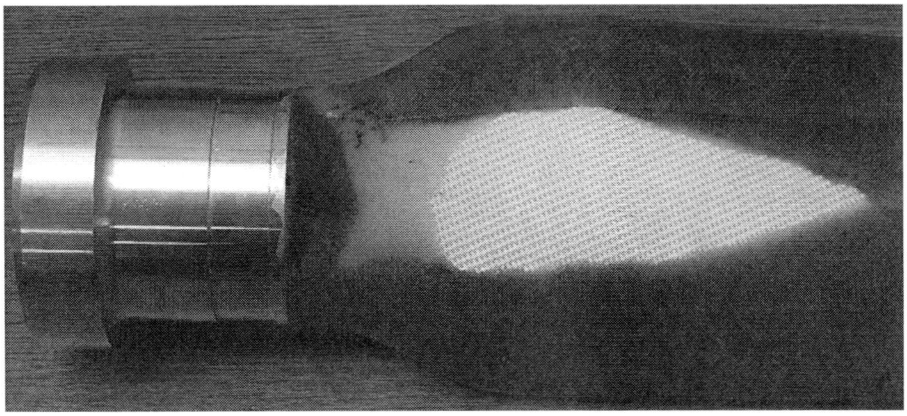

Fig. 1: An example of a typical dry spot;

Methods for determining the permeability coefficients are based on Darcy's law. The principal information is resin flow position during permeability measurement. Various methods have been used already. One of the first and well-known method is the SMARTweave™ (Sensor Mounted as Roving Threads) [1] that utilises a grid of conductors embedded into the fabric and employed for measuring of direct current conductivity at the grid node. The main disadvantage of the SMARTweave is invasiveness by wires incorporated into a final product or mounted on the mould wall still in touch with the product. Moreover, it is demanding on a plenty of wires. A similar method with lower demands on wires developed by Luthy et al. [2] utilises wires in a lineal configuration for direct current conductivity measurement. Flow position is thus continuous. Another method with similar invasiveness has been developed by Dominauskas et al. [3]. The method is based on a time-domain reflectometry, which measures the time delay between sending a high-speed electrical pulse and its echo reflected from impedance discontinuities. The main advantage is an ability to detect multiple flow fronts (dry spots). An interesting approach has chosen Lekakou et al. [4], who embedded an optical fibre core at the surface of the moulding. The flow front position and curing of an epoxy resin is measured by a loss of light. Another possibility for flow front measurement utilises pressure sensors wich was performed by Wei et al. [5]. The pressure beside the flow front is a critical parameter for permeability measurement, so their approach leads to reliable results.

Work described in this paper falls into an area of capacitive sensors. One of the interesting work has been done by Yenilmez et al. [6], who embedded a grid of opposite electrodes into the mould wall. Due to a precise implementation of the electrodes, their solution is non-invasive. The invasiveness is one of the most focused features indisputably; nevertheless, this solution has high demands on production and so costs, especially for moulds with complex geometry. This disadvantage has partially overcome by Kobayashi et al. [7], who proposed a patch-type flexible matrix capacitive sensor. Their work was probably inspired by Hegg et al. [8]. The work, most related with our, was proposed by Matsuzaki et al. [9] thanks to employing of fringing electric field capacitive sensors with interdigital electrodes. Theirs sensor configuration enables full-field flow monitoring.

Next chapter describes the development and demonstration of a monitoring system based on point fringing electric field (FEF) capacitive sensors with interdigital electrodes. The main demands on the monitoring system are the non-invasive, non-intrusive, low-cost and easy installation of sensors, and wireless. An intended usage of the monitoring system is for either permeability estimation purposes in arbitrary mould section or as a verification tool during process design. Demonstration of the monitoring system is performed on a transparent mould and is also compared with the simulation.

2 Methodology and implementation

2.1 Online monitoring system

The monitoring system is based on capacitive sensors with a fringing electric field. Capacitance changes of sensor depend on changes of dielectric property of material penetrated by sensor electric field. The sensitivity of FEF sensor decreases exponentially with the distance between the sensor surface and dielectric changes [10]. The sensor consists of several comb-shaped coplanar electrodes. An example of the electrodes arrangement is shown in Fig. 2a.

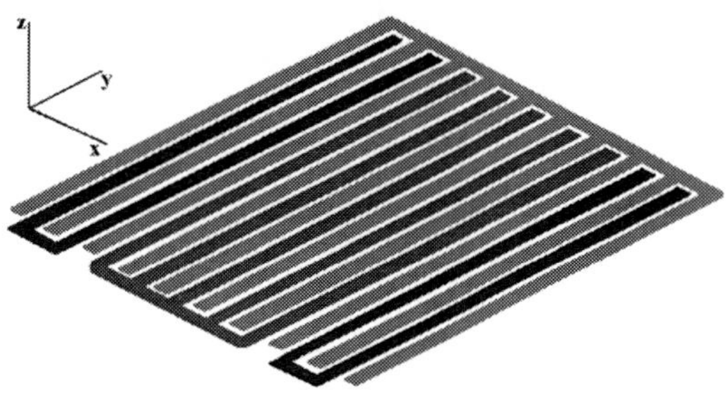

a) An arrangement of FEF sensor coplanar electrodes;

b) An example of used FEF sensor;

Fig. 2: A demonstration of sensor with coplanar electrodes;

The driving electrode (red) is connected to the harmonic voltage supply. The sensing electrode (blue) is connected to the signal processing circuit. The sensor can also be equipped with a guard electrode (in the figure marked in black) which shields, to a certain extent, the surrounding field that might cause interference. The guard electrodes also ensure that the measurement area is not affected by a deformed electric field at the sensor edges. Developed monitoring system has the side guard electrodes excluded for simplicity. Nevertheless, used sensors have included backplane guard shield. An example of used FEF sensor and their model is depicted in Fig. 2b. It is evident that FEF sensors are intended solely for non-metallic moulds. To satisfy main requirements on the monitoring system it is necessary to instal the sensors into the mould wall as close as possible to the mould surface, but still below the surface as is depicted in Fig. 3. An electrical field intensity lines in Fig. 3 (marked by 6) has a pattern typical for interdigital electrodes. As it is indicated by attenuated field line thickness, sensed area is penetrated by the field lines with lower intensity, and so sensor features lower sensitivity. Moreover, due to low cost and easy sensor installation, the sensitivity may differ rapidly among sensors. The development of the monitoring system was commenced by author's previous work [11], which via pioneering experiments helped to find main requirements on the final system. Pioneering experiments confirmed FEF sensor sensitivity, as it became a crucial factor for next analogue circuit design.

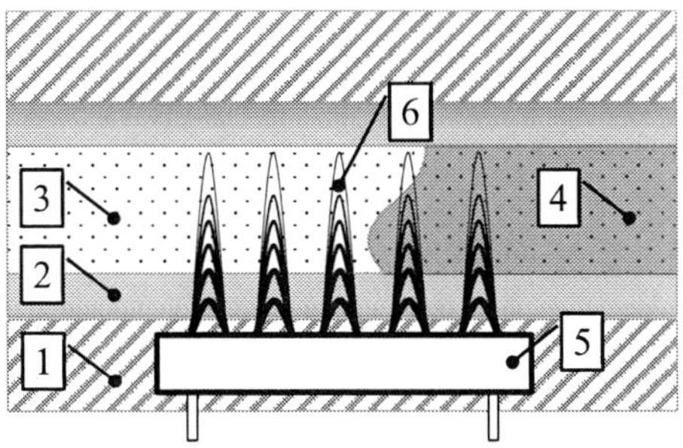

Fig. 3: Side view of FEF sensor installed in the mould wall.
(1-mould, 2-gelcoat, 3-fabric, 4-fabric&resin, 5-sensor, 6-FEF);

Developed monitoring system, depicted in block diagram in Fig. 4, is intended to instal onto the mould for better manipulation during the production process. Wireless communication is provided by radio frequency module XBee ® by Digi International. The pair of modules works in a transparent operation regime when act as a serial line replacement. Both communication directions are used. The wireless module at the mould part is connected with a microcontroller hardware serial interface. The microcontroller controls a multiplexer ADG1406 to channel selection according to preprogrammed measuring cycle in a range up to 48 channels. Sensing electrodes are connected directly to multiplexer inputs. The multiplexer output leads measured signal into analogue circuits which are the heart of the monitoring system because it converts the measured electrical capacitance of the sensor to an electrical voltage. The analogue circuits will be described in the next paragraph in detail. The measured analogue signal is digitised by 24-bit analogue-to-digital converter AD7710, which is controlled by the microcontroller via a serial peripheral interface. Because the mould part of the monitoring system is powered by a battery, the analog-to-digital converter is also used for monitoring battery condition. The microcontroller periodically reads the set of channels and send data frame into a personal computer.

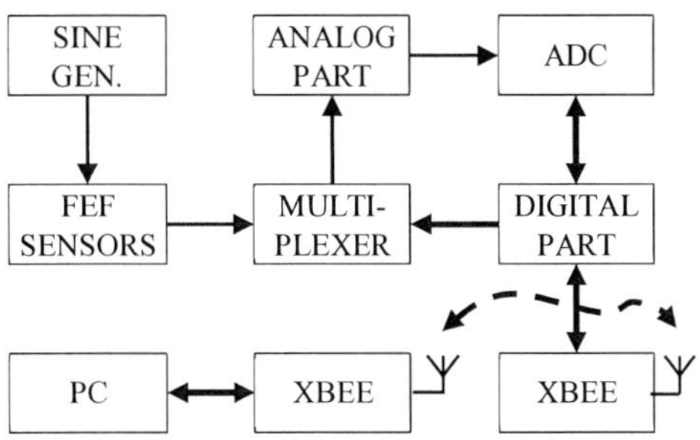

Fig. 4: Schematic diagram of main components of the monitoring system;

A user application for data processing was created by LabVIEW development system. The user application is among others able to visualise data by the help of a 3D graphic CAD model loaded in STL format.

The analogue circuits design is closely related with satisfying the main demands on the monitoring system. Above mentioned problems with sensor sensitivity leads to analogue circuits, which are able to handle with low signal-to-noise ratio. Synchronous demodulation is a well-known method for recovering these signals. Synchronous demodulation is based on multiplying the measured signal by reference signal and subsequent filtering off every harmonic component except DC component of the signal. As is depicted in the block diagram of the analogue part (Fig. 5), the synchronous demodulation is used for conditioning of both measuring and reference (harmonic voltage supply on the driving electrode) signals. These signals are consequently processed proportionally. The ratio A/B ensures a suppression of reference signal drift. Both synchronous detector and ratio implementation are part of AD698, which is here used untraditionally because it is signal conditioning subsystem for a linear variable differential transformer. One of the most important parts is the capacitance converter, which is based on AD823 wired in current-to-voltage configuration. Harmonic voltage supply with frequency 15 kHz is generated by low distortion sine generator, which is also part of the AD698.

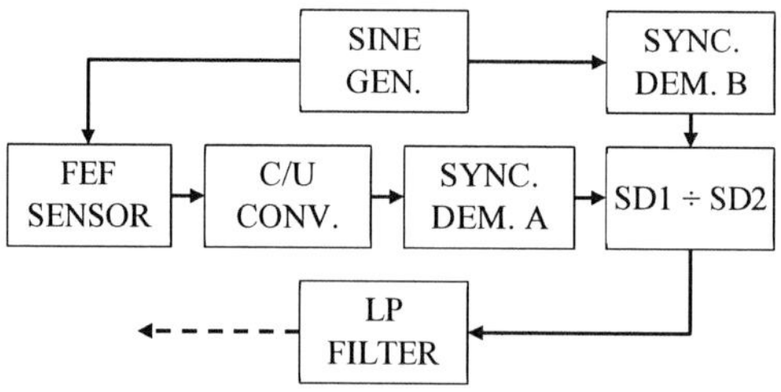

Fig. 5: Schematic diagram of analogue circuits;

2.2 Permeability estimation

The resin flows through a porous medium such as fibrous reinforcement according to Darcy's Law, which states that the flow rate of resin per unit area is proportional to the pressure gradient and inversely proportional to the viscosity of the resin. Darcy's Law is given as follows

$$V = \frac{K}{\mu} \nabla P \,, \tag{1}$$

where V is Darcy's velocity, K is permeability tensor, ∇P is a pressure gradient, and μ is a viscosity of the resin. In order to in-plane permeability measurement or estimate respectively, the thru-thickness permeability is neglected. In case of thin shell structures with isotropic fabric is possible to express in-plane permeability as follows:

$$K = \mu L^2 \frac{\varphi}{2T\Delta P} \,, \tag{2}$$

where μ is the fluid viscosity, φ is the preform porosity, and L is the distance between pressure difference ΔP. For accurate permeability measurement is necessary to measure the pressures in the place of FEF sensors. Nevertheless, hereafter system demonstration uses pressure gradients estimated from the numerical simulation.

3 System demonstration

3.1 Experimental setup

The transparent mould in the form of a fan blade with dimensions 75 x 33 cm was provided with 20 sensors monitoring system. Square sensors of 1 cm^2 area were installed into the mould wall approximately 0.5 to 1 mm below the mould surface. Whole experimental setup is depicted in Figure 6. Laminate stacking sequence of thin shell product was composed of 3 layers of glass multimat (fabric 0/90, mat, 900 g/m2) with epoxy resin Havel LH288. Moreover, the laminate was provided by flow mesh Compoflex RF 150 for increasing permeability. Two inlets and one outlet gates were equipped by runners where of 1.5 mm diameter. The resin at the inlet was freely sucked into the mould by a pressure gradient between atmospheric pressure and pressure of vacuum pump, which kept absolute pressure 15 kPa. Based on fibre volume fraction measurement, the porosity of laminate was 65 %. Resin viscosity was measured during the filling process for both, more accurate permeability estimation and numerical simulation. The filling process was recorded by a camera from the top view so that monitoring system outputs could be easily compared with reality.

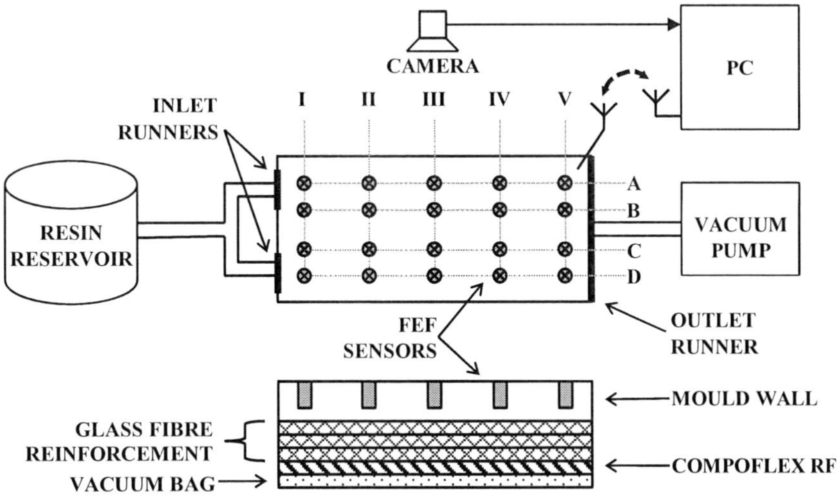

Fig. 6: Schematic setup of the experimental demonstration;

Numerical simulation was performed with the help of PAM-RTM software. Based on the above-mentioned experimental setup, the filling process was modelled as two-dimensional by neglecting the resin flow in the thickness direction. Among others, the pressure gradient was obtained from the numerical simulation for permeability estimation purpose.

3.2 Result and discussion

Based on the experiment mentioned above, permeabilities in five sections of the mould was estimated with the help of equation 2. The permeabilities and relevant parameters are summarised in Table 1.

Tab. 1: Estimated permeabilities and relevant parameters;

section	time (s)	length (m)	pressure (Pa)	viscosity (Pa·s)	permeability (m^2)
I-II	101	0.17	50549	0.24	4.4E-10
II-III	167	0.17	32463	0.21	3.6E-10
III-IV	230	0.17	24777	0.20	3.3E-10
IV-V	321	0.17	18814	0.19	2.9E-10
entire mould	959	0.75	85000	0.21	4.7E-10

As is mentioned in Table 1, the permeability estimation was performed in the sections I-II, II-III, III-IV, IV-V, and within the whole mould. The filing times of the sections were evaluated from relative changes in the capacitance (see Fig. 7) measured and stored by the monitoring system. Relevant pressures gradients in the sections were estimated from the numerical simulation, whereas pressure gradient in the entire mould was given by the vacuum pump and atmospheric pressure.

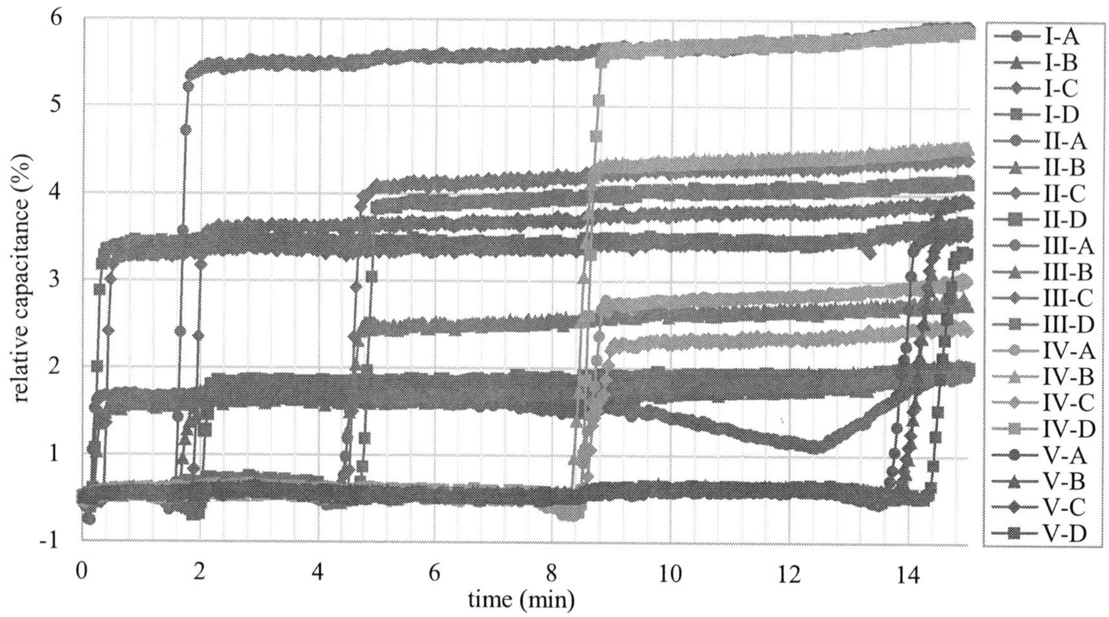

Fig. 7: The sensor relative capacitance changes during filling process;

The permeability was also estimated by best fitting the simulated and measured filling times in the relevant sections. Thus founded permeability was 4.75E-10 m^2, and it was used for numerical simulation in the next comparison. In Table 2 it is possible to see the comparison between real filling process from a camera, the graphical output from the monitoring system, and numerical simulation. It is clearly seen that simulation matches real filling process very well. Only a little difference can be found at the beginning of the filling process. The permeability founded by best time fitting is in good accordance with the permeability estimated from the entire mould filling time. Other permeability estimates are affected probably by pressure gradients, whose value was roughly estimated from the numerical simulation.

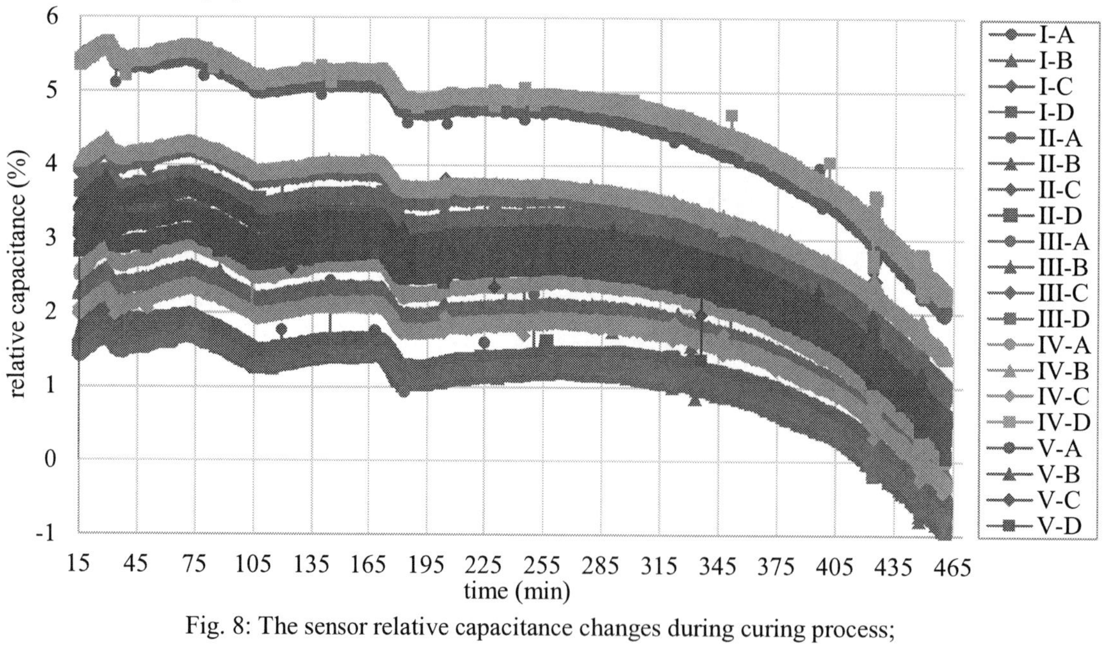

Fig. 8: The sensor relative capacitance changes during curing process;

The curing process was also monitored. Epoxy resin contains considerable polar groups that lead to the intense capacitive response [12]. The concentration of the polar groups changes by the epoxy resin crosslinking during its curing. The curing is possible to see in Figure 8. Because used capacitive sensors are sensitive not only to resin but also to fabric and its stacking sequence, it is necessary to verify more rigorously the relation between capacitive response and curing process before utilisation the monitoring system for cure monitoring at this sensor setup.

Tab. 2: Comparison of flow front progress;

time [s]	camera	monitoring system	simulation
0		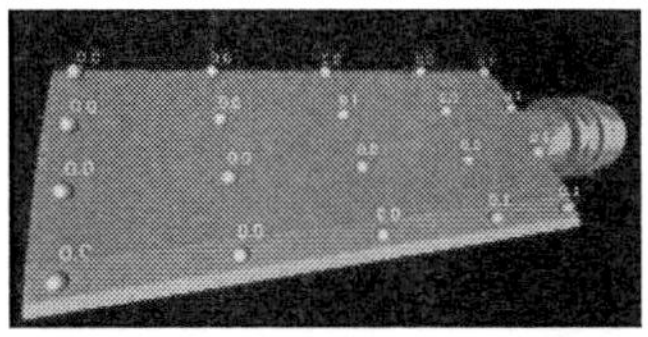	
28		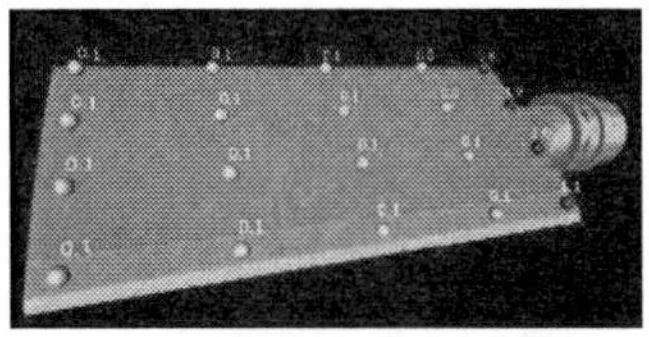	
129		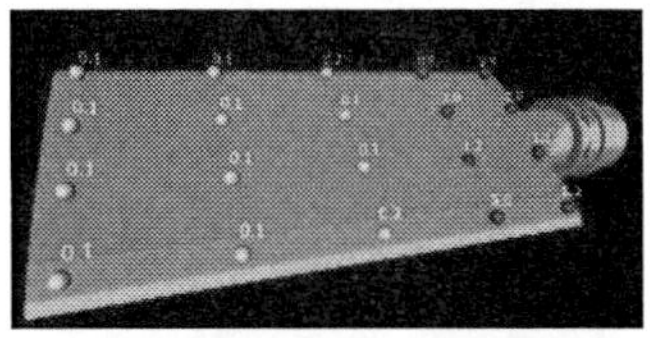	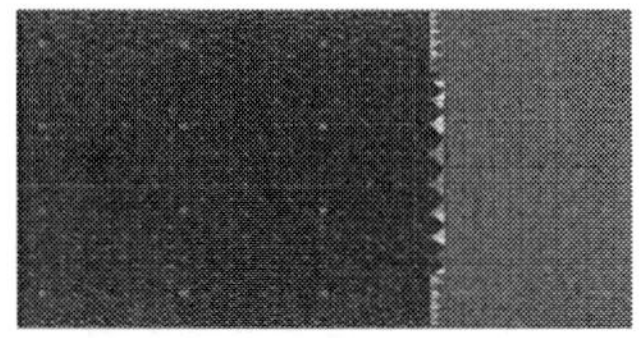
296		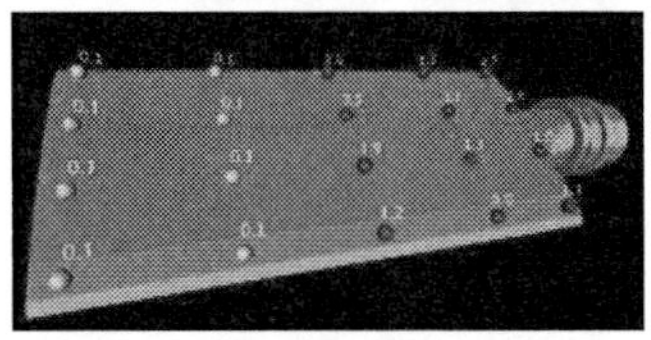	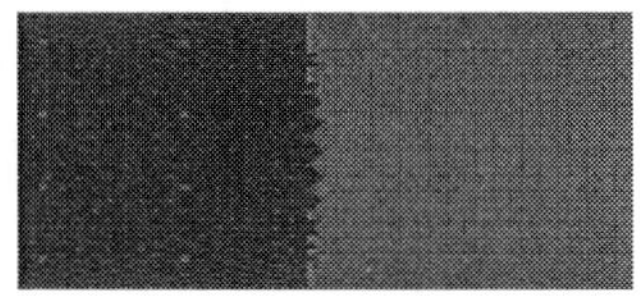
526		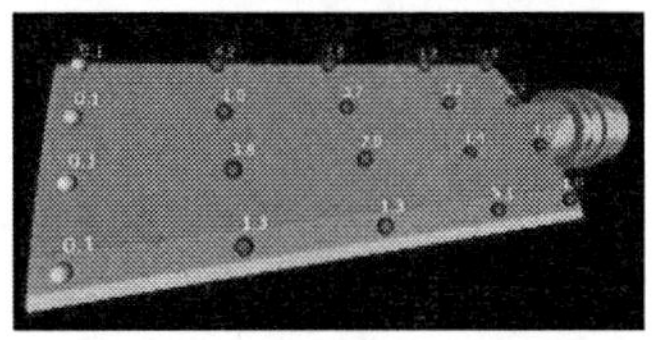	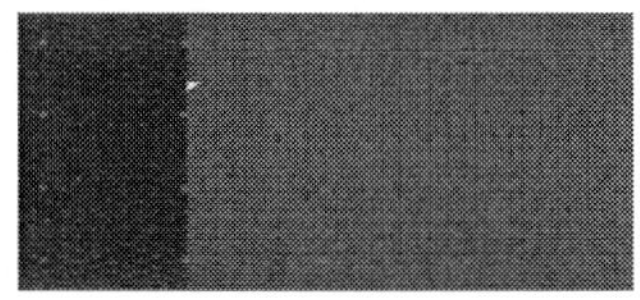
847		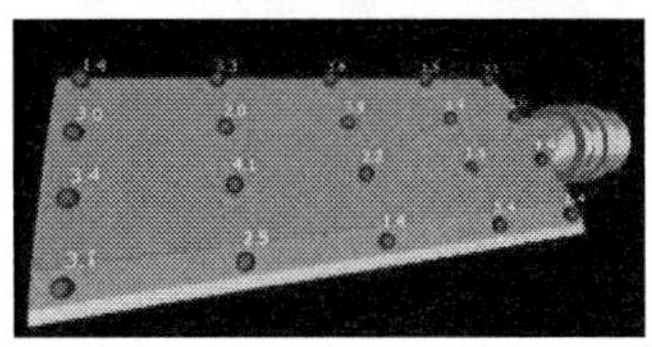	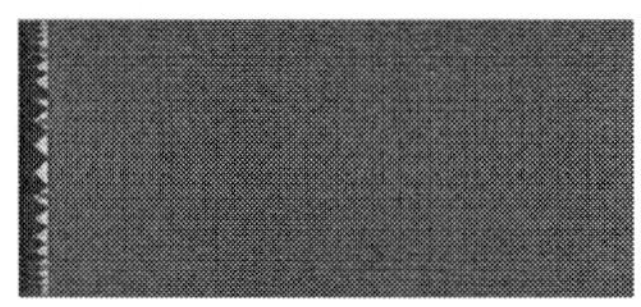

4 Conclusion

The on-line monitoring system was developed and experimentally demonstrated on the permittivity estimation. Thanks to the signal conditioning with synchronous demodulation has accessed the utilisation of capacitive sensors with the fringing electric field. The sensor configuration allows non-invasive and non-intrusive measuring the presence of the resin. Due to non-invasiveness and non-intrusion of the monitoring system, the permeability can be estimated directly in the full-scale production mould with no affect to moulding quality. Nevertheless, future work will be focused on enriching of monitoring system by pressure measurement for permittivity estimation improvement. As was shown, the graphical output from the on-line monitoring system can be easily used as the tool for verification of the process design, whereas possibility to store capacitance data can be used for permeability estimation during post-processing.

Acknowledgement

This work was performed with the financial institutional support from government budget through the Ministry of Industry and Trade of the Czech Republic.

References

[1] G. Lebrun, R. Gauvin, K. N. Kendall, Experimental investigation of resin temperature and pressure during filling and curring in a flat steel RTM mould, Composites Part A: Applied Science and Manufacturing 27 (1996) 347–356.

[2] T. Luthy, P. Ermanni, Linear direct current sensing system for flow monitoring in Liquid Composite Moulding, Composites Part A: Applied Science and Manufacturing 33 (2002) 385–397.

[3] A. Dominauskas, D. Heider, J. W. Gillespie Jr, Electric time-domain reflectometry sensor for online flow sensing in liquid composite molding processing, Composites Part A: Applied Science and Manufacturing 34 (2003) 67–74.

[4] C. Lekakou et al., Optical fibre sensor for monitoring flow and resin curing in composites manufacturing, Composites Part A: Applied Science and Manufacturing 37 (2006) 934–938.

[5] B.-J. Wei et al., Online estimation and monitoring of local permeability in resin transfer molding, Polym. Compos. 37 (2016) 1249–1258.

[6] B. Yenilmez, E. Murat Sozer, A grid of dielectric sensors to monitor mold filling and resin cure in resin transfer molding, Composites Part A: Applied Science and Manufacturing 40 (2009) 476–489.

[7] S. Kobayashi, R. Matsuzaki, and A. Todoroki, Multipoint cure monitoring of CFRP laminates using a flexible matrix sensor, Composites Science and Technology 69(2009) 378–384.

[8] M. C. Hegg et al., Remote Monitoring of Resin Transfer Molding Processes by Distributed Dielectric Sensors, Journal of Composite Materials 39 (2005) 1519–1539.

[9] R. Matsuzaki et al., Full-field monitoring of resin flow using an area-sensor array in a VaRTM process, Composites Part A: Applied Science and Manufacturing 42 (2011) 550–559.

[10] X. B. Li et al., Design of multichannel fringing electric field sensors for imaging. Part I. General design principles, in Conference Record of the 2004 IEEE International Symposium on Electrical Insulation (2004), 406–409.

[11] J. Cagan, F. Martaus, Monitoring of RTM production process, Czech Aerospace Proceedings 2 (2012).

[12] P. Hedvig, Dielectric spectroscopy of polymers, Budapest, Akadémiai Kiadó, 1977.

Influence of Cutting Conditions on Surface Integrity after Progressive Grinding of Blade Roots from Inconel 713LC

J. Čapek [1], K. Kolařík [1,a], Z. Pitrmuc [2], N. Ganev [1], L. Beránek [2], K. Trojan [1], J. Kyncl [2], J. Šimota [2]

[1] *Department of Solid State Engineering, Faculty of Nuclear Sciences and Physical Engineering, Czech Technical University in Prague, Trojanova 13, 120 00 Prague 2, Czech Republic*

[2] *Department of Machining, Process Planning and Metrology, Faculty of Mechanical Engineering, Czech Technical University in Prague, Technická 4, 166 07 Prague 6, Czech Republic*

[a] *kamil.kolarik@email.cz*

Abstract: The goal of this article is to describe changes of properties after progressive grinding of blade roots from Inconel 713LC. Especially, surface residual stresses were studied as a function of different grinding conditions and microstructure of bulk material. Residual stresses of surface layers were determined by using X-ray diffraction techniques.

Keywords: grinding; blade root; Inconel 713LC; residual stresses; cracks.

1 Introduction

Nickel-based superalloys are often used in automotive and aerospace industry as turbine blades for its excellent properties. Nevertheless, very stringent requirements on surface integrity specifically, residual stresses (RS) and microstructure are required, especially for blade roots because of its anchoring [1] and for service life under dynamic loads [2].

Therefore, the suitable selection of grinding conditions and adequate cooling are very important to prevent grinding burns. Otherwise, "thermal" RS are created due to stronger influence of accumulated heat more than the impact of plastic deformation. These unfavorable RS can result in crack initialization and its further propagation. Service life is not limited only by mechanical and thermal influences but also by the microstructure of the bulk material. The particles of ceramic inclusions, which can form in the material, are very hard and brittle; therefore, potential areas of the initialization of cracks are created by casting or ensuing solidification [3].

Using X-ray diffraction (XRD) techniques, the area with accumulation of unfavorable tensile RS of γ and γ' phases on a macroscale can be localized. Since the precipitates (γ' phase) are too hard to be plastically deformed, the local strain is very heterogeneous on a microscale – in the matrix channels, i.e. γ phase [4].

2 Experiment

The tested samples (blades) were made from nickel-based superalloy Inconel 713LC. The nominal chemical composition (in wt.%) of the as-cast samples was as follows: 12.0 Cr, 5.7 Al, 4.6 Mo, 2.0 Nb, 0.7 Ti, 0.2 Fe, 0.1 Zr, 0.08 Co, 0.05 C, 0.013 B and balance of Ni.

Surfaces of both sides, namely A (ground after wheel dressing) and B (without wheel dressing), of turbine blade roots with dimension 52 × 15 mm^2 were face ground. Further research will focus on the selection of grinding conditions for profile grinding. The grinding conditions were as followed: cutting speed v_c = 20 and 35 m/s, feed f = 450 and 600 mm/min, number of cuts N = 2 and 3, and final depth of cut (finishing) a_p = 0.05 and 0.07 mm. Together, 11 combinations of grinding conditions were used. Grinding wheel marked as *560038 STRATO ULTRA SU33A 542 GG 11 VB1* with dimensions 500×20×203.2 mm^3 was used. Grinding wheel is based on aluminum oxide grains a high porosity bond, suitable for creep feed grinding and grinding of blade roots.

Cooling fluid *Quakercool 710 LF* was used in concentration of 5 %. Cooling system enabled to bring enough amount fluid through tangential nozzle under the pressure of 10 bar.

Using Design of experiment, 3 areas were analyzed by XRD on each side of root, namely *I, II, III* and *IV, V, VI* on the side A and B, respectively, see Fig. 1. Longitudinal direction was parallel to the grinding direction and perpendicular to the axis of blades, see Fig. 1.

PROTO iXRD COMBO diffractometer in ω-goniometer set-up was used to measure lattice deformations using manganese radiation. The diffraction lines *{311}* of γ and γ' phases were measured in order to obtain macroscopic residual stresses [4]. Rachinger's method and Absolute peak method were used for determination of the diffraction angles $2\theta^{311}$ of the diffraction lines $K\alpha_1$. X-ray elastic constants $\frac{1}{2}s_2 = 6.57$ TPa^{-1}, $s_1 = -1.56\ TPa^{-1}$ for the residual stress determination using $\sin^2\psi$ method were calculated from components of stiffness tensor ($C_{11} = 234.6$ GPa; $C_{12} = 145.4$ GPa; $C_{44} = 126.2$ GPa) using Hill elastic model [5]. The cylinder collimator with Ø 2 mm was inserted into the primary beam path. The average value of the penetration depth of manganese radiation into the nickel-based material was approx. 4–5 μm [6].

In accordance with standard ČSN EN ISO 3452-1, the capillary tests were performed. Surface macro-etching was employed to reveal grain size and orientation obtained during casting. Also differentiation of intercrystalline and transcrystalline cracks was enabled. Metallographic samples were prepared in areas of capillary indications. Metallographic cross-sections were observed both in polished and etched (Marble's) condition.

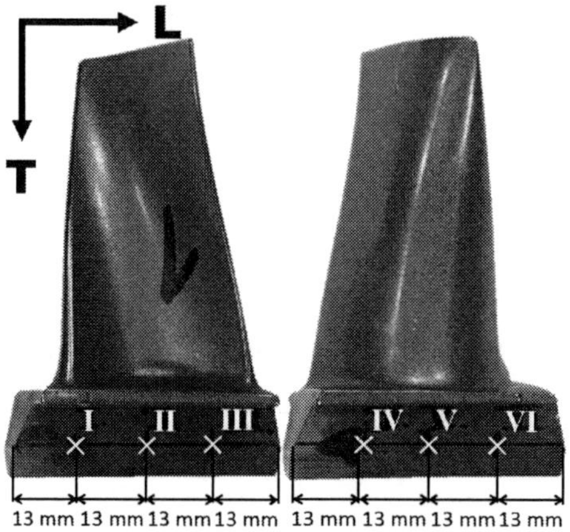

Fig. 1: Experimental sample (blade) with marked analyzed areas *I-III* and *IV-VI* on the side of root A and B, respectively.

3 Results and discussions

3.1 Microstructure and defectoscopy

Measured blade roots had homogenous values of RS on the ground surfaces areas *I-III* or *IV-VI* except for the roots which are depicted in Figs. 2-4. For the purpose of finding the reason of inequalities of these RS, the capillary tests, the metallographic cross sections and electro-etching for microstructure visibility were performed, see Figs. 5-8.

On the surface of several blade roots, the grinding burns were found, see Fig. 5. The grinding burns resulted in decrease of compressive RS values which are depicted in Fig. 2. The grinding burns were caused by insufficient cooling during grinding and probably by grain size and its orientation. The grinding burns were primarily detected on the edges of roots, as a result of a smaller amount of material for heat distribution[1], see Fig. 5. The anisotropy in polycrystalline samples is generally reduced more in comparison with single crystal, because of the distribution of crystal orientation. It means that in the case of coarse-grained material, there may be a grain which is oriented in a direction with smaller thermal conductivity parallel to the heat flow [7]. These grains may cause grinding burns.

[1] The thermal conductivity of Inconel is approx. 4× smaller in comparison with steel.

The sharp decrease of compressive RS on the surfaces for several blades was explained by metallographic cross sections obtained by capillary test through which, the cracks were found, see Figs. 3, 6 and 7. Regardless of the compressive RS in the surface layer 4-5 microns thick of ground roots, it is possible to expect that there are tensile RS in the deeper subsurface layers [8]. These RS have a negative effect on the functional characteristics of the blade roots and probably could cause cracks.

The rest of inhomogeneity were probably caused by coarse-grained microstructure of bulk material, see Figs. 4 and 8, which enables to predict that slip directions of several grains are parallel to the primary grinding force. Therefore, even smaller force could cause the slip and thereby smaller compressive RS are determined.

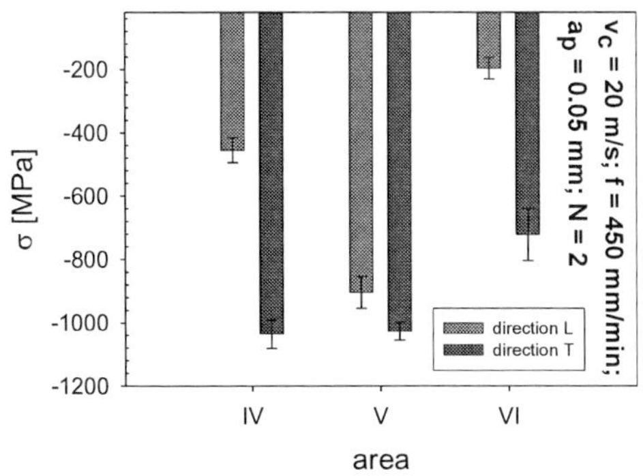

Fig. 2: Distribution of RS on the surface of the side B affected by grinding burns.

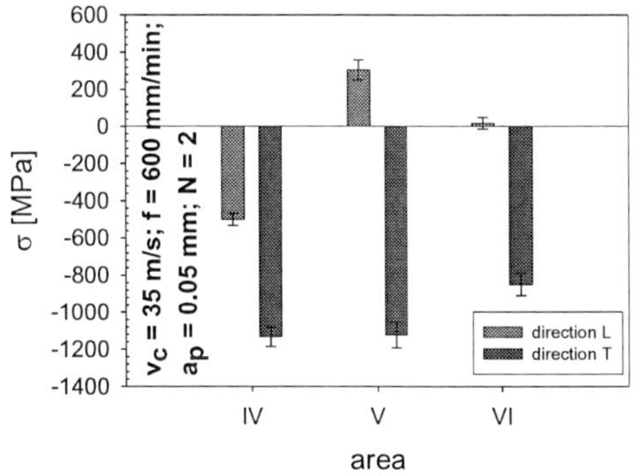

Fig. 3: Distribution of RS on the surface of the side B affected by presence of crack.

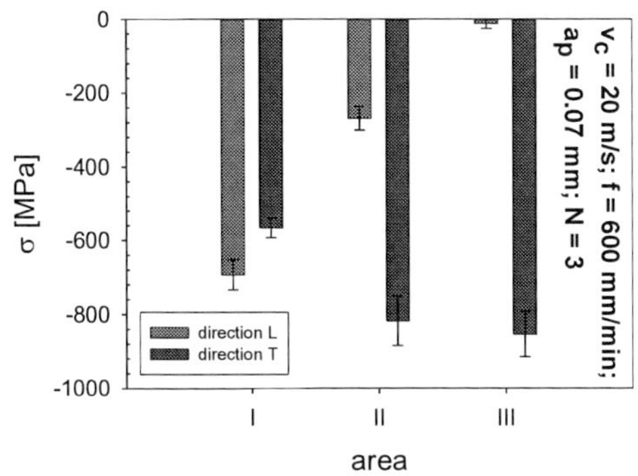

Fig. 4: Distribution of RS on the surface of the side A affected by coarse-grained bulk material.

Fig. 5: Root surface of selected blade with the grinding burn.

Fig. 6: Blade root with line indications of defects determined by capillary testing.

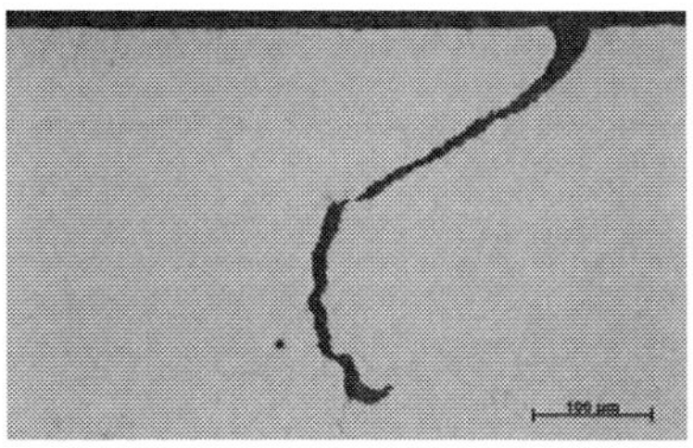

Fig. 7: Metallographic cross section of the line indication (crack) from Fig. 6.

Fig. 8: Fine-grained and coarse-grained microstructure of bulk material of selected samples.

3.2 Influence of grinding conditions

For the purpose to find the RS dependence on the grinding conditions, RS were averaged from three areas on each side of blade roots separately, see Tab. 1. Surface RS of various grinding conditions obtained by XRD are depicted in Figs. 9-11.

The values of RS depending on the final depth of cut a_p are in Fig. 9. A clear trend for both the analyzed sides and directions is observed, specifically, compressive RS increased in most cases with the depth of the cut increased. Greater plastic deformation (hardening) of material increased of mentioned RS in the surface layer 4-5 microns in thickness.

With increasing cutting speed v_c, the values of compressive RS had an increasing trend too, see Fig. 10. Greater hardening of ground surface was probably the reason of this trend.

Finally, with increasing number of cut[2], the compressive RS were higher. Various values of hardening were caused by different depths of cut after roughing.

From Tab. 1, it is possible to state that RS were for all the samples higher on the side B in comparison with RS on the side A in the transversal direction *T*, i.e. direction perpendicular to the grinding direction. In this case, the different values of RS were caused by wheel dressing.

According to the general assumptions of grinding theory [8], higher depth of final cut, faster cutting speed lead to greater increase of the material temperature which leads to tensile RS. The results of this article dealt with thin surface layer of compressive RS are inconsistent with the mentioned grinding theory. Nevertheless, with increasing of the subsurface depth, the tensile RS are predicted to the detriment of compressive RS.

Tab. 1: The average values of RS depending on grinding conditions.

Grinding conditions				side A				side B			
v_c [m/s]	f [mm/min]	N [-]	a_p [mm]	$<\sigma_L>$ [MPa]	$\Delta\sigma_L$ [MPa]	$<\sigma_T>$ [MPa]	$\Delta\sigma_T$ [MPa]	$<\sigma_L>$ [MPa]	$\Delta\sigma_L$ [MPa]	$<\sigma_T>$ [MPa]	$\Delta\sigma_T$ [MPa]
35	600	2	0.05	**-465**	17	**-648**	27	**-289**	18	**-930**	23
35	450	2	0.07	**-502**	14	**-730**	19	**-393**	21	**-858**	22
35	600	3	0.05	**-820**	14	**-776**	15	**-520**	19	**-947**	19
35	600	2	0.07	**-581**	20	**-804**	18	**-479**	18	**-910**	18
20	450	3	0.07	**-402**	15	**-780**	23	**-480**	17	**-1075**	21
20	600	3	0.07	**-306**	14	**-752**	24	**-413**	17	**-920**	24
20	600	2	0.07	**-384**	17	**-641**	24	**-84**	23	**-600**	28
20	600	3	0.05	**-304**	17	**-729**	23	**-368**	18	**-982**	25
35	450	3	0.05	**-588**	18	**-835**	24	**-633**	14	**-991**	23
20	450	2	0.05	**-258**	22	**-712**	24	**-393**	28	**-871**	24
35	450	3	0.07	**-554**	15	**-887**	15	**-637**	20	**-1125**	18

[2] Samples were ground by two cuts (roughing and finishing) or using three cuts (roughing, semi-finishing and finishing).

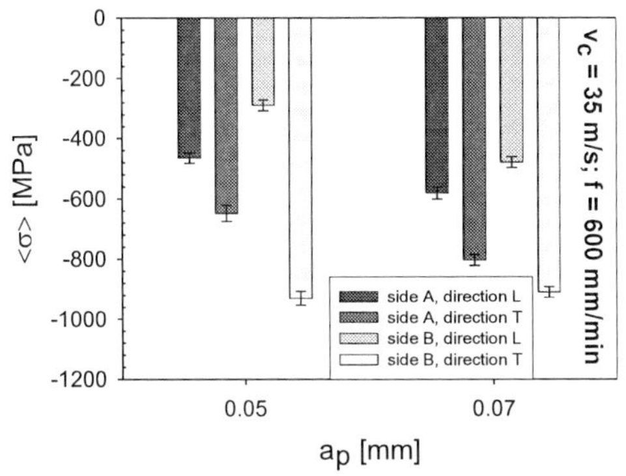

Fig. 9: Average RS on the surface of the both sides and directions depending on the depth of the final cut a_p.

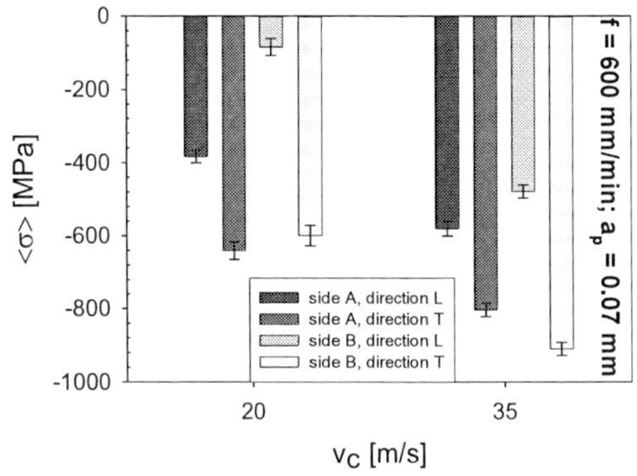

Fig. 10: Average RS on the surface of the both sides and directions depending on the cutting speed v_c.

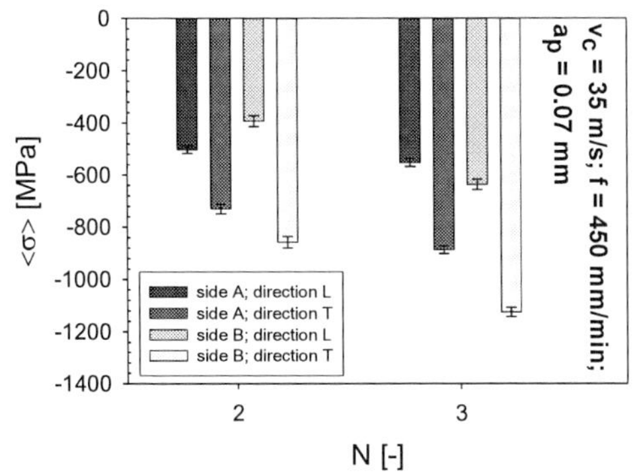

Fig. 11: Average RS on the surface of the both sides and directions depending on the number of cut N.

4 Conclusion

On all ground surfaces of the investigated blade roots, except surfaces with cracks, compressive RS were observed, see Tab. 1. The greater depth of cut, preceding the final cut, led to higher value of macroscopic compressive RS. It is possible to pronounce that greater final depth of cut and higher cutting speeds led to higher compressive RS. Examples of determined results of RS were presented in Figs. 9-11. The drops in compressive RS on the surfaces of several blades were explained by cracks, grinding burns and coarse-grained microstructure of bulk material, see Figs. 6-8.

Acknowledgement

This work was supported by the governmental funding of Technological Agency of Czech Republic – project number TA04010600.

References

[1] W. Österle, P. X. Li, Mechanical and thermal response of a nickel-base superalloy upon grinding with high removal rates, Materials Science and Engineering: A 238.2 (1997) 357-366.

[2] K. Kolařík et al., Non-Destructive Inspection of Surface Integrity in Milled Turbine Blades of Inconel 738LC, Applied Mechanics and Materials 486 (2014) 9-15.

[3] J. Čapek et al., Grinding of Inconel 713 Superalloy for Gas Turbines, Manufacturing Technology 16.1 (2016) 38-45.

[4] J. Li, R. P. Wahi, Investigation of γ/γ' lattice mismatch in the polycrystalline nickel-base superalloy IN738LC, Acta Metallurgica et Materialia 43 (1995) 507-517.

[5] R. Hill, The elastic behaviour of a crystalline aggregate, in proc.: Proceedings of the Physical Society: A 65.5 (1952) 349.

[6] J. Čapek, Z. Pala, Auxiliary programs for diffraction experiments, Materials Structure 22 (2015) 78-81.

[7] Z. Pala, N. Ganev, J. Drahokoupil, A. Sveshnikov, Surface Layers' Real Structure of Metals Exposed to Inhomogeneous Thermal Fields and Plastic Deformation, Solid State Phenomena 163 (2010) 59-63.

[8] F. Neckář, I. Kvasnička, Vybrané statě z úběru materiálu, Praha, ČVUT, 2011.

Evaluation of Depth and Profile of Surface Longitudinal Cracks and their Growth in DN300 High Pressure Pipeline

I. Černý [1,a], J. Sís [1]

[1] *SVÚM a.s., Tovární 2053, 25088 Čelákovice, Czech Republic*
[a] *Ivo.Cerny@seznam.cz*

Abstract: In an operated DN300 high pressure pipeline, crack-like defects were detected by internal inspection methods. The pipeline section could not be therefore further operated without detailed and careful evaluation of its safety and reliability. Several meters long section of the pipeline was taken out from ground for the purpose of experimental laboratory evaluation of the damage extent. Magnetic defectoscopy confirmed an existence of numerous distinct indications of longitudinal defects. Methodology of measurement of depth of these cracks was proposed using direct current potential drop (DCPD) method. An experimental calibration dependence of potentials on crack depth and shape was evaluated in laboratory. Precision limits of the DCPD method being used in this specific case are discussed in the paper, particularly considering inclined shape of the cracks. Results are related to actual crack depth values and their inclination evaluated by metallographic method.

Keywords: high pressure pipeline; crack-like defect; DCPD method; crack depth.

1 Introduction

Natural-gas pipeline accidents mostly result in major damage to buildings and other constructions located not only nearby but also quite far away and often put a lot of people in danger of injury or death. Such crashes are almost always connected with enormous material and financial damages. Therefore, safety and reliability management of high pressure gas pipelines is one of the most important issues for the operators. If damage to the pipeline is detected, such pipeline sections can no more be operated as long as safety and reliability is not confirmed on the basis of comprehensive theoretical and experimental evaluation.

In one of DN300 high pressure pipelines in Czech Republic, crack-like defects were detected. Several meters long section of the pipeline was taken out from ground for the purpose of experimental laboratory evaluation of the damage extent followed by high pressure fatigue test and test to rupture. One of the important points to be evaluated was, whether the cracks are stable or grow during the tests.

The only possibility to evaluate the crack growth was to use indirect methods of measurement of their depth. As there have been many years experience with DCPD method of crack measurement (e.g. [1-3]), it was decided to explore possibilities and conditions of its use for this specific case of longitudinal cracks in pipe. Note that the DCPD method is particularly suitable for measurement of cracks in laboratory specimens. As far as pipes are concerned, it can be used with a high precision for transversal cracks. In both cases, theoretical calibration curves can be computed and used with high precision. This does not, however, concern longitudinal cracks. when it is practically impossible to reach homogeneous direct current field in the crack area, which is a fundamental condition for using the analytical calibration. Therefore, experimental calibration had to be carried out, dedicated for the specific pipe section of specific dimensions.

2 Experimental Programme

Section of the pipe with no NDT indications, i.e. probably no cracks was used for the experimental calibration. 60 mm long longitudinal line was indicated on the pipe surface and at its centre, row of five electrodes was spot welded to the pipe surface in perpendicular direction. Two electrodes near the line, at the distance 5 mm from the line were used for measurement of potential V(a) on crack, the third electrode located at the distance 10 mm from one of the potential electrodes measured reference potential Vref and two fairly thick outside electrodes, at distance 20 mm from the nearest potential electrode, served as the direct current source. After measurement of initial value of the potentials V(a) / Vref, which was close to ideal ratio 1, artificial crack was modeled by thin gradual cuts do depth with surface length 60 mm. Position of the electrodes and cutting procedure are shown in Fig. 1.

Besides experimental measurement of the potential dependence on crack depth, modified analytical calibration curve based on Johnson's formula [5] according to [1-3] was calculated for fictive values of specimen width W = 10 mm, which corresponded to pipe wall thickness.

Fig. 1: Modelling of crack by cutting with potential measurement

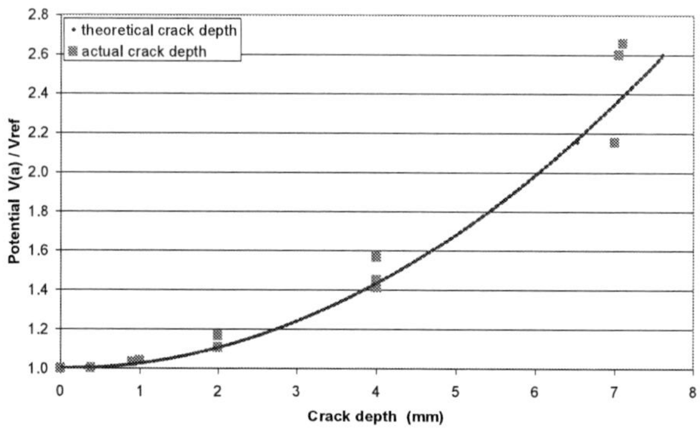

Fig. 2: Analytical calibration curve and points of experimental measurement

3 Results and Discussion

The calculated calibration curve with experimentally measured points is shown in Fig. 2. The results were very promising, because in spite of that the direct current field was not homogeneous near the crack area, results indicated that for the first approximation, the calculated curve could be used. The agreement between the analytical curve and experimental points was particularly good for the case that the depth was not constant along all the 60 mm surface length but was maximum in the crack centre and gradually decreased to its ends, approximately according to parabolic dependence. If the depth was constant along the whole crack length, the potential ratio V(a) / Vref was slightly higher than the calculated curve and on the contrary, if crack depth decreased to its ends linearly, then the points were bellow the curve.

In the second step, the DCPD method was verified on actual NDT indications of cracks in the pipe section, with typical surface length between 40 mm and 100 mm, sometimes even more. Electrodes were spot welded to the surface near four crack indications Nos. 2, 4, 12 and 25 in the same way and distances like during the experimental calibration. Measured potentials were recalculated to crack depths using the analytical curve in Fig. 2. The calculated values corresponded sequentially to depths 1.30 mm, 0 mm (actually negative value, which is nonsense), 2.17 mm and 1.60 mm. Then the pipe was sectioned and metallographical cuts perpendicular to the cracks in their centres were prepared.

The metallographical examination provided surprising results. The most important was the fact that practically all crack or crack-like defects were considerably inclined – Fig. 3. Some of them were in fact subsurface defects almost parallel to the pipe surface. Crack No. 4 was even curved back to the surface, so its maximum depth was not coincident with its tip – Fig. 3 b. Note that the selected cracks corresponded to the strongest NDT indications, as shown in Fig. 3 e, f.

Both values of crack length and depth were evaluated from the cuts in Fig. 3 and were correlated with calculated depth values from the DCPD measurement using the analytical calibration curve in Fig. 2. The results were much less encouraging than in case of the previous calibration with the artificial longitudinal radial cut. The comparison is in Table 2, dimensions are in mm.

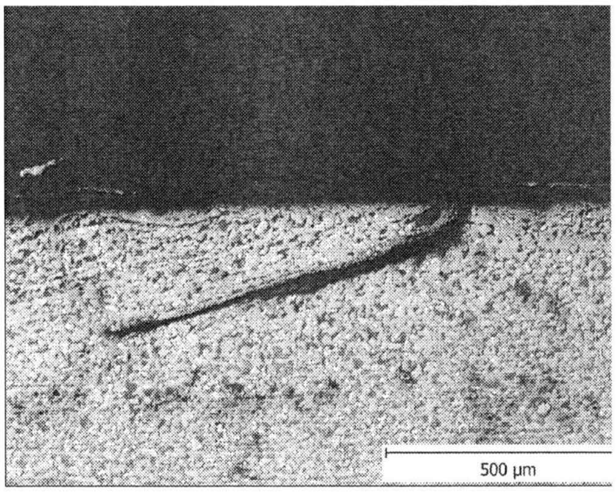

a) crack No. 2

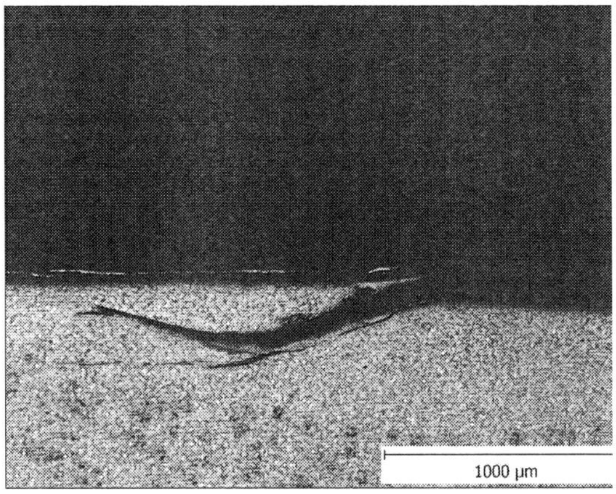

b) crack No. 4

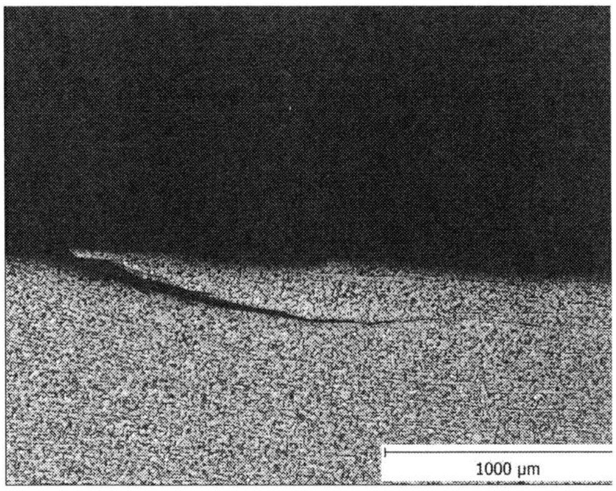

c) crack No. 12

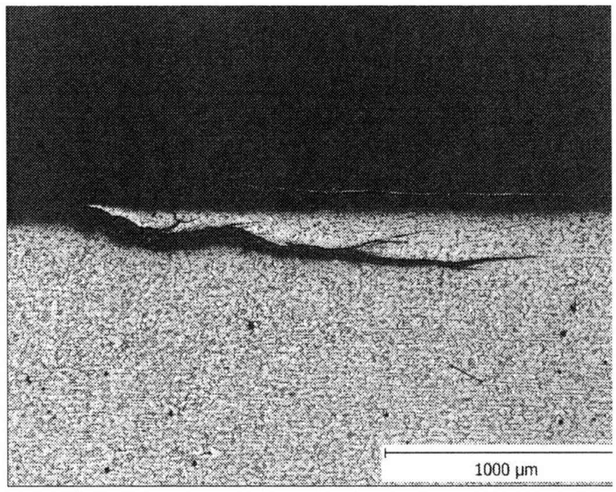

d) crack No. 25

e) NDT indication No. 12

f) NDT indication No. 25

Fig. 3: Cracks in cross section metallographical cuts and NDT surface indications of selected cracks: a) No. 2, b) No. 4, c) No. 12, d) No.25, e) indication No. 15, f) indication No. 25

In the next step, the measured potentials were plotted against values of crack depth and crack length, respectively, with the aim to try to explain the unusual an unexpected results. Individual values of the potential V(a) and also values of the ratio V(a) / Vref were plotted in the diagrams in Fig. 4.

Tab. 2: Actual dimensions of the cracks with calculated depths

Crack No.	Actual depth	Calculated depth	Actual length	Crack No.	Actual depth	Calculated depth	Actual length
2	0.25	1.30	0.75	12	0.26	2.17	1.56
4	0.35	0.00	1.52	25	0.26	1.60	1.95

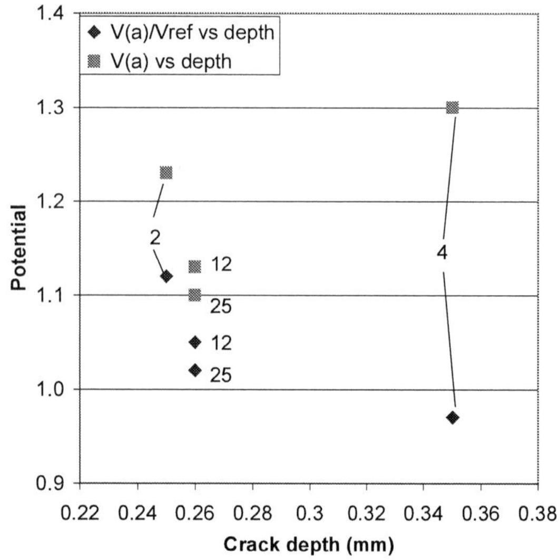

a) measured potentials V(a)/Vref and V(a), respectively, as dependence on crack depth

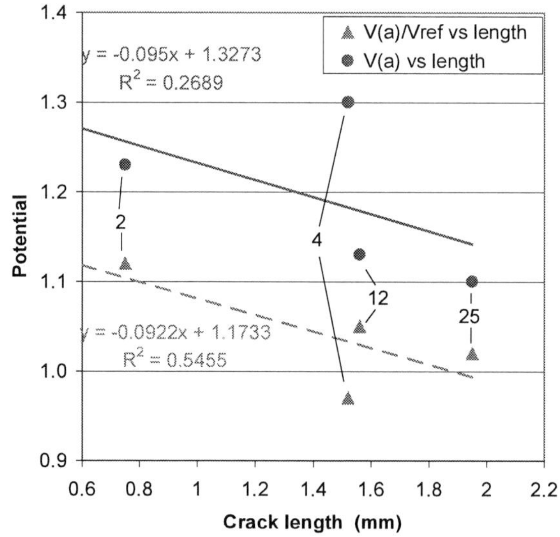

b) measured potentials V(a)/Vref and V(a), respectively, as dependence on crack length

Fig. 4: Dependence of potential ratio V(a) / Vref and pure V(a) on depth and length of cracks

It follows from Table 2 and Fig. 4 that there is no reasonable agreement between actual and calculated values of crack depth, respectively. There is even no meaningful agreement between potential V(a) or potential ratio V(a) / Vref on either crack depth or crack length. On the contrary, the potentials have a decreasing tendency with increasing crack length – Fig. 4b, which seems to be more or less absurd. There is, however, a reasonable explanation: The strong inclination of the cracks causes a kind of potential shadow on the crack side in the direction of inclination resulting in lower measured values. The shadow is the higher the crack is longer and so, measured potentials are inversely related to the length of strongly inclined cracks.

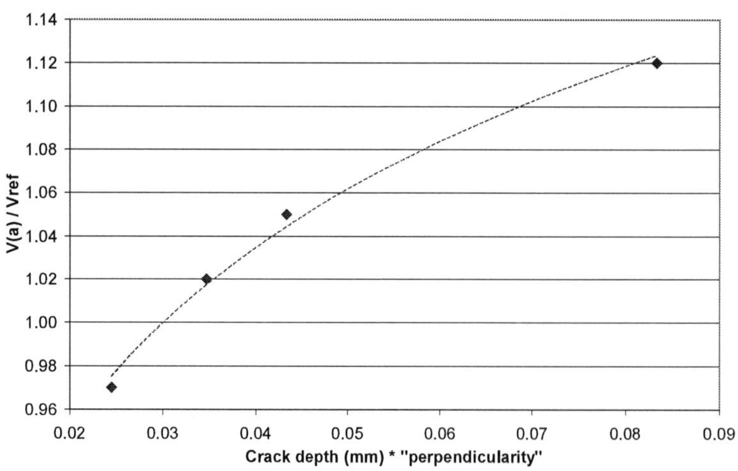

Fig. 5: Potential ratio V(a) / Vref was plotted against crack depth multiplied by "crack perpendicularity"

Another attempt was made to achieve a reasonable relationship between crack dimensions and measured potentials, namely introducing a parameter called "crack perpendicularity", defined as ratio of crack depth to its length. It is evident that such ratio equals one for ideally perpendicular crack. When actual crack depth values were multiplied by this parameter and measured potential ratio V(a) / Vref was plotted against these values, a reasonable diagram was obtained – Fig. 5, with a logarithmic dependence. More experimental points would be, however, needed to make fairly general conclusions.

The very different bahaviour of DCPD potential measurement on the inclined cracks in the pipe can also be distinctly shown in another diagram – Fig. 6, where the first part of the analytical calculated calibration curve is shown together with actual potential measurement, ratios V(a) / Vref, carried out on the crack modelled by cutting in exactly radial direction, i.e. with no inclination. The agreement is very good. On the contrary, the same V(a) / Vref values evaluated on the inclined real defects not only do not agree with the theoretical curve, but have even opposite slope.

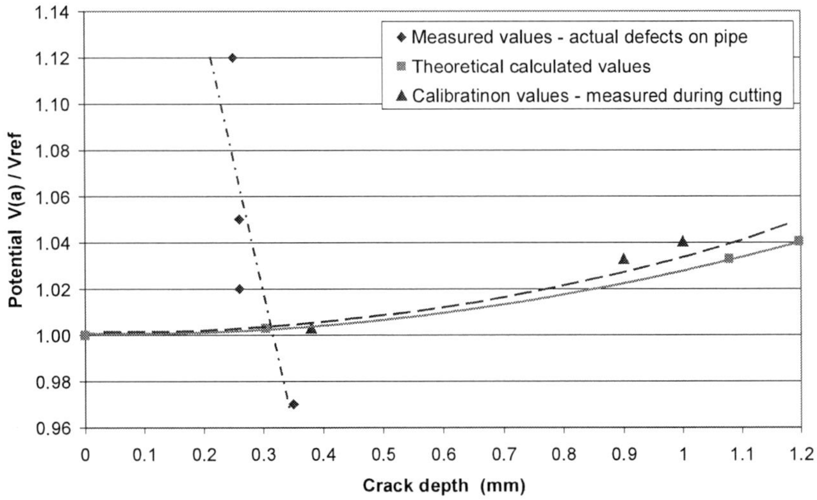

Fig. 6: Potential ratio V(a) / Vref was plotted against crack depth multiplied by "crack perpendicularity"

It also should be pointed out that measured potential, more exactly potential ratio is affected by important fact, whether the reference electrode is placed in the direction of the crack inclination or at the opposite side of the crack. In this work, it was quite probably in the direction of inclination, but not sure. Another comment concerns a possibility to calculate potential field near crack of any shape and dimensions by resolving the relevant Laplace equation using numerical methods, which is, however, very complicated and time consuming task.

4 Application of DCPD Measurement during Pipe Pressure Tests

The DCPD method was eventually used for measurement of possible crack length changes of actual cracks in the pipe during pressure cycle induced fatigue crack growth assessment and during static pressure test to failure, however with the reservation about low precision of absolute crack depth assessment. For variously inclined cracks the method can only be used for assessment of occurrence of crack increments during tests.

DCPD measurement was performed near four defects with most distinct NDT indications Nos. 11, 17, 10 and 14. The configuration of the electrodes was the same as in case of previous calibration and measurement, just the distance of current electrodes was slightly higher, approximately 100 mm from the nearest potential electrode. The pipe section was loaded by internal pressure, namely fatigue loading between zero and maximum service pressure to 1200 cycles followed by static test to failure. Results are shown in Fig. 7, where measurements Nos. 1 – 8 correspond sequentially to the following conditions: 1 – initial measurement, 2 – 1 cycle, 3 – 200 cycles, 4 – 1000 cycles, 5 – 1200 cycles, 6 – static loading to 12.5 MPa, 7 – static loading to 24.9 MPa and 8 – 34 MPa just before failure.

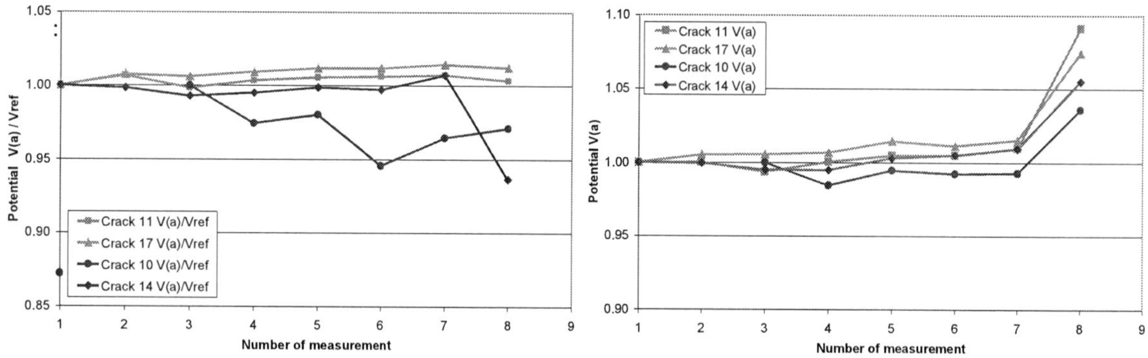

a) V(a) / Vref values during pipe pressure test

b) values of pure V(a) during pipe pressure test

Fig. 7: Application of DCPD measurement on four selected cracks during pressure cycle induced fatigue test and static pressure test to failure

Results in Fig. 7 confirm that some changes of the potentials, at least quite minor occurred on all of the four defects. In spite of the fact that measurement of reference potential in standard situations increases stability and self-consistence of the method, in this specific case, measurement of pure V(a) values provides more reasonable information, namely increasing tendency of the potential values. Note that in the cases of strongly inclined cracks, increased internal pressure may result in better contact between fracture surfaces and eventually in slight decrease of the measured values. However, this concerns not only pure potentials V(a) but also potentials Vref. It is the reason why the values in Fig. 7 b are considerably more consistent than those in Fig. 7 a. It looks that the most important changes occurred at the end of fatigue loading stage. This indicates that a kind of fatigue growth of the defects occurred. However, as far as the last strong increase of the potentials V(a) is concerned, it can be connected to the overall plastic deformation before final failure rather than growth of corresponding cracks.

5 Conclusions

The main results of the work aimed at exploration and verification of possibilities to use DCPD method, successfully being used for exact measurement of crack length and growth in laboratory specimens and several components, for measurement of depth and growth of longitudinal cracks in pipes during internal pressure induced fatigue and static tests can be summarised as follows:

- In case of the specific configuration of the potential and current electrodes, the method can not only be used for a rough assessment of crack occurrence and growth, but even analytical calibration curve based on the Johnson´s formula is applicable with quite high degree of accuracy.
- Actual existing crack in the assessed pipe section were substantially inclined, which strongly complicated application of the method for exact measurement. In such case, the use of method can be considered for rough assessment of crack existence and its growth, more or less from the qualitative viewpoint only.
- The method applied to measurement of actual indications during pipe fatigue test and static test to failure provided useful information about changes of the character of the defects. The main changes concerned last stages of the fatigue loading and static loading before failure. On the other hand, the changes could concern not only actual crack growth, but also contact conditions between the fracture surfaces.

Acknowledgement

Support of the Ministry of Education, Youth and Sports of the Czech Republic is acknowledged – grant MSM25797000.

References

[1] I. Černý, V. Linhart, J. Minárik, P. Hulík, Modification of DC Potential Method for Short Specimens, in Proc. of the 2nd Int. Conf. on Computer-Aided Assessment and Control: Localized Damage, edited by M.H. Aliabadi, D.J. Cartwright and H. Nisitani, Computational Mechanics Publications Ltd., Southampton, UK (1992) pp. 61-76.

[2] I. Černý, The use of DCPD method for measurement of growth of cracks in large components at normal and elevated temperatures, Eng. Fract. Mech. Vol. 71, No. 4-6 (2004) pp. 837-848.

[3] I. Černý, Measurement of Subcritical Growth of Defects in Large Components of Nuclear Power Plants at Elevated Temperatures, Int. J. Pres. Ves. Pip., Vol. 78, 11-12 (2001) pp. 893-902.

[4] H.H. Johnson, Calibrating the Electrical Potential Method for Studying Slow Crack Growth, Mater. Res. Stand., Vol. 5, 9 (1965) pp. 442-445.

Allowed Stress Range Prediction of Fatigue Loading for Steel Alloy in Steam Turbines

M. J. CERNY[1,a]

[1] *Czech Technical University in Prague, Klokner Institute, Solinova 7, 166 08 Prague 6, Czech Republic*
[a] *cerny@hppro.klok.cvut.cz*

Abstract: Corrosion pitting is responsible for the initiation of fatigue cracks in alloy steel used in steam turbines. In the paper the criteria for prediction of allowed stress range of cyclic loading will be given.

Keywords: corrosion; fatigue; cracks; steel alloy; steam turbines.

1 Introduction

Knowledge of corrosion pitting is basic for the nucleation and propagation of fatigue cracks in steel alloy of steam turbine blades [1]. For quantifying pitting-induced corrosion fatigue, various models have been proposed to characterize the corrosion fatigue crack nucleation.

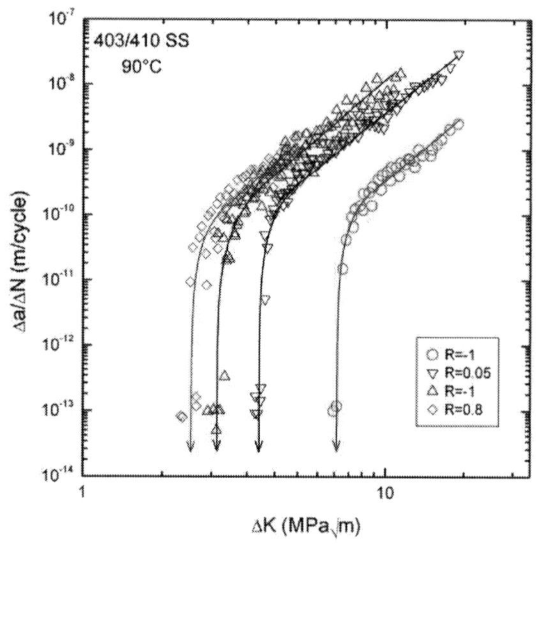

Fig. 1: Fatigue Crack Growth Rate (FCGR)

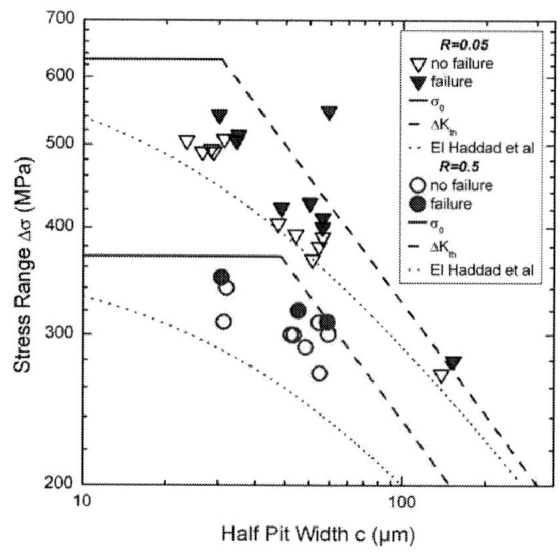

Fig. 2: Kitagawa Diagram (90°C Air at R=0.05, 0.5)

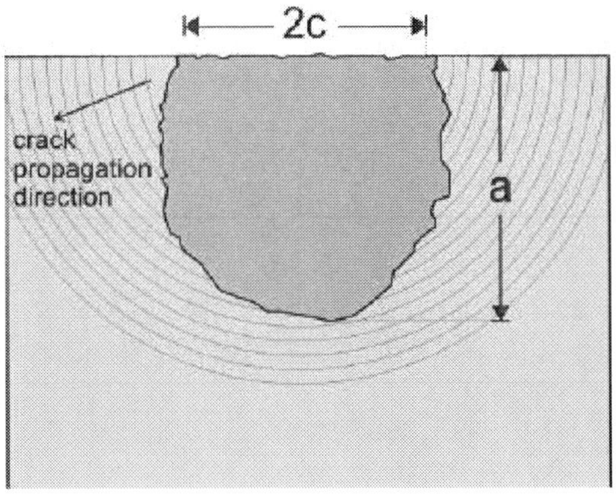

Fig. 3: Schematic Illustration of Pit Geometry

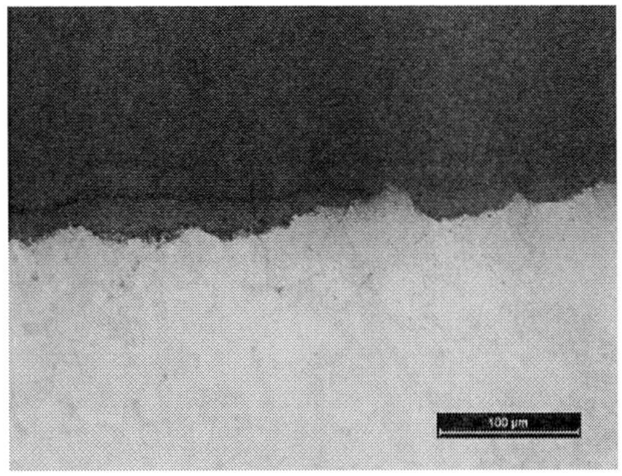

Fig. 4: True Geometry (steam 650°C,6700 hrs)

2 Fracture Mechanics and Kitagawa- Takahashi Diagram

To evaluate the influence of corrosion pits on fatigue, linear elastic fracture mechanics is used. It has been observed, that most pits have roughly a semielliptical shape with the width at the surface, 2c, and the pit depth, a (Fig.3). Investigations show that the pits can be treated as semi-circular surface cracks with a geometry described above. With this approximation and that the pit size is small in terms of crack behavior, the stress intensity factor of pits ΔK , FCGR diagram (Fig.1) and finally a Kitagawa- Takahashi (K-T) diagram (Fig.2) can be created for 403/410 12% Cr martensitic steel [3]. The K-T diagram correlates pit-to-crack fatigue data. The data can be obtained for specific values of R (ratio of minimum to maximum stress for combined steady and cyclic stress). As a result, K-T diagrams can be used for assessment of the fatigue limit and life-time of corroded parts of steam turbines.

A key parameter referred to frequently in this paper is R, which is the ratio of minimum to maximum stress in a combined steady and cyclic stress environment, as shown in Equation 2.

$$R \equiv \frac{\sigma_m - \sigma_a}{\sigma_m + \sigma_a}$$

Where:

σ_m is the steady (mean) stress, and
σ_a is the cyclic (alternating) stress.
Note that the stress range ($\Delta\sigma$) is equal to 2x the cyclic stress amplitude.

The description of true pit geometry has been published by author earlier [2]. The max. and min. depths of pits and their densities have been found for areas shown at Fig.2. Table 1 comprises the parameters of corrosion evaluated for part of turbine packing ETU23 (alloy X22CrMoV12-1).

A FCGR line can be drawn through the crack growth data. This characteristic shape is consistent with data for many materials. The three-component straight line is defined as follows:

- ***Region 1:*** This is the slow growth portion of the curve. The fatigue crack threshold (ΔK_{th}), corresponds to the stress intensity factor range, below which cracks do not propagate.
- ***Region 2:*** This is the power law growth region (also known as the "Paris" region).
- ***Region 3:*** Rapid unstable crack growth just prior to failure.

$$\Delta K \equiv Y \Delta \sigma \sqrt{\pi c}$$

where:

ΔK is the cyclic stress intensity factor, MPa$\sqrt{}$m,
$\Delta \sigma$ is the cyclic stress range, MPa,
c is the instantaneous crack size, meters and
Y is a geometry factor (for this case Y = 0.65).

The values shown are for illustrative purposes to demonstrate Region 1 behavior.

During this research, the concern has been to identify the conditions so that the risk of initiating a crack (Region 1) is quantified. The left portion of Fig.1 shows that for a small value of cyclic stress intensity factor ($\Delta K < \Delta K_{th}$), there would be no growth of an incipient fatigue crack. When the cyclic stress intensity factor is maintained below the threshold value, then a crack will not initiate or grow.

Table 1 Parameters of pitting determined experimentally

					Depth [μm]			
Area	**No of Pits**	**Σ L i [mm]**	**Density**	**Oxide [%]**	**Max**	**Min**	**Mean**	**St. Deviation**
1	6	1.5567	3.85	79.00	35.39	13.06	25.95	6.997
2	31	3.9507	7.85	66.74	36.39	5.65	21.58	7.825
3	17	0.7627	22.29	12.09	29.05	5.80	19.73	5.802
4	15	0.8248	18.19	33.35	30.00	14.52	20.78	5.079
5	18	0.8534	21.09	28.19	28.08	7.40	19.05	5.113
6	3	2.8381	1.06	99.22	32.88	29.19	30.81	1.540
7	42	5.3321	7.88	64.00	39.71	4.35	21.91	8.932
8	29	2.4345	11.91	41.68	36.76	9.67	22.65	7.453

As a result, from Kitagawa- Takahashi diagram we get - for area 4 and R= 0.5 - stress range is 280 MPa, what is 47% of material yield strength. This stress range can be understood as limit for stresses induced by loading in steam turbine. Allowed stress range will be different, as the K-T diagram is valid actually only for 90°C air.

3. Conclusion

In the presented paper the prediction of allowed stress range of cyclic loading with consideration of cracks due to corrosion – fatigue in steam turbine blades is described. The prediction is based on description of true pit geometry. The method follows the Kitagawa- Takahashi diagram.

Acknowledgement

The presented work has been supported by grant No TE01020068.

References

[1] Černý M.: Damage and Localized Corrosion in Steam Turbines, Conference EAN 2014, Proceedings

[2] Černý M.: Localized Corrosion in Alloy Steel of Steam Turbines, Applied Mechanics and Materials, vol.827 (2016), Trans Tech Publications, doi:10.428/www.scientific.net/AMM.827.177

[3] Salzman R., Gandy D., Rieger N., Schönbauer B., Tchegg S., Zhou S., Turnbull A.: Corrosion-fatigue Prediction Methodology for Steam Turbine Blades, Energy-tech, December 2013

Bending of Sandwich Beam with Visco-elastic Core

M. J. CERNY [1,a], P. SLAPAK [1]

[1] *Czech Technical University in Prague, Klokner Institute, Solinova 7, 166 08 Prague 6, Czech Republic*

[a] *cerny@hppro.klok.cvut.cz*

Abstract: An analysis of viscoelastic bending of sandwich beam, consisting of elastic bearing layers and viscoelastic core will be given. The analysis is based on Volterra integral equations with exponential kernels and Volterra resolvent operators. Theory is verified by experimental data from three- point bending.

Keywords: viscoelasticity; sandwich beam; light core.

1 Introduction

Polymer foam-core sandwich panels and beams are increasingly used for load-bearing components in buildings. Sandwich panels and beams are offering a high stiffness per unit weight. But such structures from polymers creep at room temperature, what is limiting their use in structural applications. In this study, we combine the viscoelastic model for foam creep with the analysis of deflection of a sandwich beam to develop an expression for the creep of a sandwich beam with a polymer foam core. The results are compared with data from a series of tests on sandwich beams with polymer based composite and rigid foam core.

2 Experimental evaluating of creep for the core

For mechanical testing has been delivered a sandwich plate with bearing layers from composite, reinforced by glass fibres and polyester matrix. First, the foam creep has been tested on cuboid testing samples, which were loaded in compression by constant load in a set-up developed and manufactured in Klokner Institute. Transversal displacements were measured by LVDTs and strains by strain gauges. Load and unload application time was at least 24 hrs. The course of strain for 24h loading and 24 h unloading is shown on Fig.1.

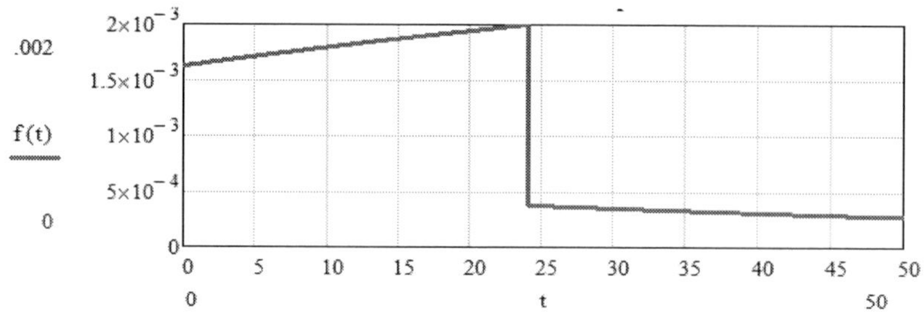

Fig. 1 Strain vs time in compression of sandwich core

Theoretical course of strain vs time according to Poynting- Thomson model has relationship

$$\varepsilon = \frac{\sigma_0}{E_1} + \frac{\sigma_0}{E_2}\left(1 - e^{-\frac{E_2}{K}t}\right)$$

and strain is depending on three material parameters E, which can be evaluated by collocation method. For the material in Fig.1 the following values have been found:
E1=24.615 N/m2, E2=45.714 N/m2, K=1961.0 N/m2.hr

3 Experimental evaluating of vertical deflection at tree-point bending

Testing samples were put into a set-up for three point testing and loaded by a constant load 300, 600 and 800 N. Distance between supports was 200 mm, length of samples 300 mm, width 50 mm. Vertical deflection has been measured by LVDT sensors and data acquisition system National Instruments NI 1052 in program environment LabWindows. Measured values have been evaluated by MS Excel and compared with theoretical values. The comparison of results shows good agreement (< 10%). Results of long-term testing of sandwich beam are shown in Fig.2.

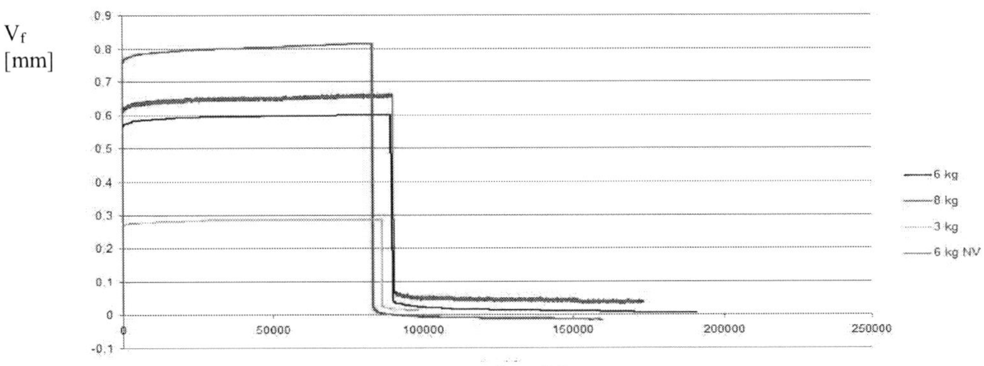

Fig. 2 Vertical displacement vs time in bending of sandwich beam

4 Formulation and solution of the problem

Paper deals with integral equations Volterra and kernels mostly of exponential type. Further, for simplicity, a symmetrical structure of the simply supported beam is considered. The load $q(x,t)$ is acting in direction of beam thickness z according to relation $q(x,t) = g(x)\,\Phi(t)$ where $\Phi(t)$ is Heaviside function.

Below, for the simplicity, the symmetric structure of the simply supported beam is considered.

Let us treat by beam core, where the shear stress τ is of main importance and define $\tau = \varphi_1(x,t)$, $\sigma_z = \varphi_2(x,t) - z\varphi_{1,x}$ where $\varphi_2 = 0$ can be considered. The relations satisfy the equilibrium condition in z direction z. Further

$u_{z,z} = \dfrac{\hat{C}}{E_c}\left(-z\,\varphi_{1,x}\right)$, $u_{x,z} + u_{z,x} = \dfrac{\tilde{C}}{G_c}\varphi_1$, where time integral operators are of form

$C = 1 + \lambda\ C^*(\alpha)$, and parameters λ, α are changing in dependence on adopted mechanical model. After integration we obtain

$$u_x = z\left(\frac{\widetilde{C}}{G_c}\varphi_1 - w_{,x}\right) + \frac{z^3}{6}\frac{\hat{C}}{E_c}\varphi_{1,xx}$$

On this basis the similar relations for bearing layers are derived. Here we will define the integral operators S, S^{-1} for a model with structural equation $E - (E/K)$.

Decisive component of stress in bearing layer is σ_x . Relation $\sigma_x = E\,S\,\varepsilon_x$ is valid and after some arrangement

$$\sigma_x = E\,S\left[\pm s\left(\frac{\widetilde{C}}{G_c}\varphi_{1,x} - \varphi_{3,xx}\right) \mp \frac{s^3}{6}\frac{\hat{C}}{E_c}\varphi_{1,xxx} - z\,w_{,xx}\right]$$

kde $S = I - \dfrac{E}{K}E^*\left(-\dfrac{2E}{K}\right)$ $\qquad S^{-1} = I + \dfrac{E}{K}E^*\left(-\dfrac{E}{K}\right)$

More advantageous formulation we can get on basis of generalized function $\omega(x,t)$, where we consider

$$\varphi_1 = s_0\,S\,\omega_{,xxx}\,, \qquad w = -\frac{\omega}{E\,r} + \frac{s\,S\,C_2}{G_c}\omega_{,xx} - \frac{s^3\,S\,C_1}{3E_c}\omega_{,xxxx}$$

Now one basic equation is satisfied identically and the second is transformed into form

$L\,\omega = -q$, where operator L is expressed by relation

$$L = \frac{1}{6}S\frac{d^4}{dx^4}\left[12s_0{}^2 + r^2 - \frac{r^3\,s\,E\,S}{G_c}\frac{d^2}{dx^2}\left(C_2 - \frac{s^2\,G_c\,C_1}{3E_c}\frac{d^2}{dx^2}\right)\right]$$

Boundary conditions for simple support are given by relations

$$\frac{d^{2k}}{dx^{2k}}\omega = 0 \qquad k = 0,1,2,3$$

At solving the problem for load $q(x,t) = g(x)\,\Phi(t)$ we will apply Fourier expansion of functions $g(x,t)$ a $\omega(x,t)$.

$$g(x) = \sum_m g_m \sin\frac{m\pi x}{l}\,, \qquad \omega(x,t) = \sum_m \omega_m\,\Phi(t)\sin\frac{m\,\pi\,x}{l} \qquad (m = 1,2,3,\ldots)$$

By substitution into the basic equation we obtain

$$\omega_m = \frac{6\dfrac{l^4}{m^4\pi^4}g_m}{12\,s_0{}^2 + r^2 + \dfrac{r^3\,s}{G_c}E\,S\dfrac{m^2\pi^2}{l^2}\left(C_2 + \dfrac{s^2}{3}\dfrac{G_c}{E_c}\dfrac{m^2\pi^2}{l^2}C_1\right)}\;\frac{1}{S\,\Phi(t)}$$

As illustration we give expression for vertical displacement for the middle of the beam ($x = \frac{l}{2}$).

$$w = -\sum_m\left[\frac{1}{E\,r} + \frac{s}{G_c}m^2\frac{\pi^2}{l^2}S\left(C_2 + m^2\frac{\pi^2}{l^2}\frac{s^2}{3}\frac{G_c}{E_c}C_1\right)\right]\Phi(t)\,\omega_m\,\sin\frac{m\,\pi}{2}$$

$m = 1,2,3, \ldots.$

After substitution we obtain the relation

$$w = -\frac{6l}{\pi^4 E}\sum_m g_m \frac{1}{m^4}\frac{\frac{l}{r}+\frac{s}{l}\frac{E}{G_c}m^2\pi^2 S\left(C_2+\frac{s^2}{3}\frac{G_c}{E_c}\frac{m^2\pi^2}{l^2}C_1\right)}{\frac{12s_0^{\ 2}+r^2}{l^2}+\frac{r^3 s}{l^4}\frac{E}{G_c}m^2\pi^2 S\left(C_2+\frac{s^2}{3}\frac{G_c}{E_c}\frac{m^2\pi^2}{l^2}C_1\right)}\frac{\Phi(t)}{S\Phi(t)}\sin\frac{m\pi}{2}$$

For concentrated load we can introduce a generalized function $\omega(x,t)$ instead of functions $\varphi_1(x,t)$, $w(x,t)$ by substitution $\omega_{,xxx} = -2s_0\,\varphi_1$ and further

$$w = \frac{\omega}{2Ers_0^{\ 2}} - \frac{\tilde{C}}{2G_c s_0}\omega_{,xx} - \frac{s_0\hat{C}}{12E_c}\omega_{,xxxx}$$

By applying the boundary conditions $\omega = \omega'' = 0$ we receive for the middle of the beam

$$w\left(\frac{l}{2},t\right) = \frac{\overline{P}l^3}{48E.2rs_0^{\ 2}} + \frac{\tilde{C}\overline{P}l}{8G_c s_0}$$

Operator $\tilde{C}$ is of the form $\tilde{C} = 1 + \lambda\ C^*(\alpha)$. Parameters λ, α are changed according to adopted mechanical model. For the Poynting-Thomsonův model (linear solid) and arrangement $G_c - G_x/K_2$ it will be valid

$$\tilde{C}\,\Phi(t) = \Phi(t) + \frac{G_c}{K_2}\int_0^t e^{-\frac{G_x}{K_2}(t-\tau)}\Phi(\tau)\,d\tau = 1 + \frac{G_c}{G_x}\left(1-e^{-\frac{G_x}{K_2}t}\right)$$

Relation $\tilde{C}\,\Phi(t)$ will be substituted into the above equation:

$$w\left(\frac{l}{2},t\right) = \frac{\overline{P}l^3}{48E.2rs_0^{\ 2}} + \frac{\tilde{C}\overline{P}l}{8G_c s_0} = \frac{Pl}{8bs_0}\left(\frac{l^2}{12Ers_0}+1+\frac{G_c}{G_x}(1-e^{-\frac{G_x}{K_2}t})\right)$$

Conclusion

Experimental evaluation of the analysis of bending for sandwich beams of a symmetrical structure with elastic bearing layers and viscoelastic core is given. Three point testing by a constant load has shown good agreement with theoretical values.

Acknowledgement

The presented work has been supported by grant No 8301600J100.

References

[1] Plantema, F, J., 1966, Sandwich Construction: The Bending and Buckling of Sandwich Beams, Plates, and Shells, Wiley and Sons, New York.

[2] Cerny M.J., Slapak P., 2012, Behavior of Composite Sandwich Beam in Bending, Proceedings, ECCM15, Venice

[3] Dundrova V., Kovarik V., Slapak P., 1970, Biegungstheorie der Sandwich- Platten, Academia-Verlag.

[4] Vinson J.R., 1999, Sandwich Structures of Isotropic and Composite Materials, Technomic.

The Electrobus Skeleton Parts – Experimental and Numerical Testing

K. Doubrava[1,*], C. Novotný[1]

[1] *Czech Technical University in Prague, Faculty of Mechanical Engineering,*
Department of Mechanics, Biomechanics and Mechatronics,
Technická 4, Praha 6, Czech Republic
* *Karel.Doubrava@fs.cvut.cz*

Abstract: The paper describes the experimental tests and numerical simulations of important parts of the skeleton of an electric bus, such as roof and side pillars and details of the windshield. The research was conducted static and cyclic bending tests sandwich samples whose composition corresponded to perform sandwich roofs. It was conducted static test load roof segment and material identification of windscreen. The obtained data was used for the verification of numerical simulations carried out at the FEM model of the car.

Keywords: Electrobus; Sandwich Roof; 4PB Test; Fatigue Test.

1 Introduction

Use of new materials in the construction of many products is motivated by an effort to increase their usefulness. Sandwich structures with various combinations of core and faces are used in transport industry [1]. Reduction of total weight of electric bus is the main goal, which allows among others to increase the useful range of the electrobus with the same electric equipment. This was solved as part of the project coordinated in cooperation with the firm VISION CONSULTING AUTOMOTIVE and the manufacturer SOR Libchavy. Efforts to achieve efficient and lightweight design can be divided into several steps, choice of sandwich roofs and select the appropriate version, the question of the use of better performing materials with traditional construction, verification of the final solution numerical and experimental manner on the roof segment. The main criterion is then fulfilling rollover testing at the lowest weight of the skeleton of the vehicle.

2 Tests of basic structure

2.1 Static bending test of sandwich specimen

With regard to previous research [2] final combination of sandwich structure consisting of a core of PVC foam and covers made of aluminium alloy was chosen. Beam specimens that option was made by sandwiches technology prototype production, which later proved as problematic factor.

Sandwich beams were subjected to four-point bending test see Fig. 1. All types of samples had the same core – PVC foam with density ca 60 $kg{\cdot}m^{-3}$. Specimens Foam A and Foam F faces were made of steel, specimens Foam D [2] faces were made of stainless steel. Top face of specimen Foam E were made of stainless steel and bottom were made of dural alloy. Both specimen Foam G faces were made of dural alloy. Results can be compared even with earlier variants see Fig. 2. Stiffness of various specimens corresponds faces materials. Specimen Foam A and Foam F are stiffer than specimens Foam D, F and G. Results static tests were affected by higher variance of strength caused by different types of failures when some samples were poorly glued.

Model sandwich roof will be used for calculations of behaviour of the bus body using FEM software ABAQUS. The PVC foam was modelled as an isotropic material with linear 8 node elements. Adhesive joint between the core and the faces was modelled using cohesive elements. Numerical model of 4PB tests verify the suitability of the selected elements and material properties. Static bending tests of sandwiches were numerically simulated by FEM.

Fig. 1: Static test of 4PB of sandwiches.

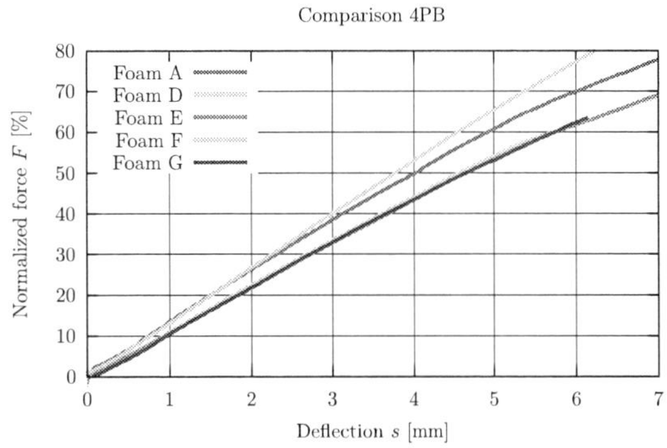

Fig. 2: Comparison results of static test.

2.2 Fatigue bending test of sandwich specimen

Based on the a static test the initial level of fatigue tests had been set, but the results of fatigue tests were affected by different types of failures the cause of the uneven quality of sandwiches prototype production. Scheme of the fatigue test is in Fig. 3. The results of fatigue tests show greater variance similar to the static test, again caused by apparently non-constant production quality of prototype specimens.

Fig. 3: Fatigue test of 4PB of sandwiches.

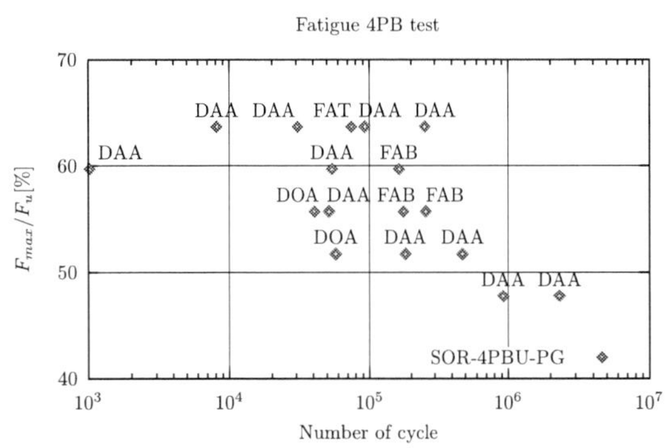

Fig. 4: Results of fatigue test.

The results of fatigue experiments are plotted in the graph in Fig. 4. Types of defects are marked in accordance with standard ASTM C393 [3], when F in the first place represents facing failure D in first place represents skin to core delamination. Second code is failure area, when A means at load bar and O means outside gauge. Third code is failure location, when A means core–facing bond, T means top facing and B bottom facing. Examples of failures can be seen in Fig. 5 and 6.

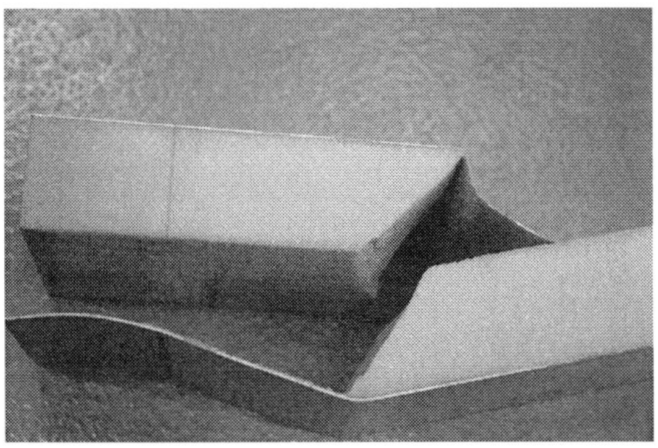

Fig. 5: Example of DOA failure.

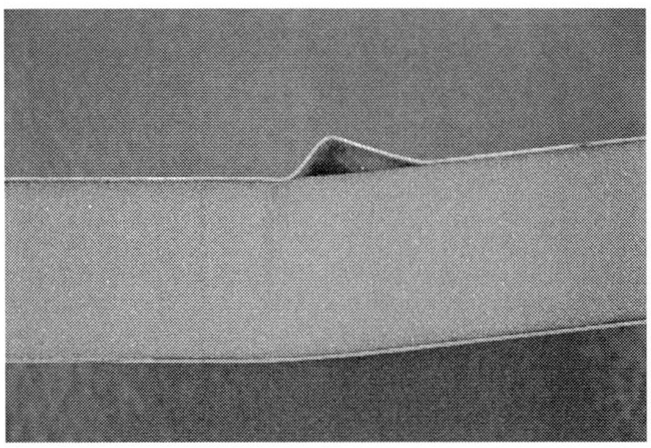

Fig. 6: Example of DAA failure.

2.3 Bending test of duplex profile

The side pillar is one of the other elements of the structure affecting the safety of the vehicle rollover accident. Three point bending tests (see Fig. 7) the connection side and window (scheme see Fig. 8) pillar were made to verify the numerical calculations. Different combinations used sections and variations of structural modifications with respect to higher loading capacity from loss of stability of the structure were investigated.

Fig. 7: 3PB test of duplex profile.

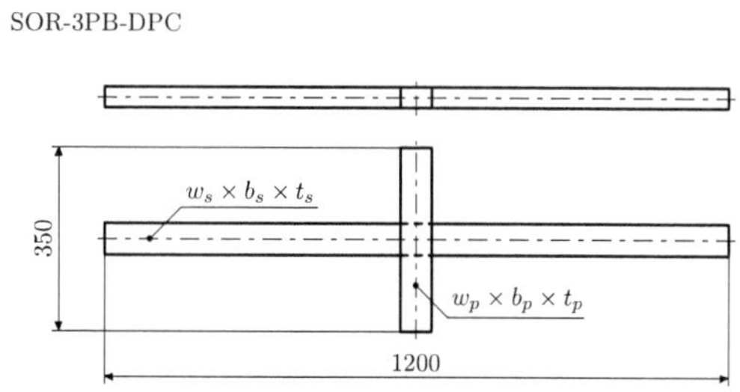

Fig. 8: Test specimen scheme.

2.4 Windscreen tests

Windscreen material properties is important parameter for tuning the numerical calculation of the electric bus rollover test. To verify the properties of glass in the numerical model tests of windscreen were made. The flexural properties were obtained by four points bending test see Fig. 9. The test was numerically simulated with good agreement between experiment and numerical calculation.

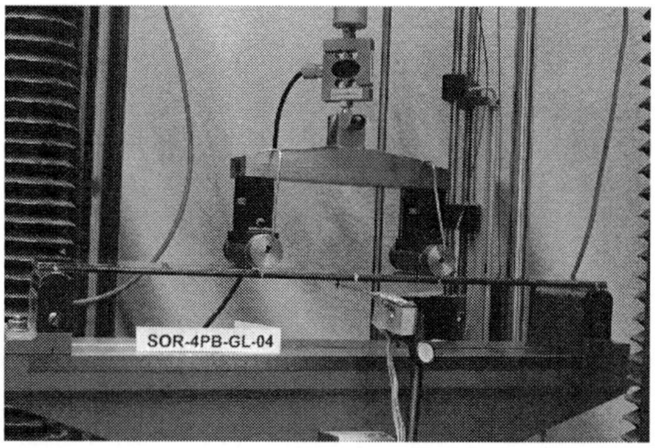

Fig. 9: Bending test of windscreen specimen

Fig. 10: Shear test of windscreen.

Shearing test was designed to simulate shearing load of glass based on the shear tests of composite panels see Fig. 10. Polyurethane adhesive was used to connect glass with loading fixture. The rate of cure adhesives emerged as a major influence on the results of shear tests, the tentative results show a large variance. Research in this area will continue.

3 Test of electrobus roof segment

To assess the resistance of the entire system, experiment of roof and side pillars was designed, when was loaded roof segment with a length of 700 mm with side pillars. The segment was loaded by a hydraulic actuator, when the force and the displacement of the piston rod were monitored. Loading force direction corresponded to the initial direction of the contact force during the bus rollover test. Segment after the test is shown in Fig. 12. Two type of specimen's were tested, specimens Segment-01–03 have not overlap panel between roof and side profile, specimens Segment-04–06 were complemented with overlap in this location.

Numerical model tests of roof segment was created and the results obtained correspond to experiment, when first FEM model corresponds to situation without overlap and second model was designed with overlap. Results of roof segments experiments and numerical simulation were plotted in Fig. 11. It can bee seen, that overlapping has no significant effect on the strength of the segment. Plastic hinge pillar was the result of test, sandwich roof carry introduced load without significant damage.

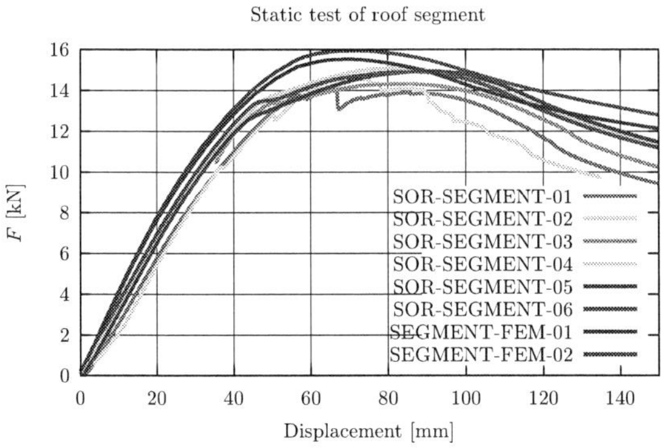

Fig. 11: Roof segment test – comparing experimental and numerical results.

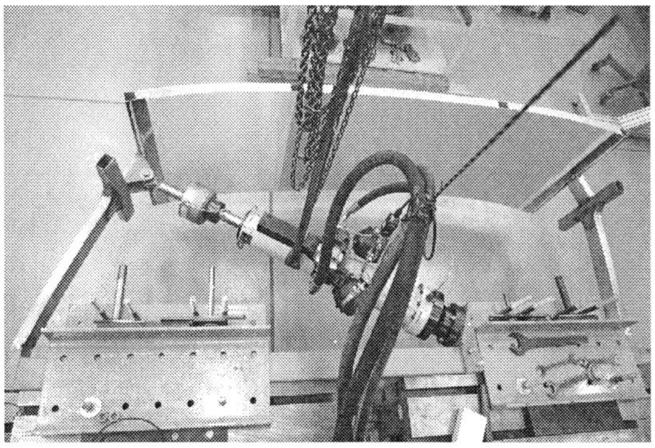

Fig. 12: Electrobus roof segment experiment.

4 Conclusion

Experiments and numerical calculation of basic elements were carried out. It was confirmed carrying capacity of the roof designed variants bus and validated numerical solution useful for further development. An important result of testing is needed to control technology in production to achieve roofs represent a consistent properties with regard to fatigue loading.

Acknowledgement

This work was supported by the projects TA CR: TH01021172 – Research and development uni fied series of high-capacity electric bus and MIT CR: FR-TI4/349 – Research and development of the skeleton of the bus from unconventional materials.

References

[1] Fojtl, L. - et al.: Evaluation of Mechanical Performance and Failure Modes of Sandwich Structures for Transport Industry . In Experimental Stress Analysis 2015. Prague: Czech Technical University, 2015, p. 94-97. ISBN 978-80-01-05734-6.

[2] Doubrava, K. - Novotný, C.: Static and Fatigue Testing Sandwich Segment of the Bus Roof. In Experimental Stress Analysis 2015. Prague: Czech Technical University, 2015, p. 66-69.

[3] ASTM International: C393/C393M 11 Standard Test Method for Core Shear Properties of Sandwich Constructions by Beam Flexure [online], [cit. 2012-11-29], Accessible from URL: <http://enterprise.astm.org/SUBSCRIPTION/NewValidateSubscription.cgi?C393/C393M-HTML>.

[4] Doubrava, K.: Static and Cyclic Tests of Sandwich Parts. [Report]. 2015. 12105/2015/26. 26 p. (in Czech).

[5] Doubrava, K.: Bending and shear tests of the windscreen and bending tests reinforced profiles [Report]. 2015. 12105/2015/18. 20 p. (in Czech).

Test and Simulation of Energy Absorption System

R. Doubrava [1,a]

[1] *Výzkumný a zkušební letecký ústav, a.s.*
[a] *doubrava@vzlu.cz*

Abstract: An energy absorbing device is a dedicated structural member to absorb an optimum amount of energy from different types of degradation in the constituent materials in an accidental event. Indeed, the structures of a plane or a car must absorb the kinetic energy introduced during a crash for the human does not suffer from acceleration exceeding the allowable physiological threshold. This paper utilized advanced nonlinear finite element modelling for simulating tests performed by the VZLU to understand and analyses loading behaviors and parameters that influence the failure of metal material. The research was focusing on application of control failure for energy absorption system design and application. This study shows the possibility of applying realistic boundary conditions using a simplified numerical model and transferability to the complex problem analysis. This paper contributes to the region of scale test design for verification of numerical analysis of complex engineering problems such as analysis of emergency cases.

Keywords: impact; FE; energy absorption.

Nomenclature:

A	:	Yield stress (Johnson Cook material model)
B	:	Strain hardening modulus (Johnson Cook material model)
C	:	Strain rate coefficient (Johnson Cook material model)
E	:	Young's modulus
n	:	Strain rate coefficient (Johnson Cook material model)
d1,d2,d3	:	Triaxial stress state (Johnson Cook damage parameter)
d4	:	High strain rate stress state (Johnson Cook damage parameter)
d5	:	High temperature environment (Johnson Cook damage parameter)
FE	:	Finite element
ε_0	:	Strain rate factor
ρ	:	Density
ν	:	Poisson ratio

1 Introduction

Simulations are useful support tests carried out on scale models. The applications of these results to real situations depend on the appropriateness of the assumptions used to simplify the models and evaluations of their influence. For numerical simulations, solvers that produce explicit solutions are most commonly used. The solutions to these simulations require a numerical convergence of the variables and are dependent on the quality and the size element of the finite elements used in the model. To incorporate certain parameters, e.g., material type or stiffness, it is often necessary to discretize alternative models.

2 Problem Formulation

In aircraft, the forces involved when a collision takes place are enormous, and will produce permanent deformations, cracks, local buckling, collapse and tearing of the plane structure. The objective of this work is to find and validate the properties of a specified material to be used when performing a FE-analysis considering large deformation and ductile fracture of the material. The main objective is to make simplifications to the input data in order to simplify the modelling procedure and concentrate studies in a simply models to be able to execute it in the real model.

2.1 Design of energy absorption system

The absorber system is design for application in aircraft passengers seat (see figure). The aim of this system is absorbing high energy during emergency lending. The figure 1 shows 3D model of passenger seat and detail of energy system application.

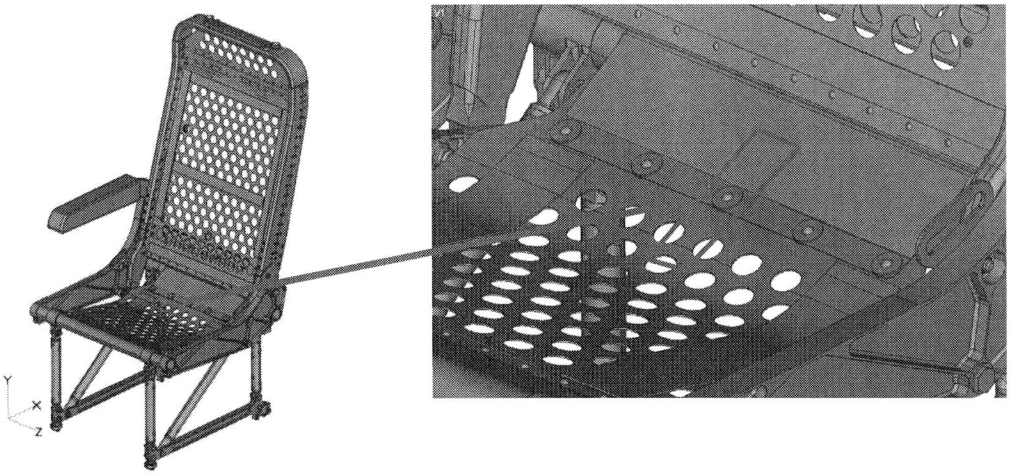

a) 3D model of passenger seat b) Selected area with energy absorption system

Fig. 1: Passenger seat with application of energy absorption system

The figure 2 shows proposed simplify model of energy absorption system for test and numerical verification.

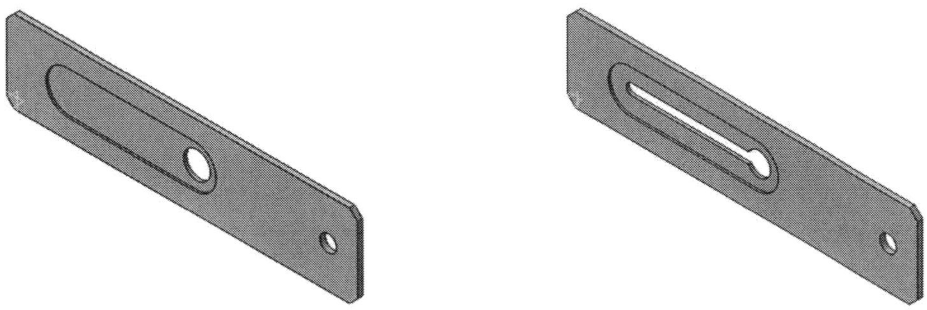

a) 3D model b) Detail of 3D model

Fig. 2: Energy absorption simplify model

2.2 Test

The test consisted to apply a tension created by a drop of weight and analyses the strain contained inside the sample and also measured the evolution of the force in relation to the time of the experiment. The schema of test is on the figure3.

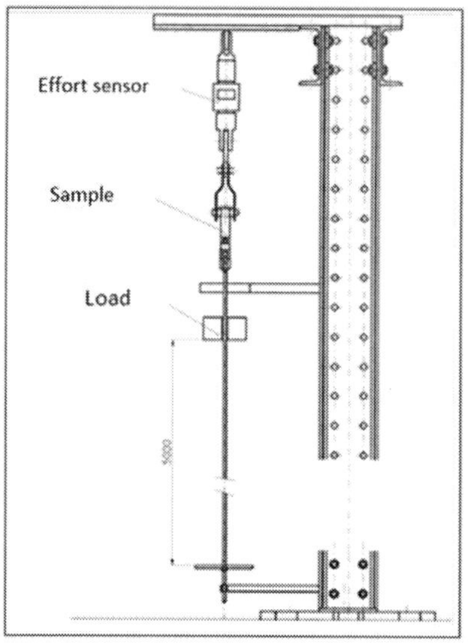

Fig. 3: Drop test set up

The results from test were analyzed from point of view reaction force measurement and plastic deformation (figure 4).

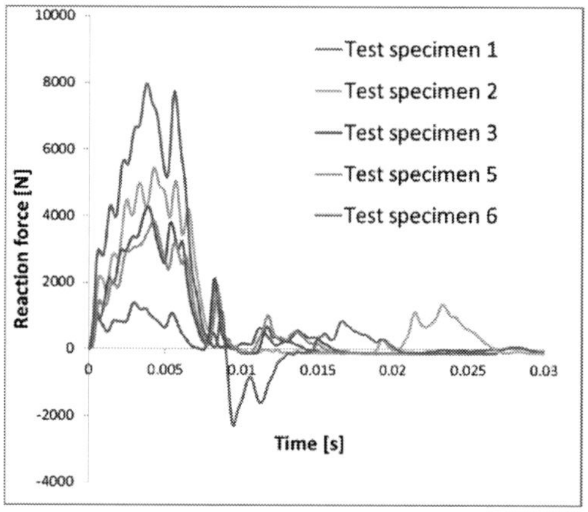

a) Reaction force results

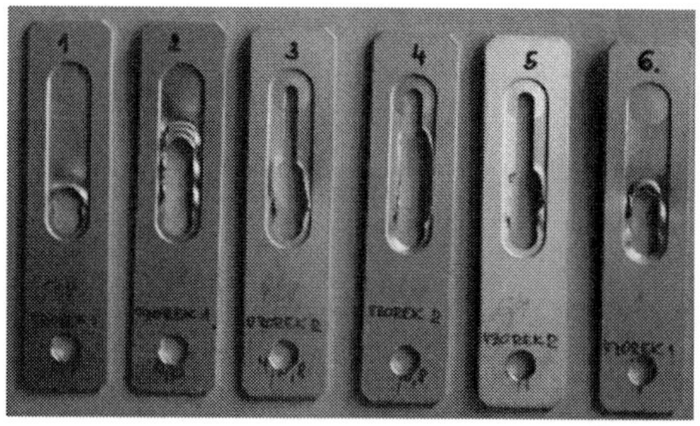

b) Deformed shape of test specimen

Fig. 4: Results from test (force vs time and deformed shape)

2.3 FE model of test

Models of test specimen and drop test set up were computed in ABAQUS code. The model was built from 2D 3D elements with Al alloy 2024 T351 isotropic material properties. The figure 5 and figure 6 shows proposed 2D a 3D FE model.

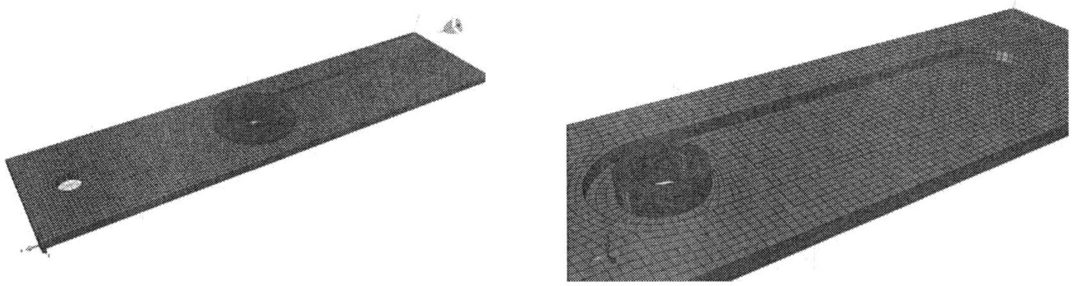

a) 3D model b) Detail of 3D model

Fig. 5: 3D FE model

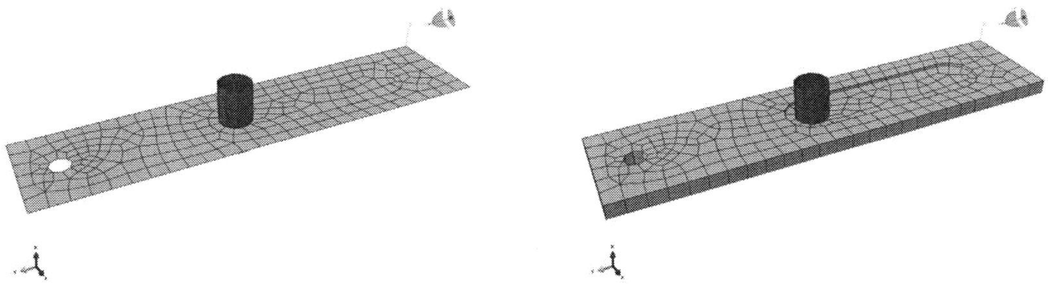

a) 2D model b) 2D model with thickness visualization

Fig. 6: 3D FE model

A Johnson-Cook material model was used for the body elements of the proposed energy absorption system and is an accumulative-damage model that accounts for the loading history, which is represented by the strain-to-fracture parameter expressed as a function of the strain rate, temperature, and stress. This model is an instantaneous failure model, which means that no strength or stiffness remains after failure. Failure was also dynamically modelled based on the equivalent plastic strain at individual element integration points. Failure was assumed to occur when the damage parameter exceeded 1. The parameters of the incorporated Al alloy material model are listed in table 1.

Table 1: Properties of Johnson Cook material model of 2024 T351 Al alloy [2]

Parameter	ρ [t/mm^3]	E [GPa]	ν [1]	A [MPa]	B [MPa]	n [1]	C [1]	ε_0 [s^{-1}]	d1 [1]	d2 [1]	d3 [1]	d4 [1]	d5 [1]
Value	2.77E-09	73.1	0.33	352	440	0.42	0.0083	1	0.13	0.13	-1.5	0.011	0

The result from FE simulation was compared with test results from point of view reaction force measurement and deformed shape of test specimen. The figure 7 shows example of the comparison between test and simulation.

.

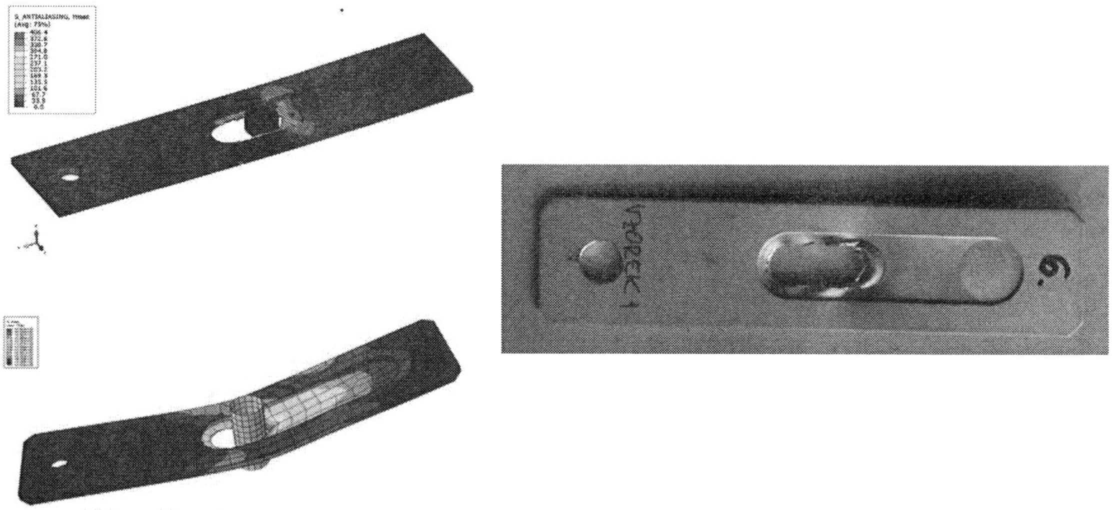

a) Results from FE analysis on 3D and 2D model

b) Results from test

Fig. 7: Comparison between FE simulation and test from point of view deformed shape

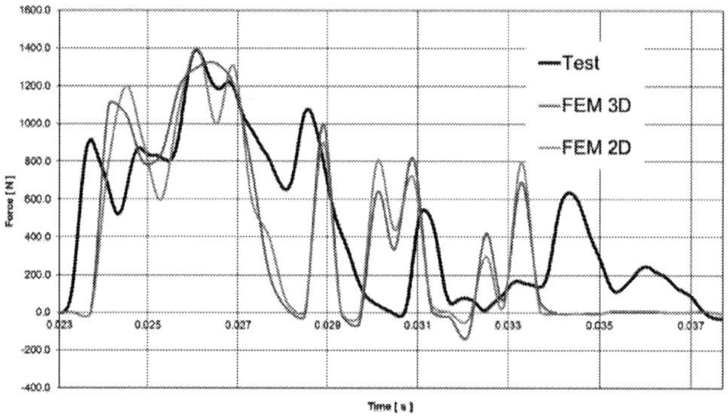

Fig. 8: Reaction force vs time

The set-up from 2D FE model was used for simulation of real structure with help to minimize mesh size especially from point of view explicit FE analysis.

The figure 9 shows summary of energy absorption system design and verification on the base of simplify model for test and simulation.

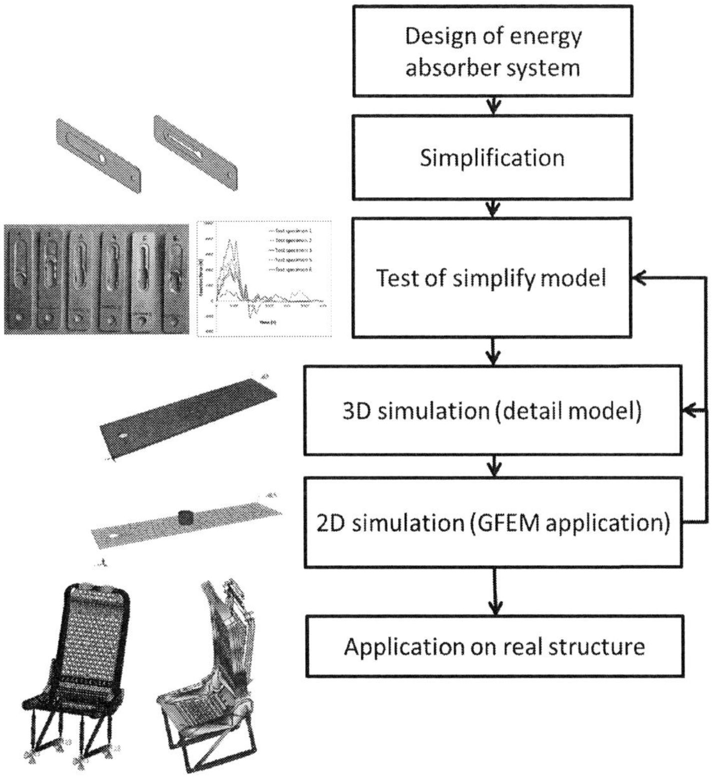

Fig. 8: Shema of energy absorption system design and optimization

3 Conclusion

This study showed the possibility of applying simplification of FE model (from 3D to 2D) by verification of test results. The simplified FE model verified by test is possible apply on real global structure simulation and optimization.

Acknowledgement

This study received financial support through the research organization development program of the Ministry of Industry and Trade of the Czech Republic.

References

[1] V. Strnad, B. Mezihorák: Protocol from dynamic test of seat plate, Protocol VZLU P-12/3220/04 (2004).

[2] W. Wang: Simulation of Hard Projectile Impact on Friction Stir Welded Plate, MSc Thesis, Cranfield University (2011).

Modernization of a vibrating machine

T. Drexler [1,a], L. Limberg [1,b],

[1] *The Faculty of Mechanical Engineering, The University of West Bohemia, Univerzitní 8, 306 14 Pilsen, CZ*
[a] *drexler@kks.zcu.cz,* [b] *limbi@kks.zcu.cz*

Abstract: This paper deals with a structural modification of parts of a vibrating machine and subsequent verification of the functionality of individual components and the whole unit. A new concept of driving parts was developed as a part of the machine modernization. It allows to improve the functionality and extends the capabilities of the machine settings. Because an intervention in the driving parts was considerable a re-design of bearings and drive shafts had to be carried out.

Keywords: Vibrator; bearings; centrifugal force.

1 Introduction

The machine which was modernized is a vibrator for technological purposes. This device is used during manufacturing of building blocks which are compacted while shaking. The device is a part of a loading machine which fills a mold with concrete and compacts it to its shape. The function of the device is based on forced rotation of eccentric weights placed on a rotating shaft. This system is put into motion by an electric motor through a V-belt. The vibrating machine consist of a pair of vibrating housings on a common base. The original design was working properly but was outdated and difficult to operate. The main requirement of this modernization was to enable force effect changing of the vibrator without a need of stopping.

2 The original design

The original design of one vibrating housing consisted of two eccentric weights on one shaft. This shaft was mounted on four bearings (two radial ball bearings and two radial roller bearings) which were placed at both ends of the shaft. Two eccentrics were located between the bearings. This type provided the compact size of bearings as the force distribution was favorable. This original solution did not satisfy the terms of maintenance during changing of the vibrator force. The force effect of the vibrator could be changed by mutual rotation of two eccentrics but only at a standstill of the machine and after removing of related covers.

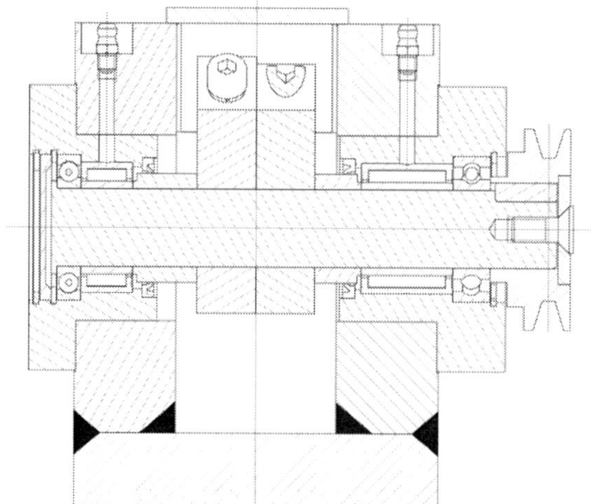

Fig. 1: The original design

3 The upgraded design

The original solution was sufficient in terms of functionality but not in terms of efficiency. It was necessary to change the design to allow adjustment of the intensity of vibration without stopping the machine. The ideal way to achieve this requirement was to split each vibrating housing into two parts so that the eccentrics can be continuously mutually turned. Vibration intensity is then given by a phase rotation of the eccentrics in each housing. This design change necessitated the addition of a second driving servo-motor and adding of a second drive train with a toothed belt. The vibrator is put into operation with no vibrating effect and the vibration can be turned on and off as needed by rotating of one eccentric weight about a desired angle.
The shaft was divided into two shafts during upgrade. Each shaft carries its own eccentric and is supported by two bearings. Force effects acting on the shaft bearings after the split are substantially more unfavorable which is reflected in size of the bearing. The position of bearings had to be modified as well.

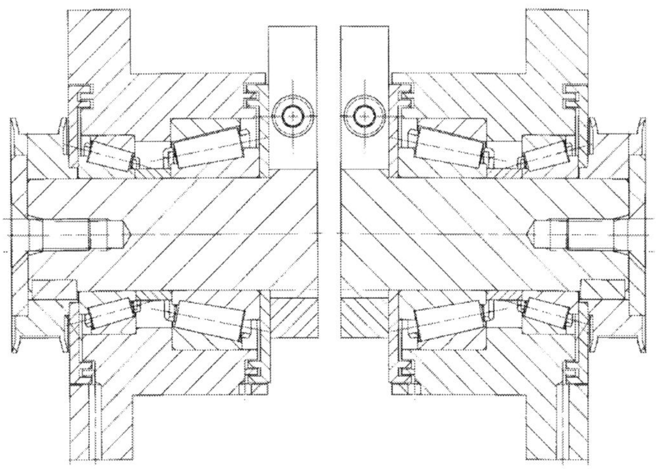

Fig. 2: The upgraded design

3.1 Inertial moments

The first requirement was to choose a right servo-motor by the starting torque to reach machine's operating speed. Each of the drive branches includes a pair of shaft with eccentric cams, two idler pulley and one drive pulley. Effect of individual components was examined based on the inertia and rotating speed of those geared parts.

Tab 1. : Effects of inertial moments

Component	Percentage effect	Inertia moment of components
Motor	*2%*	$408\ mm^2 \,.\, kg$
Pulley 26t	*12%*	$222\ mm^2 \,.\, kg$
Pulley 56t	*27%*	$4637\ mm^2 \,.\, kg$
Shaft + Eccentric weight	*59%*	$1089\ mm^2 \,.\, kg$

Total time dependence of starting torque could be examined after considering effects of all rotating parts.

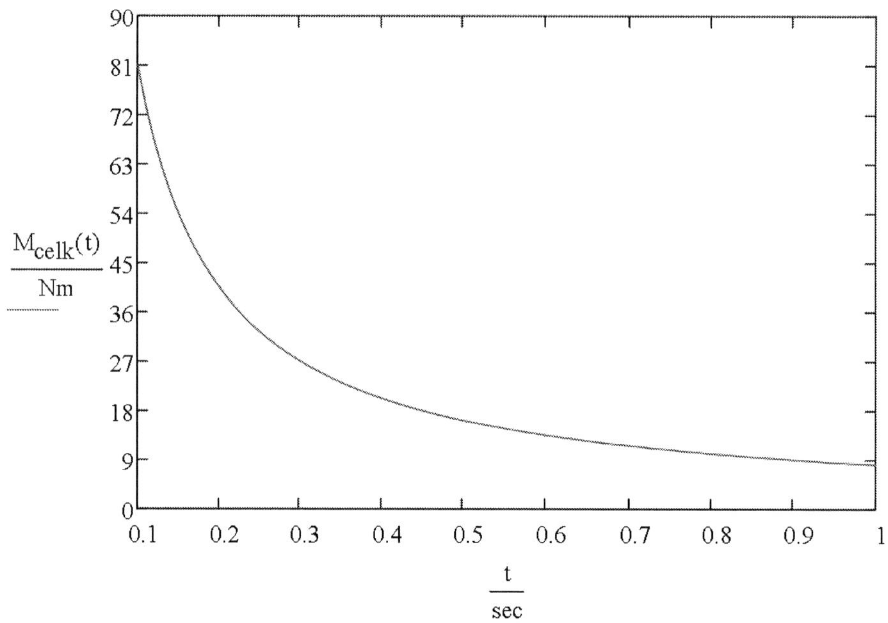

Fig. 1: Time dependency of starting torque

As the force effect of the vibrator is changed by rotation of one eccentric weight it was necessary to verify this motor torque demands as well. This torque is again dependent on the amount of time needed to make this adjustment. This dependence is described in the following graph.

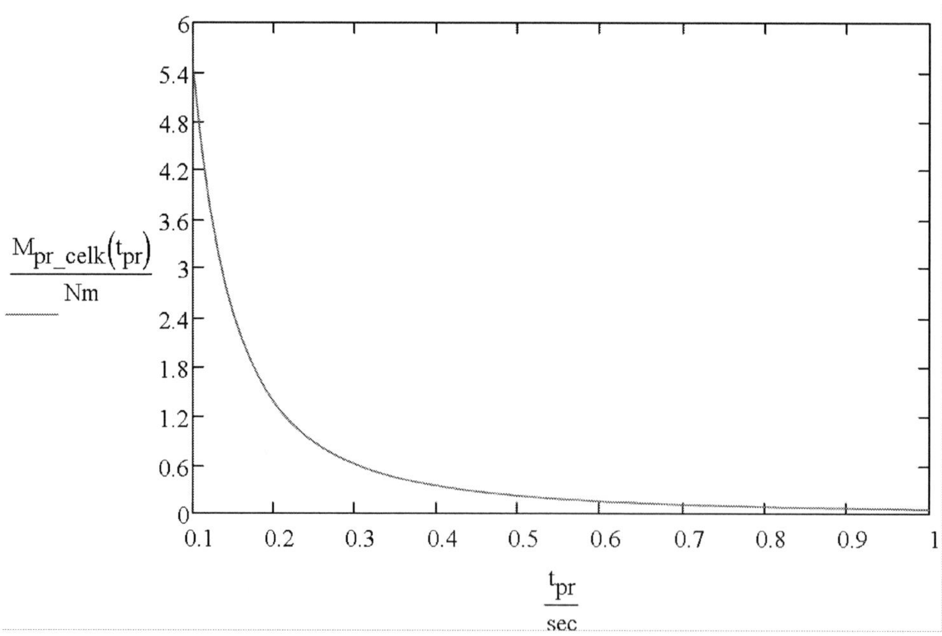

Fig. 2: Time dependency of mutual rotation torque

3.2 Centrifugal forces

Another requirement was to design appropriate bearings. Since each of the shafts is a beam on two supports it was necessary to determine bearing reactions. These reactions depend on the centrifugal force (Fd) generated by the rotation of the eccentric weights.

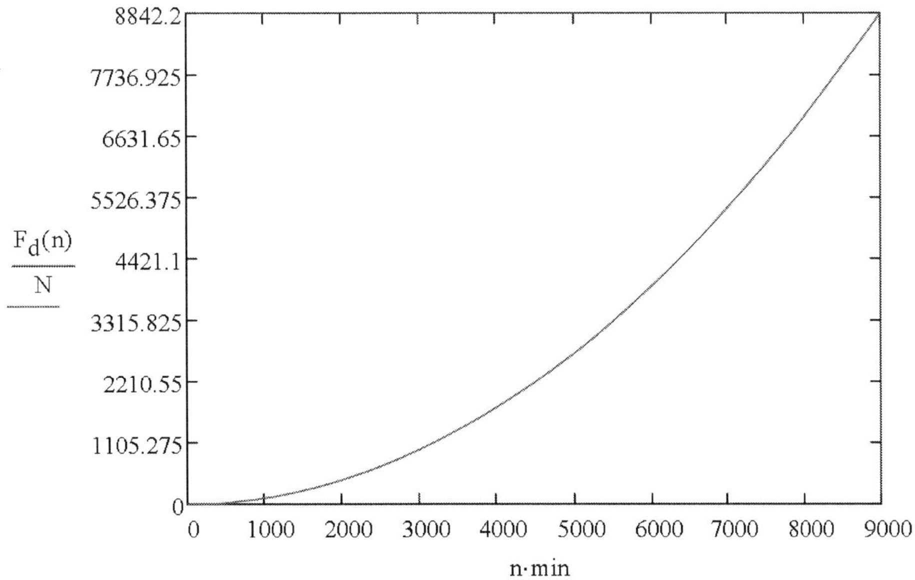

Fig. 3: Revolutions dependency of centrifugal force

The machine operates at three different states which are 9000 1000 and 5000 revolutions per minute. As the shaft is supported beam, the centrifugal force (Fd) generates reactions in the bearings. The table below contains calculated reactions in the bearings for each state.

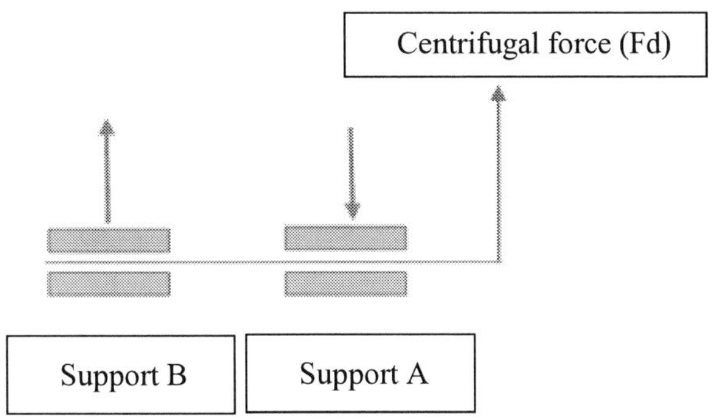

Fig. 4: Shaft support scheme

Tab 2: Reaction forces

Operation state	Reaction force (support A)	Reaction force (support B)
State 1 (Fd = 8 842 N)	*13 008,41 N*	*4 166,21 N*
State 2 (Fd = 109 N)	*428,68 N*	*319,51 N*
State 3 (Fd = 2 729 N)	*4 609,74 N*	*1 880,66 N*

Compared to the original solution, the new concept with the split shafts has disadvantage in terms of force effects and the bearings had to be adjusted from the original proportions to higher dimensional ranges. Bearing type had to be changed to the taper bearing which is characterized by reducing the force ratios by moving the shaft support towards the eccentric weights.

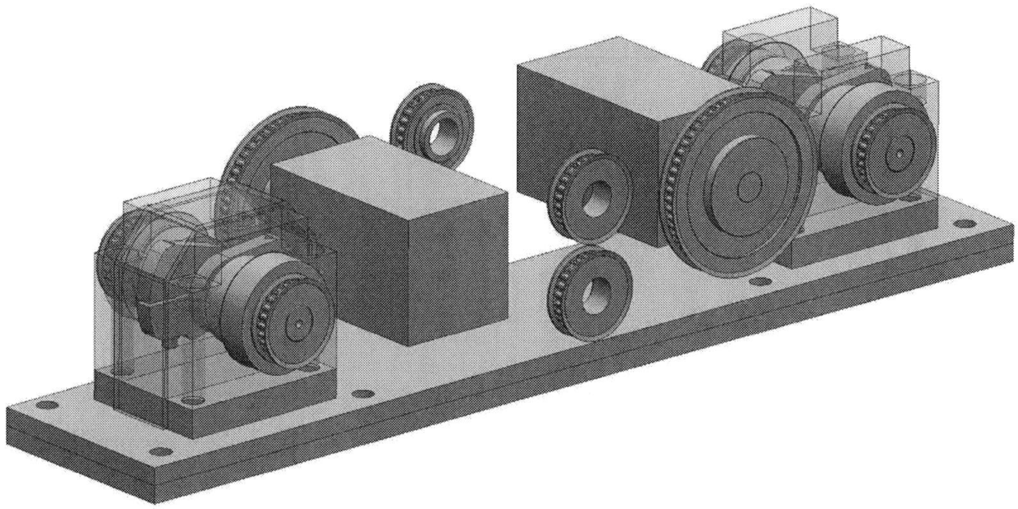

Fig. 5 : Design structure of improved device

4 Conclusion

The modernization of this vibrator was made by dividing driven shafts to separate controllable groups which extended the functionality of the device. This extension is especially advantageous in terms of service but also in terms of technological processes. The force effect of the vibrator can now be changed while running and quickly adapted to the actual requirements of a vibrated product. The new design meets all the requirements and can be safely put into operation.

References

[1] SKF Group.: Hlavní katalog SKF (6000 CS), 2007

[2] HOSNEDL, Stanislav a Jaroslav KRÁTKÝ. *Příručka strojního inženýra: obecné strojní části*. Vyd. 1. Praha: Computer Press, 1999. Edice strojaře. ISBN 80-7226-055-3.

[3] LEINVEBER, Jan a Pavel VÁVRA. *Strojnické tabulky: pomocná učebnice pro školy technického zaměření*. 1. vyd. Úvaly: Albra, 2003. ISBN 80-86490-74-2.

[4] BROUSIL, Jaromír SLAVÍK, Jaromír, ZEMAN, Vladimír, : Dynamika /1. vyd. Praha : 1989. ISBN 80-03-00164-1.

Loading of the Axle Gearbox for Metro During Drive Simulation

M. Dub[1,*], F. Lopot[1,2], O. Berka[1], V. Dynybyl[1]

[1] *Czech Technical University in Prague, Faculty of Mechanical Engineering, Department of Designing and Machine Components, Technická 4, 166 07, Prague 6, Czech Republic*

[2] *Charles University in Prague, Faculty of Physical Education and Sport, Department of Anatomy and Biomechanics, Jose Martího 31, 162 52, Prague 6, Czech Republic*

[*] *martin.dub@fs.cvut.cz*

Abstract: During working of the rail vehicle the parts of the drive are dynamically loaded by the input torque on one side and by the roughness of the rail track on the other side. The axle gearbox which is used in partly sprung drive of the metro unit is placed in the test rig for dynamic loading in company Wikov MGI a.s. The stress states of the gearbox and the hinge are investigated during the dynamic loading by means of the strain gauge measurement. Reached results are verified by the analytical calculations and finite element analysis.

Keywords: Strain Gauge; Axle Gearbox; Hinge; Rail Vehicle; Partly Sprung Drive; Drive Simulation; FEM; FEA.

1 Introduction

During working of the rail vehicle the parts of the drive are dynamically loaded by the input torque on one side and by the roughness of the rail track on the other side. Those additional loads can be significant. It is necessary to describe the dynamic loads to design appropriate parts correctly. There are various methods to reach the quantities of those loads such as analytical calculations or finite element method (FEM). It is convenient to verify calculated results by means of experimental measurement to confirm the correctness of the assumptions.

This paper deals with the measurement during the testing of the axle gearbox for metro unit. There is used the partly sprung drive, which is consisted of asynchronous motor and one-stage gearbox. The axle gearbox is embedded on the wheelset axle. Moreover, it is supported by means of the gearbox hinge, which is there to catch the reaction force caused by the torque of the wheelset. The gearbox hinge can be conveniently used for the identification of the loads, especially of the torque [1], [2], [3].

Fig. 1: The test rig with the one-stage axle gearbox.

2 Test Rig

The gearbox is placed on the test rig for dynamic testing of rail vehicle gearboxes in company Wikov MGI a.s., see Fig. 1. This test rig enables long–term runtime testing of the various types of gearboxes for rail vehicles, including testing of gears, [4]. It is possible to load the gearbox both by the torque in the closed loop and by the simulation of dynamic shocks. Those are generated by the set of hydraulic cylinders, which are installed in the triaxial arrangement, for more information see [5] and [6].

3 Measurement

The aim of the measurement is to verify quantities calculated by usage of FEM or analytically and also describe the behaviour of the tested gearbox during torque loading and during the dynamic shocks.

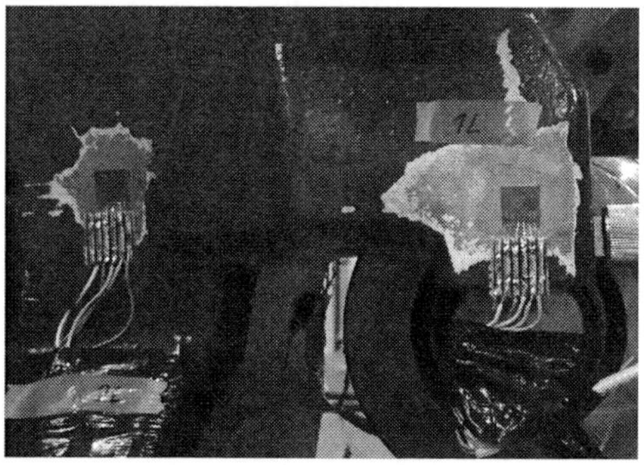

Fig. 2: Strain gauge rosettes installed on the gear box.

Fig. 3: Uniaxial strain gauges installed on the gearbox hinge.

3.1 Hinge Measurement

The gearbox is supported by the angled hinge, which is connected to the test rig frame, see Fig. 3. The hinge has to take the reaction force generated by the torque of the wheelset, but it has to allow some specific movements of the gearbox as well. The hinge is embedded into the test rig frame and into the gearbox by means of rubber–metal joints.

Four uniaxial strain gauges are used for the strain and stress measurement of the hinge. Those strain gauges were installed on each side of the rectangular cross–section of the hinge. Two of them are placed in plane perpendicular to the axle axis (strain gauge 5 and 6) and two are placed in the plane parallel to the axle axis (strain gauge 7 and 8), see Fig. 6. Both the hinge and the gearbox were tested in various load regimes. Those regimes consists of the combination of torque loading and dynamic shocks in chosen directions. In Fig. 4 and 5, there can be seen that strain gauges 7 and 8 are completely unloaded. Further, the stress level due to the torque loading remains unchanged in both types of loading. From the Fig. 5 can be analysed the response to the shock loading.

3.2 Gearbox Measurement

The cast gear box is loaded through the bearing loads and also by the reaction force of the hinge. To investigate desired strains and stresses the strain gauge measurement is designed. There are used strain gauge rosettes because of complex stress state of the gear box. The reduced stress could be calculated from obtained stresses afterwards. The strain gauge rosette installed on the gear box is shown in Fig. 2. The temperature measurement of the gearbox during load test is also carried out. This measurement was carried out to compare obtained data with results from FEA. That has been done internally in the company and the results are not available.

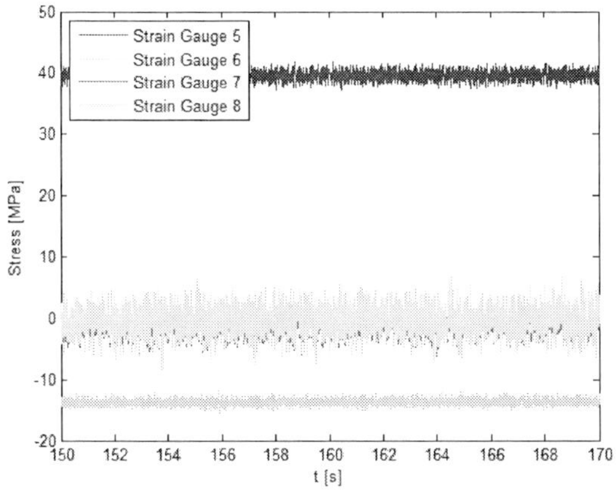

Fig. 4: Stress response of the hinge to the pure torque loading.

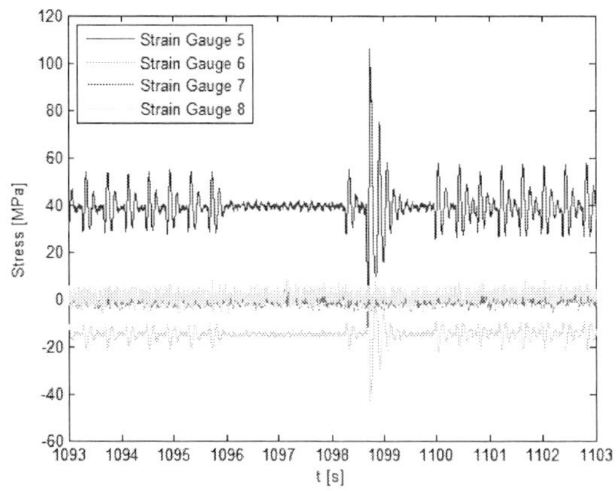

Fig. 5: Stress response of the hinge to the combination of torque loading and shocks.

4 FEA Verification

As a verification of measured stresses the FEM calculation is carried out. From known parameters of the gearbox power, torque, etc. and available 3D CAD model of the hinge the FEM model can be created. The reaction force in the hinge can be calculated analytically. For the one–stage gearbox (two gears) the reaction force S can be calculated according to the 1, [2].

$$S = \frac{M_M\,(i_c + 1) - J\ddot{\psi}}{n} \tag{1}$$

M_M is the torque of the asynchronous motor, J is the moment of inertia of the gearbox without the big gear, $\ddot{\psi}$ is angular acceleration of the gearbox sway round the axle of the wheelset (i.e. the second derivative of the rotation angle of the gearbox round the axle of wheelset), n is the perpendicular distance from the hinge to the axle (output shaft axis), see Fig. 6, and i_c is the total gear ratio of the gearbox. If the dynamic contribution to the reaction force S is neglected, the desired value of the reaction force S for the torque loading test regime can be calculated as follows:

$$S = \frac{M_M\,(i_c + 1)}{n} = \frac{1.8e6\,(5.\bar{5} + 1)}{406.5} = 29028\ \mathrm{N} \tag{2}$$

This force is used as a loading force in the FEM. The comparison to measured data is carried out for the pure torque test regime (Fig. 4) and it can be seen in Tab. 1. Desired quantities are read in two test points SG 5 and SG 6, which are designated in Fig. 6. The FEA results can be seen in Fig. 7. On this figure there is shown von Mises stress distribution, although for the comparison with measured data the stress in strain gauge longitudinal direction is used.

Tab. 1: The comparison of measured and FEA results – uniaxial stress in the longitudinal direction of the strain gauges.

Test Point	Measurement [MPa]	FEA [MPa]	Ratio [-]
SG 5	40.5	41.2	1.017
SG 6	-14.6	-15.1	1.034

From the Tab. 1 there can be seen that the results from the FEA are in very good agreement with the measurement. Certain error is caused by the accuracy of test points positioning in FEA according to the experiment.

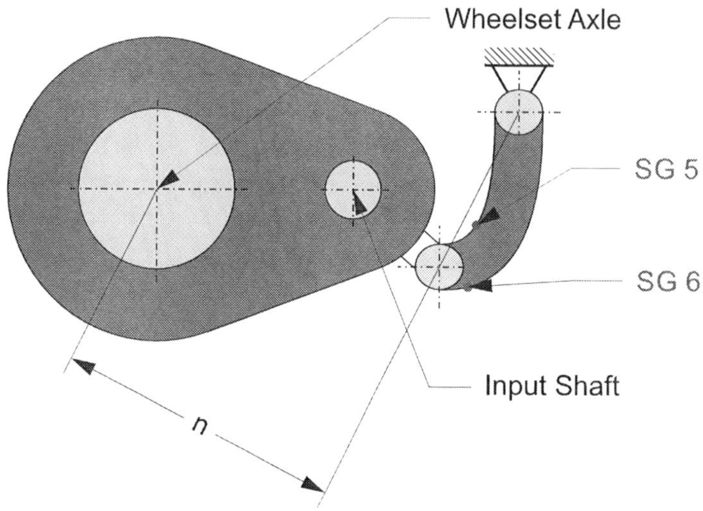

Fig. 6: Scheme of the partly sprung drive, strain gauges location.

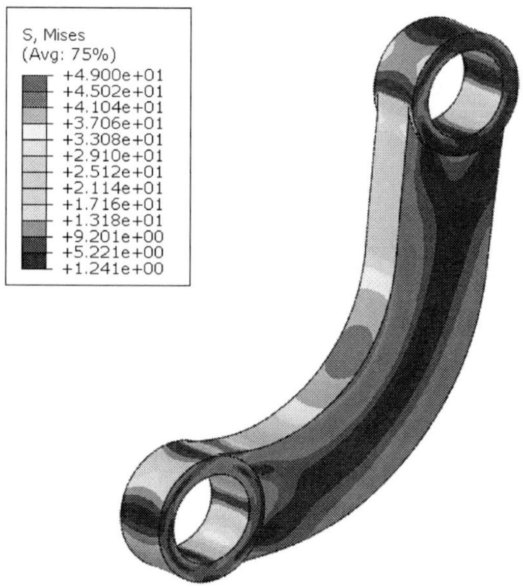

Fig. 7: Von Mises stress distribution in the gearbox hinge.

5 Conclusion

This measurement was carried out on the test rig for testing rail vehicles gearboxes. The response of the gearbox for rail vehicle and especially the hinge to the torque load and to the dynamic shocks was analysed. Those loads should simulate the real loading during working of the rail vehicle. The gearbox hinge proves to be useful and convenient object for the observation and analysis of the driving conditions and torque loading even during real drive. Loading by the torque was verified by the analytical calculation and FEA. Obtained results are in very good agreement with the measured data.

Acknowledgement

This work arose during industrial cooperation with Wikov MGI a.s. The project was supported by Czech Ministry of Industry and Trade in a program TIP: FR-TI3/261.

References

[1] J. Kolář, Design of a Wheelset Drive, Transactions on Electrical Engineering, Vol. 4 (2015), 11-19.

[2] M. Dub, etal., Analysis of a Partly Sprung Drive, Transactions on Electrical Engineering, Vol. 4 (2015), 5-10.

[3] M. Dub, etal., Experimental Analysis of Stress State and Motion of Tram Gearbox Hinge, in proc.: EAN 2014 - 52nd International Conference on Experimental Stress Analysis (2014), P. Polach, VZU Plzen, Marianske Lazne.

[4] O. Berka, etal., Mesh Quality Monitoring in Evolventn Teeth in Gearboxes under Pperation -Signal Identification, in proc.: XXI IMEKO World Congress "Measurement in Research and Industry" (2015), J. Holub, CTU in Prague, Prague, 1582-1586.

[5] K. Petr, V. Dynybyl, J. Chmelař, Test Rig for Gearboxes of Rail Vehicles, Modern Methods of Construction Design, Springer International Publishing AG (2014) 169-175, doi: 10.1007/978-3-319-05203-8_24.

[6] K. Petr, V. Dynybyl, Methods of testing gearboxes of rail vehicles, Applied Mechanics and Materials, Uetikon-Zurich: Trans Tech Publications (2014) 277-282, doi: 10.4028/www.scientific.net/AMM.486.277.

Development of Structural Health Monitoring System for Ultralight Aircraft

M. Dvorak [1,a], J. Demuth [2], M. Stepanek [3], N. Schmidova [1] and M. Ruzicka [1]

[1] *Czech Technical University in Prague, Faculty of Mechanical Engineering, Technicka 4, 166 07 Prague 6, Czech Republic*

[2] *Safibra, s.r.o., Cernokostelecka 1621, 251 01 Ricany, Czech Republic*

[2] *Phoenix Air, s.r.o., Lochmanova 64, 56201 Usti nad Orlici, Czech Republic*

[a] *milan.dvorak@fs.cvut.cz*

Abstract: Development of Structural Health Monitoring (SHM) system based on a flight data acquisition device is described in the article. System consists of a fiber optic interrogation device, analog flight data sensors and optical Fiber Bragg Grating (FBG) sensors, which are embedded in main structural joints of wings and in stabilisers.

Keywords: FBG sensor; SHM; composite; aircraft; flight measurement.

1 Introduction

The main objective of our research is a development of system for monitoring of flight loads and flight parameters, such as speed, altitude, acceleration etc. Measured flight data will be then used for design of suitable SHM model for critical parts of airframe. Data acquisition device and related sensors are mounted in the Phoenix Air S-LSA Motorglider (see Fig. 1).

The airplane has all-composite construction, structural parts are bonded together using the epoxy adhesive. Adhesive joints are critical points of technology procedure. During the operation of wing skins assembly there is no direct control of bonded joint. Airframe damage during flight because of bad quality of bonding was observed in the past. Some of the manufacturers check quality of final assembly using the small cameras, but there is no method used for in-flight measurement of adhesive joints performance/health.

Fig. 1: Phoenix Air S-LSA Motorglider [1].

We have already shown earlier that FBG sensors embedded in the glued joints or carbon/epoxy spar caps are suitable for the monitoring of critical places of the wing structure during the torsional and bending stiffness tests [2, 3]. Results obtained from the FBG sensors were fully comparable to readings from the electrical strain gauges and were in good agreement with the results from analytical and numerical wing loads models. Moreover, it was shown in [4] that optical fibers in the glued joints don't lower the static shear strength or fatigue performance of such joint.

2 Description of the Monitoring System

In-flight data acquisition system consists of FBG interrogator Safibra FBGuard (modified to measure both optical and analog electric signals) and various sensors. System is designed to measure common flight loads, not taxiing or flutter loads, thus at least 5 Hz scanning frequency is required.

2.1 Sensors for the Strain Analysis of Airframe

The strain monitoring sensor system is based on chains of optical FBG sensors (multiple strain sensing areas on a single optical fiber). The following parts of the airplane are equipped with the sensors:

- *vertical stabiliser* (4 surface-mounted strain FBGs),
- *horizontal stabiliser* (4 embedded strain FBGs),
- *left wing* (26 embedded strain FBGs, 2 surface mounted temperature FBGs, conf. A),
- *right wing* (52 embedded strain FBGs, 4 surface mounted temperature FBGs, conf. A; 10 embedded strain FBGs, conf. B).

Configuration of the FBG sensors in the wings is schematically pictured in the Fig. 2 (top view). Positions of FBG chains inside the wing spar cross-section are pictured in the Fig. 3.

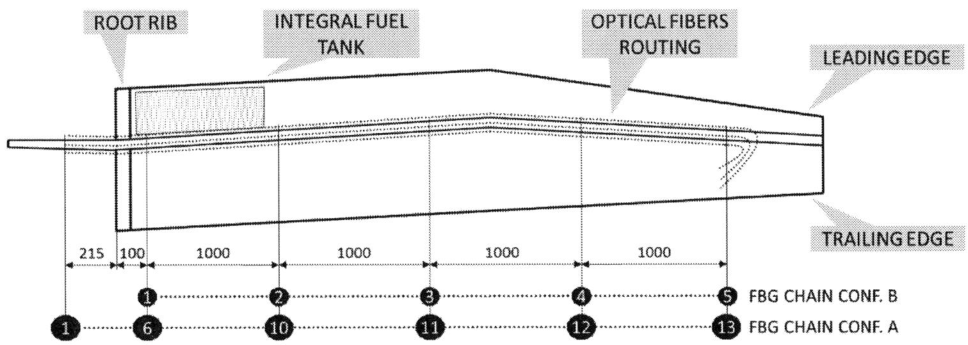

Fig. 2: FBG sensors configuration in the wing (top view).

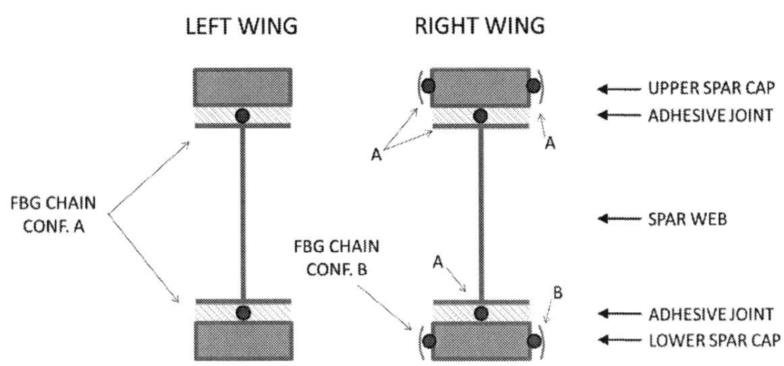

Fig. 3: FBG sensors configuration in the wing spars (cross-section).

Optical fibers, which were placed on the sides of the spar caps, were covered with the thin glass fabric for additional protection (see Fig. 3 – right wing spar cross-section). Optical fibers, which were embedded into the adhesive joints, were at first attached to the surface of the spar caps (using the fast curing adhesive, see Fig. 4) and then buried into the layer of epoxy adhesive. All optical fibers or cabling routes inside the wings were protected using the plastic tubes and boxes made from hard foam (to protect ingress points of the fibers, see Fig. 5).

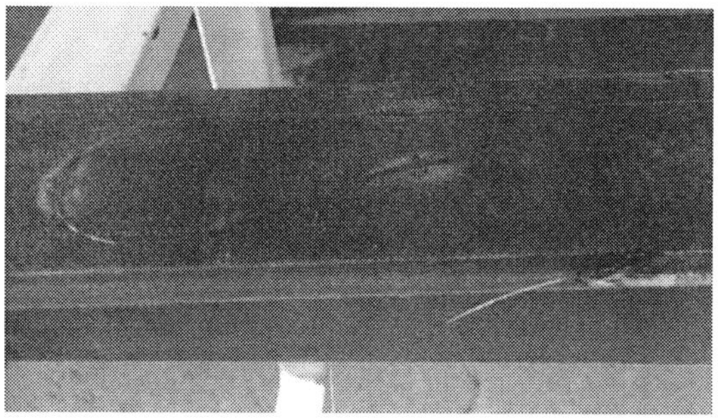

Fig. 4: Detail of optical fiber on the spar cap, just before application of epoxy adhesive.

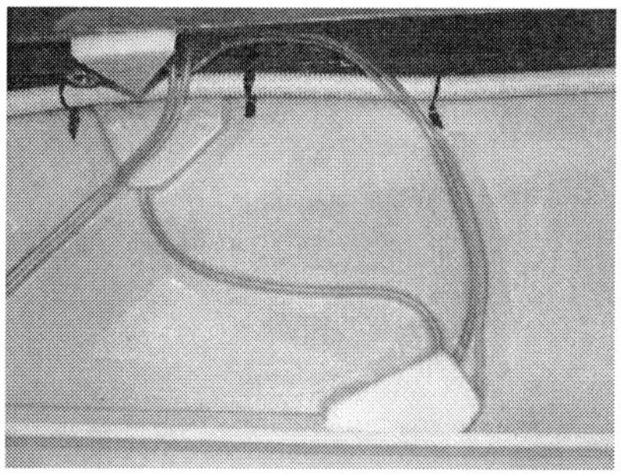

Fig. 5: Detail of protection of optical fibers and cablings inside the wing.

2.2 Sensors for the Measurement of Flight Parameters

Strain readings from the FBG sensors alone provide only partial amount of information about the airframe behavior during the service. Flight parameters such as acceleration, speed, deflection of control surfaces etc. are needed for detailed analysis of the flight conditions/loads. The following flight parameters are measured:

- *altitude* (absolute pressure sensor connected to a pitot-static system),
- *speed* (differential pressure sensor connected to the pitot-static system),
- *engine speed* (pulse counter and frequency to voltage converter connected to an ignition coil),
- *pressure in the engine intake* (absolute pressure sensor),
- *acceleration in the airplane center of gravity* (3-axis accelerometer),
- *acceleration in the empennage* (3-axis accelerometer),
- *deflection of ailerons* (linear/single-turn potentiometer),
- *deflection of an elevator* (linear/single-turn potentiometer),
- *deflection of a rudder* (linear/single-turn potentiometer),
- *deflection of flaps* (four positions, linear/single-turn potentiometer),
- *deflection of airbrakes* (linear/single-turn potentiometer).

2.3 Prototype of Flight Data Acquisition Device

The modified Safibra FBGuard 1550 FAST interrogation device was used as a basis for flight measurement system. It is equipped with 8 optical channels (250 Hz scanning frequency per channel) and 16 analog input channels (5-10 Hz per channel). All channels are synchronized.

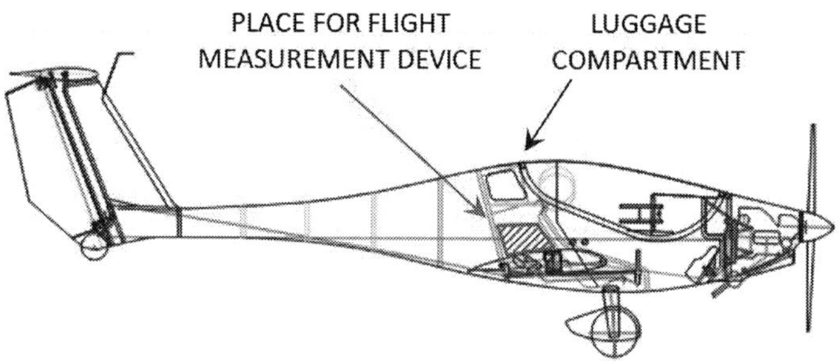

Fig. 6: Position of data acquisition device in the fuselage (side view).

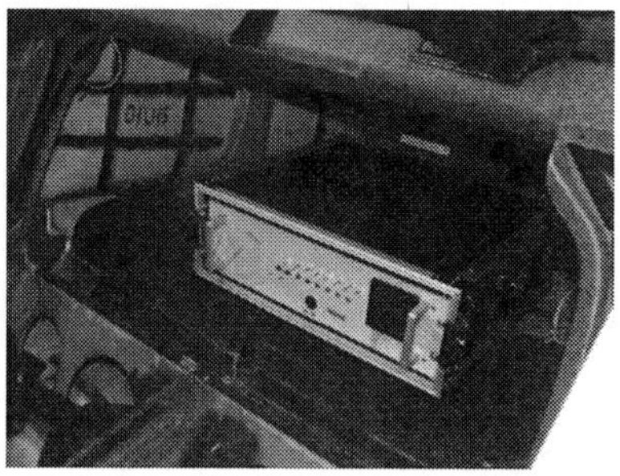

Fig. 7: Detail of device inside the fuselage (behind the seats).

Device is placed under the luggage compartment (see Fig. 6) and it is fixed to structural parts of the fuselage (Fig. 7). Electric power supply is provided by the airplane electrical system. No connected PC is needed, measured data are stored inside the device.

3 Conclusion

Currently, the main parts of the system have passed laboratory tests and they are being installed into the airplane. Load time histories obtained from the flight measurements will be then used for finite element model refinement. In the next steps, local stress & strain histories will be derived from the model for the critical places. This will allow reduction of strain sensors, that are really needed for sufficient flight load measurement, also this will provide data for possible fatigue analysis of the bonded joints.

Acknowledgement

The authors would like to thank the Technology Agency of the Czech Republic for supporting this research with project no. TA04031450.

References

[1] Retrieved from: http://www.phoenixairusa.com/

[2] M. Dvorak, M. Kabrt, M. Ruzicka, The use of Fiber Bragg Grating Sensors During the Static Load Test of a Composite Wing Structure. In Proceedings of the 51st Conference on Experimental Stress Analysis. Usti nad Labem: FVTM, UJEP, 2013, ISBN 978-80-7414-579-7.

[3] M. Dvorak, M. Ruzicka, M. Kabrt, Design of Embedded FBG Sensor System for Ultralight Aircraft Wing Monitoring. In Structural Health Monitoring 2015. Lancaster, Pennsylvania: DEStech Publications, Inc., 2015, p. 1665-1672. ISBN 978-1-60595-275-8.

[4] M. Dvorak, M. Ruzicka, Influence of Embedded Fiber Optic Sensor on Performance of Single-lap Bonded Joint. In 52nd conference on experimental stress analysis 2014. Research and Testing Institute Plzen, 2014, ISBN 978-80-231-0377-6.

The anisotropy influence on the thickness of deep drawn cup

M. Fedorko [1,a], M. Urbánek [1], M. Rund [1]

[1] *COMTES FHT, Czech Republic*

[a] *mikulas.fedorko@comtesfht.cz*

Abstract:

Manufacture of press-formed parts often involves deep-drawing operations. Deep drawing, however, can be deemed an industrial branch in its own right. Today, many experimental as well as numerical methods are available for designing and optimizing deep drawing operations. The best option, however, is to combine both approaches. The present paper describes one such investigation. Here, experiment and numerical simulation were used for mapping the impact of anisotropy on thickness variation in a spherical-shaped drawn part of DC01 steel.

Variation in the sheet thickness was measured on the drawn part using two cameras, and evaluated by means of digital image correlation provided by the ARAMIS software from the company GOM. The forming experiment was carried out on an INOVA 200 kN servohydraulic testing machine in which the force vs. piston displacement curve was recorded. This experimental arrangement was then numerically simulated using the AUTOFORM software. For the purpose of this simulation, a series of mechanical tests was conducted to obtain a description of the experimental material of 1.5 mm thickness. Material models were constructed from the tests data, describing the work-hardening curve, the impact of anisotropy and the forming limit diagram. Specifically, these tests included tensile tests, the Nakajima test and stacked test which were carried out to determine materials data for individual models. Finally, the actual sheet thickness values were found by sectioning the spherical drawn cup and by measuring under a NIKON optical microscope.

The variations in thickness along defined lines were compared with the data from digital image correlation, numerical simulations, and measurements on the sectioned cup. The above-described experimental programme is suitable for calibrating a material model for any computational software which can correctly solve a deep-drawing problem.

Keywords: cup drawing; anisotropy; sheet metal forming; finite element simulation.

1 Introduction

The automotive industry is a sector with the strictest demands on lightweighting and highest pressure on prices. Consequently, various tools are employed by producers of deep-drawn parts for predicting the part's behaviour before the actual manufacturing operation. Generally, the available options are either simple trials on small-size parts or computer simulations. With intricate press-formed parts, the best choice is to combine both of these options.

Authors of this paper have measured fundamental properties of DC01 sheet, constructed a material model for computation with the aid of the AutoForm software, and ran a computer simulation of drawing of a spherical cup. The final steps of this investigation involved validation of the simulation by an actual test of drawing and measuring a spherical cup in a mechanical testing laboratory. The test was performed on a servohydraulic machine and the cup deformation was measured using digital image correlation.

Describing the deep drawing process by means of numerical simulation poses a complex non-linear problem. Deep-drawing grades of sheet tend to be anisotropic, exhibit a defined work-hardening behaviour, and their mathematical description requires up to 10 parameters. Using AutoForm, a special deep-drawing simulation software, parameters of a real-world test can be easily input into the simulation for evaluation. In this study, the amount of thinning in the sheet at the end of the process was compared between a computer simulation and an actual test.

2 Experimental material and method

In order to construct the material model of the DC01 steel, a set of mechanical tests had to be carried out. Anisotropy of the material was first mapped using a series of tensile tests on specimens oriented in various directions (0°, 45°, and 90°) with respect to the sheet rolling direction. Additional material parameters were found by conducting and evaluating stack tests. The outcome of these tests was a curve which describes the anisotropy of material properties.

2.1 Material model

The material model is highly dependent on accurate descriptions of the material´s mechanical behaviour, the selected constitutive model as well as the input data used for its calibration. Therefore, trying to create the appropriate description for the DC01 steel sheet (Tab 1.), the following material model was selected to describe as accurately as possible the material behaviour based on the hardening law and the yield criteria.

Tab. 1: Chemical composition steel DC01

element	C	Mn	P	S	N	Al
%	0.12	0.6	0.045	0.045	0.003	0.04

2.1.1 Yield criteria

The Von Mises is the most widely used yield criteria, but it is only describing isotropic metallic materials. One of the most implemented yield criteria is the one proposed in 1948 by Hill [7], that is the first orthotropic yield criterion.

In order to obtain a balance between the accuracy of the model and parameter identification costs, most of the advanced yield criteria are using a set of seven or eight experimental values. Usually, six input values are used for parameter identification: the three yield stresses (σ_0, σ_{45}, σ_{90}) and the three *r*-values (ratio between the strain in the width and the thickness) obtained from uniaxial tensile tests (performed with orientations of 0°, 45° and 90° to the rolling direction), calculated using the formula:

$$r = \frac{\varepsilon_w}{\varepsilon_t} = \frac{ln(\frac{w}{w_0})}{ln(\frac{t}{t_0})} \tag{1}$$

However, these six experimental values are insufficient to calibrate recent advanced yield criteria as BBC2008 [8]. The additional values needed, are commonly the balanced biaxial stress (σ_b) (measured from bulge test or equibiaxial tensile test) and the balanced strain ratio r_b-value (gained from stack test and the equibiaxial tensile test)[9].

For the definition of our material model, we will use the balanced biaxial strain ratio (r_b) obtained from the stack test and the biaxial test. The balanced biaxial strain ratio r_b-value can be calculated comparing the strain in two directions (rolling direction and transverse to rolling direction) and using the formula:

$$r_b = \frac{\varepsilon_{TD}}{\varepsilon_{RD}} \tag{2}$$

2.1.2 Hardening law

In order to define the hardening behaviour of the metallic sheet we decided to use the Swift hardening law. In the model of Swift law, three material constants, strength coefficient (K), strain-hardening exponent (n) and initial strain (ε_0) should be estimated by curve fitting the measured true stress-strain data before necking to the following equation:

$$\sigma = K(\varepsilon_0 + \varepsilon)^n \tag{3}$$

2.2 Uniaxial tensile test

The first step of the evaluation process has been the measurement of the tensile test in three directions (0°, 45° and 90°) related to the rolling direction (Fig.1). Three tensile tests have been performed for each direction using the universal testing machine Zwick Roel 250kN. The strain measurements of the tests have been performed by Digital Image Correlation (DIC) method [10] with the use of ARAMIS system [11]. The system is using two cameras that enable optical 3D strain measurements. The elastic and plastic behaviour of the material was obtained after the evaluation of the tests.

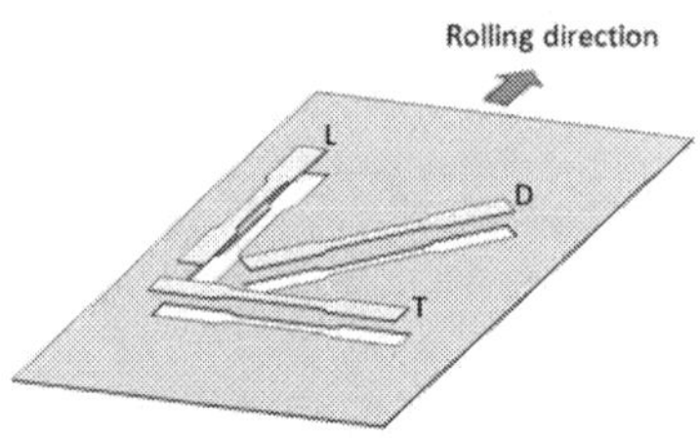

Fig.1 Sample directions related to the rolling direction

The mechanical properties evaluated from the uniaxial tensile test are the engineering strain (e) and engineering stress (s), these properties are obtained directly from measurements:

$$e = \frac{\Delta L}{L_0}, \qquad s = \frac{F}{A_0} \tag{4}$$

Finally the flow stress behaviour has been obtained rcalculating the true strain (ε) and true stress (σ). These parameters calculated directly from the engineering parameters before the necking and with an extrapolation after it.

$$\varepsilon = ln(1 + e), \qquad \sigma = s(1 + e) \tag{5}$$

We calculate these parameters for each direction (0°, 45° and 90°) to obtain the flow stress curves before necking, the flow stress beyond necking point can be extrapolated to fit the force-displacement curve. The flow stress curves are shown in the Fig.2:

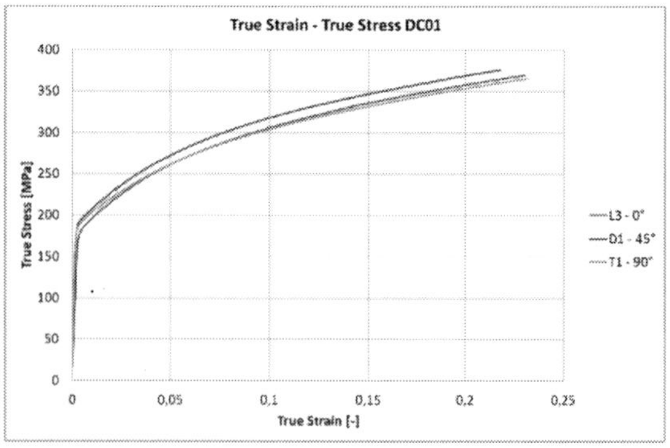

Fig.2 Flow stress curves for each direction

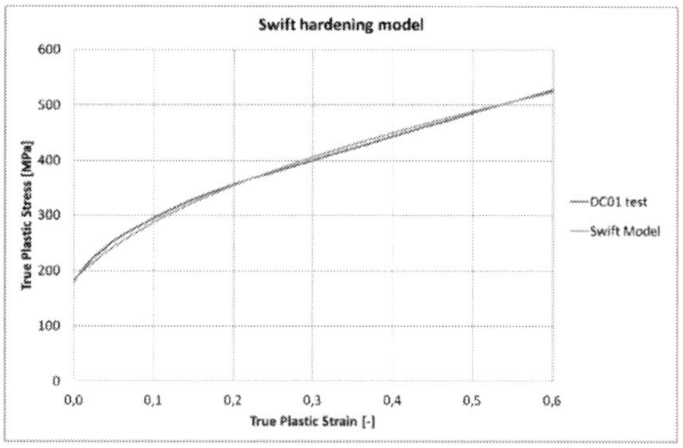

Fig.3 Swift model fitted to the tensile test

After the calculation of the true stress-strain we fitted the constants for the Swift hardening law for the rolling direction curve (0°), obtaining the following parameters:

Tab. 2: Swift model parameters

K	ε_0	n
625.1	0.005	0.412

The anisotropy parameters have been calculated after the initialization of the straining and before the necking of the specimen. The results obtained after the evaluation of the uniaxial tensile tests are:

Tab. 3: Anisotropy parameters gained from uniaxial tensile tests

σ_0[MPa]	σ_{45}[MPa]	σ_{90}[MPa]	r_0[-]	r_{45}[-]	r_{90}[-]
178.7	192.7	184.8	1.78	1.25	2.15

2.3 Stack test

To complete the characterization of the anisotropy model it is necessary to obtain another parameter in order to create a material model more reliable for advanced yield criteria. This parameter is the balanced biaxial strain ratio r_b-value that is obtained from the disk compression test (stack test), the bulge test or the biaxial test.

Fig.4 Compression stack test sample

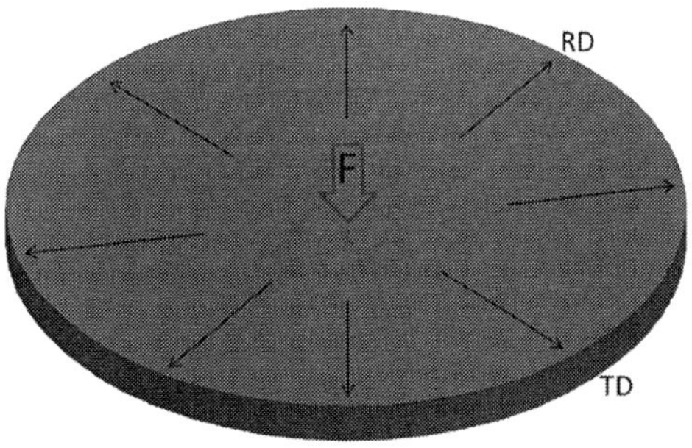

Fig.5 Compression and force and strain distribution during stack test

To obtain this parameter the stack test has been performed, using disks of 25mm diameter and stacking a total of 10 disks [12]. To conserve the anisotropy properties, the disks were piled considering the rolling direction. To reduce the friction between the stacked specimen and the compression plates of the tool, a Teflon-foil acting as a solid lubricant is placed at each end of the specimen.

During the test, the force was gained from the universal testing machine Zwick Roell 250kN and the strain and displacement values were gathered by another DIC method using MERCURY system [13]. Two independent cameras were arranged with an angle of 90° to each other, one of them in the rolling direction and the other one perpendicular to the rolling direction. With this setup it was possible to cover more than the half lateral surface and to measure the strain in the rolling direction and in the transversal direction. Using the formula previously described we have obtained the value $r_b = 0.951$ gathering the strain values of the central disks of the pile, where the strain field is homogeneous.

In order to calculate the yield locus for the DC01 steel sheet, the BBC2008 yield criterion was used [8]. The material parameters used as in put in the BBC2008 identification procedure [15] are the previously calculated:

- Three Yield stresses for each direction (σ_0, σ_{45}, σ_{90})
- Three r-values (r_0,r_{45},r_{90})
- Balanced biaxial stress (r_b)

The calculated BBC2008 yield locus is shown in the figure 6a. In our case was used 7 parameters BBC2008 yield criterion plotted on the yield locus graph (fig. 6a) [11]. Based on the results, a forming limit curve as a function of ε_1 and ε_2, true strain values on the sheet surface, was constructed, as illustrated in Fig. 6b [11]. Functions obtained in this manner were input into numerical simulations of the spherical cut drawing process in the AUTOFORM software.

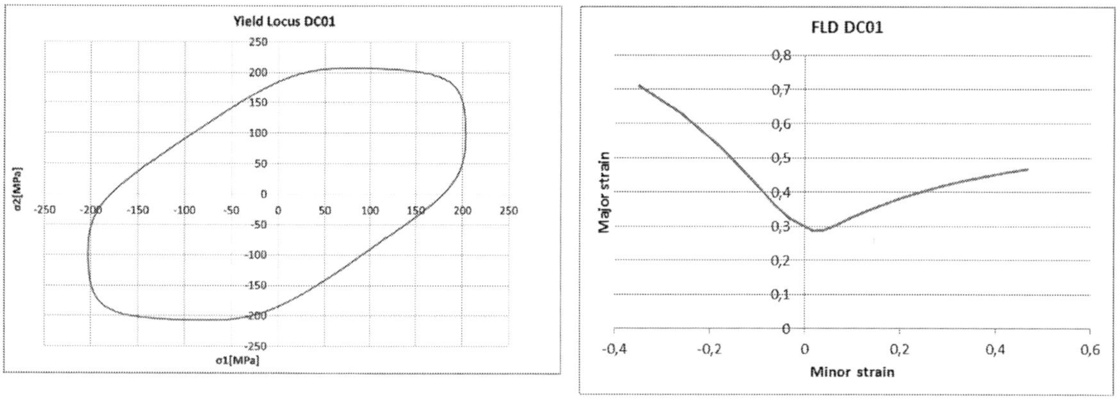

a) Yield surface b) Forming limit diagram

Fig. 6: Material properties of DC01.

3 Verification measurement

The test was carried out in an INOVA 200 kN hydraulic testing machine. The purpose was to assess the effects of lubrication between the spherical punch and the sheet. In the present case, no lubricant was used, which led to tearing away from the top of the spherical cap. By means of a spherical-end male punch of 100 mm diameter, the initial 200 mm-diameter blank was drawn at a velocity of 2 mm/s.

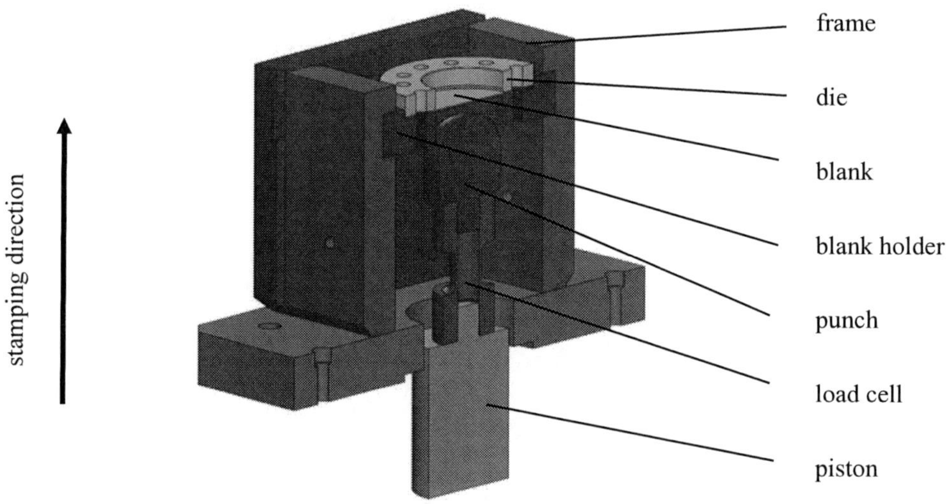

Fig. 7: Configuration of tool and blank before forming

The main outcome of the test was validation of the computational model which focused on the thinning of the sheet upon the test. During the test, the forming force was recorded by means of a load cell, and the cup shape was monitored using the ARAMIS digital image correlation system from the GOM company. Tearing in the sheet occurred at a 45° angle with respect to the rolling direction.

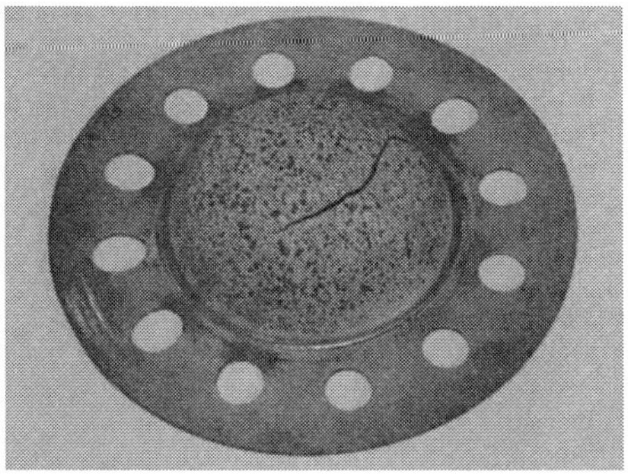

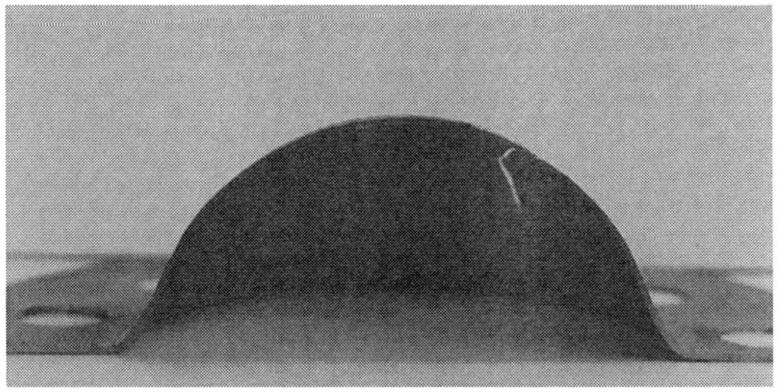

Fig. 8: Photo of sample after test

Fig. 9: Cut sample

4 FEM simulation in AUTOFORM

The AutoForm software was developed for modelling deep drawing processes. It relies on the finite element method and uses no other than shell elements. In the program, the mathematical descriptions of materials properties are tailored to deep drawing simulations. The data and the graphs listed above (Figs. 2, 3, and 6) were used as inputs into the numerical simulation of the sheet behaviour. Drawing on hands-on experience and expertise from sheet-forming industry, the program is intended mainly for the automotive sector. Every step in the sequence can be analyzed and optimized separately as an independent solution. The program features a built-in function which provides a full-scale solution for the entire manufacturing process. Default variables evaluated by the program include the sheet thickness, % thinning, wrinkling, and other values associated with deep drawing.

The input into the AutoForm simulation was a CAD file with the geometry of the Nakajima test fixture shown in Figure 7. Only a single-action draw was simulated, where the positions of the punch, the die and the blank holder are precisely defined. In this test, the blank was bolted down firmly. Therefore, the blank holder in the model was fixed. The velocity of punch was set in accordance with the completed validation test. Results of the computer simulation show the thinning of the spherical cup, as well as the indication of thinning in the graph.

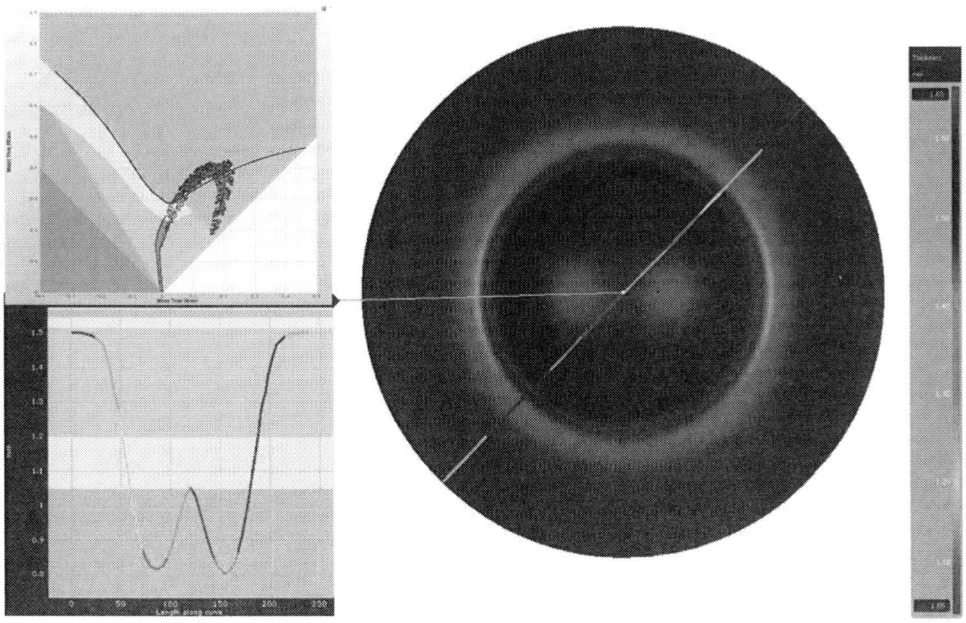

Fig. 10: Thinning in the simulated drawn cup

5 Comparison and discussion

The key outcomes of these simulations were the shape of the drawn part and thickness variations along defined cross-sections. Other variables which were compared between the actual test and the numerical simulations included the behavior of the drawing process, the force required for drawing the part and the impact of anisotropy on strain distribution.

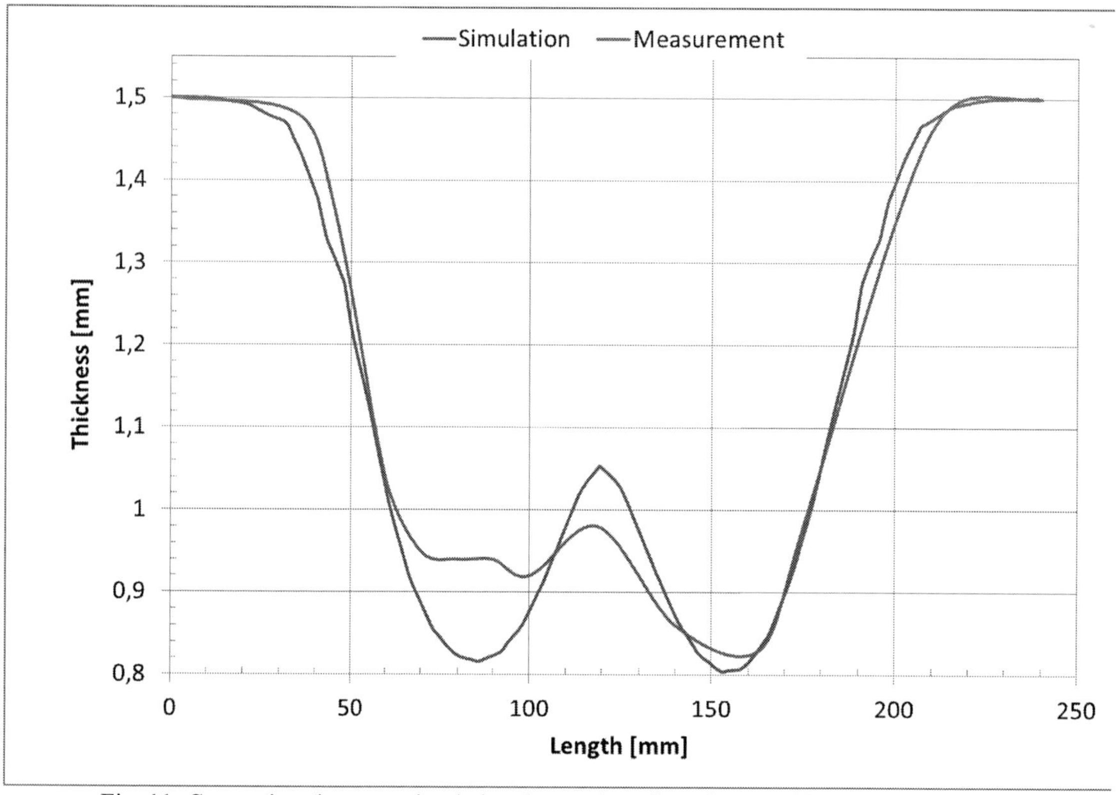

Fig. 11: Comparison between simulation data and readings from verification measurement

Figure 11 compares the thickness variation in the drawn part along a line across the tear. The simulation indicates uniform thinning in the sheet, whereas the validation test shows led to a non-symmetric profile caused by tearing. In the FLD diagram from the simulation (Fig. 10), the limit curve was exceeded at the top of the drawn part, which corresponds to the initiation of tearing in the validation test.

As evidenced by the results, the material model which had been constructed using data obtained from basic mechanical tests can be used for simulating drawn parts of relatively intricate shapes.

6 Conclusion

This paper evaluates sheet thickness variation in a spherical drawn part of DC01 deep-drawing steel. Based on mechanical testing data, a material model was constructed and input into numerical simulations in AUTOFORM software. Real-part measurements and numerical simulation data were compared. The study produced a material model for DC01 steel which describes the anisotropy of properties and its effect on sheet thickness variation in Nakajima drawing test.

Acknowledgement

This study was created by project Development of West-Bohemian Centre of Materials and Metallurgy No.: LO1412, financed by the MEYS of the Czech Republic.

References

[1] Vegter H., An Y.: Mechanical Testing for Modeling of the material behavior in forming simulation, Numisheet 2008, Interlaken, Switzerland.

[2] ČSN ISO 16842.

[3] Deng N., Kuwabara T., Korjolis Y.P.: Cruciform specimen design and verification for constitutive identification of anisotropic sheets, Experimental Mechanics (2015).

[4] Zidane I., Guines D., Leotoing L., Ragneau E.: Development of an in-plane biaxial test for limit curve (FLC) characterization of metallic sheets, Measurement Science and Technology, 21,(2010) 1-11

[5]Liu W., Guines D., Leotoing L., Ragneau E.: Identification of sheet metal hardening for large strains with an in-plane biaxial tensile test and dedicated cross specimen, International Journal of Mechanics Sciences, 101-102 (2015), 387-398

[6] Mitukiewicz G., Glokowski M.: Cruciform specimen to obtain higher plastic deformation in a gauge region, Journal of Materials Processing Technology, 227 (2015) 11-15.

[7] Hill R., A theory of the yielding flow of anisotropic metals, Proc. Roy. Soc. A193 (1948) 281-297

[8] Comsa D.S., Banabic D., Plane/stress zield criterion for highly-anisotropic sheet metals, Numisheet 2008, ed. P. Hora, Interlaken, 2008, p.43-48.

[9] Khalfallah A., Alves J.L., Oliveira M.C., Menezes L.F., Influence of the characteristics of the experimental data set used to identify anisotropy parameters. Simulation Modelling Practice and Theory, 53(2015)15-44

[10] Sutton, M., Digital and Image Correlation: Principle Developments and Applications for Parameter Estimation, University of South Carolina. http://www.gdr2519.cnrs.fr/ecole2011/ET_2011/M_Sutton.pdf

[11] Folgar Ribadas H., Urbánek M., Procházka R., Džugan J., Experiences with cruciform sample to characterize anisotropy plasticity and hardening for steel sheet, IRF2016, Porto, Portugal

[12] Merklein M., Kuppert A., A method for the layer compression test considering the anisotropic material behaviour. Int J Mater Form, (2009) Vol.2 Suppl 1: 483-486. DOI 10.1007/s12289-009-0592-8

[14] MARC-MENTAT, MSC Software Corporation, 2015

[15] Jurco P., Banabic D., A user-friendly programme for calculating Forming Limit Diagrams, Techinical University of Cluj-Napoca, Daicoviciu 15, Cluj-Napoca, Romania

[16] Prantl A., Urbanek M., Prochazka R., Dzugan J., Evaluation of influence strain path on FLD diagram, International Deep Drawing Research Group, IDDRG 2016, Linz, Austria

Design of Measurement System for Rotary Swaging Force Measurement

F. Fojtík [1,2,a], M. Fusek [1,2]

[1] *Department of Applied Mechanics, VŠB – Technical University of Ostrava, Czech Republic*
[2] *IT4Innovations, VŠB – Technical University of Ostrava, Czech Republic*
[a] *frantisek.fojtik@vsb.cz*

Abstract: This article describes a design of the measurement system for rotary swaging force measurement. Rotary swaging in its present form is an up-to-date way of material forming. Rotary swaging is an incremental process of material forming for the purpose of reducing and forming raw material cross-sections and can be used, for example, in the production of various rotary and non-rotary forged iron products, rods, wires, sensor detectors or composite materials. However, force measurement is no standard part of this production technology. In this article, the first version of the measurement system design is introduced. The force measurement has been conducted in the case of two raw products in different deformation and reinforcement stages of materials. The obtained results and swaging force progress are also commented here.

Keywords: Rotary swaging; Force measurement; Force sensor; Signal transmission.

1 Introduction

Rotary swaging in its present form is an up-to-date way of material forming, which unlike other technologies of forming has the following advantages. The main advantages of rotary swaging are the high-quality surface, dimensional accuracy, and productivity of the production output of forged iron products. It is usually not necessary to perform subsequent machining or other surface treatment. In this sense, this technology can be applied in, for example, production of various rotary as well as non-rotary forged iron products, rods, wires, sensor detectors, or composite materials, e.g. copper-clad steel composite conductors used as conductors in automotive and aviation industry[1, 2, 3, 9]. However, force measurement is no standard part of this production technology. The determination of required material deformation stage and optimization of shapes and sets of metal liners is usually based on the existing acquired experience and lately also on MKP analyses based on tested load capacity of the forging machine [4]. In cooperation with the Department of Materials Forming, Faculty of Metallurgy and Materials Engineering, in finding new research solutions, the demand for a measurement system allowing real-time recording and demonstrating swaging force has arisen.

2 Rotary Swaging Description and Measurement System Design

In reality, there are several forging machine arrangements. They can roughly be divided on the basis of the number of the forging mechanism parts rotating inter se. For the purpose of our experiments, the forging machine arrangement (Fig. 1) has been used. Raw product (1) is deformed into the required final shape by forging segments (2), which perform a radial movement up and down, i.e. perpendicular to the axis of rotation, in rapid succession one after another. There can be two to six forging segments. The force required for the raw product deformation is induced by dynamic impacts occurring between a forging metal liner (3) and rollers (6), which are placed in the cage (5). The rollers roll against the supporting ring (7), which is static on the basis of the selected constructional arrangement, or it can also perform rotary movement along with the cage. The swaging segment and metal liner are placed in a carrier (4) and they rotate around the axis of rotation together. The respective planes of the metal liners are profiled into the required shape of the final forged piece. The raw product is let in the space between metal liners in the direction of the axis of rotation. Using this constructional arrangement with the induced radial forces, the raw product is deformed into the required shape. The longitudinal feed of the raw product can be either manual or smooth using a feeder. In the workplace of the authors, the prototype of the measurement system suitable for measuring rotary swaging force [5, 6] has been designed and further constructed. The system consists of a tensometric force sensor, which replaces a selected control member of the forging system [7]. The sensor is connected to the suitable apparatus, which is installed on the carrier (4), which it performs the rotary movement with. Using wireless transfer, the apparatus communicates with the computer installed out of the working area of the forging machine. A suitable graphic interface is installed in the computer and provides wireless communication with the selected apparatus and records the measured data and processes them. The graphic and communication interface has been developed in LabVIREW [8] program. The apparatus location on the forging machine is shown in Fig. 2.

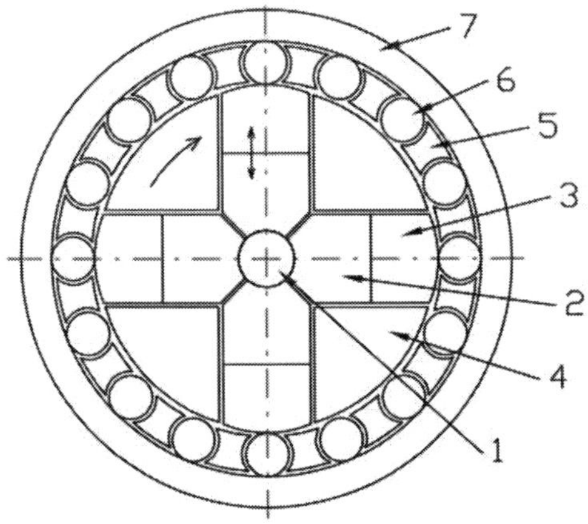

Fig. 1: Rotary swaging principle.

Fig. 2: Forging machine with the installed measurement system.

3 Experimental Results

In order to test the designed measurement system, the raw product with a circular section of 50mm in diameter and 100 mm of length has been selected.

The final forged piece will also have the circular section of 40 mm in diameter. The selected material reduction comes from used forging segments supplied by their producer and these were then fit on the forging machine before installing the measurement system. The raw product was made of spring steel. The record of the forging force progress in time during the whole length of operation of the forging machine is shown in Fig. 3.

The longitudinal feed of the raw product into the space between the forging segments was manual, and thus not smooth. Therefore, the forging force progress shown in Figure 3 also corresponds to that. The forging force progress has several characteristic intervals. The time record in the interval of 0 – 32s corresponds to the rest mode of the machine. The time record in the interval of 32 – 80s shows the forging force at the moment of the carrier approaching with forging parts in selected revolutions and in idle run without the raw product. The forging force value here is around 100kN. From the interval of 80s, the raw product starts to be pressed into the working space of the rotating forging segments. There is an apparent rapid increase in forging force up to the value of about 645kN. A slight decrease of this force corresponds to lower speed of the longitudinal feed of the raw product into the working space of the forging machine. In the interval of 280s, the process of swaging the raw product was finished and the forged piece left the working space of the forging machine. The record of the interval of 280 – 400s corresponds to the idle run of the machine without the forging process and coasting of the machine as a result of momentum at the time of turning the machine off. From the interval of 400s until the end of recording, the machine is in the rest mode.

The force curve in Figure 3 shows the resulting envelopes of force, which burns maximum values of the measured force in each cycle of the machine. The recording of the force in time has been conducted at the frequency of 2000 kicks of the forging segment per minute. In Fig. 4, the real progress or curve of the forging force is shown in time during swaging, which provides foundation for generating the envelope of the forging force curve shown in Fig. 3.

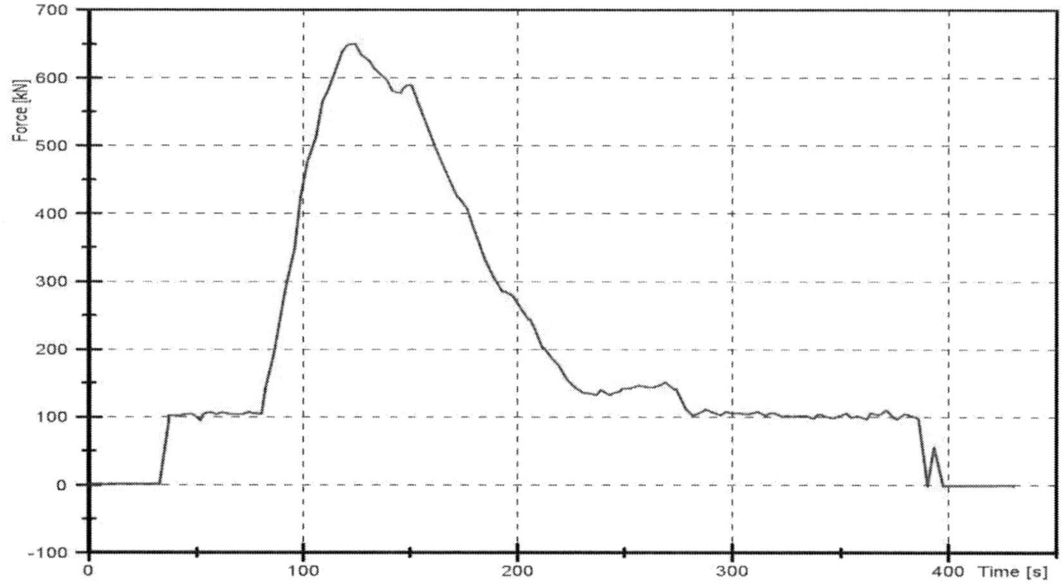

Fig. 3: Record of the forging force progress in time during the whole length of operation of the forging machine.

The resulting forged piece with the diameter of 40mm was significantly reinforced by the deformation during swaging. Thus treated, the forged piece was again put into the feeder of the forging machine. In the forging machine, the forging segments were replaced so as the resulting forged piece also had circular section with the diameter of 30mm. The record of the forging force curve in time during the whole length of operation of the forging machine is shown in Fig. 5.

Maximum force measured during this swaging was 735kN. The curve of the forging force has similar character as in the previous case including the machine operation stage without applying swaging load.

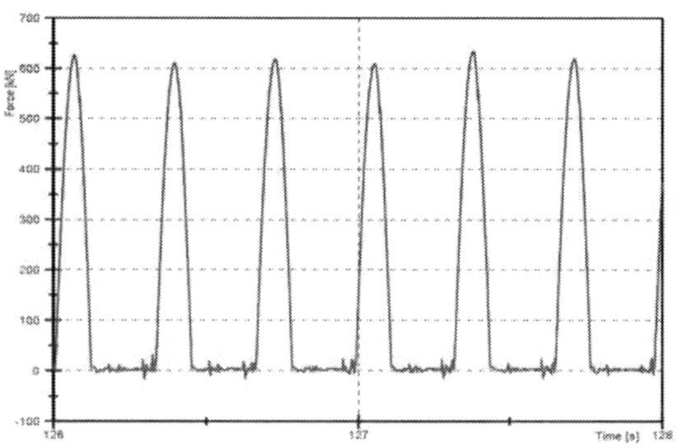

Fig. 4: Cut of the real forging force curve in time during forging (swaging).

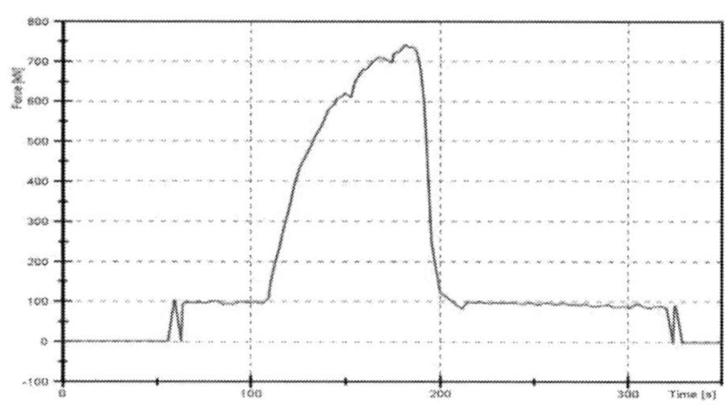

Fig. 5: Record of the forging force curve of the second raw product.

4 Conclusion

The presented article shows Design of Measurement System for Rotary Swaging Force, which is the up-to-date way of very accurate raw products forming. The prototype of the measurement system was designed and installed onto the forging machine. The measurement system consists of a special force sensor, measuring apparatus, and a computer with tailor-made graphic and communication interface. This interface is further used for setting up the measurement system in order to record and evaluate obtained results from the forging force measurement. Communication between the computer and the measuring apparatus has been performed using wireless data transfer.

For verification of the measurement, two experiments have been conducted. The progress of the forging force is demonstrated in Figures 3 and 5. The measurement results in the envelope of the forging force curve and its shape. The maximum value of the forging force is dependent on the speed and size of the conducted deformation during the rotary swaging and also on the inner stress of the raw product. In the first case, the measurement result was compared to the results obtained from MKP simulation. The maximum force here is in compliance in the order of units. The measurement results and acquired experience will lead to upgrading the constructional version of the measurement system applied. The measurement system is protected by a patent with the registration number: 305725.

Acknowledgement

This work was supported by The Ministry of Education, Youth and Sports from the National Programme of Sustainability (NPU II) project „IT4Innovations excellence in science - LQ1602“, by the Czech Science Foundation (GA15-18274S) and by the Specific Research (SP2016/145). The support is gratefully acknowledged.

References

[1] R. Kocich, A. Macháčková, L. Kunčická, F. Fojtík, Fabrication and characterization of cold-swaged multilayered Al–Cu clad composites, Mater. Des. 71 (2015) 36–47, http://dx.doi.org/10.1016/j.matdes.2015.01.008.

[2] R. Kocich, L. Kunčická, C. F. Davis, T. C. Lowe, I. Szurman, A. Macháčková, Deformation behavior of multilayered Al-Cu clad composite during cold-swaging, Mater. Des. 90 (2016) 379–388.

[3] T. Nakagawa, T. Amano, Compacting of high density P/M bars by rotary swaging, JSTP 17 (1976) 209–215.

[4] J.P. Domblesky, R. Shivpuri, Development and validation of a finite-element model for multiple-pass radial forging, J. Mater. Process. Technol. 55 (1995) 432–441.

[5] A. S. Khan, X. Wang, Strain Measurements and Stress Analysis, New Jersey, Prentice Hall, 2001.

[6] P. Macura, Experimental Methods in Elasticity and Plasticity, Ostrava, VŠB-TUO, 2001, in Czech.

[7] ŞTEFĂNESCU, Dan-Mihai. Handbook of force transducers: principles and components. Berlin: Springer, 2011, xxix, 612 s. ISBN 978-3-642-18295-2.

[8] Retrieved from: www.nationalinstruments.com

[9] Retrieved from: http://www.hmp.de/cms/upload/Rotary%20swaging%20GB.pdf

Experimental Determination of Stress Fields near Notches by Reflection Photoelasticity

P. Frankovský [1,a], F. Trebuňa [1], O. Ostertag [1], F. Šimčák [1], I. Delyová [1], J. Kostka [1]

[1] *Technical University of Košice, Faculty of Mechanical Engineering, Košice, Slovakia*
[a] *peter.frankovsky@tuke.sk*

Abstract: The paper aims at the analysis of stress fields which occur while applying loads to a photoelastically coated notched sample. The analysis was done by means of reflection photoelasticity using polariscope LF/Z-2 and digital video camera. On the notched sample subject to loads by eccentric tension we determined the differences of principal normal stresses in specified point at loads from 3 kN to 12 kN. In addition, we determined the value of principal normal stress at the edge of the sample during load of 12 kN which was later compared with numerical solution in programme SolidWorks. On the photoelastically coated sample we observed maximum elongation of the coating, too.

Keywords: reflection photoelasticity; isoclinic fringes; isochromatic fringes; notch; stress field.

1 Introduction

It is crucial for design engineers to have the knowledge of stress fields of loaded components, structure knots and structural systems. At the same time, it is important to identify areas of extreme stresses properly. Such knowledge enables design engineers to optimize technical, design as well as economical features of components, structure knots and structural systems. Experimental methods are therefore helpful when carrying out stress analyses. These methods can be classified as contact methods or contactless methods. The contactless group of methods includes the optical experimental method of photoelasticity which uses polarized light to analyze the stress state. This is a progressive way of stress assessment. PhotoStress is the method of reflection photoelasticity and is one of the methods which use the phenomenon of birefringence which occurs when loading an optically sensitive material on the surface of some component. This method enables us to establish a picture of a stress field as well as gradient of stresses which occur on the surface of an examined structural component. Reflection photoelasticity is especially suitable for stress analysis of real structural components. For identification of direction and magnitude of principal normal stresses we use isoclinic and isochromatic fringes [1, 2].

Compared to other experimental methods in mechanics, PhotoStress method has the following advantageous features [3, 4]:

- visual overview of stress and strain fields throughout the whole photoelastically coated surface and, in this way, the method offers information on critical stress concentrations around holes, notches and other areas of possible failures as well as information on maximum or minimum stressed areas etc.;
- universal application which enables us to determine stresses and deformations in any point of photoelastic coating applied in laboratory conditions or directly in the field;
- non-destructive impact on surface – the experiment can be carried out repeatedly with changing load parameters;
- wide range of measurements – the method enables us to measure individual parameters and determine differences and individual principal strains and stresses, residual stresses etc.

2 Specification of sample

The determination of stress fields around the notch was done by means of reflection photoelasticity method on a steel sample made of steel X70 of 12 mm. The shape, dimensions and load of the sample are depicted in Fig. 1. The sample was cut from a steel plate with photoelastic coating PS-1A [1]. The cutting technology used is called Waterjet Cutting. Steel X70 has only small volume of carbon and its bainitic microstructure secures its high toughness and ductility level.

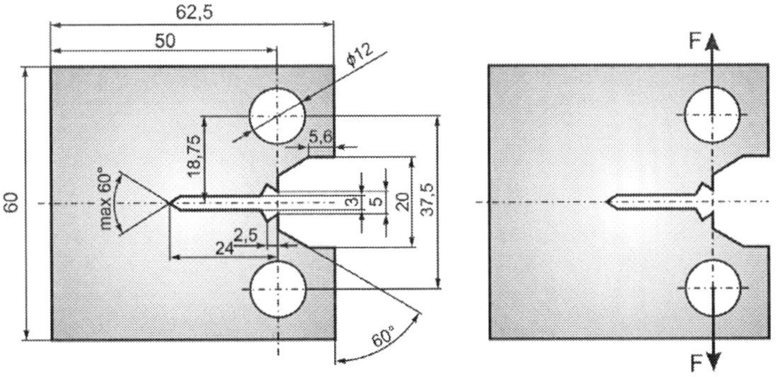

Fig. 1: Shape, dimensions and load of analyzed notched sample.

Material characteristics of the steel X70 are listed in Tab. 1 [4].

Tab. 1: Material characteristics of steel X70

Parameters	*Indication*	*Unit*	*Value*
Modulus of elasticity	E	MPa	$2.14.10^5$
Proof strength	$R_{p0.2}$	MPa	575
Proof strength	R_{eH}	MPa	586
Tensile strength	R_m	MPa	650
Elongation	A_{10}	%	24
Poisson's ratio	μ	-	0.3

Photoelastic coating PS-1A of 2.05 mm was chosen while following systematic selection methodology [9, 10]. The coating was applied to the surface of the notched sample by two-component adhesive PC-1. Photoelastic coating PS-1A is highly sensitive and hence allows us to use it in elastic as well as in elastic-plastic areas. It is delivered as a plane plate with reflective layer.

Material characteristics of the PS-1A photoelastic coating used are listed in Tab. 2 [2].

Tab. 2: Material characteristics of photoelastic coating PS-1A

Parameters	*Indication*	*Unit*	*Value*
Strain-optic coefficient	K	-	0.150
Modulus of elasticity	E	MPa	2500
Poisson's ratio	μ	-	0.38
Elongation	A	%	5
Maximum usable temperature	t	°C	150

Photoelastic coating applied to the analyzed notched sample (Fig. 2) transforms a part of the load and, simultaneously, strengthens the sample. As a result, deformations of the sample are lower than if no coating had been applied [5, 6].

Fig. 2: Analyzed sample with photoelastic coating.

For the above-stated purposes a correction of reinforcement effect of the photoelastic coating was done through correction coefficient for reinforcement while considering plane stress [1]

$$C_{PS} = 1 + E^{*}t^{*}, \tag{1}$$

where $E^{*} = \frac{E_c}{E}$ is the relation of Young's modulus of photoelastic coating and Young's modulus of the analyzed sample;

$t^{*} = \frac{t_c}{t}$ is the relation of coating thickness and thickness of the analyzed sample.

After calculations the value of correction coefficient, by which the order of isochromatic fringes is multiplied, is 1.0122.

3 Load specification

Experimental measurement was done on a tearing machine Zwick 1387 (Fig. 3) while loading the sample with eccentric tension. In order to fulfil the conditions of static loading the movement of the machine jaws was set to 0.5 mm/min [6].

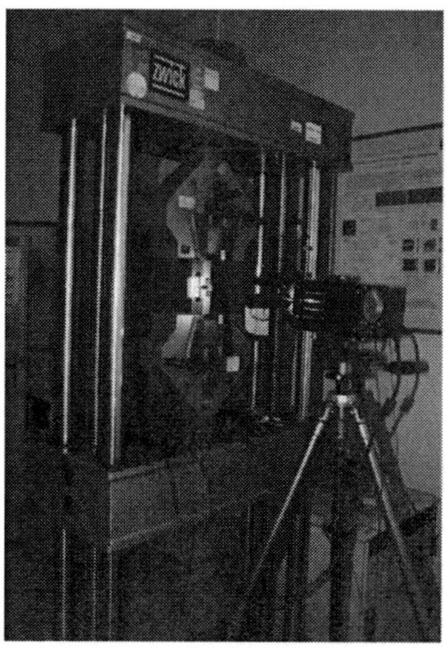

Fig. 3: Tearing machine Zwick 1387 and reflection polariscope LF/Z-2.

Reflection polariscope LF/Z-2 with usual source of white light (Fig. 3) was used to observe isoclinic and isochromatic fringes which occur when loading the notched sample. Compensator 832 was used to read order values of isochromatic fringes in pre-selected points. The compensator was based on the principle of null balance. Digital video camera was used to record isochromatic fringes during gradual loading of the sample.

The measurement of notch opening was done with a strain gauge and a converter (Fig. 4) which assessed the notch opening in mm [6] on the basis of electric voltage calculation. Every 1.3 s the converter registered the set of 8 000 values of electric voltage within 300 ms.

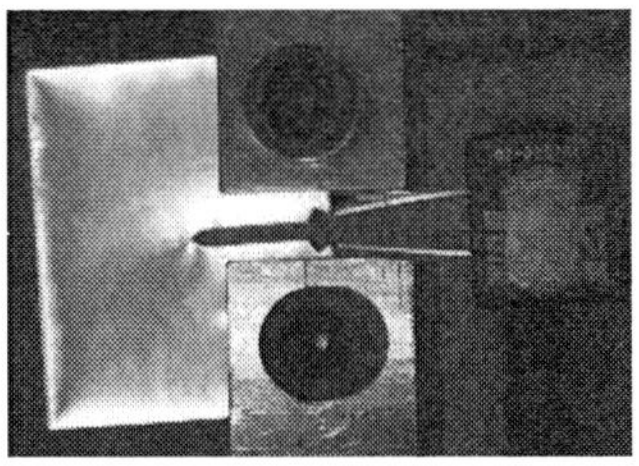

Fig. 4: Notch opening sensor.

A computer programme was written for the assessment of recorded data. The programme synchronized forces and notch openings in a time line. In order to synchronize measured data it was needed to launch the tearing machine and the converter simultaneously [6].

The relation of force F in kN to elongation f in mm was a direct outcome of the tearing machine software (Fig. 5).

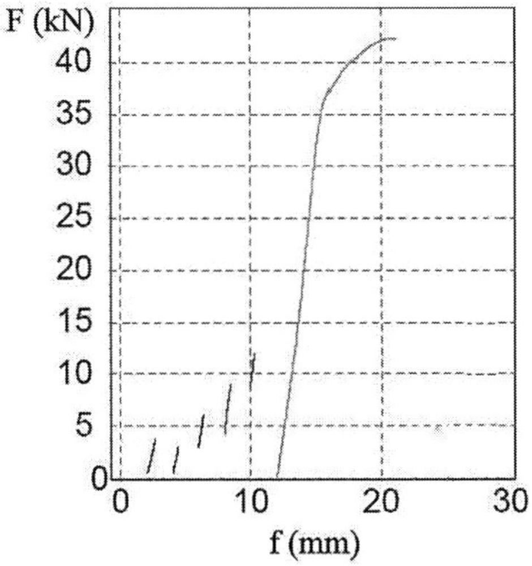

Fig. 5: Notch opening in relation to loading force.

The relation in Fig. 5 consists of a number of curves. The light green curve represents force distribution while loading the sample until crack. The notch opening increased without increasing force. The cause of this effect is that after fastening the sample in the device both surfaces had to fit firmly and adjust to each other [6]. The short curves in the initial part of the diagram (red, blue, brown, violet, dark green) represent force distributions during analysis of isoclinic and isochromatic fields while applying loads 3 kN, 4.5 kN, 6 kN, 9kN and 12 kN.

For the sample under examination were determined the following relations [6]:

- notch opening in relation to time (Fig. 6a);
- force in relation to time (Fig. 6b);
- force in relation to notch opening (Fig. 7).

Fig. 6a, Fig. 6b and Fig. 7 show us a little bounce of distribution at 38 kN. This bounce is the result of a crack in photoelastic coating.

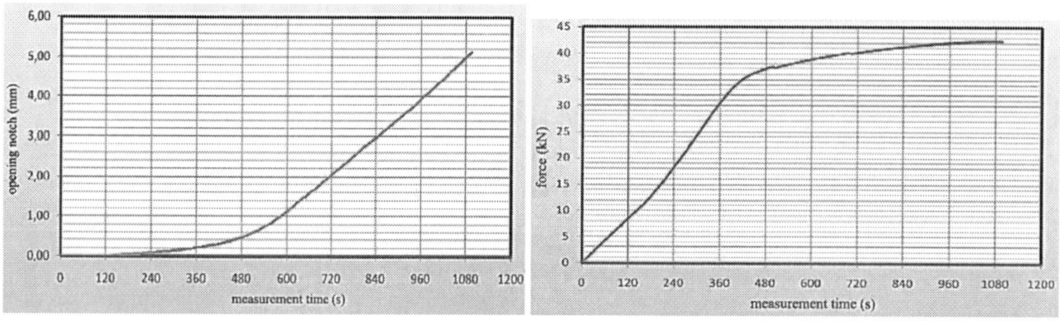

a) b)

Fig. 6: a) Notch openings in relation to time; b) Loading force in relation to time.

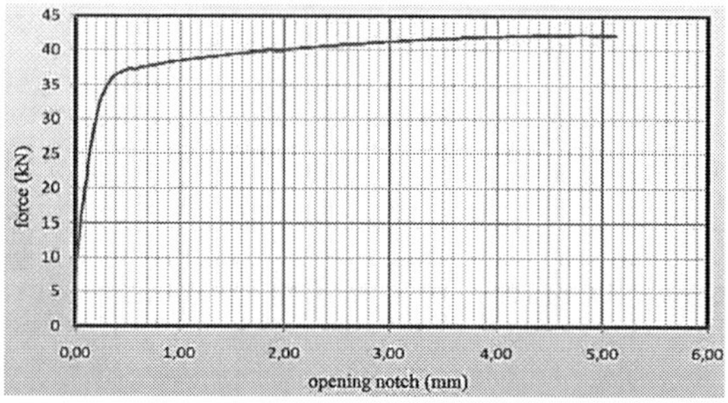

Fig. 7: Loading force in relation to notch opening.

4 Isoclinic and isochromatic fringes on the notched sample

Fig. 8 depicts isoclinic fringes on the notched sample with angle parameter α 0° to 90° and 10° increase. Isoclinic fringes were gained by reflection polariscope LF/Z-2 at plane polarized light. During examination, the analyzed notched sample was loaded by eccentric torsional force of 3 kN. Isoclinic fringes are areas in which directions of principal normal stresses are equal [6-8].

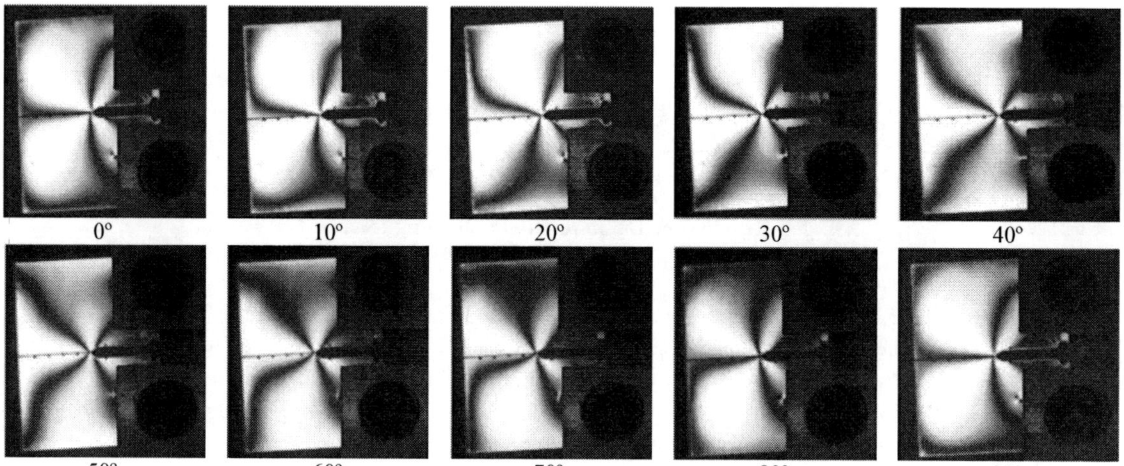

0° 10° 20° 30° 40°

50° 60° 70° 80° 90°

Fig. 8: Isoclinic fringes on the notched sample under examination.

Fig. 9 depicts isochromatic fringes during gradual loading of examined notched sample with force from 0 to 12 kN. Isoclinic fringes were observed by reflection polariscope LF/Z-2 at circularly polarized light. At loads 12 kN and higher it is possible to observe stress gradient around notch root. The high gradient is a result of high stress concentration in the point of crack occurrence [6, 9, 10].

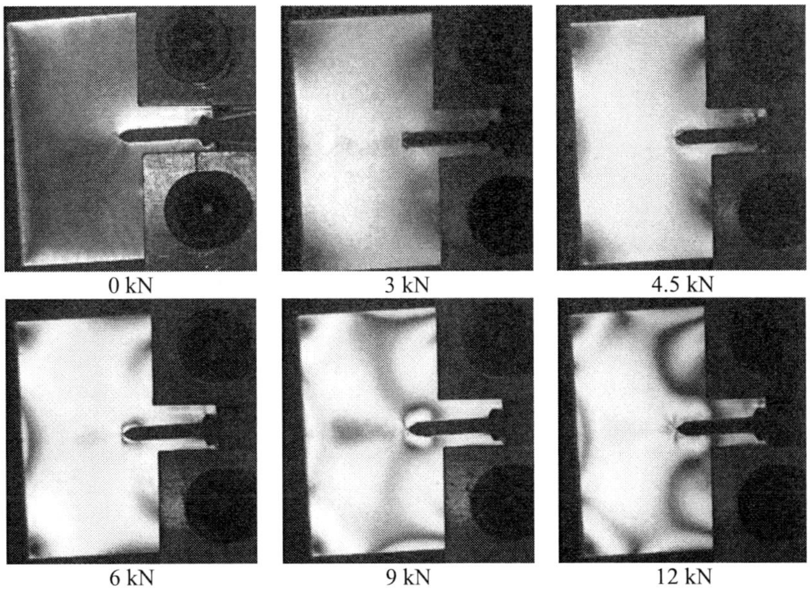

Fig. 9: Analysed sample with photoelastic coating PS-1A.

5 Principle stress analysis

Determination of principal normal stresses' difference was performed in point 1 (Fig. 10) at loading forces 3 kN, 4 kN, 6 kN, 9 kN and 12 kN.

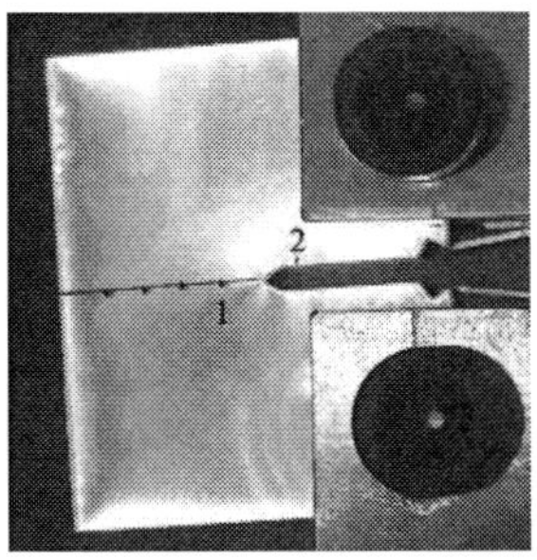

Fig. 10: Fringe order measurement points.

At the above-stated loading force values we used compensator 832 to determine the isochromatic fringe order *N*. Differences of principal normal stresses in point 1 at individual loads were determined by means of the following relation:

$$\sigma_1 - \sigma_2 = \frac{E}{1+\mu} f \cdot N, \tag{2}$$

where E is the Young's modulus of sample material, μ is Poisson's ratio of sample material and f is fringe constant of photoelastic coating.

Listed in Tab. 3 are isochromatic fringe orders *N* at point 1 under loading force f 3 kN, 4 kN, 6 kN, 9 kN and 12 kN and differences of principal normal stresses $\sigma_1 - \sigma_2$ corrected by correction coefficient *CPS* and derived from the relation (2) [1, 4].

Tab. 3: Fringe orders and difference of principal normal stresses at individual loads

Measurement	F (kN)	N (-)	$\sigma_1 - \sigma_2$ (MPa)
1	0	0	0
2	3	0.28	37.39
3	4.5	0,47	62.77
4	6	0,62	82.80
5	9	0,97	129.54
6	12	1,27	169.61

As it is known from the theoretical background, biaxial stress state requires additional measurement to determine individual values of principal normal stresses. For these purposes we use separation methods. However, there are cases when the equation (2) can be transformed to the following form:

$$\sigma_1 = \frac{E}{1+\mu} f \cdot N. \tag{3}$$

One of these cases requires free unloaded edge of an object subject to analysis. In each point of such edge we find one principal normal stress, which is non-zero tangential to the edge, and another zero normal stress. In an experimental way we used a compensator with null balance to specify the isochromatic fringe order in point 2 on the notched sample at loading force 12 kN (Fig. 10). Its value was 1.57. Using the relation (3) we subsequently determined the value of principal normal stress $\sigma_1 = 209.67$ MPa.

Numerical stress analysis was done in programme SolidWorks (Fig. 11) in order to verify previously gained value of principal normal stress σ_1 in point 2. Principal normal stress in point 2 determined by computer simulation has the value 200.2 MPa. The difference between principal normal stress σ_1 gained in an experimental way and numerically is 3.6%. As a result, we can state that the experimental measurement was relatively precise. This way of verification is easily applicable in everyday practice since the majority of structural elements include free edges [11].

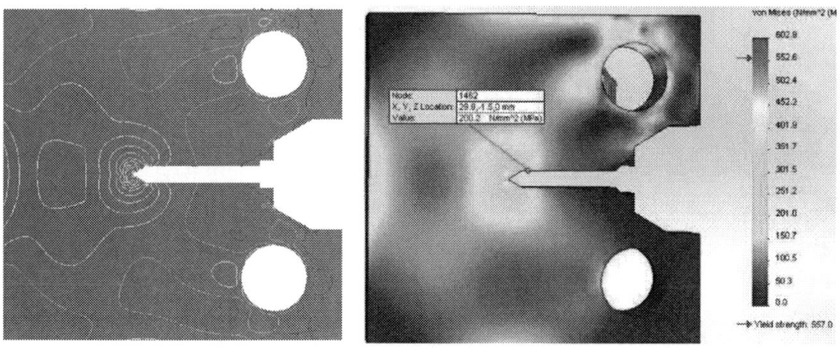

Fig. 11: Field of reduced stresses according to von Misses.

6 Verification of endurance limit of photoelastic coating

As regards tensibility of used photoelastic coating PS-1A, the manufacturer states the value of 5%. This implies that photoelastic coating is able to transform 5% of relative elongation value. Examinations were carried out on the photoelastically coated notched sample regarding the value of relative elongation at which the failure of photoelastic coating occurs [12, 13].

The failure of photoelastic coating on the notched sample occurred at loading force 38 kN and relative elongation 4% (Fig. 12). The fracture was clearly visible immediately while the crack was accompanied with sound effect [6, 14].

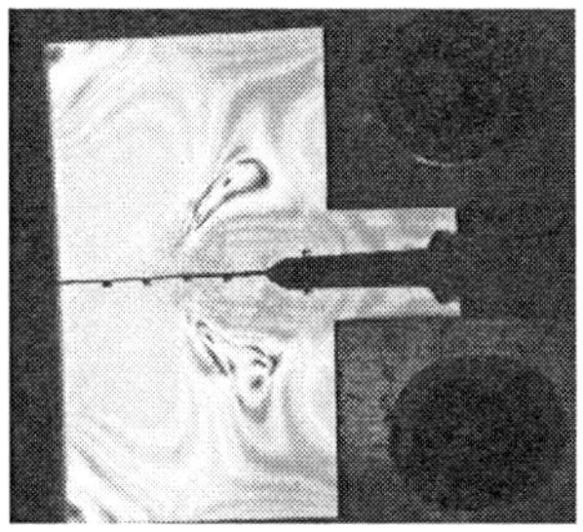

Fig. 12: Failure of photoelastic coating.

For the completeness of this issue we analysed video recording which was recorded by HD video camera and cut to individual shots of JPEG format. Fig. 13 depicts shots which are representations of characteristic moments throughout the analysis. It is obvious from Fig. 13 that stress gradient around the notch is high when the photoelastically coated sample is gradually loaded. Considering the shot No. 13 and higher it is obvious that photoelastic coating loses its ability to transform deformations and, at the same time, the ability to display isochromatic fringes. Fig. 13 no. 17 depicts the first moment when failure of photoelastic coating occurred. The shot No. 19 shows the crack spreading throughout the whole width of the sample, whereas the shot No. 20 depicts the coating peeled off under applied stresses and deformations.

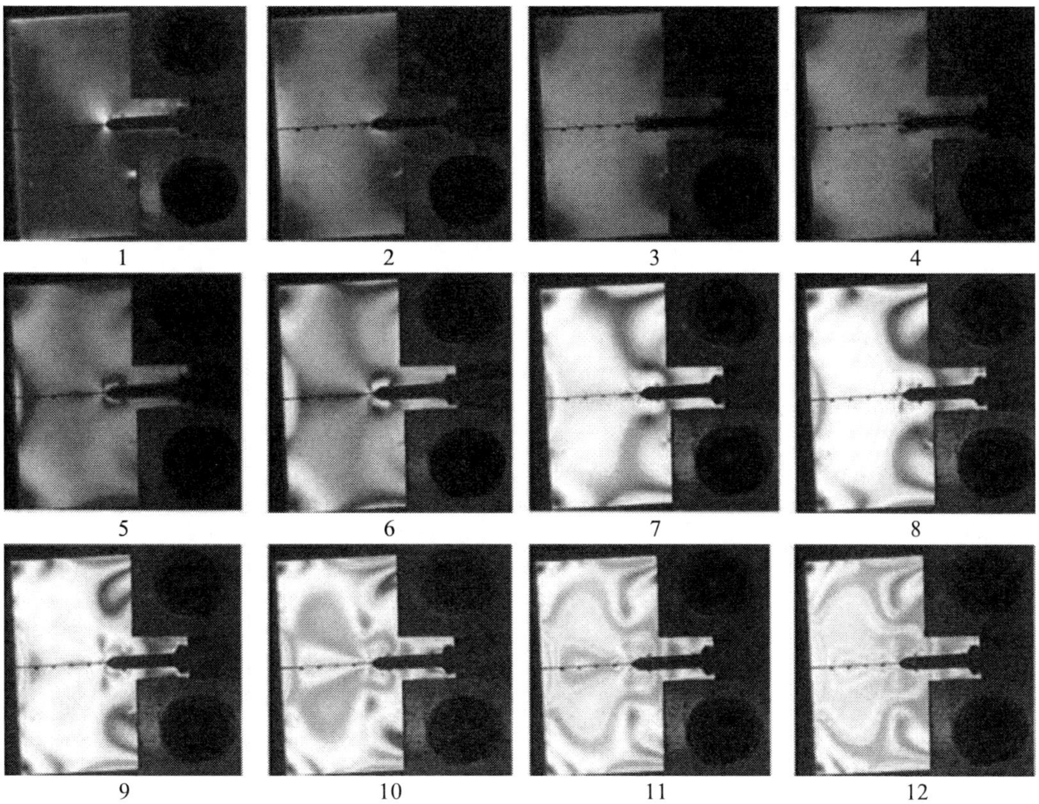

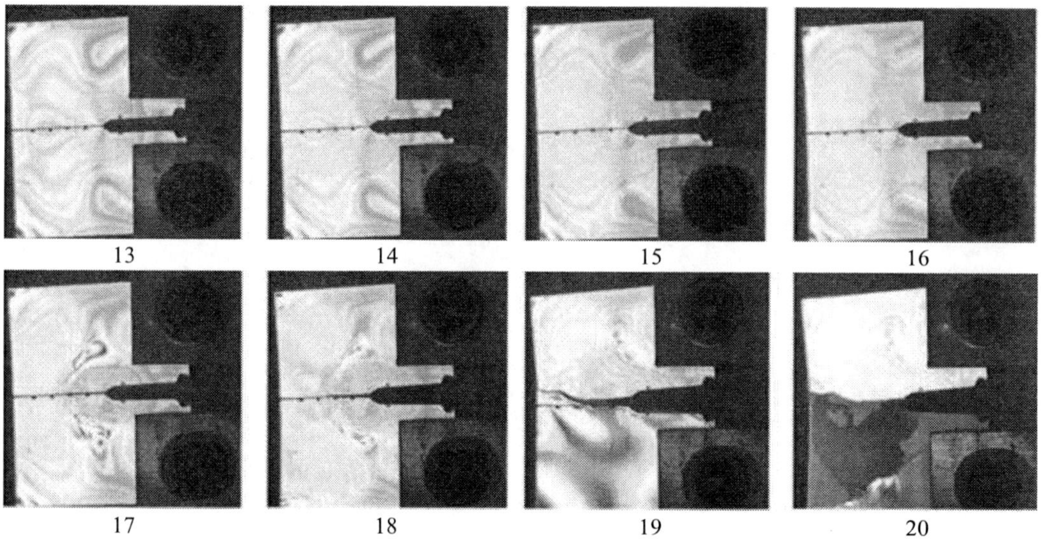

Fig. 13: Analysed sample with photoelastic coating PS-1A.

Fig. 14 depicts damaged photoelastic coating PS-1A with the notched sample and relative elongation 4%.

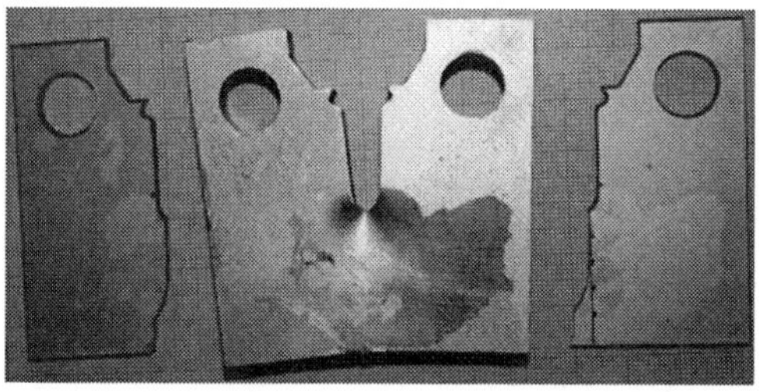

Fig. 14: Damaged photoelastic coating PS-1A with the notched sample and relative elongation 4%.

7 Conclusion

Using reflection photoelasticity method we were able to determine the difference of principal normal stresses in a particular point at various values of loading force. The correctness of experimentally gained data was verified through determination of one of principal normal stresses on the free edge of photoelastically coated notched sample. This measurement was verified by means of numerical simulation in programme SolidWorks. Additionally we examined the maximum elongation of the applied photoelastic coating PS-1A of the sample. The value provided by the manufacturer is 5%. Through measurements we finally determined that maximum elongation of the photoelastic coating PS-1A of 2.05 mm is 4%. This difference might have been caused by the structure of bonded joint.

Acknowledgement

This work was supported by the grand projects VEGA no. 1/0393/14, 1/0872/16 and project ITMS 26220120060.

References

[1] F. Trebuňa, F. Šimčák, Handbook of Experimental Mechanics (in Slovak), Košice, TypoPress, 2007.

[2] F. Trebuňa, J. Jadlovský, P. Frankovský, M. Pástor, Automation in the method PhotoStress (in Slovak), Košice, Technical University of Košice, 2012.

[3] A.S. Kobayashi, Handbook on Experimental Mechanics, Seattle, Society for Experimental Mechanics, 1993.

[4] M. Šmida, Using of PhotoStress method for identification of possible failures of elements in mechanical devices. Diploma thesis. TUKE, Košice, 2013.

[5] F. Trebuňa et al., The use of optical methods in experimental mechanics 2 (in Slovak), Košice, Technical University of Košice, 2015.

[6] F. Trebuňa et al., The use of optical methods in experimental mechanics 1 (in Slovak), Košice, Technical University of Košice, 2014.

[7] M. R. Ayatollahi, M. Nejati, Experimental evaluation of stress field around the sharp notches using photoelasticity. Materials & Design, 2 (2011), pp. 561-569.

[8] M. N. Pacey, M. N. James, E. A. Patterson, A new photoelastic model for studying fatigue crack closure. Experimental mechanics, 1 (2005), pp. 42-52.

[9] F. Fojtík, P., Ferfecki, Z. Paška, Computer Aided Evaluation of the Stress Tensor in the Two-Dimensional Photoelasticity, Applied Mechanics and Materials, 827 (2016), pp. 181-184.

[10] D. Post, Photoelastic stress analysis for an edge crack in a tensile field. Proc. SESA, 1 (1954), pp. 99-116.

[11] J. E. Akin, Finite element analysis concepts: via SolidWorks, 2010, World Scientific.

[12] W. B. Bradley, A. S. Kobayashi, Fracture dynamics—a photoelastic investigation. Engineering Fracture Mechanics, 3 (1971), 317-332.

[13] M. Ramulu, A. S. Kobayashi, Dynamic crack curving—a photoelastic evaluation. Experimental Mechanics, 1 (1983), pp. 1-9.

[14] C. A. Sciammarella, F. M. Sciammarella, Experimental mechanics of solids. John Wiley & Sons, 2012.

Testing the Mechanical Reliability of Plastic Bottles for Drinking Beverages

K. Frydrýšek [1,a], F. Sejda [1,b], V. Bajtek [1,c], L. Dvořák [2,d], Š. Michenková [1,e]

[1] *VŠB–Technical University of Ostrava, Faculty of Mechanical Engineering, Department of Applied Mechanics, 17. listopadu 15/2172, 708 33 Ostrava, Czech Republic*

[2] *VŠB–Technical University of Ostrava, Faculty of Mechanical Engineering, Department of Hydromechanics and Hydraulic Equipment, 17. listopadu 15/2172, 708 33 Ostrava, Czech Republic*

[a] *karel.frydrysek@vsb.cz;* [b] *frantisek.sejda@vsb.cz;* [c] *vojtech.bajtek@vsb.cz;* [d] *lukas.dvorak@vsb.cz;* [e] *sarka.michenkova@vsb.cz*

Abstract: In cooperation with industry, testing was conducted to determine the reliability (i.e. applicability in regular use by people) of "sports caps" on plastic bottles. These bottles are popular all over the world and comply with health standards. They are designed to contain various types of popular beverages. The maximum overpressures for situations of liquid leakages were measured (i.e. volumes of carbonated or uncarbonated beverages, sealing, stress analyses etc.). For the laboratory testing at the VŠB–Technical University of Ostrava, a custom-made hydraulic/pneumatic system was built, based on a PR 15 sensor with accessories (Hyrotechnik Lumburg Multisystem 5050 measurement system, Orlík 1500 W compressor, tubes, etc.). The experiments were statistically evaluated (in the form of histograms) and the sealing reliability of the bottle + "sports cap" closure system was assessed. The presented results are important for future numerical modelling, design improvements and other investigations. Stresses and their variabilities were calculated according to Laplace's theory of thin shells.

Keywords: Sealing; Reliability; Plastic Bottle Closure; Overpressures; Measurement; Hydraulic/Pneumatic System, Leakage, Stresses, Statistics, Histograms

1 Introduction

In cooperation with the VŠB–Technical University of Ostrava (Ostrava, Czech Republic), the manufacturer required an investigation of the sealing, strength and reliability of the given closure system for plastic bottles. The bottles have the popular type of closure known as a "sports cap", and they are designed for drinking still (uncarbonated) or carbonated beverages (water, Coca-Cola etc.); see Fig. 1. The bottles were produced in two versions: ones without a hygienic spout cover and the others with this cover.

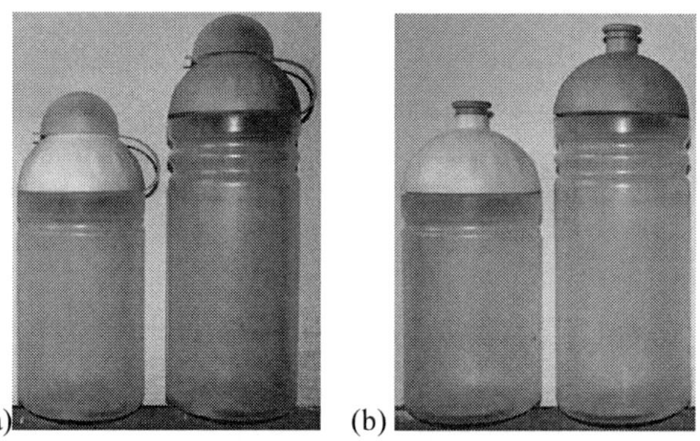

(a) (b)

Fig. 1: Bottles with "sports caps" (a) without a hygienic spout cover, (b) with a hygienic spout cover.

The sealing of the bottles was tested under conditions simulating their use in ordinary conditions (i.e. with new and pre-used "sports caps"); see Fig. 2. The measurements were conducted on randomly selected bottles supplied by the manufacturer; these were numbered in ascending order. The selection was made from a set of 28 bottles. The bottles were divided into 4 types (7 × 0.75 l with hygienic spout cover, 7 × 0.75 l without hygienic spout cover, 7 × 0.5 l with hygienic spout cover and 7 × 0.5 l without hygienic spout cover); see Fig. 1.

The aim of the study was thus to test and statistically evaluate possible undesirable leakages of beverages from plastic bottles with the "sports cap"-type closure; such leaks cause dissatisfaction among the producer's customers. The results of the testing and their statistical evaluation (histograms) are original outputs and are used by the producer for ongoing research and development purposes. The study also determines the variance of stress values in the plastic bottle wall according to Laplace's theory of thin shells.

Fig. 2: Measurement equipment setup and positioning of the bottles (containing liquid).

2 Experiments

It was necessary to conduct experiments in order to meet the requirements of the producer – i.e. to verify the sealing reliability of the bottles when containing carbonated or still (uncarbonated) beverages. As a first step it was necessary to determine the overpressure in the bottle caused by the release of carbon dioxide from the beverage. An initial measurement was therefore conducted: a beverage was poured into the bottle, then the bottle was closed and shaken. After shaking, the overpressure was measured using a pressure sensor. The initial measurement showed that the overpressure inside the bottle reached approx. 1.2 bar (depending on the type of beverage and its volume inside the bottle). This was the first conclusion; it enabled us to determine the pressure interval occurring within the bottles.

It was then necessary to determine the types and methods of testing to be used. It was decided that the following tests would be conducted:

1) Identification of leakage points using pressurized air, with the bottle positioned underwater,
2) Measurement of pressure after opening the bottle closure,
3) Decrease in air pressure over a 24-hour period,
4) Determination of the extent of liquid leakage caused by released carbon dioxide.

The sealing test was conducted by a semi-destructive method as follows. A hole was drilled in the lid of the bottle, and a screw-fixture was inserted into the hole to enable the tube to be connected. First, the seal was tested underwater using pressurized air; this enables any points of air escape to be easily identified; see Fig. 2. The measurement setup is shown in Fig. 3(a). The pressure regulator was set to the required air pressure, and the bottle was pressurized underwater through the directional control valve. The design of the measurement system was based on measurements of static characteristics of pneumatic elements [1].

This test showed that in most cases leakages occurred via the bottle cap. However, in some bottles there were leakages via the screw-thread which serves to attach the cap to the bottle.

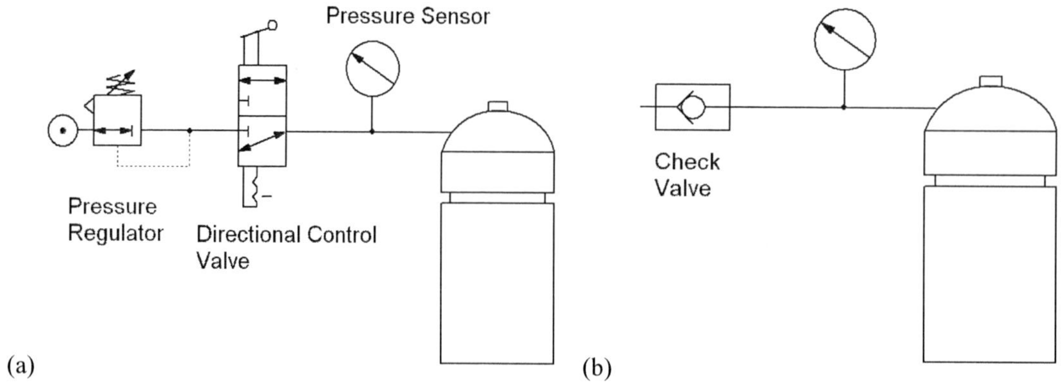

Fig. 3: (a) Setup of bottle sealing test, (b) Setup of check valve connection.

The same setup was used when determining the overpressure required to cause the spontaneous opening of the bottle cap, which was in the unlocked position. The test was conducted on both new and pre-used bottle caps. In this case, the pressure regulation valve was used to gradually increase the overpressure inside the bottle until the air pressure caused the cap to burst open (i.e. an unpleasant occurrence for the user). The overpressure at the moment of opening was determined from records taken using the Hydrotechnik M5050 measurement system. A PR 15 sensor was used, with a range of –1 to 6 bar. The mean overpressures required for spontaneous opening of an unlocked bottle cap are 1.115 bar (for a new, unused cap) and 0.889 bar (a pre-used bottle cap). The results are presented in section 3.

Measurements were also conducted to determine pressure loss from inside the bottle over a 24-hour period. In this case, the bottle was pressurized and then disconnected from the pressurized air source. In order to prevent air loss via the tube, a check valve was added to the measurement setup; see Fig. 3(b). After disconnection from the air source, measurement of the overpressure inside the bottle commenced; see Fig. 4. The results show that once the overpressure inside the bottle has decreased to approx. 0.9 bar, it ceases to decrease any further, and no further leakage occurs.

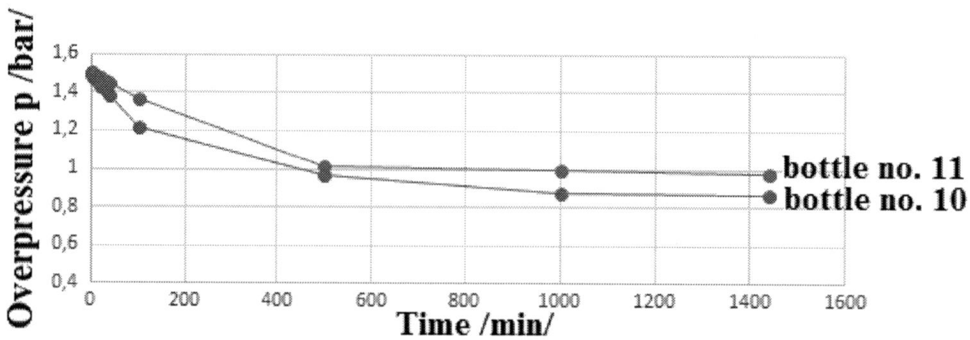

Fig 4: Graph of overpressure decrease (leakage from bottle) – results of 24-hour measurement.

The final measurement determined the quantity of liquid leaked from the bottle. A 0.75 l bottle was partially filled with 0.5 l of a selected carbonated beverage (of the Coca-Cola type). The bottle was then closed, fitted with a pressure sensor, and weighed. The same process was applied to 0.5 l bottles, into which 0.35 l of the beverage was poured. The bottles were shaken, releasing carbon dioxide. The overpressure of the gas was almost 1 bar in all cases. The bottles were then placed so that the bottle neck was positioned below the beverage surface level. In some cases, the overpressure of the released carbon dioxide caused leakage of the liquid (probability of leakage is 50%). After an hour, the bottles were re-weighed and the quantity of leaked liquid was thus determined.

All measurements were conducted at normal temperatures, i.e. 20° C.

3 Results

The measurements were conducted in order to test the reliability of the bottles for normal use. The measurements determined the overpressure p in the bottle (see Fig. 5 and Tab. 1), the decrease in overpressure during the monitored period, and the quantity of liquid leaked over a certain period. The results indicate that the type of liquid inside the bottle has a major influence on leakage. There was no liquid leakage from bottles containing still (uncarbonated) water. The quantity of liquid leakage increases in bottles containing carbonated mineral water or beverages of the Coca-Cola type. Leakage of liquid from the bottles thus only occurred in cases where there was increased internal overpressure p. The largest percentage of leaks occurred via the rubber "sports cap" closure.

Liquid	Bottle no.	Bottle volume /l/	Volume of leakage /ml/	Point of leakage	Max. overpressure /bar/	Min. overpressure /bar/	Mean overpressure /bar/
Coca-Cola	12	0.75	0	-	0.803	0.787	0.782
Coca-Cola	13	0.5	8	cap	0.875	0.822	0.822
Mattoni mineral water	18	0.75	4	screw-fitting	0.935	0.897	-
Mattoni mineral water	19	0.5	0	-	0.961	0.918	-
Kofola carbonated soft drink	14	0.75	0	-	0.734	0.702	0.701
Kofola carbonated soft drink	15	0.5	2	cap	0.77	0.742	0.734
Mattoni mineral water	16	0.75	0	-	0.526	0.52	0.524
Coca-Cola	17	0.5	2	cap	1.042	1.037	1.038
Water	20	0.5	0	-			
Coca-Cola	21	0.5	0	-			
Water	22	0.75	0	-			
Coca-Cola	23	0.75	2	cap			

Tab 1: Results of measurements for bottles containing carbonated and uncarbonated beverages

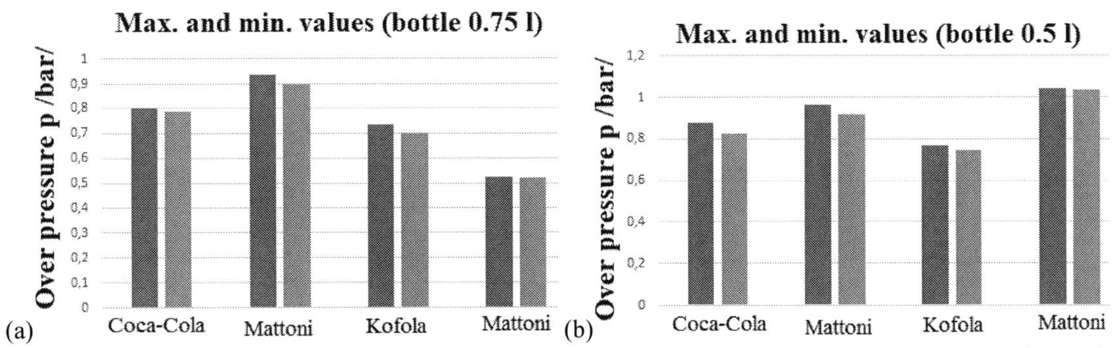

Fig 5: Plastic bottles with the "sports cap" closure (max. and min. overpressure when containing carbonated beverages).

There was a noticeable difference in the sealing reliability between new and pre-used bottles; on the pre-used bottles, the "sports cap" was manually opened and closed 500×. Leakage from these bottles occurs at much lower pressures. A completely new bottle suffers minimal (zero) liquid leakage. Another factor influencing sealing is the force and thoroughness with which the bottle cap is closed (i.e. it may not be closed properly – the influence of the human factor). Further factors include the way in which the bottle is used (opening with teeth instead of hands, dropping the bottle, etc. – these are consequences of "normal" human behaviour) and material degradation (the mechanical properties of rubber and plastics change over time). The older the bottle (i.e. the more wear it has incurred through use), the greater the probability of liquid leakage. Based on the test results, it can be stated that a bottle of this type can be safely used (i.e. with zero liquid leakage) for uncarbonated beverages (e.g. still water, fruit juice, etc.). If the bottle contains a carbonated beverage (carbonated mineral water, beverages of the Coca-Cola type, etc.), there is a 50% chance of leakage from new bottles. If the bottle is older, and its cap has incurred wear, it is no longer suitable for carbonated beverages (probability of leakage is almost 100%).

The results shown in Tab. 1 and Fig. 5 enable the determination of meridian stress σ_m and tangential stress σ_t in the bottle wall; see [3]. At a sufficient distance from the base and the cap (approx. at the mid-point of the bottle's height), Laplace's theory states

$$\sigma_m = \frac{pD}{4t}, \quad \sigma_t = \frac{pD}{2t}, \tag{1}$$

where D is the mean bottle diameter and t is the thickness of the bottle wall. The stresses are given in Tab. 2 (a 0.75 l bottle) and Tab. 3 (a 0.5 l bottle).

Beverage	Meridian stress σ_m /MPa/		Tangential stress σ_t /MPa/	
	MAX	MIN	MAX	MIN
Coca-Cola	2,54	2,44	5,08	4,87
Mattoni	1,99	1,91	3,99	3,81
Kofola	2,18	2,14	4,36	4,28

Tab. 2 Stress in the wall of a closed plastic bottle (0.75 l) with the “sports cap” closure.

Beverage	Meridian stress σ_m /MPa/		Tangential stress σ_t /MPa/	
	MAX	MIN	MAX	MIN
Coca-Cola	2,61	2,49	5,22	4,99
Mattoni	2,83	2,82	5,66	5,63
Kofola	2,09	2,02	4,18	4,03

Tab. 3 Stress in the wall of a closed plastic bottle (0.5 l) with the “sports cap” closure.

The results can be statistically processed e.g. via histograms, producing a suitable basis for future research and development of plastic bottle systems including the “sports cap” closure. Fig. 6 shows the histogram of the probable volume of liquid leakage (the vertical axis is frequency, the horizontal axis the leakage volume) presented discretely or continuously.

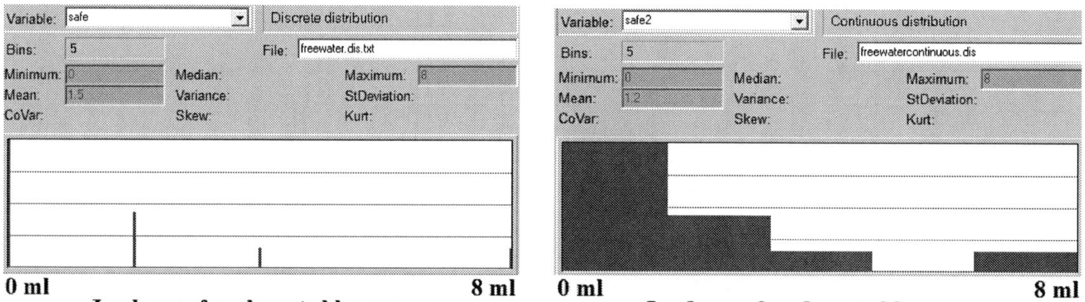

(a) **Leakage of carbonated beverages** (b) **Leakage of carbonated beverages**

Fig. 6: Plastic bottle with a “sports cap” closure (statistical evaluation of the volumes of liquid leakage for carbonated beverages – (a) discrete histogram, (b) continuous histogram).

4 Conclusion

Various types of plastic bottles with “sports cap” closures are internationally popular. However, the literature still lacks even basic studies investigating these systems. For this reason, the results presented in this paper are highly original.

The results indicate that the type of liquid inside the bottle has an unexpectedly large influence on its sealing reliability. No leakage was recorded in bottles containing still (uncarbonated) water. The quantity of leakage increases if the bottles contain carbonated mineral water or beverages of the Coca-Cola type; in such cases the reliability of the closure is 50%. Leakage thus occurs in bottles with increased internal overpressure. The largest proportion of leakages occurred via the rubber seal of the bottles.

Histograms of liquid leakage or stress variability in the bottle wall are important statistical outputs which can be used in future research and development of plastic bottles for beverage consumption.

The calculated stresses and their variance (minimum and maximum values) in the bottle wall were determined via a simple and elegant method using Laplace's theory of thin shells.

The results enable us to draw the general conclusion that an older bottle with a worn "sports cap" closure no longer gives a reliable seal, thus potentially causing damage (e.g. leakage may damage personal property, food items, clothing etc.). In older bottles there is an almost 100% probability of liquid leakage. In newer bottles (with a non-worn "sports cap" closure) the probability of leakage when containing carbonated beverages is 50% (only if the "sports cap" is properly closed).

The results indicate that plastic bottles with "sports cap" closures are unsuitable for carbonated beverages.

In the future, further testing could be conducted on different bottles; this may include cyclical loading (e.g. according to [4]) or more detailed solutions applying a probability-based method [5] or a complex numerical method [6].

References

[1] FOJTÁŠEK, K, DVOŘÁK, L. Measurement of Static Characteristics of Pneumatic Motors with Elastic Working Elements, In *Transactions of the VŠB – Technical University of Ostrava, Mechanical Series*, vol. LVIII, no. 1, 2012, pp. 53 – 56, ISSN 1210-0471

[2] FRYDRÝŠEK, K., DVOŘÁK, L., BAJTEK, V., MICHENKOVÁ, Š., SEJDA, F. *Measurement Report about the Measurement of Leaking of Bottles (Zpráva z měření - testování těsnosti lahví)*, VŠB–Technical University of Ostrava, Department of Applied Mechanics, Ostrava, 2014, pp. 1-31, written in Czech.

[3] TIMOSHENKO, S. WOINOWSKY-KRIEGER, S. *Theory of plates and shells*. 2nd ed., reissued. New York: McGraw-Hill, 1987. ISBN 0070647798.

[4] HALAMA, R., SEDLAK, J., ŠOFER, M., *Phenomenological Modelling of Cyclic Plasticity*. Numerical Modelling, InTech, Croatia, 2012, pp. 329-354, ISBN 978-953-51-0219-9.

[5] LOKAJ, A., VAVRUŠOVA, K., Application of laboratory tests results of dowel joints in cement-splinter boards VELOX into the fully probabilistic methods (SBRA method). *APPLIED MECHANICS AND CIVIL ENGINEERING*, Book Series: Applied Mechanics and Materials, Vol. 137, 2012, ISSN: 1660-9336, pp. 95-99.

[6] JANČO, R., KOVÁČOVÁ, M., Using Packages Mechanics Explorers 'Beam' in Program Mathematica for Solution of Euler-Bernoulli Beams. *APLIMAT 2005 - 4TH INTERNATIONAL CONFERENCE, PT II*, ISBN:978-80-969264-2-8, Bratislava, Slovakia, 2015, pp. 197-203.

Specimen design for low-cycle fatigue experiments under large strain amplitude loading

J. Fumfera [1,a], R. Procházka [2]

[1] *Czech Technical University in Prague, Faculty of Mechanical Engineering, Department of Mechanics, Biomechanics and Mechatronics, Technická 4, 166 07 Praha 6, Czech Republic*

[2] *COMTES FHT a.s., Průmyslová 995, 334 41 Dobřany, Czech Republic*

[a] *jaromir.fumfera@fs.cvut.cz*

Abstract: Specimen design for uniaxial low-cycle fatigue test of 08Ch18N10T austenitic steel under large strain amplitude (up to 3%) loading conditions is presented. Design phase FEM analysis including post-buckling analysis with inclusion of non-linear material plasticity is presented. New axisymmetric specimen design based on meridian section of elliptical shape is introduced. Some details about experimental equipment and experimental results are presented.

Keywords: Low-Cycle Fatigue, Large Strain Amplitude Experiments, Specimen Design

1 Introduction

Low-cycle fatigue (LCF) is a part of fatigue phenomenon, where loading implies higher nominal stresses than yield stress. Maximum number of cycles to failure for common steel-like materials is usually thousands of cycles or less [1]. Due to large stresses and strains in material, effects like post-buckling and additional bend can occur. The higher the loading amplitude is, the more obvious are these effects and specimen design must reflect these phenomena.

LCF experiments are usually strain-controlled. Deformations are measured for example by extensometer attached to a specimen's surface and applied load is feedback controlled by required value of extensometer's displacement. This type of controlling induces strain field in specimen. Usually there is required specific value of total strain amplitude in specific point or area. For 1D tests, bar specimens with uniform-gage test section (hereinafter "uniform-gage specimen") or specimen with variable longitudinal test section (i.e. hourglass specimens) are used [2]. Test are performed on different strain amplitude levels of total strain in test section.

Typical low-cycle fatigue life of 08Ch18N10T austenitic stainless steel presented in this paper usually shows initial cyclic hardening followed by saturation of cyclic deformation curve (stress-strain response of next cycle is almost the same as previous one). For larger strain amplitudes, cyclic hardening (or sometimes softening) can occur for some types of materials. Last stage shows cyclic softening due to degradation of material properties and crack growth up to final failure [3].

2 Specimens Design

2.1 Original specimen design

For the first series of experiments, uniform-gage specimen was used (see Fig. 1). Ex-post analysis of results shows that almost all uniform-gage specimens fail outside the test section in radius notch (that also means outside extensometer tips that control loading).

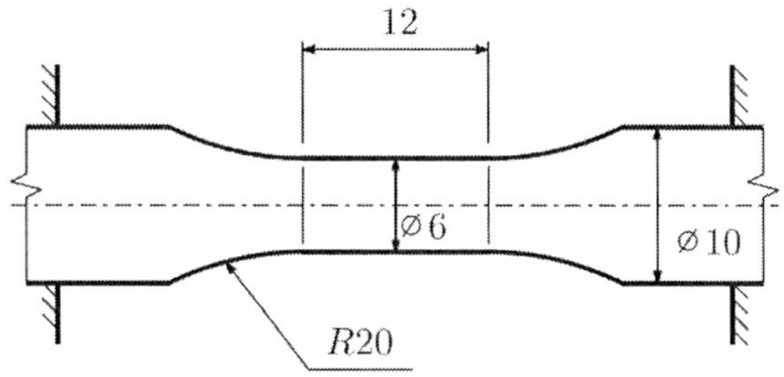

Fig. 1: Original uniform-gage specimen

Data analysis also shows different history of material hardening than expected. For almost all specimens, there is no material cyclic softening in the final stage (see Fig. 2). That indicates different type and cause of failure than expected. Finally, data analysis shows loss of stability control due to post buckling effect of several specimens under the highest strain amplitudes. Due to all these reasons, it was decided to propose new specimen design.

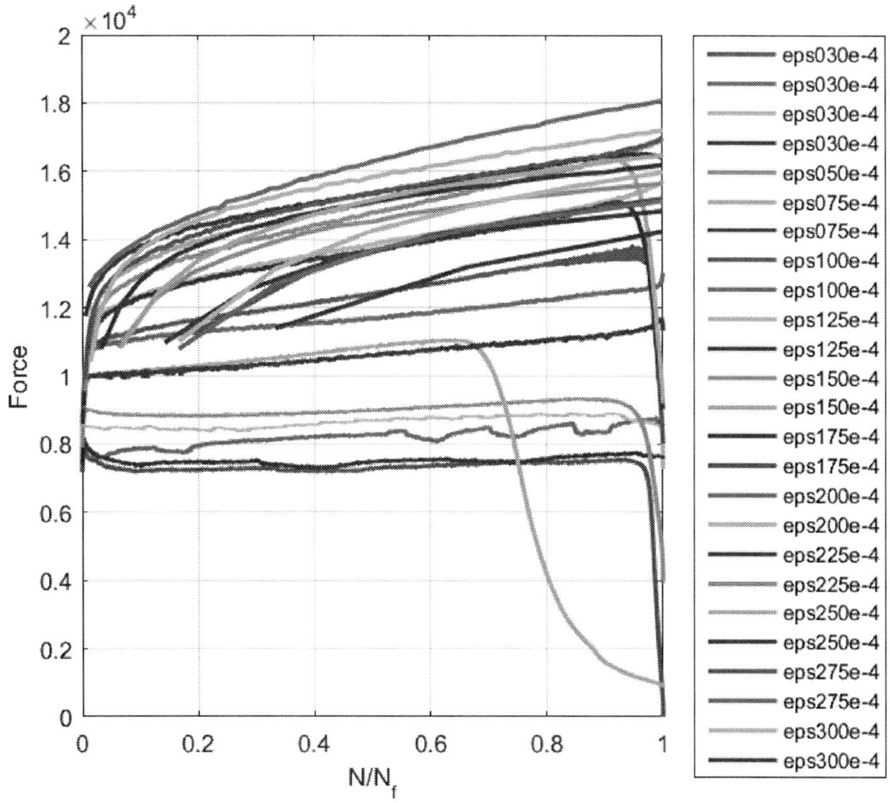

Fig. 2: Force progress during fatigue life for original uniform-gage specimen. N_f represents number of cycles at failure. See no-cyclic softening in the end of the fatigue life for almost all specimens.

2.2 New specimen design

Two candidates for new specimen design was proposed, both with variable longitudinal-section. This type of geometry localizes deformation into one cross-section (in the narrowest part) and also should be more post buckling-resistant. First specimen has circular longitudinal section (and is usually called hourglass specimen, see Fig. 3a), second one has elliptical longitudinal section (see Fig. 3b). Both specimens' dimensions are limited by experimental equipment and material cast.

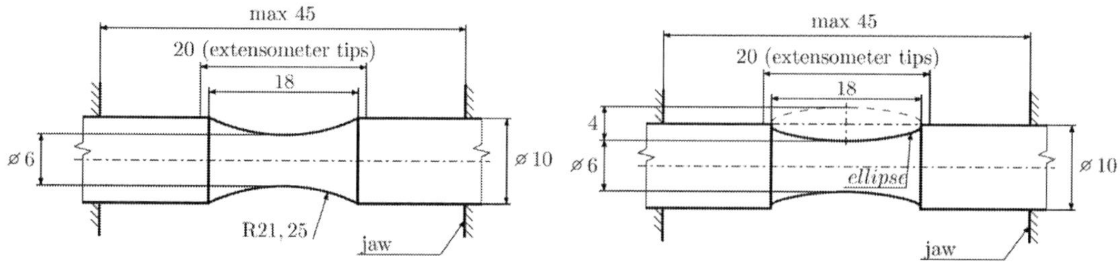

a) Hourglass specimen

b) Elliptical geometry specimen

Fig. 3: New specimens suitable for LCF experiments under large strain range.

3 Design Phase Analyses

Series of Finite element method (FEM) simulation in FEM software Abaqus was done to analyze suitability and loading conditions of both newly proposed geometries.

3.1 Post-Buckling Analysis

Research team was not looking for the real cause[1] of loss of stability control due to post-buckling effect. Problems with stability control occurs only on the highest loaded specimens (for total strain amplitude $\epsilon_a = 2{,}75\% - 3\%$). Proposed post-buckling analysis should only estimate whether new specimen design is more post-buckling resistant than the original one.

Proposed procedure consists of two steps. First, specimen's geometrical imperfection, modelled as the first eigenmode (of sine shape, see Fig. 4), as maximum manufacturing tolerance was imported into FEM model as well as maximum allowed axis misalignment of experimental device. In second step, specimen FEM model is loaded by forced displacement in axial direction. Criterion for determination of the limit reaction force is chosen as transversal displacement of the specimen by $0{,}5\ mm$. Material model includes non-linear plasticity behavior and is approximated from static tensile curve.

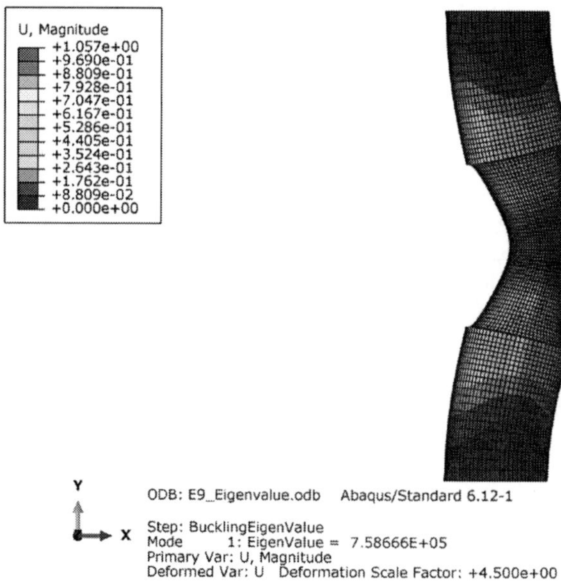

Fig. 4: Scaled initial imperfection of specimen.

---

[1] Origin of imperfection can be technological, caused by some non-isotropical material behavior, or caused by crack growth on one side of specimen...

3.2 Strain Field Analysis

Unlike the uniform-gage specimen, dependency between axial total strain amplitude and extensometer's displacement for specimen with variable longitudinal section is not trivial. Strain field analysis finds relationship between displacement of the extensometer's attached points and value of total strain amplitude on the surface of specimen in the narrowest point of specimen (point with highest deformation level, where crack usually starts to grow).

FEM simulation on the number of different load levels is performed and total axial strain amplitude value as a function of extensometer displacement is determined.

3.3 Triaxiality Factor Analysis

Triaxiality factor (hereinafter triaxiality) is dimensionless parameter defined as

$$TF = \frac{\sigma_{hydro}}{\sigma_{Mises}} \qquad (1)$$

where σ_{hydro} is hydrostatic stress and σ_{Mises} is Mises equivalent stress (also known as HMH stress). It characterises stress field and for example the value of triaxiality for pure tension is $TF = \frac{1}{3}$. The purpose of this type of analysis is to check whether the stress field in analyzed variable geometries of specimens is approximately comparable with smooth specimen.

4 Analyses results

Results of post-buckling analysis are in Tab 1. It shows that proposed geometries can handle about 20% higher load, so they are significantly more post-buckling resistant. Again - it would be good to mention that these results should only be taken as a relative comparison of post-buckling resistance of presented geometries rather than exact procedure of determination of a limit loading.

Tab. 1: Post-buckling analysis results

Specimen Geometry	Limit Force [kN]
Uniform-gage (original)	17700
Hourglass	22700
Elliptical	22000

Strain field analysis of both specimen shows slightly non-linear dependency between extensometer's displacement and total strain amplitude value (see Fig. 5, Fig. 6).

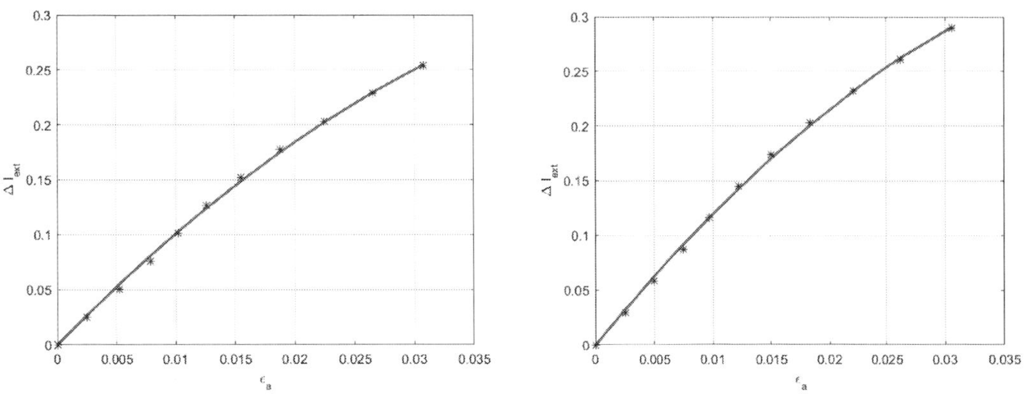

a) Hourglass specimen b) Elliptical geometry specimen

Fig. 5: Dependency of extensometer's displacement on axial strain amplitude.

Results of triaxiality analysis are in Fig. 6 and show very good agreement with the original uniform-gage specimen. The elliptical geometry shows values of triaxiality slightly closer to the original one.

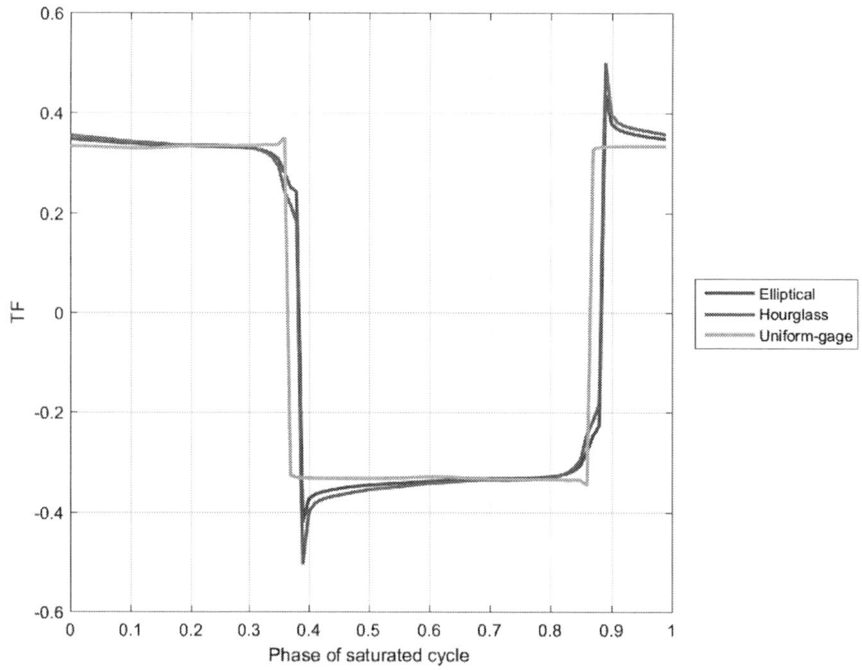

Fig. 6: Progress of triaxiality (TF) during saturated cycle.

All analyses show, that both proposed geometries should be suitable for LCF experiments even for large strain amplitude loading. Finally, it was decided to use elliptical geometry, because it's stress triaxiality parameter closer to the uniform-gage specimen.

5 Experiment

5.1 Experiment description

Tests were carried out by means of a computer-controlled electromechanical Mayes machine with 100 kN capacity employing the LabControl software tool. The maximum stroke is 150mm. The base is not equipped with MTS Series 609 Alignment Fixture to reduce bending strains by improving concentric. Therefore, the test facility was adjusted for alignment fixture assembly which is mounted between the load cell and the load frame. Fatigue specimens were placed in the 647.25A side-loading hydraulic wedge grips (vee-shaped wedges). Constant strain amplitudes were controlled by the extensometer MTS model 634.25 with initial gauge length of 10mm with a 50% measuring range, in case of prismatic specimens, and length of 20mm with a 20% measuring range, in case of elliptical geometry specimens. The MTS model 650.03 extensometer calibrator was employed to get high accuracy at the lower strain levels where the data is more important. The calibration of the extensometer was done before and after each test program in accordance with ASTM E606 standard [1].

Fatigue tests were carried out on round and elliptical geometry specimens at constant strain amplitudes from 0.5% up to 3% of total deformation at ambient temperature. The load ratio of -1 was applied. The frequency of testing relates directly to the strain level at which the specimen was loaded. To keep a strain rate (0.002s-1) factor negligible and constant, the triangular waveform of the loading was used along with low loading frequencies. To reduce the influence of knife edges of contact extensometer, which is directly mounted on the specimen surface in gauge length area, the dull edges were used.

Each test was terminated after the specimen was separated. Fatigue tests were evaluated after the tensile force amplitude dropped 20% from the maximum of the force response.

Furthermore, the attention is paid to keep the fracture surface of the specimen in good condition for additional fractography investigations. Consequently, the crosshead software limits had to be set to avoid damaging fatigue crack surfaces which may lead to make the observation impossible.

Fig. 6: Experimental setup includes extensometer and hydraulic wedge grips.

5.2 Experiment Results

Experimental results confirm, that elliptical geometry has no tendency to post-buckle and force progress during fatigue life is in accordance with expectations (see Fig. 7).

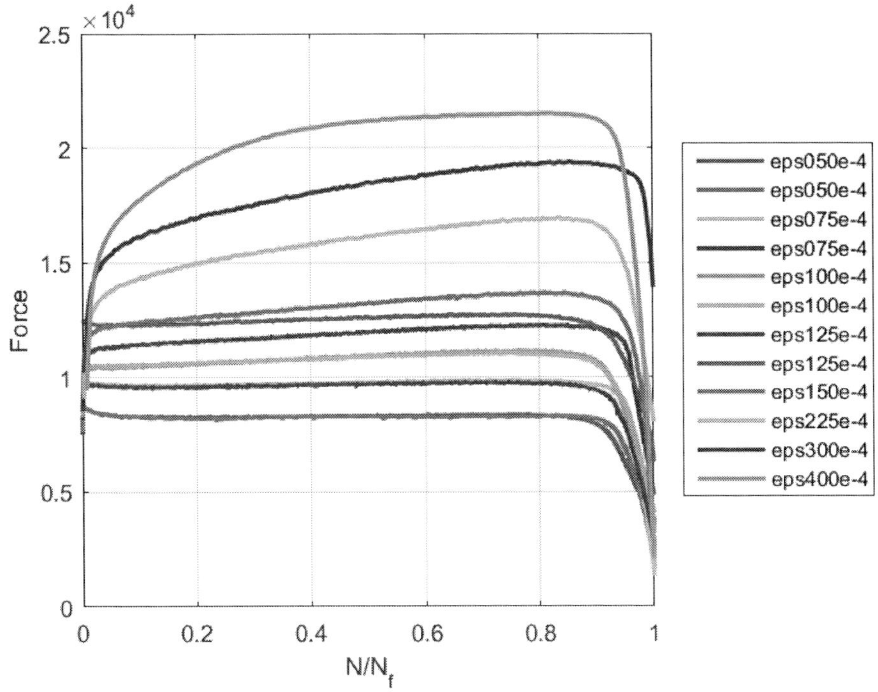

Fig. 7: Force progress during fatigue life for elliptical geometry specimen.

6 Conclusion

Unsuitability of originally used uniform-gage specimens was briefly analyzed. Two new geometry designs of specimens' candidates were proposed. Post-buckling, strain field and triaxiality analysis was done. Elliptical geometry specimen was chosen because of better triaxiality factor progress. Subsequent experimental procedure was described. Experimental results proved buckling resistance of new elliptical geometry design and force progress during fatigue life in accordance with expectations.

Acknowledgement

The authors would like to acknowledge support from the Technology Agency of the Czech Republic, grant No. TA04020806, and support from the Grant Agency of the Czech Technical University in Prague, grant No. SGS15/187/OHK2/3T/12.

References

[1] Lemaitre, J. and Desmorat, Rodrigue. Engineering Damage Mechanics. New York: Springer, 2005. ISBN 3-540-21503-4.

[2] American Society for Testing and Materials. Standard practice for strain-controlled fatigue testing, E606-04. ASM International: The Materials Information Society; 2004.

[3] Fekete, B.: Isothermal and thermal–mechanical fatigue of VVER-440 reactor pressure vessel steels. Journal of Nuclear Materials 464, 2015.

[4] Fumfera, J.: Návrh vzorků a parametrů zatežování. Zpráva ČVUT FS č. 12105/15/02, 2015.

Measurement and Evaluation of Spatial Motion of Washer Extractor

M. Fusek[1,2,a], R. Halama[1,2], M. Marek[2], F. Fojtík[1,2], Z. Poruba[1]

[1] *Department of Applied mechanics, Faculty of Mechanical Engineering, VŠB-Technical University of Ostrava, Czech Republic*

[2] *National Supercomputing Centre IT4Innovations, VŠB-Technical University of Ostrava, Czech Republic*

[a] *martin.fusek@vsb.cz*

Abstract: This paper describes the measurement stand for the determination of the spatial motion of the rigid structure. As the example, the drum of the washing unit can be given. The measurement system uses 2D laser sensors to obtain the instantaneous position of the point at the device investigated. The motion of any point can be then calculated on the base of known body stiffness. The measurement method is sufficiently accurate and makes possible the cheap and simple evaluation of the motion or vibrations of the construction.

Keywords: measurement; spatial motion; washer extractor; laser distance sensor.

1 Introduction

This paper builds on contributions [1, 2] which deal with the measurement of the motion of the washing unit using six one-component position sensors. The newly designed system is based on 2D laser sensors. The measurement is primarily focused on the determination of the drum motion of the washing unit. However, this method can be used for measurement of any other rigid object.

2 Description of the Measurement System

The measurement and evaluation is realized based on two main components - hardware and software.

The first one utilizes two 2D laser distance sensors (Position 4, Fig. 1) placed at a specific angle and distance toward the auxiliary measurement jig (Position 2, Fig. 1), which is attached by magnets to the object being investigated (Position 1, Fig. 1). The proper placement of laser distance sensors is guaranteed through the auxiliary bracket with the fixed space for these sensors (Position 5, Fig. 1). The auxiliary bracket also provides the possibility to adjust the height (Position 6, Fig. 1). The whole scheme allows fast installation and start of the measurement process. The hardware equipment provides the power for sensors and realizes the transfer of data collected during the measurement to a service PC.

The second part contains the measurement and evaluation software. The measurement software (Fig. 2a) provides information at the connection of laser sensors and allows the user to adjust precisely the mode of their operation, filtering and smoothing of profile data in every image. Furthermore, it includes a function to control and input the measured data into a file, including runtime control with information about the iterative processing of individual images from the measurement.

The evaluation software (Fig. 2b) analyses and displays the spatial motion of the body based on the data acquired and stored via the measurement software. With this program, the user can directly see the basic data and individual images from the measurement or perform the complete analysis of the motion of the body in both 2D and 3D visualizations.

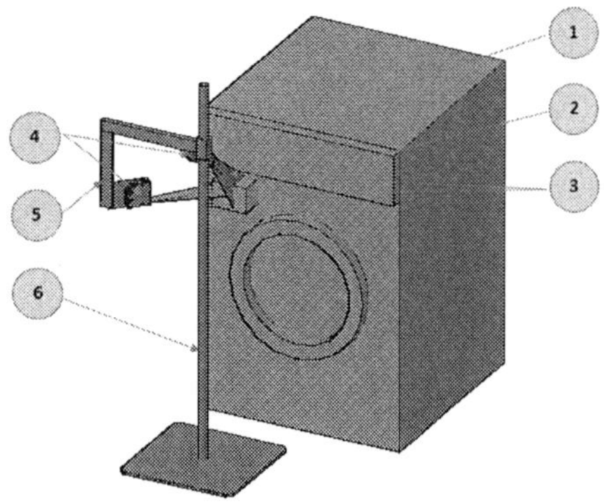

a) Stand scheme

b) Real configuration

Fig. 1: Measurement stand for evaluation of spatial motion.

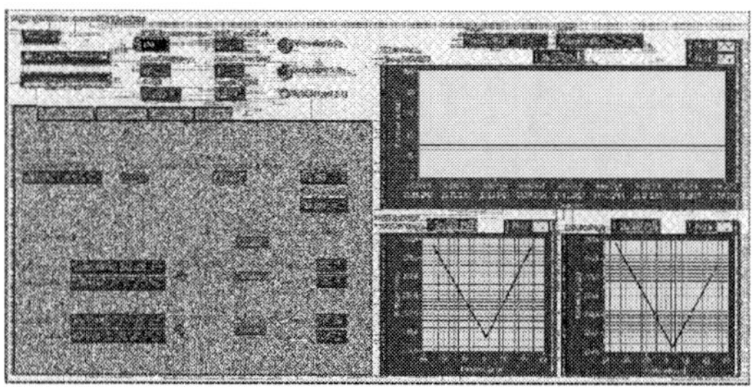

a) Measurement software

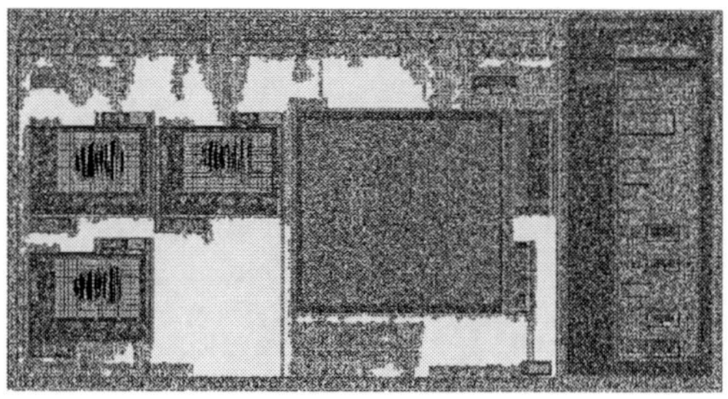

b) Evaluation software

Fig. 2: Software for measurement and evaluation of spatial motion.

3 Measurement

The measurement system was developed and tested in laboratory conditions where very good results were achieved. In order to verify the functionality, the stand was equipped with a series of additional laser one-dimensional sensors. Subsequently, the calculated results were compared with those from auxiliary sensors.

In addition to laboratory measurements, the measurements in practical conditions in the factory was realized.

In order to design the chassis of the washing unit, it is necessary to know the motion of individual points of the washing drum. The inner space of the washing machine is filled with other members such as driving unit, mechanical members (for example vibration and impact dampers), electrical wiring, control systems and emergency members. Thus, the motion of the washing drum is limited. The shape of the chassis is fully influenced by exterior design of the washing machine given by the appearance. Thus, it is obvious that knowledge of the motion of individual drum points in all modes of the washing cycle is essential.

The following table (Tab. 1) summarizes results obtained from measurements in practical conditions on two types of washing units. The evaluation software makes possible to approximate the trajectory of any point at the washing drum. In Table 1, amplitudes of the point (located at the front of the washing unit at its upper left corner) in individual directions are recorded.

The column "Mode" in Table 1 represents the washing program (short 13 min., long 40 min.). The column "Filling" represents the contain of the washing drum (jeans – washing machine was filled by jeans cloths, rubber – washing machine was filled by rubber segments simulating the 15% imbalance, stop in the column "Notice" indicates the emergency state – see bellow).

Tab. 1: Measurement results.

Measurement	Washing machine	Mode	Filling	Δx [mm]	Δy [mm]	Δz [mm]	Notice
1	1	short	jeans	13,5	4,9	9,9	
2	1	short	jeans	48,3	8,7	17,36	stop
3	1	short	jeans	17,3	4,7	12,37	
4	1	short	jeans	6,9	4,7	8,96	
5	1	short	jeans	25,3	4,7	16,55	stop
6	1	long	jeans	28,2	6,2	15,52	
7	2	short	rubber	23,9	4,9	14,6	
8	2	short	rubber	24,1	4,9	14,4	
9	2	short	rubber	24,1	5,0	14,5	

Records of the measurement 3 on the washing unit 1 is depicted in following figures. Again, the results correspond to the front of the washing unit at its upper left corner. The motion depicted in 3D visualization is obvious in Fig. 3. Individual graphs represent the motion in individual axes x, y, z with respect to the main coordinate system of the measuring stand. Motions can be easily converted to any coordinate system chosen.

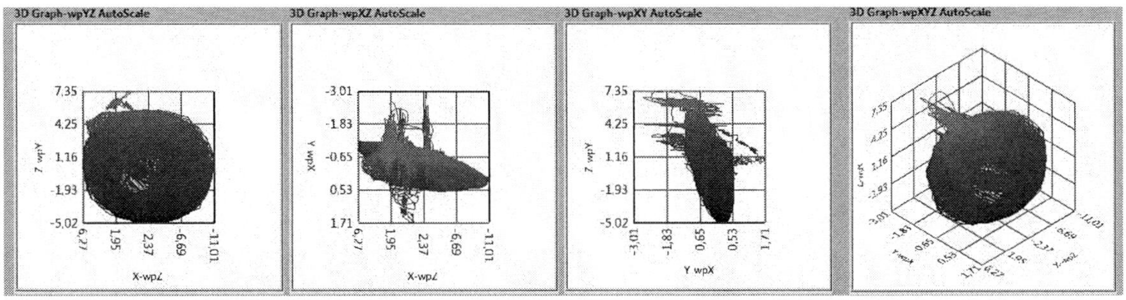

Fig. 3: Motion during the whole washing cycle in 3D visualization (measurement nr. 3).

Figure 4 represents the motion during the whole washing cycle. The significant shift of the washing drum can be seen at the beginning of the record. This shift is related to the filling of the drum and needs to be taken into the account. Then, the washing process proceeds. At the end of the washing process it is obvious that the washing unit does not return to its initial position. The initial sag does not appear during testing with the rubber imbalance. It is necessary to realize that simulation with clothes (in our case with jeans at 50% of maximal load) gives truer image than an experiment with mere imbalance. On the other hand, there are uncertain inputs concerning the repeatability and potential numerical simulation. Further, also other types of filling need to be used for more complex description of the behavior.

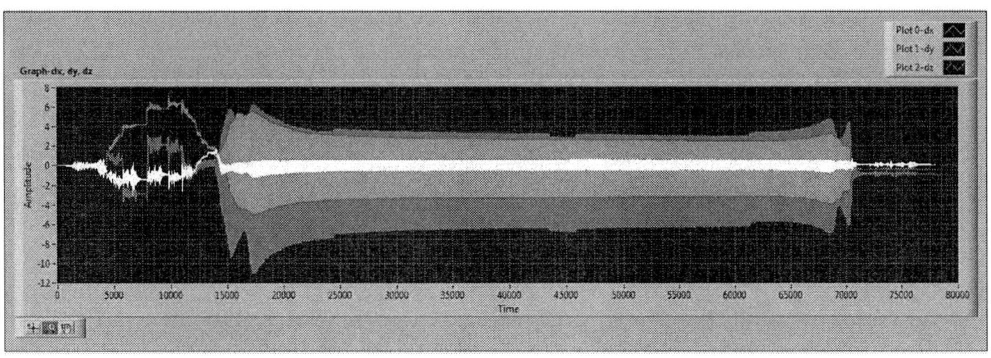

Fig. 4: Motion during the whole washing cycle (measurement nr. 3).

The emergency switch was activated during the measurement nr. 2. This switch guards the deflection of the washing drum so that both the damage of the devices located in the inside of the washing unit frame and collision between the drum and housing of the washing unit is avoided. After the guarding sensor is activated, the chosen washing program is interrupted, speed of the drum is reduced so that the cloth adopts more uniform position and after approx. four minutes the original washing program proceeds. Such case represents the emergency state which is not desirable. The emergency state was activated also during measurement nr. 5 – see Fig. 5. Results obtained by measurement make possible a better understanding of the motion parameters of the washing unit and allow to optimize the inner arrangement of the washing unit. The information obtained can also help during setting up the simulation models. The software for evaluation was positively accepted by the company designers and is being tested at present time.

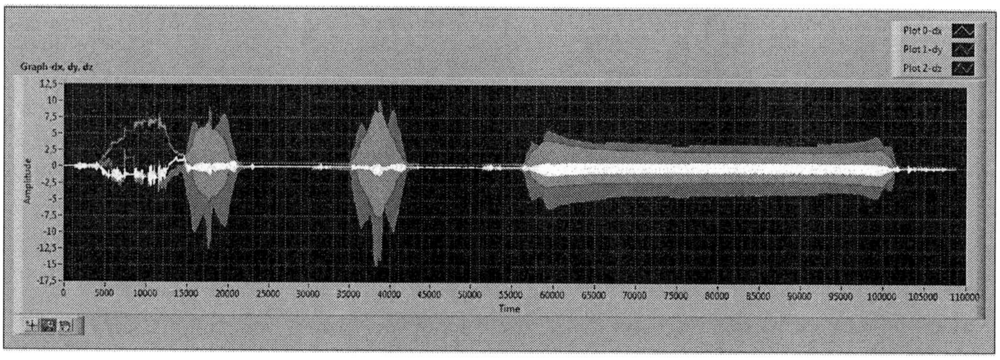

Fig. 5: Motion during the whole washing cycle with emergency switch (measurement nr. 5).

4 Conclusion

This contribution describes the laser system designed for the measurement of the spatial motion of (sufficiently) rigid object, for instance the drum of washing unit. It is the improved stand related to the original measurement system which is consisting of six one-component sensors of motion – see [1, 2]. The great benefit of the new measuring stand is significantly faster manipulation and measurement preparation. The contribution deals with the measurement performed in the factory designing and producing washing units. However, the method developed is not limited only to the measurements of washing drums. It can be used for measurement of any object engaging the spatial motion.

Acknowledgement

This work was supported by The Ministry of Education, Youth and Sports from the National Programme of Sustainability (NPU II) project „IT4Innovations excellence in science - LQ1602“. This work was also supported by project CZ.1.05/3.1.00/14.0318 Pre-seed activities of VSB-TUO II – Engineering (Individual activity 6). This work was also supported by the Grant Agency of the Czech Republic (GACR), project No. 15-18274S. Authors would like to appreciate the assistance of Alliance Laundry CE s.r.o. Pribor company during the preparation and realization of experiments throughout the article preparation.

References

[1] Halama, R. et al., Real Time Measurement and Evaluation of Washer Extractor Vibrations, in proc: EAN 2012: 50th Annual Conference on Experimental Stress Analysis (2012), ed. M. Růžička, K. Doubrava, Z. Horák, Prague: Czech Technical University in Prague, Faculty of Mechanical Engineering, Tábor, Czech Republic, 129-134. ISBN 978-80-01-05060-6

[2] Halama, R., Fusek, M., Fojtík, F., An Industrial Application of a Method for Spatial Motion Measurement. In proc: Experimental Stress Analysis (2010). ed. P. Šmíd, P. Horváth, M. Hrabovský. Olomouc: Palacky University, Velké Losiny, Czech Republic, 75-80. ISBN 978-80-244-2533-7.

Cyclic Hardening Behavior of Stainless Steel 316L - Experiments and Modeling

R. Halama [1,a], D. Ličková [1], J.J.R. Morales [2]

[1]Department of Applied Mechanics, Faculty of Mechanical Engineering & National Supercomputing Centre IT4Innovations, VŠB - TU Ostrava, 17. listopadu 15, 70833 Ostrava, Czech Republic
[2] Universidad De Castilla - La Mancha, Calle Altagracia, 50, 13071 Ciudad Real, Cdad. Real, Spain
[a] radim.halama@vsb.cz

Abstract: This paper shows main results of research emphasized on stress-strain behavior of the 316L stainless steel under proportional as well as non-proportional loading. The 316L steel shows significant additional hardening due to non-proportional loading. The results from all realized fatigue tests including the number of cycles to crack initiation and cyclic hardening/softening curves are presented. Two ways of fatigue parameters estimation are applied with comparable results. Important conclusions for the cyclic plasticity modeling considering investigated material are also discussed with emphasize on observed Non-Masing behavior and strain range dependent cyclic hardening.

Keywords: multiaxial fatigue; additional hardening; cyclic plasticity; 316L stainless steel.

1 Introduction

In the classical theory of plasticity the non-proportional loading is defined as a type of loading when principal stress directions rotate. The intensive research in the field of cyclic plasticity in last 30 years have shown that in case of non-proportional loading the cyclic hardening/softening behavior is more significant than in case of equivalent proportional loading [1]. Cyclic plasticity models based on phenomenological approach have to be very robust to describe cyclic hardening properties correctly under proportional as well as non-proportional loading [2]. The situation becomes more complicated in the case of stainless steels, because their transient behavior is strain range dependent [3]. Moreover, such cyclic plasticity model contains a large number of parameters which need to be estimated using a large number of experimental data from fatigue tests. This paper is focused on the stress-strain behavior investigation of the 316L stainless steel emphasizing the additional hardening behavior and the strain range dependent cyclic hardening.

2 Experimental study

All experiments were conducted on the 316L stainless steel. The electro-servo-hydraulic testing machine LabControl 100kN/1000Nm placed at the Faculty of Mechanical Engineering of the VŠB - Technical university of Ostrava was used for all cases.

The EPSILON extensometer - type 3550 with 25.4mm gauge length was used to measure/control the axial and shear strain simultaneously. For biaxial testing the specimen has tubular active part with outer and inner diameter of 12.5 mm and 10 mm respectively. The uniaxial fatigue tests were realized on solid round-type specimen with diameter of 5 mm. All low-cycle fatigue tests were realized on smooth specimens under strain control with zero mean strain during the cycle. The strain rate of about $6{\cdot}10^{-3}$ s^{-1} was considered in all tests during cycling. The number of cycles realized before crack initiation was determined in all tests based on the decrease of force (moment) amplitude of twenty percent in comparison to the saturated values. All experiments were performed at room temperature. The loading paths considered in this study are presented schematically in Fig.1. Eight different cases were realized. These paths were considered: uniaxial (case A), torsional (B), proportional (C), square (D), rhombic (E), two blocks (F), circular (G) and elliptical (H). The elliptical path shape corresponds to the 45° out of phase fatigue test.

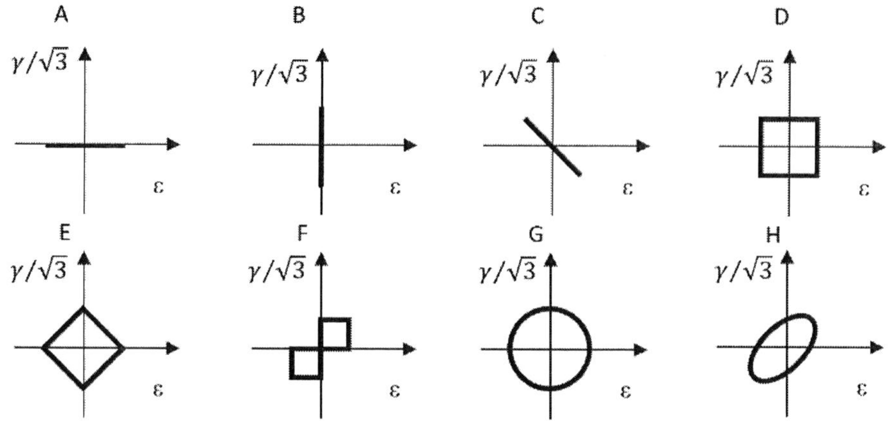

Fig. 1: Strain path shapes in particular cases

2.1 Uniaxial fatigue tests

In total nine low-cycle fatigue tests with constant amplitude of total strain were performed. The results of tests are summarized for all levels of loading in the Tab.1. Stable hysteresis loops evaluated in a half of fatigue life are displayed in the graph at the Fig.2. The Non-Masing behavior of the SS316L material is obvious from the Fig.2. It can be caused by a primary creep even under room temperature, which is well known phenomenon of stainless steels.

Tab. 1: Results of uniaxial fatigue tests.

Total strain amplitude ε_{at} [1]	Stress amplitude σ_a [MPa]	Number of cycles to failure N_f [1]
0.02433	627	24
0.01752	564	56
0.01349	501	133
0.00989	466	594
0.00699	424	897
0.00488	380	5700
0.00395	362	8630
0.00293	336	124450
0.00265	322	536512

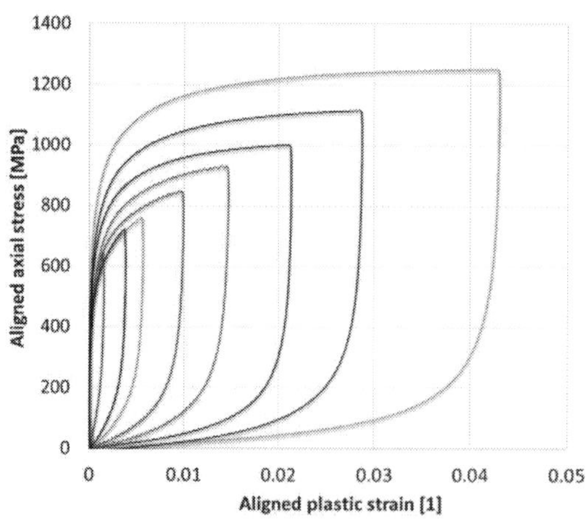

Fig. 2: Saturated hysteresis loops of SS316L alligned to the bottom peak

2.1.1 Uniaxial stress-strain behavior

The evolution of stress amplitude during cycling for selected levels of strain amplitude can be seen in the Fig.3. The investigated material reveals very rapid cyclic hardening in initial cycles followed by slow process of the cyclic softening.

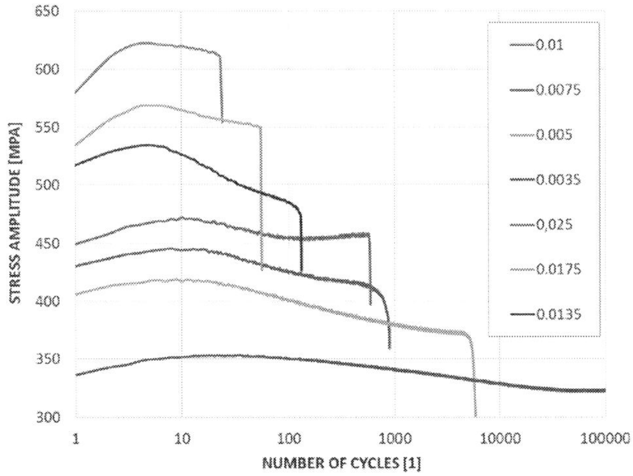

Fig. 3: Cyclic hardening/softening curves of SS316L for certain strain amplitudes

2.1.2 Fatigue parameters estimation methods

For evaluation of tests of low cycle fatigue in tension-compression, the Manson-Coffin-Basquin equation [4] is usually applied. It illustrates the relationship between the amplitude of total strain ε_{at} and number of cycles to crack initiation N_f

$$\varepsilon_{at} = \varepsilon_{ae} + \varepsilon_{ap} = \frac{\sigma'_f}{E}\left(2N_f\right)^b + \varepsilon'_f\left(2N_f\right)^c, \quad (1)$$

where ε_{ae} is the amplitude of elastic strain, ε_{ap} is the amplitude of plastic strain, ε'_f is the fatigue ductility coefficient, c is the fatigue ductility exponent, σ'_f is the fatigue strength coefficient and b is called as the fatigue strength exponent. The Young modulus was evaluated as a dynamic elastic modulus from all uniaxial stable hysteresis loops E=169785MPa. Amplitude of the total strain can be also expressed by the Ramberg-Osgood equation [5] as a function of stress amplitude in the case of tension-compression

$$\varepsilon_{at} = \varepsilon_{ae} + \varepsilon_{ap} = \frac{\sigma_a}{E} + \left(\frac{\sigma_a}{K'}\right)^{n'}, \quad (2)$$

where K' is the cyclic strength coefficient and n' is the cyclic strain hardening exponent.

Six fatigue constants have been usually determined by the conventional method based on approximation of the experimentally obtained data by a line (three independent linear regressions). Niesłony [6] has proposed a new method for identifying fatigue constants, the so-called 3D method. The main advantage of this method is that it ensures the compatibility of six fatigue constants. The method is based on the approximation by a straight line in xyz space, where $x = log(\varepsilon_{ap})$, $y = log(\sigma_a)$, $z = log(2N_f)$. The regression line is determined by the direction r (r_1, r_2, r_3) and a point P (x_P, y_P, z_P), see Fig.4. Based on the direction vector coordinates and coordinates of point P the six fatigue coefficients are calculated as follows

$$n' = \frac{r_2}{r_1}, \quad (3)$$

$$K' = 10^{y_P - x_P n'}, \quad (4)$$

$$c = \frac{r_1}{r_3}, \quad (5)$$

$$\varepsilon'_f = 10^{x_P - z_P c}, \quad (6)$$

$$b = \frac{r_2}{r_3}, \quad (7)$$

$$\sigma'_f = 10^{y_P - z_P b}. \quad (8)$$

When solving the problem, 4 regressions are made. The first regression determines the equation of a plane in space and the other three are used to calculate regression lines in the planes $\mathbf{x} - \mathbf{y}$, $\mathbf{y} - \mathbf{z}$ and $\mathbf{z} - \mathbf{x}$, where $\mathbf{x} = \boldsymbol{log(\varepsilon_{ap})}$, $\mathbf{y} = \boldsymbol{log(\sigma_a)}$, $\mathbf{z} = \boldsymbol{log(2N_f)}$. Subsequently, two regressions with the highest coefficient of determination are selected. It is obvious that at least one of the selected regressions is a line in plane, i.e. it is necessary to establish a plane which is given by the straight line and which is, at the same time, perpendicular to the plane in which it lies. In order to make methods more comfortable for use, there was developed algorithm in MATLAB.

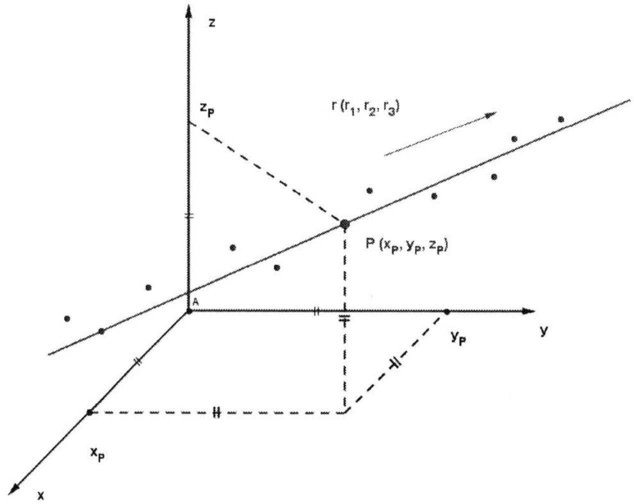

Fig. 4: Points in the 3D space with regression line determined by point P and directional vector r

2.1.3 Results of fatigue parameters estimation

Both methods (conventional and 3D method) were applied to obtain fatigue parameters of SS316L material. The resulting regression dependencies for the stainless steel 316L based on uniaxial data are shown as output from the MATLAB program in Fig.5. Values of fatigue constants obtained after application of the 3D and the conventional methods are presented in Tab. 2.

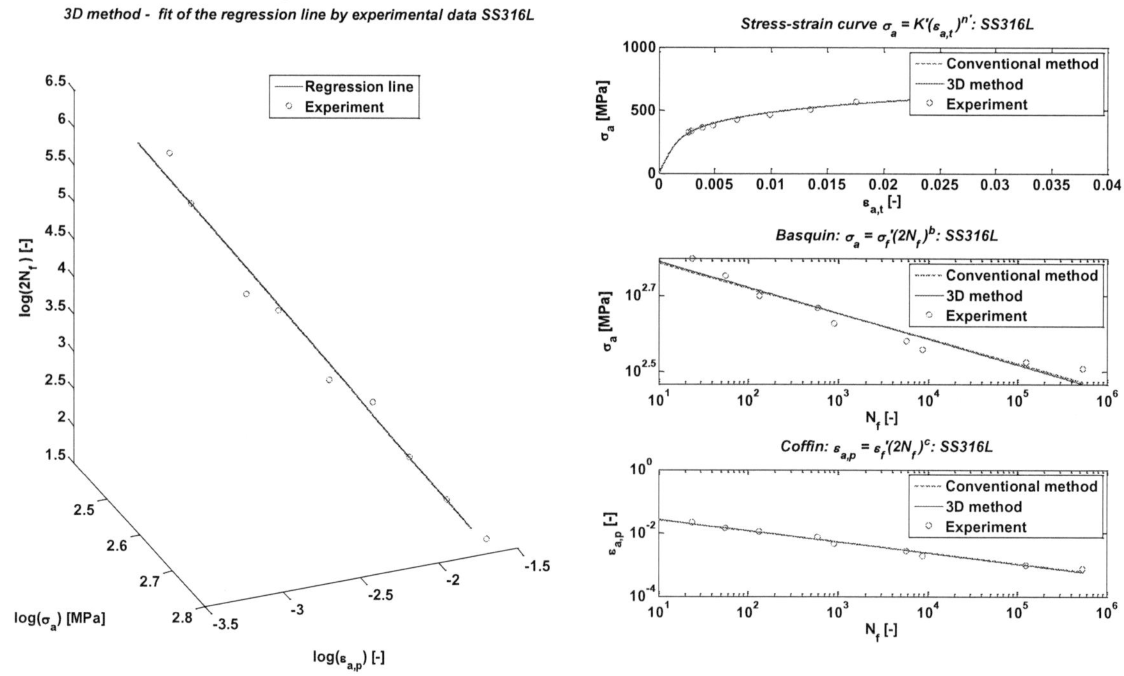

Fig.5: Graphical representation of 3D method application and its comparison with conventional method for the stainless steel 316L – tension-compression

Tab. 2: Results of 3D method and conventional method application.

Parameter	Conventional method	3D method
ε'_f	0.070	0.074
c	-0.343	-0.350
σ'_f [MPa]	742	755
b	-0.066	-0.068
K' [MPa]	1251	1251
n'	0.194	0.194

2.2 Biaxial fatigue tests

Totally 10 biaxial fatigue tests were realized. In order to compare individual cases the amplitude of equivalent strain was calculated for each test as the minimal radius of the circumscribed circle in the diagram axial strain ε - equivalent value of shear strain ($\gamma/\sqrt{3}$). Similarly the amplitude of equivalent stress was calculated for each test as the minimal radius of the circumscribed circle in the diagram axial stress σ - efective shear stress ($\sqrt{3}\tau$). Main results of all biaxial tests are summarized in the Tab.3.

Tab. 3: Results of biaxial fatigue tests.

Spec. number	Path type	Axial strain amplit. ε_a [1]	Axial stress amplit. σ_a [MPa]	Eff. shear strain amplit. $\gamma_a/\sqrt{3}$ [1]	Eff. shear stress amplit. $\sqrt{3}\tau_a$ [MPa]	Equiv. strain amplit. ε_a^{eqv} [1]	Equiv. stress amplit. σ_a^{eqv} [MPa]	Number of cycles to failure N_f [1]
B11	torsion	0	0	0.0042	422	0.0042	422	6200
B1	torsion	0	0	0.0077	477	0.0077	477	1099
B10	torsion	0	0	0.0151	569	0.0151	569	416
B2	proport.	0.0055	348	0.0056	345	0.0078	490	818
B3	rhombic	0.0055	513	0.0056	554	0.0056	534	618
B4	circle	0.0075	689	0.0076	742	0.0075	715	434
B5	elliptic	0.0075	594	0.0076	649	0.0099	880	548
B7	2 blocks	0.0052	543	0.0052	551	0.0074	561	180
B8	square	0.0044	578	0.0047	617	0.0067	660	480
B6	square	0.0076	728	0.0076	780	0.0107	1067	108

2.2.1 Non-proportional hardening

Additional hardening of the material can be evaluated by the comparison of the amplitudes for proportional cases and non-proportional cases. Equivalent stress amplitudes for different loading paths are shown in the Fig.6.

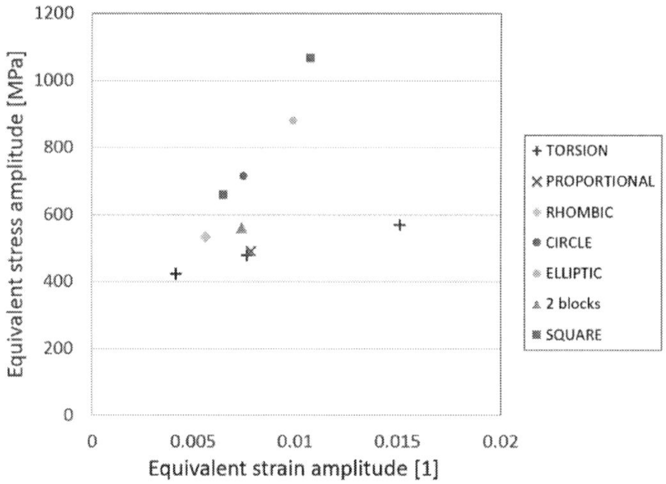

Fig. 6: Stress-strain diagram for different biaxial cases

The influence of strain path shape can be apparent from the Fig.7. All cases considered in the Fig.7 correspond to the same equivalent strain amplitude value of about 0.786 percent.

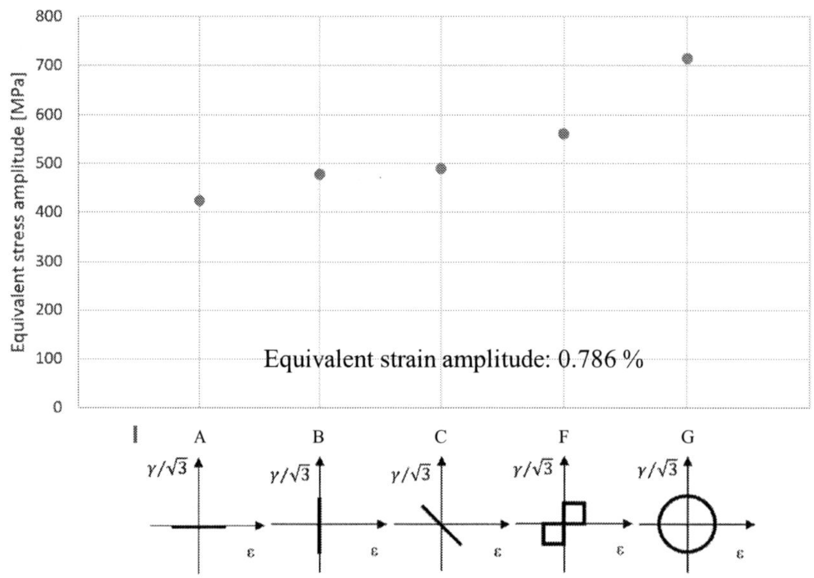

Fig. 7: Comparison of equivalent stress amplitude for different loading paths.

3 Discussion of results

The results of fatigue tests presented in the Fig.7 were obtained for a half life cycle. The smallest value of the equivalent stress amplitude is that of the uniaxial case. The slightly higher values were observed under torsion and proportional tension-torsion loading tests. Stronger additional hardening than in previous cases corresponds to the two block loading path, what is clear also from the Fig.6. The most significant additional hardening was observed under non-proportional 90° out-of-phase loading (about 69 per cent) and for square strain path shape, see Fig.6. The number of cycles to failure in the case of highly non-proportional loading is almost six times less than the number of cycles to failure under uniaxial loading with the same equivalent strain amplitude. A large amount of plastic work is accumulated, which ultimately leads to the lifetime shortening.

4 Conclusion

The main results from the fatigue tests realized on SS316L material under uniaxial as well as biaxial loading were presented. The cyclic hardening/softening curves shows similar behaviour on particular levels of strain amplitude. Very important conclusion of the experiments is, that really significant Non-Masing behaviour occurs. Regarding this fact a cyclic plasticity model with memory surface have to be used to describe the cyclic hardening/softening behaviour correctly. Each experimentally realized case has been simulated by the modified AbdelKarim-Ohno model [3] in the finite element program ANSYS 15.0. These results will be presented in a future paper.

Acknowledgement

This work was supported by The Ministry of Education, Youth and Sports from the National Programme of Sustainability (NPU II) project „IT4Innovations excellence in science - LQ1602", the project of specific research of Ministry of Education, Youth and Sports of the Czech Republic under No. SP2016/145 and by the Grant Agency of the Czech Republic (GACR), project No. 15-18274S.

References

[1] R. Halama, J. Sedlák, M. Šofer, Phenomenological Modelling of Cyclic Plasticity, in: P. Miidla (Ed.), Numerical Modelling, InTech, Rijeka, 2012, pp. 329-354.

[2] R. Halama, M. Šofer, F. Fojtík, Choice and Calibration of Cyclic Plasticity Model with Regard to Subsequent Fatigue Analysis. *Engineering Mechanics* 19 (2012) 87-97.

[3] R. Halama, F. Fojtík, A. Markopoulos, Memorization and Other Transient Effects of ST52 Steel and Its FE Description, *Applied Mechanics and Materials* 486 (2013) 48-53.

[4] Bäumel A. Seeger T. Material data for cyclic loading. supplement 1. materials science monographs. 61. Amsterdam: Elsevier Science. 1990.

[5] Ramberg W. Osgood WR. *Description of stress–strain curves by three parameters. Technical note no. 902.* Washington (DC): National Advisory Committee for Aeronautics. 1943.

[6] Niesłony A. El Dsoki Ch. Kaufmann H. Krug P. New method for evaluation of the Manson–Coffin–Basquin and Ramberg–Osgood equations with respect to compatibility. *International Journal of Fatigue* 30 (2008) 1967–77.

Numerical and experimental analysis of composite nacelle behaviour under different boundary conditions

HAVLÍK Vojtěch [1,a*], KAFKA Václav [1,b], HRON Robin [1], MLCH Ivan [1]

[1] Aerospace Research and Test Establishment (VZLU), Strength of Structures Department, Beranovych 130, 199 05 Prague – Letnany, Czech Republic

[a] vojtech.havlik@vzlu.cz, [b] kafka@vzlu.cz

Abstract: In order to proof composite part design the compressive stiffness test of engine nacelle demonstrator was carried out in VZLU. The nacelle was made from monolithic glass epoxy laminate and sandwich glass epoxy composite with honeycomb core. The aim of this article is to harmonize numerical boundary conditions with the real boundary condition of the nacelle test fixture. Identifying of the real boundary conditions will help to increase the precision of numerical model prediction capability. Deflection of nacelle near the place of loading was used to compare how the finite element method analysis results match with the values measured from the test.

Keywords: composite, honeycomb, nacelle, boundary condition, deformation shape.

1 Introduction

One of the ESPOSA project aims was to obtain most suitable composite materials and processes for small gas turbine engines nacelle. Nacelle demonstrator was made of glass epoxy laminate combined with honeycomb sandwich. The test was designed to control quality and strength of the tested nacelle.

2 Test description

Test arrangement is shown on the figure1. The nacelle fixture was designed to correspond maximally with real joint conditions on the airplane. Nacelle was supported on all free edges. Test was divided into two sub-tests with different loading places - see Fig. 4. First loading was up to force 350N and loading was applied in the rear area where the nacelle is made from sandwich composite. Second loading was up to force 200 N and loading was applied in the front area where the nacelle is made from monolithic glass epoxy laminate.

Fig. 1: Test arrangement

Fig. 2: Loading support foot

3 FEM model description

The aim of FE model is to harmonize numerical boundary conditions with the real boundary conditions during test. FE model was created in Patran. Due to assumption of geometrical nonlinearity the calculations in Nastran were carried out using nonlinear solver Sol 106. Material nonlinearities were not taken into account.

Loading in the place of monolithic glass epoxy laminate (Load 2) leads to bigger deflection and geometrical nonlinearities. Boundary condition influence on deformation shape is also bigger in this case. That's why only results of the model for load 2 will be taken into account in the following chapters.

3.1 Materials

Material characteristics used for defining ACG glass-epoxy prepreg and for A1-48-5 Honeycomb are shown in the table1. Datas for ACG glass-epoxy were taken from [2]. Datas for Hexcel honeycomb were taken from [3]

Tab. 1: Material data.

	E_{11} [MPa]	E_{22} [MPa]	E_{33} [MPa]	G_{12} [MPa]	G_{23} [MPa]	G_{13} [MPa]	μ_{12} [-]
ACG glass-epoxy prepreg	23400	23400	8760	3700	1100	1100	0,2
HEXCEL A1-48-5 Honeycomb	25	40	140	-	-	-	-

Each of composite material is shown on the figure 3 with different color. Material angle 0° is oriented in direction of X axis on figure 3. Plane xz is symmetry plane of the nacelle. Axis Z is oriented vertically and in the same direction as loading force.

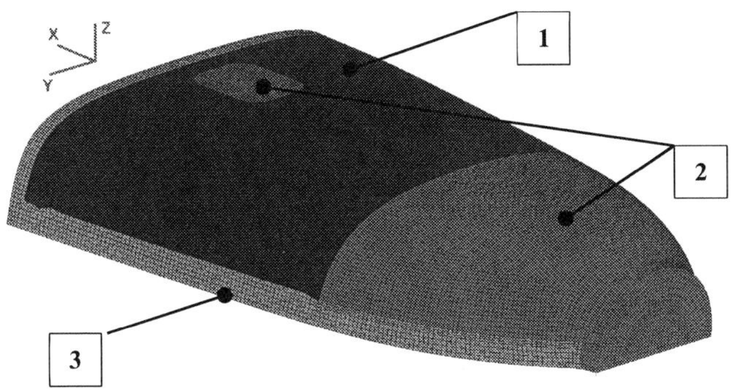

Fig. 3: Positions of monolithic laminate and sandwich composite

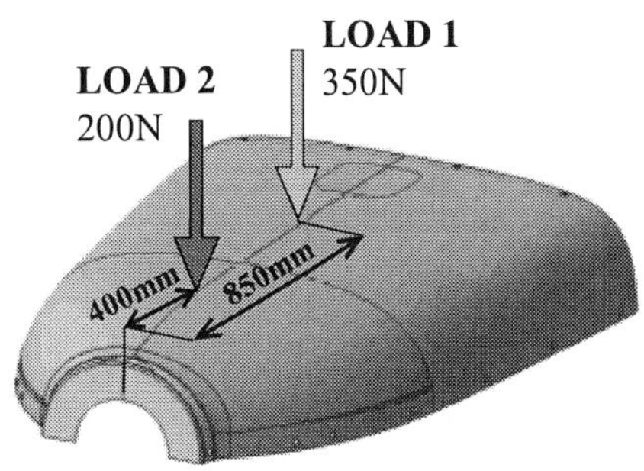

Fig. 4: Positions of loading

Sandwich composite structure made of glass epoxy-prepreg with honeycomb core is placed in area 1. Areas 2 and 3 represent monolithic laminate made of glass-epoxy prepreg. The orientation of each ply is written in the following list:

- 1 - sandwich composite - [45°/0°/Honeycomb/45°]
- 2 - monolithic laminate - [45°/0°/0°/45°]
- 3 - monolithic laminate - [45°/0°/-45°/0°/0°/-45°/0°/45°]

3.2 Boundary conditions and loading

There were two variants of modeling the boundary conditions. In the first case are all free edges of the nacelle constrained in all 6 degrees of freedom.

In the second case the gap elements were used on the free edges of the nacelle. In nonlinear solver gaps interact only in cases when they are in compressive stress state so in this way they substitute contact with nacelle fixture .Gaps enable translations of free edges only in the same directions as support fixture.

Due to avoid puncture of the material in place of loading the nacelle was tested with loading support foot 60x60mm.

4 Results comparison

Figures 7 and 8 show nacelle deformation shapes for different boundary condition. Each picture represents result for loading by vertical force 200N. To find the best boundary condition the deformation shapes from FEM analysis will be compared with the test deformation shape. Photo of the test deformation shape is shown on figure 9.

4.1 Deformation shapes

Figure 7 shows deformation shape of nacelle with free edges constrained in all 6 degrees of freedom. Figure 8 shows deformation shape of nacelle with gap elements on free edges.

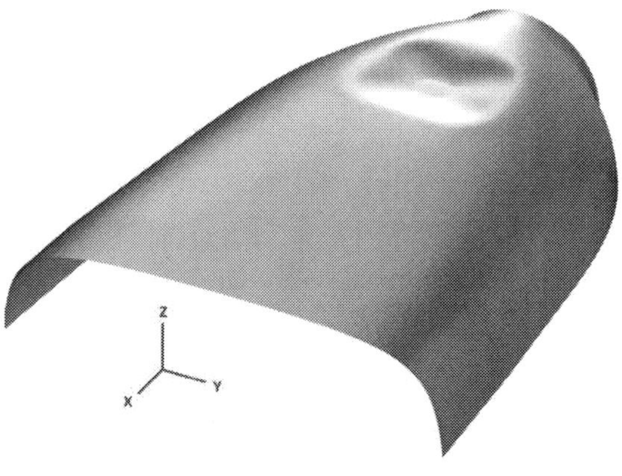

Fig. 7: Nacelle with constrained all 6 degrees of freedom on free edges

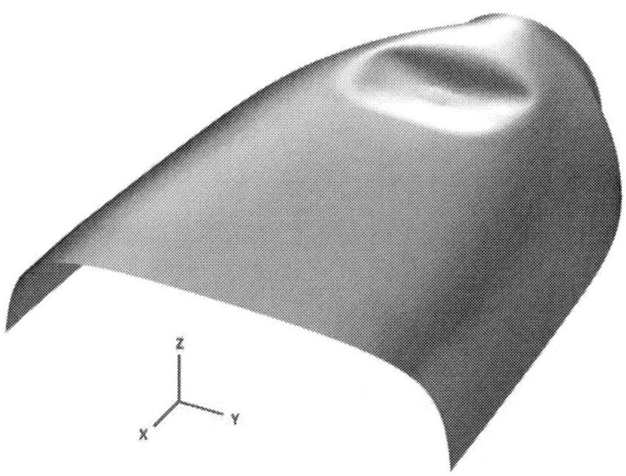

Fig. 8: Deformation shape of nacelle with gap elements on free edges

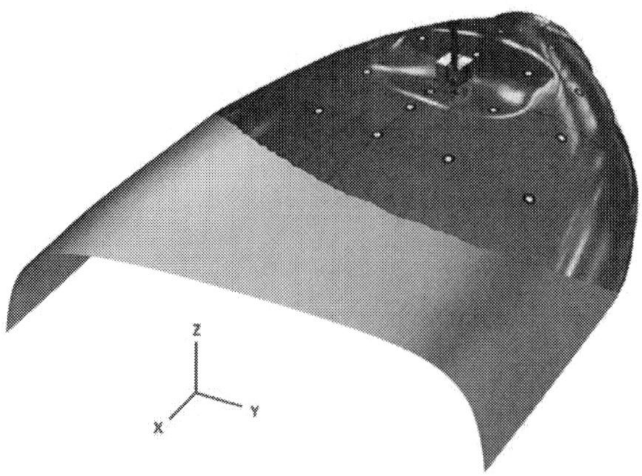

Fig. 9: Deformation shape of nacelle during test

Comparing results with the photo from the test on figure 9 the best fit in the deformation shape is with figure 8. That's why the boundary condition with gap elements on free edges was chosen as the best way to model the nacelle test from deformation shape point of view.

4.2 Deflection of the nacelle

To compare FEM results with some measurable parameter the deflection of nacelle was chosen. The deflection was measured near the place where the load was applied. Comparison of deflection from FEM and test is shown on figure 10. The deflection in the monolithic structure for 200 N was during test 29,43 mm. The FE model shows in the same place deflection 22 mm. Difference between test results and FEM analysis may be caused by manufacturing of the nacelle. That's why the thickness of nacelle can be in some areas slightly different from the thickness which was modeled in FEM model.

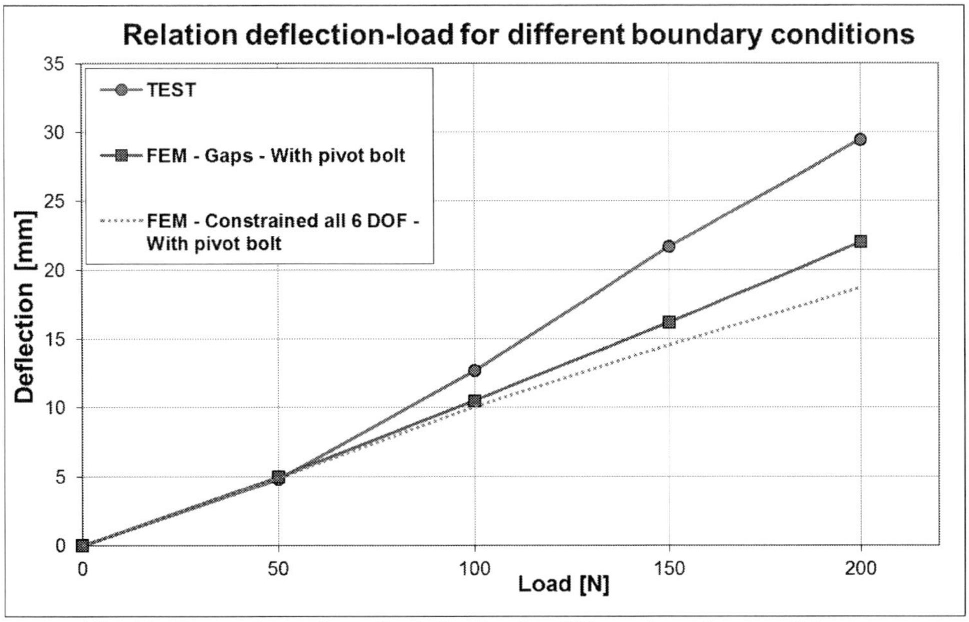

Fig. 10: Comparison of nacelle deflection for different boundary conditions

5 Conclusion

As it was shown in this article correct simulation of the boundary conditions has significant influence on the deformation shape. In the case of nacelle compressive test using of boundary conditions modeled by gap elements correspond better with reality then simple constraining of nacelle free edges. Identifying of the real boundary conditions helps to increase the precision of numerical model prediction

Acknowledgement

The research has received funding from the European Union's 7FRPfor Research, Technological Development and Demonstration within ESPOSA (Integrated approach to efficient propulsion and related aircraft systems for small-size aircraft) project.

References

[1] V. Kafka, Report on test run on sub-scale component BE1, ESPOSA Deliverable WP6.5/D65.25 (2015)

[2] D. Szeląg, A. Osmęda, Detailed FEM analysis BE1, ESPOSA Deliverable D6.5.7 ILOT (2014)

[3] Retrieved from: http://www.hexcel.com/Resources/DataSheets/Honeycomb-Data-Sheets/A1A10_eu.pdf

Complement of Calibration of Measuring Element Using FEA

J. Hoidekr [1,a], P. Malý [1], F. Starý [1], P. Mossóczy [1],V. Dynybyl [1]

[1] *CTU in Prague, Faculty of Mechanical Engineering, Department of Designing and Machine Components, Technická 4, 166 07 Prague 6, Czech Republic*

[a] *jan.hoidekr@fs.cvut.cz*

Abstract: This paper focuses on complement of calibration of measuring element directly integrated into industry machine using finite elements method analysis. It describes a method to convert calibrated load by concentrated force into real industry load by pressure. It is based on similarity and comparison of curves of deformation for both these cases of load.

Keywords: Calibration; Strain Gauge; FEM analysis.

1 Introduction

In industry, there are lot of requests to measure stress, deformation, pressure, force, velocity, etc. For the experimental measurement are often used special sensors with strain gauges. The sensors are often directly integrated into industry machines, e.g. [1, 2].

In this case, there was a request to calibrate measuring element with given geometry, where strain gauges are used. The measuring element is directly integrated into industry machine, so there was no possibility to change the geometry. The measuring element is loaded with constant pressure in the industry machine, but there was no way to precisely simulate constant pressure during the calibration.

Hence, a method has been devised how to convert calibrated load by concentrated force into load by pressure using FEA. This method is based on defining curves of deformation for both cases of loads and their comparison. This will allow us to obtain a constant of transformation of load by concentrated force to load by pressure.

2 The Geometry of the Measuring Element

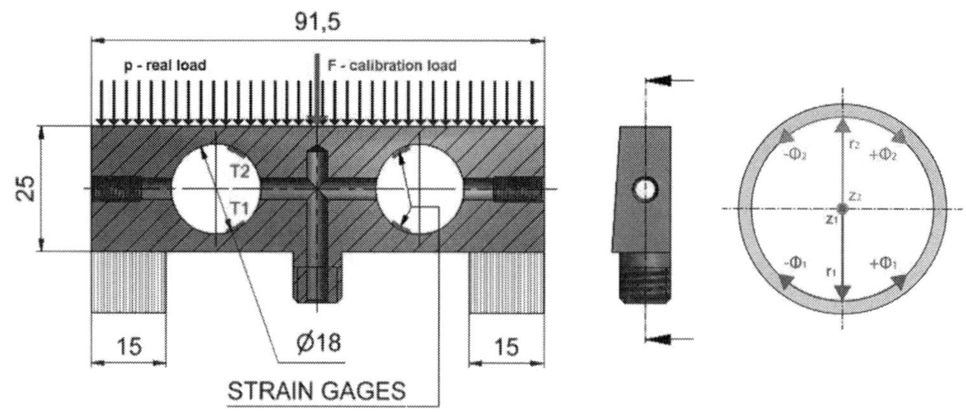

Fig. 1: The geometry of the measuring element with illustrated load, boundary conditions, position of strain gauges and definition of local cylindrical coordinate systems.

The two defined local cylindrical coordinate systems on measuring element are shown in Fig. 1, named C_1 (r_1, Φ_1, z_1) and C_2 (r_2, Φ_2, z_2). Results of FEA are shown just in these coordinate systems which corresponds to deformation of strain gauges placed on circumference of circular holes.

3 Defining Curves of Deformation Using FEA

A very fine mesh in the area of strain gauges was designed using C3D8R elements in Abaqus software [3]. The mesh is shown in Fig. 2. There are 16 elements in the 'z' axis direction and 148 elements in the 'Φ' axis direction. Nodes are numbered in the (i, j) coordinate system, where i = {1, 2, …, 149} and j = {1, 2, …, 17}. After analysing the data, the values of relative deformation of circumference of the hole for both cases of load, in every node N(i, j), were found.

The results were constant by the 'z_1' and 'z_2' axis in C_1 and C_2 systems. Hence, an average of the results was taken. The average was taken only in strain gauge area, which is located between nodes j = 6 and j = 12. Averaged result by the 'i' axis then corresponds to the result by the 'Φ' axis. This is described in Eq. (1) and shown in Fig. 2, where the surface of the hole is developed. The result is a function of one variable and can be shown in 2D graph. This is how the curves of deformation are defined.

$$\varepsilon_{(\Phi_1,\Phi_2)} \approx \varepsilon_{(i)} = \sum_{j=6}^{j=12} \bar{\varepsilon}_{(i,j)} \tag{1}$$

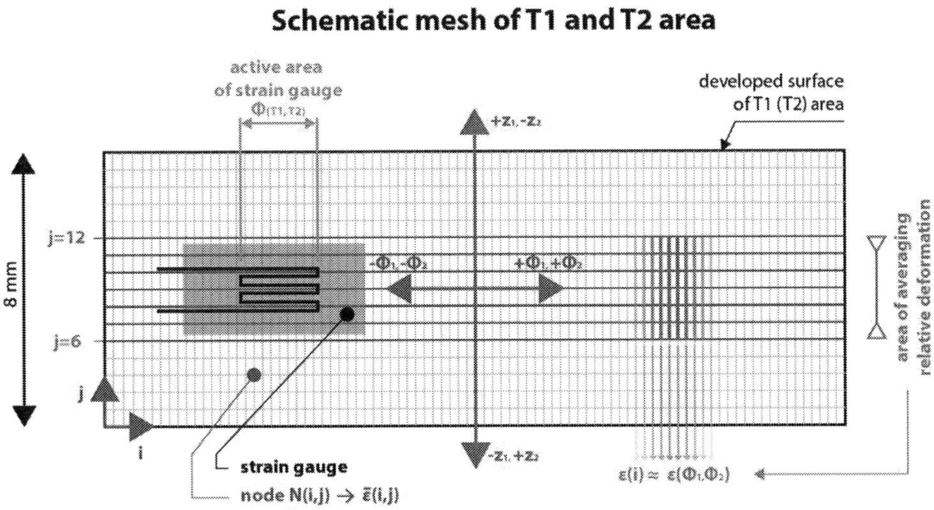

Fig. 2: The mesh of T1 and T2 area, where the strain gauges are located.

The exact position of strain gauges was measured using the optical method. The averaging of the curve of deformation in the zone of measured real position of active area of the strain gauge should be theoretically equal to the value measured by the strain gauge. These relations are shown mathematically in Eq. (2) and Eq. (3).

$$\varepsilon_{(T1,T2)}^{force} = \frac{1}{\Phi_{(T1,T2)}} \cdot \int_{(T1,T2)} \varepsilon_{(\Phi_1,\Phi_2)}^{force} \cdot d\Phi \tag{2}$$

$$\varepsilon_{(T1,T2)}^{pressure} = \frac{1}{\Phi_{(T1,T2)}} \cdot \int_{(T1,T2)} \varepsilon_{(\Phi_1,\Phi_2)}^{pressure} \cdot d\Phi \tag{3}$$

$$R = \frac{\varepsilon_{(T1,T2)}^{pressure}}{\varepsilon_{(T1,T2)}^{force}} \tag{4}$$

Based on the similarity of these curves of deformation for both cases of load, the ratio R between the integrated values of the curves of deformation for both loads is the constant that should be used while converting calibrated load into real industry load. This relation is shown mathematically in Eq. (4).

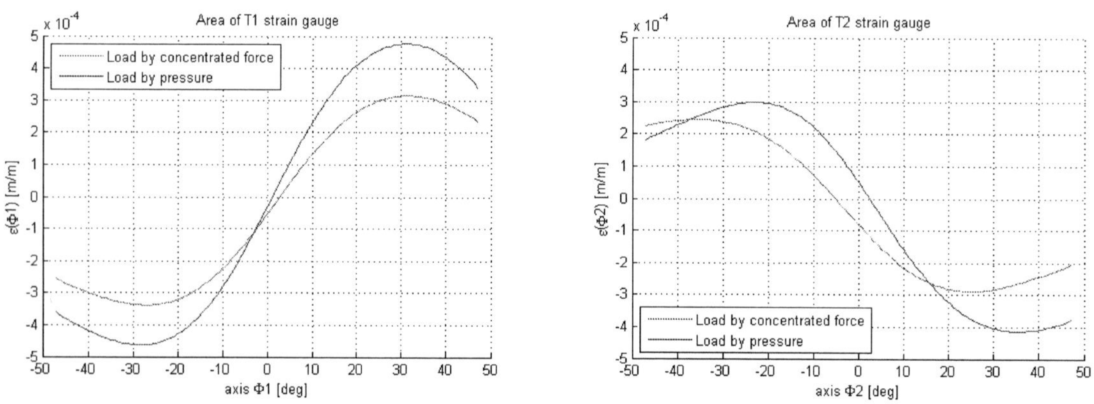

Fig. 3: The curves of deformation in T1 and T2 area for load by concentrated force and load by pressure.

4 Converting loads

The analysis was divided into 10 equal steps, given in Tab. 1. The load by concentrated force from 0 to 3.0 kN with step 0.3 kN was used during calibration. For load by pressure the values from 0 to 10 MPa with step 1 MPa were used. Graphs of curves of deformation are always shown for the loads in the same step only. The ratio between them does not change for any step. It is because of linearity of the model. For example, in Fig. 3 curves made by loads from step 10 are shown.

Tab. 1: Table of conversion of loads for FEA divided by the steps.

step [-]	1	2	3	4	5	6	7	8	9	10
load by force [kN]	0.3	0.6	0.9	1.2	1.5	1.8	2.1	2.4	2.7	3.0
load by pressure [MPa]	1	2	3	4	5	6	7	8	9	10

The ratio between values of loads in Tab. 1 for every step is fixed and its value is 0.3. The relation of conversion from load by concentrated force to load by pressure can be set with knowledge of Tab. 1. The ratio R obtained from the analysis will also be used. This is mathematically described in Eq. (5).

$$p_{meassured}\ [MPa] \approx \frac{1}{0.3} \cdot R \cdot F_{calibrated}\ [kN] \tag{5}$$

5 Strain gauges

Linear HBM strain gauges type 1-LY11-1.5/120 were used. In total 4 pieces were used, 2 for every circular hole of measuring element. This theory is given only for one half of the measuring element. This is the reason why only two areas of strain gauges T1 and T2 are mentioned.

This theory can also be analogically used for the second half of the measuring element because all the model is symmetric by the plane where the concentrated force is applied.

Every strain gauge is connected in quarter-bridge circuit, so the data from each strain gauge is taken separately.

6 Conclusion

The proposed method for calibration of the given measuring element takes advantage of the simplicity of the loading by a concentrated force during the calibration. The FEA based on the similarity of curves of deformation in strain gauges areas and their comparison gives the sufficient method how to link the calibration and the real measurement of the pressure load.

References

[1] O. Berka, M. Dub, F. Lopot, Experimental analysis of gear loading in planetary transmission, EAN 2014 - 52nd International Conference on Experimental Stress Analysis (2014)

[2] M. Dub et al., Experimental analysis of stress state and motion of tram gearbox hinge, EAN 2014 - 52nd International Conference on Experimental Stress Analysis (2014)

[3] ABAQUS (2011) 'ABAQUS Documentation', Dassault Systèmes.

Compressed Brick Masonry Columns Strengthened by Textile Reinforced Concrete

O. Holčapek [1,a], F. Vogel [1], P. Reiterman [1]

[1] *Czech Technical University in Prague, Faculty of Civil Engineering, Experimental Centre, Thákurova 7, 166 29, Prague 6, Czech Republic*

[a] *ondrej.holcapek@fsv.cvut.cz*

Abstract: The aim of presented article is to describe the influence TRC's slim layer applied on compressed brick masonry columns. Simple application of TRC is not the only benefit of this solution. This solution stabilizes analysed columns and due to the confined higher maximum load-bearing capacity is achieved. Three tested columns, which two of them were strengthened, had height 1040 mm and plant dimension 300 x 300 mm. Burnt bricks 290 x 140 x 65 mm with declare compressed strength 30.0 MPa were joint by lime-cement mortar with compressed strength 5.0 MPa. Layer of approximately 10 mm of high strength cement matrix containing alkali-resistant glass textile reinforcement was applied around the columns. Compressive strength 100 MPa and flexural strength 15.7 MPa characterize used cement matrix, while AR-glass textile is characterized by 585 g/m^2 and windows' size was 5.5 x 4.5 mm. The strengthening increased maximum force to 109 %, resp. 148 %, in compare to non-strengthened column.

Keywords: Masonry, compressed columns, strengthening, TRC, AR-glass textile reinforcement.

1 Introduction

Process of reparation, renovation, strengthening and stabilization of existing load-bearing structures has long history and development; used methods and materials reflected actual state of knowledge. The steel banding has been used for strengthening compressed load-bearing elements (concrete and reinforced concrete columns, masonry pillars) or bending load-bearing elements (wooden beams, reinforced concrete beams). In addition to steel the concrete, reinforced concrete or reinforced cement mortal are proven approaches of masonry's strengthening. Strengthening of existing masonry columns takes place due to several reasons, mainly when the permanent load rise up (when the purpose of building is change) or when the masonry is breached (due aggressive environment, long-term overload, missing maintenance). More widespread use of newly developed composite materials (high performance concrete, glass, basalt or carbon composites) leads to their application in several areas, including strengthening. The wrapping of columns and pillars by GFRP (glass fibres reinforced polymer) or CFRP (carbon fibres reinforced polymer) increases their load-bearing capacity in compression [1]. The radius of column's corner affects the final achieved strength [2]. Two times higher ductility showed reinforced concrete column strengthened by 40 mm of UHPFRC than traditional jacket from reinforced concrete (RC) and thickness of 60 mm [3]. The 350 x 350 x 3000 mm columns were strengthened in real condition (reconstruction of 80 years old factory). It should be noted, that the maximum load in kN of traditionally reinforced column was over 40% higher, than in the case of RC jacket. Benefit of this solution lies in higher resistance to earthquake action. Very promising material for strengthening masonry structures is textile reinforced concrete (TRC), which combines fine ground cement matrix and textile fabrics from high strength materials (carbon or glass). Several article deals with strengthening reinforced concrete by TRC [4] but masonry application is still not sufficient described; tensile characteristics of history masonry, in the context of earthquake, strengthened by TRC were investigated in [5]. There are two approaches to strengthening – wrapping, that maximizes the load-bearing capacity in compression or additional layer of concrete or cement mortar, which increases the cross-section that resist load action. The load transfer from structure has to be provided on the additional layer.

2 Description of Performed Experiments

Performed experimental program deals with the testing of bricks masonry columns strengthened by slim layer of textile reinforced concrete, where the textile fabrics were applied in two different ways. Columns were tested at 28 days after strengthening process by static load test.

2.1 Characteristics of Used Masonry

Following part describes properties of lime-cement mortal, brick's units, masonry and textile reinforced concrete. Standard burnt bricks with dimension 290 x 140 x 90 mm were used; compressive strength of three specimens was tested (f_u) and the normalized mean compressive strength was calculated according to Eq. (1). Lime-cement prescription mortal with volume ratio 1:3:5 (cement, lime, sand 0/4 mm) was used and the mechanical properties (f_m – compressive strength of mortal) were tested according to [6] on specimens 40 x 40 x 160 mm. Final characteristic strength of masonry was calculated according to equation (3). All these equations are available in [7].

$$f_b = \delta \cdot \eta \cdot f_u \quad (1)$$
$$f_b = 0.77 \cdot 1.0 \cdot 39.5 = 30.42\, MPa \quad (2)$$
$$f_K = K \cdot f_b^{0.65} \cdot f_m^{0.25} \quad (3)$$
$$f_K = 0.4 \cdot 30.42^{0.65} \cdot 5^{0.25} = 5.51\, MPa \quad (4)$$

f_b – normalized mean compressive strength of masonry unit (brick)
f_u – average compressive strength of masonry unit
δ – coefficient of masonry unit shape (0.77)
η – coefficient of conditioning (equal to 1.0 for 6% humidity of masonry unit)
f_k – characteristic compressive strength of masonry
f_m – characteristic compressive strength of mortal
K = 0.4 (longitudinal joint, group 1, according to [7] CSN EN 1996-1-1)

2.2 Characteristics of Used TRC

TRC consists of two main parts – textile reinforcement and cement matrix. The cement matrix can be generally characterized as high performance concrete. Used cement matrix has been verified in several applications. The binder and filler consists of cement CEM I 42.5R, microsilica, silica powder ST6, silica aggregates in fractions 0.1/0.6, 0.3/0.8 and 0.6/1.2 mm. Due to the superplasticizer SVC 1035 the low water to cement 0.35 has been achieved. The compressive strength of this composite is 43.3 MPa at the age of 24 hours and 102.2 MPa at the age of 28 days [8], measured on cube specimens (100 x 100 x 100 mm). The flexural strength at the age of 28 days is 15.7 MPa.

The textile reinforcement was made by Adfors Saint-Gobain Company. This textile reinforcement can be used in concrete elements, wall reinforcement etc. The textile reinforcement has a lot of advantages and is made of glass fibres with high tensile strength. The textile reinforcement has alkali resistance surface due to safe use in concrete elements [9]. The textile reinforcement R 585 A 101, which is used in this experimental program, has basis weight 585 g/m2 and window's size 5.5 x 4.5 mm. The window's size of textile limits the maximum grain size of used cement matrix, in this case the maximum grain was 1.2 mm.

2.3 Strengthening Process

Two different ways of alkali-resistant glass textile reinforcement application were tested (in the angle of 45° and parallel to columns dimension). Two layers of glass-fibres textiles ADFORS R585 were applied on each side of columns. The scheme of application of textile reinforcement can be seen on Fig. 1 and Fig. 2. Fig. 1 shows first type of the textile reinforcement application and Fig. 2 shows second type of the textile reinforcement application. The strengthening process can be observed on Fig. 3. The strengthening process started after one month after columns' production. Due to the character of bricks and their porosity, the surface was pre-wetted by spraying of water. First layer of fine ground cement matrix was applied on the surface of pre-wetted column and inside the concave masonry joints. The textiles were pressed into the layer of fine ground cement matrix. Slim layer of cement matrix finished the surface and creased the final surface. Columns were wrapped by foil to avoid the evaporation of water and eliminated shrinkage and creation of cracks.

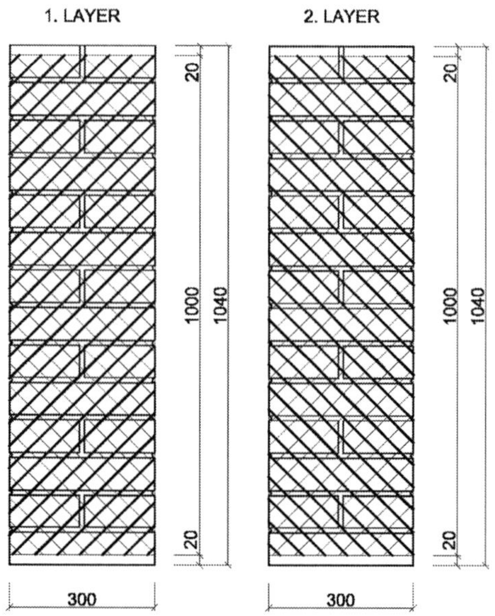

Fig. 1: First type of textile´s reinforcement application.

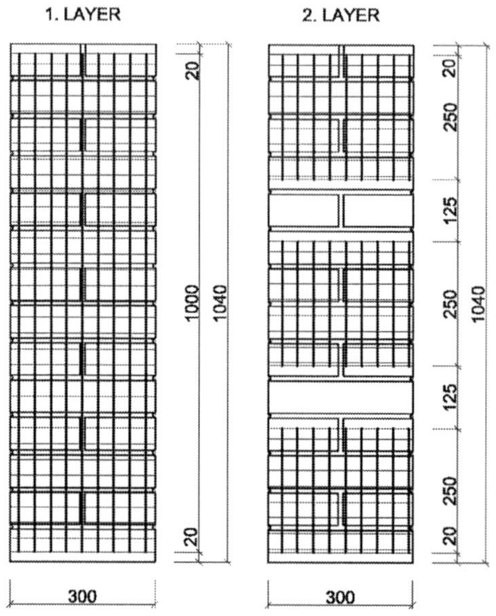

Fig. 2: Second type of textile´s reinforcement application.

2.4 Static Load Testing

Columns with dimension 300 x 300 x 1040 mm were tested at the age of 60 days (28 days after strengthening process). The hydraulic loading press INOVA 2000 was used for testing. The testing was controlled by linear increase of vertical deformation of loading plate (0.004 mm/sec). The evaluation of vertical force and the crosshead of the loading plate of the hydraulic press were recorded in time. The tested column in the loading press is shown in the Fig. 4.

Fig. 3: Application of TRC on brick masonry column´s surface.

Fig. 4: Tested column in the loading press.

3 Results and Discussion

The obtained basic results are written in next Tab. 1. The table contains maximum value of vertical force in the second column, a ratio maximum force to characteristic force calculated from standards in the third column and a ratio maximum force to maximum force of non-strengthened column in the fourth column.

The maximal calculated force according standard CSN EN 1996-1-1 of non-strengthened masonry column is equal to 459.9 kN. The maximal force of the tested non-strengthened column reached 135 % of this value, which is 669.1 kN. First strengthened column reached 727.3 kN (147 % of the characteristic force) and second strengthened column reached 991.5 kN (200 % of the characteristic force). This experimental program was designed as a pilot testing to verify basic assumptions of strengthening by TRC. These results of two strengthened columns are spread due to not ideal technological approach in this pilot experimental program. First column strengthening was not geometrical great and this fact caused spread in the results.

We can see better resistance of strengthened columns, if we compare non-strengthened column and two strengthened columns. First column reached 109 % of maximal force on non-strengthened column and second column reached 148 % of maximal force on non-strengthened column.

Tab. 1: Obtained results from experimental program.

Column	Maximum force - F_i	F_i / F_k	F_i / F_{ref}
	[kN]	[%]	[%]
According to CSN EN 1996-1-1	F_i = 495.9	100	74
Non-strengthened column	F_{ref} = 669.1	135	100
Strengthened column (45°)	727.3	147	109
Strengthened column (0°)	991.5	200	148

Fig. 5: Tested strengthened column.

Fig. 6: Detail of collapsed strengthening layer and column after performed test.

The collapse of the non-strengthened column was regular like it was expected. The similar failure mode was developed on both strengthened columns. The collapse of the strengthened columns was caused by failure of the TRC layer in corner of the top of the column in both cases. The collapsed column it can be seen on Fig. 5 and detail of the collapsed corner it can be seen on Fig. 6.

The loading force and displacement of the load cell was recorded during testing. From obtained data for both strengthened columns is made a chart in Fig. 7. There is load cell displacement on the horizontal axis and force on the vertical axis. The linear increase of the force can be seen between 3.5 and 7 mm displacement. After linear area started cracking of the TRC strengthening around upper corner of the columns. The peaks of the curves indicate failure of the columns, which was caused by breaking the TRC layer on the upper corners of the columns.

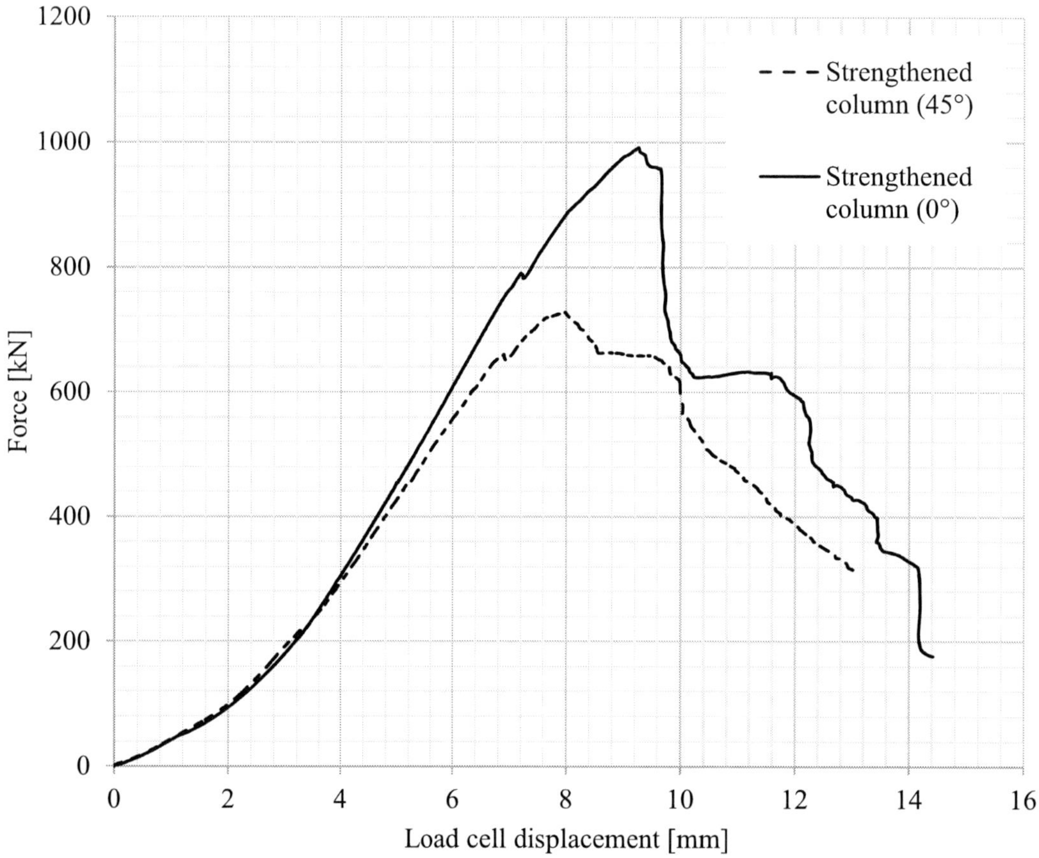

Fig. 4: Force x displacement chart of strengthened columns.

4 Conclusion and Further Research

The aim of this article is to describe pilot experimental testing of the masonry columns strengthened by thin layer of the textile reinforced concrete. Totally three columns with dimensions 300 x 300 x 1040 were made for the experimental program. Two of them were reinforced by two layers' of glass fibres textile reinforcement and cement matrix. The testing was controlled by linear increase of vertical deformation of loading plate (0.004 mm/sec). The maximum force of strengthened columns achieved 109 %, resp. 148 %, in comparison with non-strengthened column. These results are spread due to not ideal technological approach in this pilot experimental program. This experimental program was created for confirmation basic premises of strengthening of masonry by TRC. Anyway from the results we can conclude, that application of TRC seems to be a progressive solution for strengthening of compressed brick masonry columns and interesting alternative to conventional solution (reinforced concrete wrapping, application of CRF, GRF, etc.).

Acknowledgement

This research work was financially supported by SGS16/199/OHK1/3T/11 – Multicomponent binder systems for composite building materials, which is gratefully acknowledge.

References

[1] J. Witzany, R. Zigler, Stabilization and strengthening of historic building´s stone masonry columns, Advanced Materials Research 923 (2014) 93-96. doi: 10.4028/www.scientific.net/AMR.923.93

[2] J. Kubát, J. Karas, The Corner Radius Influence on the Deformation Behavior of Brick Pillars Strengthened by CFRP Confinement, Advanced Materials Research 1122 (2015) 273-2779. doi:10.4028/www.scientific.net/AMR.1122.273

[2] D. Rosignoli et al., Structural Reinforcement and Seismic Retrofitting with UHPFRCC Special Formulation Jacketing of 1930 Building RC Structural Elements, The Fourth International FIB Congress 2014, Mumbai, 125-129.

[3] V. Mechtcherine, Novel cement-based composites for the strengthening and repair of concrete structures, Construction and Building Materials 41 (2013) 365-373. doi:10.1016/j.conbuildmat.2012.11.117

[4] A. D'Ambrisi, L. Feo, F. Focacci, Experimental and analytical investigation of bond between Carbon-FRCM materials and masonry, Composites: Part B 46 (2013) 15-20. doi:10.1016/j.compositesb.2012.10.018

[6] CSN EN 196-1 Methods of testing cement – Part 1: Determination of strength (2005)

[7] CSN EN 1996-1-1. Eurocode 6: design of masonry structures – Part 1-1: General rules for reinforced and unreinforced masonry structures (2007).

[8] O. Holčapek, F. Vogel, P. Konvalinka, Analysis of mechanical properties of hydrothermally cured high strength matrix for textile reinforced concrete, Acta Polytechnica 55 (2015) 313-318. doi:http://dx.doi.org/10.14311/AP.2015.55.0313

[9] Retrieved from: http://www.sg-adfors.com

Proposal of Strain Field Calculation Procedure in 2D Digital Image Correlation

R. Huňady [1,a], M. Hagara [1], P. Pavelka [1]

[1] *Technical University of Košice, Faculty of Mechanical Engineering, Slovakia*
[a] *robert.hunady@tuke.sk*

Abstract: The calculation of strains in digital image correlation is traditionally based on the method of affinity deformation of image elements (facets). Amount of strain of each facet is determined based on analysis of their local curvatures. To obtain relevant results, deformation gradients should be included in the solution. It requires the use of smoothing filters. Filtering level largely affects the resulting fields of deformation. The article describes the procedure of the calculation of strains that is based on the algorithms used in the finite element method and properly combined with result smoothing process. The aim of the article is to assess the efficiency and accuracy of the presented procedure.

Keywords: digital image correlation, strain calculation, smoothing.

1 Introduction

Digital image correlation method is a contactless method that uses precise cameras for the measurement of surface displacements and strains. In the case of 2D DIC, when only one camera is used to measurement, deformations are analyzed in the plane parallel to image plane of CMOS sensor (Fig. 1). 3D correlation systems use minimally two cameras arranged stereoscopically. The basic principle of digital image correlation consists in the correlation of images captured during loading. The correlation runs by small unique image elements, called facets. The uniqueness of the facets is achieved by creating a high contrast random speckle pattern on the surface of the object being analyzed. Information on the displacements is then obtained by correlation of corresponding facets in the state before and after loading of the object [1].

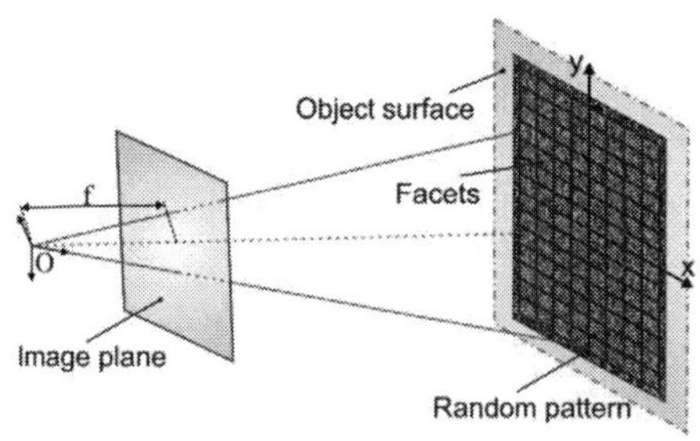

Fig. 1 Principle of 2D digital image correlation.

Displacement of any point of one facet can be described on the basis of shape functions of the first order which take into account affinity deformation [1]:

$$\xi_1(x_i, y_j) = u + \frac{\partial u}{\partial x}\Delta x + \frac{\partial u}{\partial y}\Delta y, \qquad \eta_1(x_i, y_j) = v + \frac{\partial v}{\partial x}\Delta x + \frac{\partial v}{\partial y}\Delta y,$$

where Δx a Δy represent distance of point (x_i, y_j) to the center of the given facet. If the displacement vectors of all points are known, it is possible to calculate strain of each facet. This calculation is based on analysis of the local curvature of facets. It practically means that strain of any facet is independent on other facets. To obtain relevant strain fields, deformation gradients have to be included in the solution. It requires the use of smoothing filters [2, 3]. As Fig.2 shows, filtering level largely affects the resulting strain fields. Therefore, a correct adjustment of filter is very important. However, a determination of a correct value is difficult. This is why the new procedure based on a combination of FEM algorithm and special smoothing filter was proposed.

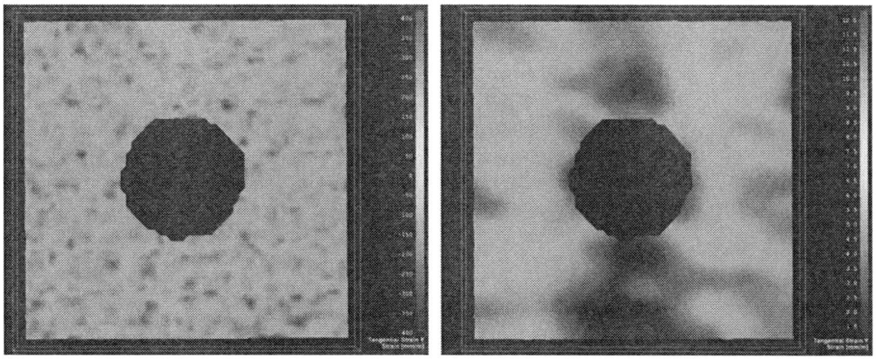

Fig. 2 The strain field before and after the smoothing of results.

2 Proposed procedure

The proposed procedure for the determination of deformation fields in DIC method has three basic steps: 1) the calculation of strain on elements, 2) determination of averaged nodal strain values, 3) smoothing of results. To calculation of deformation parameters, procedure uses the shape functions of planar triangular finite elements with three nodes [4]:

$$N_i(x,y) = a_i + b_i x + c_i y,$$

where a_i, b_i, c_i are the coefficients of bilinear polynomials. Components of strains can be calculated as

$$\boldsymbol{\varepsilon} = \mathbf{Bd}$$

where $\mathbf{d}$ is a vector of nodal displacements of the given element, $\mathbf{B}$ is a matrix of partial derivatives of the shape functions

$$\mathbf{B} = [\mathbf{B}_1 \ \mathbf{B}_2 \ \mathbf{B}_3], \quad \mathbf{B}_i = \begin{bmatrix} b_i & 0 \\ 0 & c_i \\ c_i & b_i \end{bmatrix}, \ i = 1,2,3.$$

Note that nodal displacements are determined by correlation software. After the conversion of element values (Fig. 3a) to the nodal values (Fig. 3b), results are smoothed (Fig. 3c). The procedure uses smoothing filter that replaces value in the node i by mean value of nodal values of its surrounding. This surrounding area is defined by a square whose width can be varied depending on the settings of correlation parameters (especially facet size). Smoothing algorithm can be described as

$$\widetilde{\mathbf{E}} = \boldsymbol{\Sigma}\mathbf{E}\boldsymbol{\Sigma}^T,$$

where $\mathbf{E}$ and $\widetilde{\mathbf{E}}$ are matrices of nodal strain values before and after the smoothing, $\boldsymbol{\Sigma}$ is sparse band matrix whose band width is equal to twice the square width.

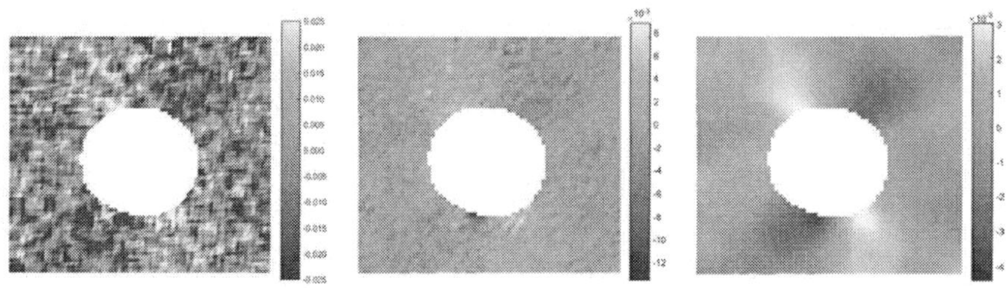

a) b) c)

Fig. 3 Shear strain fields around the hole: a) strains on elements, b) averaged nodal strains, c) smoothed nodal values.

3 Calibration process

Calibration process was performed to determine a relation between a degree of smoothing and a facet size. The object of a calibration measurement was the flat planar sample of 30 mm wide made of aluminium sheet of a thickness of 2 mm. The sample was loaded statically in a tension machine in the longitudinal direction (Fig. 4a). Strain in this direction is therefore dominant. In an ideal case, the strain values are the same at all points. Due to spatial and correlation errors, this is not simple to achieve in reality, therefore average value $\bar{\varepsilon}_{yy}$ of a smoothed strain field (Fig. 4b) was considered in a calibration. Reference value ε_{ref} was calculated on the basis of maximal tension force $F_{\max}$ in the last loading step:

$$\varepsilon_{ref} = \frac{F_{\max}}{EA} = \frac{3600\text{N}}{69 \cdot 10^3 \text{MPa} \cdot 60\text{mm}^2} = 8.696 \cdot 10^{-4},$$

where E is Young's modulus and A is the cross-sectional area of the sample.

a) 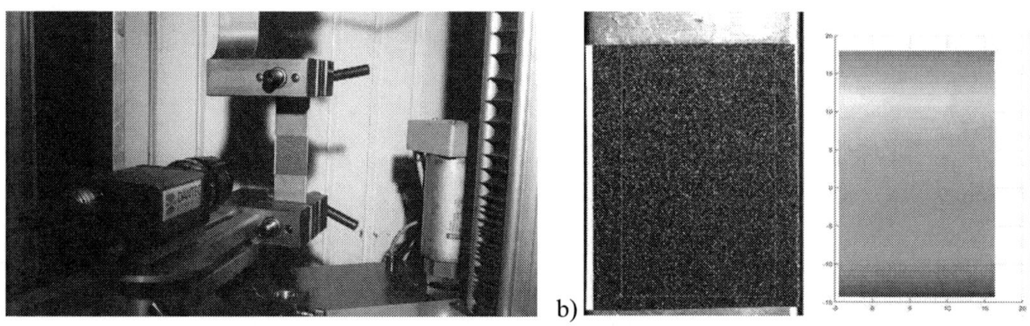 b)

Fig. 4 a) Calibration measurement, b) Camera view and smoothed strain field.

System Dantec Dynamics Q-400 with one 5 Mpx sensor was used for measurement. Nodal displacements were evaluated in software Istra4D for different facet size. Correlation files were subsequently imported to Matlab for the post-processing corresponding to the procedure described above.

It is clear that average strain value was determined for different correlation files and different settings of smoothing filter, but still for the same measurement. The results were related to reference value, as can be seen in Fig. 5a, where blue squares represent average strain values for given facet size and degree of smoothing. Intersection points of approximated blue lines and reference value line define a relation between a degree of smoothing and a facet size. Trend curve that is the result of this calibration process is shown in Fig. 5b and is valid only for conditions of the performed calibration measurement. Fig. 5 shows that the degree of smoothing increases non-linearly with decreasing facet size.

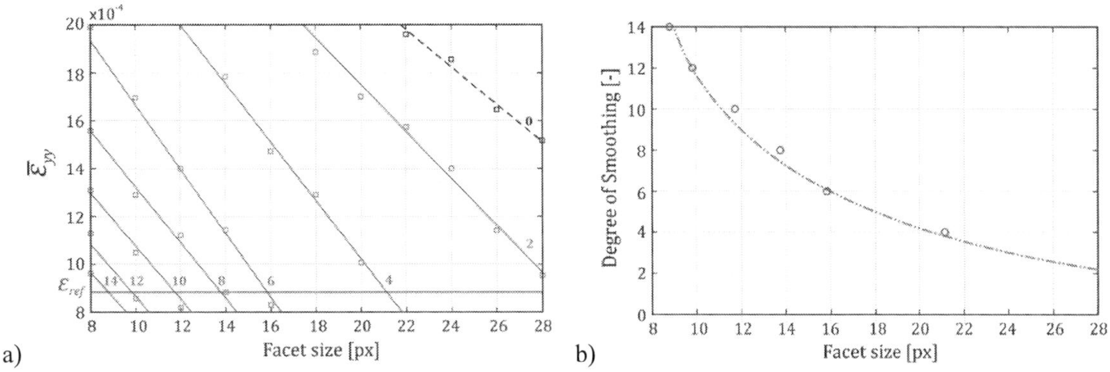

Fig. 5 a) Average strain values, b) Trend curve.

4 Experiment

Experimental measurement was carried out in the identical conditions as calibration process. The object of measurement was the flat sample with a 3 mm diameter hole placed in the center. Sample material and dimensions were the same as in the previous case. The sample was loaded by tension as is shown in Fig. 6. Maximal force in the last step was 3600 N.

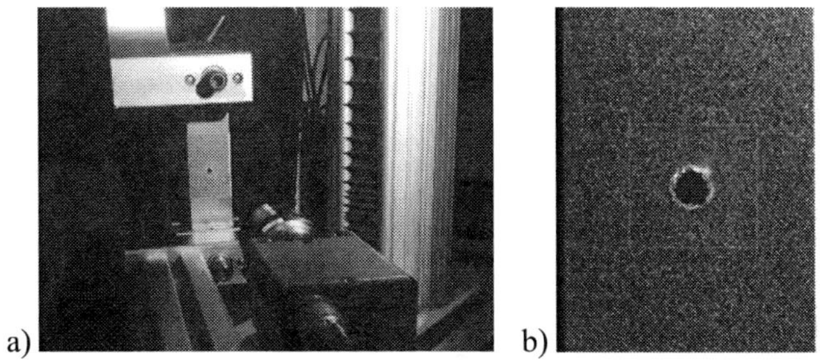

Fig. 6 a) Measurement configuration, b) Camera view and evaluated area.

The aim of experiment was to investigate deformations around the hole. Strain fields were determined using the proposed algorithm. Degree of smoothing for the given facet size was defined using the curve obtained on the basis of calibration. For an example, Fig. 7 shows the resulting fields of strain ε_{yy} for two different sizes of facets.

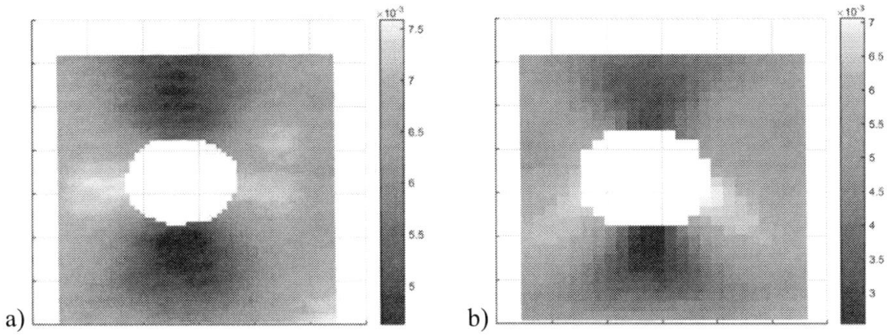

Fig. 7 Strain field ε_{yy} around the hole, a) facet size: 12px, degree of smoothing: 9, b) facet size: 28px, degree of smoothing: 2.

Fig. 7 shows that the resolution of resulting strain field primarily depends on a facet size. If the bigger facets are being used, the information about deformation is averaged over the bigger area during the image correlation process. In addition, part of the information near the geometrical boundary is definitely lost. If the smaller facets are being used, the systematical correlation errors increase due to a bigger noise. In that case, the adequate stochastic speckle pattern has to be used, i.e. every facet must contain characteristic part of a pattern with sufficient sharpness. The mentioned aspects are typical for digital image correlation method.

Conclusion

The results of experiment presented in the paper indicate that the introduced procedure is very well applicable in DIC method. The necessary part of this procedure is the calibration process to determine the correct smoothing parameter. As was shown, a degree of smoothing is primarily dependent on a facet size. The future work will be focused on verification of the procedure by using a conventional experimental method, e.g. Photostress®. With respect to a number of practical problems associated with 2D DIC method, the authors will focus their efforts on an application of proposed procedure for 3D measurement.

Acknowledgement

This work has been accomplished under the projects VEGA 1/0393/14 "Analýza príčin porúch prvkov mechanických sústav kvantifikáciou polí deformácií a napätí" and ITMS: 26220220141 "Výskum modulov pre inteligentné robotické systémy".

References

[1] F. Trebuňa, R. Huňady, M. Hagara: Experimental Method of Mechanics – Digital Image Correlation (in Slovak), TU v Košiciach, 2015.

[2] M. Hagara, R. Huňady: Q-STRESS v.1.0–a Tool for Determination of Stress Fields Using Digital Image Correlation Systems. In: Procedia Engineering, Vol. 96, 2014, pp. 136-142.

[3] M. Hagara, R. Huňady, F. Trebuňa: Stress Analysis Performed in the Near Surrounding of Small Hole by a Digital Image Correlation Method. In: Acta Mechanica Slovaca, Vol. 18, No. 3-4, 2014, pp. 74-81.

[4] V. Ivančo, R. Vodička: Numerical Methods of Body Mechanics and Selected Applications (in Slovak), TU v Košiciach, 2012.

Bearing Race and Roller Line Contact Analysis in 2D

J. Chmelar[1,*], V. Dynybyl[1]

[1] *Czech Technical Uiversity in Prague, FME, Department of Designing and Machine Components*
[*] *Jakub.Chmelar@fs.cvut.cz*

Abstract: Bearing race and roller contact analysis was performed. Comparison of Hertz analytical with fem elasto-plastic solution gives very similar results of contact pressure profile and sub-surface stress development in elastic domain. When material yield occurs results become different. Johnson-Cook's constitutive material model was used for elasto-plastic material behaviour.

Keywords: Bearing; Contact; Johnson-Cook; Hertz; Subsurface stress

1 Introduction

Dominant loading mode of bearing is contact between bearing race and rolling elements. Elemental load depends on bearing inner geometry, amount of rolling elements, radial clearance, external load character and supporting structure stiffness. Identification of elemental load could be estimated using standard ISO/TS 16 281: 2008 [1], that provides bearing analysis method based on bearing mathematical model or with use of FEM bearing modelling, that is more complicated and require more time resources. Bearing reliability is dependent on bearing roller and race contact properties mainly pressure, lubrication and material. Induced sub-surface stress influences material behaviour and is responsible for sub-surface initiated damage.

2 Contact definition

Analysis is focused on line contact modelling in plane (2D). The contact is defined by parameters listed in Fig 1 - Contact bodies curvature - R_i , Material of bodies represented by Young's Modulus - E_i, Poisson's constant - ν_i and acting force per millimetre of length - P. The contact area is considered as rectangle with infinite length and width according to Fig. 2

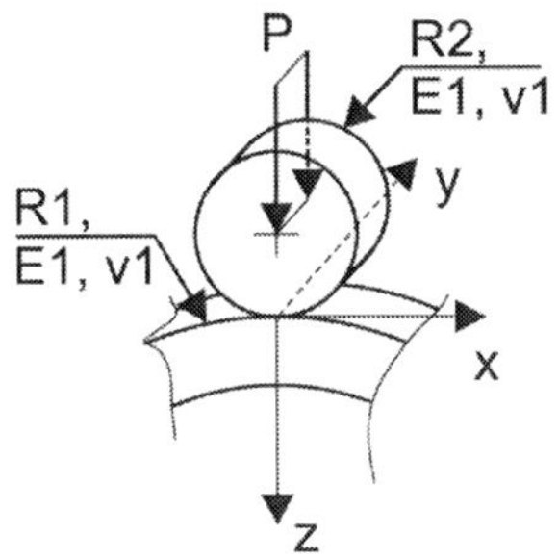

Fig. 1: Hertz line contact model definition.

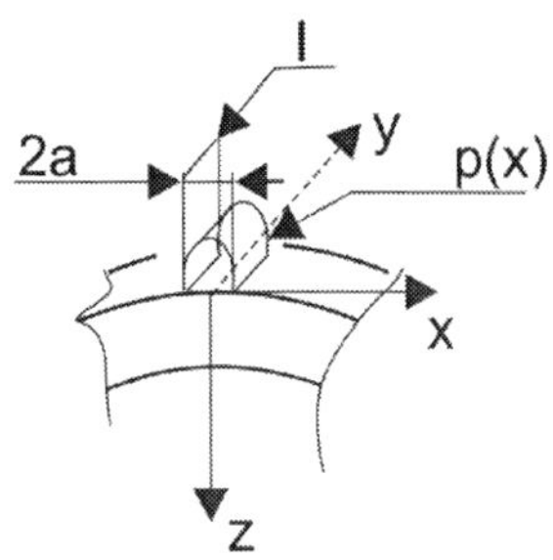

Fig. 2: Hertz line contact pressure description.

2.1 Hertz line contact analytical model

When two cylindrical non-conforming, smooth elastic bodies are forced into contact, the area that is formed in contact spot is assumed to be an infinite rectangle [2] with width accoding to Eq. (1)

$$a = \sqrt{\frac{4PR}{\pi E^*}} \tag{1}$$

Induced contact pressure distribution, assumed to be constant in direction of y coordinate Fig. 2, has parabolic shape described by equation Eq. (2) where the phrase in front of square root is maximal pressure [2].

$$p\left(x\right)=\frac{2P}{\pi a^{2}}\sqrt{\left(a^{2}-x^{2}\right)} \tag{2}$$

In Eq. (1) and Eq. (2) are used following substitutions: mutual contact bodies curvature R and mutual Young modulus E^{*}. They are calculated according to equations Eq. (3) and Eq. (4) respectively.

$$\frac{1}{E^{*}}=\frac{1-\nu_{1}}{E_{1}}+\frac{1-\nu_{2}}{E_{2}} \tag{3}$$

$$\frac{1}{R}=\frac{1}{R_{1}}+\frac{1}{R_{2}} \tag{4}$$

Hertz theory is according to [2] defined with following assumptions: (i) The contact area width is small comparing to size of bodies in contact. (ii) The contact area width is small comparing to curvature of both contact bodies. (iii) Elastic - linear material behaviour. (iv) Surfaces are smooth and frictionless. (v) Plane strain conditions considered. (vi) Load is constant in off plane direction.

2.1.1 Sub-surface stress conditions

Contact of bodies induce stress response in material. The stresses could be described using potential theory [2]. For contact centre, they were simplified to following form Eq. (5) - (7). Orthogonal stresses below contact centre match principal stresses, that makes it analysis straight forward.

$$\sigma_{1}=\sigma_{x}=-p_{0}\left[\left(2-\left(\frac{z^{2}}{a^{2}}+1\right)^{-1}\right)\sqrt{\frac{z^{2}}{a^{2}}+1}-2\left|\frac{z}{a}\right|\right] \tag{5}$$

$$\sigma_{2}=\sigma_{y}=-2p_{0}\nu\left[\sqrt{\frac{z^{2}}{a^{2}}+1}-\left|\frac{z}{a}\right|\right] \tag{6}$$

$$\sigma_{3}=\sigma_{z}=-p_{0}\left[\sqrt{\frac{z^{2}}{a^{2}}+1}\right]^{-1} \tag{7}$$

2.1.2 Material Yield prediction

Material Yield sub-surface could be predicted according to available yield criterion. There are two widely used when considering rolling contact fatigue [3]. Max. shear stress according to Tresca (8) and distortional energy criterion Von Mises Eq. (9). The σ_{Y} is material yield strength in pure tension.

$$\tau_{max}=\left|\frac{\sigma_{1}-\sigma_{3}}{2}\right|>\frac{1}{2}\sigma_{Y} \tag{8}$$

$$\sigma_{vm}=\sqrt{\frac{1}{2}\left[\left(\sigma_{1}-\sigma_{2}\right)^{2}+\left(\sigma_{1}-\sigma_{3}\right)^{2}+\left(\sigma_{2}-\sigma_{3}\right)^{2}\right]}>\sigma_{Y} \tag{9}$$

Material yield occures when limiting stress according to chosen criterion is exceeded. [2]. It could be proven by combining Eq. (5) - (7) to (8) or (9) respectively contact pressure for material yield could be calculated. The resulting expressions are summarized in Tab. 1.

2.2 FEM Model

FEM model of contact was assembled in Abaqus to support and enhance Hertz contact theory results in plastic domain. Standard roller bearing NU206 was used as reference geometry. Parameters are summarized in Tab. 3. The model consists of two planar bodies (3 DOF) - roller and race, forced into contact. Both bodies are driven by reference points RP1 and RP2 Fig. 3, where RP1 has constrained 2 DOFs: X - direction and RY-Rotation around Y axis. Force is applied to RP1. RP2 has all 3 DOFs constrained. Both RPs are connected with geometry by rigid beam elements.

Tab. 1: Summary of Hertz Line Contact Sub-surface stresses.

Criterion	Designation	Unit	Expression
Max. of Tresca shear stress	τ_{Tr}	$[N/mm^2]$	$0,3\,p_0$
Max. Tresca shear stress depth	$z_{\tau_{Tr}}$	$[mm]$	$0,78\,a$
Max. Von Mises stress	σ_{vm}	$[N/mm^2]$	$0,56\,p_0$
Max. Von Mises stress depth	$z_{\sigma_{vm}}$	$[mm]$	$0,7\,a$
Contact press. Tresca yield	$p_0\;Y_{Tr}$	$[N/mm^2]$	$1,67\,\sigma_Y$
Contact press. von Mises yield	$p_0\;Y_{VM}$	$[N/mm^2]$	$1,79\,\sigma_Y$

Tab. 2: Johnson-Cook material model parameters for 100Cr6 steel, 300 HV10, non-hardened. [5]

A $[N/mm^2]$	B $[N/mm^2]$	n	m
688.17	150.82	0.3362	2.7786

2.2.1 Mesh

For whole model there were used planar strain quadratic elements - CPE8R [4]. The mesh is designed as structured in area of contact and where inspected sub-surface stresses were expected. The rest of model is meshed with free mesh consisting of quad elements.

2.2.2 Material

Constitutive material model Jameson-Cook was chosen. This model is based on empirical data and provides good results for isotropic materials [4]. The model is defined by Eq. (10). The advantage of this model is that is is widely used, so parameters are well accessible. In this model A - is yield stress in 0.2 % strain offset, B - is strain hardening coefficient and n - is hardening exponent. The other coefficient are not applicable for this quasi static fem analysis since they are related to time dependent dynamic strain flow in material which is not considered in this model.

$$\sigma_o = \left[A + B\overline{\varepsilon^{pl}}^{n}\right]\left[1 + C\ln\frac{\dot{\overline{\varepsilon}}^{pl}}{\dot{\varepsilon_0}}\right]\left(1 - \widehat{\Theta}^{m}\right) \tag{10}$$

Used Johnson-Cook's material values for steel 100Cr6 in non-hardened state are summarized in 2. For elastic material behaviour Young elastic modulus E = 210GPa and ν = 0.277 was used.

2.2.3 Contact properties

The contact was modelled using finite sliding algorithm and surface-to-surface contact discretization. The contact was set up as frictionless with no stiffness penalty in normal direction.

Tab. 3: FEM roller bearing model geometry.

Bearing	hole dia. [mm]	Inner race dia. [mm]	Roller length [mm]	Roller dia. [mm]
NU206	30	37.5	10	9

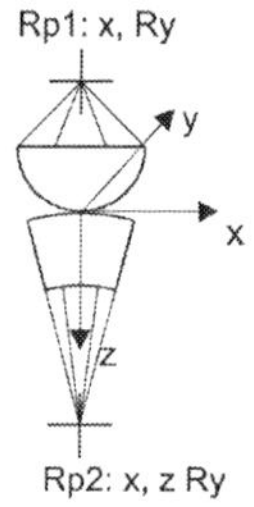

Fig. 3: Contact model with applied boundary conditions.

3 Results

There were performed calculations on FEM model for 5 different load cases expressed in contact pressure value: 0,5 GPa, 0,8 GPa, 1,3 GPa, 1,62 GPa and 2,5 GPa. Calculations were performed both for elastic and elasto plastic material model. The results are presented in following sections.

3.1 Contact pressure

Comparison of contact pressure for calculated cases are presented in Fig. 4. Calculated contact pressure (according to Tab. 1) where von Mises criterion predicts first yield in material fits the declination of line connecting discrete result points of elasto plastic material solution. Below that declination point, in elastic material model, all three solutions give similar results. In Fig. 5 contact pressure profiles of load case 1,62 GPa are compared. Elastic Hertz solution well follows FEM elastic solution. For elasto plastic material behaviour there is obvious a maximal contact pressure drop accompanied with extension of contact zone width. Assumed parabolic pressure profile of elastic solution is slightly flattered.

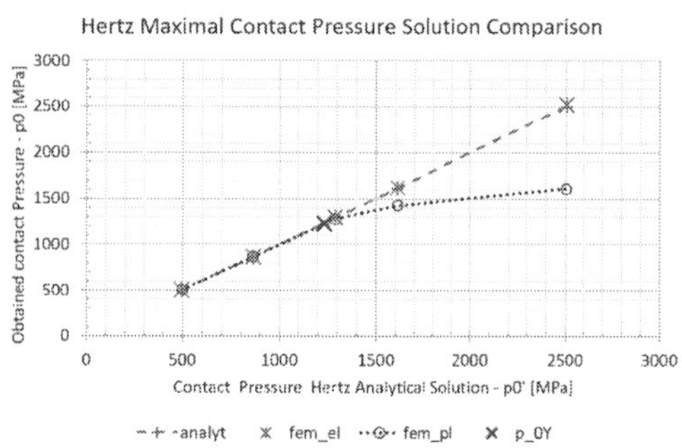

Fig. 4: Hertz contact comparison of Hertz-anal, fem-el and fem el-pl solutions in 5 load points. Green cross indicates contact pressure where first von Mises yield to occurs according to Tab. 1

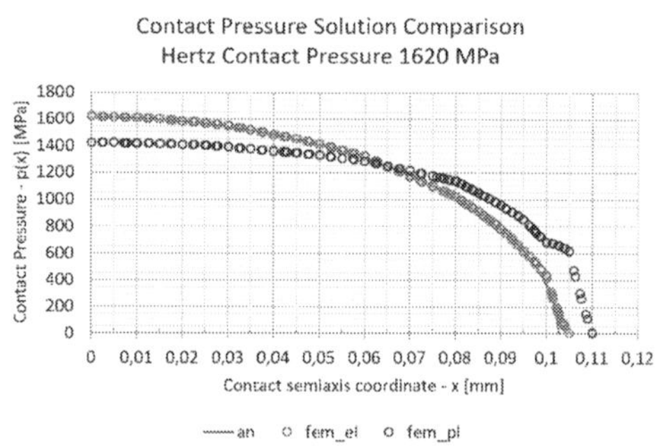

Fig. 5: Comparison of Hertz contact pressure profile for Hertz-anal, fem-el and fem el-pl solutions at cpress 1,62 GPa.

3.2 Sub-Surface stress

3.2.1 Elastic solution

Analysis of sub-surface stress field was performed. Comparison of orthogonal sub-surface principal stresses calculated according to Eq. (5) - (7) with fem solution is presented in Fig.6. Stresses and depth are normalised to maximal pressure and contact half-width respectively. It is obvious that results from both solutions with respect to small deviations caused by different solution nature agree. Presented results are valid only in elastic domain, but provides clear insight into material loading modes below contact centre. Recalculated principal stresses to Max. plane shear stress according to Tresca and von Mises stress are in Fig. 7. It can be seen that both stresses reach their maximal values in certain depth below surface. This depth could be easily estimated according to equations provided in Tab. 1.

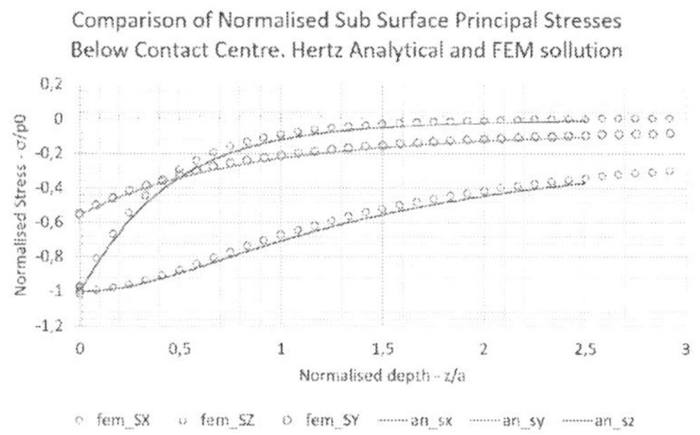

Fig. 6: Comparison of analytical and FEM principal stress development solution below centre of contact. The solution is valid only in elastic domain

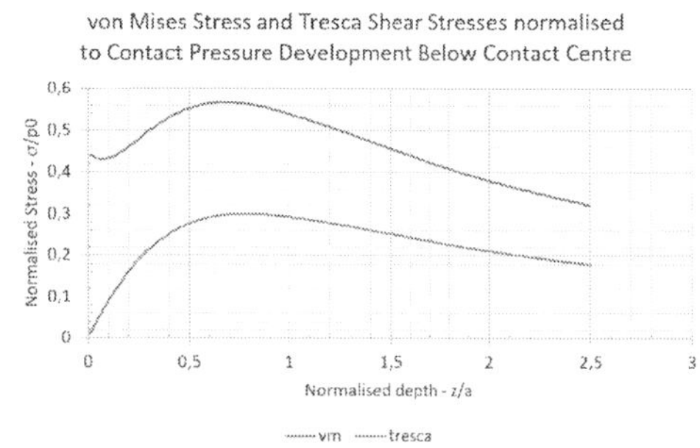

Fig. 7: Normalised development of sub-surface stresses Tresca and von Mises. Analytical solution, valid only in elastic domain.

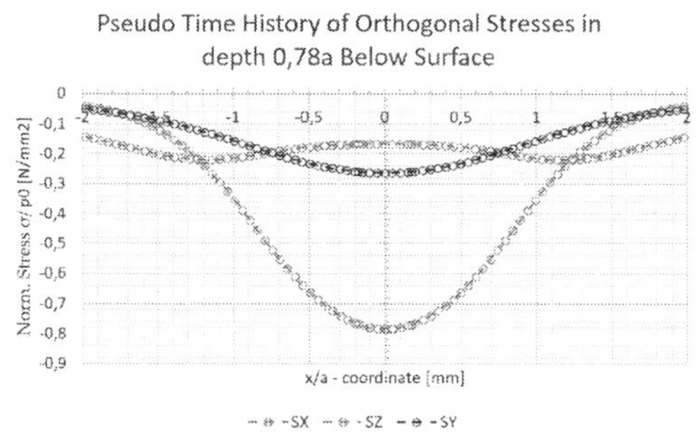

Fig. 8: History of principal stresses in depth 0.78a below surface. Point of contact at 0 point.

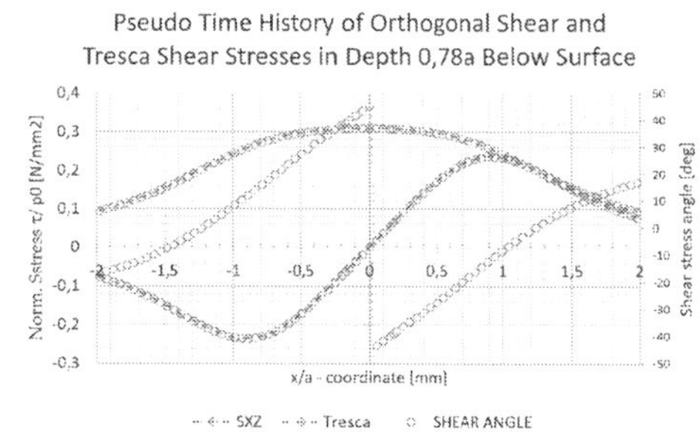

Fig. 9: History of shear stresses (Orthogonal and Tresca) in depth 0.78a below contact surface and development of Tresca shear stress angle.

The complexity of stresses below contact is well illustrated in Fig. 8 where all three orthogonal stresses are viewed. The point of contact is at zero coordinate, so when roller moves, observed point goes through all viewed stress states. It is obvious that stresses change with time unproportional. In Fig. 9 are viewed orthogonal shear stress and Tresca shear stress bellow surface during roller pass. It is apparent that below observed point the shear stress changes it's direction by 90.

3.2.2 Elasto-plastic solution

Results from load cases beyond material yield were obtained only by FEM method. Comparison of von Mises stress obtained is in Fig. 10. Material Yield cause maximal stress value drop and induce material plastic behaviour. Progression of material plasticity with respect to applied contact pressure is viewed in Fig. 11.

4 Conclusion

The analysis of roller bearing line contact was performed. Method of analytical Hertz contact stress solution and FEM were used and compared during analysis. It was proved that analytical solution provides very efficient tool for contact analysis. It is although very important to know the limits of this method that are summarized in section 2.1. FEM was used to analyse stress field below contact. The model in FEM was built as planar with plain strain elements. Johnson-Cook material model was used to model elasto plastic material behaviour as described in 2.2.2. The contact pressure solutions are presented in Fig. 4 where is obvious material yield when higher load is applied. When material yield strength is exceeded the contact area is extended and thus there is declination of contact pressure. Parabolic contact pressure profile flatters as in Fig. 5. Comparison of sub surface principal stresses below contact centre proves FEM model Fig. 6. Von Mises stress and shear stress according to Tresca are in Fig. 7. There is no shear stress on the surface, because off plane stress component is

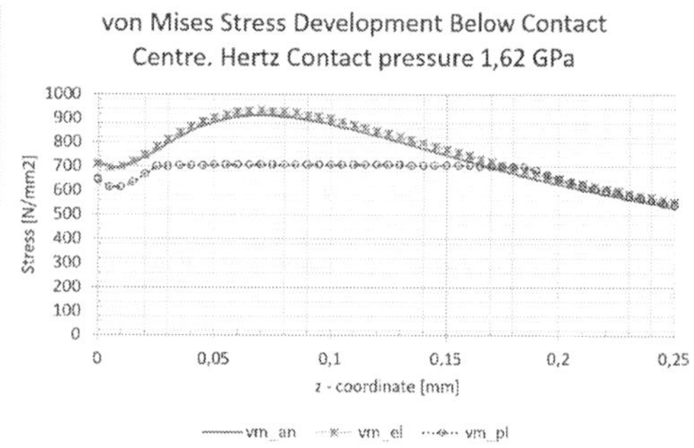

Fig. 10: Comparison of analytical and FEM principal stress development solution below centre of contact. The solution is valid only in elastic domain. $\sigma_Y = 688\ N/mm^2$.

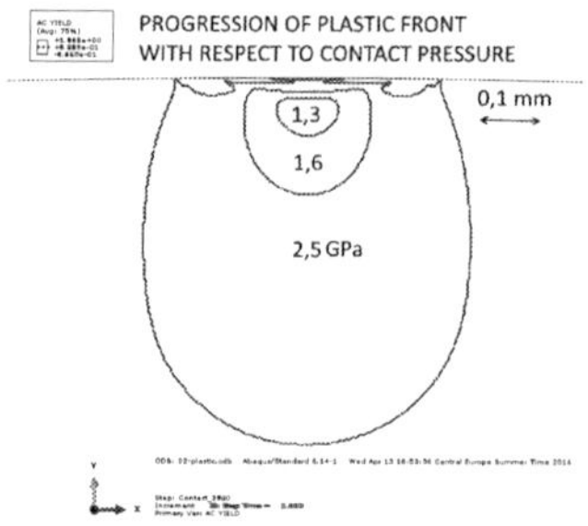

Fig. 11: Progression of material plasticity (contours delimiting areas with plastic mat. behaviour) with respect to contact pressure bellow contact centre. $\sigma_Y = 688\ N/mm^2$

not considered in Tresca. If analysed Von Mises stress, surface stress is calculated. Maximal values are for both reached below surface. Equations for quick estimation of these values are in Tab. 1. When stresses development is plotted for one point below surface during roller pass is obvious that complex and unproportional material stresses are present. See Fig. 8. Below contact centre the direction of principal Tresca shear stress changes rapidly, that is in Fig. 9 illustrated by Tresca shear stress angle development. Comparison of elastic material solution with plastic is in Fig. 10 where is obvious material yield limit on Von Mises's curve. The progression of plastic zone with loading is illustrated in Fig. 11. The first plastic point forms in location of Max. von Mises stress and grows.

References

[1] ISO 16281:2008 Rolling bearings Methods for calculating the modified reference rating life for universally loaded bearings.

[2] K. L. Johnson, Contact Mechanics, Cambridge, Cambridge University Press, 1985.

[3] Ervin V. Zaretsky, Rolling Bearing Life Prediction Theory and Application, 2013, Available electronically at `http://www.sti.nasa.gov`.

[4] ABAQUS 6.14 Online Documentation, Dassault Systmes, 2014.

G. Author et al., Title of Proceedings Paper, in proc.: Proceedings Title (2010), ed. A. Editor, Organizing Institution, Conference Location, 181–190.

[5] A. Shrot, M. Baker, Is It Possible to Identify Johnson-Cook Law Parameters From Machining Simulations?, International Journal of Material Forming (2010) Vol. 3, Suppl 1, 443–446, doi: 10.1007/s12289-010-0802-4.

Difference of Cylindrical Locks Break Resistance in the Dependence on Outdoor Temperature

J. Ivanka [1,a], P. Navrátil [2]

[1] *Tomas Bata University in Zlín, Department of Security Engineering, Nad Stráněmi 4511, Zlín, 760 05, Czech Republic*
[2] *Tomas Bata University in Zlín, Department of Process Control, Nad Stráněmi 4511, Zlín, 760 05, Czech Republic*
[a] *ivanka@fai.utb.cz*

Abstract: The present study is focuses on the interaction of temperature with cylindrical lock break resistance. It is based on an analysis of fundamental temperature and mechanical impacts on lock systems with regard to the provided protection, presenting verification and analysis of laboratory measurements and documentation of the testing methodology, describing the relevance of temperature for break resistance and emphasising the significance of temperature dependence of the material.

Keywords: break resistance; hardness; cylindrical lock; cylinder; lock systems.

1 Introduction

Lock systems have guaranteed protection of our property, privacy and security for thousands of years. They have gradually been introduced worldwide and nowadays they are used by millions of people. Since their beginning they have become a fundamental component of the access control system in the area of mechanical protection systems. A topic issue in access control is nowadays enlarging the user identification possibilities and increasing their security, sometimes achieved through changes in technology of the inner mechanisms retaining however the time-proven fundamental course of functions resulting in the targeted unlocked/locked status. The most important requirement for lock systems is their reliability, i.e. stability of the system properties. Regardless of the weather or time of the day, we expect them to retain their primary functions.

2 Analysis and Verification of the Measured Values

Cylindrical locks of the entrance door are during the year exposed to temperatures ranging approximately (for the Czech Republic conditions) from -30°C to +40°C; their material therefore must cope with the temperature difference up to 70°C (however, in reality the working temperature range is reduced by the indoor temperature in the house). These differences result in changes in material properties, which is the subject of the following section of the report. Phenomenological thermodynamics describes behaviour of a macroscopic system for a thermal energy region characterizing changes in physical values and in the system status during a dynamic development of the system energy. Static thermodynamics focuses on this issue with regard to elementary particles and micro-volumes. When choosing materials for mechanical protection systems it is necessary to determine the required properties of the material under process which will correspond to the requirements on the future security element in an appropriate way. Hardness tests mostly present their results in the form of a dimensionless number. It is important to record properties of the environment in which the tests are conducted as these may have a significant impact on the values obtained. For example hardness decreases with increasing temperature energy because the atom bonds become looser. Measurement of bending strength of the tested samples forms the major part of practical tests of the study. Strength is defined as resistance of material to external forces, such as pressure, tension, shear and bending. Natural reaction to the application of force is a change in shape. The change may be momentary or lasting and these factors divide deformation into an elastic or plastic deformation. Fractography is an analytical method of studying fracture processes indicating factors which will help us define kinetic behaviour of destruction of material in time. Fracture identification is carried out based on fracture morphology, magnitude of the total energy necessary for destruction of the sample and the extent of the plastic deformation. The laboratory measurements were carried out using devices *Zwick 1456* and *Integral 2 E*. The cooled down samples, 422 in total, were placed in a Liebherr freezer with the minimum temperature reached at -25.5°C. The measured samples were double sided cylindrical locks of the Extol Craft 65 mm and FAB 200 – 1000 types made mostly of brass plated by nickel to provide protection against corrosion. They are cylindrical locks for buildings which meet the requirements that we have for lock systems designed for security systems. The lowest material support for breaking resistance of the lock system is in the weakest point of the cylindrical lock situated at the revolving tooth and the hole for the fixing screw (about 3 mm under the screw hole at the widest point, about 2 mm above the hole and near the tooth about 14 mm of non-homogenous and non-compact material). The difference lies in choosing higher speed of stress to which the tested samples are subjected (Fig. 1). The experiment was carried out using a device made by Zwick. Product Zwick 1456 is designed for compression, tensile and bending tests of materials which can be conducted at different temperatures (within the range from -80°C to +250°C). The device is equipped with the testXpert II software providing the necessary setting options and input data, the calculating part of the measurement followed by visualization of the obtained data. One of the essential parts of Zwick 1456 enabling measurements within the temperature ranges above is a thermal chamber. It is a thermally insulated space which can be positioned so that it surrounds the measured sample. Break resistance was tested by determining bending strength. The cylindrical lock is during the bending strength test placed on two points 50 mm from each other. The third point is situated between them in the centre in the opposite direction to the support. This third acting point applies force to the weakest part of the cylindrical lock body between the both cylinders of the lock.

Fig. 1: Testing of cylindrical locks outside and inside the thermal chamber.

The measured values of the break resistance were within the range from 2630 to 2150 N. It is obvious from the values that critical stress increases as a result of increasing temperature and distribution of the applied force into the neighbouring zone of the fracture defined by plastic deformation. The increase in the critical stress in the negative region of the temperature axis due to higher elastic modulus is also of interest. The escalation of the critical stress for the specified area is very significant. The samples tested for break resistance were subjected to fractographic analysis in order to define the type of fracture resulting in the destruction of the cylindrical lock. Specifically they were specimens tested at -20°C and +40°C, i.e. the suprema and infima samples of the selected temperature interval. Substantial differences in the tested samples were already obvious from the macroscopic point of view. The analysis revealed a difference between both fractures where in the case of the cooled down sample (temperature -20°C) the style of deformation corresponded to brittle fracture while the warmed up sample (temperature +40°C) showed a process of considerable plastic deformation (Fig. 2).

a) -20°C b) +40°C

Fig. 2: A close-up of the fracture surface formed at -20°C and +40°C.

The micro-fractographic analysis was carried out using the Carl Zeiss Jena light microscope equipped with a video camera connected to a computer. The tested objects were lit by an illumination arm in order to emphasise details. The surface images for microscopic analysis were obtained at 15-fold magnification. The application of higher, 50-fold magnification, was rejected due to significant differences in height of the fracture areas surfaces which made required focusing of the image impossible (Fig. 3).

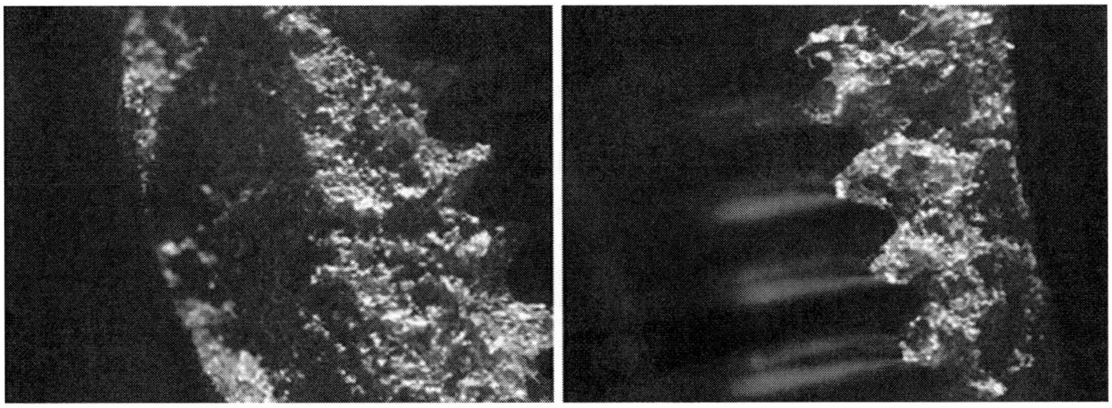

Fig. 3: Microscopic close-up of the fracture (-20°C).

Fig. 4: Microscopic close-up of the fracture (+40°C).

Lower necessary energy of the critical flow stress and higher looseness of atoms in their bonds is obvious from the warmed-up sample analysis. It can be clearly seen that the destruction progressed in the direction of the flow planes at the approximate angle of 45°. Much larger plastic deformation in the course of destruction also caused substantial changes in height on the fracture surface (Fig. 4).

Measuring the samples revealed transitional fracture behaviour (transition between brittle and ductile fracture).

Tab. 1: Selected values of break resistance in testing the product FAB 200.

temperature[°C]	temperature difference [°C]	time interval [s]	average values of break resistance [N]
+40	0	14	2630
+35	0	14	2460
+30	0	14	2340
+25	0	14	2360
+20	0.0065	14	2770
+15	0.0877	14	2300
+10	0.169	14	2130
+5	0.25	14	2210
0	0.331	14	2360
-5.5	0.42	14	2360
-10.5	0.5	14	2900
-15.5	0.583	14	2950
-20.5	0.664	14	2750

3 Conclusion

The tests of break resistance simulating attempts at breaking the cylindrical lock were conducted with homogenous thermal structure of the material. In the real conditions there is a temperature transition within the locking system between the indoor temperature in the house and the outdoor temperature. The temperature difference in question however is of negligible significance (unless temperature differences in hundreds of degrees Celsius are involved) as the characteristic of the fracture occurring in the cylindrical lock under a great pressure is perpendicular to the supremum (maximum) temperature variations and is therefore situated in the part with resumed temperature homogeneity. In conclusion, based on the measured values we can say that break resistance (measured for temperature -20°C) is temperature unstable and for the tested lock systems it reached the variation up to 15% of the reference value.

Acknowledgement

This work was supported by the Ministry of Education, Youth and Sports of the Czech Republic within the National Sustainability Programme project No. LO1303 (MSMT-7778/2014) and also by the European Regional Development Fund under the project CEBIA-Tech No. CZ.1.05/2.1.00/03.0089.

References

[1] V. Dalibor, Kovové materiály (Metal Materials), Prague, VŠCHT, 2006.

[2] J. Pokluda, Mechanické a strukturní materiálové charakteristiky (Mechanical and Structural Material Characteristics), Brno, FS VUT, 1989.

Application of Notches in Machine Elements for Measurement Sensitivity Increase

M. Janda [1,a], J. Kamenický [1], O. Berka [1], P. Malý [1], F. Lopot [1,2], V. Dynybyl [1]

[1] *CTU in Prague, Department of Designing and Machine Components, Technická 4, 166 07 Prague 6, Czech Republic*

[2] *Charles University in Prague, Faculty of Physical Education and Sport, Department of Anatomy and Biomechanics, Jose Martího 31, 162 52, Prague 6, Czech Republic*

[a]*Martin.Janda@fs.cvut.cz*

Abstract: This work is focused on the development of a method how to increase the sensitivity of measurement of machine elements using strain gauges. Many machine components are often intentionally designed as overdesigned therefore the surface deformation and the corresponding stress are difficult to measure with standard wire strain gauges. The research is carried out using mainly experimental work, the obtained results are compared with the FEM analysis.

Keywords: Strain Gauge; Notch; Stress Increase; Linearity

1 Introduction

Nowadays, measurements with strain gauges represent widespread and desired method for assessment of the stress state of machine elements because there is the possibility for the online measuring during the operation of a machine. However, there are many cases when the limitations of this method become dominant and the method cannot be used. Such case may be the measurement of machine components designed with high safety factor when the stress state during the operation is very low. Large industrial gearboxes are the examples of machine components that are designed with respect to the mechanical and economic efficiency. Application of strain gauge measurement gives information about the real efficiency of the operated machine to the customers. Moreover, service organization can get the information that helps them to improve the maintenance of the production equipment, and the producers receive important feedback for their development and production. It is also important to measure the efficiency of such large industrial gearboxes with high accuracy because for example the loss of 1 % of 6 MW gearbox is 60 kW.

Motivation of this article is to give the outline and insight how to use strain gauge measurement for such applications.

2 Theory

There are many possibilities how to increase the sensitivity of strain gauge measurement. For example the measuring equipment can be tailored. The sensitivity of equipment can be increased by usage of converter with higher bit width, or the application of semiconducting strain gauges that have 80 times higher sensitivity than wire strain gauges provides better results. Another possibility is to find the appropriate place for the strain gauge where the local increase in stress can be assumed. Also the increase of the measurement sensitivity can be expected proportionally to the local stress increase. Notches represent such places where the increase of stress arises.

From the theory of notch stress state follows that there is a small area near the notch where the increase of the stress is above the nominal value. Figure 1 shows the example of such notch created by a shaft shoulder where the stress flow is concentrated. The FEM analysis of the shaft with two shoulders, i.e. notches, and the course of shear stress along the shaft are shown in the figure 2. The increase of the stress near the notches is obvious.

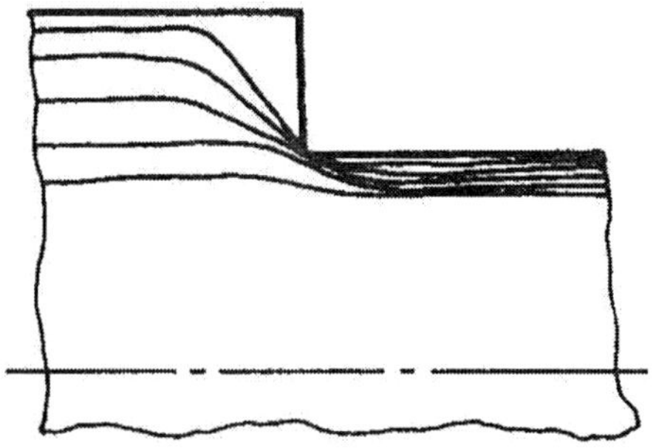

Fig. 1: Notch and the stress field

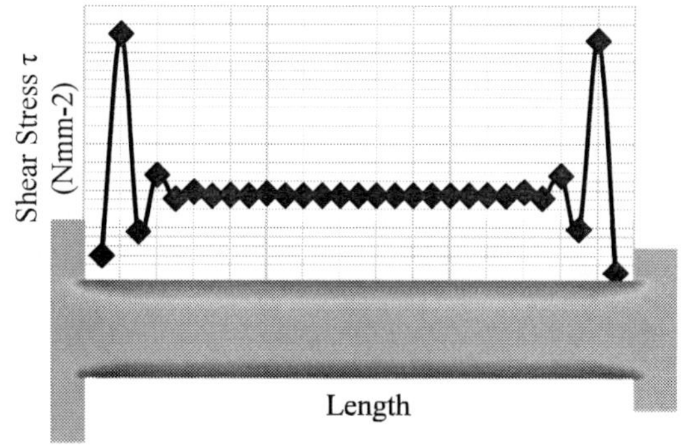

Fig. 2: Course of shear stress along the shaft

The value of the shear stress in the notch can be easily calculated from the known equation (1), where α is the stress concentration factor, τ_{nom} is the nominal shear stress and τ_{max} is the maximal value of the shear stress in the notch. Stress concentration factor α for the considered geometry can be found in known nomograms. Example of the nomogram for the shaft shoulder can be seen in the Figure 3 [2], where D and d are the shaft diameters and ρ is the notch radius.

$$\alpha = \frac{\tau_{\max}}{\tau_{nom}} \tag{1}$$

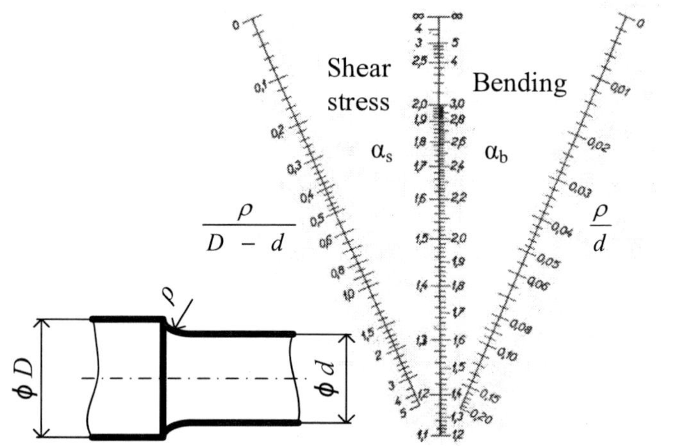

Fig. 3: Nomogram of the stress concentration factor α for the shaft shoulder

Using the stress concentration factor the value of the notch stress can be calculated, but in most cases it is not possible to place strain gauge directly to the notch. So there has to be found another place, where the stress is higher than the nominal value, which will guarantee an increase in sensitivity and where the strain gauge can be placed. To find this place a stress gradient γ can be used, which is calculated from the known equation (2).

$$\gamma = \frac{1}{\rho} + \frac{4}{D+d} \tag{2}$$

3 Experiment

The specimen of the shaft was created for the verification of the stress state in the real parts. On the shaft there were two notches/shaft shoulders. The measurement was obtained using 120hm strain gauges near the notches. The strain gauges were configured for measuring of the shear stress. The examples of measured results in both strain gauge positions can be seen in figure 4 and figure 5. The shear stress is linearly dependent proportional to the torque load.

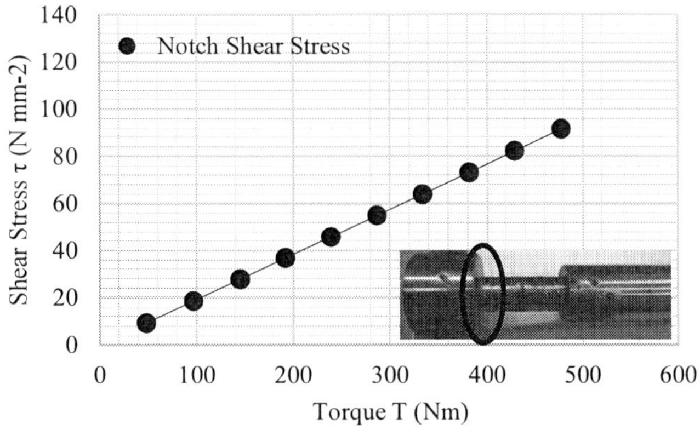

Fig. 4: Measurement of shear stress near the notch

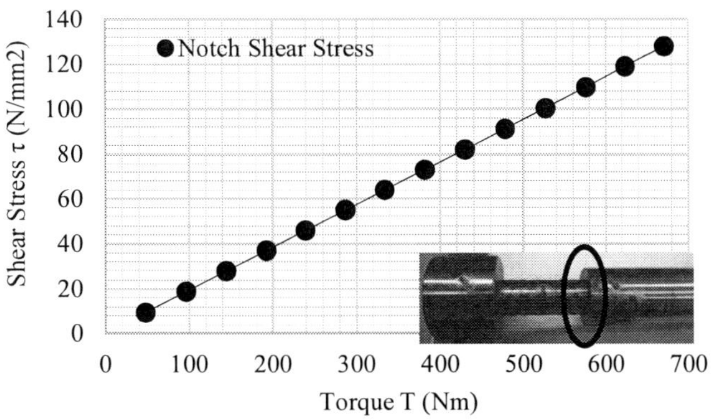

Fig. 5: Measurement of shear stress near the notch

4 Summary

The stress theory near the notch of the shaft loaded by torque was verified with help of experimental measurement and FEM analysis. It was confirmed that the increase of the stress near the notch is linearly proportional to the loading. Following research in this field is focused on the verification and application of other types of notches with respect to the increase in the sensitivity of the strain gauge measurement.

Acknowledgement

This work is supported by the grant TAČR TH01020152.

References

[1] O. Berka, M. Dub, F. Lopot et al, Analysis of satellites loading in planetary transmission, 53rd International Conference on Experimental Stress Analysis, EAN 2015; Cesky Krumlov; Czech Republic.

[2] M. Ružička, M. Hanke, M. Rost, Dynamická pevnsot a životnost, Praha, Vydavatelství ČVUT, 1992.

[3] J. Michalec et al.,Pružnost a pevnost II, Praha, Vydavatelství ČVUT, 2001

The influence of peritubular and intertubular dentin for the development of the micromechanical properties

P. Hájková [1], A. Jíra [1,a]

[1] *Czech Technical University in Prague, Faculty of Civil Engineering, Czech Republic*
[a] *jira@fsv.cvut.cz*

Abstract: This paper is devoted to micromechanical properties of a tooth in dentin part. Mechanical properties of dry dentin from the human tooth were measured by a method of nanoindentation. The indentation was performed by loading to the maximum force of 3 mN. The trend of properties was observed depending on distance from the nearest tubule, its location in a peritubular or intertubular dentin area. A huge number of indents were made during the experiment and mechanical properties of these indents were compared, especially reduced modulus and hardness.

Keywords: human tooth; dentin; nanoindentation; hardness; reduced modulus.

1 Introduction

The knowledge of microstructure and mechanical properties of biological material such as human dentin is important for understanding processes which are coming during life and for designing elements (dental implants or tooth filling).

Dentin which forms most of the crown and tooth roots consists of main matter from odontoblasts and dentin tubules with Tomes´ fibers. Diameter of tubules is 1-2,5 μm and its density is 20.000-50.000 tubules/mm2 [1]. Peritubular dentin (PTD) forms a wall around tubules (width ~ 1,0-2,5 μm) and is highly mineralized [2]. Instead of intertubular dentin (ITD) which forms material between tubules and is less mineralized with more collagen.

The importance of knowing mechanical properties of dentin led to many researches in the past. These were carried out by quasi-static and dynamic methods whose usage was limited. Regarding it, indentation started to be used. The indentation have been used to analyse hardness and Young´s modulus of human dentin since the last century [4,5]. New methods for obtaining more detailed results are tested now [6].

The research by the method of the nanoindentation enabled people to obtain micromechanical properties of segments of human dentin – ITD and PTD. Average values of the hardness and the Young´s modulus in PTD measured by different researchers have been almost the same. For example average value of the Young´s modulus ~ 29,8 GPa and 29,3 GPa [4, 7]. Much more differences have been recorded for ITD properties. There the position of indents is really important because values near enamel and a pulp can differ to 20% [4]. Similar results and differences were obtained by carrying out indents large enough to reduce influence of tubules sufficiently [8].

We present our research which deals with the trend of properties in PTD and ITD depending on the distance of the nearest tubule. Other authors mentioned decrease of properties from interface of tubules in PTD and constant value in ITD [7]. The results of our research just as [7] were conducted by the indentor with operating mode CSM and the Berkovich´s tip with the methodology of Oliver & Pharr [9].

2 Materials and Methods

2.1 Preparation of Specimens

Four specimens were prepared from four molars in total. Two of them were assigned for the measurement of human dentin properties. The cleaned dry molars were embedded in epoxy resin (Fig. 1) and cut into slices ~ 10 mm after induration. Subsequently, these samples were grinded and polished by abrasive papers and canvas (Fig. 2). We used ultrasonic cleaning in alcohol bath for disposal of fine impurity.

Plane section was transversal (indentation in longitudinal direction) which crosscut tooth roots. The position of cuts had been chosen with regard to visible defects which could cause changes in mechanical properties by reason of demineralization.

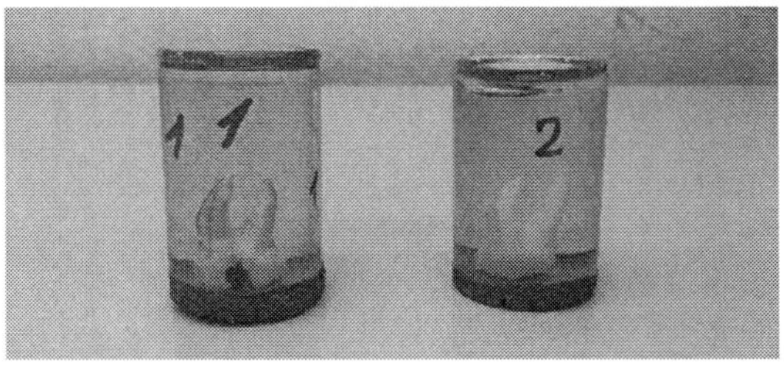

Fig. 1: The molars embedded in epoxy resin.

Fig. 2: Grinded and polished samples.

2.2 Nanoindentation

The indentation was made by the indentor CSM Instrument Nano Hardness Tester with Berkovich´s tip.

The first sample was used for obtaining an appropriate maximum force during loading. Seven matrixes were performed, each one with 25 indents. The maximum force for each matrix was different – 3, 5, 10, 15, 20, 25 and 30 mN. The penetration depth varied ~ 332 – 1163 nm. Force controlled tests were used for each cycle. The parameters of cycle: loading 120 mN/min, pause 10,0 s, unloading 120 mN/min.

Second sample which made for observing the trend of mechanical properties depending on the distance from tubule interface was loaded by the force of 3 mN. The penetration depth varied ~ 255 – 483 nm. The cycle parameters of the force controlled tests: loading 18 mN/min, pause 10,0 s, unloading 18 mN/min. The distance between indents was 4 μm regarding mutual affection.

3 Results

3.1 Test of Appropriate Maximum Force

7x25 indents were made for obtaining an appropriate maximum force during loading. Each matrix was loaded by a different force. From photos it was evaluated which indents had been located in ITD and which had interfered completely or partly with tubules. The most obvious difference in values of properties was found out if the lowest force had been used.

The hardness of ITD obtained by loading to the maximum force of 3 mN was ~ 973 MPa (Fig. 3). It was more than 30% increase of hardness compared to indents which had interfered with tubules. The differences were less obvious during usage of higher forces. It was caused mainly due to the size of indents which had interfered in ITD, PTD and tubules.

Regarding these results, the maximum force of 3 mN was chosen for next test. That could provide quality results and the penetration depth was lower so it could enable more exact location of indents in dentin – tubules, PTD or ITD.

Fig. 3: Average values of the hardness obtained during loading different forces

	H_{IT} [MPa]						
Maximum force [mN]	3	5	10	15	20	25	30
ITD	973,17	1038,86	1033,90	1022,06	1122,22	1017,97	980,18
tubules	658,15	995,61	867,87	1078,59	1057,37	983,09	975,39
difference	315,02	43,25	166,06	56,53	64,85	34,88	4,79

Fig. 4: Average values of the reduced modulus obtained during loading different forces

	E_r [GPa]						
Maximum force [mN]	3	5	10	15	20	25	30
ITD	33,90	31,80	28,40	27,26	28,66	27,50	26,72
tubules	27,10	31,50	27,57	28,25	27,78	27,56	27,10
difference	6,80	0,30	0,83	0,99	0,88	0,06	0,38

3.2 Test of Properties Trend

The indentation for obtaining the trend of properties in PTD and ITD depending on the distance from the nearest tubule interface required a huge number of indents. Two matrixes were made (8x8 and 6x6), 100 indents in total. The matrixes were located near the pulp where density of tubules was higher. According to photos, the indents in ITD were put into six sets (Fig. 5). The indents which interfered completely or partly into tubules were marked as the set 0.

Fig. 5: Putting indents into sets according to distance from tubule interface.

Set	1	2	3	4	5	6
distance from tubule interface [µm]	0 - 1	1 - 2	2 - 3	3 - 4	4 - 5	5 - 6

After evaluation of the sets, it was obvious that values of observed properties were increasing with the distance. The observed properties were the hardness (H_{IT}) and reduced modulus (E_r). It was obvious that the values of observed properties were increasing with bigger distance. The observed properties were the hardness (H_{IT}) and the reduced modulus (E_r). The first set indicated the average hardness ~ 965 MPa as the last set ~ 1222 MPa (Fig. 6, 7). It is increase more than 20%. Likewise the reduced modulus of first set ~ 25,8 GPa, the last set ~ 29,2 GPa (Fig. 6, 8).

The high standard deviation of the set 0 (the indents which interfered into tubules) is probably caused by various volume of interference in these tubules. The high values of hardness were reached if the indents had been located mainly in PTD.

Fig. 6: Average values of the hardness (H_{IT}) and the reduced modulus (E_r) of each set.

Set	0	1	2	3	4	5	6
H_{IT} [MPa]	837,77	965,58	1100,10	1162,66	1205,97	1162,17	1222,57
E_r [GPa]	24,94	25,75	27,20	27,56	28,37	28,37	29,24

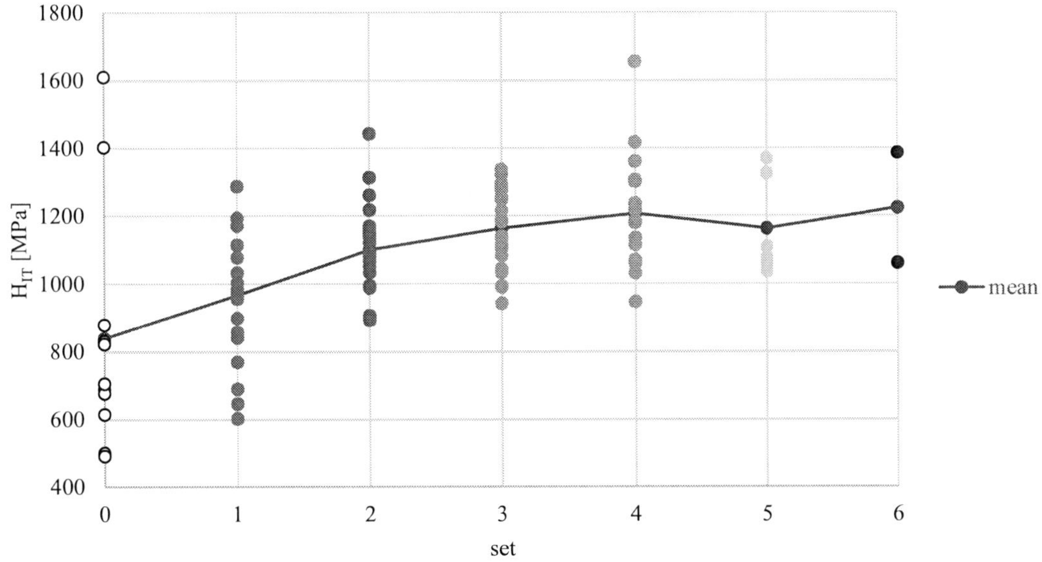

Fig. 7: The trend of the hardness in dentin part depending on the distance from the nearest tubule.

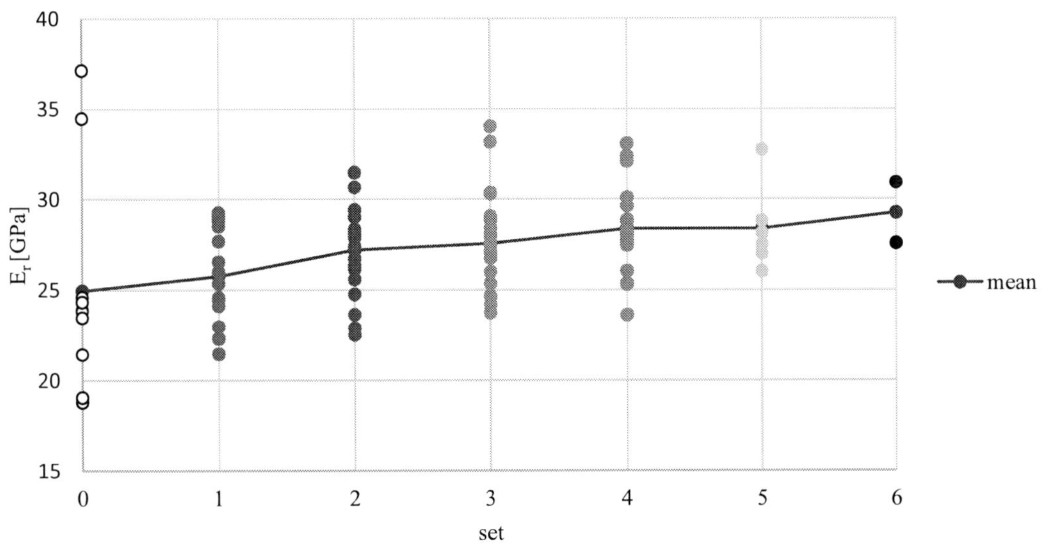

Fig. 8: The trend of the reduced modulus in dentin part depending on the distance from the nearest tubule.

The results show different trend than it has been performed by other authors so far. The authors point out decreasing trend of the observed properties in PTD and almost constant values in ITD [7]. The trend of the hardness and the reduced modulus obtained in this test can be assigned to ITD inhomogeneity and its affection by presence of tubules.

4 Conclusion

The influence of the distance from tubule interface for the development of the micromechanical properties in ITD was published in this paper. Firstly, the appropriate force for loading had to be found out to obtain optimal results and small indents. The maximum force of 3 mN was chosen for the next test. Subsequently, 100 indents were made. The increasing trend of the hardness and the reduced modulus from tubule interface was obtained. It is the different trend than have been published by other authors so far. They have denoted almost constant values in ITD. Because the measurement had been located near the pulp where the tubule density had been higher, the values of the properties were obtained only to the distance of 6 μm from the nearest tubule interface. In the next test it would be good to verify if the trend continues with the same character even further.

Acknowledgement

The financial support by the Technology Agency of the Czech Republic (TAČR project no. TA03010886) and Faculty of Civil Engineering, Czech Technical University in Prague (SGS project No. SGS14/122/OHK1/2T/11) is gratefully acknowledged.

References

[1] K. J. Anusavice, Phillip´s science of dental materials, Missouri, Elsevier, 2003.

[2] R. Garberoglio, M. Brannstrom, Scanning electron microscopic investigation of human dentinal tubules, Archives of Oral Biology 21 (1976) 355-371, doi:10.1016/S0003-9969(76)80003-9.

[3] R. G. Craig, F. A. Peyton, Elastic and mechanical properties of human dentin, Journal of dental research 37 (1958) 710-718.

[4] J. H. Kinney et al., Hardness and Young´s modulus of human peritubular and intertubular dentine, Archives of Oral Biology 41 (1996) 9-13, doi:10.1016/0003-9969(95)00109-3.

[5] N. Meredith et al., Measurements of the microhardness and Young´s modulus of human enamel and dentine using an indentation technique, Archives of Oral Biology 41 (1996) 539-545, doi:10.1016/0003-9969(96)00020-9.

[6] G. Balooch et al., Evaluation of a new modulus mapping technique to investigate microstructural features of human teeth, Journal of Biomechanics 37 (2004) 1223-1232, doi:10.1016/j.jbiomech.2003.12.012.

[7] D. Ziskind et al., Young´s Modulus of Peritubular and Intertubular Human Dentin by Nano-indentation Tests, Journal of Structural Biology 174 (2011) 23-30, doi:10.1016/j.jsb.2010.09.010.

[8] A. Jíra, J. Němeček, Nanoindentation of human tooth dentin, Key Engineering Materials 606 (2014) 133-136, doi:10.4028/www.scientific.net/KEM.606.133..

[9] W. C. Oliver, G. M. Pharr, An Improved Technique for Determining Hardness and Elastic Modulus Using Load and Displacement Sensing Indentation Experiments, Journal of Material Research 7 (1992) 1564-1583, http://dx.doi.org/10.1557/JMR.1992.1564.

Use of Short Carbon Fibres as a Reinforcing Material for the Refractory Cement-based Composite

M. Jogl [1,a]

[1] *CTU in Prague, Faculty of Civil Engineering, Experimental Centre, Thákurova 7/2077, 166 29 Prague 6, Czech Republic*

[a] *marcel.jogl@fsv.cvut.cz*

Abstract: A type of fibre-reinforcing material was developed by carbonization of high quality polyacrylonitrile precursor, which is specifically engineered. Together with resin, carbon fibres are usually manufactured into a composite. These carbon fibre components are light-weight and strong compared to parts made of metals (e.g. aluminum) or other fibre reinforced composites. For the purpose of this experiment, the attention was focused on the application of chopped, dispersed carbon fibres. These extremely strong and stiff fibres are typically 6 mm in length and due to their properties they can also be used as reinforcement for high-temperature resistant composites. This experimental program addresses the effect of the addition of short carbon fibres on mechanical properties of fibre reinforced cement-based composites exposed to high temperatures of 600 °C and 1000 °C. The proposed composites differed in doses of fibres, which are 0.0 %, 0.5 % and 2.0 % by volume.

Keywords: aluminous cement; carbon fibre; short dispersed fibre, cement-based composite, mechanical properties; high temperature.

1 Introduction

The bond and failure mechanisms in fibre-reinforced concrete are presently being investigated widely by many researchers [1, 2, 3, 4]. Fibre-reinforced composites rank to the most frequently used composites because of their suitable mechanical properties. Application of fibre reinforcement in the form of short fibres ensures increased load capacity of such developed materials [1, 5]. Carbon fibre is exceptionally versatile in the civil engineering field thanks to its high strength and flexibility. It can be used to restore existing structures just as efficiently as it can be used to reinforce brand new ones. In case of high-temperature resistant cement-based composites fibres significantly contribute to the reduction of crack initiating which accompanies volume changes caused by the structural transformation due to high temperature impact.

In this research the main goal of realized experimental was to determine the influence of carbon fibres [6] in refractory cementitious composites after exposure to high temperatures.

2 Creation of the Composites

The main component of interest in this experiment was short carbon fibres. The carbon fibre production process is relatively energy intensive, but carbon fibres popularity is growing because they have excellent strength due to its weight, see Tab. 1. In comparison with steel, the carbon material delivers twice the strength at one-fifth the weight.

Tab. 1: Properties of carbon fibres.

Properties	Specific Weight [g.cm^{-3}]	Tensile Strength [MPa]	Elastic Modulus [GPa]	Thermal Conductivity [W.m^{-1}.K^{-1}]	Coefficient of Thermal Expansion at 100 °C [K^{-1}]
	1.80	3000	500	25	0

Each experimental mixture for the manufacture of composites, shown in Tab. 2, contains as refractory binder the aluminous cement Secar®71, which due to its high content of aluminium oxide (Al_2O_3) ensure sufficient temperature resistance over 1000 °C [7, 8]. Expanded glass Liaver was applied as lightweight aggregate to reduce final bulk density of composites. Liaver is produced by recycling of waste glass by patented technology. The plasticizer Sica Viscocrete 1035 provides fluidizing effect of the fresh mixture and required workability while the low water/cement (w/c) ratio is kept.

To investigate the structural effect of carbon fibre reinforcement in cement-based composites nine prismatic specimens were tested for each proposed mixture. Composite mixtures were cast in steel forms (prisms 40×40×160 mm) and then cured at temperature 20 ± 3 °C and wet condition, the curing time lasted for 28 days.

Tab. 2: Composition of experimental specimens.

Components		Ref [kg.m^{-3}]	C0.5 [kg.m^{-3}]	C2.0 [kg.m^{-3}]
Aluminous cement	Secar®71	900	900	900
Carbon fibres	6.0 mm	0	8.75	35
Water	potable	224	224	224
Plasticizer	Sika 1035	22.75	22.75	22.75
Fireclay aggregate	0.0-1.0 mm	490	490	490
Liaver	0.25-0.5 mm	70	70	70
	0.5-1.0 mm	15	15	15
Amount of fibres	[volume %]	0.0 %	0.5 %	2.0 %

3 Measurement Methods and Results

Dried samples were divided into three groups. The first group was represented by dried specimens at temperature 105 °C. The second and the third groups were submitted to high-temperature loading up to 600 °C and 1000 °C. The required high temperature was maintained for 180 minutes for each composite. It is necessary to carry out a drying procedure and firing after cement composites maturity [9].

A series of uniaxial, deformation-controlled tests was performed to study the strength of specimens. After high-temperature loading was investigated flexural strength of the composites and on the remaining fragments was tested compressive strength. All measurements of mechanical properties were carried out according to the standard EN 196-1:2005 [10]. On the basis of these data, the mechanisms of the interaction between short dispersed fibres and the matrix of cement-based composite are discussed.

The summary results of the measured properties of the produced refractory composites are shown in Tab. 3. Table shows also an overview of the dependence of bulk density on temperature.

Tab. 3: Summarized characteristic of experimental specimens.

	Bulk Density ρ [kg.m^{-3}]			Flexural Strength f_{tm} [MPa]			Compressive Strength f_{cm} [MPa]		
	105 °C	600 °C	1000 °C	105 °C	600 °C	1000 °C	105 °C	600 °C	1000 °C
Ref	1911	1725	1721	8.9	8.2	7.6	44.5	36.1	34.4
C0.5	1868	1647	1635	14.9	12.3	10.8	52.0	41.5	37.6
C2.0	1737	1526	1504	19.4	13.8	10.4	72.3	52.5	25.2

We can observe the gradual decline of bulk density due to effect of high temperature loading, when moisture and physically bounded water is evaporated first. Increase of temperature led to further decrease of bulk density which is caused by partial chemical decomposition of hydration products.

4 Conclusion

Realized program confirmed satisfactory application of short carbon fibres as reinforcement for high-temperature resistant composites. Flexural strength f_{tm} measurement was organized as a three point test with supports distance of 100 mm and was calculated by help of the maximum reached force. Graph on Fig. 1 and Fig. 2 compares values of flexural and compressive strength according the different amount of fibres in total volume of cement composite after exposure to elevated temperature.

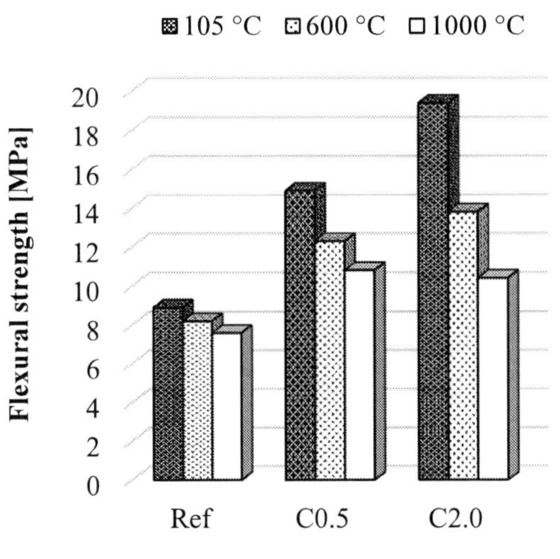

Fig. 1: Values of flexural strength.

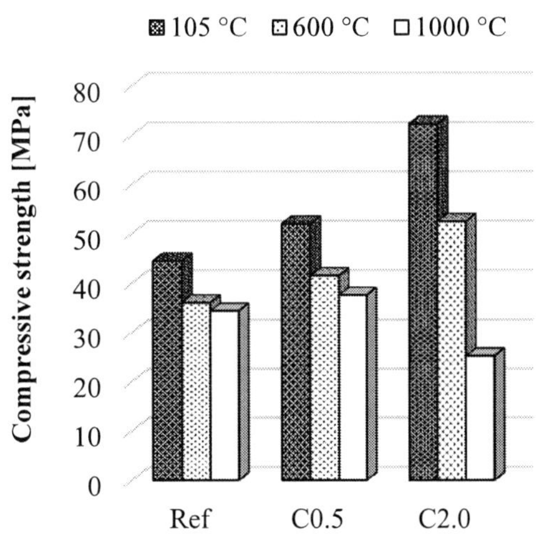

Fig. 2: Values of compressive strength.

The resulting values of Ref composites, after drying and high-temperature loading, exhibited the lowest flexural resistance of all specimens as it was expected. On the other hand, the specimens with 2.0 % of carbon fibres reached the highest values of flexural (f_{tm}) and compressive (f_{cm}) strength in experimental program after drying and exposure to 600 °C. But at the load temperature of 1000 °C occurs the most significant changes in strength. The reduction is probably caused by disruption of the interaction between the mixtures components. According to performed tests we can also determine the positive influence of fibre dosage of 0.25 % (due to mechanical properties and good workability of fresh mixture).

The optimal temperature level for using this type of composite is below 1000 °C, where the values are still sufficient. Good mechanical parameters can be observed on specimens after exposure to 600 °C. None of the thermal loading scheme did not resulted into visible cracks in aggregates or cement matrix, volume expansion, observable disintegration or any negative phenomenon know from interaction ordinary concrete and environment characterized by high temperature (typically up to 600 °C).

Acknowledgement

The authors gratefully acknowledge the support provided by the Czech Science Foundation under the project No. P104/12/0791: Fibre-Reinforced Cement Composites for High Temperature Applications.

References

[1] D.J. Kim et al., Comparative Flexural Behaviour of Hybrid Ultra High Performance Fiber Reinforced Concrete with Different Macro Fibers, Construction and Building Materials 25 (2011) 4144-4155, http://dx.doi.org/10.1016/j.conbuildmat.2011.04.051

[2] R. Barhum, V. Mechtcherine, Effect of short, dispersed glass and carbon fibres on the behavior of textile-reinforced concrete under tensile loading, Enginnering Fracture Mechanics 92 (2012) 56-71, http://dx.doi.org/10.1016/j.engfracmech.2012.06.001

[3] B. K. Kandola, M. H. Akonda, A. R. Horrocks, Fibre-reinforced glass/silicate composites: effect of fibrous reinforcement on intumescence behavior of silicate matrices as a fire barrier application, Materials and Design 86 (2015) 80-88, http://dx.doi.org/10.1016/j.matdes.2015.07.103

[4] D. Koňáková, E. Vejmelková, Reinforced Cement Composites – Effect of Hybrid Fibres on Selected Properties, Materials Science Forum 824 (2015) pp 179-183, http://dx.doi.org/10.4028/www.scientific.net/MSF.824.179

[5] S. E. Artemenko, Yu. A. Kadykova, Polymer composite materials based on carbon, basalt and glass fibres, Fiber chemistry 40 (2008) pp 37-39, http://dx.doi.org/10.1007/s10692-008-9010-0

[6] Retrieved from: http://www.tohotenax-eu.com/en/products/tenax-carbon-fiber/short-fibers.html

[7] P. Reiterman et al., Physical and Mechanical Properties of Composites Made with Aluminous Cement and Basalt Fibers Developed for High Temperature Application, Advances in Materials Science and Engineering 2015 (2015) 10, http://dx.doi.org/10.1155/2015/703029

[8] M. Čáchová et al., Basic Physical and Mechanical Properties of Composites Based on Three Different Cements, Key Engineering Materials 677 (2016) pp 186-190, http://dx.doi.org/10.4028/www.scientific.net/KEM.677.186

[9] R. Boris et al., Effect of holding temperature on properties of different types of heat-resistant concrete, Refractories and Industrial Ceramics 54 (2014) 397–400, http://dx.doi.org/10.1007/s11148-014-9619-x

[10] EN 196-1:2005 - Methods of testing cement. Determination of strength.

Fastener stiffness measurement methodology

M. Kadlec [1,a], P. Kucharský [1]

[1] *VZLU – Aerospace Research and Test Establishment, Beranových 130, Prague, Czech Republic*
[a] *kadlec@vzlu.cz*

Abstract: The stiffness of fasteners is needed for calculation of crack growth and damage tolerance parameters for large aeronautical structures. Experiments were performed on single fastener specimens made of aluminium alloy joined by aluminium rivets. The joints were loaded in pairs to prevent secondary bending and the load was introduced by increasing steps to monitor the change in stiffness with the increasing load level. Various plate thicknesses from 1.2 mm up to 6 mm and different rivet diameter combinations were analysed based on the structure of L-410 NG airplane. The paper includes the results of quasi-static testing of the riveted joints and a comparison with analytical models. The modelled stiffness was on average 16% higher than measured values. Maximum analysed difference was 44% which means that the experiment is irreplaceable.

Keywords: Aluminium alloy; fastener; stiffness; experiment; aerospace.

1 Introduction

The structure of L-410 NG airplane developed by Aircraft Industries will be certified by damage tolerance philosophy. Therefore, finite element calculations are performed to analyze fatigue crack growth and residual strength of the structure in the vicinity of joints. The large and complex structure parts cannot be efficiently modeled with individual rivet meshing and various contact parameters. Consequently, a rivet is substituted by a virtual spring with a stiffness parameter. This simplification is making the modelling much more time efficient.

2 Material and Methods

Specimens for the experiment were made of aluminum alloy D16čATV for series 1, 2, and 4, and aluminum alloy 7475 T7351 for series 3 and 5. The rivets were made of an aluminum alloy with the diameters and eventual countersink according to Tab. 1. The specimens were tested in pairs to prevent secondary bending (Fig. 1). Doublers were used to compensate the thickness of the samples that were attached to each other during the test (Fig. 2a). Each series consisted of 3 pairs of samples and the results were averaged. Loading was controlled by force that was introduced in increasing steps according to Ref. [1] as shown in Fig. 2b.

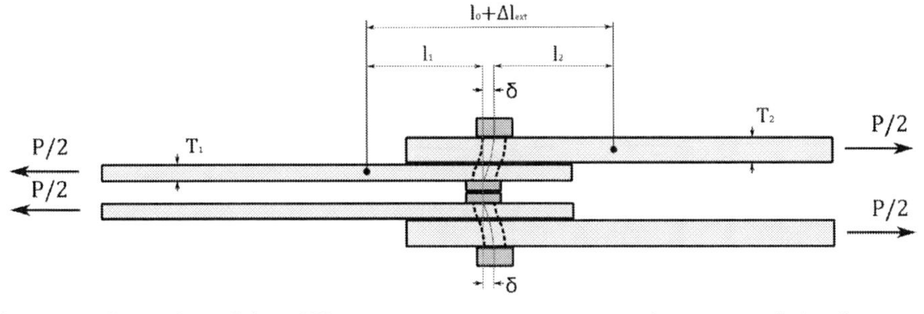

Fig. 1: Configuration of the stiffness measurement on two specimens tested simultaneously.

This type of loading can show the joint behavior even after certain rivet settling. The elastic deformation of joined parts over the 50 mm extensometer base was calculated and subtracted from the total deformation to evaluate only the local fastener flexibility. The fastener flexibility is a measure of the influence of fasteners on the flexibility of the whole joints. It plays an important role when considering the factors influencing the strength level and fatigue life of an aircraft joint. The flexibility f is reciprocal value of stiffness K:

$$flexibility\ (f) = \frac{1}{stiffness\ (K)} = \frac{\delta}{P} \tag{1}$$

in which P refers to the external force applied to a single fastener (P/2 in case of paired testing) and δ to the deflection of the joint due to the fastening.

a) Test set-up with attached extensometer.

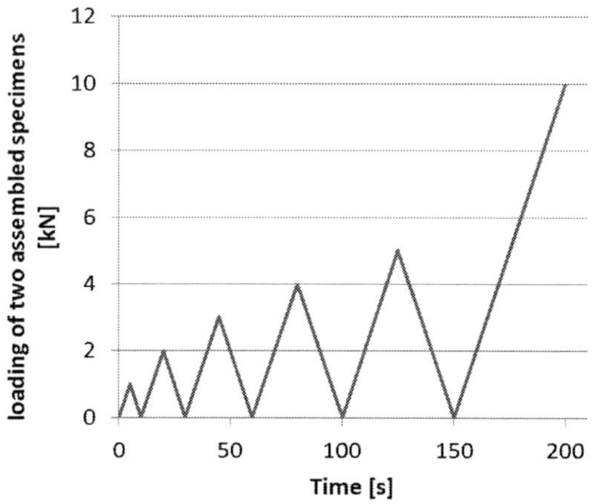

b) Introduced loading.

Fig. 1: Stiffness measurement test set-up and loading.

3 Results

In total 30 specimens were tested divided to 5 series. Typical loading curve with loops is visible in Fig. 2a. The parameters for each series of joints and the average stiffness measured for various loop load levels is included in Tab. 1. The stiffness K was evaluated for the increasing load of each loop. Generally, K was linearly decreasing with the increase of load level of loops (Fig. 2b). Series 2 had significantly higher stiffness than series 1 even for lower fastener diameter. It could be explained by the sealant effect that affect the gap between connected parts and thus cause the stiffness difference.

Tab. 1: Specimen parameters and measured average stiffness K for various load levels.

series no.	T_1 [mm]	T_2 [mm]	rivet diameter [mm]	countersink [yes/no]	average stiffness K [kN/mm]				
					0.5 kN	1 kN	1.5 kN	2 kN	2.5 kN
1	2	1.2	3.5	y	19.5	19.0	18.3	17.1	14.9
2	2	1.2	3	y	29.3	26.4	24.8	21.4	17.9
3	4	1.2	3	y	18.1	18.4	18.0	16.7	14.0
4	2	1.2	4	n	27.8	26.9	25.8	24.1	22.0
5	6	1.2	4	n	20.4	19.4	18.5	17.5	15.7

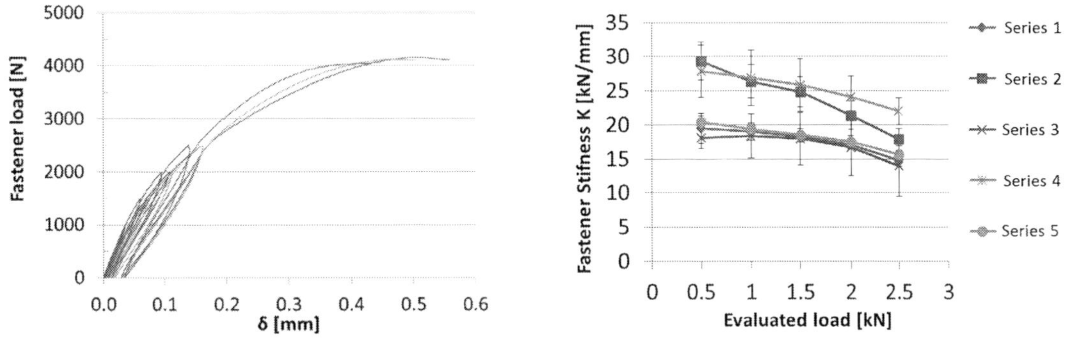

a) Typical loading curves. b) Measured stiffness based on load level.

Fig. 2: Stiffness measurement test set-up parameters and a picture of a configuration.

Fig. 3 shows typical fracture of a joint with visible dark sealant area and a fastener failed surface. The bearing was approximately 0.5 mm and the failed surface of the fasteners was smooth and flat. The sealant covered both fracture faces.

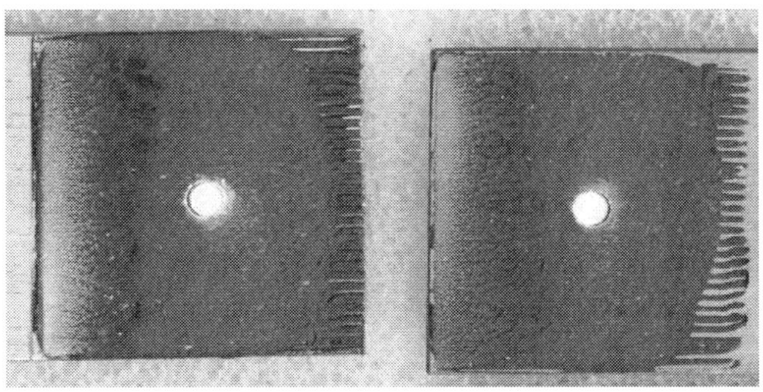

Fig. 3: Typical fracture of a joint with visible dark sealant area and a fastener failed surface.

4 Model comparison and discussion

Several models for stiffness were retrieved from Ref. [2, 3]. Five models are expressed below as equations for stiffness f where E and E_f are Young modulus of the sheet material and the fastener, respectively; D is rivet diameter and μ is Poisson number of the sheet material. The equations are not in general form but they were simplified for single shear riveted joints with same sheet material for both parts. Barrois model, that distinguishes free heads and clamped heads, was evaluated by reading values from graphs included in Ref. [4].

Boeing I:

$$f = \frac{1}{K} = \frac{1}{ED}\left[5 + 0.8\left(\frac{D}{T_1} + \frac{D}{T_2}\right)\right] \tag{2}$$

Douglas (Swift):

$$f = \frac{1}{K} = \frac{5}{DE_f} + 0.8\left(\frac{1}{T_1E} + \frac{1}{T_2E}\right) \tag{3}$$

Tate & Rosenfeld:

$$f = \frac{1}{K} = \frac{1}{T_1E_f} + \frac{1}{T_2E_f} + \frac{1}{T_1E} + \frac{1}{T_2E} + \frac{32(T_1+T_2)(1+\mu)}{9E_f\pi D^2} + \frac{8\left(T_1^3+5T_2^2T_1+5T_2T_1^2+T_1^3\right)}{5E_f\pi D^4} \tag{4}$$

Boeing II:

$$f = \frac{1}{K} = \frac{2\left(\frac{T_1}{D}\right)^{0.85}}{T_1}\left(\frac{1}{E} + \frac{3}{8E_f}\right) + \frac{2\left(\frac{T_2}{D}\right)^{0.85}}{T_2}\left(\frac{1}{E} + \frac{3}{8E_f}\right) \tag{5}$$

Huth:

$$f = \frac{1}{K} = \left(\frac{T_1+T_2}{2D}\right)^{2/5} 2.2\left(\frac{1}{T_1 E} + \frac{1}{T_2 E} + \frac{1}{2T_1 E_f} + \frac{1}{2T_2 E_f}\right) \quad (6)$$

The modelled stiffness was on average 16% higher than measured stiffness for 0.5 kN load as is shown in individual comparisons in Tab. 2. Maximum analyzed difference was 44%.

Tab. 2: Relative difference between a modelled stiffness and the measured stiffness K for 0.5 kN load.

Series	Boeing I (2)	Douglas (3)	Tate&Rosen. (4)	Boeing II (5)	Huth (6)	Barrois (clamped)	Barrois (free)
1	32%	32%	13%	36%	13%	48%	-37%
2	-11%	-11%	-43%	-1%	-39%	9%	-91%
3	36%	36%	-6%	45%	15%	44%	15%
4	10%	10%	-18%	11%	-18%	28%	-88%
5	42%	42%	-3%	46%	10%	48%	26%
Median	32%	32%	-6%	36%	10%	44%	-37%

These significant differences were also reported in Ref. [2] where the measured and modeled stiffness differed by factor of 2 up to 10. This can be explained by the fact that the equations were developed internally in commercial companies where various geometry, technology and materials were used. Also the test set-up and method of data evaluation is not known for the models. In terms of fatigue crack modelling, the modeled stiffness, that was generally higher than measured, is not on the safe side. This is because the crack faces would be hold open in the stiffer configuration and the modelled crack growth would be slower than in reality. Tate & Rosenfeld model seems to be the most suitable for this case.

5 Conclusion

To conclude, the results of quasi-static testing of riveted joints showed that the modelled stiffness was on average 16% higher than measured values that is not on the safe side considering fatigue crack growth modelling. Maximum analysed difference was 44% which means that the experiment is irreplaceable for precise stiffness related calculations needed for aerospace structures.

Acknowledgement

This work was funded by the Technology Agency of the Czech Republic, project no. TE02000032.

References

[1] H. Huth, Zum Einfluss der Nietnachgiebigkwit mehrreihiger Nietverbindungen auf di Lastübertragungs- und Lebensdauervorhersage [disertation thesis], Darmstadt, Fraunhofer - Institut für Betriebsfestigkeit (LBF), 1984.

[2] P. Adamík, Experimentální stanovení tuhosti nýtových spojů a jejich modelování metodou konečných prvků [dissertation thesis], Brno, VUT, 2009.

[3] N. R. Valencia, Fastener Flexibility, Ingeniería Aeronáutica, 2007. Retrieved from: http://tinyurl.com/zh47zaj.

[4] W. Barrois, Stresses and displacements due to load transfer by fasteners in structural assemblies, Engineering fracture mechanics 10 (1978) 115-176.

Probabilistic approach to the assessment of composite steel-concrete structure

P. Kesl [1]

[1]*University of West Bohemia in Pilsen, Faculty od Applied Sciences, Department of Mechanics, Univerzitní 8, 306 14 Plzeň, CZECH REPUBLIC, kesl@kme.zcu.cz*

Abstract: Probabilistic approach to the assessment of composite steel-concrete structure of the building DT Plzeň. A brace of symmetrical "I" ross- section of steel S 235, C20/25 concrete. To assess the structure, we performed simulation with the use of the SBRA probability Method by Anthill software and with subsequent comparison ofultimate load results with percentage of use of most stressed part of the segment determined through calculation by Fine-EC-EC4 program.

Keywords: Reliability of structures; steel-concrete; composite; probability of failure; design probability; simulation; defining a limit state; Simulation-Based Reliability Assessment (SBRA Method).

1 Introduction

For the composite steel and concrete ceiling construction in the building of the Technology Centre - DT Pilsen a symmetrical rolled profile IPE 220-240 of steel grade S 235 and reinforced concrete slabs C20/25-25/30-XC, class reinforcement B550B was used. On this structure calculation is carried out according to DIN-EN-1991 report profile DIN-EN-1994 inner strength Fine-EC-2D, 3D, and subsequent optimization probabilistic method SBRA-program Anthill with the probability of structural failure Pf (i) ceiling segment (Fig.1) in used.

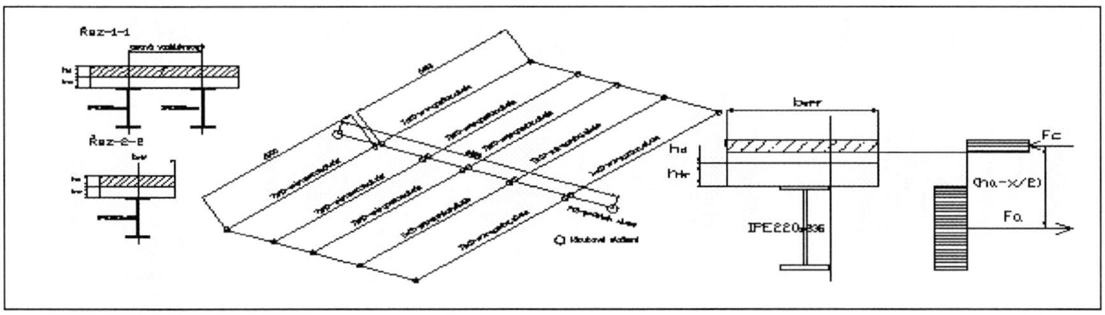

Fig. 1: Static scheme ceiling of composite structures, TS (i), P (i)

2 Assessing structures

The composite rod of symmetric cross section as a ceiling segment (Fig. 1) of rolled IPE 220- 240 of steel grade S 235, and the concrete slab of concrete C20/25-25/30-XC ,class reinforcement B550B, the load width of 2000 mm and span of 6000 mm was analyzed for different load conditions (ZS, S, G, W, Q, A (i)) i = 1 to 6). Decisive load combinations was determined by Fine-EC program and the structural response for each load was analyzed for the TS-component (i), P(i) in accordance with a combination of theory I. , II. regulations.

2.1 Equations - Model in Anthill-SBRA method:

$$V_{smyk} = V_{ed} / V_{rd}, \quad (1) \qquad V_{rd} = \frac{A_V \; A\,\mathrm{var} \dfrac{F_y \,\mathrm{var}}{\sqrt{3}}}{10} \quad (2)$$

V_{ed} $[N]$ .... design combination of shear forces

V_{rd} $[N]$ .... resistance of Shear

$$M_{ohyb_{adol}} = M_{ed} / M_{Pl\,rd\,a}, \quad (3) \qquad M_{Pl\,rd\,a} = A \; A_{\mathrm{var}} \; f_{yd} \; z \; r \left(10^{-3}\right) [Nm] \quad (4)$$

$$Ohyb_{,adol} = \sigma_{,adol} / f_{yd} \quad (5) \qquad \sigma_{,adol} = \left(\left(M_{ed} / I_y\right) z_{dol}\right)\left(10^{-3}\right) \quad (6)$$

M_{ed} $[Nm]$ .... design combination of bending moment

$M_{Pl\,rd\,a}$ $[Nm]$ .... the reduced bending resistance in cross-section

Ohyb,$_{adol}$ [%] .... utilization of the cross-section under the action of bending moment

M_{ed} $[Nm]$ .... design combination of bending moment

σ_{adol} $[MPa]$.............. educed resistance sectional flexural

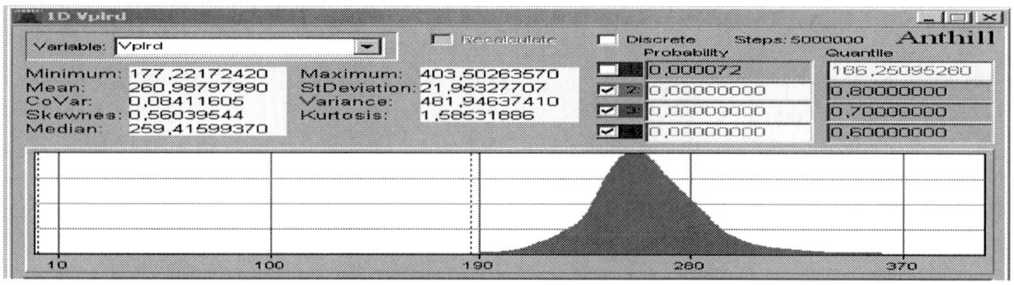

Fig.2: Shear capacity sectional *Vpl rd [N]*

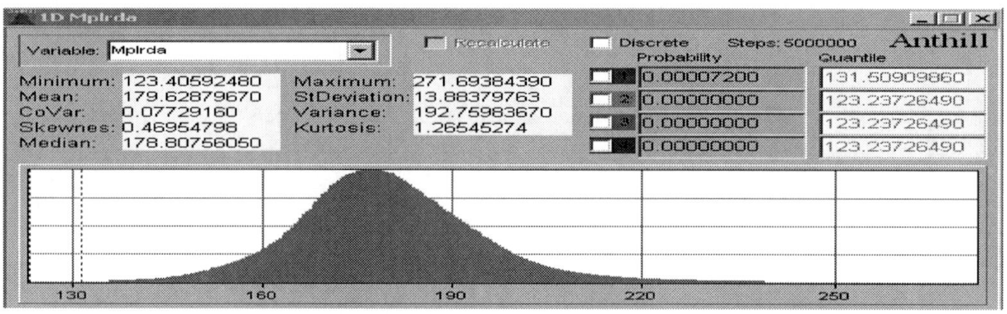

Fig.3: The reduced bending resistance in cross-section $M_{Pl\,rd\,a}$ $[Nm]$, $M_{Pl\,rd\,c}$ $[Nm]$

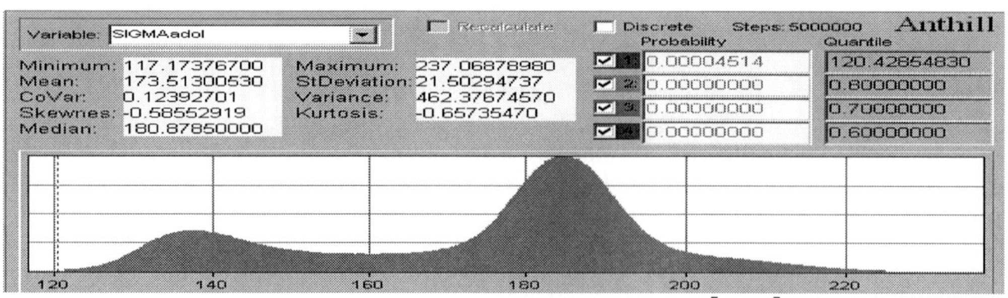

Fig.4: The resulting reduced stress σ_{adol}, $\sigma_{c,hor}$ $[MPa]$

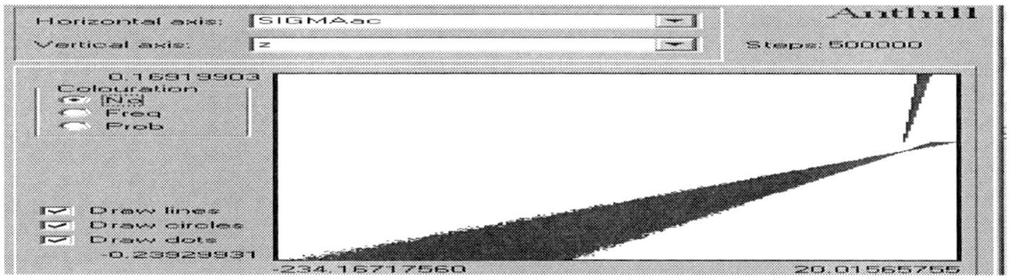

Fig.5: Tension in cross - elasticity [MPa]

The determined value for the probability of failure Pfd(i) = 0.000072 are: according to the model diagram - plasticity (Fig.3) the torque is 131.51 kNm, and according to the model diagram – elasticity (Fig.4) the stress is 173,513 MPa.

2.2 Assessment of the cross section on the probability of failure - Pf(i) - (SF(i))

For the simulation interval and determining the failure probability Pf (i) or SF (i) of the steel-concrete profile a set of calculations (series 1, series 2, series 3) was performed for the different number of simulation steps: 500.000, 1.000.000, 2.000.000, 5.000.000, 6.000.000 cycles. Category 4 is the design life, life of 50 years, the consequences class CC2, maintenance IL2, these values are Pfd(i) (MSU) = 7.2 (10^{-5}) Pfd(i) (MSP) = 6.7 (10^{-2}). The failure probability on the structure is in the interval with the number of cycles 5.000.000 is Pf(i)= (from 1,12(10^{-5}) to 1,375(10^{-5})) (Fig.6), or the use of the cross-section of 94% -97% (Fig.7).

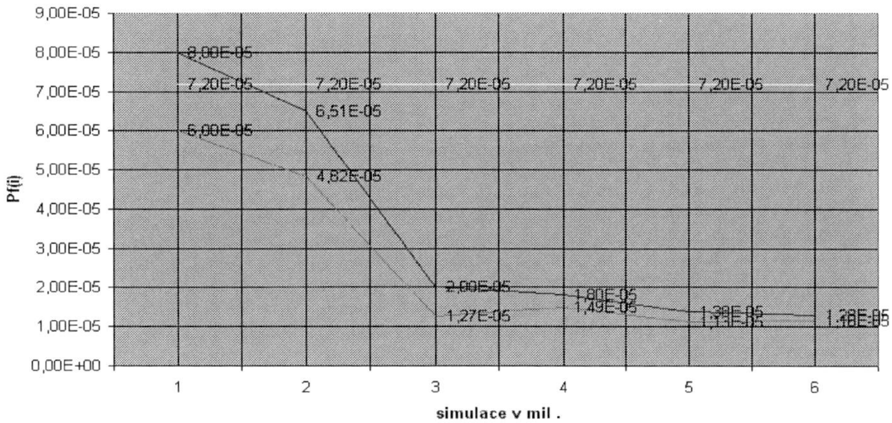

Fig.6: The course of the failure probability Pf (i), SF(i) the number of simulations

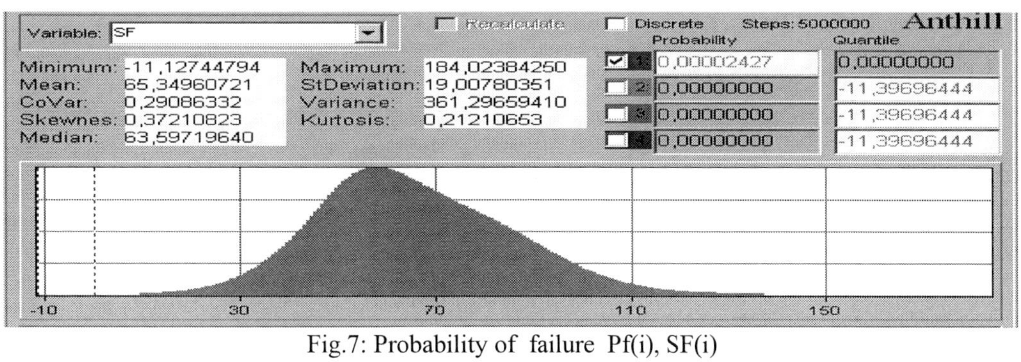

Fig.7: Probability of failure Pf(i), SF(i)

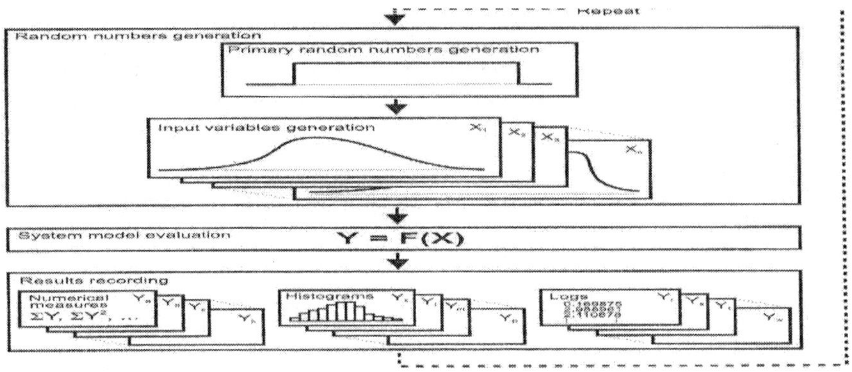

Fig. 8. Simulation model of solved beam

3 Conclusion

Summary and conclusions - evaluation of results obtained using SBRA and EC4. For the comparison of results obtained by DIN-EN-1994 and SBRA methods (fig.2,3,4,5,6,7) follows that using probabilistic SBRA method leads to savings of material in extremely stressed ceiling segment. This conclusion is achieved by a different understanding and approach to the calculation and design, which is based primarily on the probability theory and simulation (Fig.8). Taking into account the influences acting on the structure, external and internal conditions and structure usage savings in material and finance are obvious (Tab. 1).

Tab. 1: Comparison of results extremely stressed segment

Equations	Results obtained by calculating		Difference in %.
	DIN-EN-1994	SBRA Anthill	
Vpl,rd	48,43%	45,50%	+2,93%
Mpla,Rd, Mplc,Rd, plasticity	88,00%	75,00% ÷ 80,00%	+8,00% ÷ +13,00%
$\sigma_{a,dol}$ $\sigma_{c,hor}$, elasticity	96,00%	89,00% ÷ 90,00%	+6,00% ÷ +7,00%
probability of failure Pf(i)	Pfd(i)=0,000072	(1,12*10-5 ÷ 1,375*10-5)	Probability of failure < Pfd(i)
% Utilization, fault - is Pf (i)	-	(94,00% ÷ 97,00%)	Values at the extreme limits
% Utilization, fault - not Pf (i)	Pf(i)=0	(cca 90,00%)	Recommended value assessment.

Acknowledgement

The research work was supported by the Project SGS-2016-038.

References

[1] P. Marek, M. Guštar, T. Anagnost, (1995): Simulation Based Reliability Assessment for Structural Engineers,. CRC, Boca Raton, Florida, U.S.A., ISBN 0-8493-8286-9.

[2] P. Marek, J. Brozzetti, M. Guštar, P. Tikalsky, E. Editors (2003): Probabilistic Assessment of Structures using Monte Carlo Simulation. Basics,Exercises, Software,Second edition, ITAM CAS, Praha.

Effect of Shielding Gas on Residual Stresses and Mechanical Properties of Laser Welding of S355 Steel

K. Kolařík [1], N. Ganev [2,a], K. Trojan [1], J. Sís [3], S. Němeček [4]

[1] *CTU in Prague, Department of Solid State Engineering, Faculty of Nuclear Sciences and Physical Engineering, Trojanova 13, 120 00 Prague 2, Czech Republic*

[2] *TU of Liberec, Institute for Nanomaterials, Advanced Technologies and Innovation, Studentská 1402/2, 461 17 Liberec 1, Czech Republic*

[3] *SVÚM a.s., Strength Department, Tovární 2053, 250 88 Čelákovice, Czech Republic*

[4] *Matex PM, s.r.o., Morseova 1126/5, 301 00 Plzeň, Czech Republic*

[a] *nikolaj.ganev@fjfi.cvut.cz*

Abstract: The aim of this article is to assess the impact of the behaviour of the molten bath, penetration geometry, mechanical properties and mainly residual stress surface distribution of laser welded joints. In the case of classic fusion welding an exothermic or endothermic effect of the used shielding gas can be observed according to its dissociation energy. The presented experiments verify this behaviour also for laser welding, which could affect much higher rate of melting and cooling in compassion with other classic arc welding methods.

Keywords: laser welding; shielding gas; S355 steel sheets; XRD; residual stresses.

1 Introduction

While in the case of conventional arc welding methods effects of shielding gases are already described, literature for laser welding is not adequately documented. Firstly, the parameters of fibre diode laser welding were optimized to obtain stable keyhole. After that the aim of the research was to describe the influence of classic shielding gases taken from arc welding methods to the width of the weld and heat affected zone (HAZ). The effect of gases on the resulting surface residual stresses and possible changes in mechanical properties, which have a significant effect on the fatigue life under dynamic fatigue stress [1], was also investigated. Another aim was to describe whether the higher cost of shielding gases provides sufficient added value of more expensive compounds. From the records of speed camera during welding, see Fig. 1, different plasma formations were demonstrated. The smallest plume of plasma and the highest stability was observed using He, which has a higher thermal conductivity and the highest ionization potential [2]. To avoid loss of the power input, the minimum content of 30% He is needed in the gas mixture for the elimination of the formation of plasma plum above the welding surface for laser technologies according to the literature [3]. The use of pure argon as a shielding gas in arc methods leads to a reduction in penetration and degradation of weld quality due to the low thermal conductivity of argon [4]. Gas mixture using active components CO_2, O_2 are used mainly in arc methods for reducing the surface tension of molten metal due to formation of oxide surface layer, which stabilizes the arc [4, 5]. The impact of individual gases and gas mixtures are not fully understood for fibre diode laser welding yet.

2 Experimental

Altogether, six welded plates $150 \times 200 \times 4.6$ mm^3 were prepared from S355J2 steel sheets using a 3.2 kW laser with 0.5 m/min beam speed and optimized welding parameters. One side of the samples was ground before welding and the other side was not machined. Different shielding gas was used for all samples, i.e Ar, C12X2 (12 % CO_2, 2 % O_2, 86 % Ar), He3H1 (3 % He, 1 % H_2, 96 % Ar), C6X1 (6 % CO_2, 1 % O_2, 93 % Ar), He50 (50 % He, 50 % Ar), He. The X-ray diffraction (XRD) measurements were performed by a PROTO iXRD COMBO diffractometer with ω-goniometer and *{211}* diffraction line of α-Fe was measured by CrKα radiation. Residual stresses (RS) were calculated from the lattice deformation by using generalized Hook's law. RS were obtained in both the directions, i.e. perpendicular and parallel to the weld. Aside from stresses, macro-photos of welds macrostructure and heat affected zone (HAZ) were taken. Furthermore, welds were subjected to impact test and tensile test in SVÚM, a.s. Pictures of welding process with different shielding gases from speed camera were taken in the company MATEX PM.

3 Results

Selected pictures of welding process with different shielding gases from speed camera are demonstrated in Fig. 1. Macrostructure photographs of laser welded joints using different shielding gases are presented in Fig. 2. Surface residual stresses obtained by XRD on all six samples are depicted in Figs. 3 and 4.

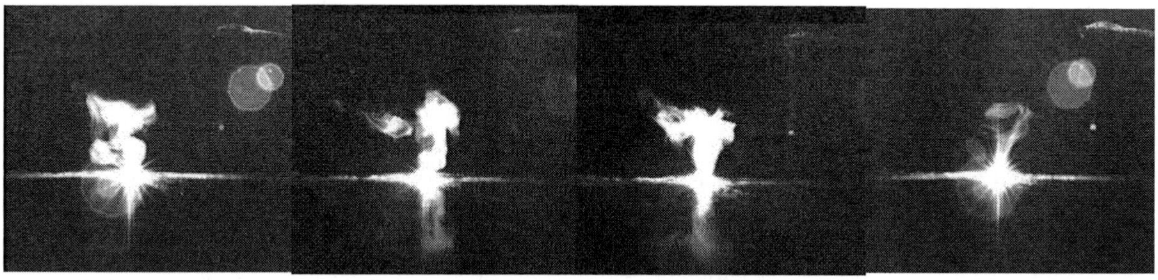

a) sh. gas Ar b) sh. gas C12X2 c) sh. gas He3H1 d) sh. gas He

Fig. 1: Records of welding process with different shielding gases from speed camera.

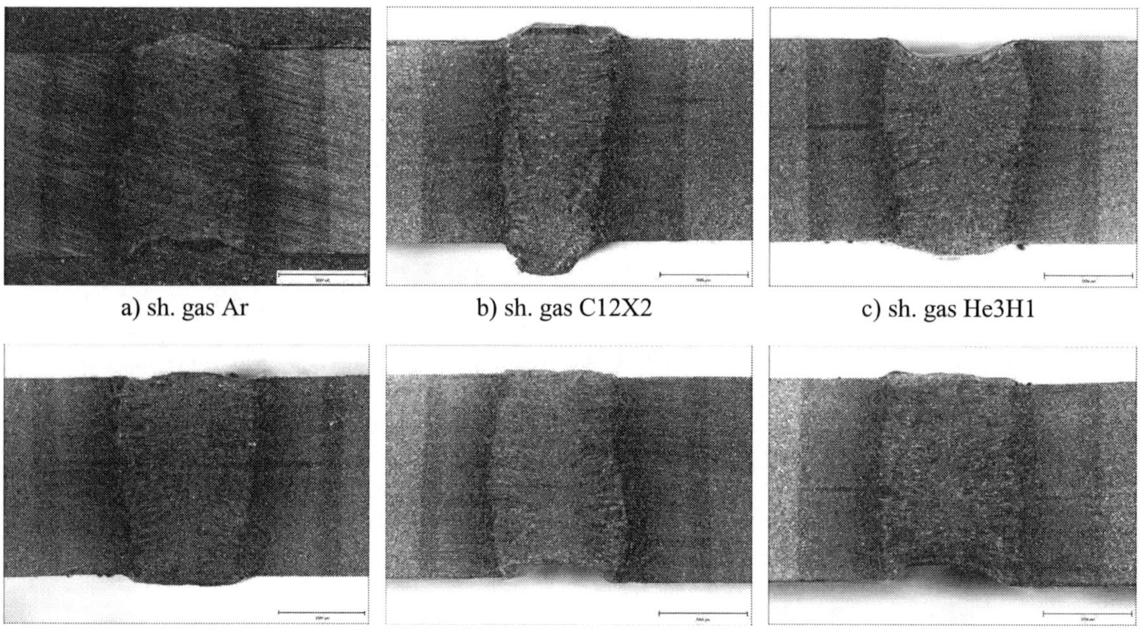

a) sh. gas Ar b) sh. gas C12X2 c) sh. gas He3H1

d) sh. gas C6X1 e) sh. gas He50 f) sh. gas He

Fig. 2: Macrostructure photographs of laser welded joints using different shielding gases; ground sides are at the bottom of the welds.

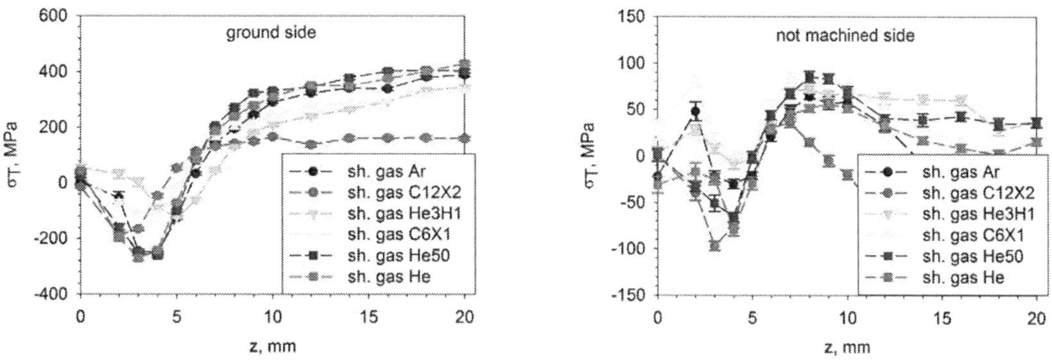

a) ground side b) not machined side

Fig. 3: Comparison of surface residual stresses gradients (thickness of the irradiated area approx. 4 μm) in the direction perpendicular to the weld for both sides of samples.

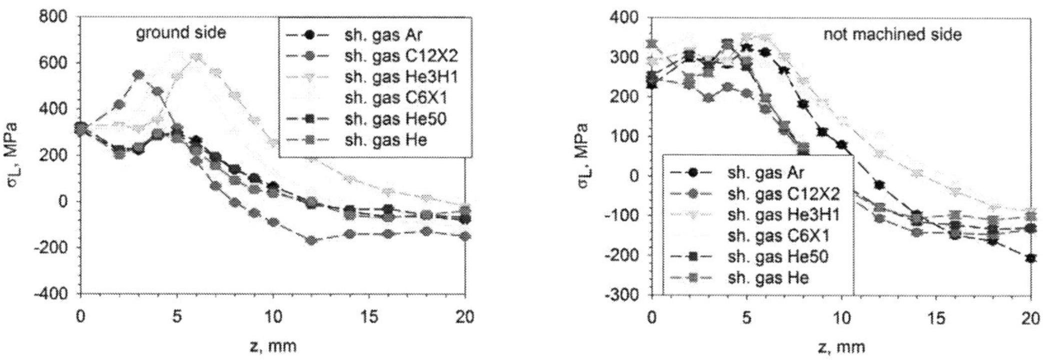

a) ground side b) not machined side

Fig. 4: Comparison of surface residual stresses gradients (thickness of the irradiated area approx. 4 μm) in the direction parallel to the weld for both sides of samples.

4 Discussion

All used shielding gases meet the minimum requirements of weld quality in terms of notch toughness and hardness. The results of the impact bending tests show that regarding strength currently the most widely used shielding argon gas appears to be the best. There was not significant effect of shielding gas to the width of welds and HAZ from the macrostructure photographs of laser welded joints (see Fig. 2). But there was some change in the actual geometry of the weld itself, which is reflected at the position of maximum of RS observed on both the analysed surfaces. A greater difference in the width of the welds would probably occur when higher laser power input is used, which could highlight different ionization potential of the individual gases.

The gas C12X2 reported narrowest HAZ according to macrostructure photographs (Fig. 2) and values of RS (Figs. 3 and 4). Furthermore, the gas C12X2 reported the best result of fatigue life test. Not machined side for gas C12X2 also has the highest observed surface compressive residual stress. In contrast, sample welded using shielding gas C6X1, which contains half of the content of active components, reported according to Fig. 2 wider fusion zone and HAZ. Surface residual stresses of the sample C6X1 are in all cases qualitatively higher compared to sample C12X2, see Figs. 3 and 4. The sample welded using shielding gas C6X1 also exhibits lower values of the fatigue life. Using gas with hydrogen has led to a similar course of the RS as the gas C6X1 but resistance to fatigue damage was unfavourable. Hydrogen is generally not recommended for ferritic steels, since it causes hydrogen cracks in the welds [5].

Ground sides in comparison with not machined surfaces have different gradients of surface RS in the HAZ but RS have a similar pattern in the fusion zone, see Figs. 3 and 4. The growth of macroscopic RS in the direction parallel to the weld in the HAZ and in the direction perpendicular to the weld outside the HAZ compared to the not machined side is caused by grinding technology. During interaction of the cutting tool (grinding wheel), tensile stresses are formed. Due to the superposition of RS resulting from grinding and substantial contraction of the cooling plate along the weld, the tensile RS on the machined side of the HAZ in the direction parallel to the weld are formed and the compressive stresses are formed in the second direction. On the not machined side of the samples, of course, the superposition is not observed. The macroscopic surface RS in the direction perpendicular to the weld on not machined sides are in comparison with the ground side qualitatively lower and their values are in the range from –100 to +100 MPa. This may be important when the cracks are initializing during the fatigue life tests, which were performed in SVÚM.

The macroscopic residual stresses observed in direction parallel to the weld are on the not machined surface for all the gases similar and on the ground side is a noticeable drop in surface residual stresses for gases He, He50 and Ar. However, gas mixtures using hydrogen or a low content of the active components exhibit unfavourable results of RS at the interface of the HAZ and the base material. Shielding gas mixtures using helium (He, He50) showed on the not machined side in direction perpendicular to the weld similar progression of surface RS in the fusion zone and HAZ as when is used only pure Ar as a shielding gas.

Consequently, further experiments are necessary to describe different process of formation of surface residual stresses and the associated different results of fatigue life for the shielding gas C12X2. Observable difference in weld geometry, see Fig. 2, and the results of surface RS and fatigue life for gas C12X2 is probably due to sufficient positive effects of active components in shielding gas. These components sufficiently modify the surface tension of the molten metal and thus completely penetrated weld was formed without an unfavourable undercut and also without a high overlap.

5 Conclusion

The effects of different shielding gases for laser welding of S355 steel plates to residual stresses and mechanical properties were investigated. Generally speaking, the most suitable distribution of residual stresses and the best fatigue life were achieved with a shielding gas C12X2. On the contrary, the unsatisfactory results showed the joint welded using gas He3H1. The use of pure inert gases or their combination showed similar macrostructure photographs, results of RS and mechanical properties.

Acknowledgement

This research was was carried out in the frame of the research projects LO1201 of the Ministry of Education, Youth and Sports in the framework of the targeted support of the "National Programme for Sustainability I" and by the Grant Agency of the Czech Technical University in Prague, grant No. SGS16/245/OHK4/3T/14. The authors thank MATEX PM for supplying the laser welds.

References

[1] Pan, Lingyun, et al., Welding residual stress impact on fatigue life of a welded structure, Welding in the World 57.5 (2013) 685–691.

[2] Quintino, L., et al., Gas shielding in fibre laser welding of high strength pipeline steel, Science and Technology of Welding and Joining 16.5 (2011) 399–404.

[3] Tani, G., et al., The influence of shielding gas in hybrid LASER–MIG welding, Applied Surface Science 253.19 (2007) 8050–8053.

[4] Wang, L. L., et al., Effects of shielding gas composition on arc profile and molten pool dynamics in gas metal arc welding of steels, Journal of Physics D: Applied Physics 47.46 (2014) 465202.

[5] Kah P., Martikainen J., Influence of shielding gases in the welding of metals, The International Journal of Advanced Manufacturing Technology 64 (2013) 1411–1421.

Impact Force Response of Foams and Rubbers

R. Kottner[1,*], T. Mandys[2], J. Bartošek[1], J. Kocáb[2]

[1] *European Centre of Excellence, NTIS - New Technologies for Information Society, Faculty of Applied Sciences, University of West Bohemia, Univerzitní 8, 306 14 Plzeň, Czech Republic*

[2] *Department of Mechanics, Faculty of Applied Sciences, University of West Bohemia, Univerzitní 8, 306 14 Plzeň, Czech Republic*

* *kottner@kme.zcu.cz*

Abstract: Expanded polystyrene, open-closed cell foam, rubber based on synthetic isoprene and butadiene elastomers, and cork/rubber composite were tested in this work. The aim was to obtain the impact force response which can be further used either for an immediate comparison of the material energy absorbing properties or for the validation of material models. Impact tests were performed in a drop tower designed by the authors. To obtain more data for the validation, both displacement and force responses on the impact were investigated.

Keywords: drop test; open-closed cell foam; rubber; EPS; cork.

1 Introduction

Energy absorbing materials have been used in many applications: From packaging of goods to military devices, from civil to aerospace engineering [2]. Producing optimal products requires primarily the utilization of appropriate materials. Certainly, the products must also have optimal shape and dimensions. To find the optimal geometry properties, accurate material models are often necessary [3], [6].

Four materials were tested in this work:

- Expanded polystyrene (EPS), which is a crushable foam commonly used in personal protective equipment [1].
- Open-closed cell foam, which is a modern energy absorbing material having strain rate sensitivity (produced e.g. by D3O®).
- Rubber based on synthetic isoprene and butadiene elastomers (hardness 80 ShA), which is used e.g. as a damping element in rubber sprung wheels [4].
- Cork/rubber composite, which is used usually as the core material in constrained damping layer constructions (sandwich panels) [5], [8]

The aim was to obtain the impact force response which can be further used either for an immediate comparison of the material energy absorbing properties or for the validation of material models.

2 Experimental methodology

The impact tests were performed in a drop tower designed by the authors (Fig. 1). Spherical (s) and flat (f) steel impactors having weight of 2.4 kg were used. Geometry of the spherical impactor is shown in Fig. 2, the flat impactor had a square surface with side length 30 mm. The impactors reached impact velocity of 2 m/s or 3 m/s.

Rectangular samples measured 20×20 mm and had a height (thickness) of 9 mm in case of the cork/rubber composite and 10 mm in all other cases. Density and designation of the materials is listed in Tab. 1. The samples were exposed to drops repeatedly (four times in identical impact velocity). In order to distinguish test conditions in following results, a sample designation consisting of 4 symbols was used. The first symbol denotes material, the second the type of the impactor, the third the impact velocity in meters per second, and

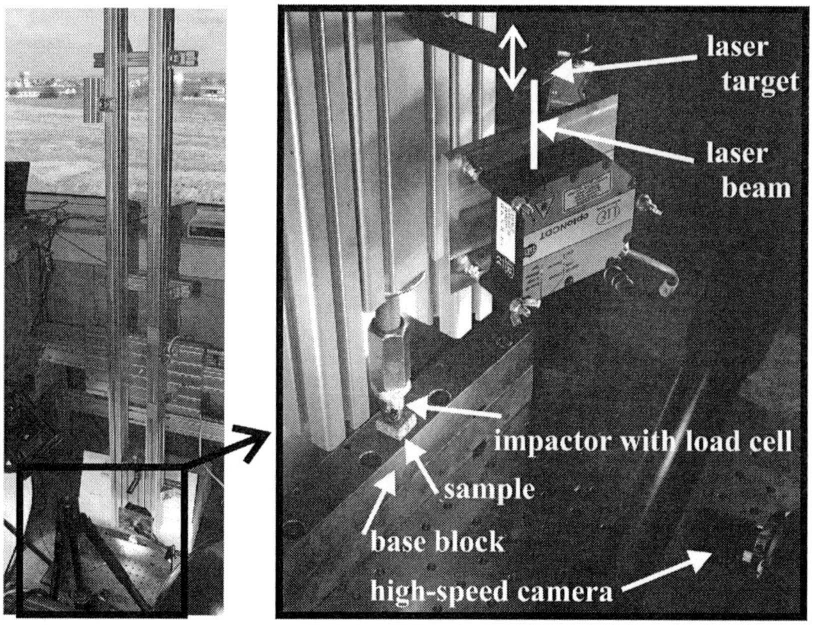

Fig. 1: Drop test.

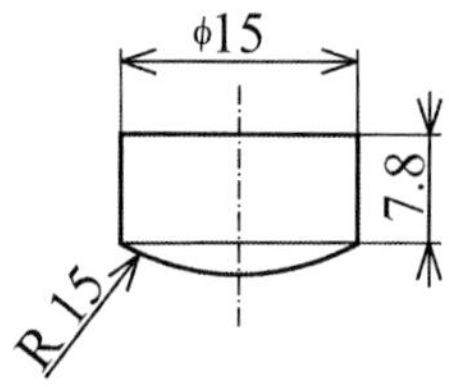

Fig. 2: Geometry of spherical impactor [mm].

the fourth the ordinal number of impact for given impact velocity (e.g. R-s-2-1 is a rubber sample that was tested using the spherical impactor and this sample was exposed to the first impact having velocity of 2 m/s).

To obtain more data for the validation, both displacement and force responses on the impact were investigated. Micro-Epsilon optoNCDT 2300-50 laser and KISTLER 9712B load cell were used. Data from these devices were supplemented with images from high-speed camera OLYMPUS i-SPEED 2.

Tab. 1: Designation and density of tested materials.

material	expanded polystyrene	open-closed cell foam	rubber	cork/rubber composite
designation	E	F	R	C
density [kg/m^3]	100	500	1200	800

3 Experimental results

Fig. 3–5 show displacement and force responses on first impacts. When an impactor touched the steel base block (sample height was zero), the displacement indicated using the laser was zero. Maximum sample compression (Fig. 6–11) and the sample height after the impact can be read from the displacement responses. Moreover, the displacement responses detected rebounds of an impactor within the analyzed time period (0.5 s). When the rebound height (RH) was higher than 38 and 46 mm in case of flat and spherical impactors, respectively, the displacements exceeding those values were not registered due to the laser measuring range. The most important information in the force responses are the peak values. Fig. 12 demonstrates these maximum forces

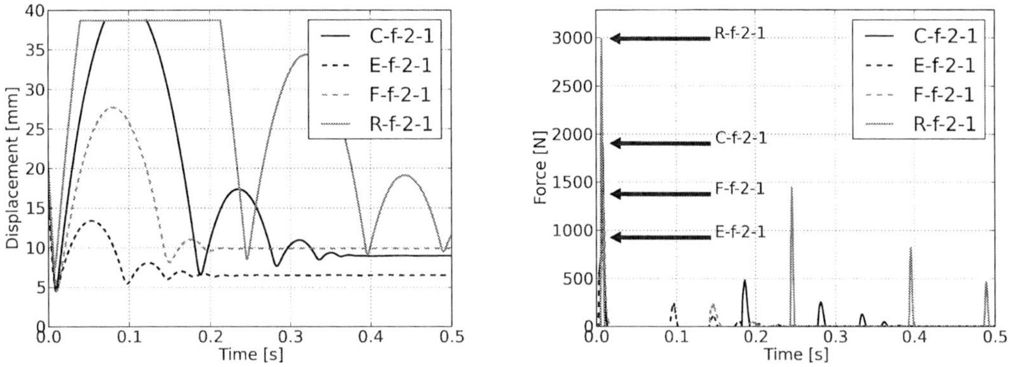

(a) Displacement response

(b) Force response

Fig. 3: Responses on impacts of flat impactor for velocity of 2 m/s.

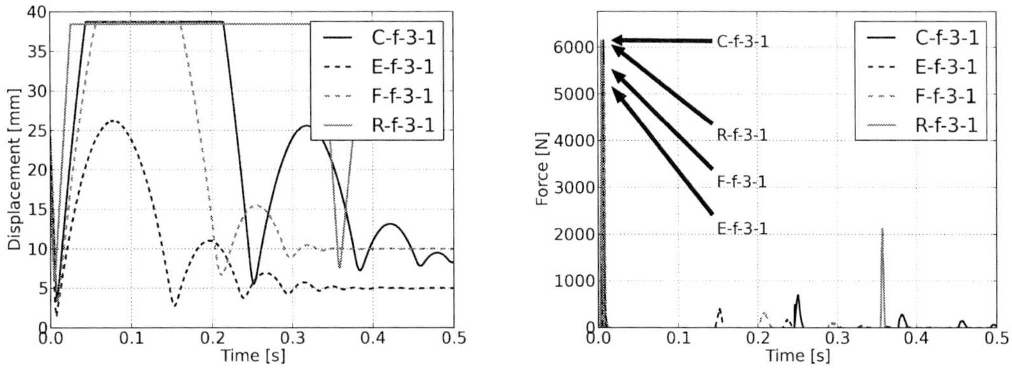

(a) Displacement response

(b) Force response

Fig. 4: Responses on impacts of flat impactor for velocity of 3 m/s.

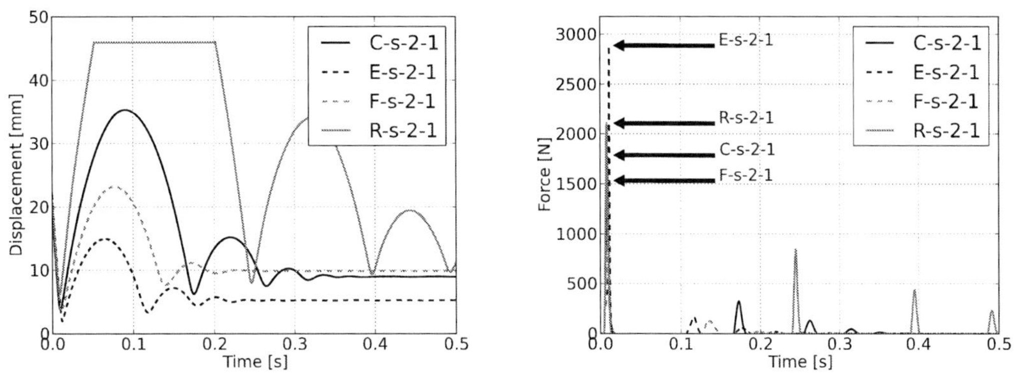

(a) Displacement response

(b) Force response

Fig. 5: Responses on impacts of spherical impactor for velocity of 2 m/s.

(a) Velocity of 2 m/s (C-s-2-1)

(b) Velocity of 3 m/s (C-s-3-1)

Fig. 6: Maximum compression of C samples using spherical impactor.

(a) Velocity of 2 m/s (F-s-2-1)

(b) Velocity of 3 m/s (F-s-3-1)

Fig. 7: Maximum compression of F samples using spherical impactor.

(a) Velocity of 2 m/s (R-s-2-1)

(b) Velocity of 3 m/s (R-s-3-1)

Fig. 8: Maximum compression of R samples using spherical impactor.

(a) Before impact (b) Maximum compression

Fig. 9: Test of E sample using spherical impactor for velocity of 2 m/s (E-s-2-1).

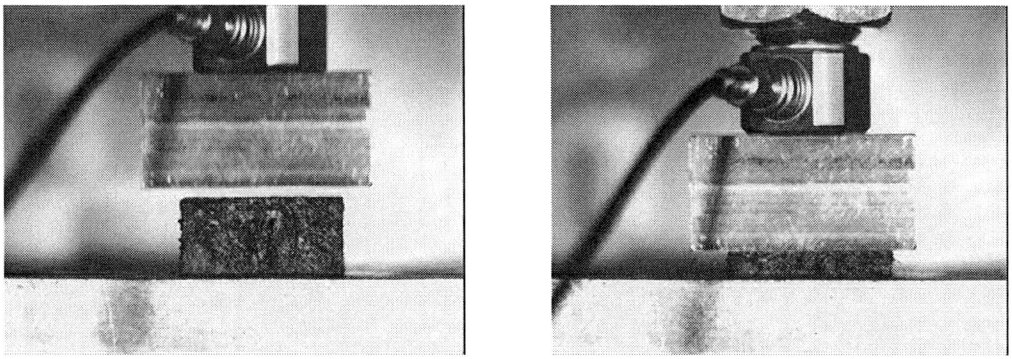

(a) Before impact (b) Maximum compression

Fig. 10: Test of E sample using flat impactor for velocity of 2 m/s (E-s-2-1).

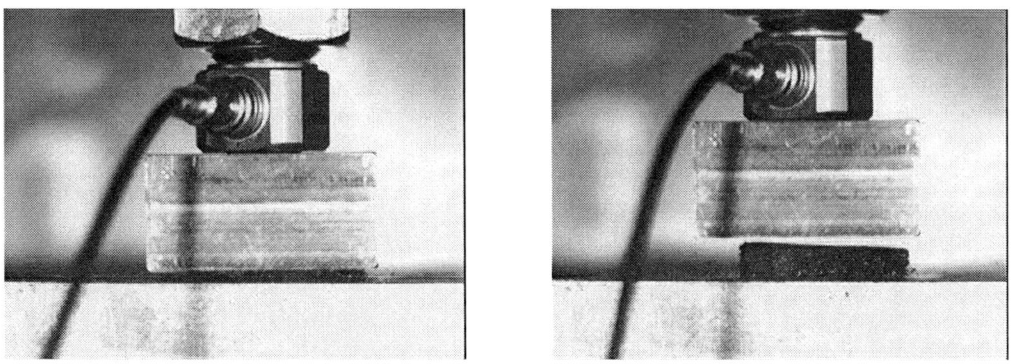

(a) Maximum compression (b) Sample after first impactor rebound

Fig. 11: Test of E sample using flat impactor for velocity of 3 m/s (E-f-3-1).

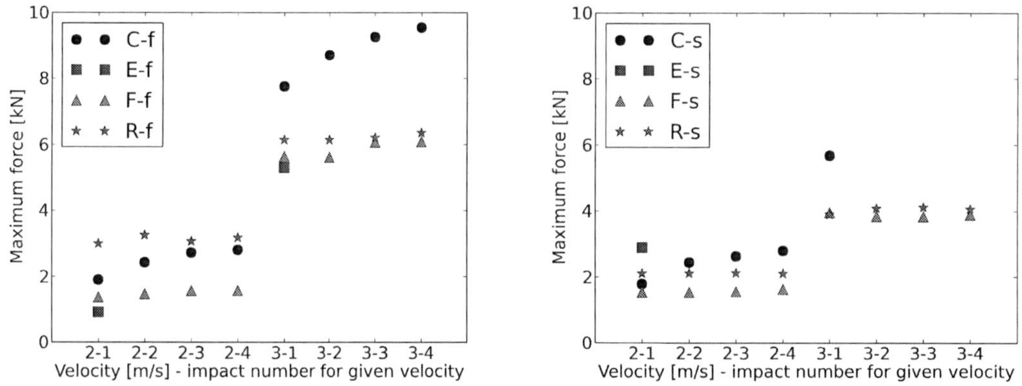

(a) Flat impactor

(b) Spherical impactor

Fig. 12: Maximum force (MF) dependence on impact repetition.

(a) E sample for velocity of 2 m/s (E-s-2-1)

(b) C sample for velocity of 3 m/s (C-s-3-1)

Fig. 13: Samples after test using spherical impactor.

(MF) in case of repeated impacts (excepting EPS; unimpacted E samples were used for each test of EPS).

Lowest RH and MF values were determined in the E-f-2-1 test (see Fig. 3 and 12). EPS exhibited lowest RH and MF values also in the E-f-3-1 test in comparison to other tested materials impacted using the flat impactor and having velocity of 3 m/s. However, when the spherical impactor was used, the MF value of the E-s-2-1 test was the highest (Fig. 5 and 12). The reason for that is that the impact loading was concentrated (diameter of the impactor was too small). The sample thickness between the spherical impactor and the base block in time of the maximum compression (see Fig. 9) was only 1 mm (see Fig. 5(a)). Therefore, E samples were not tested using the spherical impactor for velocity of 3 m/s (danger of the load cell damage). The E sample thickness between the impactor and the base block in time of the maximum compression was also very small when the flat impactor was used (see Fig. 4(a)) and 11(a)), however, the distribution of the loading in the sample was significantly more uniform. Significant plastic deformations of E samples after the tests are obvious from Fig. 11(b) and 13(a). Since majority of the shock-absorbing capacity was exhausted [7], E samples were not exposed to drops repeatedly (danger of the load cell damage).

In case of the spherical impactor (Fig. 5), the lowest MF value was determined in the test of open-closed cell foam. RH and MF values were the second lowest in F-f-2-1 and F-f-3-1 tests (Fig. 3 and Fig. 4). Considering repeated impacts (Fig. 12), F samples responses in all four drops for a given velocity were nearly identical.

The highest values of RH and MF were determined in case of rubber except tests where the impact loading was concentrated or where samples were exposed to drops repeatedly. When R samples were exposed to the impacts repeatedly in tests having velocity of 3 m/s (Fig. 12), the MF had nearly identical values as those of the F samples.

In case of first impacts of the cork/rubber composite, RH and MF values were lower than in case of rubber (Fig. 12). Shock-absorbing properties of C samples in repeated impacts were reduced by damage which occurred in the material (see Fig. 13(b)). When an unimpacted C sample was impacted in the C-f-3-1 test (without any previous test having velocity of 2 m/s), the MF value was 6146 N, which is nearly identical MF value as in the case of other tested materials. Similarly, when an unimpacted C sample was impacted in the C-s-3-1 test, the MF value was 4552 N. In contrast to it, when C samples had been already exposed to four drops having velocity of 2 m/s, the MF values in C-f-3-1 and C-s-3-1 tests were 7758 N and 5683 N, respectively (Fig. 12).

4 Conclusions

Displacement and force responses on the impact of two types of impactors were investigated. Results of the four tested materials (considering used samples thickness) showed that the maximum reduction of impact forces can be achieved using expanded polystyrene despite having the lowest density. However, the impact loading of the EPS sample must be sufficiently distributed into sample volume. Moreover, EPS is not suitable for repeated impacts. In contrast to that, open-closed cell foam and rubber based on synthetic isoprene and butadiene elastomers are suitable for repeated impacts. Significantly lower values of impact forces were determined in tests of open-closed cell foam in comparison to rubber testes when the flat impactor were used and impact velocity was 2 m/s. Cork/rubber composite can be exposed to impacts repeatedly, however, its shock-absorbing properties worsen with each impact. When responses on only first impact were compared, cork/rubber composite samples had better or similar shock-absorbing properties as rubber samples despite having smaller thickness.

Acknowledgement

R. Kottner and J. Bartošek were supported by the project LO1506 of the Czech Ministry of Education,Youth and Sports. T. Mandys and J. Kocáb were supported by the grant project SGS-2016-038.

References

[1] D. H. Blanco et al., Design of Protective Equipment, MRTN-CT-2009-ESR11 MYMOSA, Deliverable 3.2, 2010.

[2] F.A.O. Fernandes et al., Comparing the mechanical performance of synthetic and natural cellular materials, Materials and Design 82 (2015) 335–341, doi: 10.1016/j.matdes.2015.06.004.

[3] F.A.O. Fernandes, R.J.S. Pascoal, R.J. Alves de Sousa, Modelling impact response of agglomerated cork, Materials and Design 58 (2014) 499–507, doi: 10.1016/j.matdes.2014.02.011.

[4] R. Kottner, V. Laš, Numerical Smulation of Behaviour of the Sprung Wheel Rubber Segments, New Rail Trends 15 (2007) 12–16.

[5] Materials Description & Properies retrieved from: http://www.corkhouse.co.il/WEB/8888/NSF/Web/8030/ACM-87.pdf

[6] M.A. Forero Rueda, L. Cui, M.D. Gilchrist, Optimisation of energy absorbing liner for equestrian helmets. Part I: Layered foam liner, Materials and Design 30 (2009) 3405–3413, doi: 10.1016/j.matdes.2009.03.037.

[7] F.M. Shuaeib et al., A new motorcycle helmet liner material: The finite element simulation and design of experiment optimization, Materials and Design 28 (2007) 182–195, doi: 10.1016/j.matdes.2005.04.015.

[8] V. Lašová, J. Vacík, R. Kottner, Investigation of dynamic properties of hybrid laminate structure, Procedia Engineering 48 (2012) 358–366, doi: 10.1016/j.proeng.2012.09.526.

Punch test of breast implant

M. Kovar [1,a], J. Sobotka[2], L. Capek [1],

[1] *Department of Applied Mechanics, Technical University of Liberec*
[2] *Department of Engineering Technology, Technical University of Liberec*
[a] *marek.kovar@tul.cz*

Abstract: The treatment of breast cancer is one of the most significant medical problem. The patient undergoes ablation after diagnosing breast cancer. Another procedure of treatment consists of radiotherapy and chemotherapy. These procedures are very hard to patient psychological state of mind. But the most psychological problem is loss of the breast. The surgeon wants to return the breast for the patient as soon as possible. The field of reconstructive surgery is still open for new approaches from different branches. Tension in resulted scar can be found out by numerical model before the surgery and the surgeon can find better approach for the surgery. The essential task for the customized virtual surgery, the material model of the breast implant is needed. There is no established model for this implant. The aim of this contribution is to perform a punch test in application of the breast implant.

Keywords: silicone gel, FEM, material model, breast implant.

1 Introduction

Virtual customized surgery is one of the trends in the field of clinical biomechanics. This approach enables to predict the surgery result, so that many problems can be avoided before real surgery. The reconstruction of a breast after ablation is one of the clinical applications. The essential task for this virtual surgery is to find out the material model of the breast implant. This contribution is focused on a punch test done on the standardized breast implant.

This work has two parts. The first one is an experiment in the laboratory. Second part is about finding suitable material model and its material constants.

2 Materials and methods

The experimental sample was the round breast implant *McGhan* TSLP235g (textured shell surface, soft touch, low profile plus, size range 235 g).The implant was put on the plane desk and compressed by hemispherical punch with a diameter 25 mm. The velocity movement of the punch was 30 mm/min with total displacement 25 mm. All processes of measurement were captured by an optical system ARAMIS (GOM) for taking true strain on the sample. The frame rate was set on three images per second. The experiment was repeated five times after material relaxation. This measurement is seen on Fig. 1.

The MSC.Marc software (*MSC.Software Ltd., the Czech Republic*) was used for numerical analyses of the breast implant. The material identification was done by material data fitting function. The most suitable material model was chosen Yeoh one. The strain energy function is written in terms of the principal stretches as

$$W = \sum_{i=1}^{3} C_i (I_1 - 3)^i \qquad (1)$$

where C_i are material constants, K is the initial bulk modulus. The original model proposed by Yeoh had a cubic form with only I_1 dependence and is applicable to purely incompressible materials. The initial constants from the material data fitting function are seen in Tab. 1.

Fig. 1: Experiment measurement.

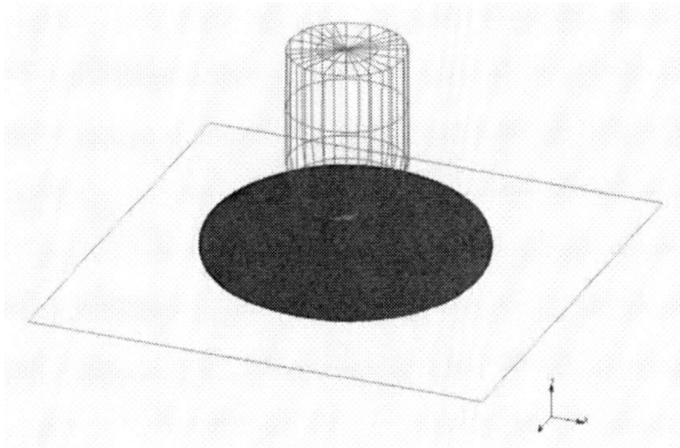

Fig. 2: FEM model.

Tab. 1: Initial material [MPa]

C10	C20	C30
0.00157439	0.000117081	4.3026×10^{-7}

The real geometry of the breast implant was taken by 3D scanner NextEngine before measurement. This scanner based on a laser beam technology. The geometry from 3D scanning is not complete due to light reflexing during scanning. The scanned surface was repaired in Geomagic software. The volume model of the implant was imported into FEMAP preprocessor software. The meshing of the model was done by 96 887 tetrahedral elements. The preprocessing of the model was done in MSC.MARC too, Fig. 2. The material model was defined as above. The contact between the punch and the breast was defined as with friction Coulomb bilinear friction model. The friction coefficient was set to 0.5 values. The punch and plane desk are defined as a rigid body.

3 Results

The comparison of force displacement curve from FEM analysis and the experimental one can be seen on figure 3.

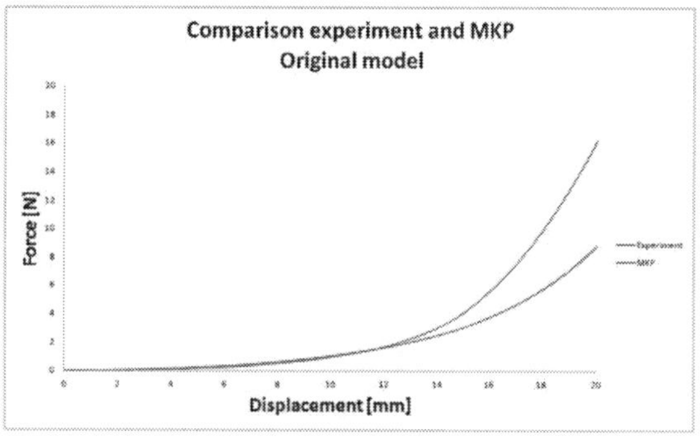

Fig. 3: Initial material data fitting

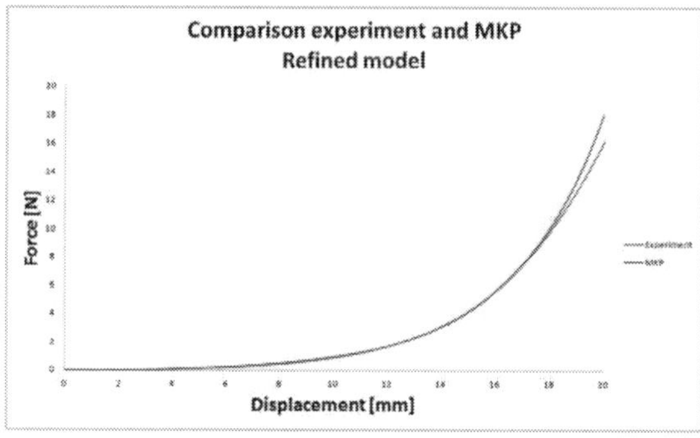

Fig. 4: Updated material data fitting

In case of initial material data fitting the error between experiment and FEM model was very significant. The value of error reached up to 60 %. Updated material constants are in Tab.2. Comparing experiment with FEM model after optimization is on the picture 4. The new value of the error is less than 15 %. This inaccuracy is supposed to be sufficient, because in real operation with silicone breast implants such large deformations are not expected.

Tab. 2: Updated [MPa]

C10	C20	C30
0.000174393	0.000467081	4.3026×10^{-7}

Comparing force displacement curve of the punch is just one part of the task in case of structural large deformation. In this part of the work the frozen deformed FE mesh was compared with data from ARAMIS system. The inspection was done in software Geomagic, Fig. 5. There is only rhomboid part of full screen where the best results are. Geometry accuracy between real model and FEM results are on the picture 6. Length differences are from 0.293 mm to -0.558 mm.

Fig. 5: Showing captures part in experiment.

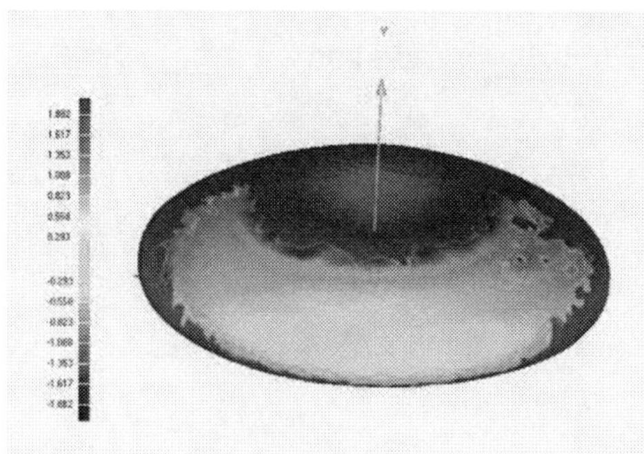

Fig. 6: Comparing geometry from experiment and FEM.

4 Conclusion

Modelling of breast augmentation after ablation due to a tumor is interesting not from the research point of view, but mainly from the social demand of the task. The essential task for this virtual surgery is the material model used for the breast implant. This contribution is focused on a punch test done on the standardized breast implant. It was shown, that punch test can be used as an experimental approach, and nevertheless the standardized material models are not very applicable for the silicon breast implant. The next goal of the work must be development of the more accurate material model, focused on a viscoelastic behavior. Found material model is in a good agreement with only a round implant. If this material model was used in a punch test of the anatomical shape implant, result was not usable. The highest simplification of the task can be seen in creating the FEM model as an one object. In reality, the silicone implant is composed from two parts. One part is a silicone cover and second part is silicone gel infill. In future work will divided breast implant into two parts and create a material model for each part separately. This approach can be supposed to be a more accurate for any type of a breast implant.

Acknowledgement

This publication was written at the Technical University of Liberec as part of the project "Research on advanced composites materials, polymeric materials, development and simulation of mechanical and mechatronic systems" with the support of the Specific University Research Grant, as provided by the Ministry of Education, Youth and Sports of the Czech Republic in the year 2016.

Stress analysis of the winding head frame for the production of the composites

R. Kovář [1,a], P. Kulhavý [1], A. Lufinka [1]

[1]*Technical University of Liberec, Centre for nanomaterials, advanced technologies and innovation, Bendlova 1409/2, 460 01 Liberec 1, Czech Republic*

[a]*radovan.kovar@tul.cz*

Abstract: New technology for production of composite materials is to wrap the plastic core using fibers with specific properties. Subsequently wrapped core is poured using polymer matrix and thereby allow it to obtain parts with excellent mechanical properties with low weight. For the production of such parts, the winding head is used. The supporting frame of this head is divided and has a C-shaped. Two rotary wheels, which are fixed through bearings in the paths of the frame, carry fibers wound on coils across aluminum rings. Due to the tension that occurs when the fibers are pulled during winding, some deformation of the frame does exists in the area, where the frame is. This work shows the behavior of the frame during the winding process.

Keywords: aluminum frame; tension; winding process

1 Introduction

Currently there is growing production of composite materials having improved mechanical properties in compare with conventionally used materials. For composite frames a method of winding a plastic core using carbon fibers can be used. For this production process construction of winding head is designed. It is composed in basic of an aluminum frame and two rotary rings which are fitted with coils. On the coils are wound fibers. The rotating wheels are mounted in the frame by means of bearings. During winding the fibers are fixated on the surface of the core. By dragging cores through the head a force in the fibers is generated, which is transmitted to the head structure as shown in Fig. 1. The force is caught in the construction on an aluminum frame. Deformation of the frame depends on value of the force in combination with speed of rotation of the rotating rings.

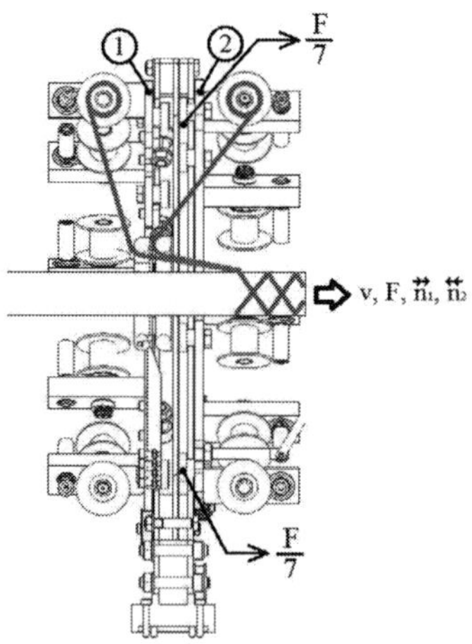

Fig. 1: Scheme of the winding head.

During the process of winding the core is pulled through the center of the winding head under force of 100 N. This force is given by dragging fibers over aluminum rings and brake resistance in the coils. Through the fibers, the force comes to the rotary wheel. Rotary wheels have speed 36 RPM and they have different direction of rotation. Each wheel is fitted with seven bearings, which are fixed in the paths located in support frame. This leads to deformation of the supporting frame, which is exposed to a cyclic stress. The supporting frame is divided and his shape looks like number C. During the process bearings go across the space in the frame and they do an impact on his edges, as shown in Fig. 2. At this moment the shape of the supporting frame is burdened by stress in direction of head axis.

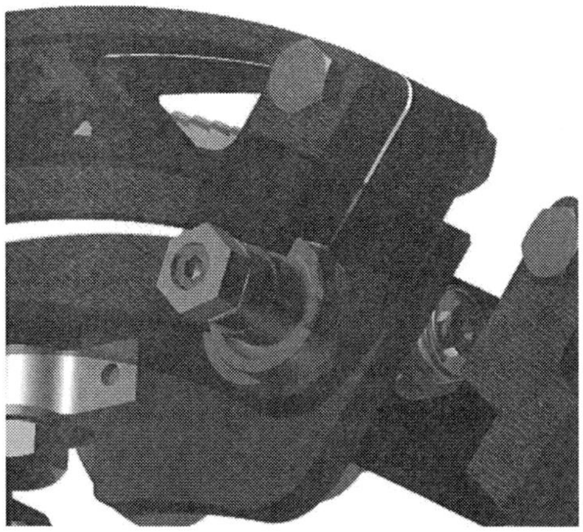

Fig. 2: View on the bearing making impact into the groove of the frame.

2 Material and methods

2.1 FEM analysis of the base frame

Based on the CAD geometry a FEM model has been built [1, 2, 3]. Result of the analysis shows, how the frame behaves during the process. Frame arms are loaded by cyclic bending and vibration. Vibrations caused by impact of bearings on the frame with frequency 0, 2 Hz. Results of FEM analysis are shown Fig. 3 and Fig. 4. Fem analysis showed the places where the greatest deformations occur during the winding process. Frame arms are loaded by the cyclic tension in the direction of base frame axis. This arms are created, when the base frame is gonna be divided. In this case it is possible to wrap closed frames. Areas, where it is possible to find the highest values of deformation, are placed at the begininnig of the arms. Diameter of the inner ring is 300 mm. So when the bearings act on the ends of arms, it is possible to measure behavior of both arms in the direction of base frame axis.

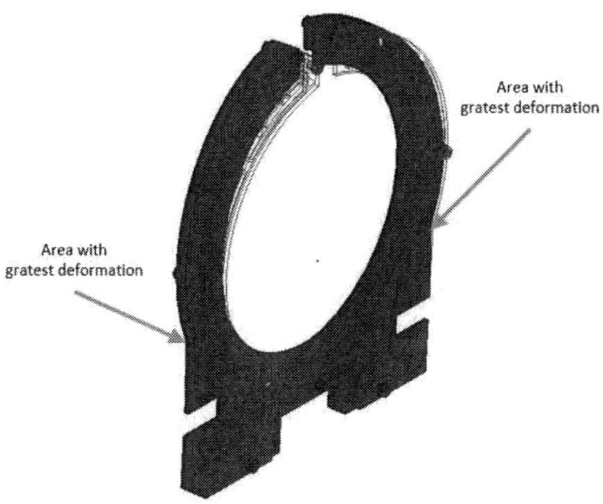

Fig. 3: Isometric view of the model.

Fig. 4: Top view of the model.

During the process, arms of the base frame are loaded by bending stress in the direction of the wrapped core motion. Bending is invoked by braking coils rotation. The size of friction indicates the tension in layed fibers. Fibers transmit power to the two rotating wheels, which are caught on the base frame. Base frame has a sandwich construction type and is composed of six aluminum sheets. Combining of aluminum sheets give solid structure with two arms, forming a circle. The circle is divided, however, giving rise to the possibility of increased stress in both arms. Arms are mostly strained at their ends, where they are bend. At the same time there are impacts of bearings into ends of base frame arms. This reality is given by divided design of the base frame. During the process, bearings are going in orbit paths created in the base frame. Where the frame is divided, bearings are going into a free space. When the bearing try to return back into path, it occurs an impact. This activity creates impulses in a bending stress. Bending stress is given by Eq. (1), Eq. (2).

$$\sigma = \frac{M_o}{W_o} \tag{1}$$

$$M_o = F.r, W_o = \frac{h.b^2}{6} \tag{2}$$

Where F is a force caused by the coils braking system, r is length of arms, W_o is section modulus in bending.

2.2 Experiment

According to the simulation results an experiment was compiled. Experiment include measuring of the base frame deformation during the winding process. This indicates show optimal settings of components having an influents on the winding process.

To determine the deformation value of the base frame, an experiment was performed. Strain values were determined using two strain gauges, which have been glued to place with the occurrence of maximum deformation [4], as shown in Fig. 5. These places were found using FEM simulation. For testing two strain gauges of the type 3/120LY41 were used. Properties list of the strain gauge show: sensitivity +0.2%, nominal resistance $120\Omega \pm 0.3\%$, gage factor tolerance $2.02 \pm 1\%$. Strain gauge measures the elongation by Eq. (3).

$$\varepsilon = \frac{\Delta l}{l} \tag{3}$$

Where l is the length of the sample. By the elongation value it is possible to calculate the size of bending stress using Hook's law. The experiment was performed during the real winding process for 5 seconds. This period represents the displacement of the head in the direction of its own axis in 1 meter. During that time, execute each rotary wheel 3 turns.

Fig. 5: View on strain gauges glued on the base frame.

3 Results

The output from the strain gauges are the values of elongation. With Hook's law these values can be converted to values at the bend. This makes it possible to observe the behavior of the base frame arms during the winding process. The graphs placed below shows the tension at locations where the strain gauges are glued. Bending of both arms are in the same direction. Because each strain gauge was stuck on the other side of the frame, they are in one graph negative values (pressure) and the second positive (tensile). The average value of elongation in the graph showed in Fig. 6 is -22. The average value of elongation in the graph showed in Fig. 7 is 5. From the results it is seen that both arms are not stressed equally. This may cause gravity. Because the head is during the winding process in a vertical position, first arm is above the second one. The upper arm is used only as a guide for bearings. The lower arm bears the weight of rotating wheels. Bearings thus increasing rigidity of lower arm. In both graphs there are shown some kinds of jumps that are created due to impact bearings on the frame body. During the three turns of ration wheel, it is seen in one arm 21 impacts from the graph. In two arms it is therefore absorbing 42 impacts. To optimize the operation of the winding head would be advisable to colligate frequency of bearings impacts on both rotary wheels. Technically it is possible, because the spacing between each bearing on both rotating wheel are provided identically. Then, the frame structure will absorbed only half number of impacts.

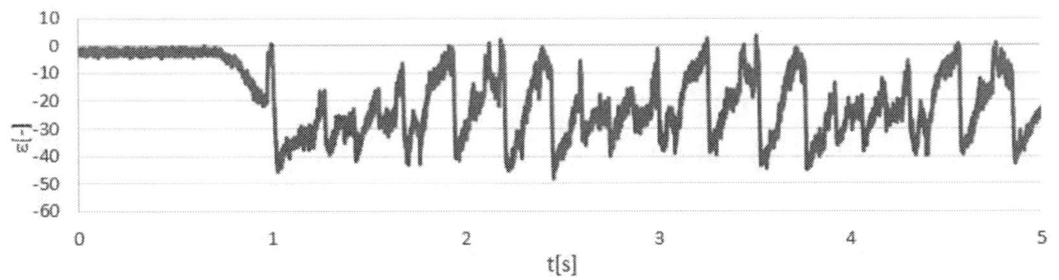

Fig. 6: Strain gage on the upper arm (pressure).

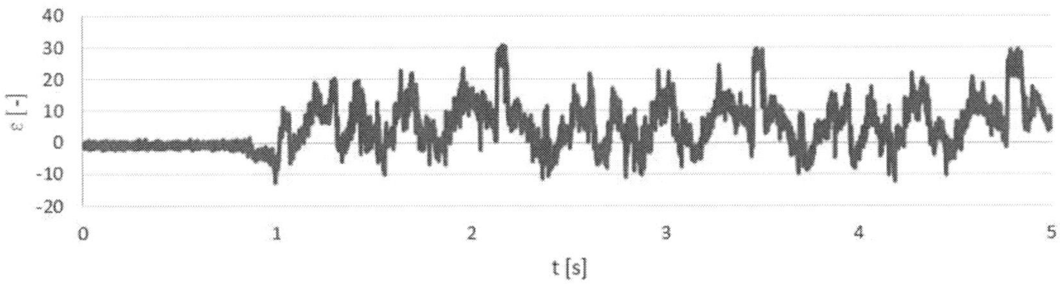

Fig. 7: Strain gage on the lower arm (tensile).

Acknowledgement

The results of this project LO1201 were obtained with co-funding from the Ministry of Education, Youth and Sports as part of targeted support from the "Národní program udržitelnosti I" programme.

References

[1] B. Teplý, S. Šmiřák, Pružnost a plasticita II, Brno, CERM, 2000

[2] V. Kolář et al., Výpočet plošných a prostorových konstrukcí metodou konečných prvků, Praha, SNTL, 1979

[3] V. Kolář, I. Němec, V. Kanický, FEM Principy a praxe metody konečných prvků, Praha, Computer Press, 1997

[4] P. Macura, Experimentální metody v pružnosti a plasticitě, Ostrava, Skripta VŠB – TU Ostrava, 2001

Force Gauge for Measuring Actual Force during Measuring Bucket Calibration

P. Křibala [1, a], J. Kamenický [1], V. Dynybyl [1]

[1] *Czech Technical University in Prague, Department of Designing and Machine Components, Czech Republic*

[a] *petr.kribala@fs.cvut.cz*

Abstract: The article describes design of force gauge for experimental calibration of measuring bucket. The force gauge uses semiconductor strain gauges for measuring actual calibration force loading measuring tooth on the test stand. The force gauge was tested on test stand in laboratory.

Keywords: force gauge, measuring bucket, test stand, semiconductor strain gauge

1 Introduction

Test stand [1] is used for calibration of measuring bucket which measures digging forces. Test stand is shown in Fig. 1. The force gauge is used for measuring loading force on the test stand in laboratory. Each bucket tooth is loaded by hydraulic motor which presses to the arm. The force gauge with contact bolt is between the arm and the measuring tooth, it is shown in Fig. 2. At the first dimensions and shape of force gauge were designed after that shape and dimensions were modified due to FEM analysis. Final force gauge shape was tested by experimental measurement on test stand.

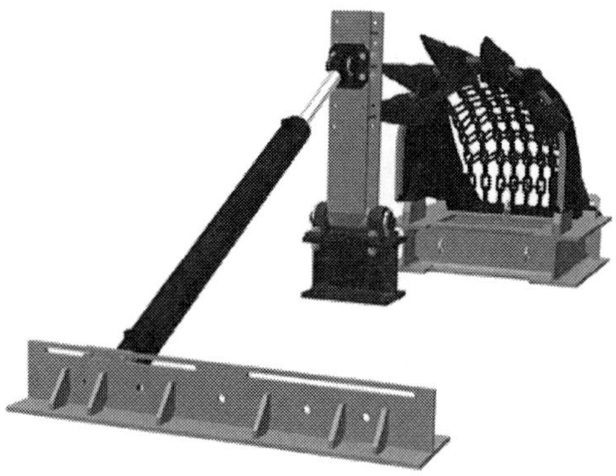

Fig. 1: Test stand for measuring bucket calibration

Fig. 2: Force gauge between measuring tooth (on the left) and arm (on the right)

2 Design and Calculation

2.1 Design

Main part of the force gauge is shank (tube) where the strain is measured on outer surface. Force gauge has inner thread on the one side and outer thread with hexagon for attach to the arm on the other side. Loaded force gauge is stressed by pressure. When the force gauge is loaded by eccentric force, the force gauge is stressed by combination of pressure and bending. Force gauge is shown in Fig. 3. Shape and dimensions were modified due to FEM analysis. Normal stress of force gauge loaded by tensional centric force is shown in Fig. 4.

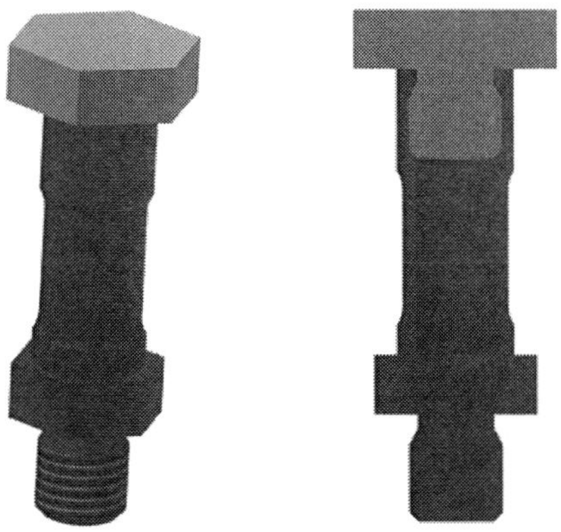

Fig. 3: Force gauge with contact bolt

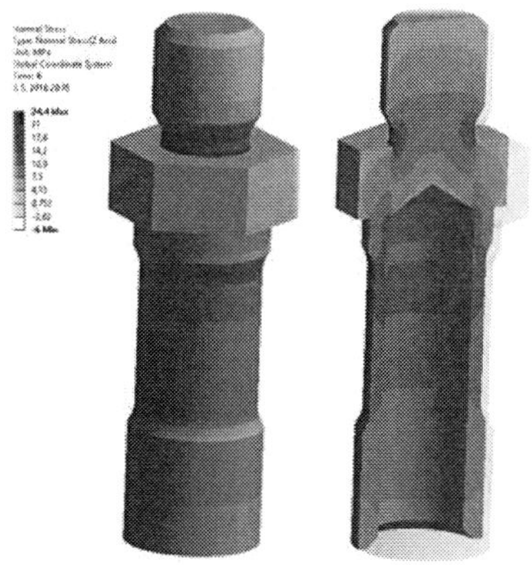

Fig. 4: Normal stress of force gauge loaded by tensional centric force

2.2 Strain gauges

On the force gauge there are four semiconductor strain gauges with support, which are bonded on shank parallel with axis of force gauge. Strain gauges are bonded on shank turned about 90 degrees. Force gauge can compensate bending stress. Semiconductor strain gauge with support is shown in Fig. 5. The force gauge with semiconductor strain gauges and wires is shown in Fig. 6.

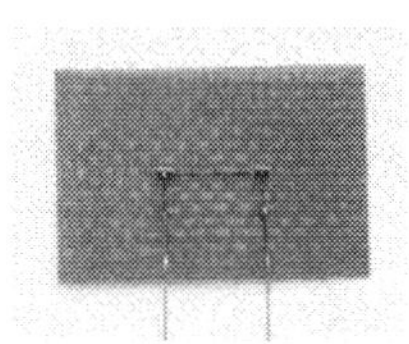

Fig. 5: Semiconductor strain gauge with support [2]

Fig. 6: Force gauge with semiconductor strain gauges and wires

Strain gauges are used for quarter bridges. 60 x higher threshold sensitivity of semiconductor strain gauge (compared with wire strain gauges) makes possible measurement of strain in metals beginning with a value as low as a millionth of one millimeter per meter. [3].

2.3 Experimental measurement

Measurement was carried out in laboratory at CTU in Prague. There were used laptop with measuring PCI Express card National Instruments and measuring device PXIe - 1073 with module PXIe - 4330 (both from National Instruments). Force gauge was attached on the frame of test stand. The rod was screwed into the force gauge. Weights were successively put on the rod and stressed the force gauge. The force gauge was tensioned in this measurement. Test stand for force gauge is shown in Fig. 7 and in Fig. 8.

Fig. 7: The test stand for force gauge - unloaded

Fig. 8: The test stand for force gauge – loaded by weights

Force gauge attachment and its weight loading caused stress by combination of tension and bending. It is possible to calculate tension stress and bending stress because of using four strain gauge. Bending stress isn't mentioned in results because it isn't important.

2.4 Results

Results are normal stress calculated in FEM and measured normal stress depending on loading weight which is shown in Fig. 9.

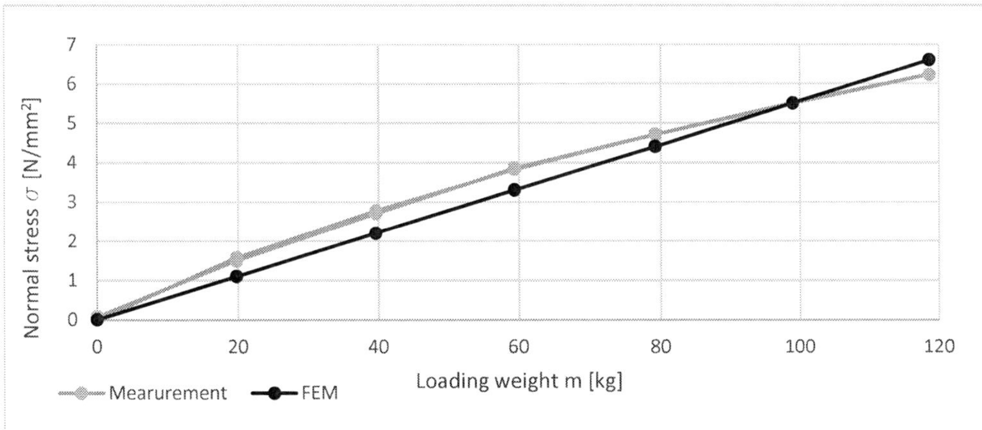

Fig. 9: Normal stress dependency on loading weight

3 Conclusion

The force gauge for calibration of the measuring bucket teeth on the test stand was designed. Dimensions and shape of force gauge was modified according to results of analytical and FEM calculations. The force gauge uses four semiconductor strain gauges for measuring strain during calibration on the test stand. It was discovered by measurement, that the force gauge has moderately nonlinear characteristics. For that reason additional measurement will be done or the force gauge shape will be modified.

Acknowledgement

This project has been supported by Czech Technical University in Prague grant No. SGS14/134/OHK2/2T/12.

References

[1] KŘÍBALA, P., V. DYNYBYL a J. KŘIČKA. Method of Experimental Measurement of a Bucket in Laboratory Conditions. In: *The latest methods of construction design.* Springer International Publishing, 2016, p. 433-436. DOI: 10.1007/978-3-319-22762-7_65. ISBN 978-3-319-22761-0

[2] Retrieved from: http://www.vtsz.cz/images/podlozka-small.jpg

[3] VTS Zlín - strain gauges. 2010. http://www.vtsz.cz/en/strain-gauges.php

Vibration response of composite structures

P. Kulhavý [1,a], J. Petřík [1], P. Srb [1], P. Lepšík [1]

[1]*Technical University of Liberec, Faculty of Mechanical Engineering, Studentská 2, 461 17 Liberec 1, Czech Republic*

[a]*Petr.Kulhavy@tul.cz*

Abstract: Nowadays development of composite materials, optimization of their properties, manufacturing processes and still expanding fields of using devotes constantly through all industry sectors. Pre-saturated fabrics known as prepreg have an indispensable position among the long fiber based composite materials. Their application seems to be very similar to the conventional methods, but in fact they have some specifications especially in view on their manufacturing process. In this article is with using an experiment and numerical model based on ACP pre- and post-processor evaluated the vibration transmission of this way created materials. These were mainly found options by changing the orientation of the individual material layers influence the shape of vibration transmission function and modal characteristics. At reduction of the vibration we are usually trying to improve parameters of created devices including their durability and eliminate resonance at the operating frequency range of the device. The excessive vibrations could significantly disrupt e.g. passengers and driver and this has an impact not only to their comfort but even safety. In real life, of course, we could find countless fiber directions and layers combinations for each applications. However, a four layers material with 2 constant and 2 variables plies was chosen as the default concept of this work. An expected dependence of the plies orientation onto a shape of the transfer function and even swapping of some MOD characters (bending, torsion and their combinations) have been found.

Keywords: composite material; modal analysis; comparison; vibration.

1 Introduction

The use of composite materials has been still growing especially because of their potential to target on a variety material properties to suit actual needs. A material composed of several elements is nowadays called as a composite when each of components still retain its self autonomy. It means that they are never chemically merged or dissolved, even though that in their interaction to environment they looks as a one independent part [1]. Long fiber composites offer generally the greatest flexibility for design specification. As well as knowledge of the basic mechanical properties of laminated materials is for us important to know their vibrational response, dumping characteristics and possibility of their modification by varying composition of the individual plies. Another important feature is the ability to specifically target the resulting synergistic effect - it means the considerable anisotropy of the materials considering for example a direction of the real load as can be seen in Fig. 1. To ensure high reliability of the structures, the actual behaviors of the laminated composite parts must be accurately predicted and carefully monitored. According to [2, 3] the material properties determined from standard specimens tested in laboratory may often significantly deviate from components manufactured in factory.

Reduction of production costs, improving quality and using of modern CAD technologies are the key points in the future of composite parts.

1.1 Material

Carbon fibers pre-impregnated with resin are the main constituent of so called prepregs. These are called "composites of the first degree" which has a dispersion of the solid phase. Used material is an Epoxy Carbon UD Prepreg with thickness of 0.2 mm. For this unidirectional material is the longitudinal Young's modulus in comparison with the transversal significantly lower. With the "rule of mixture" is possible to estimate the longitudinal modulus of a unidirectional long fiber composite with formula (1):

$$E_L = E_f E v_f + E_m v_m \qquad (1)$$

Where E is a Young modulus appropriate to the uniaxial usually quasi-static stress; v_i is a constituent volume fraction and the index L, m and f mean longitudinal, matrix and fiber respectively [2].

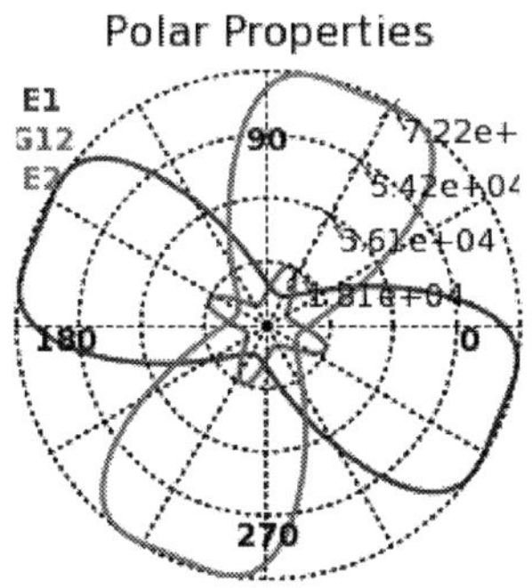

Fig. 1: The basic material properties

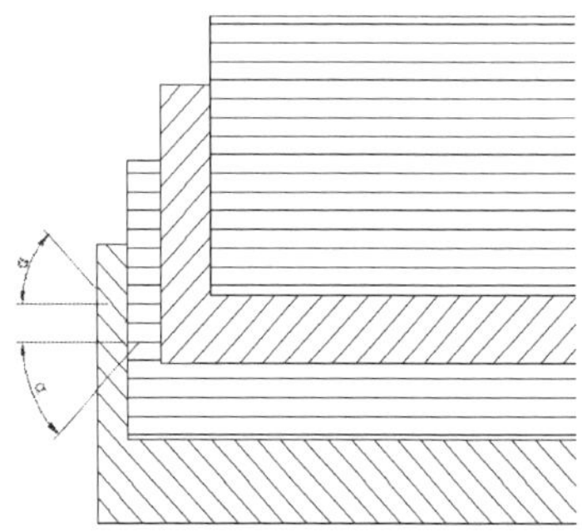

Fig. 2: One of the tested plies layout (0°/45°/0°/-45°)

Samples of size 120 x 40 mm arranged according to Fig.2 and Tab. 1 have been produced from a prepreg sheets and subsequently cured under the pressure and temperature of 150 °C for 15 minutes. The thickness of individual layers after their finishing was 0.16, 0.29, 0.4 and 0.52 mm.

Tab. 1 Composition of the tested materials

Sample nr.	Plies - angle α °			
1	0	0	0	0
2	0	20	0	-20
3	0	45	0	-45
4	0	60	0	-60
5	0	90	0	-90

1.2 Modal analysis

Current destructive mechanical tests usually based on static loading are slow and expensive. As a non-destructive alternative for the classical tensile test can be applied a method based on using of resonant frequencies when the composite body is from one side vibrationally excited and on the second scanned. On the basis of transfer functions can be determined the modulus of elasticity and size of the Poisson constant. According to [4] theories of Lagrange, Euler-Bernoulli, Rayleigh and Timoshenko based on a vibration of thin beams are in this area mostly used.

Modal vibration tests are usually based on the measurement of modal frequencies, damping factors and the mode shapes on the specimens or parts [5]. They have potential quickly and enough precisely provide the basis of rapid and inexpensive characterization of elastic and viscoelastic properties of the tested materials. Generally, the damping in metal structures is low, which results in high amplitudes of the vibrations. The damping of the fiber reinforced composite materials is generally higher and significantly depends on the constitution of materials.

According to Gibson [6], an empirical relation for modal frequencies of a flat cantilever beam is (2).

$$f_n = \frac{(\lambda_n L)^2}{2\pi^2}\left(\frac{EI}{\rho A}\right)^2 \tag{2}$$

Where fn = frequency of n-th mode (Hz), E = modulus of beam material, I = moment of inertia of beam about its neutral axis, ρ mass density of beam material, A = cross sectional area of beam, λ_n = eigenvalue for n-th mode which depends on boundary conditions, L = beam length.

Laminated composite plates have been analyzed by many authors [2, 3, 6]. It is well documented that shear deformation can have significantly influence on the natural frequencies of such plates. The damping property of this kind of plate is usually expressed in terms of the modal loss factor η calculated for each vibration mode of the plate. The loss factor (3), a measure of internal damping, is calculated by applying the half-power bandwidth.

$$\eta = \frac{\Delta f}{fn} \tag{3}$$

2 Experiment

The aim of the experiment is to assess the impact of the individual plies direction of the composite material on to the overall transfer function. To a console the composite plate was mounted with using the clamping jaws (Fig. 3, 4) fixed on three bolts - at the sides and one in the middle [7]. It was necessary to provide polished clamping surfaces for perfect fixing of the sample and to avoid possible undesirable vibrations. Accelerometer Acc1 was mounted near to the jaws and at the free end of the composite sheet was mounted the second accelerometer Acc2. The acceleration arising at Acc1 and its response in Acc2 depending on time for an angle of 20 ° and 60 ° are shown in Fig. 5. The excitation force and its response converted to the frequency spectrum and calculated ratio of these two functions is called the transfer function. In this function doesn't depend on the type of excitation. It is possible to excite harmoniously, randomly or by an impulse and the results of one type of excitation may be used for predicting response of the structure at a different type.

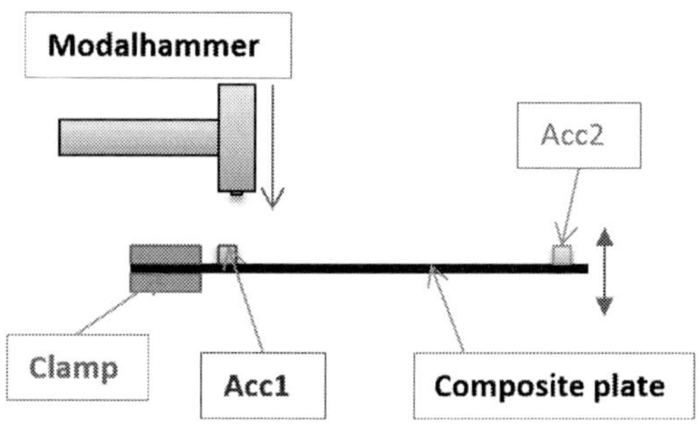

Fig. 3: Scheme of experiment

Fig. 4: Photo of the real device

In our case a modal hammer has been used. Another possibilities of excitation in similar cases could be using of hydraulic systems due to the possibility of a smooth speed control, shock absorbing and movement without "steps" typical e.g. for pneumatic systems due to the compressibility of the medium. According to the analyzed dynamic response, these methods can be subdivided into modal analysis, frequency domain, time domain and impedance domain. According to [9, 8] the excitation and the response signals are digitalized and processed by an analyzer of signals. The damping characteristics of the beams are deduced from the Fourier transform of the response to an input impulse by fitting this experimental response with the analytical response of the beam using the Ritz method.

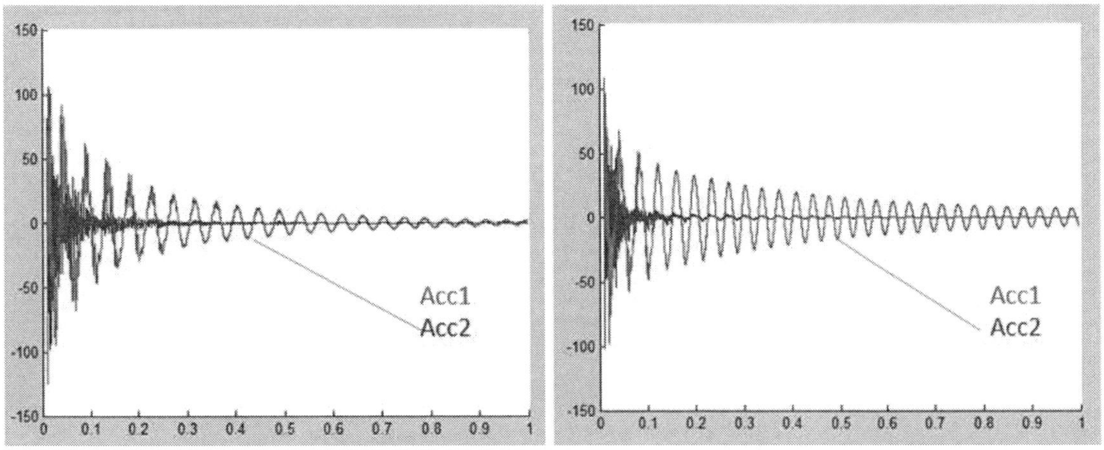

Fig. 5: Acceleration response to the excitation: 0/60/0/-60 (left) and sample 0/20/0/-20 (right)

3 The model section

Based on the CAD geometry and measured material characteristics [1] with using the ACP preprocessor a FEM model (Fig. 6) has been built [9]. The simulation itself was distributed into 2 parts - Modal analysis and frequency response. On the resulting visualizations (Fig. 7), are quite clearly see the individual natural frequencies and example of their swapping from bending to torsion simultaneously with changing the fibers direction.

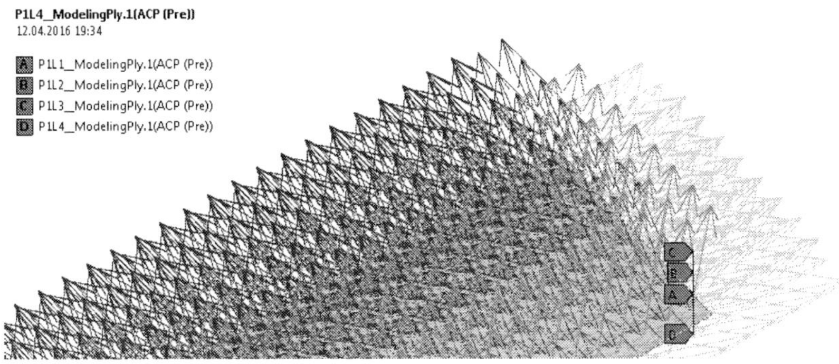

Fig. 6: Directional vectors of the individual plies

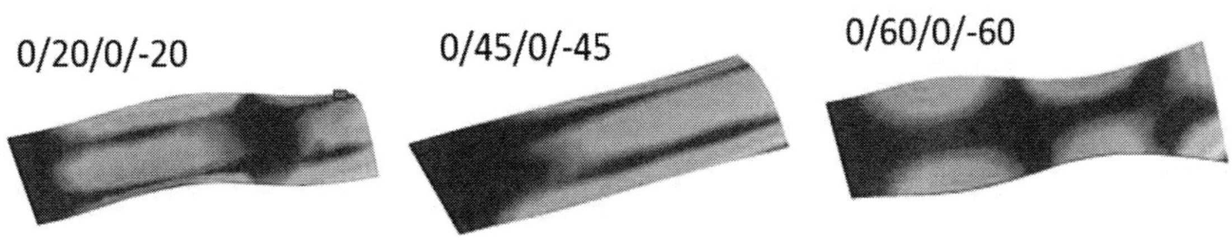

Fig. 7: Example of modifying shape of the fourth mode by changing direction of two from 4 plies

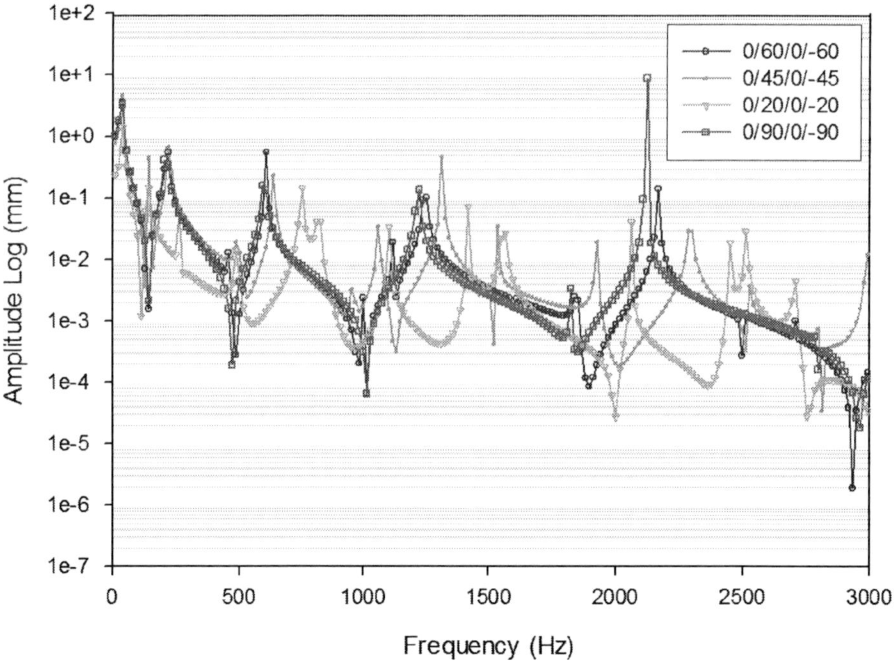

Fig. 8: Transfer function for 4 different plies angles – experimental

The material model was defined as a volume and presented by the necessary tensile and shear moduli of elasticity for each direction, density and Poisson's ratio. The resulting transfer characteristics for various angles of individual plies direction solved in ACP Ansys we can see in Fig. 8 above. Finite element analysis allows us to derive the different strain energies stored in the material directions [9] of the constituents of composite materials, and next, the energy dissipated by damping in the materials and the composite structure can be obtained as a function of the strain energies and the damping coefficients associated to the different energies stored in the material directions [10]. A Finite element model using layer-wise theory [11]. It is assumed that the same displacement distribution in the individual layers is capable of representing displacement discontinuity conditions at interfaces between layers.

4 Results

The experiment is a quick way how to find out the basic modal and transmission characteristics. Then, primarily based on the practical knowledge, it is possible to determine which modes are significant and what are just some e.g. sidebands. The numerical model could give us clear overview of the modes shape, however, their real size and relevance is necessary to verify. As is shown below in Tab. 2 for the selected case of ply orientation 0/45/0/-45° are the numerically and experimentally found eigenvalues quite similar. Significant differences are primarily at low frequencies and perhaps just there is the place where the all imperfections of manufacturing process, mass of the accelerometers and idealization in the material model will be reflected.

Tab. 2: Overview of the found eigenfrequencies [Hz].

Model	35	143	222	640	1061	1314	1536	1925	2295	2295	2988
Experiment	29	144	173	648	1012	1237	1429	1894	2241	2241	3107
Deviation	17%	-1%	22%	-1%	5%	6%	7%	2%	2%	2%	-4%

5 Conclusion

In this presented work, the experimental and numerical determination of the own frequencies and transient response for prepreg plates with 4 layers were carried out. Two layers (1st and 3rd from the top) were oriented still in the longitudinal direction, for the 2nd the angle was variable in range 0; 20; 45; 60 and 90° and the last remaining was invert to the 2nd. It has been found dependency of the transition function, its shape and particle models including their size on to the individual plies orientation. On the basis of the experimentally obtained material data [1, 7] the numerical model has been compiled. Using this model was possible to thoroughly track the process, changes in the shape of individual modes and the response of the whole structure. Results of model simulations and experiment are in good agreement. For composite parts will be error of the simulation every time bigger than for e.g. steel parts. This is caused by many parameters and conditions, that we have to consider, during the part creation (grease, pressure, temperature, imperfect vacuum, real thickness of plies) and this fact, whether we like it or not, will almost always significantly affect the final mechanical parameters. As would be expected and as was confirmed by this work, even if the weight is still the same, the fiber orientation has significantly influence to the overall rigidity. Changing orientation of the fibers we can shift the individual peaks and modify the transmission, even is possible to change character of own frequencies, as for example, change bending to torsion. The presented results are important particularly for the possibility of finding the response of the whole structure due to changes in the angle of individual layers. This could help us in our concepts to shift the modes or avoid developed devices to operate in the critical frequency areas.

Acknowledgement

This publication was written at the Technical University of Liberec as part of the project "Innovation of technical systems structures with the use of composite materials" with the support of the Specific University Research Grant, as provided by the Ministry of Education, Youth and Sports of the Czech Republic in the year 2016.

References

[1] Kulhavy P., Srb P. Syrovatkova M. *Methods for determining the tensile properties of a fiber composite rods*. In: Book of Proceedings of 56th International Conference of Machine Design Departments. Slovak University of Agriculture in Nitra, 2015

[2] Lesieutre G. A., *Modeling Frequency-Dependent Longintudinal Dynamic Behavior of Linear Voscoelastic Long Fiber Composites*, Journal of composite materials, *vol. 28 no. 18* 1994

[3] Wang W.T., Kam T.Y., *Material characterization of laminated composite plates via static testing,* Composite Structures Vol. 50, Issue 4, Elsevier, 2000

[4] Vozkova, P., *Víceúrovňové modelování mechanických vlastností textilních kompozitů*, Disertation thesis, Technical University of Liberec, Liberec, 2008

[5] Society of Automotive Engineers, *Test procedures for automotive structural composite materials*. Sufrace vehicle standard, Automotive Composites Consortium, 1995

[6] Gibson R.F., *Modal vibration response measurements for characterization of composite materials and structures*, Elsevier, Composites Science and Technology 60, 2000

[7] Petrik, J., Kulhavy, P. Pasek, M. Sasek, J. *Vibration and Noise Reduction Analysis of Sheet Metal Structures with Damping Layer,* Applied Mechanics and Materials, Vol 732, 2014

[8] Whitney JM. *Shear correction factors for orthotropic laminates under static load*. Journal of Appl Mech; 40:302-4, 1973

[9] Zou Y., Tong L. and Steven G.P., *Vibration based model dependent damage identification and healt monitoring for composite structures*. Journal of Sound and vibration 230, 2000

[10] Petru M.; Syrovatkova M.; Martinec T.; Lepšík P.: *Analysis of Changes in the Surface Quality of a UD Prepregs Composite due to Mechanical Loading*, Material Science Forum, Vol.818, 2015

[11] Barbero E. J. and Reddy J. N. *International Journal of Solids and Structure* 28.Modelling of delamination in composite laminates using a layerwise plate theory, 1991

Ultrasound gel as suitable tool for simulation of the fiber orientation in the fiber reinforced concrete

R. Lovichová[1,a], J. Fornůsek[1], L. Soukupová[2], J. Valentin[2]

[1]*Experimental Centre, Faculty of Civil Engineering, Czech Technical University in Prague, Thákurova 7, Prague, 166 29, Czech Republic*

[2]*Department of Road Structures, Faculty of Civil Engineering, Czech Technical University in Prague, Thákurova 7, Prague, 166 29, Czech Republic*

[a]*romana.lovichova@fsv.cvut.cz*

Abstract: Presented paper is focused on the creation of transparent viscous liquid substance based on the ultrasound gel to replace fresh mixture of high-strength concrete within some costly experiments. The description of the viscous behavior of both materials and optimisation of ultrasound gel are shown. Two different methods for determining and measuring the viscosity of selected materials are described. The final solution of ultrasound gel is validated by these experimental methods and it is conveniently used for further experiments where an orientation of steel fibers was examined.

Keywords: high performance concrete; ultrasound gel; fiber; orientation; viscosity; viscometer.

1 Introduction

Main aim of this research is to find a way to simply imitate an expensive X-ray analysis of fiber reinforced samples and to verify if there is influence of pouring the containing steel fibers into the formwork [1]. The flow table test was selected as a method for comparison of consistency and workability of fresh concrete mixture and ultrasound gel solution, the rotational viscometer was used for measuring of dynamic viscosity of the both materials. Measured data from these tests were successively applied in the optimization of a transparent gel and its further use when the gel with addition of steel fibers was poured into a special formwork.

2 Materials

Ultra-High Performance Concrete (UHPC) containing fine-grained aggregate up to fraction 1 mm was used for our experiments. The mixture was developed and created by Experimental center of the Czech Technical University in Prague and tested during previous laboratory experiments. It was proved that the fine aggregate cementitious composites are suitable for high performance concrete mixtures [2]. According to this UHPC mixture an aqueous solution of ultrasound gel was prepared. The viscous gel is a transparent and odorless and it is commonly used for medical purposes. Producer confirmed thixotropic behavior of the gel. The fresh concrete and an ultrasound gel are both viscous and thixotropic. However, materials are similar in some characteristics and different in others. The created gel differ in its transparent look, which was our intention. Viscosity and thixotropy have a significant effect on fiber orientation, particularly due to the behavior changes during the action of external forces. The viscosity values of high performance concrete range from 68 Pa·s to 530 Pa·s [3] and the viscosity values of used ultrasound gel lie between 60 Pa·s and 80 Pa·s. Less viscous concrete has the fibers oriented more than concrete with higher viscosity. It also affects the deployment of the fibers in the specimen. Viscosity is an important parameter, segregation may occur if viscosity is not sufficient [4]. These findings of the investigated materials were used as input parameters to create a substance of similar consistency as fresh concrete [5].

Selection of the type of fibers is limited by place and application purpose. The most suitable are high modulus fibers that can be successfully mixed with cement paste or concrete to produce material with the desired structural characteristics. Steel fibers fulfill these requirements and proves to be a very appropriate choice for interaction with concrete especially with UHPC. The length of fiber cross-sectional shape, tensile strength, ductility and interaction are the main factors which determine the final properties of the fibers in concrete [6]. A steel fibers we used are commonly added to the tested UHPC mixture. A high strength steel microfibers with straight shape, length $L_f = 13$ mm and diameter $d = 0.18$ mm and tensile strength of 2400 MPa was used as the dispersed reinforcement in this experimental programme. Microfibers were dosed by volume to the mixture. The dose of the fibers into this UHPFRC mixture is usually 1.5 %. The same amount of steel fibers was applied in an experiment with modified ultrasound gel solution where the orientation of fibers in transparent formwork was investigated.

3 Experimental programme

The purpose of the experiments was to design and create a transparent substance that effectively replaces a fresh mix of UHPC during filling into the formwork. Another aim was to demonstrate the effect of the pouring of ultra-high performance fiber-reinforced concrete (UHPFRC) on a final fiber orientation. And also its influence on the mechanical properties of the specimen.

3.1 Flow table test

The flow table test apparatus was used for optimisation of the consistency and workability of ultrasound gel in accordance with fresh mixture of UHPC. The fresh mixture of UHPC and a pure ultrasound gel were tested at first. Three tests with a fresh UHPC mixture was performed, all with practically same result, the measured diameters of the spillage circles were 16 cm. After the first flow table tests with a pure gel was clear that it is too dense without any water addition. Therefore doses of water was added, each futher test was performed with the addition of a 10 g of water until the diameter of spillage for ultrasound gel solution correspond with the diameter for the fresh mixture of UHPC. Once the final ratio of ultrasound gel and water was found, two more control measurements with a modified ultrasound gel solution followed. An identical result was evaluated for all three tests, the measured diameters of the spillage circles were again 16 cm. The spillage of both investigated materials has been documented after the flow table test, see Fig. 1.

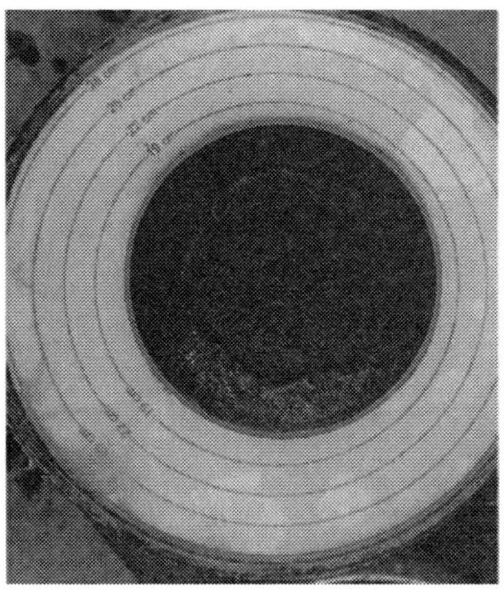

a) Fresh mixture of UHPC

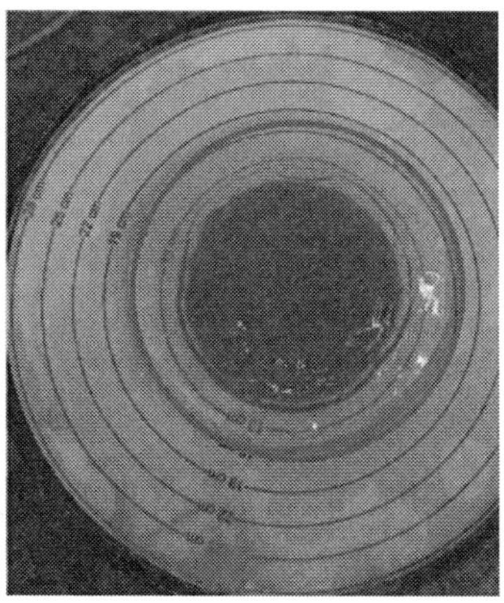

b) Ultrasound gel solution

Fig. 1: Investigated materials after flow table test.

3.2 Rotational viscometer

A rotational viscometer Brookfield DV-II+Pro was selected for practical use in further experimental and computational cases after summarization of the different possibilities of viscosity measurement and suitability of individual methods. The viscometer is based on the principle of rotation of the spindle immersed in the sample with the aid of a calibrated spring. The viscous properties of the fluid indicates the degree of reflection of the spring which is measured by rotary encoder. A spindle "RV05" from a basic set of spindles was utilized for measurement and Rheocalc software was used during the testing for data acquisition and analysis. The comparison and measured values of both materials are described on the Fig 2. The dynamic viscosity of the fresh mixture of cement (FMC) decreases from 300 Pa·s to 10 Pa·s with increasing speed of rotation per minute from 0.5 to 25 and for the ultrasound gel solution (UGS) it was from 400 Pa·s to 10 Pa·s with increasing speed of rotation per minute from 0.5 to 35. These results give a good confidence to use the ultrasound gel solution in the upcoming experiments.

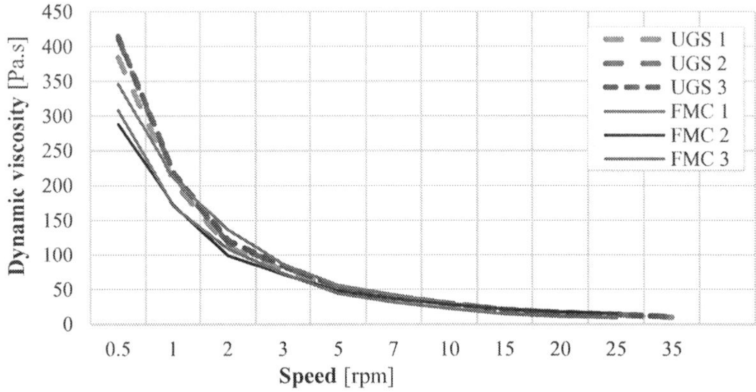

Fig. 2: The measurement of dynamic viscosity of the ultrasound gel solution (UGS) and the fresh mixture of cement (FMC), depending on the number of spindle revolutions per minute.

3.3 Simulation of pouring mixture

Based on the findings and results described above an experiment with an ultrasound gel containing fibers was performed. The created gel mixture was mixed with steel microfibres of the same dose as is usual to add into the UHPFRC mixture. This gel mixture was then poured into a special transparent formworks in the horizontal and vertical direction. An orientation of fibers was observed and studied due to a pouring of mixture and an impact of the formwork walls. See the photos bellow where the formworks filled with fiber ultrasound gel are shown (Fig. 3).

a) Filled horizontal formwork b) Filled vertical formwork

Fig. 3: The orientation of fibers in the matrix of the ultrasound gel after pouring into the transparent formwork.

The fibers in a horizontal formwork are oriented preferentially perpendicular to the direction of the pouring – horizontally (Fig 3 a). There is the visible influence of the formwork walls in their vicinity on the orientation of the fibers. This phenomenon is particularly noticeable in side view where the fibers follow a boundaries of the form on the bottom and on the sides (Fig 3 a – bottom). Mostly random fiber orientation was in the vertical formwork (Fig 3 b). The orientation of the fibers was more in vertical direction along the walls of formwork and more horizontal in the middle of thickness of the prism and close to the bottom. Fiber orientation was affected by formwork walls almost across the entire cross-section due to the small width of formwork.

The simulation of storing the fresh mixture of concrete replaced by transparent material with same consistency proved an influence of arrangement and orientation of steel fibers in a fresh mixture of high-strength concrete.

4 Conclusion

The optimal consistency of the ultrasound gel solution was found with the aid of the flow table test. The final ratio of the ultrasound gel and water was modified to 1:0,566. The dynamic viscosity of the materials was measured on the rotational viscometer Brookfield with spindle "RV05". The measured values are shown on the Fig. 2 and they confirm comparable dynamic viscosity of tested materials. The dynamic viscosity of investigated materials decreases from a values about 350 Pa·s to 10 Pa·s with increasing speed of rotation per minute from 0.5 to 25 (35).

The assumption was that the fibers are oriented perpendicular to the pouring direction of the mixture. And the assumption has been proven by our experiments. The fibers are oriented in the horizontal direction when a liquid composite is also poured into the horizontal formwork. The orientation of the fibers was also affected by walls of formwork especially when the width of formwork was smaller.

Acknowledgement

The authors gratefully acknowledge the financial support from the Grant Agency of the Czech Technical University in Prague, the grant No. SGS15/060/OHK1/1T/11.

References

[1] J. Fornůsek, M. Tvarog. Characteristics of Fiber Reinforced Cementitious Composites in Dependence on Casting Direction. In: Applied Mechanics and Materials 732: 377-380 (2015).

[2] R. Lovichová, P. Padevět, J. Fornůsek. Changes of the Properties of Cement Paste with Fly Ash Exposed to the High Temperature. In: Key Engineering Materials 677: 138-143 (2016).

[3] C. Hu, F. Larrard. The rheology of fresh high-performance concrete. In: Cement and Concrete Research 26(2):283-294 (1996).

[4] P. Stähli, R. Custer, J. G. M. van Mier. On flow properties, fibre distribution, fibre orientation and flexural behaviour of FRC. In: Materials and Structures 41:189-196 (2008). ISSN 1359-5997.

[5] B. Zhou, Y. Uchida, Fiber orientation in ultra high performance fiber reinforced concrete and its visualization. In proc.: VIII International Conference on Fracture Mechanics of Concrete and Concrete Structures (2013) paper 160, Toledo (Spain).

[6] M. Beddar. Fibre-reinforced concrete. In: Concrete 38(4) 47 (2004). ISSN 0010-5317.

Identification of the assembly line drive chain force load

A. Lufinka [1,a]

[1] *Technical University of Liberec, Studentská 2, 46117 Liberec, Czech Republic*
[a] *ales.lufinka@tul.cz*

Abstract: The small digger assembly line is composed from the series of transport trolleys, which are slowly pulled around the mounting posts. All these trolleys are moved together by one towing chain. Increasing the number of trolleys was required during the assembly line upgrade, however a power unit including the towing chain should remain as originally. Therefore it was necessary to determine whether the chain will not be overloaded with a higher number of the trolleys. Methodology for the identification of the towing chain force load is described in this article.

Keywords: assembly line; towing chain; force load; measuring; strain gauge sensor.

1 Introduction

The small digger assembly line is composed from the series of transport trolleys, which are slowly pulled around the mounting posts. All these trolleys are moved together by one towing chain. The real assembly line and its functional principle are shown in the Fig. 1.

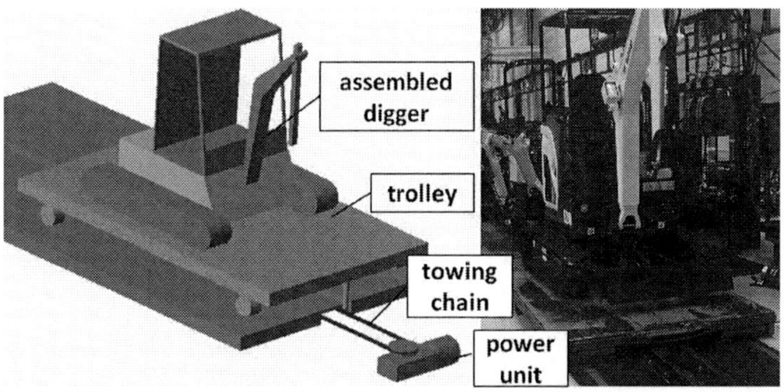

Fig. 1: The real assembly line (right) and its functional principle (left).

Increasing the number of trolleys (from 6 to 10) was required during the assembly line upgrade, however the power unit including the towing chain should remain as originally. Therefore it was necessary to determine whether the chain and the power unit will not be overloaded with a higher number of the trolleys.

The towing force can be theoretically calculated from the wheel load and the rolling resistance coefficient by the next simple equation (see [1]):

$$F_t = c_{rr} F_n \tag{1}$$

where F_t is the towing force, c_{rr} is the rolling resistance coefficient and F_n is the normal force (see Fig. 2)

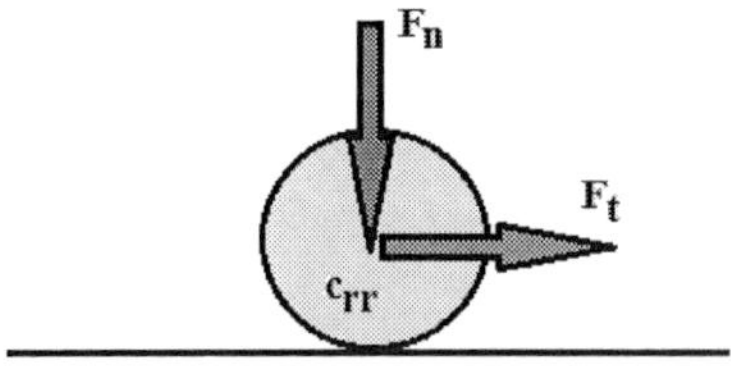

Fig. 2: The towing force theoretical calculation.

The rolling resistance coefficient depends on the floor and the wheel material and its values are published for the most used material pairs [2]. Obviously the bearing properties, the wheel shape and the dimension must be reflected for the wheels that are really used, however the towing force calculation is the same. The wheel producer publishes the rolling resistance coefficient (often called as a traction coefficient) for an every wheel and any floor type and it is used for the real towing force calculation.
So this approach was used for the assembly line towing force calculation. The trolleys have wheels with plastic surface and run on the steel track. Their producer specified the traction coefficient value 0.01828. The load distribution for the old and the new assembly line is shown in the Fig. 3 and the towing force theoretical calculation results are summarized for the both line variants in the Tab. 1.

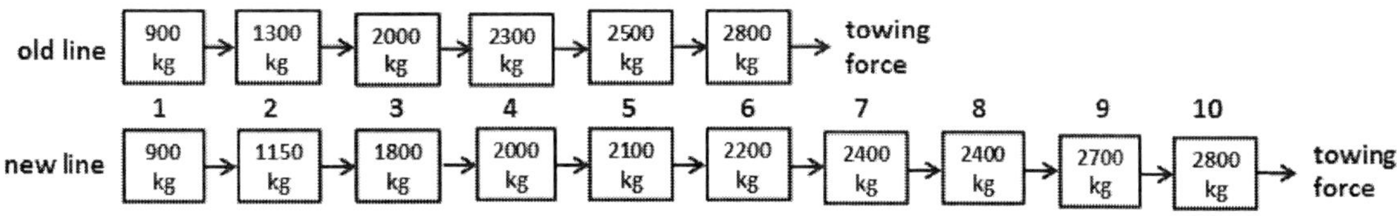

Fig. 3: The old and the new assembly line load distribution.

Tab. 1: The towing force theoretical calculation.

traction coefficient 0,01828		mounting post										total vertical load [kN]	total towing force [kN]
		1	2	3	4	5	6	7	8	9	10		
trolley weight [kg]		300	300	300	300	300	300	300	300	300	300		
digger weight [kg]	original line	900	1300	2000	2300	2500	2800	x	x	x	x	133,4	**2,4**
	new line	900	1150	1800	2000	2100	2200	2400	2400	2700	2800	230,0	**4,2**

The real towing force value can be different from its theoretical calculation because the traction coefficient value can be affected over the time by the wheels wear out, dirt, uneven floor, etc. Therefore the assembly line producer needed to verify whether the real force corresponds to its theoretical calculation and the power unit and the towing chain will not be overloaded after the line upgrade.

2 The real towing force measurement

Because the towing chain is located in a very narrow groove, the standard force sensor could not be used for the towing force measuring. Therefore the force was measured by the strain gauge sensors glued to one part of the towing chain. This part was first weakened for the sensitivity increasing and two identical strain gauge sensors HBM 1-LY41-6/350 [3] were glued to the inner chain surfaces (see Fig. 4). Each strain gauge was connected separately as a quarter-bridge, two signals were used for the force measuring. The ambient temperature change during a short term measurement was not expected, therefore the temperature non-compensated quarter-bridges could be used.

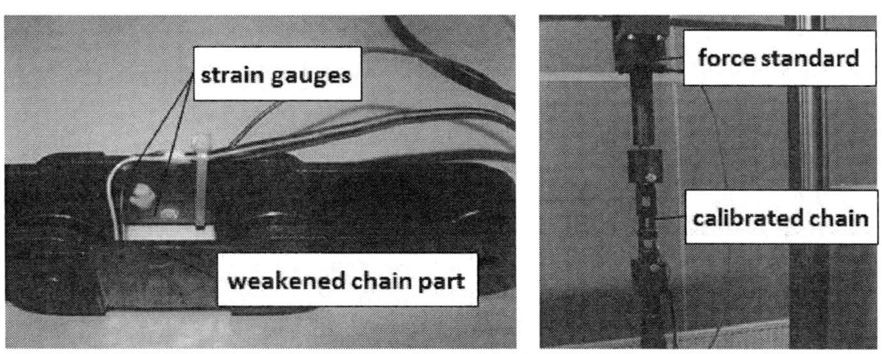

Fig. 4: The chain part with the strain gauges (left) and its calibration (right).

The strain gauge signals were calibrated to the force value using the laboratory calibration system and the total force was calculated as their average. Possible both chain sides uneven loading was eliminated with this. All this was done before the real force measurement at the Technical University of Liberec laboratory.

Finally, this part of the chain was included to the assembly line towing chain (see Fig. 5) and the real force was measured during the assembly line trolleys movement. Two measurements were planned initially (first before and second after the line upgrade) because the line producer wanted to compare these two operating statuses. Third measurement had to be added later because the towing chain tensioning mechanism failed after some assembly line operation time and the cause had to be detected.

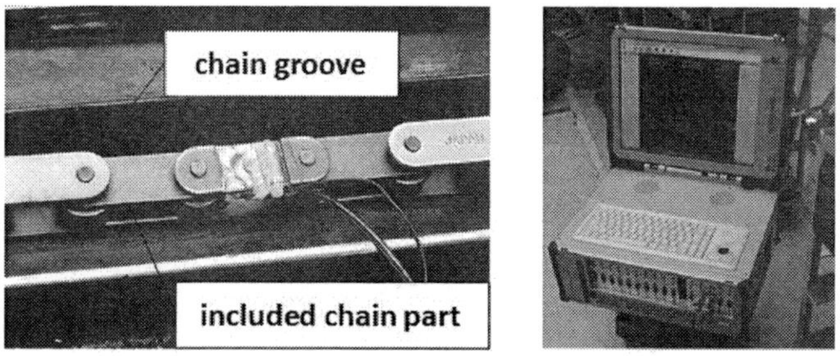

Fig. 5: The real towing force measurement.

3 Results

All the measurements were done in the same way. The assembly line movement was stopped, its towing chain was discontinued and the measuring chain segment was included. Then the line traction was launched and a displacement of approximately one meter was measured. All the trolleys were standard loaded by the assembled diggers but no production operation was done during the first two measurements.

The real towing force measured waveforms (see Fig. 6) slightly fluctuate due to the imperfect and dirty floor but their mean values in a steady state practically correspond with the calculated values (2,4kN for 6 and 4,2kN for 10 trolleys). Because the traction engine is controlled by the frequency converter, the line start is smooth, without the towing force peaks.

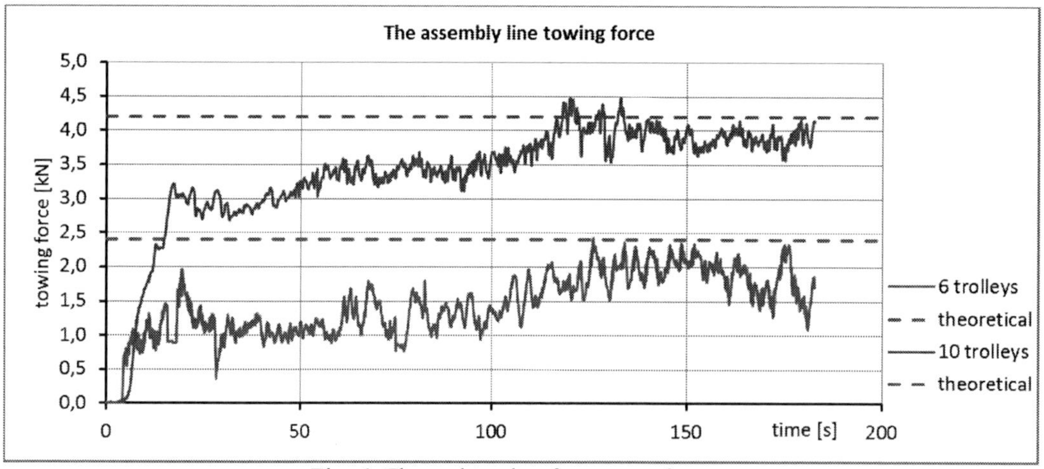

Fig. 6: The real towing force waveforms.

The first two measurement results were clear, the assembly line components are not overloaded by the trolley number increasing, the real towing force value corresponds to its theoretical calculation. The assembly line was put into operation on this conclusion basis. But the towing chain tensioning mechanism unexpectedly failed after some time and it was in a contradiction with the previous results. The entire production process was then analyzed and probable cause of the failure was determined. A new checking operation was added in the production line. All the digger functions began to be tested at the station no.9 before the digger final covering. Its hydraulics systems were activated and all the digger movements were tested. Obviously, the digger was still on the trolley, which was attached and moved by the towing chain, and all the assembly line traction system was loaded by the moving mass inertial forces. Verification measurement before the production line using was done without this checking operation, the trolleys were loaded only by the static mass. The assembly line producer did not foresee this situation because the diggers were previously tested outside the assembly line, after their final completion. Therefore a new measurement of the towing force was done during the normal line operation, including the digger function test. The result is shown in the Fig. 7.

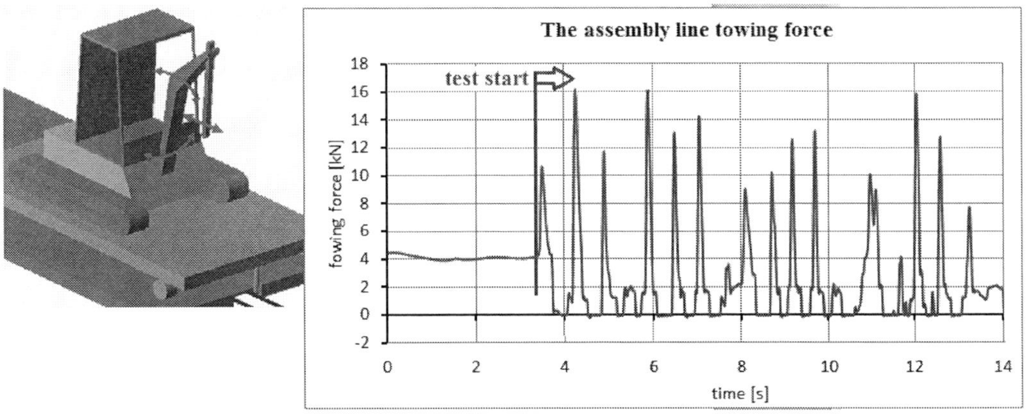

Fig. 7: The towing force waveform during the digger function test.

The measured waveform shows that the towing force before the function test corresponds with the assumption value 4.2kN but the very big peaks are caused by inertial forces during the digger test. Their maximal values increased up to 16kN. The inertial forces dynamical effect is four times greater than the static loading. The towing chain is not overloaded (its maximum allowed force is still much bigger than the force peaks) but its tensioning mechanism was not designed for this dynamical loading and it is approximately twice overloaded by the dynamical force peaks. Therefore the failure assumption was confirmed by this additional measurement.

4 Conclusion

All the measurements showed that the static tensile force corresponds to the theoretical values and the assembly line design was made correctly. The tensioning mechanism failure was caused by adding of a new checking operation which was not initially considered. There are two possible ways to solve this problem and neither of them is an easy one. The first option – return to the outside assembly line testing – decreases a productivity, the second way - with the "on trolley" digger testing – requires the assembly line modification (the tensioning mechanism redesign or the trolley disconnecting and braking during the digger test). The final decision depends on the assembly line producer and it is beyond a scope of this article. This paper is focused only on the towing force measuring methodology and the presentation of this measurement results.

Acknowledgement

The results of this project LO1201 were obtained with co-funding from the Ministry of Education, Youth and Sports as part of the targeted support from the "Národní program udržitelnosti I" programme.

References

[1] R.C. Hibbeler, Engineering Mechanics: Statics & Dynamics, Pearson, 2007, pp. 441–442.

[2] J. Leinveber, P.Vávra, Strojnické tabulky, Úvaly, Albra – educational publishers, 2008.

[3] Retrieved from: http://www.hbm.com/en/0364/strain-gauges-for-stress-analysis

Mechanical Description of The Clinched Joint

Pavel Malý [1,a], František Lopot [1], Jiří Sojka [1]

[1] *CTU in Prague, Department of Designing and Machine Components, Technická 4, 166 07 Prague 6, Czech Republic*

[a]*pavel.maly@fs.cvut.cz*

Abstract: There is a mechanical description of a clinched joint loaded in the perpendicular direction to the joint axis (shear strength) in this text. The description is based on the simplified mechanical scheme considering different components of load, such as shear, tensile load, bending or friction in the contact surfaces. The resulting stress state in the place of fracture is compared with the ultimate strength of the material considering also the material strain hardening that occurs during the creation of the joint.

Keywords: clinched joint; mechanical model; force analysis; shear loading; fracture.

1 Introduction

Nowadays, the need of connecting different components to create a final assembly is crucial in the industry production. Clinching represents one of the modern methods for the joining of the sheet metal plates. This method uses the plastic deformation of the basic material when the joint is created using special tool set (punch and die) – see figure 1. This technology is suitable for connecting components in many branches of production, primarily where the fast, cheap and reliable connection of thin sheet-metal plates in needed. Automotive industry, HVAC, appliance manufacturing etc. are typical examples of clinching application.

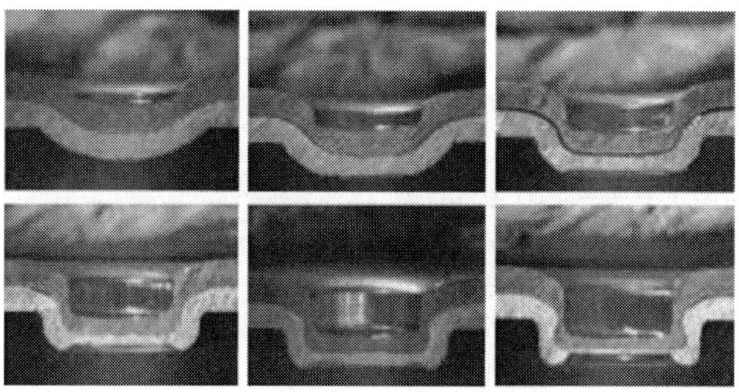

Fig. 1: TOX®-Round Joint formation, [1]

Although there are many articles describing the clinching method, process and properties that are based mainly on the FEM analysis, the new modern technologies often lack the fundamental theoretical background, therefore this article presents first insight into the basic theoretical description of the clinched joint that can be applied in a practical design of complex components and assemblies.

2 Joint and material description

The clinched joint may be loaded in two basic directions: in the direction of the axis (pull-out strength), or in the perpendicular direction to the axis of the joint (shear strength). This text focuses on the clinched joint under the shear load when the parameters were obtained for this state of loading by a single lap shear test.

There are different failure mechanisms of the clinched joint that depend firstly on the type of loading, used material or geometry of the joint. The two basic failure modes are unbuttoning of the joint and the fracture in the neck of the joint.

The shear strength of the joint (TOX®-joint with outer diameter of 10 mm) is approximately 3900 N. The clinched joint with the loading is shown in the figure 2, the joint after the fracture located in the joint neck is shown in figure 3. It is obvious from the figure that the fracture is characterized with very small accompanying plastic deformation therefore it can be supposed that the material in the joint was significantly plastically deformed during the creation of the joint.

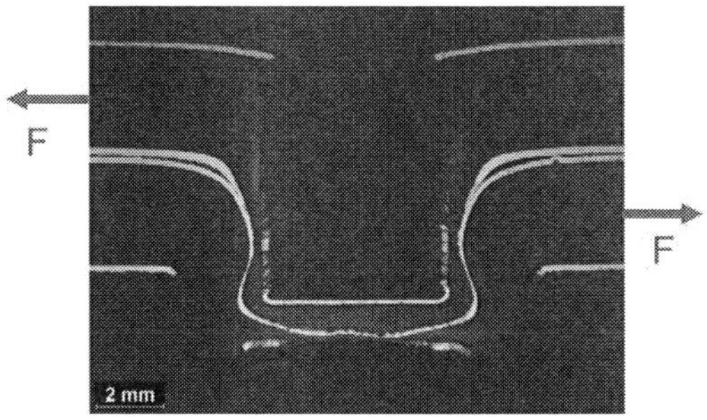

Fig. 2: Macrosection of the clinched joint

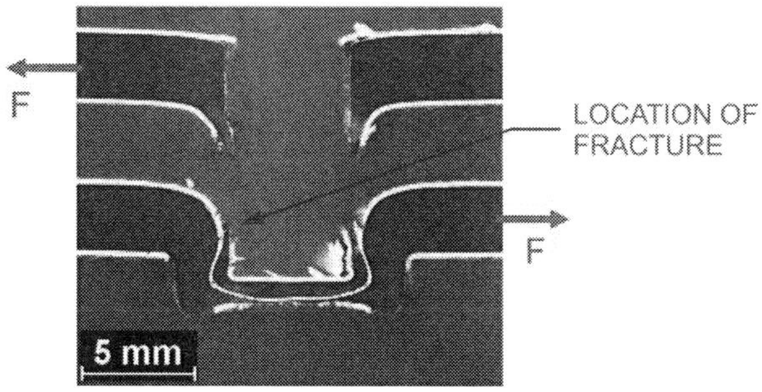

Fig. 3: Clinched joint – fracture after the single lap shear test in the joint neck

2.1 Scheme of the joint

The simplified scheme of the joint used for the mechanical description is shown in figure 4. The interlock of upper and lower sheet-metal plates is simplified by two cones and the neck. The geometry is characterized with the dimensions: outer and inner diameter of the joint neck (d, D) and the angle α. Specific dimensions were measured with usage of the stereomicroscope and are listed in the table 1.

This simplified scheme is used for the calculation of the total reduced stress in the location of the fracture. Different load components were considered for calculation of the reduced stress, so e.g. the shear load, tensile load, bending moment, friction and their combination can be used to find the optimal model of the joint.

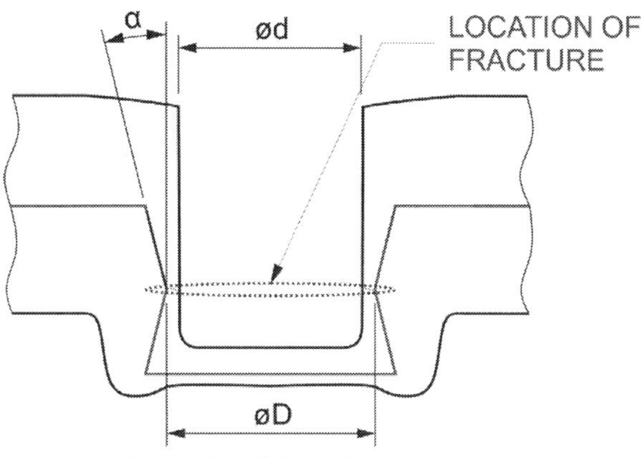

Fig. 4: Simplified scheme of the joint

Tab. 1: Dimensions of the joint.

d [mm]	D [mm]	α [deg]
5.54	6.09	~20

2.2 Material properties

The calculated reduced stress is compared with the ultimate strength of the material. The set of testing specimens was made of coated sheet-metal plates from low carbon steel DX51D (material No. 1.0226) with the tensile strength in the range of 270 to 500 N/mm^2. The detailed parameters of the used material are listed in the table 2.

Tab. 2: Material properties.

Grade		Material no.		Tensile strength R_m [N/mm^2]	Minimum elongation A_{80} [%]
DX51D		1.0226		270 - 500	22
Chemical composition (max. %)					
C	Si	Mn	P	S	Ti
0.18	0.5	1.20	0.12	0.045	0.30

Furthermore, the microhardness in the specific places of the joint was measured. The average value of hardness HV0.2 in the place of the fracture, i.e. in the joint neck of the upper sheet-metal plate, is approximately 286HV0.2.

The creation of the joint is accompanied by the plastic deformation of the base material thus the strain hardening of the material occurs. The increased tensile strength of hardened material in the joint neck was determined using hardness conversion table, e.g. CSN EN ISO 18265 Metallic materials – Conversion of hardness values. The tensile strength of plain carbon or low-alloy steels is 915 N/mm^2 for the hardness of 286HV0.2.

3 Mechanical description of the joint

The simplified mechanical description and fracture of the joint can be characterized by the stresses assumed in the location of fracture in the joint, i.e. tensile and bending stress are the stress components used for determination of the total stress that is in the material during the fracture. The shear stress in the fracture is omitted in this mechanical description of the joint because there is a contact between two sheet-metal plates in the joint during the loading.

The force components that cause the stress components during the loading of the joint in shear strength direction are shown in the figure 4.

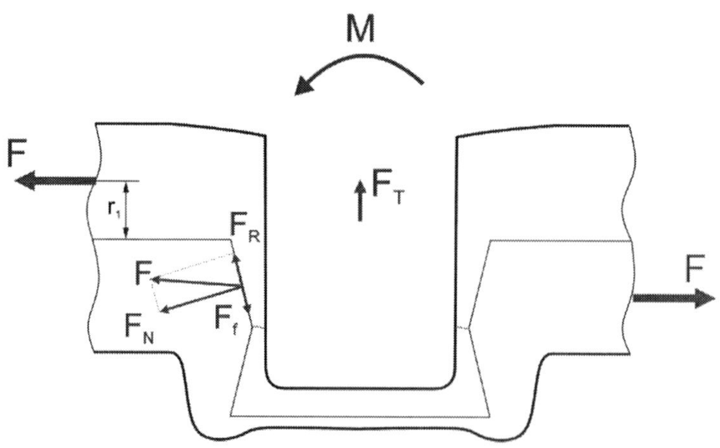

Fig. 4: Computational scheme of the joint

Force components F_N and F_R arise when considering force resolution of loading force F on the inclined surface of the cone in the joint. The normal force F_N causes the friction force F_f between upper and lower sheets in the joint that can be expressed by the static friction law ($F_f = f \cdot F_N$). The static friction coefficient f can be considered for steel-steel combination 0.05 – 0.11 for greased surfaces or 0.5 – 0.8 for dry surfaces. The total force F_T in the direction of the joint axis is then expressed by the eq. 1 and the bending moment is for simplification expressed by the eq. 2.

$$F_T = \left(F_R - F_f\right) \cdot \cos\alpha = F \cdot \sin\alpha \cdot \cos\alpha - F \cdot f \cdot \cos^2\alpha \quad (1)$$

$$M = F \cdot r_1 \quad (2)$$

The stress components, i.e. tensile and bending stress, can be calculated using the load components and cross-section characteristics (area and section modulus). The reduced stress in the location of fracture in the joint neck is calculated with help of suitable material strength theory. For this case, the maximum stress during the fracture is the sum of two above specified stress components. The resultant reduced stress can be compared with the tensile strength of the hardened material in the joint neck.

When the parameters of the joint (dimensions etc.) and experimentally obtained shear strength (force F) are used in the model, the maximum reduced stress in the joint is in the range between 800 till approximately 950 N/mm2, depending on other parameters (e.g. coefficient of friction). It is obvious that the maximum reduced stress is near the tensile strength of the material in the joint neck.

4 Conclusion

The mechanical description of the clinched joint can be used for the design and calculation of complex connection of two sheet-metal plates with more single joints, or for the computational assessment of the approximate shear strength without need of usage of the methods that can be more time consuming. This mechanical model is further subjected to the detailed specification, detailing of force analysis and verification with the help of experimental testing or FEM.

Acknowledgement

This work was supported by the industrial cooperation.

References

[1] TOX®-Product Range [online]. [cit. 2016-05-04]. Available on http://www.tox-de.com/assets/countries/EN/pdf/TOX_Product_Range_00_en.pdf

[2] J. Mucha, W. Witkowski, The clinching joints strength analysis in the aspects of changes in the forming technology and load conditions, Thin-Walled Structures 82 (2014) 55–66, http://dx.doi.org/10.1016/j.tws.2014.04.001.

[3] COPPIETERS, S.: Experimental and numerical study of clinched connections. PhD Thesis, Katholieke Universiteit Leuven – Faculty of Engineering, Leuven, Belgium, 2012.

[4] P. Groche et al., Joining by forming—A review on joint mechanisms, applications and future trends, Journal of Materials Processing Technology 214 (2014) 1972–1994

Elastic model of composite sandwich structure model with progressive damage for low-velocity impact

T. Mandys [1a], V. Laš [1], T. Kroupa [1], C. Šťádler [2], J. Bartošek [3]

[1] *University of West Bohemia in Pilsen, NTIS - New Technologies for Information Society, European Centre of Excellence, Univerzitní 8, 306 14 Pilsen, Czech Republic*

[2] *University of West Bohemia in Pilsen, Faculty of Mechanical Engineering, The Research Centre of Forming Technology - FORTECH, Univerzitní 8, 306 14 Pilsen, Czech Republic*

[3] *University of West Bohemia in Pilsen, Faculty of Applied Sciences, Department of Mechanics, Univerzitní 8, 306 14 Pilsen, Czech Republic*

[a] *tmandys@ntis.zcu.cz*

Abstract: This paper is focused on progressive failure analysis of sandwich composite beam loaded with transversely low-velocity impact. A user defined material model was used for modeling of the non-linear elastic behavior of composite skin of resulted sandwich structure. The non-linear behavior of foam core was modeled using Low-Density Foam material model. The results between numerical model and performed experiments were compared in form of deflection and contact force time dependencies and occurrence of damage too.

Keywords: Drop tester; Low-velocity impact; Sandwich structure, Fiber glass textile composite, Polymer foam core.

1 Introduction

Sandwich structures offer great potential for use in aerospace, marine and transportation and other low-bearing engineering, where the weight must be keep for any reason to a minimum value. Sandwich structure is built by two face sheets (skins) and light core. The outer skins are obviously thinner than the core. The main purpose of the core is to maintain the distance between stiffer outer skins and to transfer the shear load while skins carry compressive and tensile load. This structural arrangement have much larger bending stiffness than single solid plate made of the same total weight from the same material as outer skin only. The other benefits of sandwich panels are the excellent thermal insulation, acoustic damping, easy machining, etc. However, these sandwich panels have usually very low damage resistance and are susceptible to impact damage. The prediction and the prevention of the panel's impact damage are important in design of these panels and their applications. The sandwich structures may be subjected to randomly low-velocity impacts in every day life application such as dropped tool during repair, hit by a stone, or fall of different objects as cell phones. These low-velocity, or low-energy impacts, can cause damage on the sandwich structure that cannot be visible by naked eye, but may significantly reduce the residual strength or stiffness and affect the lifetime of structure or safety of the whole construction. In practice damage can be detected using CT-scan [1] or using structural health monitoring (SHM) [2] based on piezoelectric sensors [3]. These damage detection processes are very time-consuming and expensive. Therefore, it is necessary to ensure the safety of sandwich structures in all cases of expected loads and expected random impact loads.

Even though the non-linear behavior of composite materials is most evident in case of in-plane shear tests [4] the most reported works devoted to numerical simulations of low-velocity impacts consider the composite skin of sandwich structure as a linear orthotropic material. The experimental testing shows, that the fiber-glass fabric material behaves non-linearly in tension along wrap and weft direction and in-plane shear too [5].

2 Low-velocity impact test

The tested sandwich structure were made from 3 layers of fiber-glass fabric with the product name Aeroglass (390 g/m^2) and epoxy resin Epicote HGS LR 285. The composite laminate skins of total thickness 1.2 mm. The core was a close cell cross-linked polymer foam Airex C70.55. The wide sandwich beam of dimension 400 mm × 150 mm and overall thickness 12.5 mm was subjected to transverse low-velocity impact using drop-testing machine. The testing device enables to set the impact place on tested body and the height of impactor directly via moveable horizontal and vertical linear guides. The impactor of total weight 2.336 kg was equipped by force sensor (Kistler 9712B) that enables to record the time-force response (contact force) between the spherical head of impactor with radius 15 mm and tested body. Impacts were aimed at the centre of upper skin of tested sandwich beam. The response of sandwich structure was measured in form of deflection in three selected points simultaneously using the laser sensors. The sampling frequency of force sensor placed on head of impactor and all four laser sensor was the same, 20 kHz. The sandwich beams were simply supported on the steel stand along the shorter edges with the overlap 17.5 mm on the every side. The impact events were recorded using high-speed digital camera (Olympus i-Speed 2) with the frame rate 2000 fps. The geometry of tested sandwich beam together with the placement of selected measured points on upper skin of the beam is shown on Fig. 1.

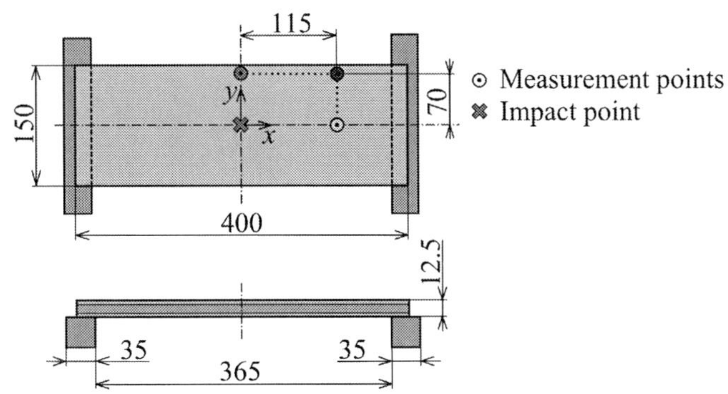

Fig. 1: Geometry of tested sandwich beam and the positions of measurement.

The range of impact velocities was varied between 2.0 and 5.0 $m \cdot s^{-1}$. The impactor was accelerated only by the gravity. The real impact velocities of impactor on sandwich beam were affected by the friction in the linear guides of drop tool apparatus, therefore the real impact velocities of impactor were measured using the fourth laser sensor. The comparison between the impact velocities given by free fall of impactor and real impact velocities of impactor affected due to friction in the linear guides is summarized in Tab. 1.

Tab. 1: Comparison between theoretical and real impact velocities subjected to tested sandwich structure.

Impact velocity $[m \cdot s^{-1}]$	Real impact velocity $[m \cdot s^{-1}]$	Real impact energy [J]
2.00	1.89	4.17
3.00	2.91	9.89
4.00	3.92	17.95
4.50	4.39	22.51
5.00	4.83	27.25

Fig. 2 shows the process of impact event on wide sandwich beam captured by high-speed camera for impact velocity 5.0 $m \cdot s^{-1}$. The foam crushing under the impactor starts at the time t = 2.5 ms after the begin of impact event (t = 0.0 s). The damage of the upper composite skin begin at the time t = 5.5 ms. This damage propagates in form of crack across the upper skin - time t = 6.0 ms. The maximal deflection of the tested beam is achieved during the impact at the time t = 15.5 ms.

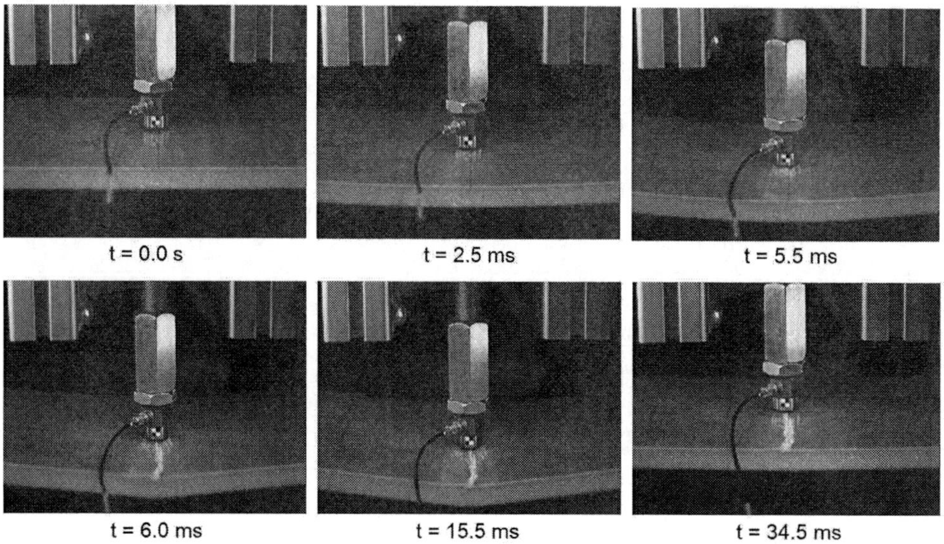

Fig. 2: Pictures of impact event on wide sandwich beam obtained high-speed camera for impact velocity 5.0 $\mathrm{m \cdot s^{-1}}$.

3 Material model and material parameters

3.1 Material model of composite skin

Sandwich composite fabric skin was modelled using a user defined orthotropic material model with the non-linear elastic behaviour. This material model was implemented into Abaqus software using VUMAT subroutine written in Fortran code. The transitions between linear and non-linear parts of stress-strain relationship are indicated using the values of deformations ε_{0i} $(i = 1, 2)$ in principal directions 1 and 2 during loading. The non-linear function with the constant asymptote was considered in the case of the shear in plane 12 [6] Eq. (6). The resulting stress-strain relationship of laminated sandwich skin is described by following equations [7]

$$\sigma_1 = C_{11} \cdot \varepsilon_1 \cdot (1 - D_{11}) + C_{12} \cdot \varepsilon_2 \cdot (1 - D_{12}) + C_{13} \cdot \varepsilon_3 \cdot (1 - D_{13}), \quad \text{for } \varepsilon_1 < \varepsilon_{01}, \tag{1}$$

$$\begin{aligned}\sigma_1 = C_{11} \cdot \left[\varepsilon_1 + \frac{A_1}{2} \cdot \left(\varepsilon_{01}^2 - \varepsilon_1^{\,2} \right) - A_1 \cdot \varepsilon_{01} \cdot \left(\varepsilon_{01} - \varepsilon_1 \right) \right] \cdot (1 - D_{11}) + C_{12} \cdot \varepsilon_2 \cdot (1 - D_{12}) + \ldots \\ C_{13} \cdot \varepsilon_3 \cdot (1 - D_{13}), \quad \text{for } \varepsilon_1 \geq \varepsilon_{01}\end{aligned} \tag{2}$$

$$\sigma_2 = C_{12} \cdot \varepsilon_1 \cdot (1 - D_{12}) + C_{22} \cdot \varepsilon_2 \cdot (1 - D_{22}) + C_{23} \cdot \varepsilon_3 \cdot (1 - D_{23}), \quad \text{for } \varepsilon_2 < \varepsilon_{02}, \tag{3}$$

$$\begin{aligned}\sigma_2 = C_{12} \cdot \varepsilon_1 \cdot (1 - D_{12}) + C_{22} \cdot \left[\varepsilon_2 + \frac{A_2}{2} \cdot \left(\varepsilon_{02}^2 - \varepsilon_2^{\,2} \right) - A_2 \cdot \varepsilon_{02} \cdot \left(\varepsilon_{02} - \varepsilon_2 \right) \right] \cdot (1 - D_{22}) + \ldots \\ C_{23} \cdot \varepsilon_3 \cdot (1 - D_{23}), \quad \text{for } \varepsilon_2 \geq \varepsilon_{02},\end{aligned} \tag{4}$$

$$\sigma_3 = C_{13} \cdot \varepsilon_1 \cdot (1 - D_{13}) + C_{23} \cdot \varepsilon_2 \cdot (1 - D_{23}) + C_{33} \cdot \varepsilon_3 \cdot (1 - D_{33}), \tag{5}$$

$$\tau_{23} = G_{23} \cdot \gamma_{23} \cdot (1 - D_{44}), \tag{6}$$

$$\tau_{13} = G_{13} \cdot \gamma_{13} \cdot (1 - D_{55}), \tag{7}$$

$$\tau_{12} = \frac{G_{12}^{0} \cdot \gamma_{12}}{\left[1 + \left(\frac{G_{12}^{0} \cdot |\gamma_{12}|}{\tau_{12}^{0}}\right)^{n_{12}}\right]^{\left(\frac{1}{n_{12}}\right)}} \cdot (1 - D_{66}), \tag{8}$$

$$C_{11} = \frac{E_1 \cdot [1 - \nu_{23} \cdot \nu_{32}]}{\Delta}, \quad C_{12} = \frac{E_1 \cdot [\nu_{21} + \nu_{23} \cdot \nu_{32}]}{\Delta}, \quad C_{13} = \frac{E_1 \cdot [\nu_{31} + \nu_{32} \cdot \nu_{21}]}{\Delta},$$

$$C_{22} = \frac{E_2 \cdot [1 - \nu_{31} \cdot \nu_{13}]}{\Delta}, \quad C_{23} = \frac{E_2 \cdot [\nu_{32} + \nu_{31} \cdot \nu_{12}]}{\Delta}, \quad C_{33} = \frac{E_3 \cdot [1 - \nu_{12} \cdot \nu_{21}]}{\Delta}, \tag{9}$$

where

$$\Delta = 1 - \nu_{12} \cdot \nu_{21} - \nu_{23} \cdot \nu_{32} - 2 \cdot \nu_{12} \cdot \nu_{23} \cdot \nu_{31}.$$

The material parameters E_1, E_2 and E_3 are Young's moduli in principal directions 1, 2 and 3; and ν_{12}, ν_{23}, ν_{31} Poisson's ratios in planes defined by principal directions 1, 2 and 3. Shear modulus in plane 23 and 13, is designated as G_{23} and G_{13}, respectively. The parameters A_1 and A_2 in Eq. (2) and (4) describe the straightening of yarns of fiber-glass fabric and the loss of stiffness in corresponding directions 1 and 2 respectively. The nonlinear behavior in case of shear in plane 12 (8) is described using initial shear modulus G_{12}^{0}, asymptotic value of shear stress τ_{12}^{0} and shape parameter n_{12}.

The maximum stress failure criterion was used to predict failure on the composite skin

$$F_{1T} = \frac{\sigma_1}{X_T}, \quad F_{1C} = \frac{|\sigma_1|}{X_C}, \quad F_{2T} = \frac{\sigma_2}{Y_T}, \quad F_{2C} = \frac{|\sigma_2|}{Y_C}, \quad F_{12} = \frac{|\tau_{12}|}{S_L}, \tag{10}$$

where X and Y are the strengths in principal directions 1 and 2, the subscripts T and C denote tension and compression and S_L is the shear strength. The values of degradation variable D are dependent on the type of occurred failure. The degradation parameter in case of shear failure ($F_{12} \geq 1.0$) before the strain γ_{12} reaches the critical value of strain and $\gamma_{12} \geq \gamma_{12}^{F}$) was implemented from [8]. The degradation parameters are presented in different kinds of occurring failure in forms

$$F_{1T} \geq 1 \Rightarrow D_{ij} = 1.0, \quad F_{2T} \geq 1 \Rightarrow D_{ij} = 1.0, \quad F_{12} \geq 1 \text{ and } \gamma_{12} \leq \gamma_{12}^{F} \Rightarrow D_{66} = 1 - e^{\left(\frac{1}{m_{12}}(F_{12})^{m_{12}}\right)},$$
$$F_{1C} \geq 1 \Rightarrow D_{11} = 0.6, \quad F_{2C} \geq 1 \Rightarrow D_{22} = 0.6, \quad F_{12} \geq 1 \text{ and } \gamma_{12} > \gamma_{12}^{F} \Rightarrow D_{ij} = 1.0. \tag{11}$$

The non-negative constant m_{12} is presented by integer and γ_{12}^{F} is the ultimate deformation, when the material is fully damaged. The Fig. 3 shows the principle of linear and non-linear material behavior in principal direction 1 together with the principle of degradation in case of unaxial loading. The principle of non-linear function with constant asymptote in case shear plane 12 and applied material degradation is presented in Fig. 4.

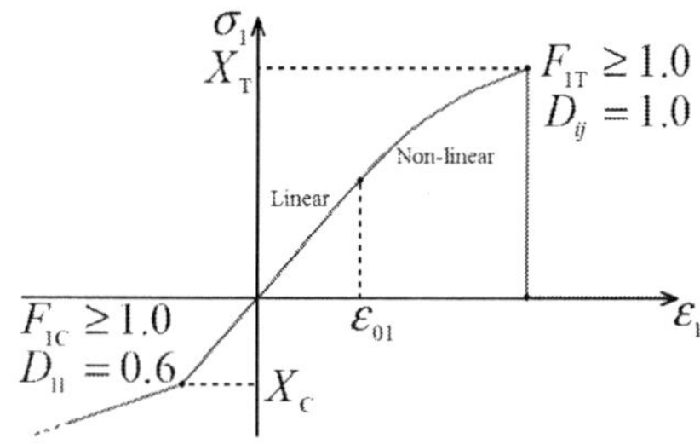

Fig. 3: The principle of linear and non-linear material behaviour together with material degradation in principal direction 1.

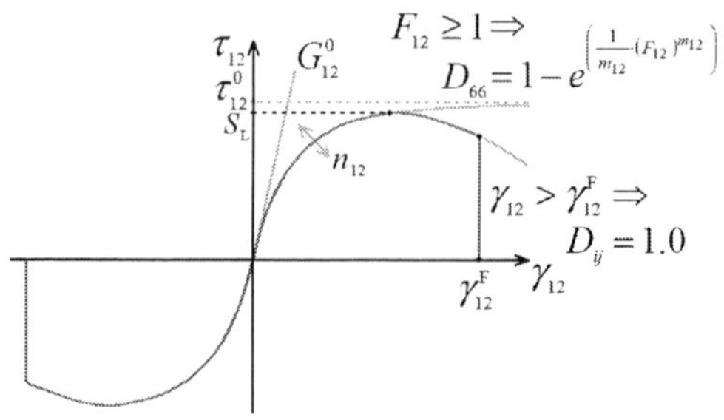

Fig. 4: The principle of material degradation of non-linear function with constant asymptote in plane 12.

Material parameters of considered material model of composite skin were identified from compressive and tensile tests using optimization process [7]. The material parameters E_3, G_{13} and G_{23} were taken from literature [9]. All material parameters of composite laminated skin of sandwich structure are summarized in Tab. 2.

Tab. 2: Material parameters of sandwich composite skin.

E_1 [GPa]	E_2 [GPa]	E_3 [GPa]	G_{13} [GPa]	G_{23} [GPa]	A_1 [–]	A_2 [–]	ε_{01} [–]	ε_{02} [–]	τ_{12}^0 [MPa]	m_{12} [–]
16.9	18.5	8.0	4.0	2.75	10.0	14.0	0.0008	0.005	39.66	5
ν_{12} [–]	ν_{23} [–]	ν_{31} [–]	ε_{12}^F [–]	G_{13}^0 [GPa]	X_T [MPa]	X_C [MPa]	Y_T [MPa]	Y_C [MPa]	S_L [MPa]	ρ_C $[\mathrm{kg\cdot m^{-3}}]$
0.337	0.337	0.28	0.32	4.96	325	65	347	67	35	1154

3.2 Material model of foam core

Foam core of the sandwich structure was modeled using the Low Density Foam model [10]. This isotropic material model for highly compressible elastomeric foams assumes the zero Poisson's ratio. Material behavior was specified via defined unaxial stress-strain curves for tension and compression. The nonlinear elastic stress-strain behavior in tension was described via curve in form [7]

$$\sigma_i(\varepsilon_i) = 92.6\cdot\ln(1+\varepsilon_i) + 198.8\cdot[\ln(1+\varepsilon_i)]^2 - 134.2\cdot[\ln(1+\varepsilon_i)]^3 - \\ 2.3\cdot[\ln(1+\varepsilon_i)]^4 + 11.8\cdot[\ln(1+\varepsilon_i)]^5, \text{ where } i = 1, 2, 3. \tag{12}$$

The Fig. 5 shows the considered compressive and tensile stress-strain relationship of foam core of sandwich structure. The material is fully damaged after reaching the tensile strength R_{mT}. The compressive behaviour of the foam core was described as an ideally elastoplastic material. The compressive stress-strain curve has three distinct ranges. When the limit for linear-elastic compressive behaviour represented by compressive strength R_{mC} is reached, the foam core starts crushing at the same constant stress. When all cells of foam are crushed, the foam structure starts to respond as compacted foam. This state is in material model represented by ultimate compressive strain ε_{U}. The response of foam is characterised after reaching ε_{U} by sharp increase in stiffness. The material parameters of foam core are summarized in Tab. 4.

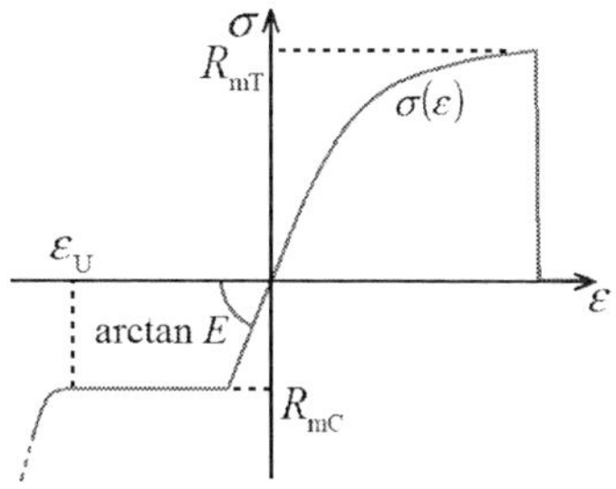

Fig. 5: Tensile and compressive stress-strain behaviour of foam core.

Tab. 4: Material parameters of sandwich foam core.

E	ν	R_{mT}	R_{mC}	ε_U	ρ_F
[MPa]	[–]	[MPa]	[MPa]	[–]	$[kg \cdot m^{-3}]$
50.0	0.0	1.5	1.2	0.53	60.0

4 Numerical simulations and results

The numerical simulations were modeled in finite element software Abaqus 6.14 using explicit solver based on central difference scheme for time integration. The finite strain theory was assumed. The finite element model was created as a fully contact problem of four bodies. The friction between bodies has been neglected. 8-node solid elements (Type C3D8) were used. The finite element model of impactor was simplified and only the head of impactor was modeled with added mass to reach its real total weight. The real impact velocities form experiment were considered (see Tab. 1). The computed values of deflection were interpolated from displacements of neighbouring nodes of finite element mesh near the measure point of laser sensor from the experiment.

The results from experiment and numerical simulation represented by deflections in measuring points and by the time dependencies of the contact force were compared for the range of t = 0 – 20 ms from the start of impact event for impact velocities in range 2 – 4.5 $m \cdot s^{-1}$. The damage area in form of spherical cap due to foam crushing was occurred for this impact velocities. The results comparison in case of impact velocity 5.0 $m \cdot s^{-1}$ is performed for the range of time t = 0 – 40 ms due to occurrence of different kind of damage in form of rupture of outer composite skin. The detail of ruptured upper composite skin at the end of impact event is shown on Fig. 6, where are compared the damage from numerical simulation and the experiment. The damage in compression in principal direction 2 is shown in case of numerical simulation. The blue area corresponds to damage in tension in principal direction 1. The comparison of deflections and contact forces dependencies on the time is shown for selected impact velocities 3, 4 and 5 $m \cdot s^{-1}$ on Fig. 7 – 9.

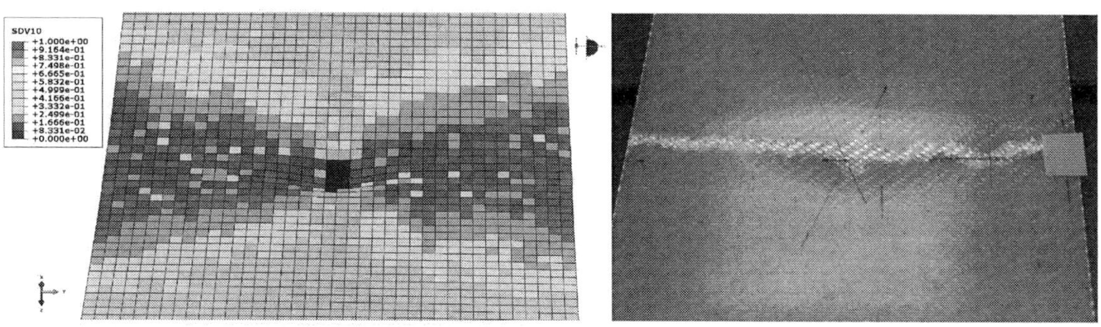

Fig. 6: The comparison of occurred damage on upper composite skin of sandwich beam from numerical simulation and experiment for impact velocity 5.0 $m \cdot s^{-1}$.

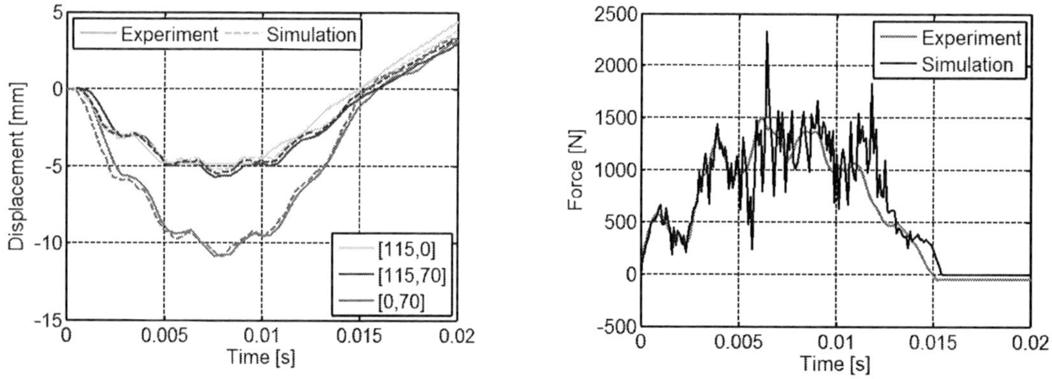

Fig. 7: Comparison of deflection (left) and contact force (right) of sandwich wide beam between numerical simulation and experiment for impact velocity 3.0 $m \cdot s^{-1}$.

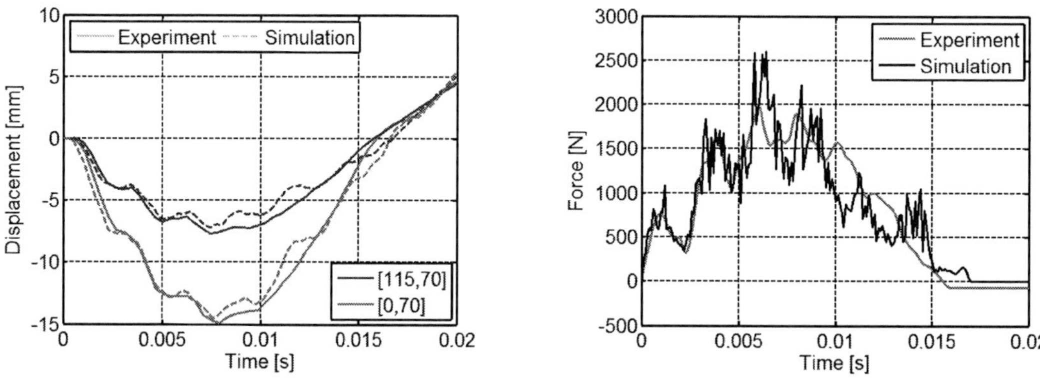

Fig. 8: Comparison of deflection (left) and contact force (right) of sandwich wide beam between numerical simulation and experiment for impact velocity 4.0 $m \cdot s^{-1}$.

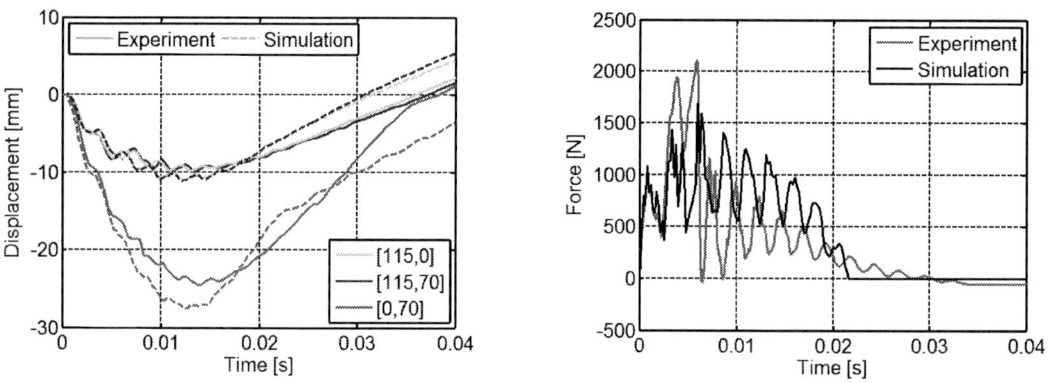

Fig. 9: Comparison of deflection (left) and contact force (right) of sandwich wide beam between numerical simulation and experiment for impact velocity 5.0 $m \cdot s^{-1}$.

5 Conclusion

The response of composite beam to low-velocity impact has been investigated experimentally and numerically. The experimental results in form of deflection and contact force time dependencies were compared with results from numerical model for all performed impact velocities. The results of contact force time dependencies do not show perfect agreement in form of occurrence of the peaks in case of numerical simulations, because the damping of contact between impactor and upper skin was neglected. The damage was compared only by the visual inspection with the resultant damage from the numerical simulations. The expected occurrence of tensile failure of lower composite skin during the impact did not occurred in case of experiment and numerical simulation too. The delamination between the skins and the foam core did not occurred in case of experimental testing, therefore thus behavior is not included in the material model because of reasonable time consumption of numerical analysis. Generally, sufficient agreement between experiments and numerical simulations was achieved.

Acknowledgement

T. Mandys, T. Kroupa and V. Laš were supported by the project LO1506 of the Czech Ministry of Education, Youth and Sports, J. Bartošek was supported by the student research project of Ministry of Education of Czech Republic No. SGS-2016-038 and C. Štádler was supported by the student research project of the Ministry of Education of the Czech Republic No. SGS-2014-022.

References

[1] Wang J., Waas A. M., Wang H., Experimental and numerical study on the low-velocity impact behavior of foam-core sandwich panels, Composite Structures (2013) vol. 96, pp. 298-311.

[2] Sadílek P., Zemčík R., Bartošek J., Mandys T., Active structural health monitoring of composite plates and sandwiches, Applied and Computional Mechanics (2013), vol. 7, pp. 183-192.

[3] Zemčík R., Laš V., Kroupa T., Bartošek J., Reconstruction of Impact on Textile Composite Plate Using Piezoelectric Sensors, Proceedings of the Ninth International Workshop on Structural Health Monitoring (2013), Stanford University, Stanford, Vol 1, pp. 393-400.

[4] Kroupa T., Krystek J., Srbová H., Janda P., Plastic behavior in shear and degradation of shear modulus of textile composite materials with simple plane wave, Experimental Stress Analysis 2011, Brno University of Technology, Brno, pp. 169-176.

[5] Kunc K., Kroupa T., Zemčík R., Krystek J., Tensile and compressive tests of textile composites and its result analysis, 22st. international conference on materials and technology, Program and book of abstracts (2014), Ljubljana, pp. 134-134.

[6] Kroupa T., Zemčík R., Klepáček J., Temperature dependence of parameters of non-linear stress-strain relationship for carbon epoxy composites, Materials and Technology (2009), vol. 43, no. 2, pp.69-72.

[7] Mandys T., Kroupa T., Laš V., Progressive failure analysis of composite sandwich beam in case of quasistatic loading, Materials and technology (2014), vol. 48, no. 4, pp. 593-597.

[8] Yen C. F., Ballistic Impact modeling of composite materials, 7th International LS-Dyna User's Conference (2006), Dearborn, Michigan.

[9] Zemčík R., Laš V., Kroupa T., Purš H., Identification of material characteristics of sandwich panels, Bulletin of Applied Mechanics (2011), vol. 26, pp. 26–30.

[10] Abaqus 6.14 Documentation, Dassault Systèmes Simulia Corp., 2014.

Numerical Modelling of Compressed Masonry Column

A. Maroušková [1,a]

[1] *Czech Technical University in Prague, Faculty of Civil Engineering, Thákurova 7/2077, Prague 6, 166 29, Czech Republic*

[a] *aneta.marouskova@fsv.cvut.cz*

Abstract: This paper deals with numerical modelling of unreinforced masonry (URM) column and masonry column reinforced (RM) by fiber reinforced polymer (FRP) wrapping. The heterogeneous model was chosen for the analysis under the assumption of linear – elastic behavior for all components simulating the state before damage. Both, the bricks and the mortar are modelled separately. The perfect adhesion was assumed between FRP external sheets and masonry support. The loading is applied through the prescribed vertical displacement at the upper surface. Stress distribution of URM and RM are mutually compared and discussed with the results obtained by experimental testing. For all simulations the commercial software package ABAQUS was used.

Keywords: finite element analysis; masonry; FRP sheet; perfect adhesion.

1 Introduction

This study deals with masonry columns with plan dimensions of 0.3 m x 0.3 m and a height of 1 m (according to the real column tested in laboratory). The column was constructed using P20 solid burnt bricks with dimensions of 0.29x0.14x0.065 m. Mortar joints with a thickness of 0.02 m were made with M2 mortar. The experimental program is part of a research project NAKI, which is being carried out at the Faculty of Civil Engineering, CTU in Prague.

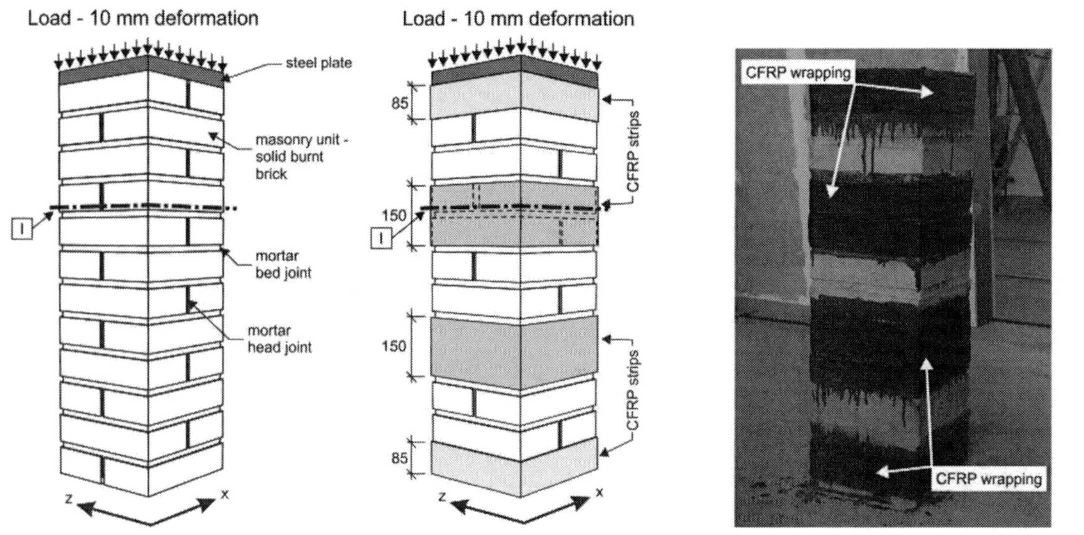

Fig. 1: Charts of URM and RM columns, photograph of real tested column.

Four external CFRP sheets (Fig. 1) were glued onto the column by means the epoxy resin – two sheets with a height of 85 mm were placed at the head and bottom of the column, another two with a height of 150 mm were placed at the column's thirds [1,2].

2 Numerical Analysis

Numerical analysis of masonry structures became one of the most significant interest for global research program in last couple years. Experimental testing enables to follow up a visual response of loading structure, whilst the numerical simulation provides a possibility of disassembling the structure and looking "inside". The obtaining further (numerical) outcomes helps to deeper understanding of structure's behavior and its failure mechanism. Also, simultaneous process of experimental and theoretical parts of research enables mutual comparison of obtained results and better evaluation of damage character.

High requirements, such as non-invasion, reversibility, same appearance keeping, on architectural heritage's protection can prevent an application of traditional strengthening and stabilization methods. FRP external sheets are one of the possible methods that is comply with these requirements. High tensile strength with low weight (thank to that no surcharge of structure appears) belongs to further advantages of FRP materials [1].

In general, modelling methods of FRP reinforcement and masonry support interface can be classified into several approaches. First approach is using zero-thickness interface elements, all the nonlinearities are concentrated along interface, while the components of masonry are usually modelled under the assumption linear-elastic behavior. An application of additional layer (layers) simulating the interface between FRP and masonry is another option. Masonry components can be assumed to behave linear-elastically or nonlinear material model can be adopted for them. The disadvantage of this method is the necessity of specifying enormous amount of materials' parameters. The assumption of perfectly adherent FRP sheet can be considered as simplifying method although complex knowledge of the issue is necessary and this method can provide correct simulation of structure's response.

2.1 The Issues of Numerical Modelling of Masonry

Masonry can be considered as heterogeneous material. The assembly of two materials with different mechanical properties (masonry units and mortar joints) causes masonry heterogeneity magnifying by the heterogeneous composition of masonry units and mortar themselves. Other effects, such as manufacturing, type of bond, damage or degradation further contribute to the heterogeneous nature of masonry. The assembly of the individual components, each of them has a huge variance of mechanical properties, together with influence of mortar joints makes from masonry numerical simulation a complex task.

Failure mechanism of concentrically compressed masonry column is accompanied by formation and growth vertical mostly tensile cracks. The behavior of unreinforced compressed masonry structures is strongly influenced by mutual interaction between masonry unit and mortar joints. The shape of solid burnt bricks can be considered as geometrically accurate, masonry has regular bond and more complaint mortar than masonry units in presented paper. The mortar's compliance to greater transverse strain causes under concentric compressive load additional tensile stresses of masonry units (solid burnt bricks). Thus it has to be taken into account that the tensile cracking of compressed masonry column with regular bond is partially caused by compressive load and partially by mutual interaction [3].

2.2 Numerical Model

The heterogeneous model was chosen for the analysis under the assumption of linear – elastic behavior for all components simulating the state before damage. Linear – elastic parameters are shown in table 1. The accurate description (geometrically and materially) of all components results in complexity of numerical model, but on the other hand allows defining contact between two materials and following up the stress's changes at interfaces. Only heterogeneous model, in which each components are modelled separately, is capable of expressing the influence of mortar joints.

The developed numerical model is composed from approx. 60000 finite elements - first order continuum (solid) elements (C3D8) for mortar and bricks discretization and linear membrane elements (M3D4) for reinforcement discretization. Maximal size of finite element's edge was a third of masonry unit's height. For brick-mortar interface, "hard" contact was assumed in normal direction and frictional behavior (with a friction coefficient of 0.6) in tangential direction. The simulation of RM was executed under the assumption of perfect adhesion between reinforcement and masonry support - masonry and composite were tied together through a standard constraint in ABAQUS program [4, 5]. The boundary conditions weren't applied directly on the masonry column, but there are two steel plates – one at the head of the column and another at the bottom of the column.

Tab. 1: Linear – elastic parameters of materials.

Material	Young Modulus E [GPa]	Poisson's coefficient ν [-]
Masonry unit	2.5	0.20
Mortar	0.5	0.15
Steel	210	0.30
FRP	140	0.33

2.3 Results

Numerical results are shown at the same spots - 1 cm over bed mortar joint, at the spot with CFRP sheet for RM column respectively. The stress distribution of RM column is significantly distinct from URM column. Maximal principal stress's trajectories of URM run along masonry units and cause transversal cracking of solid burnt bricks, whilst RM column's maximal principal stress's trajectories radically change direction as a result of CFRP wrapping. The cracking is concentrated close to masonry surface (reinforcement) and is converging at the corners of column.

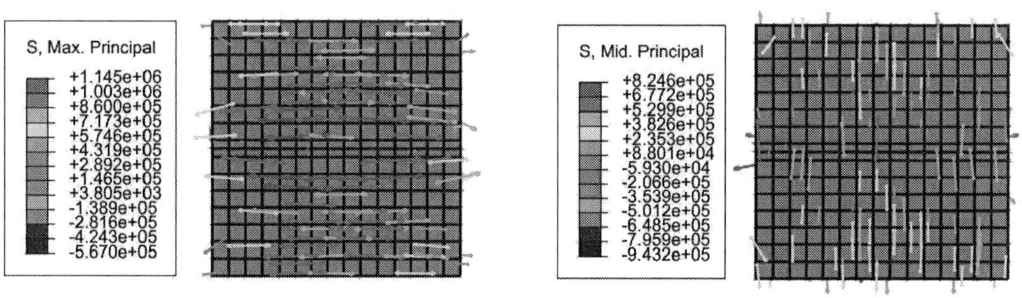

Fig. 2: Trajectory of max. and mid. principal stress along the cross section for URM column.

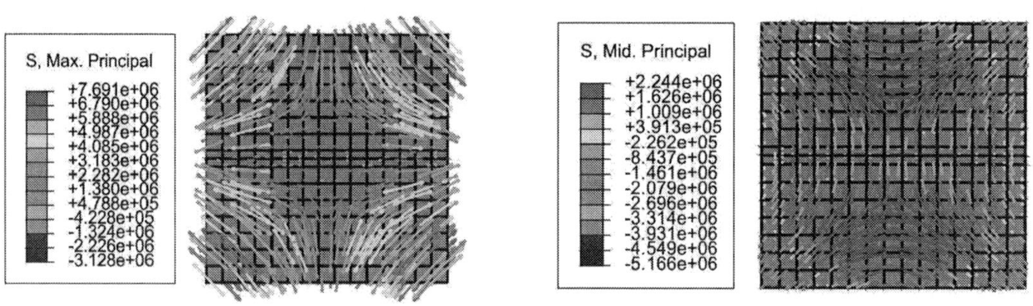

Fig. 3: Trajectory of max. and mid. principal stress along the cross for RM column by CFRP wrapping.

All laboratory tested column were disassembled after reaching the ultimate load. For both solved types – URM and RM column, the result from experimental testing clearly show cracking in direction of mid. principal stresses, normal to the direction of max. principal stresses respectively (Fig. 4).

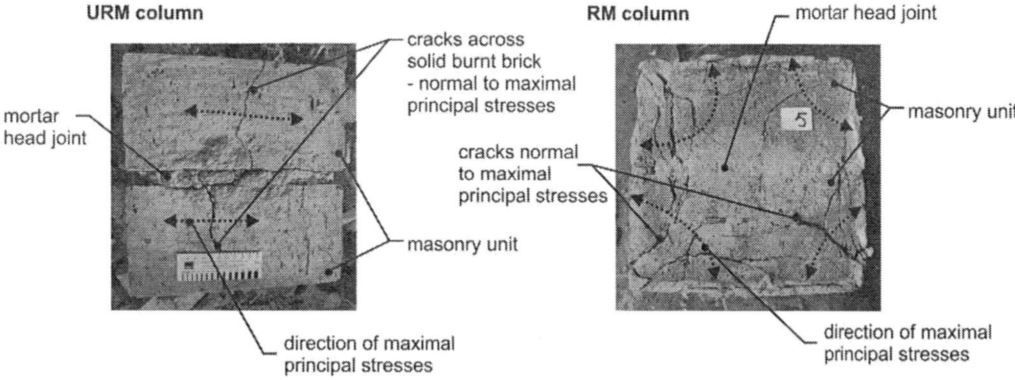

Fig. 4: Character of cracking of URM and RM column, experimentally detected.

3 Conclusion

The failure mechanism of compressed URM column is strongly influenced by mutual interaction between two materials with different mechanical properties – masonry units and mortar joints. For brick masonry, the mutual interaction causes additional transverse tensile stresses. CFRP sheets due to high tensile strength prevent the growth of horizontal deformation and vertical tensile cracks. It was shown that carbon CFRP sheets are efficient tool for increasing ultimate bearing capacity and rigidity of brick masonry [1,2]. CFRP sheets defend lateral strain and take over part of transversal tensile stresses. Reinforcement by CFRP wrapping significantly change stress distribution compared to URM column and transform failure mechanism of compressed masonry column.

Acknowledgement

The results were obtained with support from research project DF12P01OVV037 - "Progressive Non-Invasive Methods of the Stabilization, Conservation and Strengthening of Historic Structures and their Parts with fibre- and nanofibre-based Composite Materials".
The grant researcher is prof. Ing. Jiří Witzany, DrSc. dr.h.c.

References

[1] J. Witzany, R. Zigler, Stress State Analysis and Failure Mechanisms of Masonry Columns Reinforced with FRP under Concentric Compressive Load, Polymers 8 (2016), 176, doi: 10.3390/polym8050176.

[2] J. Witzany, T. Čejka, R. Zigler, Failure Mechanism of compressed short brick masonry columns confined with FRP strips, Construction and Buildings Materials 63 (2014) 180-188, doi: 10.1016/j.conbuildmat.2014.04.041.

[3] A. Maroušková, Masonry Column Reinforced by FRP Wrapping: Behavior and Numerical Analysis, Applied Mechanics and Materials 825 (2016) 27-30, doi:10.4028/www.scientific.net/AMM.825.27.

[4] M. Mrozek, D. Mrozek, A. Warwrzynek, Numerical analysis of selection of the most effective configuration of CFRP composites reinforcement of masonry specimens, Composites: Part B 70 (2015) 189-200, doi:10.1016/j.compositesb.2014.11.016.

[5] R. Fedele, G. Milani, A numerical insight into the response of masonry reinforced by FRP strips. The case of perfect adhesion, Composite Structures 92 (2010) 2345–2357, doi:10.1016/j.compstruct.2010.03.014.

Analysis and construction of measurement device

R. Martonka [1,a], V. Fliegel [1]

[1] *Technical University of Liberec, Department of Design of Machine and Mechanism, Czech Republic*
[a] *rudolf.martonka@tul.cz*

Abstract: Vibration analysis of some of the test machine equipment requiring special construction measuring devices, where the main criterion is the weight, stiffness and universally of the product. Special design must comply particular requirement of universal use, i.e. possibility of loading in different directions without disassembling equipment under test. The product should not affect his vibration measurements of machine equipment.

Keywords: analysis; dynamics; construction; measurement; vibration.

1 Introduction

Measuring device has been designed according to the requirements of a weldment. It has been suggested several variants. Subsequently, the chosen one variant for production. Measuring device was manufactured and subjected to testing. The test results are listed in the following chapter.

2 Construction and analysis

In this chapter, the authors discuss the different types of structures designed test fixtures. When structuring is primarily taken into account manufacturability in accordance with not only fit together with a source of excitation signal (hydrodynamic pulsator) but also to the actual test piece. These considerations have been proposed 3 variants, which differed from each other dominantly (Fig.1,Fig. 2 and, Fig.3). Each of these originally designed was subjected to a dynamic virtual analysis of the obtained results describing natural frequency and corresponding shapes of oscillations (Fig. 4, Fig. 5, Fig. 6).

2.1 Construction

The construction was realized three variants. The first variant of structure have been proposed from the viewpoint of ease of production and ease of installation (Fig. 1). This first option, however, almost all the natural frequency of the experiment conducted.

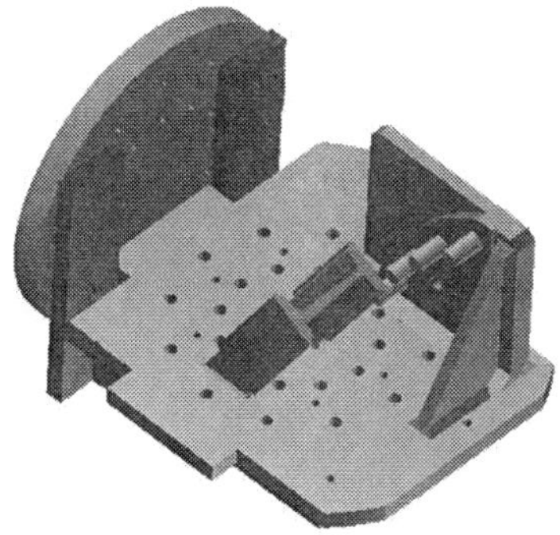

a) measuring devices

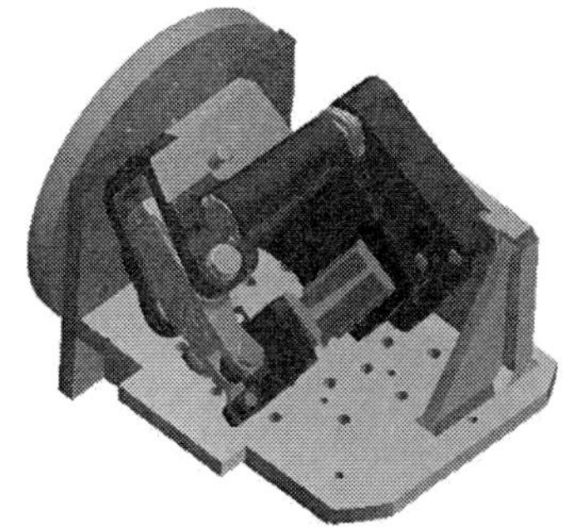

b) measuring devices by measuring machine equipment

Fig. 1: Design of measuring device – design 1.

For this variant was derived second option. The difference between these two was the addition of the reinforcing shoulders. These arms should not result in a worse fit together with an experienced sample. Positives were the reduction in weight of the total composition based on weight reduction of anchor plates and increase the stiffness of the whole structure. Said modified structure is shown in Fig. 2.

a) measuring devices

b) measuring devices by measuring machine equipment

Fig. 1: Design of measuring device – design 2.

For the previous two versions, a third variant, in which was eliminated ballast weight plates and made better placement of reinforcing shoulders. Reinforcing shoulders had a large effect of increasing the overall rigidity of elimination at reducing the total weight of the test fixture. Innovative resulting structure is shown in Fig. 3.

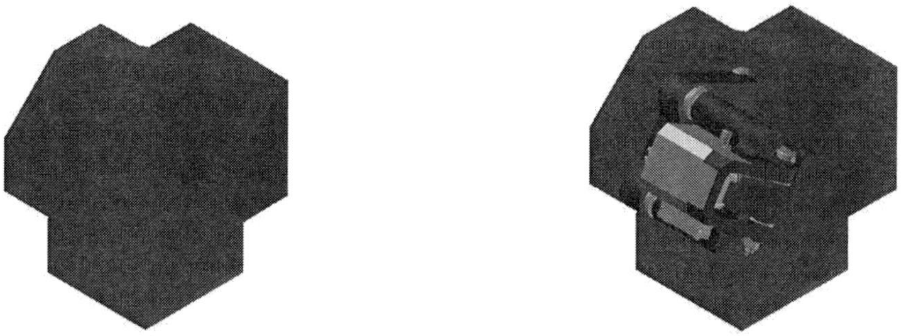

a) measuring devices

b) measuring devices by measuring machine equipment

Fig. 1: Design of measuring device – design 3.

All three types were subjected to structure a virtual analysis of the listed exemplary results shape natural frequencies with the greatest variations in the individual preparations.

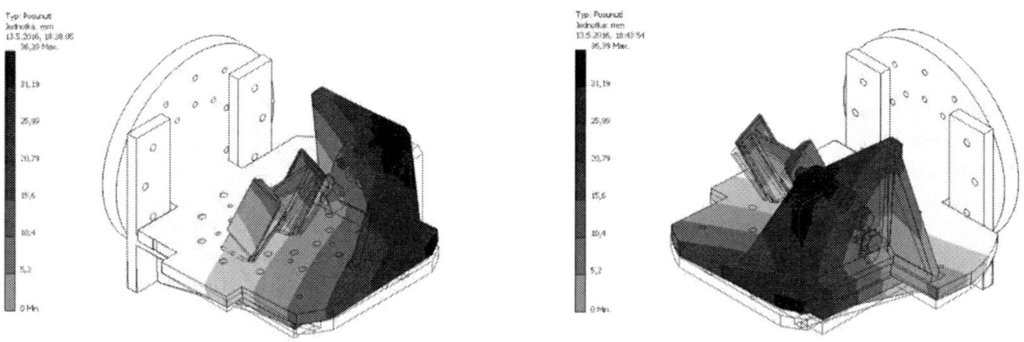

a) measuring devices

b) measuring devices by measuring machine equipment

Fig. 1: Design of measuring device – design 1.

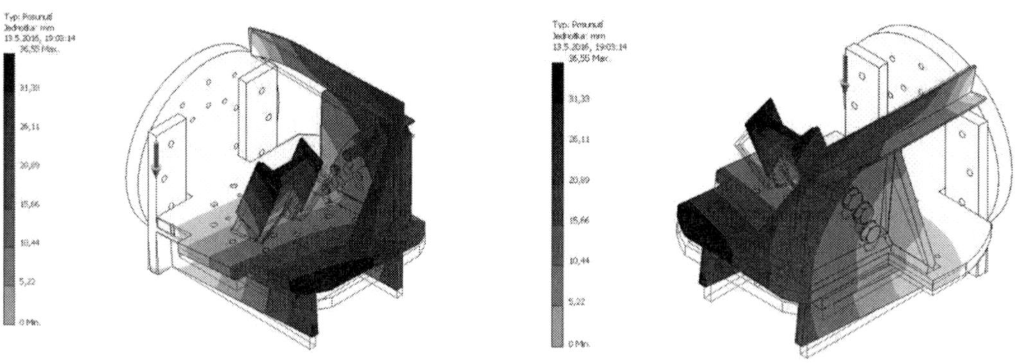

a) measuring devices

b) measuring devices by measuring machine equipment

Fig. 1: Design of measuring device – design 2.

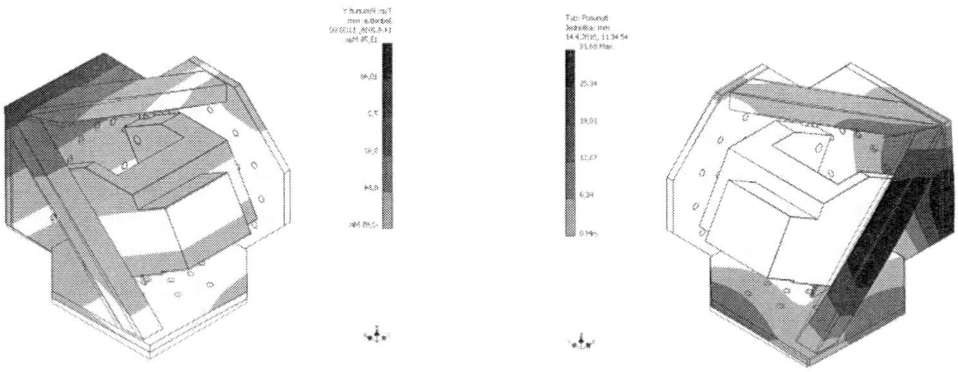

a) measuring devices

b) measuring devices by measuring machine equipment

Fig. 1: Design of measuring device – design 3.

Fig. 4: Preparation of measurements with device.

Fig. 5: Preparation of measurements with device and machine equipment.

3 Conclusion

M The latest variant of the structure was chosen as the most suitable. Drawn up according to the product of said structure was implemented and then on were made to the planned experimental tests. Real preparation is shown in fig. 7. Measured vibration values that are limited by the standard [1] are presented in an article in the Proceedings of the authors J.Petřík and A. Lufinka.

Acknowledgement

This publication was written at the Technical University of Liberec as part of the project "Innovation of products and equipment in engineering practice" with the support of the Specific University Research Grant, as provided by the Ministry of Education, Youth and Sports of the Czech Republic in the year 2016.

References

[1] ČSN EN 61373 Drážní zařízení - Zařízení drážních vozidel - Zkoušky rázy a vibracemi

Self-Heating of Filled Rubber during Cyclic Loading

B. Marvalova [1,a], I. Petrikova [1], J. Novak [1]

[1] *Dpt. of Applied Mechanics, Faculty of Mechanical Engineering, Technical University of Liberec, Studentska 2, 46117 Liberec, Czech Republic*

[a] *bohda.marvalova@tul.cz*

Abstract: The self-heating rubber behaviour is studied during a cyclical loading. Cylindrical samples of two types of filled rubber material (70 and 80 Shore A) were submitted to the compressive cyclical strain controlled loading. Temperature change was measured in the centre of samples and the dissipated energy was determined. Loading continued until the equilibrium of temperature was reached. Changes of storage and loss moduli and of loss tangent were determined by means of DMA. All these quantities first diminished in course of loading and then attained steady values.

Keywords: filled rubber; cyclical loading; self-heating; dynamical moduli.

1 Introduction

Filled rubbers have in general low heat conductivity and quite high hysteretic losses in course of cyclic loading. These features lead to considerable self-heating and to the rise of temperature which is the cause of the decrease of modulus and also of the decrease of inner damping of rubber. The fatigue endurance and the strength of rubber decline with the rising temperature as well.

The change of the temperature is the result of the competition between hysteretic self-heating and the heat dissipation to surrounding environment. The quantum of heat released during one cycle is proportional to the dissipated energy of deformation which is irreversibly lost due to the hysteresis. The dissipated energy depends on the test conditions. The heat dissipation to the surrounding depends on the gradient of temperature and on the heat transfer coefficients.

Two types of cyclic test are used in practice: strain controlled with constant strain amplitude ε_0 and stress controlled with constant stress amplitude σ_0. Since the temperature rise leads to the fall of modulus, the strain controlled test results in the decrease of the stress amplitude and, subsequently, in the decrease of hysteresis. In this case the temperature attains steady value quite soon. On the other hand, when the stress control is used, the strain amplitude increases and the dissipation of energy and heat release increase too [1].

2 Experiment

The strain controlled compressive cyclic tests were performed on filled EPDM rubbers Tab. 1 with different hardness: B samples with 70 Shore A and D samples with 80 Shore A. The samples were cylindrical of 20mm diameter and 20 mm height. The compressive preload was 4 mm, the amplitude was 0.5 mm and the frequency was 10 Hz. The sampling frequency was 1 kHz. Temperature was measured by a thermocouple in the centre of specimen and recorded every 1 min till the steady state at 16 min approximately. Every 60 seconds, 20 cycles of force versus displacement were recorded for the purpose of dynamic mechanical analysis [2]. The dynamic mechanical analysis (DMA) is well suited for the identification of the short-time range of rubber response. DMA consists of dynamic tests, in which the force response to a sinusoidal strain controlled loading is measured. DMA tests under sinusoidal tension mode were carried out on an electro-dynamic testing machine Instron ElectroPuls E3000 equipped with an environmental chamber and with WaveMatrix software. Raw test data were recorded and evaluated in the Matlab Signal Processing Toolbox.

Tab. 1: Tested samples.

sample	B	D
number	5	5
Shore A	70	80
dimensions	20/20	20/20

2.1 Cyclical Loading

Cylindrical samples are loaded by a prescribed compressive axial displacement which consists of a harmonic component superposed on a static preload. When a viscoelastic material is subjected to a sinusoidally varying strain after some initial transients the stationary stress-response will be reached in which the resulting stress is also sinusoidal, having the same angular frequency but advanced in phase by an angle δ. Then the strain lags the stress by the phase angle δ. The axial displacement $u(t)$ consists of a static pre-strain u_0 under tension which is superimposed by small sinusoidal oscillations:

$$u(t) = u_0 + \Delta u \sin(2\pi ft). \tag{1}$$

Stresses and strains are calculated with respect to the reference geometry [3] of the pre-deformed specimen

$$\varepsilon_0 = u_0 / (L_0 + u_0), \quad \Delta\varepsilon = \Delta u / (L_0 + u_0), \tag{2}$$

where L_0 is the undeformed length of the specimen. The force response $F(t)$ of the specimen is a harmonic function and can be written as:

$$F(t) = F_0 + \Delta F \sin(2\pi ft + \delta). \tag{3}$$

F_0 is the static force depending only on the pre-deformation u_0. The force amplitude ΔF and the phase angle δ depend, in general, on the pre-deformation, the frequency and the strain amplitude [4].
If the incompressibility of the rubber is assumed $A_0L_0 = A(L_0+u_0)$, where A_0 is the cross-sectional area of the undeformed specimen, we can relate the force to the cross-sectional area A of the pre-deformed specimen:

$$\sigma(t) = \frac{F(t)}{A} = \sigma_0 + \Delta\sigma\left[\cos\delta \sin(2\pi ft) + \sin\delta \cos(2\pi ft)\right]. \tag{4}$$

The dynamic stress-response $\sigma(t)$ normalized by the deformation amplitude $\Delta\varepsilon$ can be written:

$$\sigma(t) = \sigma_0 + \Delta\varepsilon\left[E'(\varepsilon_0, f, \Delta\varepsilon)\sin(2\pi ft) + E''(\varepsilon_0, f, \Delta\varepsilon)\cos(2\pi ft)\right], \tag{5}$$

where

$$E'(\varepsilon_0, f, \Delta\varepsilon) = \frac{\Delta\sigma}{\Delta\varepsilon}\cos(\delta), \tag{6}$$

and

$$E''(\varepsilon_0, f, \Delta\varepsilon) = \frac{\Delta\sigma}{\Delta\varepsilon}\sin(\delta) \tag{7}$$

are the storage and dissipation moduli respectively and δ is the phase angle. The phase angle was determined using linear viscoelastic theory by discrete Fourier transform in Matlab. In the course of loading an increase of temperature was observed due to a part of the dissipated work. The temperatures of samples increase in the course of tests and reach their steady state after 10^4 cycles approximately see Fig. 1. Area of hysteresis loop which characterises the dissipated energy decreases during the test and its inclination diminished see Fig.2. This means that the stress amplitudes and dynamic moduli decrease due to the increase of temperature. The storage modulus, the loss modulus in Fig.3 and loss angle in Fig. 4 have similar decreasing trend.

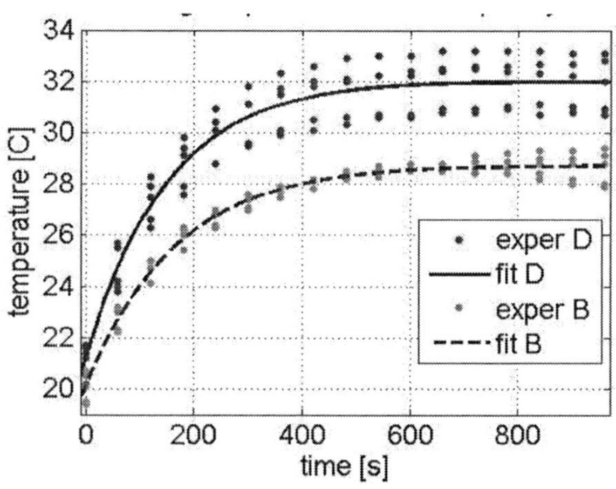

Fig. 1: Rise of temperatures during the cyclic loading at 10Hz - rubber B /70 Sh A and rubber D/80 Sh A

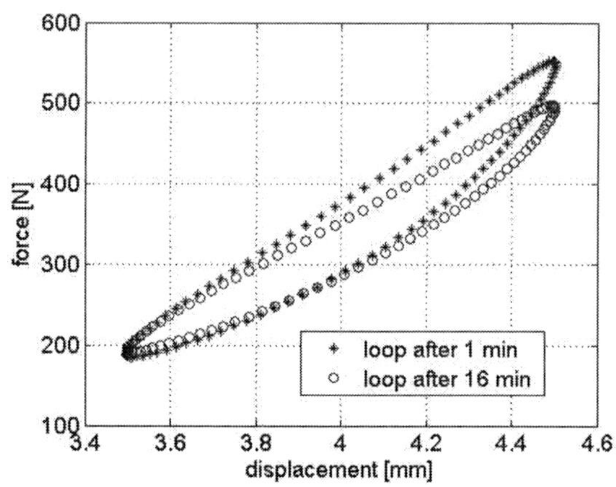

Fig. 2: Evolution of hysteresis loop during a test at 10 Hz - sample of rubber B.

2.2 Hysteretic Heating

The area of hysteresis loop corresponds to the energy dissipated in the cycle which is transformed partially into heat. The area of loop was determined in every loading cycle from the stored data. The values of dissipated energy per cycle decrease almost exponentially and reach the steady state after 10^4 cycles see Fig. 5. The steady state results from the balance in the heat generated by the dissipation of work and the heat drain to the surroundings. Studies of self-heating of polymers and rubbers [5-7] showed that only a certain part - around 55% of the dissipated work is converted to heat. The evolution of the temperature θ is governed by the heat equation

$$\rho c \frac{\partial \theta}{\partial t} - \kappa \Delta \theta = \dot{Q}, \tag{8}$$

where ρ is density, c is specific heat, κ is thermal conductivity and $\dot{Q}$ is dissipated power per unit volume.

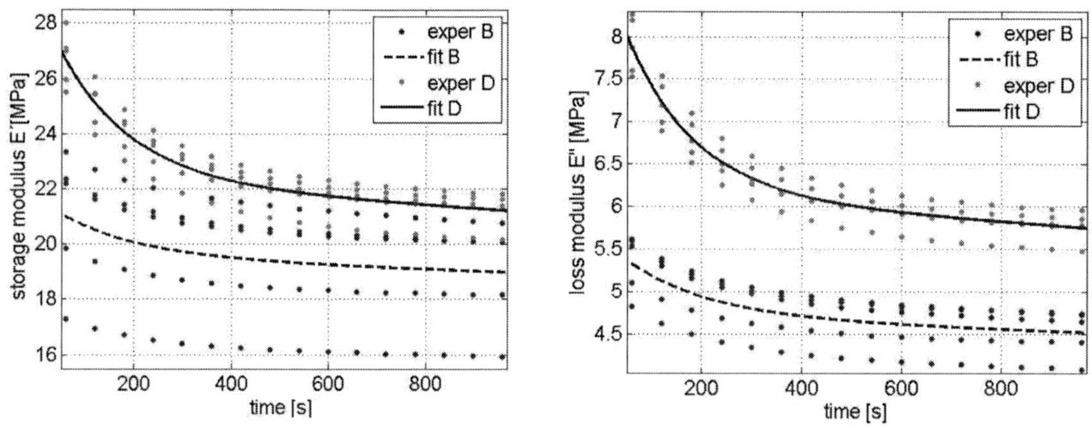

a) storage modulus

b) loss modulus

Fig. 3: Dynamical moduli decreasing due to temperature rise during cyclic compressive tests at 10 Hz.

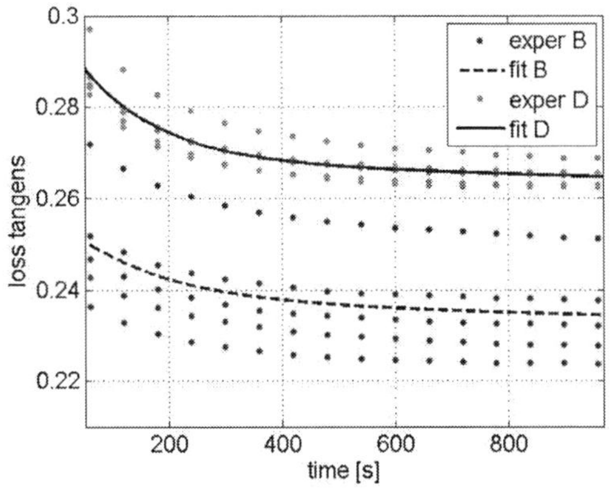

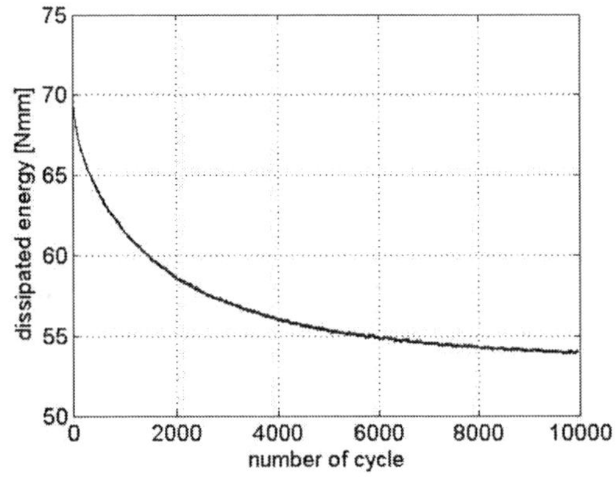

Fig. 4: The loss tangent decreasing slowly due to self-heating of rubber samples at 10 Hz .

Fig. 5: Area of hysteresis loop in every cycle at 10 Hz - the steady state after 10^4 cycles.

The thermal boundary conditions are of two types. The loading grips are coated by Teflon to reduce friction and the thermal state here can be considered as adiabatic. The heat convection should be considered at the cylindrical surface:

$$\kappa \frac{\partial \theta}{\partial r} + \lambda\left(\theta - \theta_0\right) = 0 \quad \text{at } r = R. \tag{9}$$

The evaluation of the nonlinear transient heat transfer problem will be the subject of finite element analysis. The thermo-mechanical coupling will be also considered.

2.3 Changes in Material Properties

The dynamical moduli and the loss angle as well as the area of hysteretic loops depend inversely on the rising temperature caused by self-heating see Figs.3-5. This dependency can be approximated by a linear function as can be seen in Fig. 6.

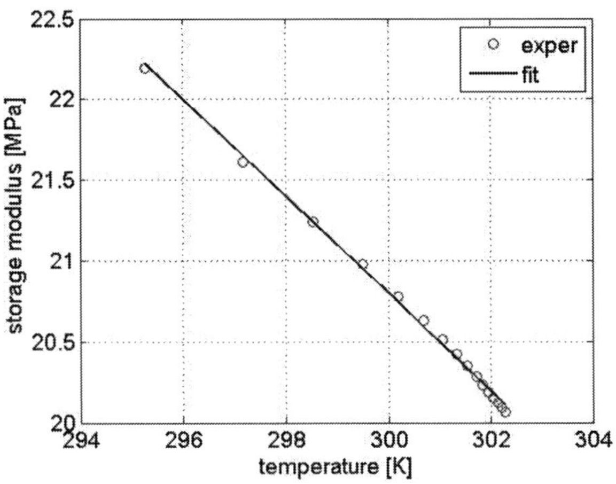

a) storage modulus

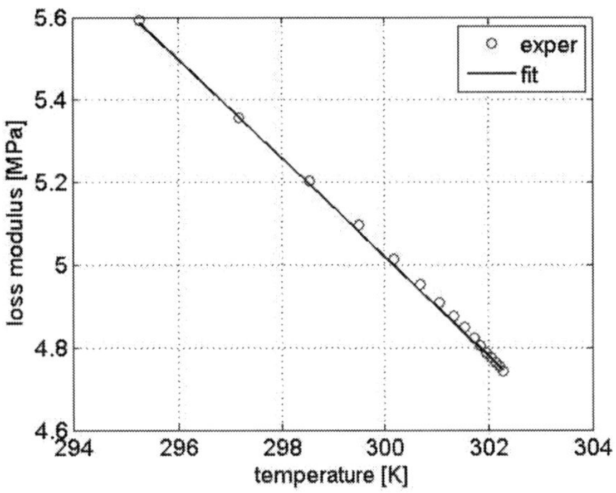

b) loss modulus

Fig. 6: Decrease in dynamical moduli of rubber B as the function of rising temperature during cyclical loading at 10 Hz.

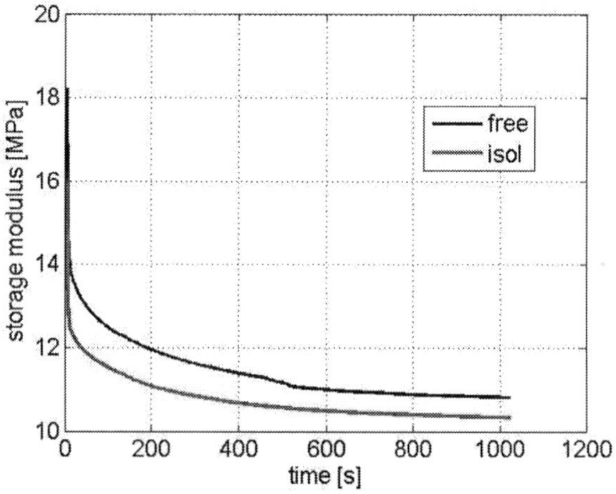

a) decreasing storage modulus - rubber B, storage modulus is slightly higher for free convection

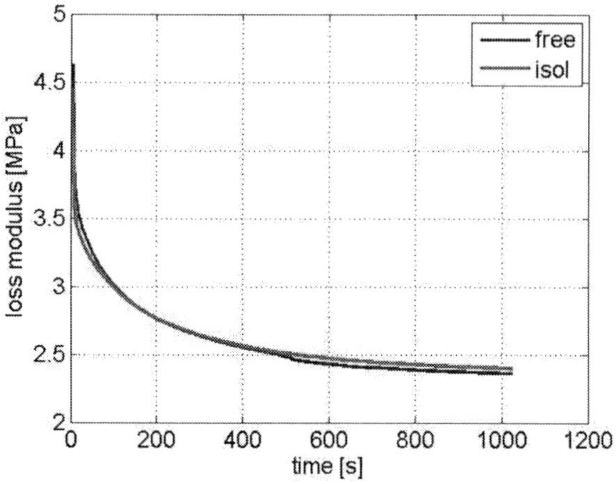

b) decreasing loss modulus - rubber B, no apparent difference in loss modulus

Fig. 7: Dynamical moduli in cyclic compressive test of rubber B at 20 Hz. Convection of heat was either free at cylindrical surface or the surface was insulated thermally.

The influence of the loading frequency was studied on the same type of cylindrical rubber samples of diameter and height 20 mm. The loading frequency was 20 Hz, the compressive prestrain u_0 was 4mm and the amplitude Δu was 0.5 mm. To determine the effect of heat transfer to the surrounding, the lateral surface of some samples was thermally isolated. The values of dynamical moduli and loss angle fall very steeply in the beginning of loading and are lower in comparison with those for the 10 Hz frequency see Fig. 7 and 8. The biggest difference is in the dissipated energy for the frequencies 10 Hz and 20 Hz but the thermal insulation did not have any significant impact on the area of hysteresis loops see Fig. 9. The temperature raised by 12 K in the free sample and by 16 K in the insulated sample.

3 Conclusion

The mechanical properties of hard rubbers (70 and 80 Shore A) and the self-heating were studied in the cyclic compressive tests with the strain control. Rubber D of higher hardness exhibits higher energy dissipation under cyclic loading. It leads to the heat generation, to the temperature rise and to decrease in the dynamic modulus and loss angle. All monitored quantities reached steady state after approximately 10^4 cycles in the strain controlled loading.

The heat generation depends on the energy dissipation due to nonlinear viscoelastic properties of rubber. The energy dissipation is proportional to the area of hysteresis loops – Lissajous curves. The area depends on the loss modulus of rubber and on temperature. The results of strain-controlled experiments show the dependence of the energy dissipation on the temperature, on frequency, on preload and on the amplitude of cyclic loading. The mechanical properties are temperature dependent.

The self-heating depends on the dissipated energy and on the heat transfer to surrounding environment. The thermal insulation leads to higher self-heating of rubber elements.

In practical engineering applications the rubber dampers are subjected to dynamic loading superimposed on a preload such as weight of engine or vehicle. Such cyclic loading is rather force controlled. In such case the heat generation and the temperature rise can attain quite high level which can significantly affect the properties of rubber damping elements and lead to a catastrophic failure.

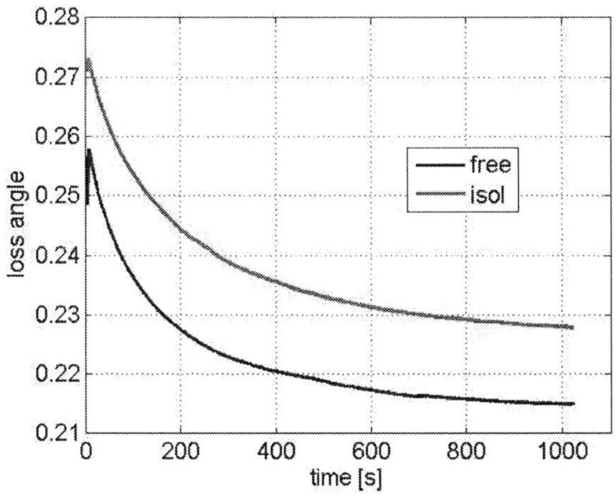

Fig. 8: Loss angle in cyclic compressive test of rubber B at 20 Hz – difference for free and prevented heat convection.

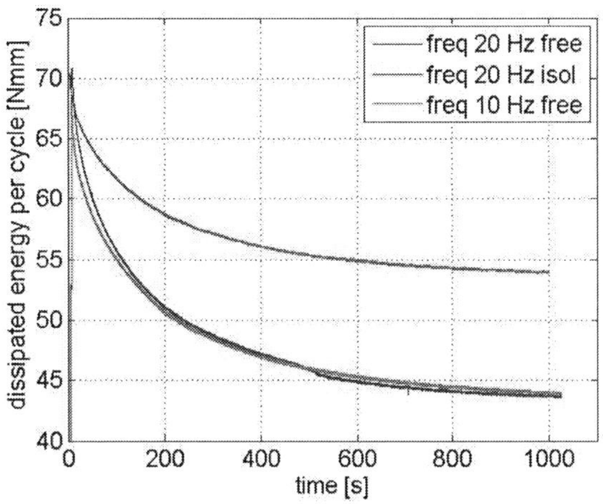

Fig. 9: Comparison of the evolution of the area of hysteresis loop in cyclic tests – rubber B.

Acknowledgement

This work was supported by national financial resources of Ministry of Education Youth and Sports of Czech Republic for specific university research (SGS 21120).

References

[1] Ratner, S.B. and Korobov, V.I., 1965. Self-heating of plastics during cyclic deformation. Polymer Mechanics, 1(3), pp.63-68.

[2] Petrikova, I., Marvalova, B. and Nhan, P.T., 2011. Influence of thermal ageing on mechanical properties of styrene-butadiene rubber. Constitutive Models for Rubber VII. Taylor and Francis, pp.77-83.

[3] Lion, A. and Kardelky, C., The Payne effect in finite viscoelasticity: constitutive modelling based on fractional derivatives and intrinsic time scales. International Journal of Plasticity, 20(7), (2004) pp.1313-1345.

[4] Höfer, P. and Lion, A., Modelling of frequency-and amplitude-dependent material properties of filler-reinforced rubber. Journal of the Mechanics and Physics of Solids, 57(3),(2009) pp.500-520.

[5] Pichon, P.G., Boutaous, M.H., Méchin, F. and Sautereau, H.,Measurement and numerical simulation of the self heating of cross-linked segmented polyurethanes under cyclic loading. European Polymer Journal, 48(4), (2012) pp.684-695.

[6] Molinari, A. and Germain, Y., Self heating and thermal failure of polymers sustaining a compressive cyclic loading. International journal of solids and structures, 33(23), (1996) pp.3439-3462.

[7] Rittel, D. and Rabin, Y., An investigation of the heat generated during cyclic loading of two glassy polymers. Part II: Thermal analysis. Mechanics of Materials, 32(3), (2000) pp.149-159.

[8] Johnson, A.R. and Chen, T.K., Approximating thermo-viscoelastic heating of largely strained solid rubber components. Computer meth. in appl. mechanics and engineering, 194(2), (2005) pp.313-325.

Experimental Verification of Modal Characteristics of Vehicle Computing Model

J. Melcer [1,a]

[1] *University of Žilina, Faculty of Civil Engineering, Slovak Republic*
[a] *jozef.melcer@fstav.uniza.sk*

Abstract: For the numerical simulation of vehicle motion along transport structures various computing models of vehicles were created. To obtaine relevant results of numerical simulation the properties of computing model must correspond to the reality. The parameters of vehicle computing model were tuned on the basis of in situ experimental test. The modal characteristics of computing model were compared with modal characteristics of real vehicle.

Keywords: experiment; numerical simulation; vehicle; computing model; modal characteristics.

1 Introduction

Solution of the problem of vehicle – road or vehicle – bridge interaction demands to create computing model of vehicle and computing model of a structure. The computing model of vehicle can be created on various levels – one two or three dimensional. For the modeling of moving load effect on pavements the three dimensional computing model of vehicle Tatra 815 was created. The properties of the model were verified by experimental test carried out on actual vehicle. The modal characteristic of computing model and modal characteristics of actual vehicle were mutually compared.

2 Vehicle computing model

The space multi-body computing model of vehicle Tatra 815 was created, Fig. 1, [1]. The model has 9 mass degrees of freedom.

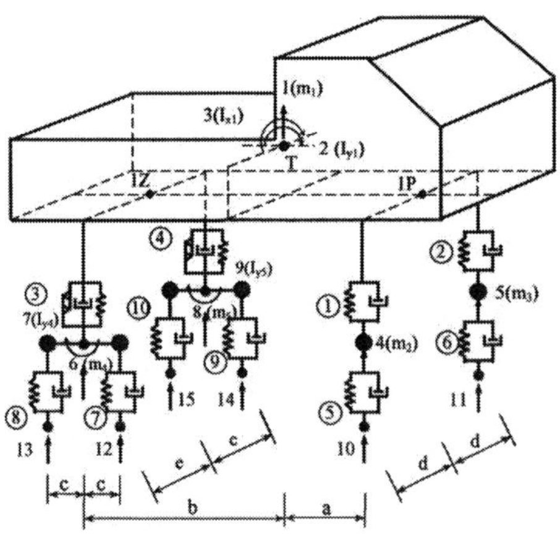

Fig. 1: Space computing model of vehicle Tatra 815.

The mass characteristics and stiffness characteristics were verify on the real vehicle. The vehicle was weighted as a whole and every axle individually. The results are put into Tab. 1.

Tab. 1: Mass of full and empty vehicle.

	m_C [kg]	m_{LF} [kg]	m_{RF} [kg]	m_{LR} [kg]	m_{RR} [kg]
full vehicle	23 250	2 840	2 830	8 905	8 675
empty vehicle	11 990	2 300	2 300	3 695	3 695
mass difference	11 260	540	530	5 210	4 980

For the determination of stiffness constants of vehicle computing model the distances of characteristic points of vehicle bed from comparative plane for empty and full vehicle were measured. The distances were measured by laser distance meter Leica DISTO^{TM} A5. Tire pressing were evaluated from digital snaps. The marking of vehicle bed characteristic points and vehicle wheels are in the Fig. 2.

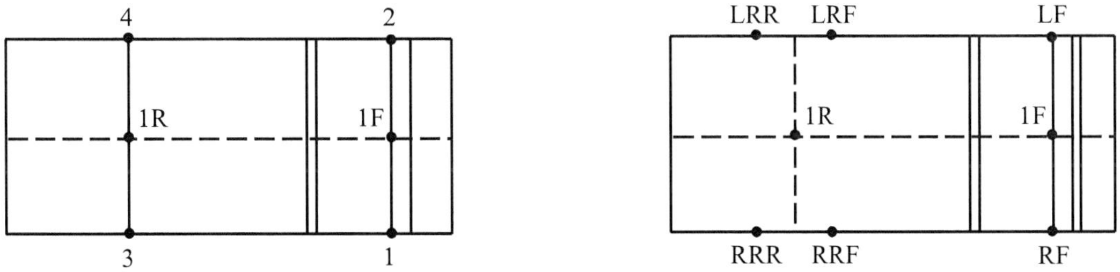

Fig. 2: Marking of vehicle characteristics points.

The distances of vehicle bed characteristic points in regard of comparative plane are put into Tab. 2 and tire pressing into Tab. 3.

Tab. 2: Distances of vehicle bed characteristic points in regard of comparative plane.

	w_1 [mm]	w_2 [mm]	w_3 [mm]	w_4 [mm]
full vehicle	1 410	1 392	1 507	1 507
empty vehicle	1 364	1 345	1 447	1 441
distance difference	46	47	63	66

Tab. 3: Tire pressing of individual wheels.

	w_{RF} [mm]	w_{LF} [mm]	w_{RRF} [mm]	w_{RRR} [mm]	w_{LRF} [mm]	w_{LRR} [mm]
tire pressing	3.2	3.4	9.7	9.8	10.2	10.3

The forces activated the displacements of vehicle bed characteristic points are put into Tab. 4 and forces activated the tire displacements of individual wheels are put into Tab. 5.

Tab. 4: Forces activated the displacements of vehicle bed characteristic points.

	F_1 [N]	F_2 [N]	F_3 [N]	F_4 [N]
force	5 300	5 400	49 800	52 100

Tab. 5: Forces activated the tire displacements of individual wheels.

	F_{RF} [N]	F_{LF} [N]	F_{RRF} [N]	F_{RRR} k_7 =	F_{LRF} [N]	F_{LRR} [N]
force	5 300	5 400	24 900	24 900	26 050	26 050

From the forces put into Tab. 4 and 5 and from displacements put into Tab. 2 and 3 the stiffness constants of individual connecting members of vehicle computing model were calculated. The stiffness constants were calculated in two variants. In variant 1 the stiffness constants were calculated from average forces and average displacements, Tab. 6. It means that the values for left and right side of vehicle are the same. In variant 2 the stiffness constants were calculated for actual values, Tab. 7.

Tab. 6: Stiffness constants for variant 1 – average values.

	$k_1 = k_2$ [N/m]	$k_3 = k_4$ [N/m]	$k_5 = k_6$ [N/m]	$k_7 = k_8 = k_9 = k_{10}$ [N/m]
stiffness constants	123 843	934 862	1 621 212	2 547 500

Tab. 7: Stiffness constants for variant 2 – actual values.

	k_1 [N/m]	k_2 [N/m]	k_3 [N/m]	k_4 [N/m]	k_5 [N/m]
stiffness constants	123 831	123 854	935 207	934 525	1 656 256
	k_6 [N/m]	k_7 [N/m]	k_8 [N/m]	k_9 [N/m]	k_{10} [N/m]
stiffness constants	1 588 235	2 567 010	2 540 816	2 553 922	2 529 126

For the variant 1 and 2 the natural frequencies and natural model were calculated.
Results for variant 1:

Diagonal mass matrix

$$\{\mathbf{m}\}_D = \{m_1, I_{y1}, I_{x1}, m_2, m_3, m_4, I_{y4}, m_5, I_{y5}\}_D =$$
$$= \{20200, 54822, 20196, 455, 455, 1070, 466, 1070, 466\}_D \quad [\text{kg, kg.m}^2].$$

Diagonal matrix of stiffness constants

$$\{\mathbf{k_i}\}_D = \{k_1, k_2, k_3, k_4, k_5, k_6, k_7, k_8, k_9, k_{10}\}_D =$$
$$= \{123843, 123843, 934862, 934862, 1621212, 1621212, 2547500, 2547500, 2547500, 2547500\}_D \quad [\text{N/m}].$$

Natural frequencies

$$\{\mathbf{f}\} = \{f_{(1)}, f_{(2)}, f_{(3)}, f_{(4)}, f_{(5)}, f_{(6)}, f_{(7)}, f_{(8)}, f_{(9)}\} =$$
$$= \{1.14,\ 1.46,\ 1.68,\ 9.86,\ 9.86,\ 10.98,\ 10.98,\ 11.96,\ 11.96\} \quad [\text{Hz}].$$

Natural modes – modal matrix

$$[\mathbf{r}[= \lfloor \{r_{(1)}\}\ \{r_{(2)}\}\ \{r_{(3)}\}\ \{r_{(4)}\}\ \{r_{(5)}\}\ \{r_{(6)}\}\ \{r_{(7)}\}\ \{r_{(8)}\}\ \{r_{(9)}\} \rfloor =$$

$$= \begin{bmatrix} -0.7589 & 0.0000 & -0.8644 & 0.0000 & 0.0022 & 0.0000 & 0.0000 & 0.0000 & -0.0118 \\ 0.5935 & 0.0000 & -0.4074 & 0.0000 & -0.0026 & 0.0000 & 0.0000 & 0.0000 & -0.0047 \\ 0.0000 & 0.9725 & 0.0000 & -0.0022 & 0.0000 & 0.0000 & 0.0000 & -0.0114 & 0.0000 \\ -0.1884 & -0.0701 & 0.0301 & -0.7071 & -0.7071 & 0.0000 & 0.0000 & -0.0017 & -0.0004 \\ -0.1884 & 0.0701 & 0.0301 & 0.7071 & -0.7071 & 0.0000 & 0.0000 & 0.0017 & -0.0004 \\ -0.0189 & -0.1489 & -0.2060 & 0.0011 & -0.0002 & 0.0000 & 0.0000 & -0.7071 & 0.7071 \\ 0.0000 & 0.0000 & 0.0000 & 0.0000 & 0.0000 & 1.0000 & 0.0000 & 0.0000 & 0.0000 \\ -0.0189 & 0.1489 & -0.2060 & -0.0011 & -0.0002 & 0.0000 & 0.0000 & 0.7071 & 0.7071 \\ 0.0000 & 0.0000 & 0.0000 & 0.0000 & 0.0000 & 0.0000 & 1.0000 & 0.0000 & 0.0000 \end{bmatrix}$$

Results for variant 2:

Diagonal mass matrix

$$\{\mathbf{m}\}_D = \{m_1, I_{y1}, I_{x1}, m_2, m_3, m_4, I_{y4}, m_5, I_{y5}\}_D =$$
$$= \{20200, 54822, 20196, 455, 455, 1070, 466, 1070, 466\}_D \quad [\text{kg, kg.m}^2].$$

Diagonal matrix of stiffness constants

$$\{\mathbf{k_i}\}_D = \{k_1, k_2, k_3, k_4, k_5, k_6, k_7, k_8, k_9, k_{10}\}_D =$$
$$= \{123831, 123854, 935207, 934525, 1656250, 1588235, 2567010, 2540816, 2553922, 2529126\}_D \quad [\text{N/m}].$$

Natural frequencies

$$\{\mathbf{f}\} = \{f_{(1)}, f_{(2)}, f_{(3)}, f_{(4)}, f_{(5)}, f_{(6)}, f_{(7)}, f_{(8)}, f_{(9)}\} =$$
$$= \{1.14,\ 1.46,\ 1.68,\ 9.76,\ 9.95,\ 10.96,\ 10.99,\ 11.95,\ 11.98\} \quad [\text{Hz}]$$

Natural modes – modal matrix

$$[\mathbf{r}] = \left[\{r_{(1)}\} \quad \{r_{(2)}\} \quad \{r_{(3)}\} \quad \{r_{(4)}\} \quad \{r_{(5)}\} \quad \{r_{(6)}\} \quad \{r_{(7)}\} \quad \{r_{(8)}\} \quad \{r_{(9)}\} \right] =$$

$$\begin{bmatrix} 0.7589 & -0.0008 & -0.8644 & -0.0016 & 0.0016 & -0.0001 & 0.0001 & -0.0070 & -0.0094 \\ -0.5935 & -0.0013 & -0.4074 & 0.0018 & -0.0018 & 0.0000 & 0.0000 & -0.0028 & -0.0037 \\ 0.0014 & -0.9725 & 0.0023 & -0.0016 & -0.0015 & -0.0001 & -0.0001 & -0.0092 & 0.0068 \\ 0.1845 & 0.0689 & 0.0293 & -0.0109 & -0.9999 & 0.0000 & 0.0000 & -0.0017 & 0.0006 \\ 0.1922 & -0.0712 & 0.0309 & 0.9999 & -0.0105 & 0.0000 & 0.0000 & 0.0010 & -0.0013 \\ 0.0186 & 0.1483 & -0.2059 & 0.0009 & 0.0005 & 0.0000 & -0.0180 & -0.1438 & 0.9887 \\ 0.0001 & 0.0011 & -0.0016 & 0.0000 & 0.0000 & -0.0001 & -0.9998 & 0.0061 & -0.0411 \\ 0.0191 & -0.1495 & -0.2060 & -0.0005 & 0.0009 & 0.0171 & 0.0000 & 0.9887 & 0.1432 \\ 0.0001 & -0.0011 & -0.0015 & 0.0000 & 0.0000 & 0.9998 & -0.0001 & -0.0390 & -0.0055 \end{bmatrix}$$

3 Experimental test

To verify the modal characteristics of vehicle computing model the in situ experimental test on vehicle was carried out. The purpose of the test was to excite vibration very closed to natural vibration. Twelve test trip of vehicle were realized during the experiment. The list of individual runs within the test is put into the Tab. 8.

Tab. 8: List of individual runs within the test.

No. of run	description of run
1	run over standard obstacle
2	run over standard obstacle
3	run over plank of dimensions 50×320 mm
4	run over plank of dimensions 50×320 mm
5	run by right wheel over plank in longitudinal direction (length 3.5 m)
6	run by right wheel over plank in longitudinal direction (length 3.5 m)
7	run by right wheel over standard obstacle and by left wheel over plank (obstacles are shift 0.9 m in longitudinal direction)
8	run by right wheel over standard obstacle and by left wheel over plank (obstacles are shift 1.2 m in longitudinal direction)
9	run to left-handed arch
10	run to left-handed arch
11	run to left-handed arch plus normal obstacle in the left track
12	run by right wheel along street curb and by left wheel along pavement

The vehicle response was registered by four accelerometers Bjuer-Kjaer BK 4508. Sensor BK1 was located on the front axle, BK2 on the rear axle of vehicle in vertical direction. Sensor BK3 was located on the bed of vehicle in horizontal direction and sensor BK4 in the front the frame of vehicle body in vertical direction. The signal was amplified, digitized by A/D interface and stored in computer. The spectral analysis of individual records was carried out. Power spectral densities (PSD) were used for evaluation of natural frequencies. The values of dominant frequencies obtained from frequency spectra calculated from individual records are put into Tab. 9. The values of numerically and experimentally obtained natural frequencies are mutually compared in the Tab. 10. The results introduced in Tab. 9 and 10 are supported by figures of power spectral densities shown in Fig. 3, 4, 5.

Tab. 9: Dominant frequencies correspond to individual sensors and individual runs.

No. of run	sensor BK1	sensor BK2	sensor BK3	sensor BK4
1	10.00	2.66	1.20;12.14	1.21
2	1.20; 9.99; 10.97; 12.18	2.92	1.45; 12.14	1.19
3	1.44; 9.99; 10.47	2.65	1.20; 1.67	1.46; 2.91
4	1.67	2.89	1.46	1.69
5	9.73; 10.97	2.91	1.43	1.46
6	9.27; 10.70; 11.95	2.92	1.46	1.19
7	9.73; 10.98	2.92	1.92	1.20
8	1.17; 9.99; 10.47	2.89	1.21; 12.40	1.20
9	9.76	3.15	1.20	1.20
10	10.00	3.13	1.44	1.70
11	1.44; 10.00; 11.70	2.91	1.19	1.44
12	10.00	2.91	1.20	1.46

Tab. 10: Numerically and experimentally obtained values of natural frequencies.

j	natural frequencies $f_{(j)}$ [Hz]		
	variant 1	variant 2	experiment
1	1.14	1.14	1.17; 1.19; 1.20; 1.21
2	1.46	1.46	1.44; 1.45; 1.46
3	1.68	1.68	1.67; 1.69; 1.70
4	9.86	9.76	9.73; 9.76; 9.99; 10.00
5	9.86	9.95	9.73; 9.76; 9.99; 10.00
6	10.98	10.96	10.47; 10.95; 10.97; 10.98
7	10.98	10.99	10.47; 10.95; 10.97; 10.98
8	11.96	11.95	11.70; 11.95; 12.14; 12.18
9	11.96	11.98	11.70; 11.95; 12.14; 12.18

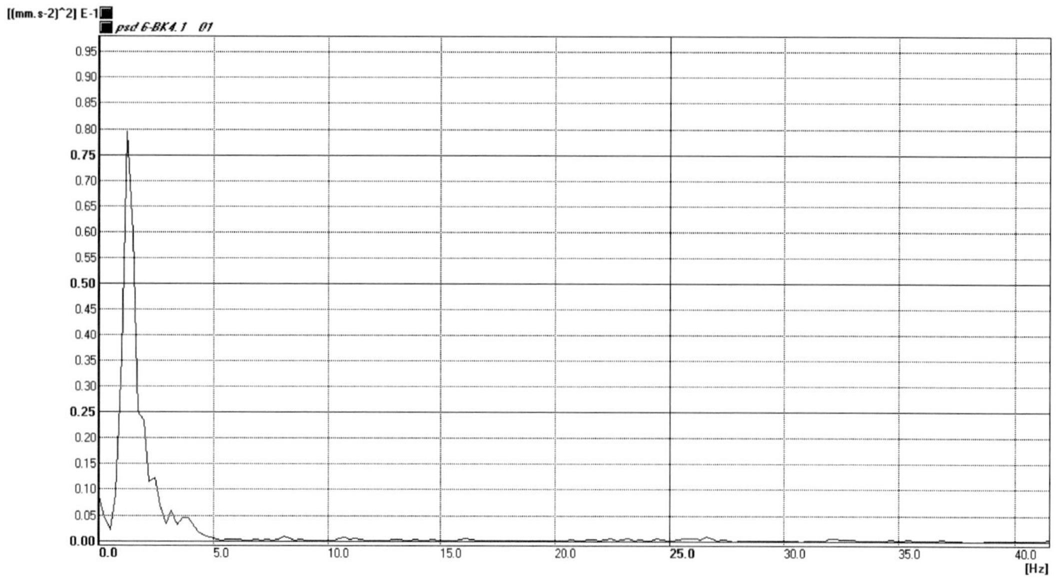

Fig. 3: PSD, run 6, sensor BK4, peak at frequency $f = 1.19$ Hz.

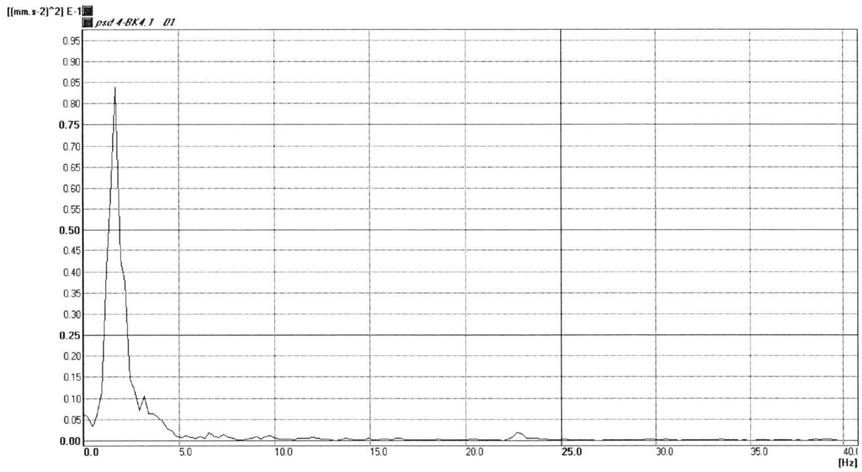

Fig. 4: PSD, run 4, sensor BK4, peak at frequency $f = 1.69$ Hz.

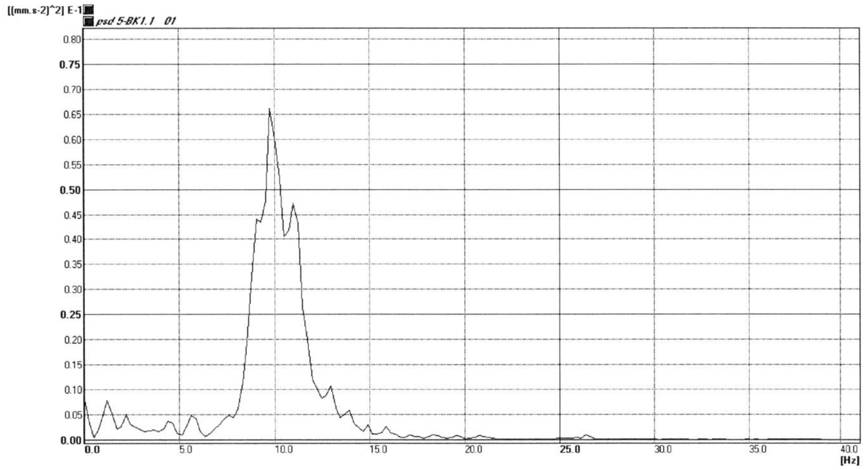

Fig. 5: PSD, run 5, sensor BK4, peak at frequency $f = 9.73$ and 10.97 Hz.

4 Conclusion

The space computational model of vehicle Tatra 815 was created as multi-body model with 9 degrees of freedom. The modal characteristics of the model were verified by in situ experimental test. The effort was to excite the vibration very closed to natural vibration and to compare experimentally obtained natural frequencies with numerically obtained ones. The results of the test confirm very good agreement between numerically and experimentally obtained values of natural frequencies. Only the frequencies in the range 2.66 – 3.15 Hz are not modeled by vehicle computing model.

Acknowledgement

The work was supported by Grant National Agency VEGA, project No. G1/0005/16.

References

[1] J. Melcer, B. - I. Martinická, Mutual comparison of FRF for various vehicle computational models, Transactions of the VŠB-Technical University of Ostrava 14 (2014) 123–130.

The effect of low temperature annealing on distribution of residual stresses of T24 steel welds

Ľ. Mráz[1], Ch. Hervoches[2], M. Vrána[2], P. Mikula[2,a], J. Kotora[3]

[1]*Welding Research Institute – Industrial Institute SR, Bratislava, Slovak Republic*
[2]*Nuclear Physics Institute ASCR, v.v.i.,25068 Řež, Czech Republic*
[3]*SES Tlmače, Slovak Republic*
[a] *mikula@ujf.cas.cz*

Abstract: The effect of low-temperature long-term heat treatment on distribution of residual stresses on the modified chromium-molybden steel of the type 7 CrMoVTi10-10 and known as T24 steel which was studied by using neutron diffraction method, is presented.

Keywords: Steel weld; residual stress; neutron diffraction; temperature anealing

1 Introduction

The modified chromium-molybden steel type 7 CrMoVTi10-10, known as T24 steel is a material that has been attracting attention for a long time [1]. This steel belongs to the group of modified CrMo low-alloyed creep resistant steels. The reasons for development of some properties of these steels is to increase the efficiency of thermal power equipment using higher media parameters in membrane walls of the super heater up to the creep zone. Post Weld Heat Treatment (PWHT) is usually applied when conventional CrMo steels are used for membrane walls. The intention is to develop and produce new steels, especially for super heater, which themselves also in an as-welded condition would achieve better properties than the parent 10CrMo9-10 steel and comparable to 9 Cr steel. However, the major objective was based on an acquisition of optimum properties of membrane walls (welded joints) without annealing, of course, it requires the determination of limits of refraining from heat treatment of welded joints in developing steels.

The leak of welds in pressure boiler systems for supercritical parameters represents an unexpected phenomenon which did not occur in conventional boilers during the development and testing. On the contrary, in certain cases, the defects, cracks, in welded joints were observed in huge components of membrane walls [2]. The same can be stated also in case of circumferential welded joints in tubes of several dimensions. The leakage of welds in T24 steel did not occur everywhere (in the region where the material of membrane walls is subject to maximum load) where this steel was used. Even workshop or on-site welds did not avoid leakage. Different opinions exist about the cause of leakage. They are briefly summarised as: material, hydrogen effects and extremely high mechanical stresses. Moreover, the consequences of potential secondary hardening of welded joints in the early stages of their operation at working temperatures have to be considered. Therefore, we would like to point out to the difficulty of the problem of crackability of welded joints in T24 steel. We assume that dealing with the above-mentioned large-sized leakage of welds in T24 steel could provide more information about this material and experience with its welding as well as the manufacture of pressure parts of this steel.

The limit for maximum hardness [3] in the vicinity of welds (HAZ) is associated with the potential risk of cold crack formation in the HAZ. The 2.25Cr1Mo steels with maximum thickness of 15 mm are generally recommended to weld at preheat temperature of 100°C when diffusible hydrogen content is below 5 ml/100 g and of 150°C when hydrogen content is from 5 to 10 ml/100 g. Due to the presence of vanadium, titanium and especially boron, these preheat temperatures can be considered as minimum ones and in some cases seem to be even insufficient.

Cold cracks, especially in the weld metal of long submerged arc welds which are oriented in transverse direction to the longitudinal weld axis, were also observed [4]. Hydrogen seems to be the most important factor for cold cracking behaviour.

Remarkable secondary hardenning for T/P24 steel after 100 hour exposure at 500°C was identified. This phenomenon is typical for vanadium hardened steels during tempering [5]. According to these results it is supposed that PWHT should be carried out in order to avoid low ductility and low toughness properties mainly during operation.

The objective of the paper is to present the effect of low-temperature long-term heat treatment on distribution of residual stresses using neutron diffraction method. The distribution of residual stresses in the vicinity of circumferential welded joints has been measured in as welded conditions and subsequently after low temperature heat treatments at 450°C for 48 hours.

2 Experimental part

2.1 Test weld preparation

Circumferential weld of two tubes made of 7 CrMoVTi10-10 creep resistant steel diameter of ∅ 38 mm and wall thickness of 6.3 mm has been prepared for this investigation. TIG process, filler material W Z CrMo2VTi (Union I P24) and shielding gas I1 have been used for welding. The heat input in the range of 8.0 to 13.0 kJ/cm, preheat temperature 150°C to 200°C and interpass temperature max. 250°C have been applied.

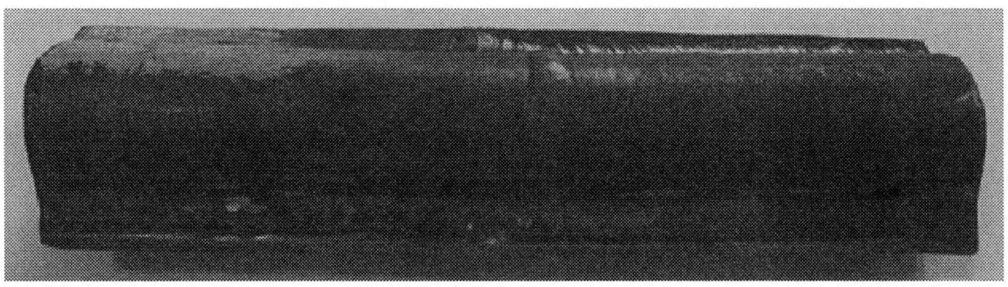

Fig. 1: Circumferential test weld of 7 CrMoVTi10-10 creep resistant steel

Hardness distribution across butt weld has been measured by Vickers method at the load of 98.1 N (HV10).

Neutron diffraction has been performed in order to identify residual stress distribution in the vicinity of the circumferential weld. Hoop, radial and axial stresses have been estimated for as/welded conditions and after long term heat treatment at the temperature 450°C for 48 hours.

2.2 Principles of the neutron diffraction method

The principle of the neutron diffraction method is quite simple. It consists in the precise determination of the d_{hkl}-spacing of particularly oriented crystal planes [6,7]. In neutron and X-ray diffraction the angular positions of the diffraction maxima are directly related to the values of the lattice spacing through the Bragg equation $2d_{hkl} \cdot \sin\theta_{hkl} = \lambda$ (d_{hkl} -lattice spacing, θ_{hkl} - Bragg angle, λ - the neutron wavelength) and thus offer a unique non-destructive technique for investigation of stress fields. When a specimen is strained elastically, the lattice spacing changes. Then, when defining the strain ε as $\varepsilon = \Delta d/d_{0,hkl}$, it is related to a change in the lattice spacing, i.e. to a component parallel to the scattering vector $\boldsymbol{Q}$ perpendicular to the reflecting set of planes.

Therefore, the knowledge of the $d_{0,hkl}$ value ($d_{0,hkl}$ is the lattice spacing of the strain-free material) is a crucial task [3]. Then by differentiation of the Bragg condition we arrive at $\varepsilon = -\cot \theta_{hkl} \cdot \Delta\theta_{hkl}$. The relation for the strain ε indicates that it gives rise to a change in the scattering angle $2\theta_{hkl}$ resulting in an angular shift $\Delta(2\theta_{hkl})$ of the peak position for a particular reflecting plane illuminated by a fixed wavelength. In such a way, the shift in the Bragg angle (relative to that of the stress-free material) permits the determination of the average lattice macrostrain over the irradiated gauge volume (see Fig. 2). The conversion of strains to stresses is carried out by means of the relation

$$\sigma_x = \frac{E_{hkl}}{(1-2\nu_{hkl})(1+\nu_{hkl})}\left[(1-\nu_{hkl})\varepsilon_x^{hkl} + \nu_{hkl}(\varepsilon_y^{hkl}, + \varepsilon_z^{hkl})\right] \qquad (1)$$

where $\varepsilon_{x,y,z}^{hkl}$ is the *x,y,z*-component of the lattice strain measured at the *hkl* crystal lattice planes, E_{hkl} and ν_{hkl} are the diffraction elastic Young modulus and diffraction Poisson ratio, respectively. For the determination of the stress tensor in this case of steel samples, three strain components should be determined as schematically shown in Fig. 3. The residual strain/stress experiment was carried out on the dedicated neutron strain/stress scanning diffractometer (http://neutron.ujf.cas.cz/en/hk4) installed at the 10 MW medium power research reactor LVR-15 (http://cvrez.cz/vyzkumna-infrastruktura/vyzkumny-reaktor-lvr-15/) and operating at the constant neutron wavelength of 0.235 nm.

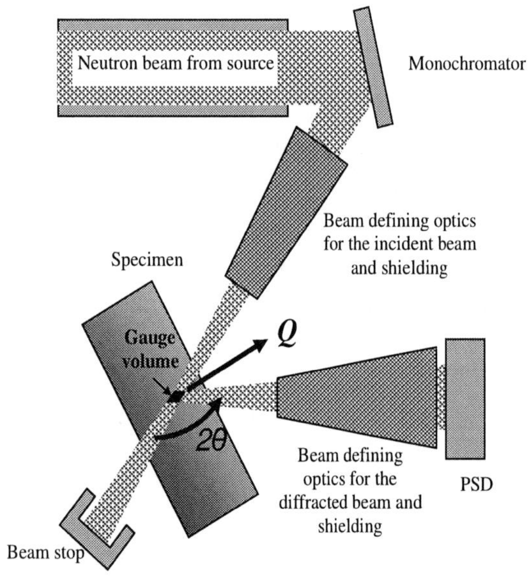

Fig. 2: Schematic illustration of a reactor source based diffractometer for strain measurement in parallel diffraction geometry

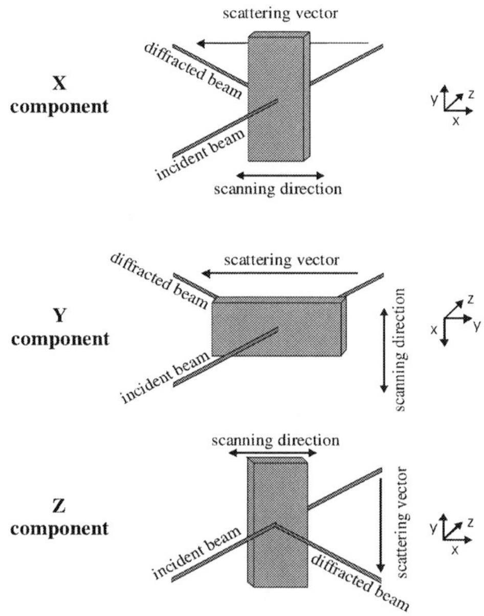

Fig. 3: Sketch of the sample setting for determination of three strain/stress components

3 Results

3.1 Hardness distribution

Hardness distribution accross the circumferential test weld prior and after low temperature heat treatment is summarized in Figure 4. No remarkable change has been noticed due to applied heat treatment.

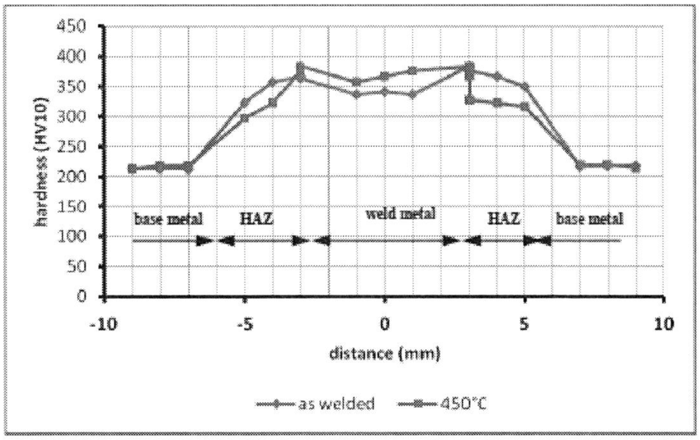

Fig. 4: Hardness distribution accross tes weld prior and after heat treatment

3.2 Neutron diffraction measurements

Axial, radial and hoop stresses have been estimated in the vicinity of the test weld (Figures 5 to 7). The results show that all stress components of weld residual stresses clear dropped after low temperature long term heat treatment.

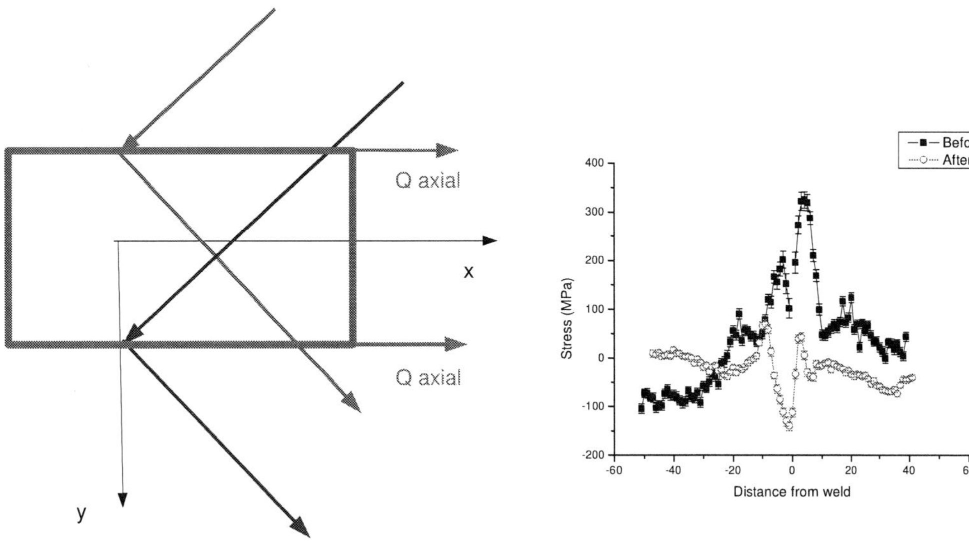

a) Scheme of the measurement b) Results

Fig. 5: Axial stresses (paralel to z-axis) in the vicinity of the test weld

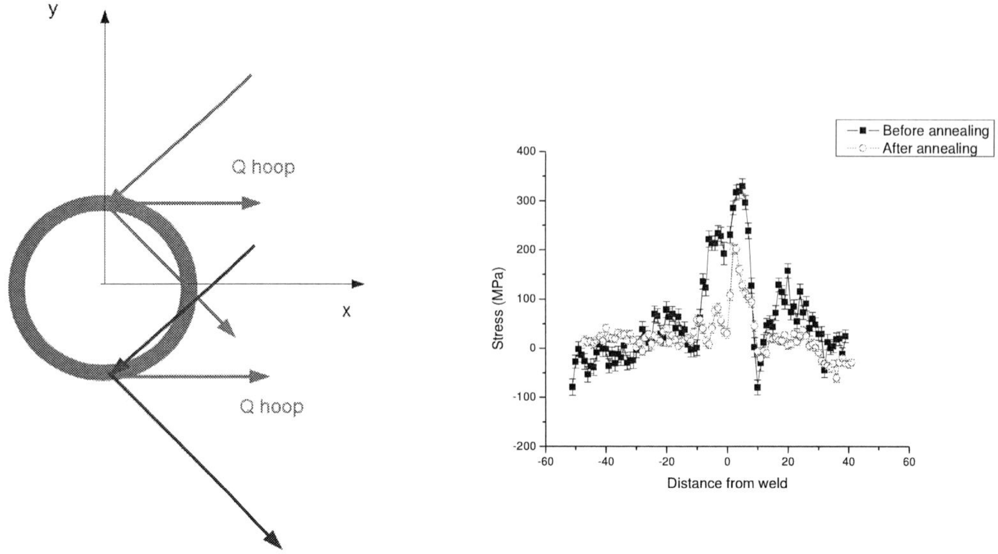

a) Scheme of the measurement b) Results

Fig. 6: Hoop (tangential) stresses in the vicinity of the test weld

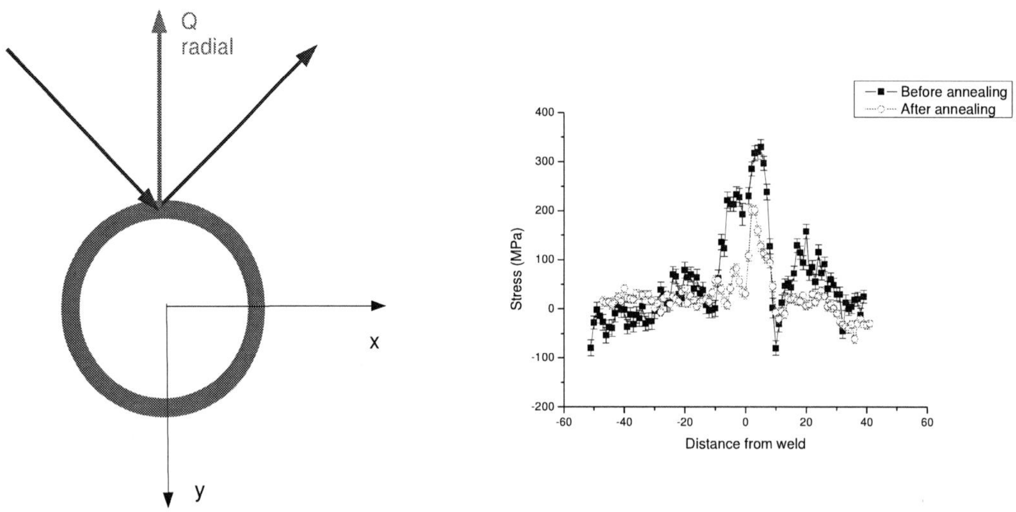

a) Sheme of the measurement b) Results

Fig. 7: Radial stresses in the vicinity of the test weld

4 Discussion

The investigation of residual stress measurement by neutron diffraction has been concentrated on the identification of the effect of low temperature heat treatment on stress distribution close to the circumferential weld of T24 steel tube. It has been recognized that the applied heat treatment did not remarkable change local properties (hardness) of the weld. Despite of this the drop of residual stresses after heat treatment compared to as welded condition is clear and remarkable. According to these results we assume that low temperature heat treatment can influence final behaviour of circumferential welds in service.

We expect that the results will contribute to the extension of knowledge about the behaviour of welded joints of the creep resistant T24 steel during the start up period of power stations.

5 Conclusions

The paper summarizes the results of neutron diffraction application for the estimation of residual stresses distribution in the circumferential weld of creep resistant T24 steel. The results show the even low temperature long term heat treatment can influence the level of residual stresses. One can suppose that this effect can play an important role in the behavior of weld during initial period of power station start up.

Acknowledgement

The related experimental studies were carried out at the CANAM infrastructure of the NPI ASCR Rez and supported by the project EU-FP7-NMI3 II: "Integrated Infrastructure Initiative for Neutron Scattering and Muon Spectroscopy", 2009-2015.

References

[1] Rohrstähle fuer moderne Hochleistungskraftwerke, Sonderdruck aus „3R International“, Heft 7/2008.

[2] VDI Nachrichten: Beizen schwächte Schwessnähte im Kraftwerk, Duisburg 23.8.2010.

[3] VdTUV-Merkblatt Dampfkessel 451-68/1 Herstellung gasdicht geschweisster Rohr wände, Ausgabe Mai 1968.

[4] Ľ. Mráz, F. Király, M. Paľo: Analysis of T24 steel welded joints – etapa IIb, october 2011, technical report 222/2000, ME095-IIb.

[5] P. Mohyla, V. Foldyna: Improvement of reliabity and creep resitance in advanced low-alloy steel, Material Science and Enginiering A 510-511 (2009) 234-237.

[6] M.T. Hutchings, A. D. Krawitz (Eds.), Measurement of residual and applied stress using neutron diffraction, NATO ASI Series, Applied Sciences 26, Kluwer Academic Publisher, 1992.

[7] V. Stelmukh, L. Edwards, J.R. Santisteban, S. Ganguly, M.E. Fitzpatrick, Weld stress mapping using neutron and synchrotron x-ray diffraction, Materials Science Forum 404-407 (2002) 599-604.

Shear Fracture - Testing Configuration for Cement Paste in Mode II

Barbora Muzikova [1,a]

[e] *CTU in Prague, Department of Mechanics, Czech Republic*
[a] *barbora.muzikova@fsv.cvut.cz*

Abstract: This article is focused on different ways of testing in mode II. There are presented three basic modes of a shear. The best way how to test the fracture mode II and determinate the critical factor K_{II} of cement paste beams is investigated here. Several methods are presented and the test based on a theory of the double-edge notched infinite plate is described in detail.

Keywords: shear fracture; mode II; experimental; fracture mechanics.

1 Introduction

The shear failure is a fundamental problem in practical engineering. Nowadays, various empirical formulae for evaluating the shear loading capacity are given. They were developed with using experimental data. In fact, this failure problem is a typical brittle fracture. Therefore, several researchers attempted to apply fracture mechanics to the shear failure for gaining a satisfactory physical model of the shear fracture.

In order to understand well the behaviour of concrete and the shear load, mode II testing needs to be performed for measuring mode II fracture toughness K_{II} and mode II fracture energy G_{IIF} of concrete materials. In practical engineering structures, it can be presented on a door-case in a building under the shear forces, the mode II fracture parameters can be directly applied to analysis. This article presents several methods for testing in the mode II and the double-edge notched infinite plate is described in detail here.

2 Mode II Tests for Concrete Materials

2.1 Fracture Modes

In general, there are three modes of the fracture failure shown in Fig. 1 - mode I opening, mode - II sliding, mode III - tearing. The combination failure in the mode I and the mode II is the main problem of testing mode II, so called a mix-mode. Each mode has its own stress factor intensity K. These factors represent the anasymptotic sphere of stress in the area of the crack mouth [2].

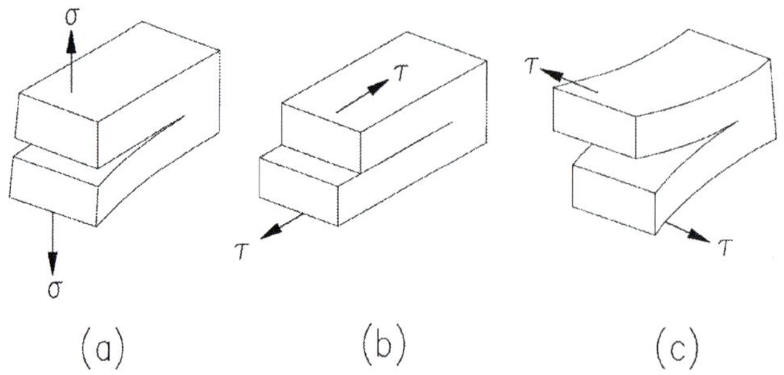

Fig.1: Modes of fracture failure.

2.2 Different Mode II Testing Configurations

In last years, many researchers paid attention on seeking mode II tests without mode I component supplement. As it turns out, not only the geometry but also the way of loading is important, equally, the geometry affecting the beam also affecting the test. Various methods were designed, some of them are shown in Fig. 2. First of all, Fig. 2a shows the theoretical pure stresses along the crack which is envisaged by testing but cannot be realized. Fig. 2b is designed by Iosipescu for metals and a welded joint, it seems to be a good solution but it came out, that the results are quite controversial and there is a thought that mode I has an influence in this test.

Fig. 2c - a push-off beam by Mattocky and Hawkins is originally presumed for testing concrete, but numerical studies have shown, that this is a mix-mode problem. The axisymmetric punch-though specimen (Fig. 2d) by Tada was used on mortar and concrete due to its easy handing, however there are large tensile stresses occured at the crack tip.

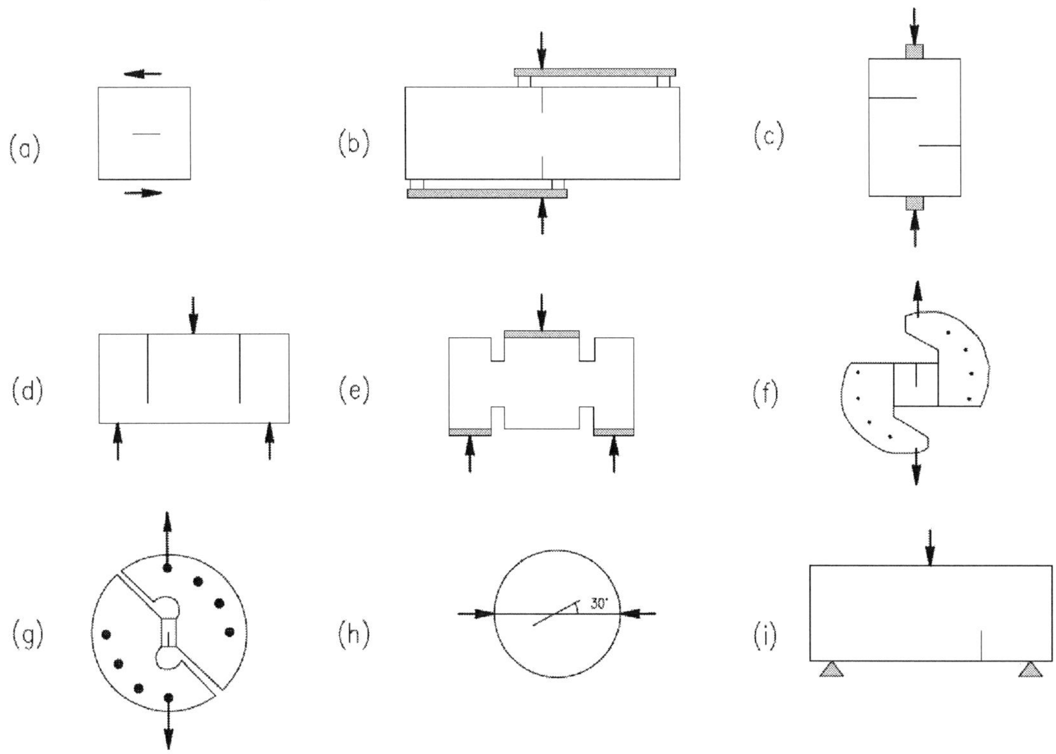

Fig. 2: Testing configuration for mode II

For a rock core testing, the cylindrical specimen with four notches (Fig. 2e) gives mixed-mode results. Fig. 2f - Although there are several methods proposed for mode II testing, none of them produces a pure mode II situation. That was the reason why Shilang Xu and W. Reinhardt carried out a new research project for an improved testing to study pure shear fracture of concrete.

Fig. 2g shows testing devices which allow various mixed combinations by rotation of the holder. Fig. 2g is not suitable for concrete, Fig. 2h designed by Irobe and Pen and finally the off-centre notched beam specimen by Jenq and Shah for studding mix-modes (Fig. 2i) [1, 3].

2.3 The Double-Edge Notched Specimen

Theoretical analysis was carried out to propose an improved testing method. The double-edge notched plate (Fig. 3) under the in-plane tensile loading seems to be appropriate, in both x-direction and y-direction the plate is infinite with unit thickness, the length of the ligament 2a is finite, but the lengths of double-edge notches are infinite. For such cases there is a formula of stress intensity factors as follows, and it is important to notice that K_I vanishes and K_{II} remains as the only one stress component.

$$K_I = 0 \quad (1)$$

$$K_{II} = \frac{\sigma}{4}\sqrt{\pi a} \quad (2)$$

The plane is still infinite in x-direction but finite in y-direction, determined by length w (Fig.3). If we take a half of the strip, the stress distribution along the symmetry axis of the strip could be assumed to be linear distribution. Using the relation between the stress intensity factor and J-integral, we can get the final formulae for K_{II} for a specimen with finite size [4, 5].

$$K_{II} = \frac{\sigma}{4}\sqrt{w} \quad (3)$$

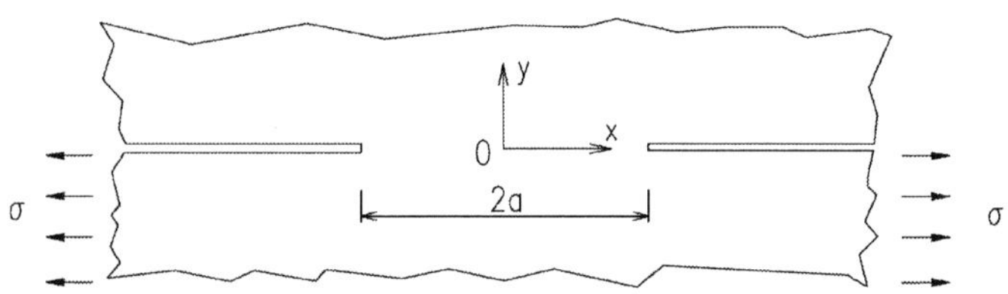

Fig. 3: The double edge-notched infinite plate.

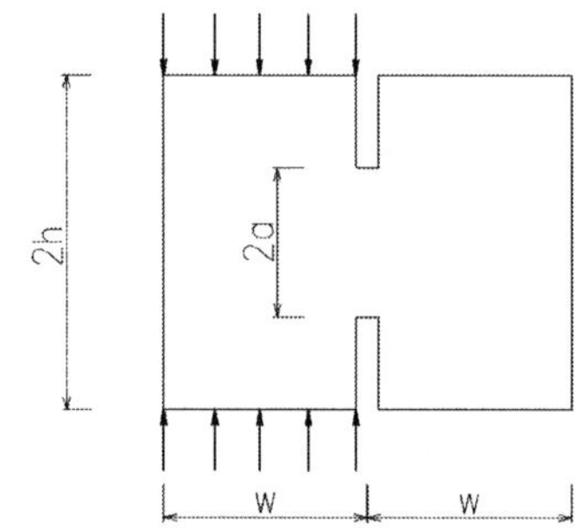

Fig. 4: The double-edge notched specimen.

And from this, the size of the specimen can be determined. It was influenced by two conditions to ensure the theory to go right, if these conditions are not executed the shear crack will not appear because the specimen will not behave as the theory of infinite plate says so, that was described above. These important two formulae are below Eq. (4) and Eq. (5).

$$w \leq \pi a \quad (4)$$

$$2a \leq h \quad (5)$$

The geometry of specimen is shown in Fig. 4 and the shear stress is determined as a stress level when shear cracks along the ligament appear. It is observable in the F-δ curve as a discontinuous zone [1, 3].

Tab. 1: The prescriptions of specimens.

Prescription	Filler [%]	Binder [%]	Binder		Water-Cement ratio
			Cement [%]	Fly ash [%]	
S1	75.0	25.0	50.0	50.0	1.0
S2	75.0	25.0	75.0	25.0	1.0
S3	75.0	25.0	100.0	0.0	0.8

2.4 The Real Test

To prove the theory into the real conditions there were made specimens from cement paste with fly ash. The prescriptions are shown in Tab. 1.

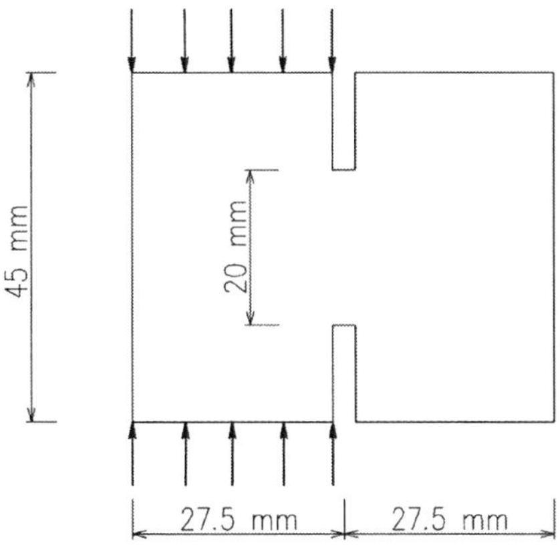

Fig. 5: Size of the testing specimen made from cement paste.

Fig. 6: The specimen in testing device.

The size of the specimen was figured out in consideration of Eq. (4) and Eq. (5). It is shown on Fig. 5. The test was carried on 55 specimens. Specimens were tested in a device for measuring compression strength. It was assumed that using this geometry and loading schema caused the concentration of shear stress along the mouth of the saw-cuts and the shear crack will go along the ligament.

It is needed to mention, that the due to the comprehensive loading on the half of the specimen, the compressive failure will appear. By using the theory of the double-edge notched specimen and executing the Eq. (4) and Eq. (5) the shear failure should appear at first. The load-displacement curve was recorded and the discontinuous zone was noticed, it is shown in Fig. 8. Also the presumed shear crack along the ligament appeared (Fig. 7). But not at all specimens are managed so, it can be seen the Fig. 9, where the factor K_{II} was determined.

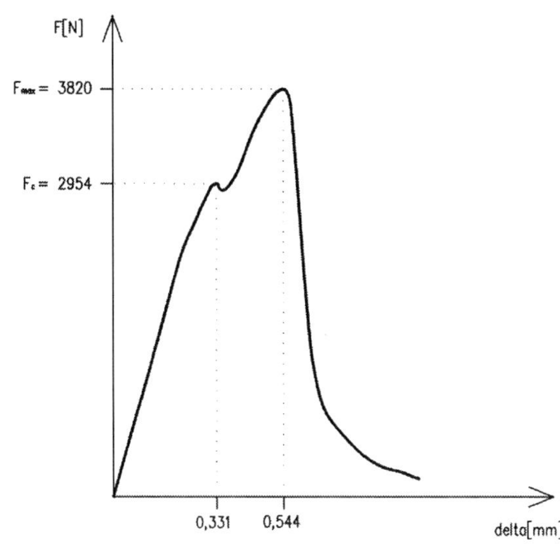

Fig. 7: Size of the testing specimen made from cement paste.

Fig. 8: The specimen in testing device.

3 Conclusion

The aim of this article is to look at the shear stress in the point of brittle mechanic, to test the shear as a mode II failure. In order to understand well the behaviour of concrete and the shear load, mode II testing needs to be performed for measuring mode II fracture toughness $K_{II.}$ There are many methods but none of them gives a result in the pure mode II here.

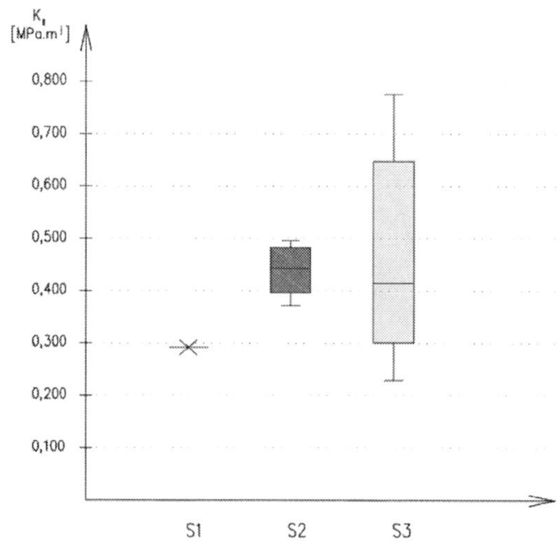

Fig.9: The K_{II} stress intensity factors.

The target of this work is also to try a new geometry and a load configuration so called the double-notched infinite plate, and there were settled conditions for the geometry and formulae for determined K_{II} of a beam based on the theory of the infinite plate in this article. This geometry and loading schema was tested on real beams from cement paste with fly ash.

The assumed shear crack and also the discontinuous zone in the load-displacement curve were observed during the real testing. To summarize, the theory of the double-edge notched specimen was performed into the real test and the antecedent progress was according to the theory of the infinite plate.

Acknowledgement

The financial support of this experiment by the Faculty of Civil Engineering, Czech Technical University in Prague (SGS project No. 14/122/OHK1/2T) is gratefully acknowledged.

References

[1] S. Shah Suredra P., S. Stuart E., O. Chenhsheng, Fracture Mechanics of Concrete, New York, John Wiley and Sons, 1995.

[2] M. Jirasek, J. Zeman, Reshape and Fracture of Materials: Reshape, Plasticity, Demage and Fracture, CTU in Prague, 2006.

[3] S. Xu, H. W. Reinhardt, Shear Fracture on the Basic of Fracture Mechanics, Otto Graf Journal, 2005.

[4] Radaj, D. and S. Zhang, Stress Intensity Factors for Spot Welds Between, Plates of Unequal Thickness, Engineering Fracture Mechanics, 1991.

[5] V Rice, J. R., A Path Independent Integral and the Approximate Analysis of Strain Concentrations by Notches and Cracks, Journal of Applied Mechanics, 1968.

Experimental and numerical analysis of glass samples

Novák. O.[1,a], Petrů M. [1], Vosáhlo J. [1]

[1] *Technical University of Liberec, Studentská 2, 46117 Liberec, Czech Republic*

[a] *novak.ondra1@seznam.cz*

Abstract: A stress analysis of material is important for a design of safe and efficient constructions. The glass is very brittle material, but its mechanical properties allow its usage as structural material. The assembling of the material models for certain type of glass can be very helpful for assessment of safety of glass constructions, but also hanging products such as chandeliers, or decorative glass.

Keywords: glass; Charpy impact test, 3-point bend test, FEM, numerical analysis

1 Introduction

Numerical analysis of glass materials is very important and helpful way for design of safe and reliable products. For assembling of suitable glass material model different parameters are required. Some of them are possible to find in the literature, but for concrete type of glass they have to be experimentally determined. The glass is very brittle material and its testing due to this characteristic could be very difficult. For example testing of Young modulus by the help of standard tensile test is practically impossible. Therefore Charpy test and 3-point bend test were involved for determination of Young and bending modulus. For both tests FEM models were assembled.

2 Experiment

Two different glass samples marked as Type 1 and Type 2 were used in the experiment. For the determination of Young modulus and flexural modulus a 3-point bend test (Fig. 1 left) in accordance with standard ČSN ISO 14125 was used. Results are shown in table 1.

$$E_{to} = \frac{4F_{max}\,L^3}{3\pi A_{max} d^4} \quad (1),$$

$$E_o = \frac{L^3\,\Delta F}{\Delta s\,\pi d^4} \quad (2).$$

Type 1						
Statistics	Amax	Fmax	σ	E_{to}	$E_{o\,(0,1mm)}$	$E_{o(0,2mm)}$
	mm	N	Mpa	Mpa	Mpa	Mpa
Average	0,21	520,50	94,19	47453,81	33405,30	35794,93
Deviation	0,02	108,89	18,12	8531,41	7729,35	7416,33
Variance	10,3%	20,9%	19,2%	18,0%	23,1%	20,7%
Type 2						
Average	0,17	464,84	72,57	42767,22	30106,18	29863,02
Deviation	0,05	193,11	23,63	5941,48	3684,57	3076,76
Variance	31,3%	41,5%	32,6%	13,9%	12,2%	10,3%

Table 1: Results of 3-point bend test

A distance between supports was set on 64 mm, a strain rate was 2 $mm.min^{-1}$. Test was done on dynamometer LabTest 2.050. Young modulus was calculated from formula (1). A bending modulus is calculated from the formula (2) for sample deformation of 0,1 and 0,2 mm.

Also Charpy test (Fig. 1 right) in accordance with standard ČSN EN ISO 179-1 was performed. The impact strength is determined from the formula (3) [1, 2].

$$H = \frac{E_c}{\frac{\pi d^2}{4}} \left[\frac{J}{mm^2} \right] \quad (3).$$

The test was performed on a device LAB TEST® CHK 50J. The distance of supports was 62 mm, a pendulum velocity was 3,7 $m.s^{-1}$. Samples were without notches. Obtained results are in table 2.

	Type 1				Type 2			
	Diameter [mm]	Cross-section [mm²]	Energy E [J]	Impact tenacity [J/mm²]	Diameter [mm]	Cross-section [mm²]	Energy E [J]	Impact tenacity [J/mm²]
Average	9,637	73,22	0,423	0,0058	8,27	54,24	0,384	0,00719
Deviation	0,634	9,70	0,063	0,0007	0,837	10,31	0,07	0,001
Variance	0,402	94,12	0,004	4,43E-07	0,701	106,37	0,005	1,08E-06

Table 2: Results of Sharpy test

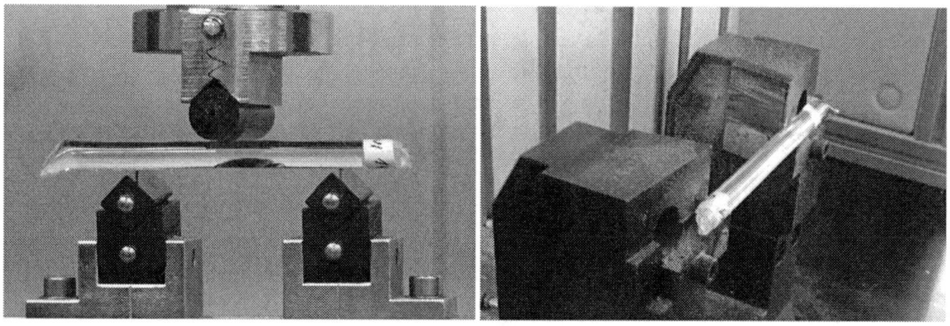

Fig. 1: Left – 3-point bend test; Right – Charpy test

The results were used for establishing of a material model of the glass material.

3 FEM model

FEM models were assembled for 3-point bend test and Charpy test. The CAD geometry was created in SolidWorks 2012 and it corresponds with real testing apparatus and samples. The geometries were imported to ANSYS software [3]. The basic geometry is created with two parts. The sample represents flexible part, supports and the indentor creates are rigid bodies. To the model contacts among the test sample, indentor and supports were introduced. A friction in contacts of individual parts was set on value 0,3. The FE mesh is created with 3D cubic elements. Boundary conditions are following: supports are fixed, the indentor in 3-point bending test is fixed, only in Z direction is allowed a displacement of 15 mm. The FE model of 3-point bending test with a mesh is shown in Fig. 2. The indentor representing Charpy pendulum can moves only in one direction with the velocity 3,7 $m.s^{-1}$. The model of Charpy test is shown in Fig. 5. FE models were optimized from point of view of material model parameters, thus model results are comparable with obtained experimental data.

3-point bend test shows stress in the structure under static loading (Fig. 3, 4), Charpy test indicates stress, energy and also failure of the glass under dynamical loading (Fig. 6, 7).

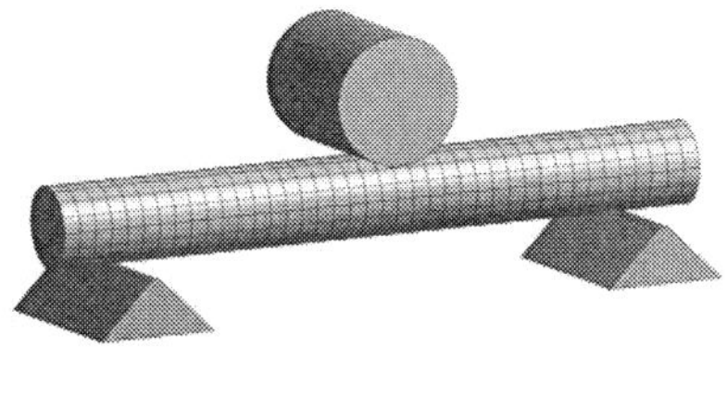

Fig. 2: Meshed geometry of 3-point bending test model

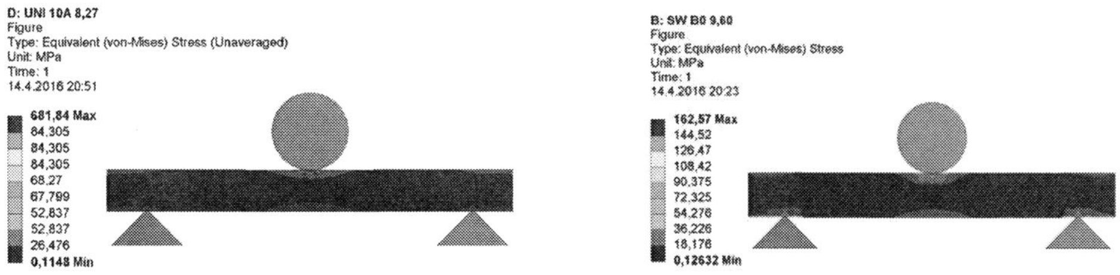

Fig. 3 and 4: Stress distribution in sample Type 1 and Type 2

Fig. 5: Meshed geometry of Sharpy test model

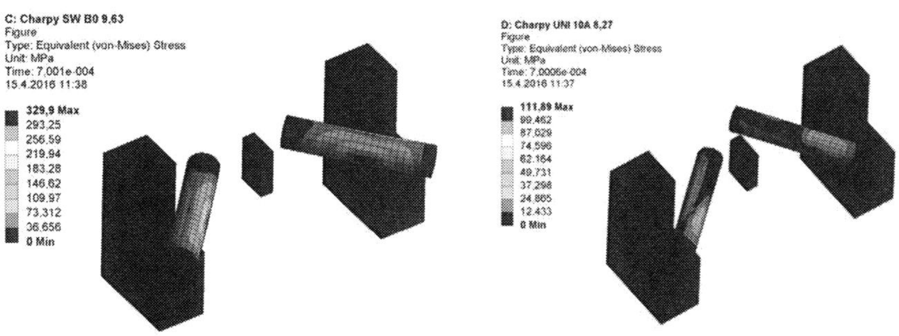

Fig. 6 and 7: Stress distribution during Sharpy test for Type 1 and Type 2 sample

4 Discussion

The model was created for a modeling of special glass products especially for decorative purposes. By the help of this material model was found for example maximal load capacity of glass bead chain that hangs in vertical direction (Fig. 8, 9). It is important to know how many beads threaded on a carrying wire can be used. It was determined by the maximal allowed stress acting on the lowest bead from the mass of another threaded beads and mass of the carrying wire. Also shape of a hole in the bead for more efficient distribution of the stress was designed by the help of this model. Another application was a determination of a tightening torque of plastic screw that fixes decorative glass plate. By the help of the model was also found optimized geometry of screw head.

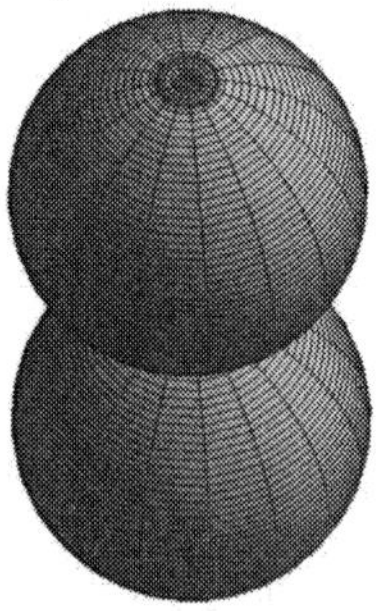

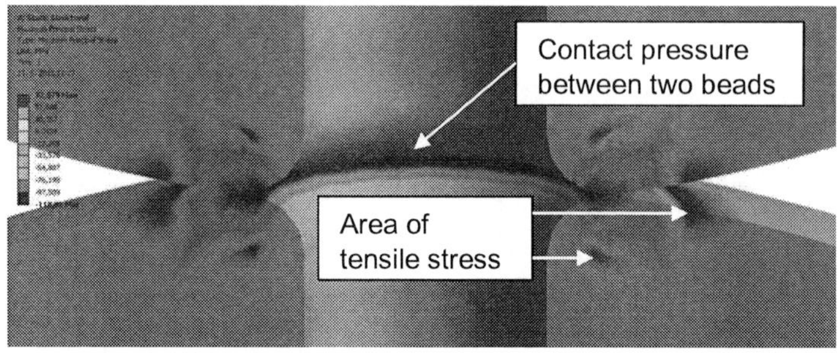

Fig. 8, 9: Model of beads pressed by the gravity force (left); a detail of contact pressure of two beads

5 Conclusion

Assembled FE models were optimized so that numerical analysis results correspond with experimental data. The models show critical stress that causes brittle fracture and crack propagation of the glass structure. Also energy during impact and propagation of stress was observed. The optimized models of these concrete types of the glass were used for stress analyses of glass products and assessment of a safety factor.

Acknowledgement

The results of this project LO1201 were obtained with co-funding from the Ministry of Education, Youth and Sports as part of targeted support from the "Národní program udržitelnosti I" program.

References

[1] *ČSN ISO 14125: Fibre-reinforced plastic composites - Determination of flexural properties*. Opr.1. Praha: Český normalizační institut, 1999.

[2] *ČSN EN ISO 179-1: Plastics - Determination of Charpy impact properties - Part 1: Non-instrumented impact test*. A1 2011. Praha: Český normalizační institut, 2001.

[3] Lee, H.: *Finite element simulations with ANSYS Workbench 13*. SDC Publications, 2011, ISBN 978-1-58503-653-0

Dependence of Compressive Strength of Rammed Earth on Used Clay Composition

T. Otcovská[1,*], P. Padevět[1]

[1] *CTU in Prague, Faculty of Civil Engineering, Department of Mechanics, Thakurova 7, Prague 6, Czech Republic*

[*] *tereza.otcovska@fsv.cvut.cz*

Abstract: This article is focuses on compressive strength of rammed earth. The mechanical properties depend on an amount of mixture water and an amount and type of clay minerals. The clay minerals play role of a binder similarly like a cement in concrete. This paper is focused on a compressive strength and its dependence on the species and quantity of clay minerals. Three earth mixtures were tested. Mechanical properties were examined on small test bodies made of the mixtures. The mechanical properties of the mixtures are compared at the end of the paper.

Keywords: Rammed Earth; Clay; Compressive Strength; Clay Mixture; Load-bearing Construction.

1 Introduction

This article focuses on a usability of unburned clay for rammed constructions where bonding component is illitic-kaolinit clay. An introduction of the article is dedicated to summary of reasons why to concern with clay construction materials in an age of modern ones. First reason is reconstruction of already built clay buildings. Clay buildings comprise considerable part of housing development in the Czech Republic and many of these building are historically valuable. Although clay structures form considerable part of traditional development there is a lack of information about the material itself [1, 2].

The second reason is conformity of clay building with principles of sustainable construction. Usage of a clay in modern constructions is an effort to apply environmentally friendly materials. Generally are these efforts connected to usage of natural materials with minimal energy demand for their preparation, usage of recyclable construction materials and usage of materials with minimal combination with synthetic substances [3, 4].

The third reason is an influence of unburned clay to human health. Main advantage of unburned clay is a quick reaction to air moisture changes and maintaining it in a level favorable to human health [5, 6].

2 Design and Manufacture of Test Bodies

The provided results concerns with clay mixtures processed by technology of ramming. A clay suitable for building constructions consists from three main components: sand (grain size 0.06-2 mm), dust particles (grain size 0.002-0.06 mm) and clays (grain size smaller than 0.002 mm). Clays are mostly composed of clay minerals which fulfill function of bonding agent similarly as a cement in a concrete. The water is added to clay mix which activates bonding ability of the clay and allows for good mixture processing. The clay minerals assure material strength and together with water are responsible for material shrinkage [6, 7].

A compressive strength of the material should rise with increasing amount of clay minerals. An amount of water in mixture may influence final strength of the material in a positive manner (better clay mix processing) or in a negative manner (higher pore count) [5, 6].

The experimental clay mix was produced by mixing of sand, clay and water. The used clay was illitic-kaolinite one (provided by the company LB MINERALS, s.r.o.). Six testing batches were produced varying by mass ratio of sand and clay (see Tab. 1). For each test batch was six test bodies produced of size $4 \times 4 \times 16$ cm and four to six test bodies of size $2 \times 2 \times 10$ cm. That bodies were created using a technology of ramming material into steel molds (see Fig. 1)

Tab. 1: Composition of the Clay Mixture Batches

set	Clay	Sand/clay ratio	Water/clay ratio
S_I	Illite, kaolinite	75/25	0.37
S_{II}	Illite, kaolinite	80/20	0.37
S_{III}	Illite, kaolinite	85/15	0.37
S_{IV}	Illite, kaolinite	75/25	0.295
S_V	Illite, kaolinite	75/25	0.335

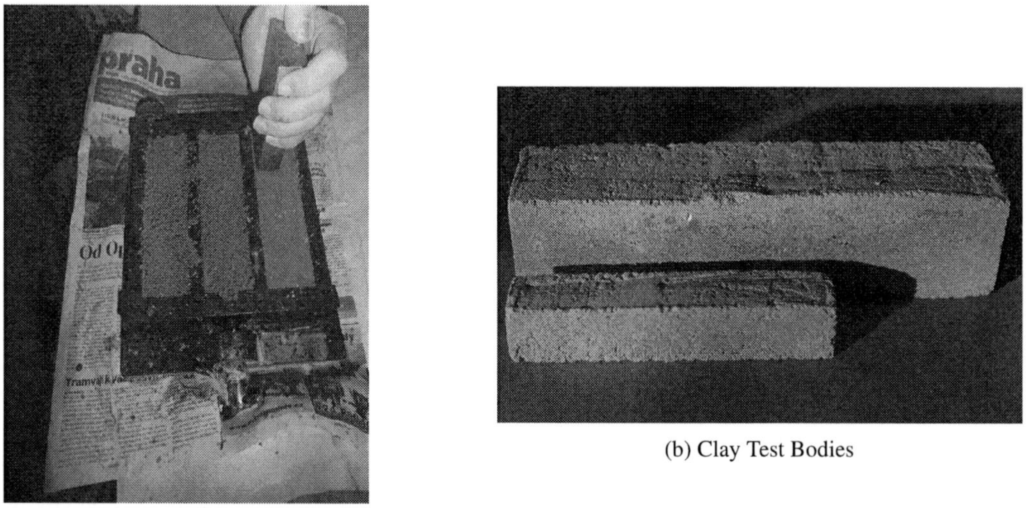

(b) Clay Test Bodies

(a) Test Bodies Ramming

Fig. 1: Manufacture of Test Bodies

3 Dependence of Compressive Strength on the Amount of Illitic-kaolinite Clay

The hypothesis that increasing amount of clay leads to increase of compressive strength was not confirmed. There were three test batches probed from which each contained six test bodies. From measured values of testing bodies compressive strength was average amount calculated (see Tab. 2).

The compressive strength dependence on an amount of a clay the average value of compressive strength 2.43 MPa for set S_I (sand/clay ratio of 75/25, w = 0.37), 2.87 MPa for set S_{II} (sand/clay ratio of 80/20, w = 0.37) and 0.59 MPa for set S_{III} (sand/clay ratio of 85/15, w = 0.37) (see Fig. 2)

Although it was not unambiguously proved that increasing amount of clay in a mix increase compressive strength, it is from measured values of test batch S_{III} clear that the compressive strength is by low amount of clay minerals influenced radically. Testing bodies from set S_{III} demonstrated high crumbliness already before beginning of compressive strength tests even though the technology and method of processing were same in all three batches.

Tab. 2: Compressive Strength of Test Bodies Dependence on the Amount of Clay in the Mixture

Set	Clay	Sand/clay ratio	Water/clay ratio	Number of test bodies	[MPa]
S_I	Illite, kaolinite	75/25	0.37	6	2.43
S_{II}	Illite, kaolinite	80/20	0.37	6	2.87
S_{III}	Illite, kaolinite	85/15	0.37	6	0.59

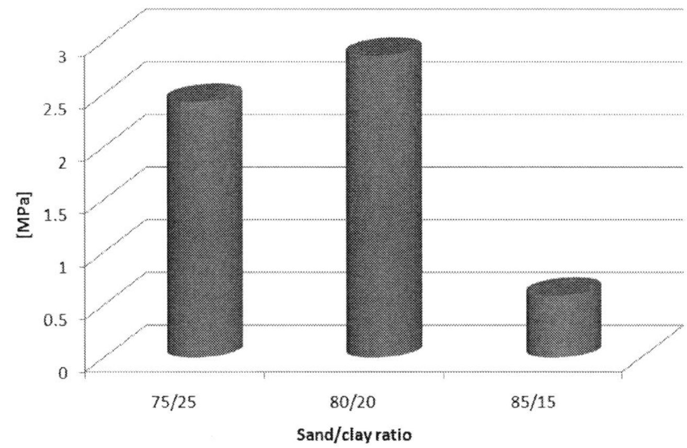

Fig. 2: Dependence of Compressive Strength on the Amount of Clay in the Mixture

Tab. 3: Compressive Strength of Test Bodies Dependence on the Amount of Mixture Water

Set	Clay	Sand/clay ratio	Water/clay ratio	Number of test bodies	[MPa]
S_I	Illite, kaolinite	75/25	0.37	6	2.43
S_{IV}	Illite, kaolinite	75/25	0.295	4	1.96
S_V	Illite, kaolinite	75/25	0.335	4	2.09

4 Dependence of Compressive Strength on the Amount of Mixture Water

The hypothesis that increasing amount of mix water leads to increase of compressive strength was confirmed. There were three test batches probed from which each contained four to six test bodies. From measured values of test bodies compressive strength was average value calculated (see Tab. 3).

The compressive strength dependence on an amount of a mixture water the average value of compressive strength 2.43 MPa for set S_I (sand/clay ratio of 75/25, w = 0.37), 1.96 MPa for set S_{IV} (sand/clay ratio of 75/25, w = 0.295) and 2.09 MPa for set S_V (sand/clay ratio of 75/25, w = 0.335) (see Fig. 3)

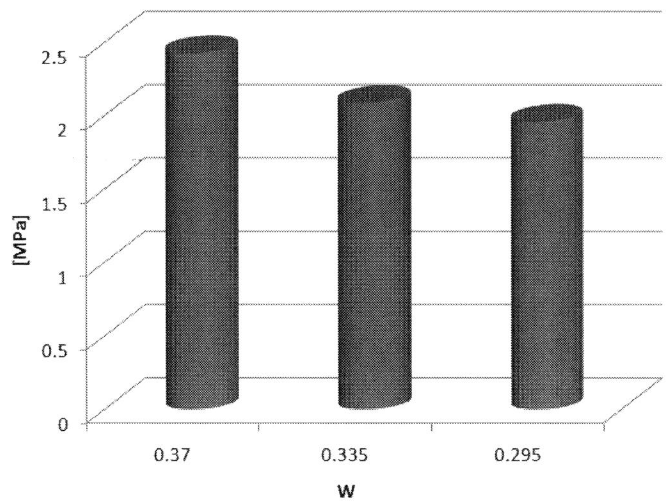

Fig. 3: Dependence of Compressive Strength on the Amount of Mixture Water

We are able to establish that increasing amount of mixture water increase compressive strength of clay material. Probable reason of this dependence is better mixture treatment and higher activation of clay minerals with mixture water increase.

However, there is a water mixture ratio point (Water Coefficient) above which the compressive strength starts to drop. All our tested bodies ware made from clay mixtures that had water coefficient much below of this critical point.

5 Conclusion

The dependability of the compressive strength on an amount of used illitic-kaolinite clay was not unambiguously proved. From measured values could be concluded that considerable change of clay amount in a mix leads to change of shrinkage and compressive strength. The hypothesis that decreasing of a clay amount decrease shrinkage during drying and compressive strength was proved for the batches with minimal and maximal amount of clay.

The compressive strength value raises with increasing of water mixture ratio. It was found from previous research that increasing amount of water mixture ratio also increase shrinkage during drying [8]. Unfortunately, these two factors are contrary to each other. It is difficult to find an optimal choice between strength and shrinkage. It is on an engineer to conclude which material parameter is more important for a given construction. Aim of the submission was to outline a sight on an area of illitic-kaolinite clay usage for load-bearing constructions from rammed unburned earth.

Acknowledgement

The financial support of this experiment by the Faculty of Civil Engineering, Czech Technical University in Prague (SGS project No. SGS16/201/OHK1/3T/11) is gratefully acknowledged.

Let us express humble acknowledgement also for LB MINERALS, s.r.o. company for free supply of material necessary for experimental measurement.

References

[1] V. Hájek, Lidová stavení: opravy a úpravy. Prague: Grada, 2001. Stavitel. ISBN 80-247-9054-8.

[2] V. Kovářů, Hliněný dům, problematika jeho památkové obnovy a využití. Zděná architektura v Čechách, na Moravě a ve Slezsku. 2004, 31 – 33.

[3] Agenda 21 pro udržitelnou výstavbu. Prague: ČVUT. 2001. CIB Report Publication 237.

[4] J. Růžička, Building structures in environmental context, mechanical physical properties of unburned bricks by impact of humidity. Brno University of Technology, Faculty of Architecture, International scientific seminar Healthy Houses, Brno, 2005. ISBN 80-214-3040-0.

[5] G. Minke, Building With Earth [online]. 2006 [vid. 18. srpen 2015]. ISBN 978-3-0346-0822-0. http://www.archive.org/details/Gernot-Minke-Building-With-Earth/

[6] I. Žabičková, Hliněné stavby [online]. Brno: Era 21, 2002 [vid. 18. srpen 2015]. ISBN 80-86517-21-7. http://www.kosmas.cz/knihy/107760/hlinene-stavby/

[7] Z. Weiss, Jílové minerály: jejich nanostruktura a využití. B.m.: Karolinum, 2005. ISBN 80-246-0868-5.

[8] T. Otcovská, P. Padevět, Microstructure of Unburned Clay and its Shrinkage During Drying, 13th Interantional Conference on New Trends in Statics and Dynamics of Buildings, Bratislava, 2015, ISBN 978-80-227-4463-8

Influence of temperature on size of creep of the cement pastes

P. Padevět[1,*], P. Bittnar[1]

[1] *Czech Technical university in Prague, Faculty of Civil Engineering, Thakurova 7, Prague 6, 166 29, Czech Republic*

[*] *pavel.padevet@fsv.cvut.cz*

Abstract: The article deals with the effect of temperature load before the test of creep on its size after finishing of measurement. Various cement paste mixtures are compared with the results obtained. The cement paste were loaded by temperature, before measurement of creep.The size of creep depending on the moisture content and water-cement ratio is compared at the end.

Keywords: cement paste; creep; shrinkage; temperature; lever mechanism.

1 Introduction

The cement paste is an essential component of concrete. Their function is a binder in concrete. The filler modifies and rather improves the properties of concrete [1]. Therefore, attention is given to the characteristics of the binder component of concrete. The properties of cement pastes are significantly influenced by the amount of mixing water. Our attention is devoted to the the creep of these materials. The temperature can change the properties of building materials. Our attention is focused on the case of load of high temperature before the experiment. Consequently, creep and shrinkage of cement pastes were measured and of them determined the final result.

2 Tested material

The preparation of the cement paste was carried out from cement CEM II and water. Water-cement ratio for the three test mixtures was 0.3, 0.4 and 0.5 [2]. The mixture No. 1 with water ratio of 0.3 was prepared by adding superplasticizer at 3 % for better processing of mixture. All mixtures were older than one month.

3 Testing procedure

The specimens were exposed to temperature of 300 °C for 2.5 hours before the test. Then it was cooled to room temperature, see Fig. 1. Firstly, the specimens was removed from the water and dried up loosely in a laboratory environment. All solids were placed in a temperature chamber and load temperature determined by the procedure after a day of drying in the laboratory. The moisture conditions were modified after the temperature load. Temperature 300 °C changes the properties of the cement paste. The question is, how temperature affects the size creep? Furthermore, the density of material and content od free water in the paste can influence on the size of creep.

The creep was measured on samples of 70 mm length, which were placed in lever arch mechanisms (LAM). The diameter of specimens was 10 mm. Specimens were split into dried and fully water saturated groups. The creep of dried cement paste was measured on the two specimens. The shrinkage of dried paste was measured on one body [3]. The same principle was applied to measurement of the water saturated bodies, two bodies to measure creep and one body to measure shrinkage [4].

The volumetric weight was determined before and after the tests. The water content in the cement pastes was slightly changed during the measurement. Creep of the dried cement pastes without the influence of of shrinkage is presented in the Fig. 2, 4 and 6. The curves of creep are significantly variable depending upon on the water cement ratio, as is evident from the pictures. Specimens were loaded with a constant load at the

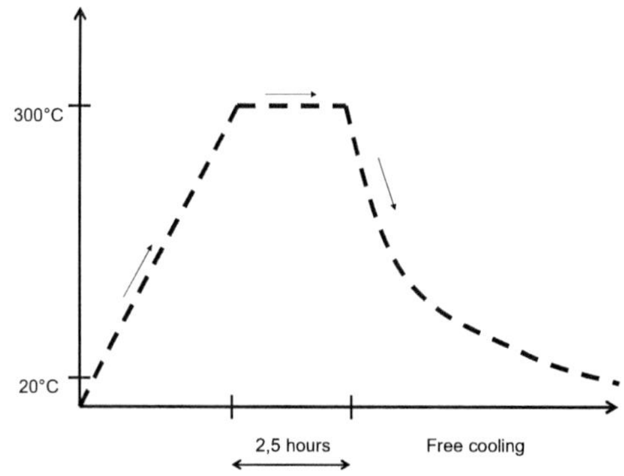

Fig. 1: Principle of temperature loading of cement paste.

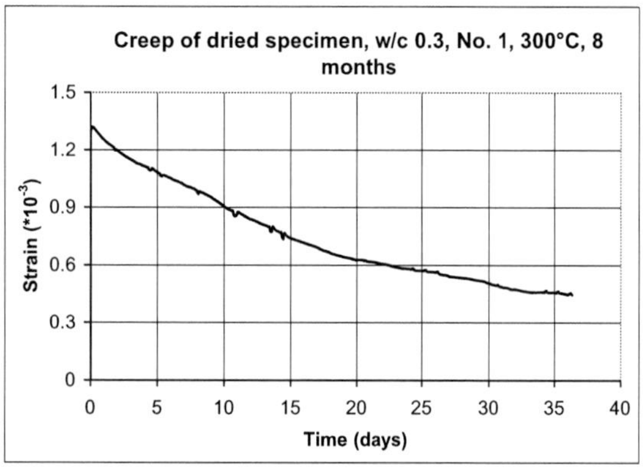

Fig. 2: Creep of dried cement paste. w/c 0.3, without influence of shrinkage.

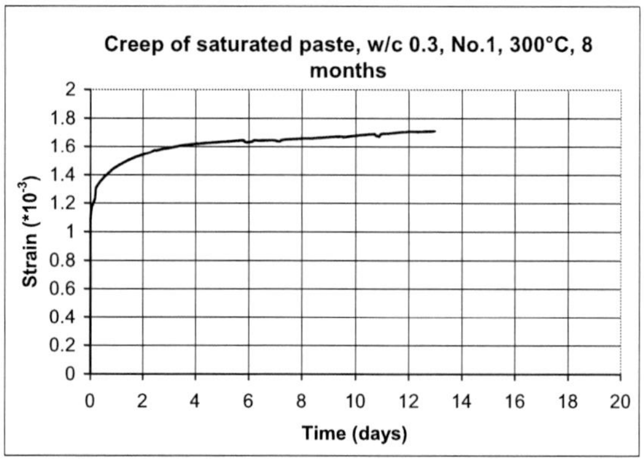

Fig. 3: Creep of water saturated cement paste w/c 0.4, without influence of shrinkage.

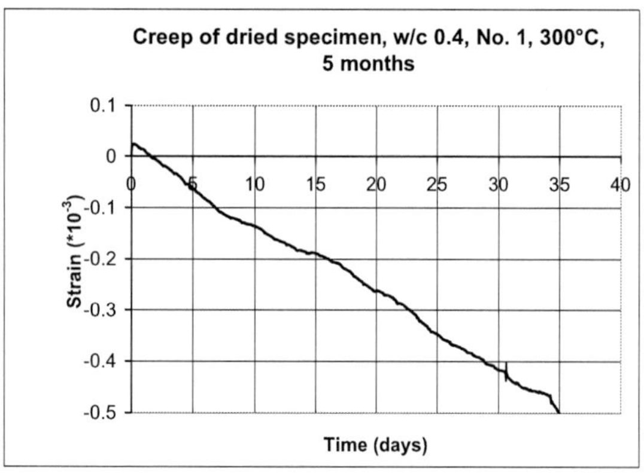

Fig. 4: Creep of dried cement paste w/c 0.4, without influence of shrinkage.

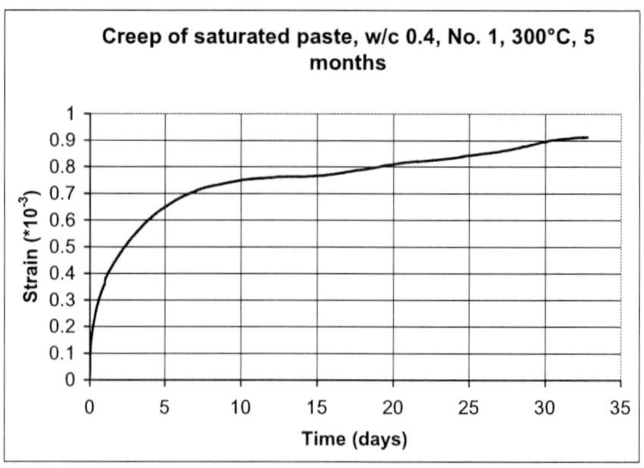

Fig. 5: Creep of water saturated cement paste w/c 0.4, without influence of shrinkage.

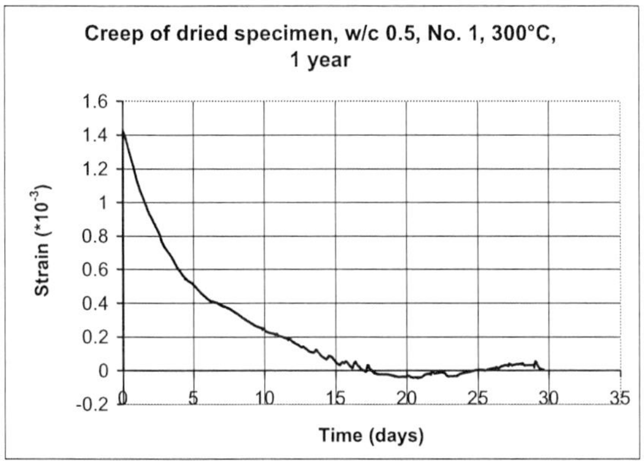

Fig. 6: Creep of dried cement paste w/c 0.5, without influence of shrinkage.

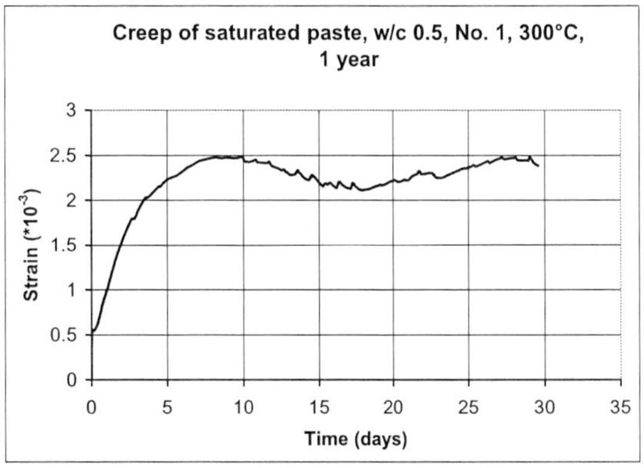

Fig. 7: Creep of water saturated cement paste w/c 0.4, without influence of shrinkage.

beginning of the test. The load was applied to the body throughout the test. The creep of dried specimens has a negative trend. This means that the length of the cylinders is not reduced, but on the contrary swelling.

The creep of water-saturated cement paste is shown in Fig. 3; 5 and 7. Presented graphs are cleaned off shrinkage. Pure creep of saturated cement pastes is shown. The creep of specimens was propagate after weight load in the lever mechanisms at the beginning of the test. There was a decrease in the length bodies as is possible see on Fig. 3; 5 and 7.

The temperature influence on the creep is negligible. This was reflected on the shape of the curves in the graphs on Fig. 6 and 7. Reducing the temperature by 2 °C caused a reduction in the rate of creep of the dried paste and the reduction in the rate of creep of water saturated paste. However, the trend of creeping dry and saturated paste is retained.

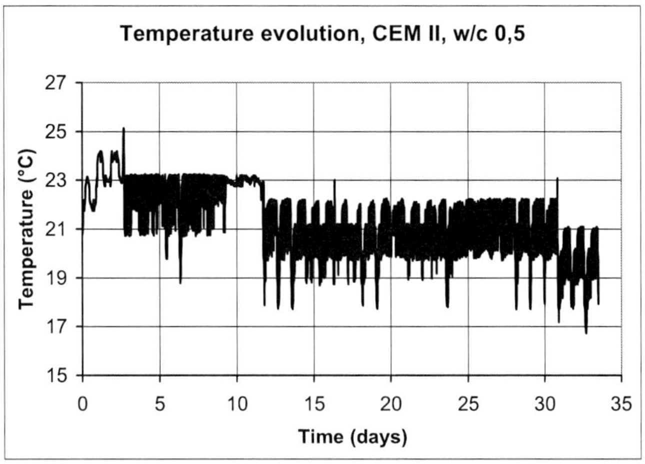

Fig. 8: Evolution of temperature for mixture with w / c ratio 0.5.

4 Results

The curves presented on Fig. 2 - 7 includes creep without shrinkage. The length of measurement was near to 30 days in all three cases. The highest rate of increase in deformation saturated cement paste is 2 days for the w / c 0.3, hereafter 6 days for the w / c 0,4 and 8 days for the w / c 0.5. Then, a rate of increase of deformation decreases. It can be concluded that extending the period in which comes moment of slowing the rate of increase of deformation and the increasing value of water-cement coefficient. The increase of deformation saturated paste is stable after 10 days testing.

Tab. 1: Results of size of creep after 25 day long measurement.

Series	w/c	Sat.	Sat.	Dried	Dried	Basic creep	Basic creep
		1	2	1	2	1	2
1	0.3	0.639*	0.564*	-0.725	-0.884	0.671	-0.094
2	0.4	0.843	0.438	-0.375	-0.411	-	-
3	0.5	1.876	2.470	-1.420	-1.508	-0.445	-0.588

* -values after 12 days (measurement terminated in 13 day).
Values in the table are in microstrains.

A different shape curves represents a dried pastes. Slowing deformation growth is not as easy to see like in the case of the saturated paste. A slowdown occurs after 15 days of testing. Cement paste with w / c 0.4 shows no slowdown. On the other hand, the cement paste with a water ratio of 0.5 shows stop the growth of deformation after 20 days. This is caused by change of environment temperature, which is shown in Fig. 8.

5 Conclusion

The results are summarized in Table 1. The data are determined to 25 day of measurement. The size of creep increases with the water content in pastes. The creep depends also on the amount of free water in the cement paste. The shrinkage of cement paste plays a significant role in the size of creep. This is evident for comparing the columns of the table Dried and Basic creep. Importance of shrinkage increases with the size of water-cement ratio. The shrinkage of very fluid mixtures (line 3 in the table) can make up more then 50 % of creep of dried cement paste.

The data listed in Table 1 relate to creep including the effect of shrinkage beyond the data of basic creep. The basic creep and creep dried paste is comparable in set 1 and 3. The difference values are caused by shrinkage. Shrinking the size of the dried material is not insignificant. It can be stated that reducing the amount of water has a positive effect on the size creep, see comparison of row 2 and 3 in the Table 1. Significance of the temperature load and water-cement ratio was reflected in the higher levels of creep and shrinkage than other materials based on cement composites [5].

Acknowledgement

This research has been supported by project SGS project No.16/201/OHK1/3T/11 Advanced from design of materials to the constructions.

References

[1] P. Padevět, T. Otcovská, O. Zobal, Influence of fly ash type on material properties of cement paste, Advanced Material research Volume 969. (2014) 212–217, doi: 10.4028/www.scientific.net/AMR.969.212.

[2] A. M. Neville, Properties of Concrete, John Wiley & Sons, (1997), ISBN 0-470-23527-6.

[3] P. Padevět, P. Bittnar, Creep of cementitous materials with addition of fly ash in time, Advanced Materials Research Volume 742. (2013) ed. 182186, doi: 10.4028/www.scientific.net/AMR.742.182.

[4] P. Padevět , Measuring of Creep of Cement Paste Specimen, Proceedings of the 2nd WSEAS International Conference on Applied Mathematics, Simulation, Modeling (ASM09), Athens, Greece, 2009, pp. 33-39.

[5] D.P. Bentz, A.S. Hansen, J.M. Guynn, Optimization of cement and fly ash particle sizes to produce sustainable concretes, Cement & Concrete Composites 33 (2011), pp. 824-831, www.elsevier.com/locate/cemconcomp.

Bearing tests of specimens manufactured from chopped C/PPS pellets

Z. Padovec[1,a], R. Sedláček[1], M. Růžička[1], M. Král[1], P. Růžička[1],

[1] *Czech Technical University in Prague, Faculty of Mechanical Engineering, Department of Mechanics, Biomechanics and Mechatronics, Technická 4, Praha 6, 166 07, Czech Republic*
[a]Zdenek.Padovec@fs.cvut.cz

Abstract: Presented article deals with description and evaluation of bearing tests of specimens manufactured from C/PPS pellets (random orientation) with titanium inserts. These inserts has two versions - cylindrical (C) and cylindrical with groove (G). Main goal of the tests is to evaluate bearing strength of the material which will be applied in the FE simulation of the real parts in construction of aircraft door and for biomedical application (part of the external fixator).

Keywords: bearing test; randomly reinforced composite; thermoplastic matrix; chopped pellets

1 Description of the Experiment

Prepared experiment was based on ASTM D5961 standard [1] for specimens with random reinforcement (randomly reinforced chopped C/PPS pellets in our case – see [4] for details) with titanium insert. Failure modes of tested specimens can be seen in Fig. 1. Specimens were manufactured in Latecoere Czech Republic. Bearing strength σ_b is defined as loading force F divided by the hole diameter D multiplied by laminate thickness t

$$\sigma^b = \frac{F}{Dt}. \tag{1}$$

Definitions of bearing strength are based on limit bearing strain values, ultimate strength, first decrease during loading or the beginning of nonlinear behavior [3]. Standard [1] defines two bearing strengths - first one is “Initial Peak Bearing Stress” which corresponds with the first decrease during loading, second one is “Offset Bearing Stress” and its computation can be described with the use of the stress/strain graph shown in Fig. 2.

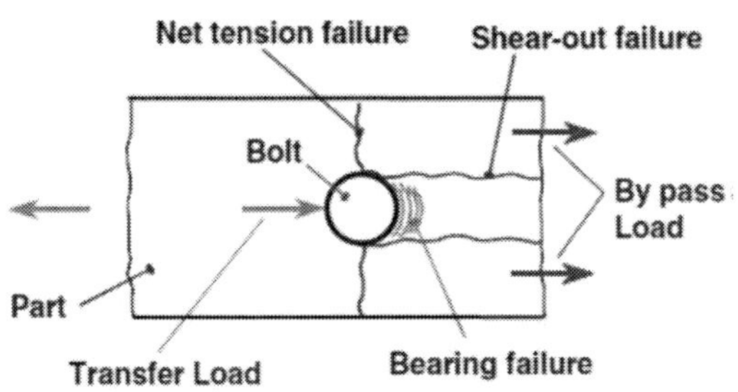

Fig. 1: Failure modes of loaded specimen [2].

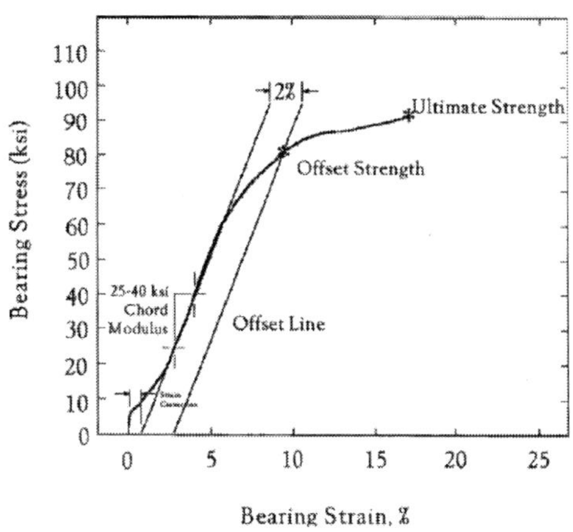

Fig. 2: Offset Bearing Stress construction [1].

Specimens were tested on TIRA 2300 machine (20 kN dynamometer) with displacement 2 mm/min and extensometer. Two sets of specimens were prepared and both bearing strengths from standard [1] were evaluated.

2 Results and Conclusions

In Tab. 1 results of the first set of specimens can be seen. Specimens with cylindrical insert cannot be evaluated because of different type then bearing failure. Second set of specimens were tested in different way (different clamping to the testing machine). Geometry of inserts is shown in Fig. 3 and 4. Both methods of clamping is shown in Fig. 5 and 6. Results from the second set can be seen in Tab. 2. From the measured values can be seen that specimens from the first set show higher average bearing strength with lower standard deviation then specimens from second set. Different configuration of experiments used during the testing of specimens from second set is better for obtaining right failure mode (right and wrong failure mode can be seen in Fig. 6 and 7). Evaluation of bearing stress can be seen in Fig. 8. Obtained results will be discussed with manufacturer and used for FE simulation of the real part manufactured with pellets (part of aircraft door construction, part of external fixator).

Tab. 1: Results from the first set of specimens.

Specimen No.	Offset bearing strength [MPa]	Initial bearing strength [MPa]
1_115_G_1	267,8	309,0
1_70_G_1	312,2	312,2
1_70_G_3	306,3	314,9
1_115_G_3	316,4	318,0
Average value	**300,7 ± 19,3**	**313,5 ± 3,3**

Tab. 2: Results from the second set of specimens.

Specimen No.	Offset bearing strength [MPa]	Initial bearing strength [MPa]
2_C_1	-	256,1
2_G_1	190,0	224,0
2_G_2	250,1	266,2
2_G_3	180,6	222,5
Average value	**206,9 ± 30,8**	**242,2 ± 19,3**

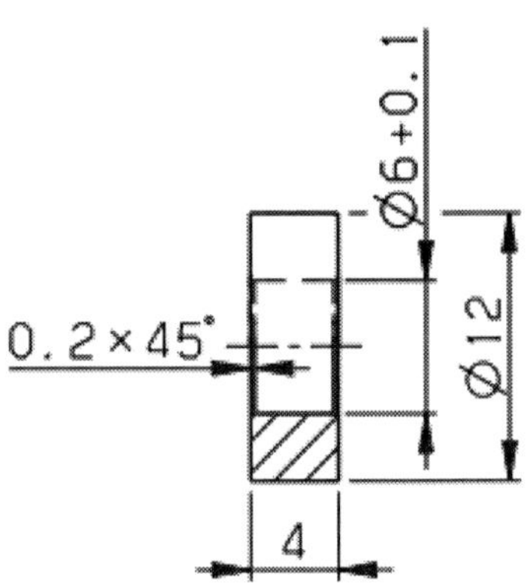

Fig. 3: Geometry of cylindrical insert.

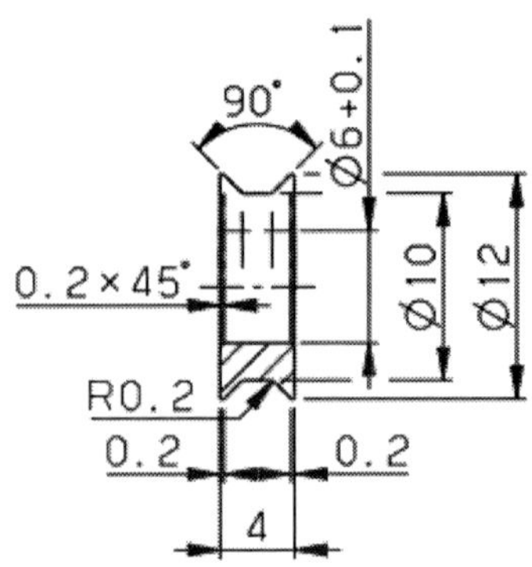

Fig. 4: Geometry of cylindrical insert with groove

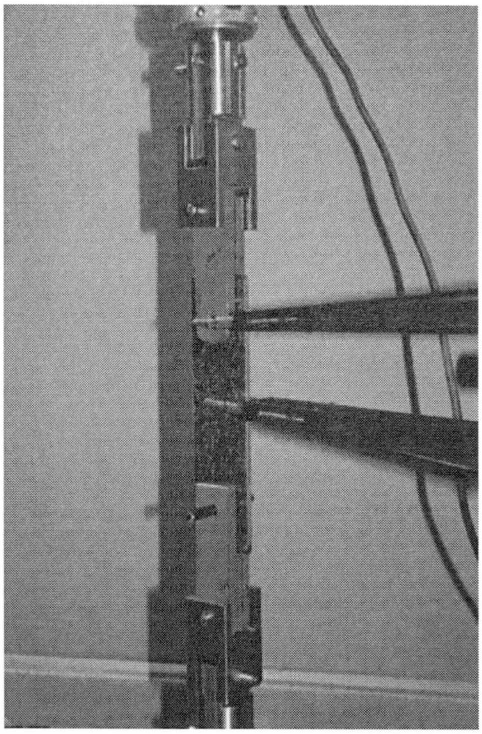

Fig. 4: Clamping of specimens from 1st set.

Fig. 5: Clamping of specimens from 2nd set.

Fig. 6: Right failure mode for bearing stress evaluation.

Fig. 7: Wrong failure mode.

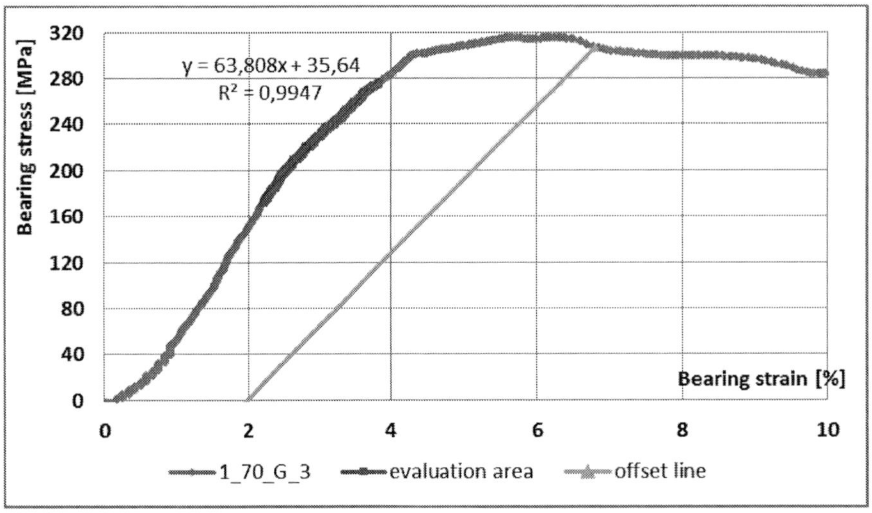

Fig. 8: Evaluation of bearing stress on 1_70_G_3 specimen.

Acknowledgement

This work has been supported by project TA03010209 of the Technological Agency of the Czech Republic and Grant Agency of the Czech Technical University in Prague, grant No. SGS15/188/OHK2/3T/12.

References

[1] ASTM D 5961 Standard Test Method for Bearing Response of Polymer Matrix Composite Laminates.

[2] B. Kolesnikov, L. Herbeck., A. Fink, CFRP/Titanium Hybrid Material for Improving Composite Bolted Joints, Composite Structures 83 (2008) 368-380, http://dx.doi.org/10.1016/j.compstruct.2007.05.010.

[3] P. P. Camanho, C. M. L. Tavares, R. de Oliveira, A. T. Marques, A. J. M. Ferreira, Increasing the Efficiency of Composite Single-Shear Lap Joints Using Bonded Inserts, Composites: Part B 36 (2005) 372-383, http://dx.doi.org/ 10.1016/j.compositesb.2005.01.007.

[4] Z. Padovec, H. Chlup, R. Sedláček, M. Král, M. Růžička, P. Růžička, Isotropy Verification of the C/PPS Samples Manufactured from Pellets by Image Processing Methods and Experiment, Materials Today: Proceedings 3 (2016) 562 – 566, http://dx.doi.org/ 10.1016/j.matpr.2016.01.091.

Phase-shifting Method and its Comparison with Procedure traditionally used in Photoelasticity

Z. Paška [1,a], F. Fojtík [1,2]

[1] *Department of Applied Mechanics, VŠB - Technical University of Ostrava, Czech Republic*
[2] *IT4Innovations, VŠB - Technical University of Ostrava, Czech Republic*
[a] *zbynek.paska.st@vsb.cz*

Abstract: The main goal of this paper is to perform the comparison of phase-shifting method with procedure, which was used in photoelasticity before advent of computers. In order to make this old procedure friendlier for use, its code was written in MATLAB software. The phase-shifting method is a procedure from digital photoelasticity. It provides isochromatic and isoclinic parameters in the full field of the specimen. In this contribution the phase shifting is carried out by three different loads acting on the specimen. The code was also written in MATLAB. Thin disc, annular discs and beam were used as models. The results are compared in certain positions on the models.

Keywords: photoelasticity; phase-shifting method; isoclinic; isochromatic; Cauchy stress tensor.

1 Introduction

Photoelasticity is an experimental method, which enables to evaluate stress state in transparent materials. Photoelasticity used polarized light and phenomenon of certain transparent materials called birefringence [1]. It is possible to solve static, dynamic tasks and problems associated with residual stresses.

This contribution is focusing on problematics of gathering data from photoelasticity measurement. There exists two types of fringe patterns, which is possible to observe. Isoclinic and isochromatic fringe patterns is representing directions of principal stresses and amount of difference of principal stresses, respectively. In work [2] was presented a procedure, which allow to obtain Cauchy stress tensor in the full field of specimen trough shear difference method. The isoclinic and isochromatic fringes were rendering manually by user and then approximated by Bézier curves. The values of isochromatic and isoclinic parameters in points between two curves were determined by the weighted arithmetic mean. This work quite well, when the density of curves is high.

In order to determine the influence of the density of interferometric fringes on results, there were used technique called phase-shifting method [3]. This procedure can provide data in the full field of the specimen. The necessity of determination of the isoclinic and isochromatic fringes manually is eliminated. There can arise some errors caused by isoclinic isochromatic interaction.

2 Phase-shifting Method

There exists many modification of this procedure. Asundi and Liu algorithm [4] was chosen for this contribution. A circular polariscope was used for this measurement (Fig. 1b). For this type of testing rig it is possible, after involving Jones' calculus, to express the light intensity as:

$$I = \mathrm{I}_0 + \mathrm{a}^2[\cos^2\beta \sin^2(\delta/2) \sin^2(2\varphi - 2\theta) + \sin^2(\beta) \cos^2(\delta/2) + \\ +0{,}5 \sin(2\beta) \sin(\delta) \sin(2\varphi - 2\theta) + \sin^2(\delta/2) \cos^2(2\theta - 2\varphi) \sin^2(\beta - 2\varphi)], \quad (1)$$

where I_0 is the background light and a is amplitude of the light vector.

The intensity of light is a function of orientations of the optic members and retardation δ introduced by the model. To calculate the isochromatic parameter δ (principal stresses difference) and isoclinic parameter θ (orientation of principal stresses) there must be considered polariscope configuration according to Tab. 1.

Tab. 1: Intensity equations for circular polariscope arrangements ($\theta = \varphi$).

configuration	φ [deg]	β [deg]	intensity equation	
1	45	90	$I_1 = \mathrm{I_b} + \mathrm{I_m}\cos\delta$	bright field
2	45	0	$I_2 = \mathrm{I_b} - \mathrm{I_m}\cos\delta$	dark field
3	90	135	$I_3 = \mathrm{I_b} - \mathrm{I_m}\sin\delta\sin 2\theta$	where
4	45	45	$I_4 = \mathrm{I_b} + \mathrm{I_m}\sin\delta\cos 2\theta$	$\mathrm{I_b} = \mathrm{I_0} + \mathrm{a}^2/2$
5	90	45	$I_5 = \mathrm{I_b} + \mathrm{I_m}\sin\delta\sin 2\theta$	$\mathrm{I_m} = \mathrm{a}^2/2$
6	45	135	$I_6 = \mathrm{I_b} - \mathrm{I_m}\sin\delta\cos 2\theta$	

From intensity equations in Tab. 1 the isoclinic and the isochromatic parameter can be expressed as:

$$\theta = (1/2)\,\mathrm{tg}^{-1}[(I_5 - I_3)/(I_4 - I_6)], \tag{2}$$

$$\delta = \mathrm{tg}^{-1}\{[(I_5 - I_3 + I_4 - I_6)/(\sin(2\theta) + \cos(2\theta))]/(I_1 - I_2)\}. \tag{3}$$

Due to tg^{-1} function and $(1/2)$ multiplier in formula (2), the isoclinic parameter is determined in range $\langle -\pi/4; \pi/4 \rangle$ and there is no information whether the angle refer to first or second principal stress. From equation (3) it can be seen, that an ambiguity of θ leads to improper determination of isochromatic parameter. It is possible to implement the three-load method to resolve this behavior and determine the photoelastic parameters in correct way.

2.1 The Three-Load Method

The model in testing rig is subjected to three different loads. For practical measurement it is useful if the difference between actual and subsequent load is constant. For loads is valid relation $\mathrm{F_1} > \mathrm{F_2} > \mathrm{F_3}$, therefore for retardation $\delta_1 > \delta_2 > \delta_3$. In regions where the retardation has been mathematically wrong determined, the relation becomes $\delta_1 < \delta_2 < \delta_3$. In any point in the stressed specimen one can find out six possible relations between retardations [4].

The main purpose of this procedure is to determine the principal stresses orientation always related to one of the major stress. If the isoclinic parameter have changed the isochromatic parameter must be recomputed according to formula (3).

3 Experiment and Results

Polariscope with sodium light was used for visualize stress state in three specimens (see Fig. 1a). Two quarter-wave plates were inserted into the polariscope. A schematic of polariscope can be seen on Fig. 1b.

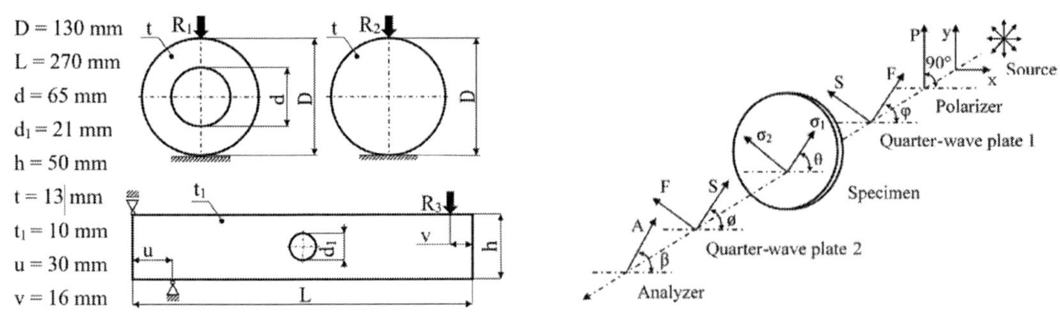

a) three investigated specimens b) orientation of optic members

Fig. 1: Models and scheme of polariscope for circular polarization.

The six intensity images for annular thin disc, disc and beam model are shown on Fig. 2. The images correspond to the maximum applied load. The eighteen images were captured by Canon EOS 60D for each specimen. The two discs were subjected to the vertical compressive force and the beam to the bending force.

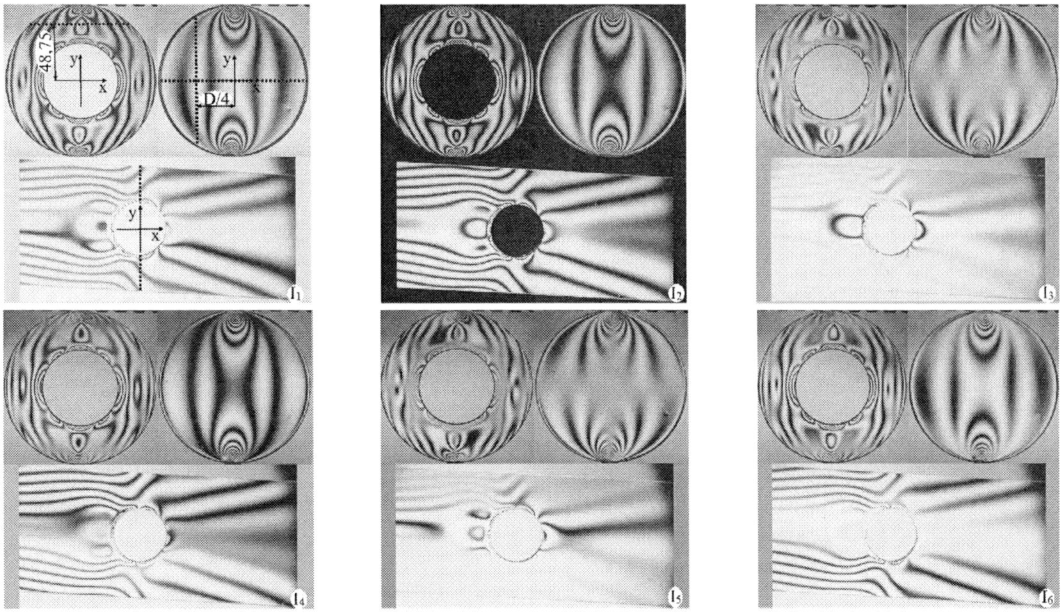

Fig. 2: Fringe patterns corresponding to polariscope arrangement mentioned in Tab. 1.

In the first three images the dashed lines denote positions where the results have been compared and they are in some meaning in symmetric positions on the models. For thin disc the vertical line lies in distance 1/4 of diameter from left. The discs are made from optically sensitive material CT 200 (material fringe value 17.4 N·mm^{-1}) and beam is made from polycarbonate (material fringe value 7 N·mm^{-1}).

3.1 Results Comparison

Results are compared on Fig. 3 and Fig. 4. No curve-fitting method was used for smoothing the data. Results on the boundary of the models can be affected by residual stresses and light scattering on the model's edge.

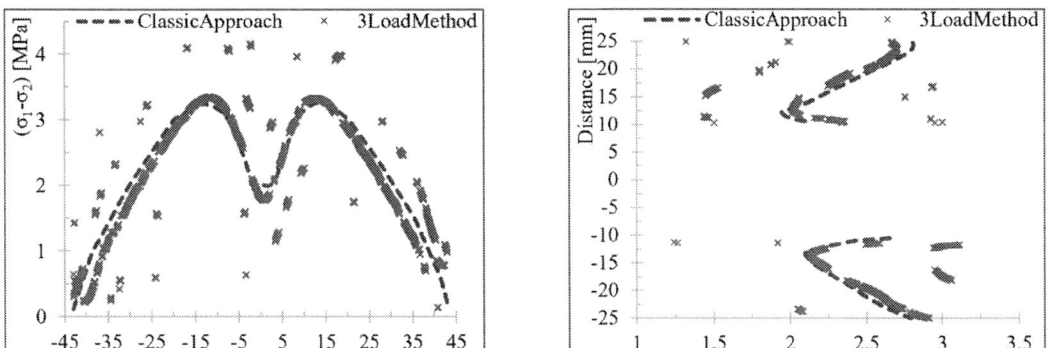

a) principal stress difference along horizontal line for thin annular disc

b) principal stress difference along vertical line for thin beam subjected to bending force

Fig. 3: Comparison of phase-shifting (Asundi's) algorithm and traditional approach for thin annular disc and thin beam.

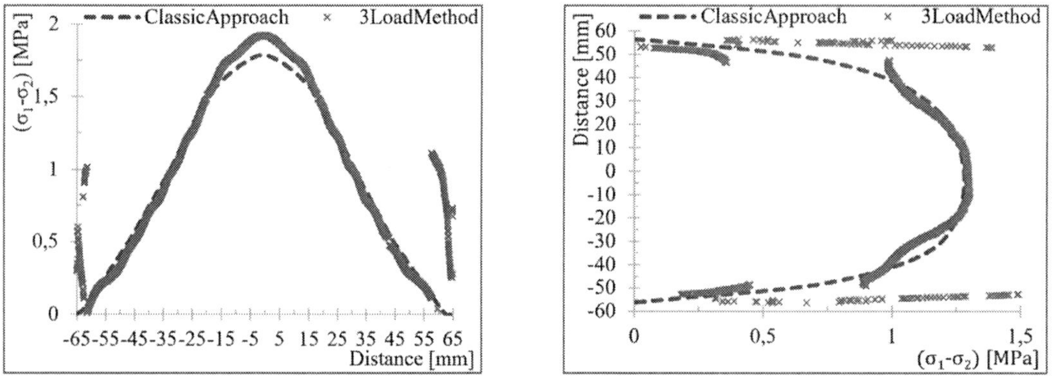

a) principal stress difference along horizontal line b) principal stress difference along vertical line

Fig. 4: Comparison of phase-shifting (Asundi's) algorithm and traditional approach for thin disc.

The Results on Fig. 3 and Fig. 4 correspond to force R_1=1240.9 [N] for thin annular disc, R_2=1678.9 [N] for thin disc and R_3=106.5 [N] for thin beam. It is apparent, that there must be implemented some post processing technique to obtain more accurate results especially on the boundary of the models. Refinement data could be used as input data for stress separation technique and then components of stress tensor can be determined.

4 Conclusion

It is apparent that there exists some differences in results obtained by two approaches. The traditional approach, which is fully computer aided, is suitable if there exists many isochromatic fringes in the field. On the other hand the phase-shifting method is more comfortable. It has clear physical and mathematical explanation, however the results in region with high density of isochromatic fringes are not good due to isochromatic-isoclinic interaction. This drawback can be eliminated by proper combination of results obtained for the three different loads. It can be proved, that with some assumption four intensity images from Tab. 1 are sufficient for calculation of photoelastic parameters.

Photoelasticity is exceptional experimental method due to its capability to directly visualise stress state in transparent models. Reflection photoelasticity is used for determination the strain distribution on surface of non-transparent parts. Solve stress condition caused by dynamic loads and residual stresses is also possible.

Acknowledgement

This work was supported by The Ministry of Education, Youth and Sports from the National Programme of Sustainability (NPU II) project „IT4Innovations excellence in science - LQ1602", by the Czech Science Foundation (GA15-18274S) and by the Specific Research (SP2016/145). The support is gratefully acknowledged.

References

[1] A. S. Khan, X. Wang, Strain Measurements and Stress Analysis, New Jersey, Prentice Hall, 2001.

[2] F. Fojtík, P. Ferfecki, Z. Paška, Computer Evaluation of the Stress Tensor in the Two-Dimensional Photoelasticity, in proc.: EAN (2015), ed. P. Padevet, Czech Technical University in Prague, Český Krumlov, 89-93.

[3] K. Ramesh, Digital Photoelasticity, Berlin, Springer Verlag, 2013.

[4] A. Asundi, MATLAB for Photomechanics, Boston, Elsevier, 2002.

Failure analysis of the moving knife holder of a shredding machine

M. Pástor [1,a], F. Trebuňa [1], F. Šimčák [1]

[1] *Technical University of Košice, Faculty of Mechanical Engineering, Department of Applied Mechanics and Mechanical Engineering, Letná 9, Košice, 04200, Slovakia*

[a] *miroslav.pastor@tuke.sk*

Abstract: The paper focuses on the analysis of the causes that led to knife holder failure. The holder is a part of the car tyre shredding equipment. Based on the evaluation of the operating conditions of the machine, examination of the condition of the knife holder after the failure, analysis of the fractured surface as well as the material properties of the examined structural elements, it was possible to draw conclusions as to the reasons for the failure.

Keywords: shredding machine, tyre, knife holder, analysis of failure.

1 Introduction

One of the options for treating used tyres is to process them into a range of rubber granules. In the first stage of the recycling process, tyres are chopped into small pieces by the shredding machine (Fig. 1), which has stationary knives and a slowly rotating rotor with moving knives (Fig. 2). Fig. 3 shows the working area of the shredding machine during its operation – tyre shredding.

Fig. 1: General view of the shredding machine with a conveyor

Fig. 2: Stationary and moving knives of the shredding machine

Fig. 3: Tyre shredding crushing in the shredding machine

The critical part of the shredder, where the breakage of the knife holder occurred, is the rotor in whose holders the moving knives are fixed with the bolts and the tongue. Their loading, wearing, subsequent grinding and adjusting are closely related to the fixed stationary knives. After approximately ten years of operation of the shredder, the moving knife holder broke. The paper presents the results of the examination of this failure and its analysis [1, 2, 3].

2 Description of the structural components of the tyre shredding machine

On the holders welded with fillet welds to the rotor of the shredding machine there are twelve cutting (moving) triangular knives (Fig.4a) arranged in two rows with six in each row (Fig.4b).

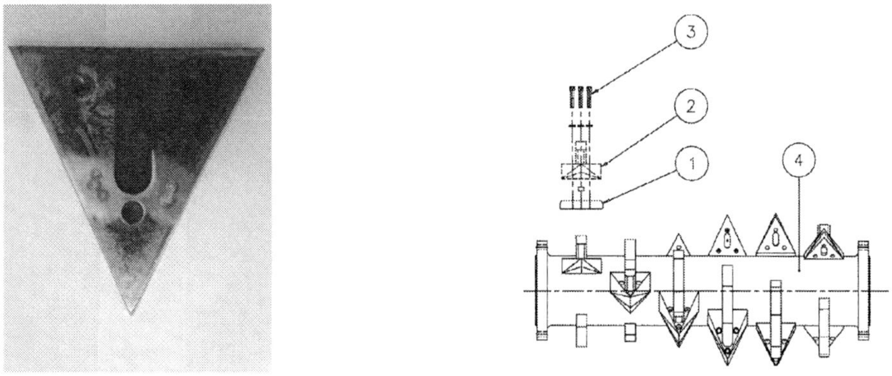

a) b)

Fig. 4: a) cutting (moving) knife, b) diagram of moving knives placement, 1 - moving knives, 2 – knife holder, 3 – bolts fixing the moving knives to the holder, 4 - rotor.

Stationary knives are fixed with bolts in the holders to the stationary parts of the machine. They are placed in one level with five full knives in the center and one half-knife at each end (Fig.2).

Tyres are shredded by the combined action of the moving and stationary knives during rotation of the machine rotor. Mechanical stress of the rotor substantially affects the sharpness of the knives‘edges as well as the size of the cutting gap between the knives. Therefore, optimal cutting conditions are provided by frequent adjusting and sharpening of the knives at the intervals specified in the instructions for the operation of the machine. The sizes and shapes of the of the stationary and moving knives are given in Fig. 5a where the cutting gap (0,2 + 0,5 mm) is adjusted by shifting the stationary knife in the grooves for bolted fixing.

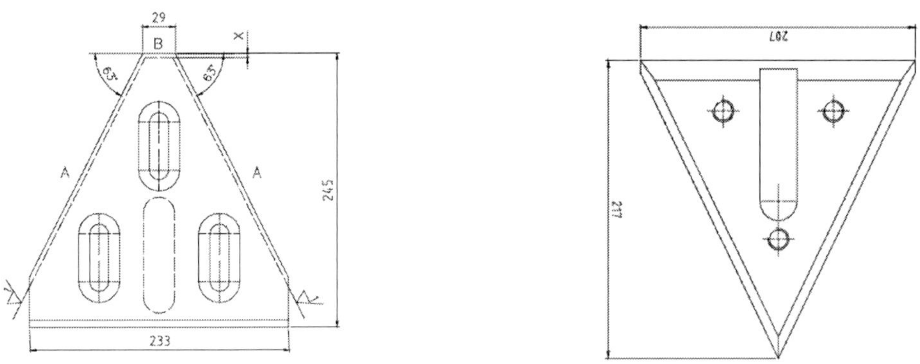

a) b)

Fig. 5: Sizes and shapes of a) stationary knife, b) moving knife.

3 Failure analysis of the moving knife holder

Moving knives are fixed by three bolts in the holders (Fig.6) which are welded with fillet welds to the rotor of the shredding machine (Fig.2).

a) b)

Fig. 6: Knife holder with a moving knife- fixed by three bolts a) top view, b) side view.

The failed weld of the moving knife holder is shown in Fig.7a. The details of the failed weld joint of the holder are given in Fig. 7b; c. Fig.8 shows the fracture surface of the holder with the moving knife of the shredding machine, Fig.9 gives a view of the fracture surface on the rotor shaft of the shredder.

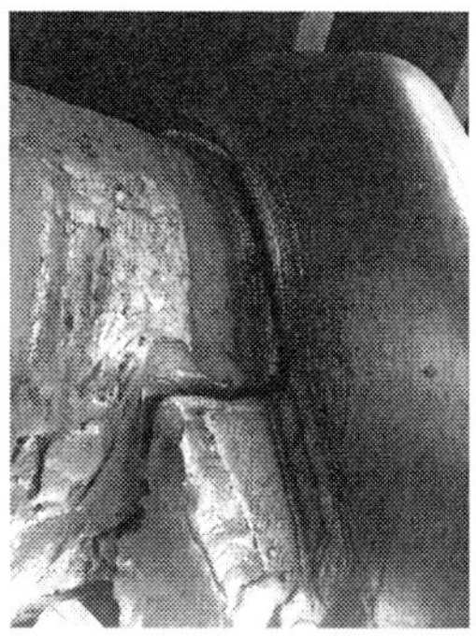

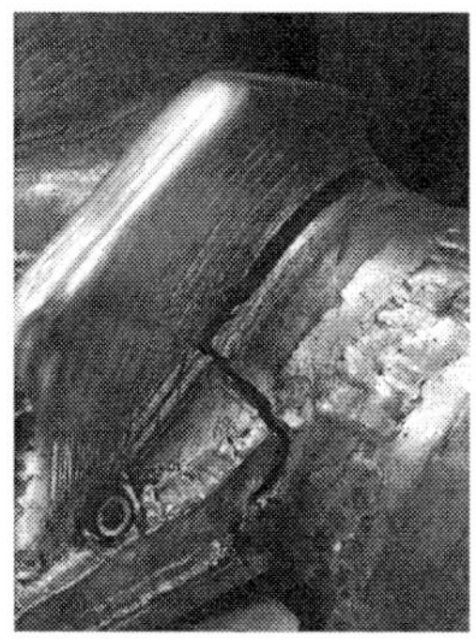

a) b) c)

Fig. 7: Failure of the welded joint of the moving knife holder a) general view , b), c) detailed view of the failed welded joint.

Fig. 8: Broken holder with moving knife

Fig. 9: Fractured surface on the shaft of the rotor

From Fig. 9 it is obvious that the fracture occurred in the location of the weld due to cyclic loading during the tyre shredding process. Samples were taken from the location of the fractured surface of the knife holder the in order to determine chemical composition and hardness as well as to perform metallographic analysis with the aim to identify material quality and assess its welding property (Fig.10a, b). In the samples, subsurface cracks were found in the locations of the welds (Fig.10c).

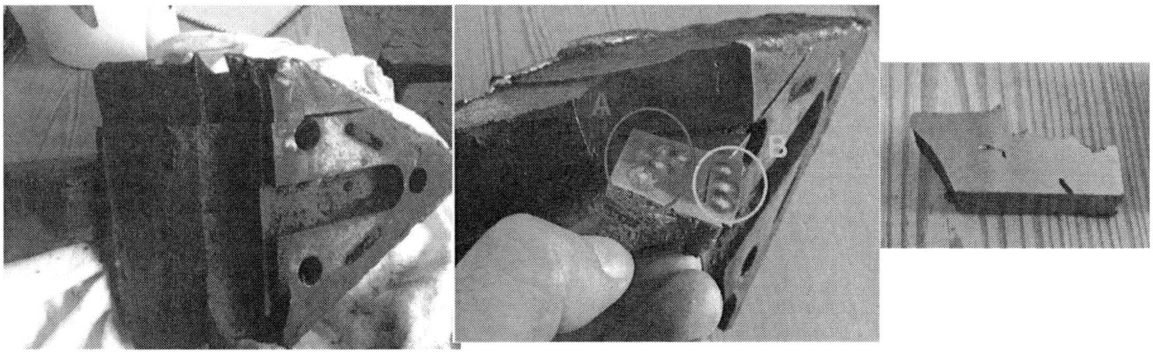

a) b) c)

Fig. 10: a, b) Locations where samples were taken and analysed, c) subsurface cracks in the samples a, b)

Chemical composition was determined by spectrometric analysis in two areas: A and B (Fig. 10b) since two different materials were used there. To adjust the moving knives, steel plates were welded to the holders to support the knives. Table 1 and Table 2 contain average values of the chemical composition obtained by measurement and chemical composition of these materials determined on the basis of the data as given in [4].

Tab. 1: Chemical composition for area A (knife holder) (wt %).

	C	Si	Mn	P	S	Cr	Mo	Ni
Determined by measurement	0,21	0,27	0,917	0,011	<0,002	0,637	0,236	0,459
Material 21NiCrMo2	0,17-0,23	0,40	0,65-0,95	Max. 0,035	Max. 0,035	0,35-0,70	0,15-0,25	0,40-0,70

Tab. 2: Chemical composition for area B (steel plate) (wt%).

	C	Si	Mn	P	S	Cu	Cr	Ni	Ti
Determined by measurement	0,146	0,48	0,581	0,11	0,002	0,091	0,055	0,143	0,02
STN 411483	max. 0,20	max. 0,55	max. 1,40	max. 0,045	max. 0,045	max. 0,30	max. 0,30	max. 0,30	max. 0,20

The hardness of the moving knife holder material was measured using testing procedure according to Vickers (STN EN ISO 6507-1) or Rockwell (STN EN ISO 6508-1). Hardness values obtained by Vickers method were in the range 263 – 287 HV. Hardness measured on the Rockwell scale was 28 HRC. The value for the tensile strength of the material was then determined to be at m $\approx$ 950 MPa based on the above-mentioned values. Mechanical properties of 21NiCrMo2 material, which are given in Table 3, show that the tensile strength of the material of the moving knife holder was about 10% higher that the upper bound of the prescribed value.

Tab. 3: Prescribed mechanical properties of 21NiCrMo2 material

E [MPa]	R_m [MPa]	A_5 [%]	σ_c [MPa]	R_e [MPa]
200.000	650 – 880	8 - 25	275 - 275	350 - 550

The place on the holder, from where the sample was taken for metallographic documentation, is shown in Fig. 11a. Metallographic analysis was performed in the cross-sections of two mutually perpendicular planes. Fig.11b shows the cuts in these planes.

a)

b)

Fig. 11: a) Sample taken for the metallographic analysis, b) cuts in two mutually perpendicular planes

Microstructure of the material is shown in Fig.12 a,b,c.

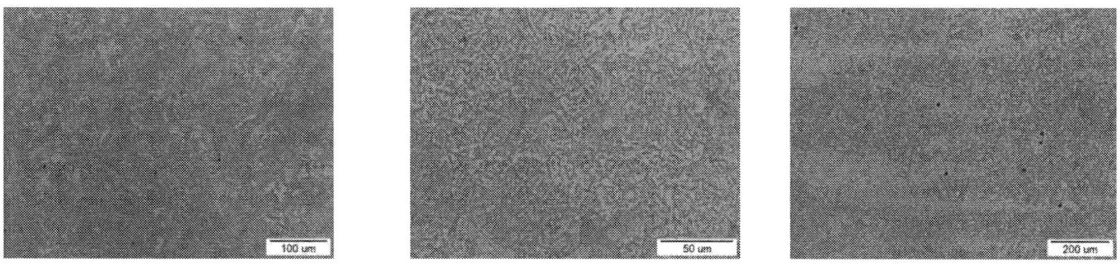

a) b)

Fig. 12: a), b), c) microstructure of the material at various scales

Based on the results of the metallographic analysis of the basic material of the knife holder, it appeared that the structure of the material in both planes was identical.

Microstructure was built by ferritic-carbidic mixture with polyhedric grains and local linelike arrangement of the carbide particles.

Weldability of the material of the holder was assessed taking into account the chemical composition measured in accordance with the procedure in STN EN 1011-2 as given in Table 1. The analysis showed that the material was weldable with preheating to 150° - 180°C.

Material of the moving knife was also assessed. Table 4 gives average values of the chemical composition determined by measurement as well as chemical composition of the corresponding quality of the material - steel STN 419 56.

Tab. 4: Chemical composition of the material of the moving knife (wt%).

	C	Si	Mn	P	S	Cr	Mo	V
Determined by measurement	0,568	0,762	0,277	<0,0 02	0,003	5,43	1,42	0,92
STN 419569	0,58-0,68	0,70-1,1	0,25-0,55	max. 0,030	Max. 0,035	4,5-5,5	0,8-1,2	0,20-0,40

The measured value of the material hardness was determined to be 53 HRC.

According to the information provided by the operator of the shredding machine, the hardness of the knives declared by the supplier was 59 HRC. This hardness for 19 569 steel is obtained by its tempering at the temperature of about 200°C [4, 5], that is why the measured hardness of 53 HRC indicated that the knife material was tempered at the significantly higher temperature.

4 Conclusion

On the basis of the analysis of the operating conditions of the used tyres shredding machine, assessment of the condition of the moving knife holder after its failure, analysis of the fractured surface and material properties of the holder and the knife, the following conclusions can be drawn:

- Based on the chemical composition it was determined that the failed material conformed to the prescribed quality– steel 21 NiCr Mo 2,
- Material hardness was 28 HRC, which corresponded to the material strength of 950 MPa,
- Microstructure was built by ferritic-carbidic mixture with polyhedric grains and local linelike arrangement of the carbide particles.
- The analysed part of the machine failed due to the repetitive mechanical stress. Fatigue failure initiated at a point of a defective, poor quality weld and stress concentration ,
- The failure could also occur because of the increased mechanical stress of the welds of the holder of the moving knife as it was not supported by the holder along its whole length (Fig.6b).
- Another cause of the failure could be attributed to improper thermal processing of the material of the moving knife that led to wear and the increase of shear force during tyre shredding.

Acknowledgement

This paper was supported by the projects of the Slovak Grant Agency, VEGA No. 1/0751/16 and No. 1/0393/14.

References

[1] T. R. Gurney, The Fatigue Strength of Transverse Fillet Welded Joints, City of Publication, Abington Publishing, ISBN 9781855730663, p. 112, 1991.

[2] L. Tong, X. Huang, F. Zhou, Y. Chen, Experimental and numerical investigations on extremely-low-cycle fatigue fracture behavior of steel welded joints, Journal of Constructional Steel Research, 119 (2016), p. 98 – 112.

[3] F. Trebuňa, M. Buršák, Medzné stavy – lomy. Grafotlač Prešov, Prešov, 2002.

[4] I. Furbacher et al., Lexikon technických materiálů. Svazek 4, Verlab Dashőfer, 2005.

[5] F. Trebuňa, F. Šimčák, Odolnosť prvkov mechanických sústav. EMILENA, Košice, 2004.

Study of the Freeze-Thaw Resistance of the Fine-Aggregate Concrete

T. Pavlů[1,2,a] and M. Šefflová [1,2]

[1]CTU in Prague, Faculty of Civil Engineering, Thakurova 7, 166 29 Prague, Czech Republic
[2]CTU in Prague, University Centre for Energy Efficient Buildings, Trinecka 1024, 273 43 Bustehrad, Czech Republic
[a]tereza.pavlu@fsv.cvut.cz

Abstract: This paper deals with possibility of use the fine fraction of recycled concrete aggregate as partial replacement of sand in concrete mixture. The main topic of this article is the study of evaluation of the durability of the fine-aggregate concrete with partial replacement of sand by fine recycled concrete. The freeze-thaw resistance was tested. The weight loss was measured after 25, 50, 75 and 100 cycles of freeze-thaw. The dependence of weight loss after freeze-thaw cycles on water absorption capacity of fine aggregate concrete were expressed. The fine recycled concrete which was used as partial replacement of sand originated from highway demolition. The replacement rate was 10 %, 20 % and 30 %. Durability properties were investigates by using cubic and prismatic specimens.

Keywords: recycled aggregate; recycled concrete; construction and demolition waste;

1 Introduction

Production and utilization of concrete is regarded as a large consumer of natural resources such as natural aggregates and non-renewable fossil fuels. Buildings consume around 40 % of the global annual stone, sand and gravel [1]. High and constantly and rapidly increasing concrete consumption causes still growing environmental impact in terms of natural resources depletion. More than 10 billion tons of concrete are produced each year [2]. Recycled C&D waste is according to standards possible to use as partial replacement of coarse natural aggregate [3]. The use of fine recycled aggregate (FRA) as partial replacement of sand has not been allowed yet. Although, the sand extraction from seaside has high environmental impact such as increasing seaside erosion, changing wave behavior, local fauna and flora ecosystems and other environmental aspects. The local ecosystems are also negatively influence during sand extraction from riverbed [4].

Cement composites are very often exposed to the effect of thermal changes. On one hand, it is the effect of cyclic temperature loading [5]. On the other hand, it is freezing and thawing during winter. The durability properties of concrete, especially freeze-thaw resistance, could be negatively influenced by water absorption of recycled aggregate concrete. The water absorption of fine aggregate concrete with partial replacement of sand is higher. The increase is between 6% and 10% in dependence on replacement rate [6]. Higher water absorption of recycled aggregates compared to natural aggregates causes higher water absorption of concrete. The dependence of water absorption of concrete on water absorption of aggregate was estimated according to results of experimental investigation [7]. Although, the durability properties of recycled aggregate concrete could be worse in comparison with conventional concrete, it is possible to use it for structural concrete elements without high requirements on durability properties.

2 Materials and Experimental program

Fine-aggregate concrete mixtures with natural sand and various replacement by fine recycled aggregate was prepared for testing freeze-thaw resistance of concrete. Freeze-thaw resistance were tested according to Czech standard CSN EN 73 1322 [8]. The same concrete mixtures were tested on water absorption by immersion and water absorption of capillarity.

2.1 Materials and mixtures

The waste concrete blocks from recycled C&D waste were crushed from fraction 32/64 mm to fraction 0/16 mm and sieved to fraction 0/4 mm. This FRA from recycled waste concrete was used as partial replacement of sand in concrete mixture. There were used Portland cement 42.5 R in mixtures and the replacement rate of sand was 10 %, 20 % and 30 %. Concrete mixtures were designed as concrete class C 30/37. All mixtures had same amount of cement and had same water-cement ratio. The mixtures are listed in Tab. 1.

Tab. 1: Concrete mixtures with different replacement of sand

Mixture Replacement rate	FNAC	FRAC 10%	FRAC 20%	FRAC 30%
NA 0/4 mm [kg]	1458	1312	1166	1020
FRA 0/4 mm [kg]	0	146	292	438
Cement CEM I 42,5 [kg]	486	486	486	486
Water [kg]	243	243	243	243
Total w/c ratio [-]	0.5	0.5	0.5	0.5

The specimens of FNAC and FRAC are shown in Fig. 1.

Fine natural aggregate concrete (FNAC)

Fine recycled aggregate concrete (FRAC)

Fig. 1: Specimens of fine natural aggregate concrete and fine recycled aggregate concrete

2.2 Experimental program

The freeze-thaw resistance was examined on beams 160 × 40 × 40 mm. The weight, dynamic modulus of elasticity, flexural strength and compressive strength on fragments were measured after 25, 50, 75 and 100 cycles of freeze-thaw. The freezing cycle takes 4 hours of freezing by air of minus 20°C and the thawing cycle takes 2 hours of thawing by water of 20°C.

Water absorption capacity of hardened concrete with different replacement rate of natural sand by FRA was tested by immersion test. Cubes 100×100×100 mm and beams 40×40×160 mm at age 28-days were after 14-days wet curing removed from water and dried in oven for 14 days. Beams 40×40×160 mm at age 28-days of hardened concrete with different replacement rate of natural sand by FRA were 3 days tested by capillarity test

3 Results and discussion

The results of freeze-thaw resistance of fine-aggregate concrete with partial replacement of sand by fine recycled aggregate were expressed by weight loss after 25, 50, 75 and 100 cycles. The freeze-thaw resistance depends on water absorption capacity of concrete. From this reason, the dependency of freeze-thaw resistance on water absorption capacity is shown in this investigation.

3.1 The freeze- thaw resistance

The results show that the freeze-thaw resistance depends on substitution ratio of natural sand. The lowest weight loss after 100 cycles of freezing and thawing, which means the higher freeze-thaw resistance, was measured for reference concrete without FRA and the highest weight loss, the lower freeze-thaw resistance, was measured for the fine aggregate concrete mixture with the highest replacement rate of sand (see Fig.2).

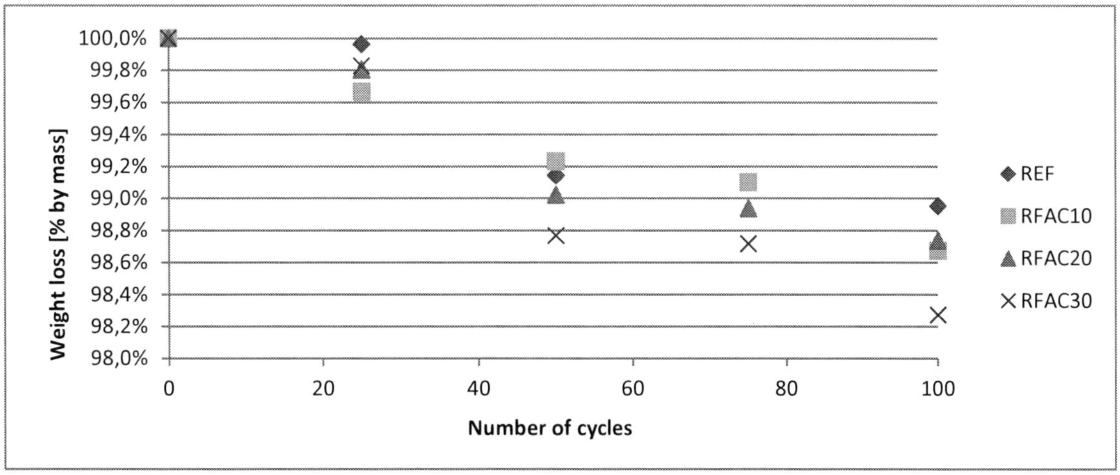

Fig. 2: Weight loss after freeze-thaw cycles of samples with different replacement rate of sand

3.2 The freeze- thaw resistance

The results show that the freeze-thaw resistance also depends on water absorption of concrete, which is naturally connected with amount of recycled aggregate in mixture and its higher water absorption. The graphs show the almost linear dependence of freeze-thaw resistance of fine aggregate concrete on water absorption by immersion and water absorption by capillarity (see Fig. 3).

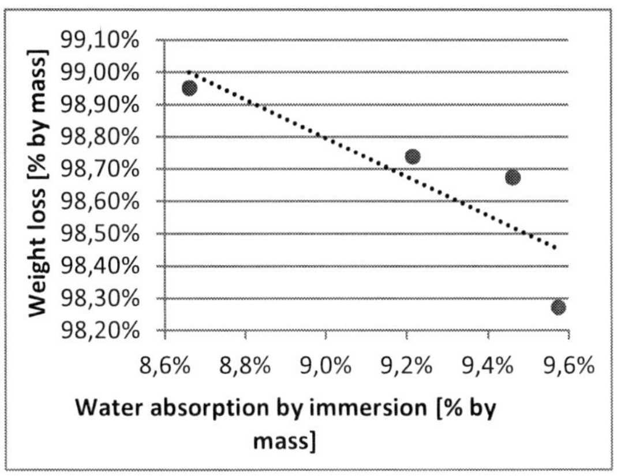

Weight loss after 100 cycles of freeze-thaw on water absorption by immersion

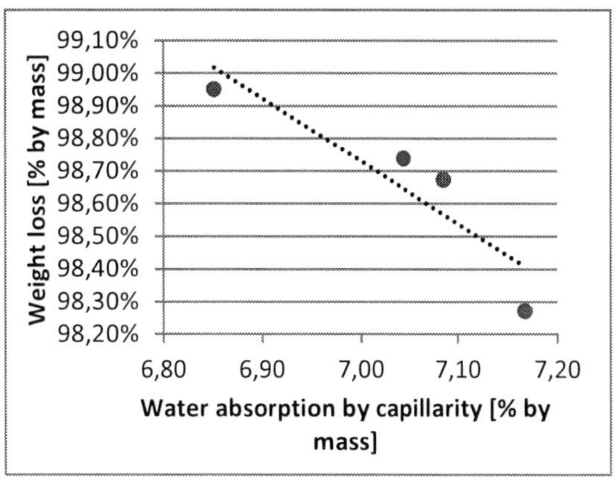

Weight loss after 100 cycles of freeze-thaw on water absorption by capillarity

Fig. 3: Dependence of weight loss after 100 cycles of freeze-thaw on water absorption capacity

4 Conclusion

In conclusion, it was proved that partial replacement of sand by FRA negatively influences durability of concrete, especially its freeze-thaw resistance. However, the weight loss of samples with FRA was less than 2 %. From this reason, it is possible to say that the weight loss of fine aggregate concrete with partial replacement of sand by fine recycled aggregate after freeze-thaw cycles is not essential. Consequently after verification this conclusion, the use of FRA is possible for concrete structures, which are influenced by freeze-thaw. Finally, is was verified that the freeze-thaw resistance is influenced by water absorption capacity of fine aggregate concrete and amount of fine recycled aggregate has impact on this characteristic. Due to the almost linear dependence of freeze-thaw resistance on water absorption capacity, the predictive model of freeze-thaw resistance of concrete could be established.

Acknowledgement

This work has been supported by the Ministry of Education, Youth and Sports within National Sustainability Programme I, project No. LO1605 and by SGS15/182/OHK1/3T/11 Verification of durability and lifespan of cement composites and recycled concrete.

References

[1] P. Joseph a S. Tretsiakova-McNally, „Sustainable Non-Metallic Building Materials," pp. 400-427, 2010.

[2] European Environment Agency, „Effectiveness of environmental taxes and charges for managing sand, gravel and rock extraction in selected EU countries," Schultz Grafisk, Copenhagen, 2008.

[3] CSN EN 206, CSN EN 206 Concrete: Specification, performance, production and conformity, Prague: CEN, 2014 (in Czech).

[4] L. Evangelista, M. Guedes, J. de Brito, A. C. Ferro a M. F. Pereira, „Physical, chemical and mineralogical properties of fine recycled aggregates made from concrete waste," Construction and Building Materials, p. 178–188, 10 April 2015.

[5] O. Holcapek, „Resistance of refractory cement composite to cyclic temperature loading," v Key Engineering Materials, Prague, 2016.

[6] M. Šefflová a T. Pavlů, „Non – destructive Testing of Concrete with Fine Recycled Aggregate," v Modern Methods of Experimental and Computational Investigations in Area of Construction, Prague, 2016.

[7] T. Pavlů, M. Šefflová „Experimental determination and estimation of water absorption capacity of hardened concrete with recycled aggregate," in EAN 2014 - 52nd International Conference on Experimental Stress Analysis, Marianske Lazne, 2014.

[8] ČSN 73 1322, ČSN 73 1322- Determination of frost resistance of concrete, Prague: Czech Standards Institute, 1969 (in Czech).

Monitoring Micro Defects of Thermally Loaded Specimens Using the Acoustic Emission Method

L. Pazdera [1,a], L. Topolar [1], L. Pařízek [2], K. Mikulasek [1], J. Smutny [1]

[1] *Brno, University of Technology, Faculty of Civil Engineering, Veveri 331/95, 602 00 Brno, Czech Republic*
[2] *Brno, University of Technology, Faculty of Chemistry Purkynova 118, 612 00 Brno, Czech Republic,*
[a] *pazdera.l@fce.vutbr.cz*

Abstract: The Acoustic Emission Method has been used to identify the thermal damage evolution in tar binders composites. Two specimen groups have been monitored. One formed by compounds based on tar while the other comprising phenolic resins and both its types, novolacs and rezoles. The signals of acoustic emission from the specimens tested in the oven at 800 degrees Celsius have been recorded during the thermal testing in order to investigate the micro change in structure.

Keywords: acoustic emission; thermally loading; micro defect; non-destructive testing; tar; phenolic resins.

1 Introduction

From the non-destructive acoustic-analysis methods the acoustic-emission method as a non-traditional method for civil engineering, was used for this experiment. The acoustic emission technique is one of the promising methods for detecting the formation of stress cracks in tests. [1] Originating in the field of structural monitoring in civil engineering, the acoustic emission testing method has the unique ability to evaluate an entire structure and locate a discontinuity as it forms and propagates. [2] It is known that, as monitoring high thermal processes is very complicated, the number of testing methods decreases with increasing temperature, which lowers the chances of getting a detailed description of the inner material structure behaviour. [3] The acoustic emission method is an applicable method that does not affect the processes occurring inside the furnace while monitoring the microscopic behaviour of the structure. [4] An acoustic emission based laboratory test was also applied to several types of asphalt materials to evaluate the low-temperature cracking performance. [5] The acoustic emission could be measured using a waveguide shaped for the signal not to change while the thermal protection of the sensor was sufficient. [6]

2 Experimental set-up

Four different types of samples were monitored under thermal loading. Each specimen of 20 °C was placed in a furnace with a loading temperature of 800 °C. The specimen was 50 mm long with a diameter of 50 mm. The experimental set-up is shown in Fig. 1. During the thermal loading, the acoustic emission activity was monitored. The acoustic emission sensor was placed on the waveguide as shown in Fig. 2. This position was chosen for its temperature to be less than 100 °C but while the detection of the acoustic emission activity is optimal.

The four types of specimens were labelled

- IS - a mixture using only tar binder (containing pitch and other products of tars distillation)
- IL - a mixture of tar binder with the highest phenolic resins content,
- KO – material similar to IL but containing another kind of resin,
- DN - composite material containing a tar binder and all phenolic resins used in IL and KO.

Fig. 1: The temperature loading of the specimen into a special oven and the position of the acoustic emission sensor on the waveguide

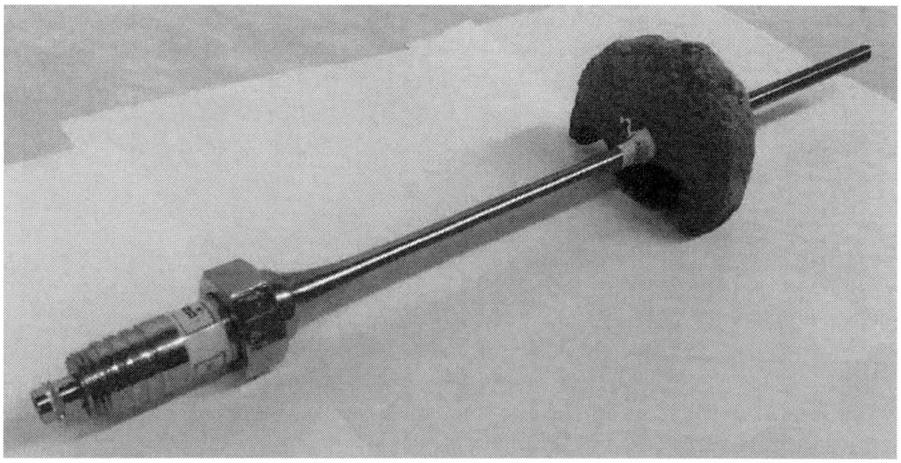

Fig. 2: The acoustic emission sensor on the waveguide

3 Results

The number of overshooting is a basic parameter of acoustic emission method. The dependence of the cumulative number of overshooting on time or temperature was used to evaluate different types of specimens. Further analyses were concerned with other acoustic emission parameters such as acoustic emission amplitude, acoustic emission energy, etc.

The acoustic emission activities during the hardening of all specimen types are shown in Figs. 3 and 4. Samples without resin (IS) took the longest time, i.e. 20 minutes to harden. The acoustic emission activities show two significant periods of changes in the mass, particularly, the first six minutes and from 8 minutes to 14 minutes. These structural changes were very slow. The specimens containing the mixture with the highest resin content (IL) change their structure very fast during the first minutes, the decline in activity follows up to the third minute when hardened. The structural changes during the hardening of similar types of specimen material (KO) are similar, there the changes occur in the first minute and the next about two minutes as the acoustic emission activity shows. There are probably two reasons for this behavior. The first one is the faster hardening of phenolic resins as compared with a tar binder. The second one is associated with the evaporation of solvents of phenolic resins, which creates a large amount of gas destroying the microstructure.

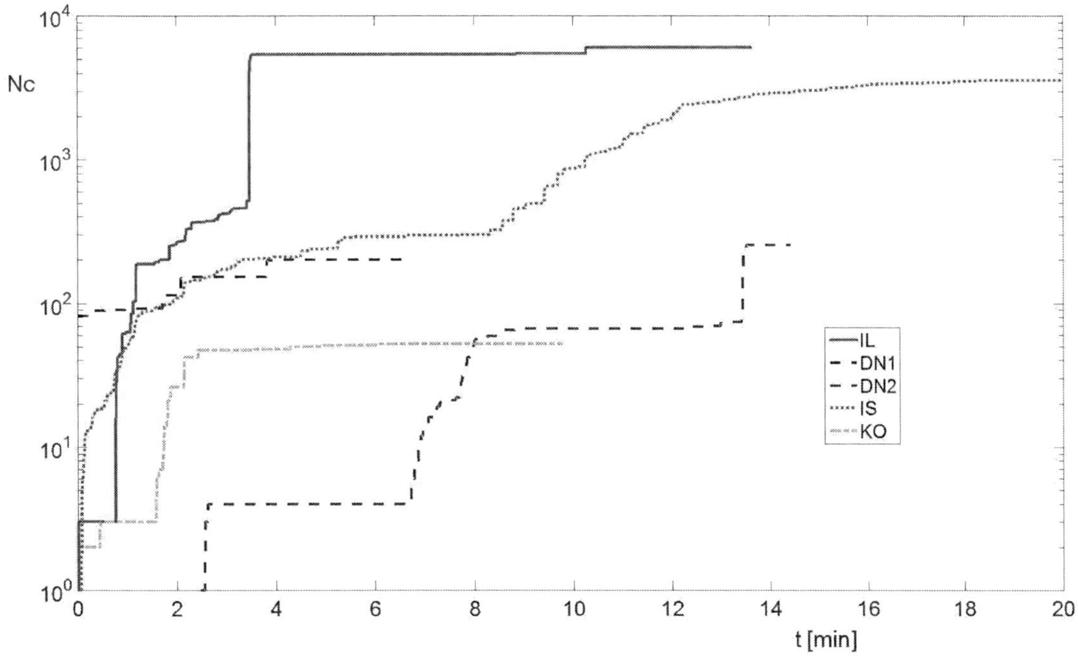

Fig. 3: The history of the cumulative acoustic emission activity of the specimens

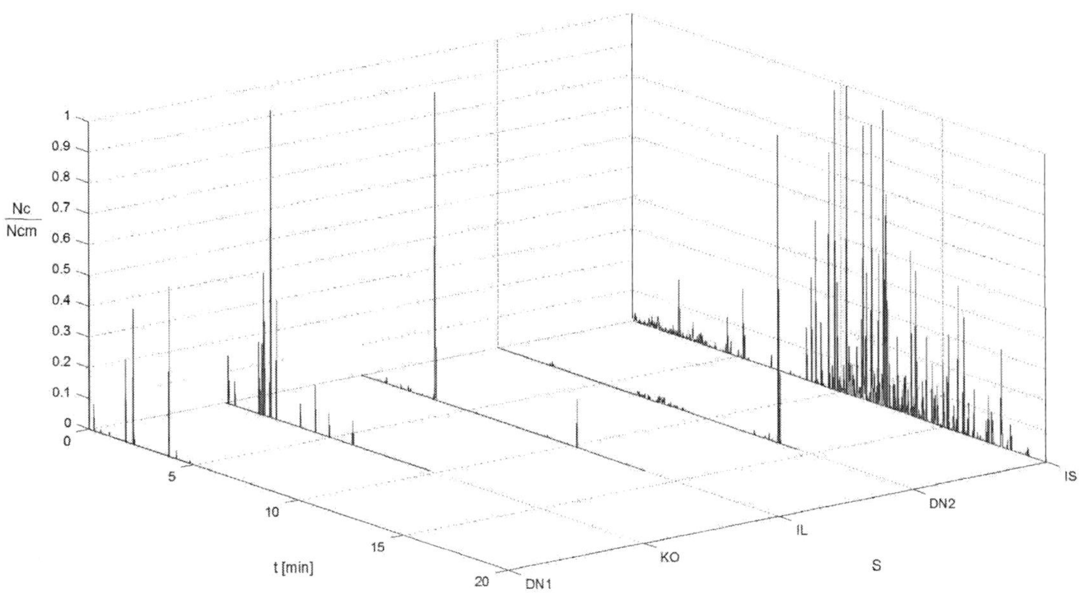

Fig. 4: The relative acoustic emission activity of specimens

Thus, in the first part (up to 5 min.), the specimen softens and settles and only in the second part (from 10 minutes) it begins to harden. As the other materials contain resin, the hardening behaviour is different. The solvents they contain are evaporated at lower temperatures, which means that the resin on the surface gets hardened fast. In the DN sample, this is given by the fact that different resins in a mixture hardening gradually, this provides ample time for a system of pores to be created through which the gas generated can escape. Thus, the material itself hardens only later when the gland varnish gets involved in the hardening, but, as compared with the IS mixture, the hardening is accelerated by resins.

4 Conclusion

Evaluation of the acoustic emission activity during thermal loading of different specimens is the focus of the paper. Acoustic emission method shows an interesting viewpoint of the thermal loading of different specimens. The chemical changes during thermal loading of specimens may be monitored with the help of the acoustic emission method. Resins mixed with tar make possible the fast hardening of the structure in a shorter time.

This method will be applied to monitoring micro changes and micro defects during thermal loading of concrete specimens. The acoustic emission method can simply be used for determining the stiffening of the matter. Therefore, when the acoustic emission activity has finished, there are no longer any chemical changes of the matter.

Acknowledgement

This paper has been worked out under the project Czech Science Foundation GACR No.16-02261S supported by Czech Science Foundation and the project No. LO1408 "AdMaS UP - Advanced Materials, Structures and Technologies", supported by Ministry of Education, Youth and Sports under the „National Sustainability Programme I" and under the project No. S-16-2967 supported by Faculty of Civil Engineering of BUT.

References

[1] M. Ohtsu, Crack classification in concrete based on acoustic emission, Construction and Building Materials, 24(12) (2010) 2339-2348
http://dx.doi.org/ 10.1016/j.conbuildmat.2010.05.004

[2] L. Topolar, P. Rypak, K. Timcakova-Samarkova, L. Pazdera, P. Rovnanik, Non-Traditional Non-Destructive Testing of the Alkali-Activated Slag Mortar During the Hardening, Materiali in Tehnologije, 50(1) (2016) 7-10
WOS:000370993000002

[3] B. Hill, B. Behnia, WG Buttlar, H Reis, Evaluation of Warm Mix Asphalt Mixtures Containing Reclaimed Asphalt Pavement through Mechanical Performance Tests and an Acoustic Emission Approach, Journal of Materials in Civil Engineering, 22(12) (2013) 1887-1897
http://dx.doi.org/ 10.1061/(ASCE)MT.1943-5533.0000757

[4] M. Kruger, C. Grosse, Event-based Data Acquisition in Wireless Sensor Networks for Structural Health Monitoring, Tm-Technisches Messen, 76(12) (2009) 568-577
http://dx.doi.org/ 10.1061/10.1524/teme.2009.0986

[5] B. Behnia, EV. Dave, WG. Buttlar, H. Reis, Characterization of embrittlement temperature of asphalt materials through implementation of acoustic emission technique, Construction and Building Materials, 111(15) (2016) 147-152
http://dx.doi.org/ 10.1016/j.conbuildmat.2016.02.105

[6] L. Pazdera, L. Topolar, J. Smutny, K. Timcakova, Nondestructive Testing of Advanced Concrete Structure during Lifetime, Advances in Materials Science and Engineering, (2015)
http://dx.doi.org/ 10.1155/2015/286469

Experimental method of measuring the efficiency of hydrophobic surface layer of concrete structures

R. Pernicová [1,a], P. Tichá [2]

[1] *Czech Technical University in Prague, Klokner Institute, Šolínova 7, 166 08 Prague 6, Czech Republic*

[2] *Czech Technical University in Prague, Faculty of civil engineering, Thákurova 7, 166 29 Prague 6, Czech Republic*

[a] *radka.pernicova@cvut.cz*

Abstract: The review of hydrophobic materials application on surfaces of concrete structures to achieve its protection again the environmental conditions is described in this paper. The article is focused on the characteristics of selected hydrophobic materials, chemical base of origin and their characteristics. The second part describes the possibility of experimental methods of measuring of new hydrophobic surface layer, particularly the quality and efficiency of their hydrophobic properties.

Keywords: hydrophobic impregnation; concrete; contact angel; surface tension; penetration.

1 Introduction

Porous building materials could absorb a lot of water due to its porous structure which may cause some negative effects on material. Temperature is one of the most significant side effects, especially below zero level. This could create cracks that can leads to deformation of integrity of the material. Another equally important negative side effect is water function as the carrier medium. Transport of ions in direction away from construction could lead to leaching of the compounds and their subsequent efflorescence on the surface [1]. In the opposite direction harmful substances such as chloride ions from de-icing salt may be carried into the material and therefore slightly damage its structure [2].

The main aim of building engineers all round the world is to provide quality isolation of the material, keep it away from humidity and avoiding the possible damage such as decrease of reliability and durability of the building constructions or its parts. Hydrophobicity of surface could be increased by using extra protective layers. This work is focused on description of characteristic of hydrophobic impregnations.

Hydrophobic impregnations have various forms of phases. The most commonly is used an emulsion, solutions, pastes and creams. An indispensable influence on quality and durability of hydrophobic product has properties of the concrete surface to which is hydrophobization applied and compliance with the technological process.

To protect outer building construction surface against external conditions, hydrophobic impregnation, impregnation or coatings could be used. Coat creates continuous film on the top of concrete that fills pores and capillaries by impregnation substance and therefore the top layer is sealed. The impregnation decrease porous characteristic of concrete and therefore strengthen its surface. As the result, capillaries and pores are filled, partly or fully. Hydrophobization of concrete generate thin layer on the surface of capillaries and pores that avoid fluid to pass through it, unlike gas and steam [3]. Pores and capillaries are internally coated but not filled.

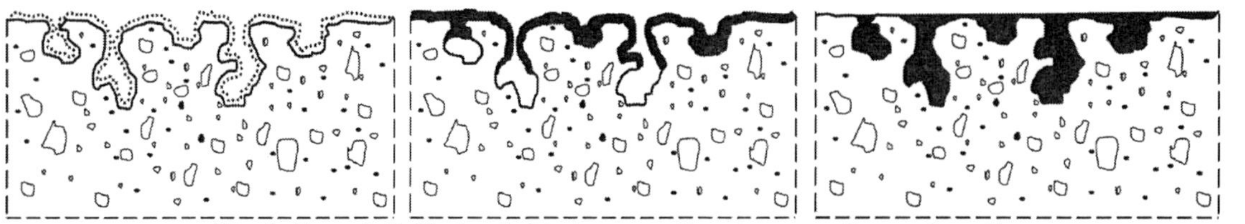

Fig. 1: From left: Hydrophobic impregnation, impregnation and coatings

2 Hydrophobic impregnation

The essential element of hydrophobic impregnations is silicon dioxide and other organic compounds that determine the final properties of the impregnation. Silica ensures adhesion to the surface because it is also inside the concrete substrate. Organic compounds remain on the surface layer are firmly embedded in the substrate and significantly alters the wettability of the surface of capillaries. Most of the hydrophobic impregnation is made by chemical base of following substances: silane, siloxane and siliconate [4] [5] [6] [7].

2.1 Silanes

Silanes have large molecules from the chemical point of view, non-hydrolyzed silanes (i.e. esters of silicic acid). Their chemical formula is SiH_4 and basic molecular bond is a C-Si. Silanes have a quality penetration properties and penetration achieves great depth. Therefore, materials treated with this substance are able to prevent the ingress of water, even threw micro-cracks in the surface of structure up to width four tenth of a millimeter. Silanes react with moisture on the silica gel which besides the hydrophobic has also the strengthening effect. Silanes have an ability to create chemical bond on the surface of quartz grains, which means that it is impossible to remove them from the surface. Anhydrous alcohols as well as aliphatic materials are used as solvents.

2.2 Siloxanes

Siloxanes are the most frequent and suitable used hydrophobic products for different kinds of building materials, out of all mentioned hydrophobic materials. It is a functional group in organosilicon chemistry with the Si–O–Si linkage. The parent siloxanes include the oligomeric and polymeric hydrides with the formulae $H(OSiH_2)_nOH$, $(OSiH_2)_n$ and have several functional groups such as silicones, alkylalkoxysiloxanes or Alkoxy polysiloxane. Siloxane products are one-component, and since the resins are soluble in white spirit, so this impregnation cannot be used for wet or damp surfaces. Siloxanes have large molecules, achieving only a little depth of penetration. Its impregnation can be renewed, because it is possible to impregnate previously hydrophobic surface. Siloxane coating is soluble in white spirit, which helps to remove old paint. However these hydrophobic impregnations represent threat to the environment and are flammable too.

The low molecular oligomeric alkylalkoxysiloxanes are mostly one-component hydrophobic impregnation that spontaneously polymerize for the presence of atmospheric moisture. They are capable of binding chemically to the surface of quartz grains, which is the reason why they are the best option for hydrophobic impregnation of silicone materials (i.e. concrete and cement products). The impregnation bond to quartz grains increase abrasion resistance of coating and at the same time prevent by coating removability. The low molecular oligomeric alkylalkoxysiloxanes are the most effective penetrate into porous materials comparing others hydrophobic impregnations. They may have either paste or cream form, in that case, are most commonly use on fresh concrete surface to achieving deeper penetration of hydrophobic impregnation.

Alkoxy polysiloxanes are solutions of polysiloxane resins transferred into an aqueous emulsion. The polysiloxanes impregnating agents are water-soluble and its hydrophobic effect creates after evaporation of water that means, they can be applied to damp surfaces. For the production of polysiloxane impregnating the emulsifiers, leach out of the polymer film, are used. That is the reason, why the final hydrophobic effect will appear after a certain time. More precisely, called polymerized siloxanes, silicones are mixed inorganic-organic polymers with the chemical formula $[R_2SiO]_n$, where R is an organic group. These materials consist of an inorganic silicon-oxygen backbone chain (-Si-O-Si-O-Si-O-) with organic side groups attached to the silicon atoms. Silicon as hydrophobic agent can be used in form micro emulsions, which create aqueous emulsion while they are mixed with water. Silicone micro emulsions have an advantage in ability to be miscible with water and therefore there is no need for organic solvent.

2.3 Siliconates

Siliconates are hydrophobic substance that reacts with carbon dioxide to form a gel of silicic acid. Chemically, they are methylsiliconate of sodium or potassium and have small polar molecules, which do not achieve significant penetration depth into the concrete structure. Incurred silicic acid gel is not soluble, therefore coating is irremovable. In the reaction siliconates with carbon dioxide sodium or potassium hydroxide is released. These new products may cause the efflorescence on surface of concrete, which appears the most often after application at lower temperature and could devaluation of concrete products as the result

3 Measurement methods

Currently, we have several measuring methods which determine level of hydrophobicity of concrete structures.

3.1 Measuring of contact angle and surface tension

Most of them measure directly or indirectly contact angle formed by tangent the surface of the liquid drop to the contact with surface of the solid phase. Hydrophobic material, which creates water-repellent layer on surface of concrete, causes an increase of water contact angle. The degree of hydrophobic efficacy is determined by the contact angle Θ formed between the tangents to the drop of water on surface with the washer.

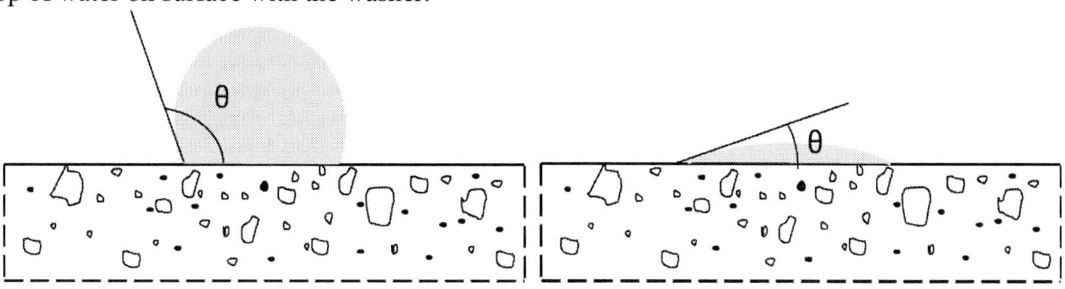

Fig. 2: Wettability of surface: hydrophobic (left) and hydrophilic (right)

There are 4 categories depending on the size of contact angle. The lower value of the contact angle means better hydrophobic properties of concrete specimens [8].

Tab. 1: Categories of hydrophobicity according the contact angle

hydrophilic	hydrophobic	super-hydrophobic	ultra-hydrophobic
$\Theta < 90°$	$120° > \Theta > 90°$	$150° > \Theta > 120°$	$\Theta > 150°$

Contact angle is one of the few directly measurable properties of the three-phase interface of a solid / liquid / gas. The balance in three-phase system describes Young's equation of equilibrium of forces on the interface.

$$\gamma_{sg} = \gamma_{sl} + \gamma_{lg} \cdot \cos \Theta \quad \text{or} \quad \cos \Theta = (\gamma_{sg} - \gamma_{sl}) / \gamma_{lg} \tag{1}$$

In this equation γ_{lg} represents the value of the interfacial tension at the interface of the two relevant phases. This equation assumes the perfectly smooth homogeneous surface [9].

Measuring the static contact angle of drop on non-porous, solid flat surface is done by the help of video cameras or goniometer. Drop of water adjacent to the solid is observed by a microscope equipped with a goniometric eyepiece allowing direct reading contact angle or evaluation later from drops image. Determination of the contact angle by this method is within 1° of accuracy. Indirect determination of contact angle is based on strain-gauge methods (capillary elevation on the vertical plate, balancing Wilhelmy plate). Based on that, detected contact angle can calculate surface energy of the material and use this information to better understand the hydrophobic behaviour [10].

3.2 Measurement of hydric properties

Basic parameters of liquid moisture transfer, namely saturated moisture content, maximum absorbency, absorption coefficient are another kind of measurement to detect the hydrophobicity level of building materials. Transport of liquid moisture was studied using the standard sorption experiment, the one-dimensional absorbency of the sample, which is based on a classic experiment of soaking from the water surface. Obtained measured data shows clearly that we can construct the dependence of the cumulative moisture content, from which we can directly determine the moisture absorption coefficient A [$kg/m^2h^{1/2}$] using the linear part of the increase in weight [11].

3.3 Measurement of penetration depth

The most important criterion of the hydrophobic impregnation durability is the penetration depth. Standard ČSN EN 1503-2 divides impregnation into two groups: I Class means penetration to a depth of 10 mm; II Class means depth more than 10 mm. Depth of impregnation is performed by informative qualitative phenolphthalein alkalinity test of concrete at borehole. Based on colorimetric test, it is in the reaction of phenolphthalein with concrete. The concrete itself has high alkalinity (pH 12-13). In contact with the impregnation concrete is neutralized and its alkalinity is significantly reduced. Main principle of the test is to immerse the borehole into the phenolphthalein solution, finding out the depth of penetration. The depth is determined by coloring concrete. The colorless solution of phenolphthalein dissolved in ethanol is turned purple at a pH greater than 9.5 and at a pH lower than 9.5 the solution stays colorless.

3.4 Conductivity measurement

Quality assessment of already applied hydrophobic impregnation on the sample surface is most commonly performed using conductivity measurement. Experimental device works on principle of measuring ac voltage of +/- 700 mV and a frequency of 20 Hz, which operates through electrodes on pre-defined surface area. In place of the impregnation failure tension is transmitted into the concrete and it is subsequent measurement on the other non-impregnated side of concrete specimen. The final value is a function of the number of defective places. The bigger is the measuring values the more damaged areas are in the hydrophobic impregnation. Escalating tension curve indicates of bad adhesion of hydrophobic impregnation to the concrete surface.

4 Conclusion

The interaction between the solid and liquid is the key role in understanding chemical and physical processes in the civil engineering, specifically in the area of interaction of building material with the external environment. The influence of UV radiation and other environmental conditions has to be taken into account, causes decay of organic ingredients and therefore hydrophobic ability is decreasing. Siloxanes are the most suitable for the most different kinds of building materials, out of all mentioned hydrophobic materials. However, it still depends on the particular material solutions and the exactly situation of the building in deciding which hydrophobic agent is best to use. Choice of an unsuitable hydrophobic product can have fatal consequences for the concrete product. Most hydrophobic impregnations are unrecoverable from the surface and they can be removed only mechanically therefore selection of suitable Impregnation is really important.

Acknowledgement

The research was supported by Grant Agency of the Czech Republic under the number GACR 15-10591S.

References

[1] R. Pernicova, Analysis of formation and testing of efflorescence on concrete elements, Advanced Materials Research Vol. 1025-1026 (2014) p. 641-644.

[2] D. Dobias, R. Pernicova, Diffusion of Chloride Ions in Ultra High Performance Concrete, Advanced Materials Research Vol. 1106 (2015) p. 21-24.

[3] J. Soucek, P. Tichá, Modeling Interior Environment of Historical Building and Determination of Critical Places for Application of Nanotextiles Protection, Advanced Materials Research, (2014), Vols. 1025-1026, p. 1141-1144.

[4] J. Vetchy, Hydrofobní impregnace betonu, MCT - Texty o betonových výrobcích, (2015) p.1-6

[5] V. Bažant, Technické použití silikonů, (1959), SNTL Praha.

[6] A. Selande, Hydrophobic impregnation of concrete structures – Effect on concrete properties, Ph.D. Thesis, (2010), Royal Institute of Technology, Stockholm, Sweden. [5] V. Bažant, Technické použití silikonů, (1959), SNTL Praha.

[7] J. Blitz, C. Little, Fundamental & Applied Aspects of Chemically Modified Surfaces, (1999), Royal Society of Chemistry, Cambridge, UK. [5] V. Bažant, Technické použití silikonů, (1959), SNTL Praha.

[8] I. Krasny, Measurement of contact angel and determination of surface free energy of plastic materials, Thesis, (2010), Tomas Bata University in Zlin, Faculty of technology, Czech Republic [5] V. Bažant, Technické použití silikonů, (1959), SNTL Praha.

[9] J.Holub, Smáčivost povrchu vůči kapalinám a druhá viskozita kapalin, Brno: Vysoké učení technické v Brně, Fakulta strojního inženýrství, (2010) pp. 38

[10] O. Chmela, Different imaging techniques for investigation of treatment effects on various substrate surfaces, Proceedings of the 21st Conference STUDENT EEICT (2015). pp. 351-355.

[11] S. Roels, J. Carmelit, H. Hens, O. Adam, H. Brocken, R. Černý, Z. Pavlík, C. Hall, K. Kumaran, L. Pel, R. Plagge, Interlaboratory Comparison of Hygric Properties of Porous Building Materials, In Journal of Thermal Envelope and Building Science, Vol. 27, (2004), pp. 307-325.

FEM Calculation of Gear Mesh on Single-stage Axle Gearbox for Trains

K. Petr [1,a]

[1] *Czech Technical University in Prague, Faculty of Mechanical Engineering, Technicka 4, 166 07 Prague 6, Czech Republic*

[a] *Karel.Petr@fs.cvut.cz*

Abstract: This article describes the methodology and calculations of gear mesh optimization by FEM for single-stage axle gearbox. Results of the calculations are used for examination of changes in the gear mesh of the tooth flank under load, particularly in term of improved uniform load distribution through a width of the tooth. The aim of the methodology is to design optimized tooth profile with application modifications of tooth flank on the basis of analysis of calculation model by FEM. The article compares several variants of modifications according to distribution of contact pressure.

Keywords: FEM; gear mesh; modifications; contact pressure; axle gearbox; gear; optimization; train; shaft; deviation; single-stage.

1 Introduction

This article describes the problem of position and shape changes of gears in the gear-mesh under load and methodology, which handles compensation for these negative phenomena. Application of the finite element method (FEM) for the calculation of gearing with respect to real shapes of individual components of the transmission system can accelerate the gear mesh optimization process [1].

The results of the calculations are used for examination of changes in the gear mesh of tooth flank under load, particularly in term of improved uniform load distribution through a width of tooth. The aim of methodology is to design optimized tooth profile with application of transverse, longitudinal and angular modifications on the basis of analysis calculation of model by FEM.

Basic technical data about single-stage axle gearbox (Fig. 1): gear ratio 3,857; nominal power 340kW; max. input torque 3528Nm; max. input shaft speed 4200min^{-1}; operating temperature -30°/+45° [2].

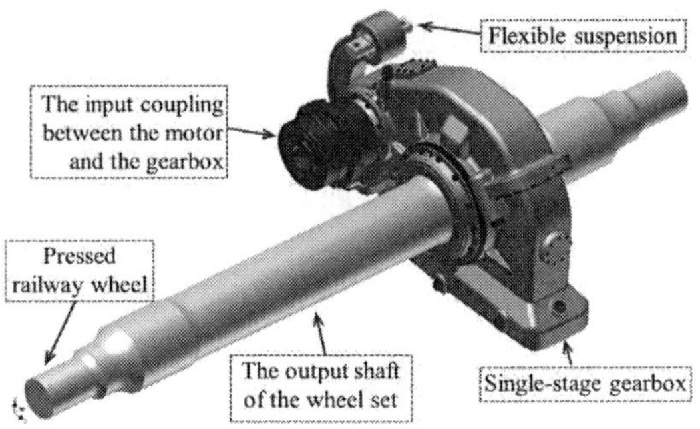

Fig. 1: Single-stage axle gearbox with coupling and wheel-set output shaft.

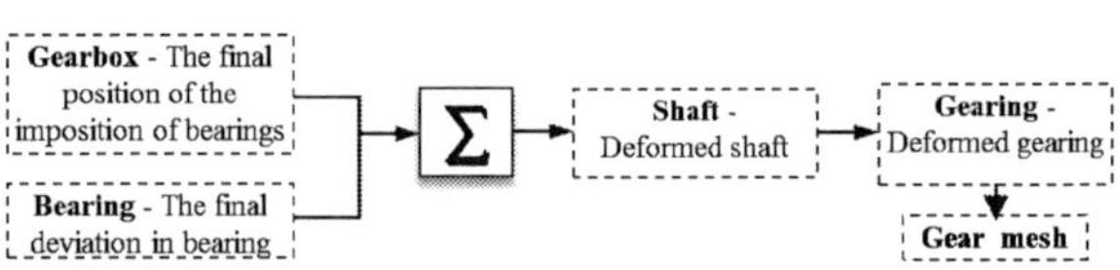

Fig. 2: Relations affecting the gear mesh.

2 The Reasons of the Inequality of Gear Mesh under Load

The final gear mesh conditions are influenced by several factors. The most important factors are deviations from the ideal shape of gears, clearance, assembly variations and transmission system parts deformation. The final position of gear mesh is heavily affected by loaded geometry of gearbox, shafts, bearings and gears. These components have fixed relations among each other and are shown in the Fig. 2.

Based on calculations by FEM for individual elements according Fig. 2, these conclusions can be made. An essential element that influences the design process is the shaft. Another important part is the gearbox, which optimized construction, must have sufficient rigidity for the position of the bearings. The part which affects the calculation on the lower level is bearing. Separate problem is the stiffness of own teeth.

3 FEM Model and Calculation of Gearbox and Shafts

After loading of CAD model into FEM connection between upper and lower part of the gearbox by the contact was made. Both parts of gearbox are bolted together. Transfer reaction forces from the shafts are applied at the center point of bearing and the transfer of the gearbox is simulated as type „Coupling" - Fig. 3.

The result of the calculation (Fig. 4) is the spatial layout of points which characterizes the entry position of the points of the shaft ($f_{xU\_deform\_i}$, $f_{zU\_deform\_i}$).

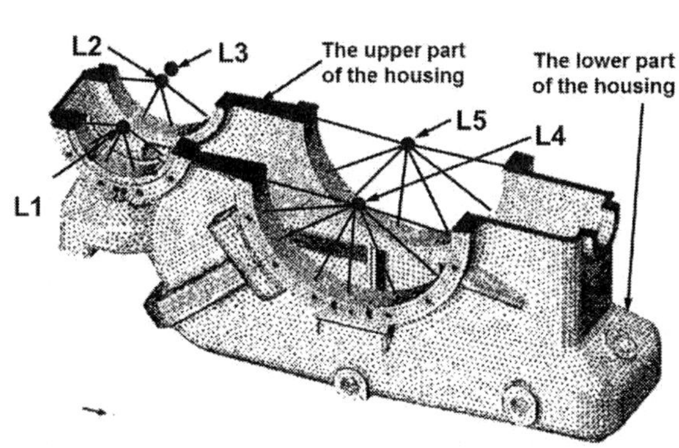

Fig. 3: FEM model of gearbox with simulation of bolts in dividing plane.

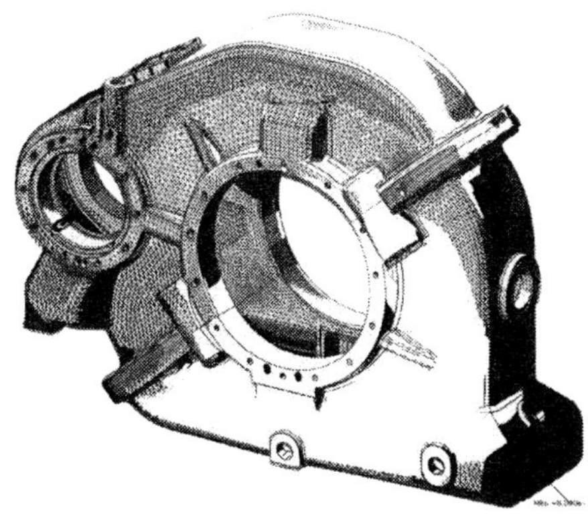

Fig. 4: Deformation of gearbox in scale.

The output shaft has locked rotation around own axis Y. On the input shaft is applied load torque 1960Nm. The calculation is performed in two steps. At first, are activated tensile forces into the bolts between the output gear and the hub, and then is applied the load torque on the input shaft.

4 Summation of all Parameters Affecting the Gear Mesh

The result is a displacement and rotation of gears. Calculation starts by displacement of points in the shaft fits. Deviations of the support and the shaft in directions X and Z are calculated according to formulas (1) for direction X and same formulas with index Z (changed X to Z) for direction Z.

$$f_{xL\_i} = f_{xU\_prod\_i} + f_{xU\_deform\_i} + \left(f_{L\_clearance\_i} + f_{L\_deform\_i}\right) \cdot \cos\varepsilon_i \tag{1}$$

where: $f_{xU\_prod\_i}$ – production deviation from the ideal position of support in the dir. X [mm];
$f_{xU\_deform\_i}$ – deviation result from the deformation of gearbox in the dir. X [mm];
$f_{L\_clearance\_i}$ – radial clearance in loaded bearing [mm];
$f_{L\_deform\_i}$ – radial deviation resulted from the deformation of the bearing „i" [mm];
ε – angle of the direction of load bearing in the plane XY [°].

In the Fig. 5 are shown representations of the new shaft positions and in Tab 1 are shown resulting values.

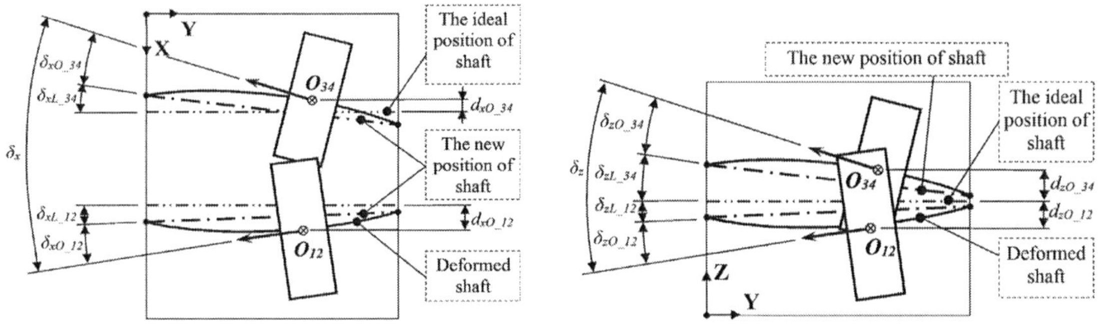

Fig. 5: Representation of total displacements and rotation of gears in the plane XZ and plane ZY.

Tab. 1: Resulting values of new midpoint position of gears.

Indication of midpoint of gears	Total Displacement [mm]		Total Rotation [°]	
	d_{xL}	d_{zL}	d_z – Plane XY	d_x – Plane ZY
O_{12}	0.02622381	-0.04098320	0.00655374	0.01529100
O_{34}	0.02077178	-0.03239230	0.00939368	0.01688300

5 FEM Model and Calculation of Gear Mesh

The solution of the contact problem is created in the gearing (3D model - gear and pinion). The contact bond among tooth flank is added to the model and meshed. The gears are connected with the center point of rotation with help of interaction type „Coupling". The subsequent rolling of the gear (pinion) and the implementation of all moments are applied through these points. Simulation of gear-mesh is performed quasi-statically (Fig. 6).

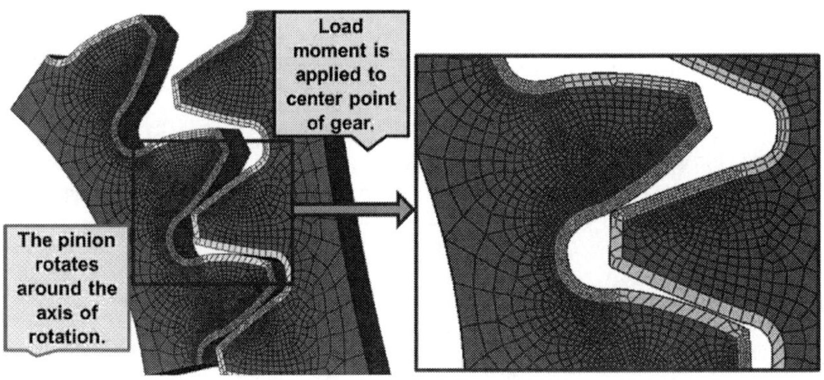

Fig. 6: FEM model of gearing (gear mesh).

6 The Variants of Modification Used for FEM Calculation

In the Tab. 2 are shown descriptions of modifications (seven variants) used for the FEM calculation of gear mesh and average and maximum values of contact pressure (FEM).

From the program KISSsoft is maximum contact stress (pressure) for pinion σ_{H_p} = 1029Nmm^{-2} and for gear σ_{H_g} = 1025Nmm^{-2} (according DIN 3990). In the Tab. 2 are shown average and maximum values of contact pressure. Contact pressures on gears for seven variants of FEM calculations are shown in Fig. 7.

Tab. 2: Description of modifications used for the FEM calculation and average and maximum values of contact pressure for each variant.

Description of modifications (variants)	Max. contact stress [Nmm^{-2}]	Average contact stress [Nmm^{-2}]
The ideal gear mesh (without modification) → **Ideal**	1038	731
Displacement and rotation (without modification) → **Def**	858	760
Transverse modification (c_{aa} = 0,015 mm) → **Trans_0.015**	993	808
Transverse (0,015mm) + longitudinal (0,009mm = c_b → pinion) modification → **Trans_Long**	865	801
Longitudinal modifications (Δβ → 0,015 mm = c_b) → **Long_0.015**	1136	725
Longitudinal modifications (Δβ → 0,009 mm = c_b) → **Long _0.009**	959	694
Longitudinal modifications (Δβ → 0,007 mm = c_b) → **Long _0.007**	899	694

7 Conclusion

Results from the application of the methodology - the calculations of the modifications with help of the finite element method confirmed the applicability in the design of longitudinal, transverse and angular modifications. Already the first step where are calculated deformation of the gearbox and shafts, provided important results for the design of angular modification. By summing the calculated deviations of deformations, manufactured deviations from the ideal shape of the bearing and clearances is calculated final position from which can be easily calculated the angle modification.

In the Fig. 7 is shown (Hertz) contact pressure on tooth flank without modification and deformation of shaft, gearbox (ideal); with respect of deformation of shaft and gearbox (def); and for five variants of modifications (with respect of deformation of shaft, gearbox, etc.).

The best solution was achieved with usage of longitudinal modification c_{bb} = 7 µm (Long_0.007). FEM results with 7 µm modification was compared in program KISSsoft and they corresponded.

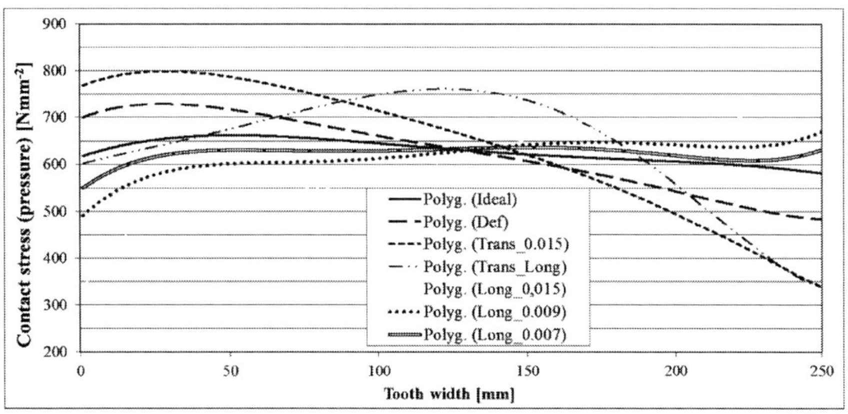

Fig. 7: Comparison of the average contact pressures on the pinion tooth flank along tooth width.

References

[1] K. Petr, P. Žák, V. Dynybyl, J. Kanaval, FEM Optimization of Gear Flanks Shape Modifications. In Proceedings of the 3rd International conference Power Transmissions '09. Thessaloniki: Sofia Publications, 2009, vol. 1, p. 93-100. ISBN 978-960-243-662-2.

[2] K. Petr, V. Dynybyl, Methods of testing gearboxes of rail vehicles, In Applied Mechanics and Materials. Uetikon-Zurich: Trans Tech Publications, 2014, p. 277-282. ISSN 1660-9336. ISBN 978-3-03785-977-3.

Measurement of Displacement of Percutaneous Lead for Spinal Cord Stimulation

I. Petrikova [1,a], B. Marvalova [1], D. Cirkl [1], V. Masopust [2]

[1] *Technical university of Liberec, Czech Republic*

[2] *Military University Hospital Prague, Czech Republic*

[a] *iva.petrikova@tul.cz*

Abstract: A Medtronic implantable neurostimulation system is indicated for spinal cord stimulation (SCS) as an aid in the management of chronic, intractable pain of the trunk and limbs. The method of anchors fixing may have a significant influence on the elimination or reduction of the SCS lead displacement. The aim of the research was an experimental examination and comparison of electrodes displacement for three different ways of fixing of the lead using anchor. The displacements were measured on pigs using optical methods. The displacement of the electrode depends greatly on the number of anchors and on the manner the lead is fixed. The fixing the lead using two anchors with a loop between them significantly reduces the total displacement of the electrode.

Keywords: measurement of displacement, digital image stereo-correlation method; image processing.

1 Introduction

The spinal cord stimulation electrodes used in the clinic are either minimally invasive cylindrical electrodes to be implanted percutaneously. The lead migration is the most common complication encountered with SCS systems. Strategies have been identified to reduce the occurrence of lead migration for SCS implantation [1, 2]. The fixing electrodes at points of their exit from the spinal space is performed using of anchors which are securely fixed to the lead and sutured with two stitches to the fascia, Fig. 1.

The aim of the research was an experimental examination and comparison of electrodes displacement for three different ways of the fixing of the lead. Three electrodes were gradually implemented into the epidural space of the pig and fixed in three different ways:

- using one anchor (Fig.1),
- using two anchors,
- using two anchors, with loop.

The set of measurements was carried out for each of fixing way. The method of fixing may have a significant influence on the elimination or reduction of the electrode feed.

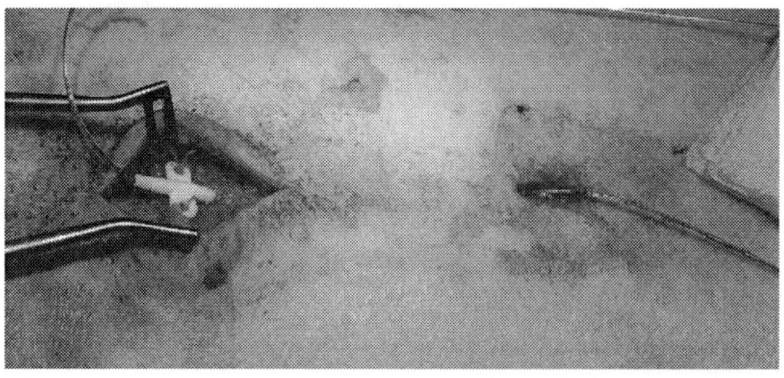

Fig. 1: The electrode fixed using one anchor.

The experimental measurements were carried out in the Laboratory of electrophysiology of the 1st Faculty of Medicine, Charles University in Prague on pigs under standard laboratory conditions. Measurements were scanned in a state without breathing.

2 Measurement of Lead Displacement

The electrode of stimulator with the connecting cable were threaded through the epidural space and led out to the surface of the body of the pig. The lead was fixed in the incision before entering into the pig's spine using one of the methods listed above. The second end of the lead was tunneled to the side of the pig and the extension of the lead connected to a rope. The rope passed over a pulley and it was loaded by a weight which caused the electrode to move. The load grew slowly from 0 to 10 N for 15 seconds. The displacement of the electrode was measured and evaluated using two different optical methods:

- Digital image stereo-correlation method,
- Digital image processing method.

2.1 Digital Image Stereo-correlation Method

The measurement and evaluation were carried out using the optical correlation device ISTRA4 Dantec Dynamics which allows for non-contact measurement of three-dimensional displacements of points on the surface of a body. One white target with a random speckle pattern of black dots was attached to the electrode and a second one was attached to the skin of pig (Fig. 2). The targets moved slowly during the loading of extension lead. The images of the two moving targets were recorded simultaneously using two cameras with a sampling frequency of 17 frames per second (Fig. 3). The recorded images were then processed in the ISTRA4 software and the 3D coordinates (x, y, z) of the target points were determined. The three components of displacement Δx, Δy a Δz are calculated as the difference coordinates of the points on the actual image and the reference image. The example of total displacement evaluation is in Fig. 4. The electrode target lies freely on the dorsal part of the pig and therefore the movement of the target is a *general spatial motion* whose approximate evaluation is only qualitative. The total displacement d of the tip of the electrode was chosen for comparison of the electrode movement in the three different ways of fixing the lead:

$$d = \sqrt{(\Delta x)^2 + (\Delta y)^2 + (\Delta z)^2} \tag{1}$$

The influence of the different ways of fixing the lead is apparent from Fig. 5 where the graphs of total displacement d during the loading are displayed. The displacement of the electrode depends greatly on the number of anchors and on the manner the lead is fixed. It is evident that fixing the lead using two anchors with a loop between them significantly reduces the total displacement of the electrode.

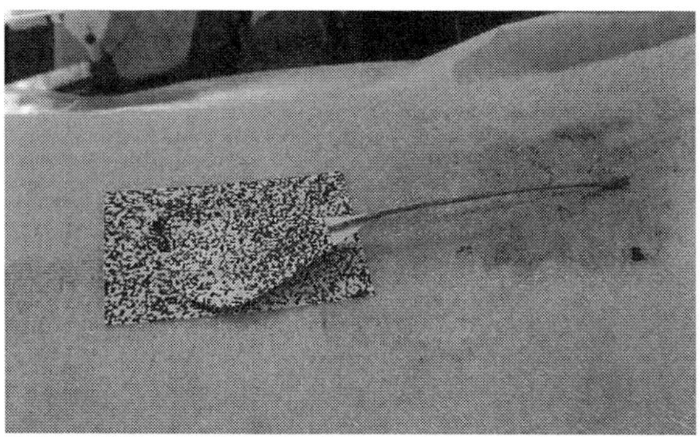

Fig. 2: Targets with random speckle pattern.

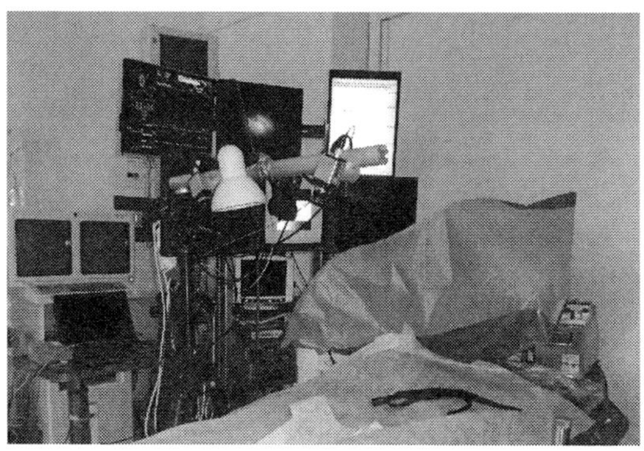

Fig. 3: Photogrammetric system with two cameras.

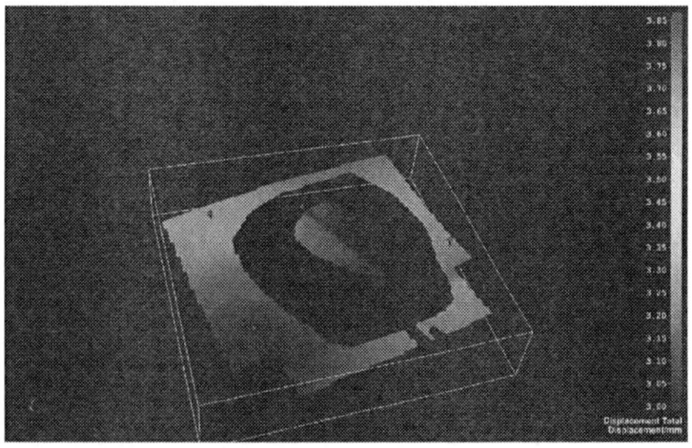

Fig. 4: The example of total displacement evaluation.

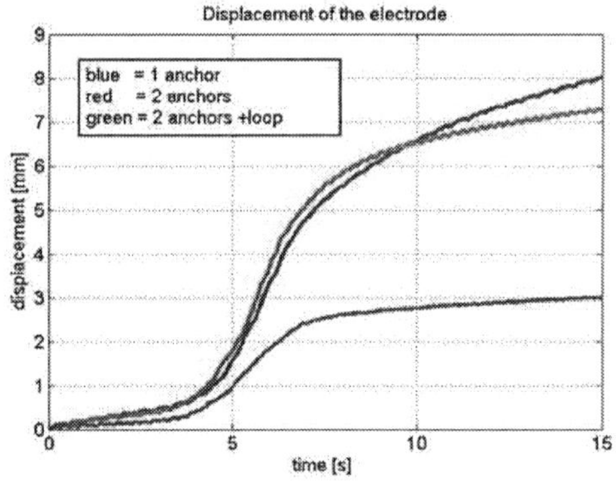

Fig. 5: Total displacement of the end the electrode during 15 seconds of loading.

2.2 Measurement of Electrode Displacement by Means of Digital Image Processing

The aim of the measurement was to precisely determine the displacement of the electrode in the longitudinal direction of the spine. A small glass tube with an internal diameter of 3 mm was sewn onto the skin of the dorsal part of the pig parallel to the spine, using two stitches. The lead was threaded through the epidural space, brought to the surface and inserted into the tube, see Fig. 6. Two targets were glued onto the surface of the tube with a distance of 100 mm between them ensuring a fixed benchmark for every photographic image. The electrode could move freely in the tube axis direction while its movement in the transverse direction was limited by the tube wall. The extension of the lead was loaded the same way as in the previous cases. In every experiment two digital images of the electrode position were recorded with a resolution of 4608 x 3456 pixels. The first image was recorded before load application when the electrode was in the starting position and the second one was recorded in steady state at the end of loading.

The pixel coordinates of both targets and of the tip of the electrode were determined in every image in Matlab Image Processing Toolbox. The coordinates correspond to places with maximum correlation coefficient between matrix A and matrix B. The matrix A contains the grayscale intensities of rectangular area around the targets or around the electrode tip. The matrix B of the same size picked from every image using detailed scanning. The correlation coefficient r is defined by

$$r = \frac{\sum_m \sum_n \left(A_{mn} - \bar{A}\right)\left(B_{mn} - \bar{B}\right)}{\sqrt{\left(\sum_m \sum_n \left(A_{mn} - \bar{A}\right)^2\right)\left(\sum_m \sum_n \left(B_{mn} - \bar{B}\right)^2\right)}}, \tag{2}$$

where $\bar{A}$ and $\bar{B}$ are the means of values in matrices A and B. The distance of targets in units of pixels in every image was used for conversion between pixels and millimetres. The targets distance was 100 mm, or about 2811 pixels, respectively, and therefore the accuracy of the measurement of the position of the electrode is higher than 0.1 mm.

The position of the electrode tip was determined in every image and the displacement of the electrode was calculated as the difference between its positions before and after loading. The results of the three sets of measurements are in Fig. 7 and in Table 1. It is evident that the way the supply lead is fixed is crucial in avoiding the displacement of the electrode due to a sudden movement of the patient.

The exploratory data analysis (EDA) was conducted within the framework of statistical analysis of the results and the parametricity was verified by Shapiro-Wilk test. Subsequently, the analysis of variance (ANOVA) was used to assess the significant differences between electrode displacements measured for three ways of fixing the electrode. It was established that the results are significantly different on the significance level $\alpha = 0.001$.

Tab. 1: Displacement of electrode for different fixation arrangements.

fixation arrangement	1 anchor	2 anchors	2 anchors and loop
number of measurements [-]	12	12	11
average displacement [mm]	15.41	9.35	0.92
standard deviation [mm]	1.41	0.46	0.48

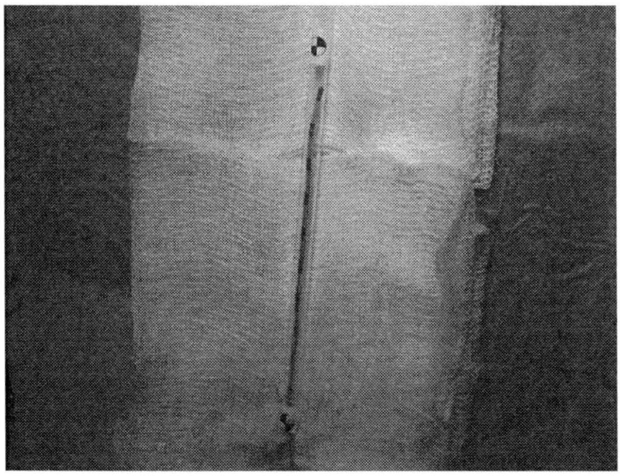

Fig. 6: Experimental setup of the measurement of electrode displacement.

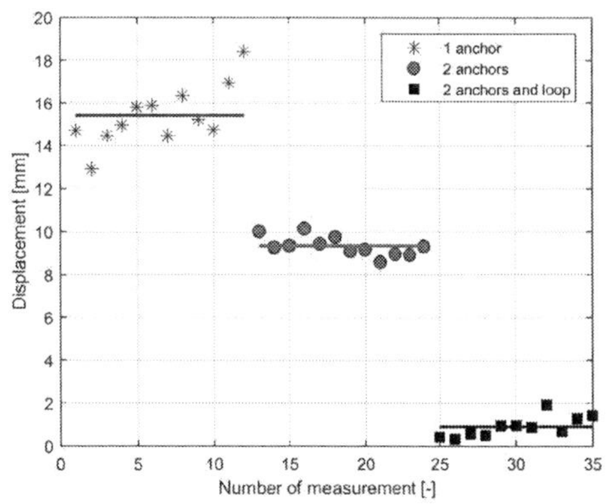

Fig. 7: Displacement of electrode for different fixation arrangements.

3 Conclusion

The results obtained by both measuring methods show the difference between the tested types of fixation.

Fixing an electrode with two anchors already considerably reduced the displacement of the electrode, and when it was fixed using two anchors and a loop the electrode barely moves as the average displacement was less than 1 mm.

Based on the statistical evaluation, it is possible to state that the differences between experimental results obtained for the three different types of fixing the lead are statistically significant.

Acknowledgement

This publication was written at the Technical University of Liberec, Faculty of Mechanical Engineering with the support of the Institutional Endowment for the Long Term Conceptual Development of Research Institutes, as provided by the Ministry of Education, Youth and Sports of the Czech Republic in the year 2016.

References

[1] A.K. Compton, B. Shah, S. M. Hayek, Spinal cord stimulation: a review, Current pain and headache reports, 16.1 (2012) 35-42, http://link.springer.com/article/10.1007/s11916-011-0238-7#/page-1.

[2] Retrieved from: https://professional.medtronic.com

Analysis of stress and deformation of the flood barrier

M. Petrů[1,a], L. Ševčík[1] , A. Lufinka[1] and T. Martinec[1]

[1]*Technical university in Liberec, Studentská 2, 46117 Liberec, Czech Republic*
[a]*michal.petru@tul.cz,*

Abstract: Increasing security protection of the population against floods has become in recent years a complex issue. It is a fact that the flood barrier should be light, easy to assemble, easy to store and also impermeable. In this article was compiled methodology for measuring and testing and analysis of flood defenses in hydrodynamic laboratory. Tests were conducted by impact with wood log on the flood barrier, which evaluated stress and strain flood barriers.

Keywords: Flood barrier, experimental measuring, numerical analysis, stress, deformation

1 Introduction

Analysis of stress and deformation of the flood barrier is Important for optimization design. Dimensioning, material properties, geometry flood barriers are very important for fast assembly, tightness and adequate mechanical properties for the protection of the population against floods. Impact with wood log into the flood barrier is one of the tests forming a methodology for testing the flood barriers [1, 2, 3].

2 Description of the solution

Experimental device parameters and settings was retained for comparison. Impactor (wood log) moves of speeds 0,2 - 2 m.s^{-1}. The whole process took away speed-camera and performed the record of measured values. Measurements were made in hydrodynamic laboratory at company JAP-Jacina (Fig.1). Measurements were performed using a system DEWETRON DEWE 5000 (Fig.2). Record the measurement data including the synchronized recording of speed camera was performed on this disk device. It was necessary to correct the force and acceleration from measured data of dynamometer and accelerometer. It was found that these forces cannot be ignored. Repeat several rounds open circuit and comparing records, we concluded that the records are identical and reproducible. Maximum deflection plates of flood barriers at the moment of impact, which was calculated as the difference signal "center" and the diameter of the signals of the two edge sensors. This calculation is based on the assumption that the extreme sensors measure the deformation of the rubber seal fins in the anchoring columns. The difference signal thus eliminated this distortion, and shows the actual deflection plates of flood barriers. From the mean values of acceleration of the impactor was calculated energy deformation by the Eq. (1).

$$E = m \int_0^{h\max} a(s)ds \quad ; \quad E_{cylinder} = \frac{1}{4}\pi d^2 \int_0^{h\max} p(z)dz \tag{1}$$

In the preceding formula represents: E - mechanical energy [J], m – mass [kg], acceleration [ms^{-2}], d – diameter of cylinder [mm], p(z) - pressure [kPa].

To determine amount of strain and the size of the stress between the wood log and plate of flood barrier were compiled numerical analysis using the finite element method in the program Ansys Workbench 14.5. Ansys uses implicit algorithm for calculations [3], where individual states of the analyzed compression are updated gradually in time t to time t+dt according to Eq. 2.

$$\delta u_{i+1} = u_{i+1}^{t+\Delta t} - u_i^{t+\Delta t} . \tag{2}$$

Where $u_i^{t+\Delta t}$ is vector of nodal displacements for ith iteration in the time $t+\Delta t$.

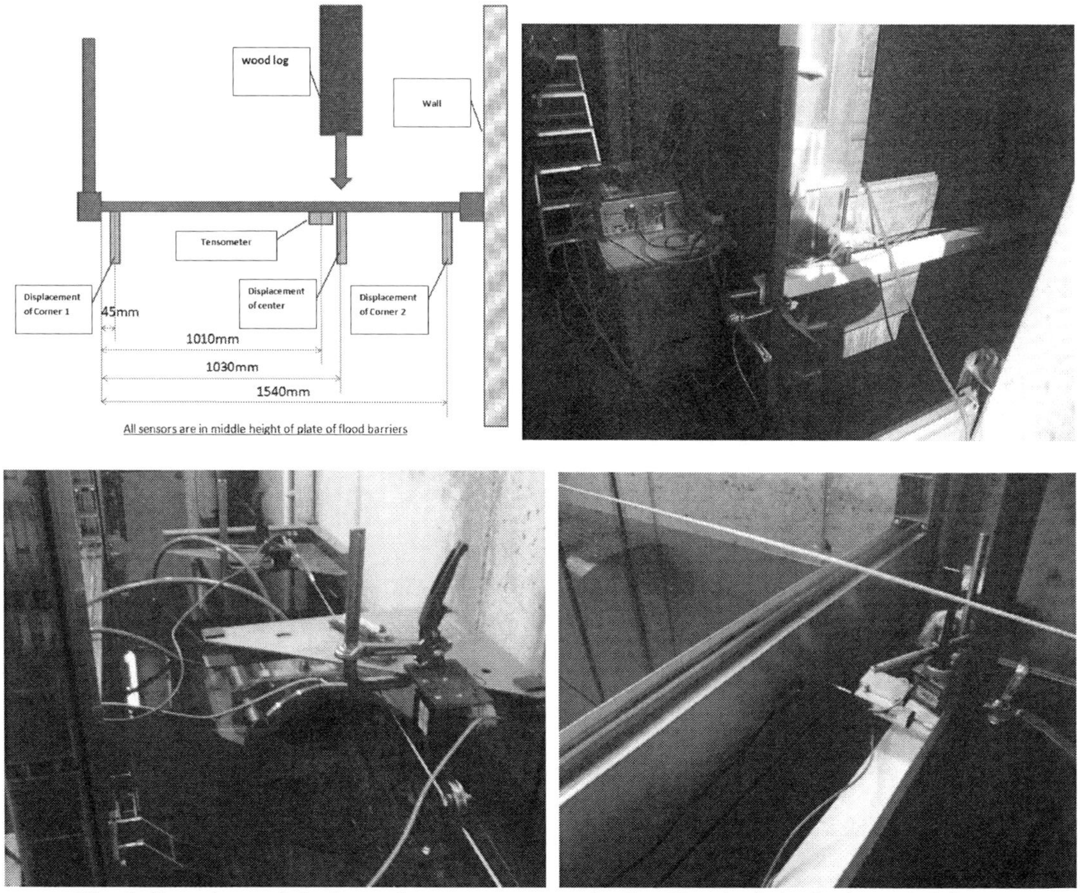

Fig. 1: Experimental device and measurement on the flood barrier in hydrodynamic laboratory

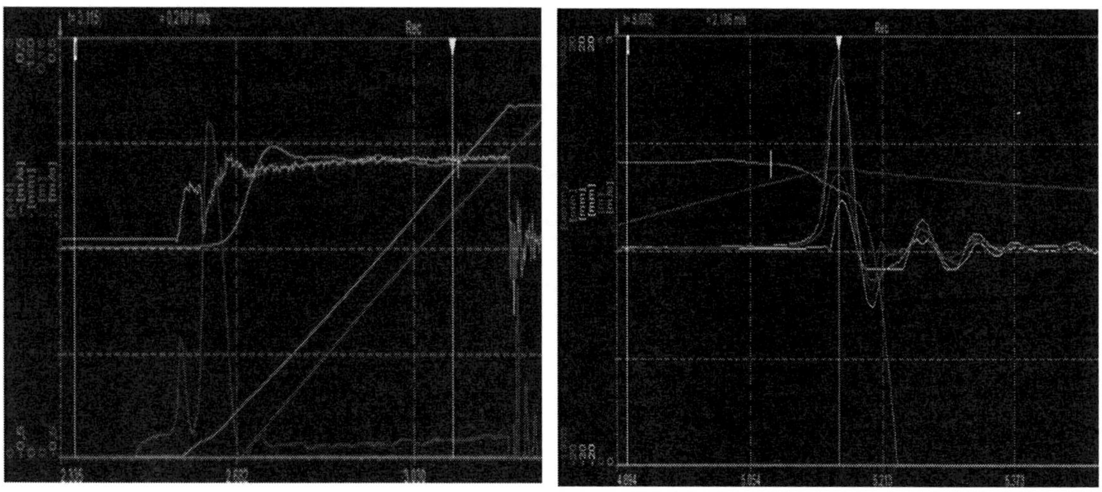

Fig. 2: Experimental data from system DEWETRON DEWE 5000

3 Results and summary

8 measurements were performed for speed of 0,1 - 2 m / s. From the measurements was evaluated the results:

- Velocity of wood log in the impact of flood barrier.
- Maximum deflection plates of flood barrier at the moment of impact. It was calculated as the difference signal "center" and the diameter of the signals of the two corners sensors (see Fig.1).

This calculation is based on the assumption that the extreme sensors measure the deformation of the rubber seal in the plate of flood barrier. The difference signal thus eliminates results distortion, and shows the actual deflection plates of flood barrier.

- Maximum value of relative elongation ε plates of flood barrier at the moment of impact.

The dependence of the measured signals at speeds aren't see clear (Fig. 3). It may be due to loss of water in a test channel due to the considerable volume of water which overflowing at impact. The water was continuously added in hydrodynamic laboratory, but the level of water between tests gradually declined. This will also reduce the impact place of wood log in plates of flood barrier. Time visual record of the impact of wood log on the plates of flood barrier is seen on the Figure 4. Figure 5 show results of deformation and stress in plates of flood barrier. Results on the Fig.5 were calculated by using FEM simulation of impact of wood log on the flood barrier according a schema on the Figure 1, above left. All results are in the table 1.

Tab. 1: Results of measurement of displacement and elongation of plate of flood barrier

Measurement	Velocity [m/s]	Center [mm]	Wall [mm]	Corner n.2 [mm]	Displacement [mm]	Elongation [um/m]
1	0,8	6,0	2,5	2,5	3,5	614
2	1,1	8,3	3,3	3,3	5,0	818
3	1,0	7,7	3,1	3,1	4,6	780
4	1,4	10,1	3,9	3,9	6,2	944
5	1,5	12,1	5,4	3,5	7,7	888
6	1,9	13,5	5,3	3,3	9,2	846
7	1,95	15,9	7,3	4,3	10,1	1138
8	2,01	17,5	7,0	5,3	11,4	1331

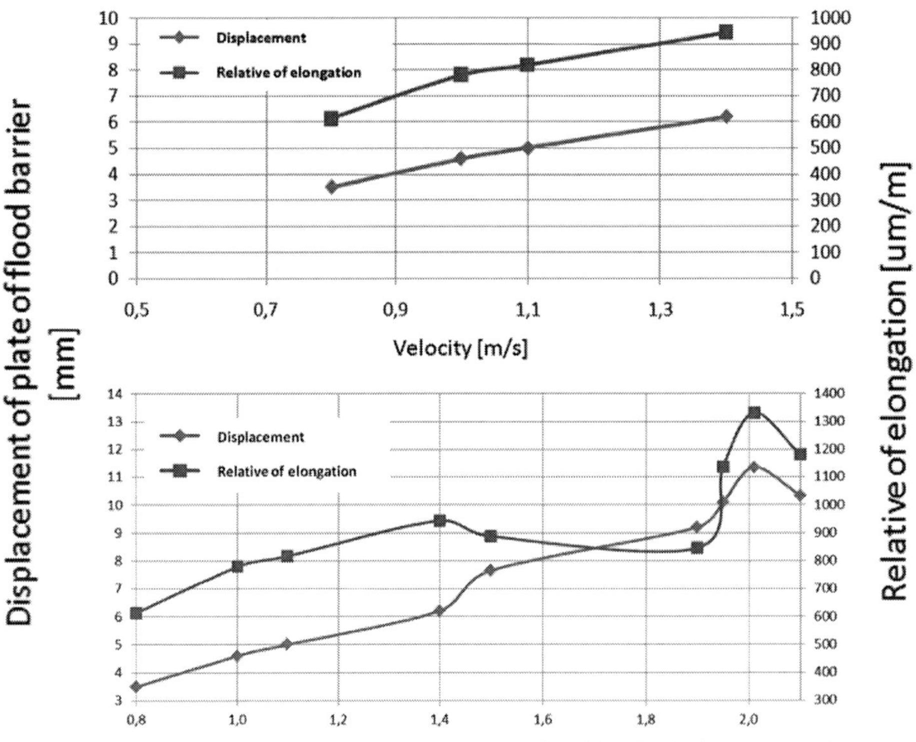

Fig. 3: Experimental device and measurement on the flood barrier in hydrodynamic laboratory

Fig. 4: Time response of detail impact of wood log on the plate of flood barrier

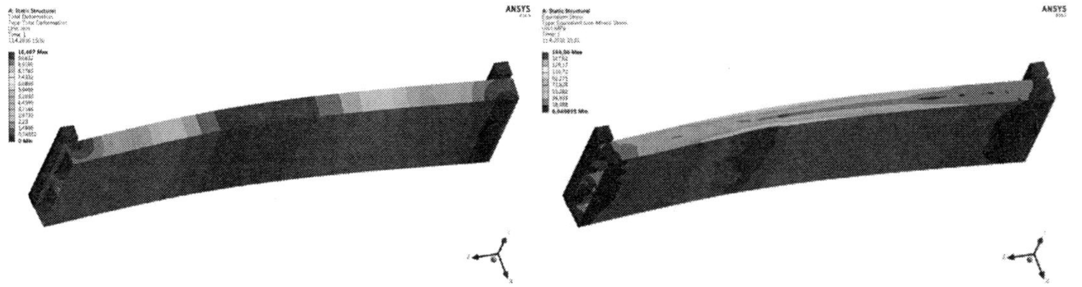

Fig. 5: FEM analysis: Deformation and stress of flood barrier

4 Conclusion

The tests and measurements of the current flood barrier suggest that are adequately dimensioned for impact of wood log. This has been demonstrated by measurement, FEM calculations and operations during the floods themselves. Of all the results show that all monitored signals are highly dependent on the speed of wood logs and place of impact in plate of flood barrier. In creating a methodology for conducting these tests will therefore need to strictly define the layout of the experiment, resp. correct placing sensors and clearly define how the impact load, thus the water level, a speed and angle of impact.

Acknowledgement

The results of this project VI20152018005 were obtained with co-funding from the Ministry of the Interior of the Czech republic as part of targeted support from the project "Vývoj protipovodňových systémů pro zvýšení ochrany obyvatelstva a infrastruktury".

References

[1] Approval Standard for Flood Abatement Equipment, Class Number 2510, FM Approvals LLC. (2013) [https://www.fmglobal.com/assets/pdf/fmapprovals/2510.pdf]

[2] Muttray, M., Oumeraci, H. Theoretical and experimental study on wave damping inside a rubble mound break water. Coast. Eng. 52 (8), 709–725. (2005)

[3] D. Malcangio, M. Torresi, G. Pascazio and A.F.Petrillo. Numerical Simulations of Water Wave Propagation by Volume of Fluid Approach, 34th IAHR World Congress - Balance and Uncertainty 26 June - 1 July 2011, Brisbane, Australia.

Experimental and numerical analysis for study a stress of new type lifting platform construction for car relocation

M.Petrů[1,a], J.Vosáhlo [1], O.Novak [1]

[1] *Technical University of Liberec, Studentská 2, 46117 Liberec, Czech Republic*
[a]*michal.petru@tul.cz*

Abstract: This article studies the mechanical loading of a new type of lift platforms and it is connectedto study of other types of lifting platforms, which has been published previously. For a stress assessment which acts in individual parts of a platform lift for moving of cars dynamic analysis and experimental studies were performed. The analysis of loads shows that immediate loading ofthe platform by bending momentumcauses that a construction of platform has to be suitably fixed. First experimental measurements were carried out on the real platform for obtaining of trajectory, velocity and acceleration during movement. Subsequently, the model analysis of the lifting platform using a four point Runge-Kutta method was assembled. Depending on the dynamic motion the stress distribution in the critical parts of platformwas investigated.The model analyses was shown that a step loading during acceleration and deceleration of the platforms leads to a dynamic impacts that areunsafefrom point of view of long term usage. The consequence of said consist in repeated overloading of pivots, bearings, rollers and other parts. Also the dynamic stress may be accompanied by the resonances. Therefore the design of lift platform has to beoptimized to ensure adequate safety.

Keywords: Lift platform; stress,experiment; numerical analysis

1 Introduction

Currently, lifting platforms in large production lines are used especially in the automotive industry. They provide the desired logistics and layout of the production process. This paper deals with a new type of platform lift, which is used for moving the car from the upper level to the lower level of the production hall. This platform moves from the top position to bottom position for 10 seconds approximately. For study stress from dynamic moving were made experimental and numerical analysis.Numerical analysis of dynamic load of the construction nodes of the platform was based on CAD data. For model dynamic analysis of platform were used all important parts and components that contribute to a creation of dynamic loading, such as base of platform, stubs, rollers, etc. similarly to the previous papers [1,2].

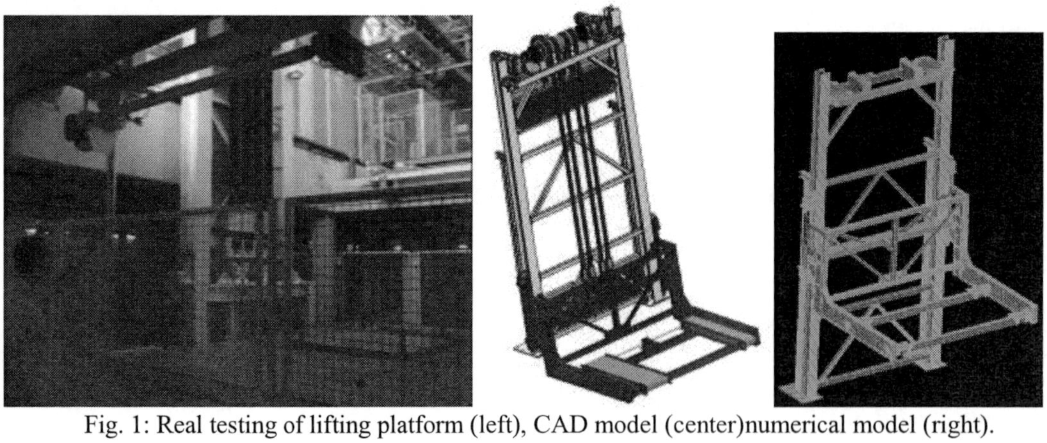

Fig. 1: Real testing of lifting platform (left), CAD model (center)numerical model (right).

2 Description of the solution

The platform moves on the given trajectory by the certain velocity. On a beginning and the end of the moving a step change of velocity appears (it is connected with extreme acceleration). It was therefore necessary for an evaluation of the trajectory to measure the velocity and acceleration of the real platform during operation of platform directly in the factory. Starting position for measurement was always in the highest and lowest position of the lifting platform (Fig.2). In between of these positions the acceleration and position were measured. The platform was loaded by the car during downward movement, and during upward movement it was without load that is corresponding with real use. Used biaxial accelerometer Analog Devices ADXL 203EB measured acceleration and static acceleration (gravity). At the beginning of each measurement the static acceleration (gravity = 9.81 ms^{-2}) was reset to zero, therefore in the vertical direction the acceleration was equal to 0. Thus results of acceleration measuring show immediate acceleration of the platform without gravity acceleration. Displacement was measured with the cable reverse sensor Micro Epsilon-WDS-4000-P115-CA-P and recorded with the data logger Dewetron DEWE 5000 with software Dewesoft 7.03.

Fig. 2:Real testing of the lifting platform in Skoda Auto.

For study a dynamical stress of new type lifting platform was made FEM simulation. Dynamic analysis was performed using the numerical solver using 4th order Runge-Kutta method (1).

$$x_{n+1} = x_n + \sum_{i=1}^{p} \psi_i k_i, \quad k_i = f(t + \alpha_i h, y_n + h \sum_{j=1}^{i-1} \beta_{ij} k_j \qquad (1)$$

Where the individual coefficients are calculated so that the method order p correspond to Taylor polynomial function x (t) for 4 orders.

Assembling of numerical model for dynamic analysis was performed in the following steps:In the pre-processor was introduced and defined local coordinate systems X, Y, Z belonging to individual parts of the platform, which are not identical with the global coordinate systems of whole construction. In the next step akinematics of the individual parts, gravitational acceleration 9,81m.s^{-2} anda load of the platform that is done by a car mass(1600 kg) were defined. The load is located in centre point of the platform. A fixation of the platform and defining of the load is shown schematically in Fig. 3.Bonds are defined for individual coordinate systems. If the global coordinate system will be used, the dynamic analysis cannot be executed. Among surfaces are set certain clearance.The defining of joints, bonds and force load from the weight of the car and the counterweight from the driving belts in the model is shown in Figure 4.In the last step material parameters for dynamic analysis were defined. Design of a new lifting platform consists of steel materials (isotropic properties, material density 7850 kg.m^{-3}, E = 210000 MPa, μ = 0.3). For dynamic analysis it is necessary defining not only the density of the material but also a volume, mass and moments of inertia with respect to the individual axes of coordinate system.These parameters are based on the real geometry of the components. It ensures required accuracy of the results. Dynamical analyses were done for different movements.

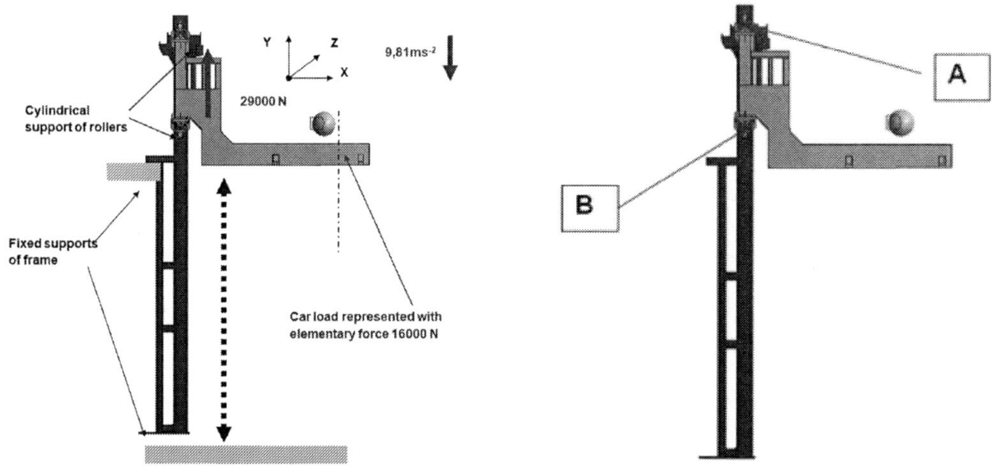

Fig. 3:Joints and boundary conditions for dynamical analyses.

For a dynamic study of platform were introduced parameters in accordance with technical documentation and previous measurements.

<u>Study 1:</u> The lifting part of the platform moves vertically from 0 to 4300 mm with constant velocity 12 m/min (0,2m.s^{-1})
<u>Study 2:</u>conditions are same as in previous case, only acceleration +0,1m.s^{-2}up to maximal velocity 0,2 m.s^{-1} is introduced
<u>Study 3:</u>conditions are same as in previous case, applied acceleration is +0,4m.s^{-2}
Dynamical analyses were applied by the help of Runge-Kutta iterative methodof order 4.
The results of simulation were used for evaluation of roller force that loading stubs and bearings.For this purpose two areas of interest will be studied (A – BFig. 3)

3 Results and discussion

Dynamical loading of the lifting part of platform during acceleration +1,1g (dynamical simulation –Study 2 causes maximal step load on the roller 7 202,7 N that is distributed on stubs and bearings. The value of maximal bending stress in the stub is 72,46 MPa. It corresponds to safety factor k_2=2,82. If the acceleration will be equal to +1,4g (Study 3) the force acting on roller will be 21150 N and bending stress in the stub will be 212,17 MPa. Then safety factor will be k_3=0,96 only. The acceleration +1,4gwas measured in the case of previous platform design. All results are put in Table 1 and in the Fig.4. These loads and consequently bending stress can be reduced by modifying of the platform design. It is evident that platform has unsuitably large overhang, which increases the bending stress.The article deals with the measurement and numerical analysis of dynamically loaded construction nodes of a new type lifting platform. From the results it is evident that the main problem consists in the acceleration and deceleration of the platform. In the construction oscillation and shock load are formed in carrier parts and together with the bending moment they cause high loads accompanied by the formation of high stress in critical locations (Fig.5).

Tab.1:Results of numerical analysis of loaded construction nodes of the platform

Study	Force on roller A [N]	Force on roller B [N]
1	5269	5157
2	7142,5	7202,7
3	19835	21150

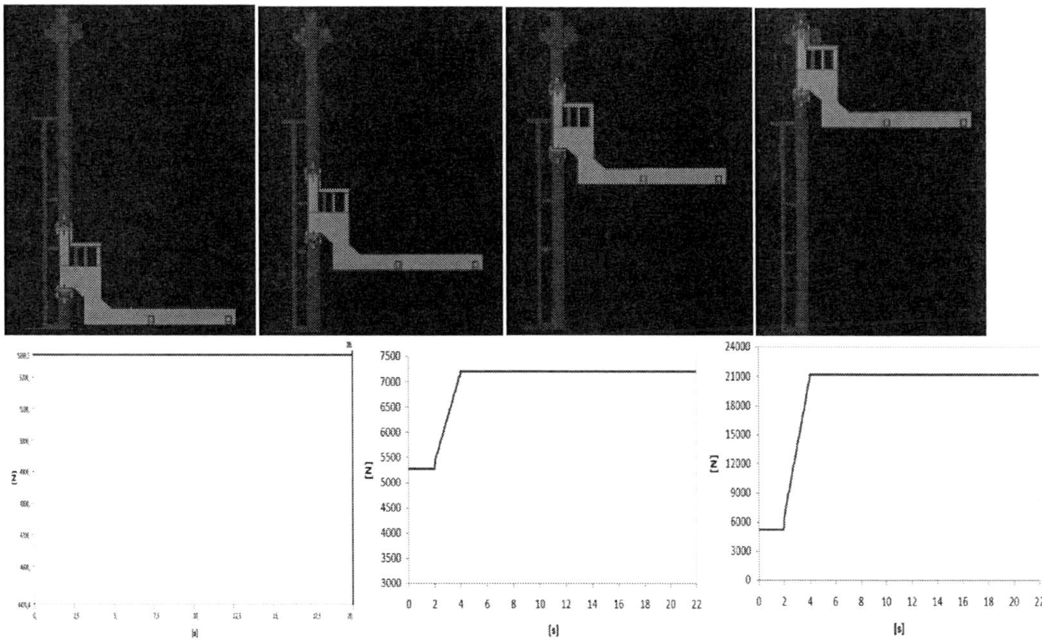

Fig .4:Results FEM: Time response of lifting platform Regal Scout (above), Force for study 1-3 (below)

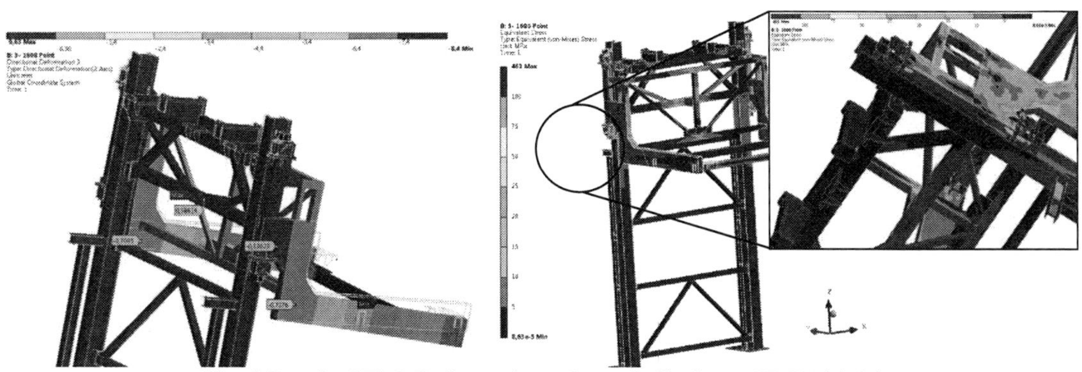

Fig .5:Results FEM: Deformation of arm (left), Stress HMH (right)

Acknowledgement

The results of this project LO1201 were obtained with co-funding from the Ministry of Education, Youth and Sports as part of targeted support from the "Národní program udržitelnosti I" program.

References

[1] Petru, M.; Lepšík, P.; Novák, L.; Lufinka, A. Measurement and numericalanalysisof lifting platformconstructionfor car relocation, EAN 2014, 2 – 5.6. 2014,Marianské Lázně, Czech Republic

[2] Petru, M. et al.Dynamical analysisof lifting platformconstructionfor car relocation, Lecture Notes in MechanicalEngineering,Volume 16, 2014, p. 517-524.

Tests of the railway components

J. Petřík [1,a], A. Lufinka [1,b]

[1] *Technical University of Liberec, Czech Republic*
[a] *jan.petrik@tul.cz*
[b] *ales.lufinka@tul.cz*

Abstract: The standards for the railway components are divided to three mounting types which include the vibration and shock tests. The levels of vibration and shock amplitudes depend on the mounting type. The highest levels are defined for the axle mounting components like brakes. The claims for the test conditions are mandatory for the testing platform development and the mounting design of the tested component which simulates the real locations. There is necessary to check the modal analysis of the mounting design cause of the Eigen modes of the fixed railway component inside the standardized vibration spectrum.

Keywords: railway, vibration, shock.

1 Introduction

The levels of the test loading are divided according to the mounting location like axle, chassis and cabin. The highest levels are defined for the axle mounting where the acceleration RMS of the white noise signal are close to the 150 m/s^2 with peaks like Normal distribution according to 3*RMS. Shock amplitude is defined as half of the sinus period with amplitude 1000 m/s^2 with 6ms duration [1,2]. The testing component is the brake system made by DAKO-CZ. We cooperate with DAKO-CZ since last year in term of durability testing and mechanical design. The mass of the brake component is 50kg. The mounting design must correspond with the real conditions and the mass limit of the whole tested system (testing brake with mounting) is 100 kg. These limitations are defined cause of the used hydraulic system with maximum oil flow of the oil unit. The optimization of the mounting system is necessary and virtual modal analysis is used.

2 Durability tests description

The vibration test signal is the pseudo stochastic signal like white noise. The prescribed spectrum is defined as the frequency spectrum of the Power Spectral density from 20 to 100 Hz and with 6dB decrease till 500 Hz which corresponds with mass of the brake. The vibration durability test takes 5 hours in each global axis (total 15 hours). Before the durability test was made the additional measurement to analyze the floating caliper modes of the brake system. The acceleration sensor is mounted to the piston of the exciter and represents the input mechanical vibration to the testing component. The testing measurement with the additional accelerometer sensor mounted to the railway brake is described on Figure 1.

The vibration control is based on the frequency domain iteration methodology described in [3]. The displacement feedback is used with acceleration control signal. The disadvantage is that is when the exciter system increases its temperature and friction, oil, etc. behavior occur the different response. There is allowed to stop the test cause of the high exciter temperature by the standard [1, 2]. The frequency analysis of the additional acceleration is used for the comparison with virtual modal analysis. There is mandatory that the modes are outside of the testing vibration spectrum. If there is not possible to fulfil this condition the control system needs to suppress the modes of the testing component with the mounting construction.

The shock test represents the prescribed acceleration signal for the piston of the exciter. The shock tests by using hydraulic system are difficult to control cause of the oil injection delay which is around 2ms. The acceleration peak is effective to perform with the displacement reversing of the piston. The servo valve opening is necessary to check during the test cause of the stability during 100% servo opening.

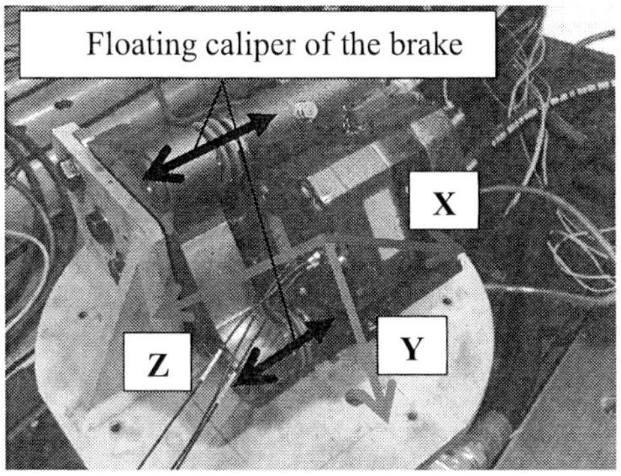

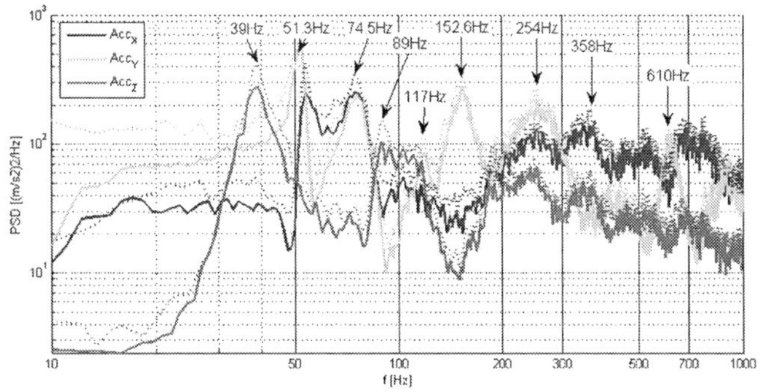

Fig. 1 a) Brake DAKO-CZ with acc. sensor b) Frequency modes of the additional acc. sensor

3 Optimization

The mounting design for the vertical tests needs to be optimized for the transverse and longitudinal direction tests. The hydraulic system does not use the vibration table and the testing component is mounted to the rotated piston flange in the tested direction. The vertical test mounting design is used for the modal analysis of the floating caliper of the brake system. The Eigen modes of the mounting system have to be outside the Eigen modes of the floating caliper (up to 150 Hz described on Fig. 2). The first two Eigen modes of the vertical mounting design analyzed by the virtual modal analysis are 187 Hz (real mode 190 Hz) and 369 Hz (real mode 343 Hz) and fulfil this condition (Figure 3a and 3b).

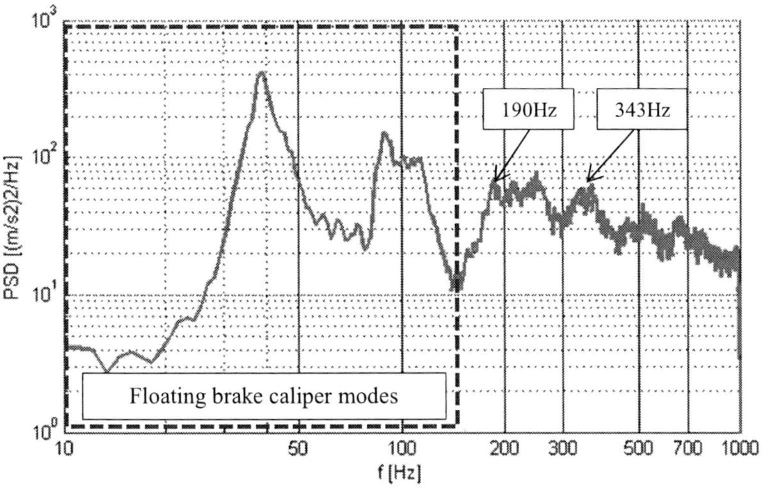

Fig. 2. Analysis of the transversal acceleration spectrum (AccZ) of the brake system

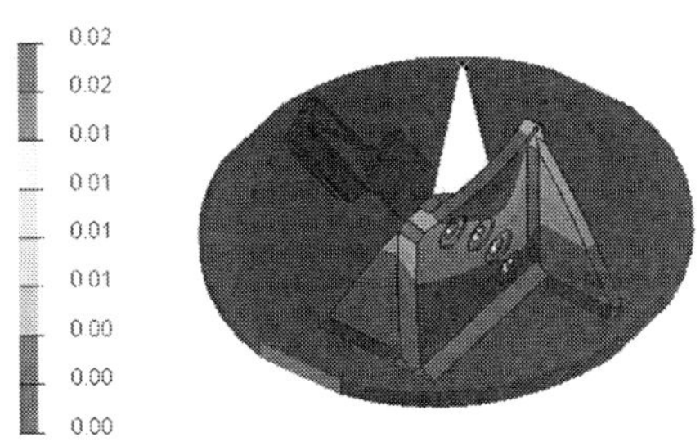

Fig. 3 a) First Eigen mode - 187 Hz b) Second Eigen mode - 369 Hz

The universal mounting design was developed and optimized by using the virtual modal analysis (Fig. 4 and Tab. 1). The initial conditions are the mass limit 100kg and Eigen modes higher than 150Hz. The optimization of the components thickness and profiles helps to predict the real behavior. The brake system was simplified in the model to rigid body with prescribed mass and inertia moments. The floating caliper was not included into the model.

Mode	f [Hz]
1	230
2	376
3	448
4	532
5	683
6	754
7	890
8	924
9	974

Tab. 1 a) Result and prescribe vibration limits

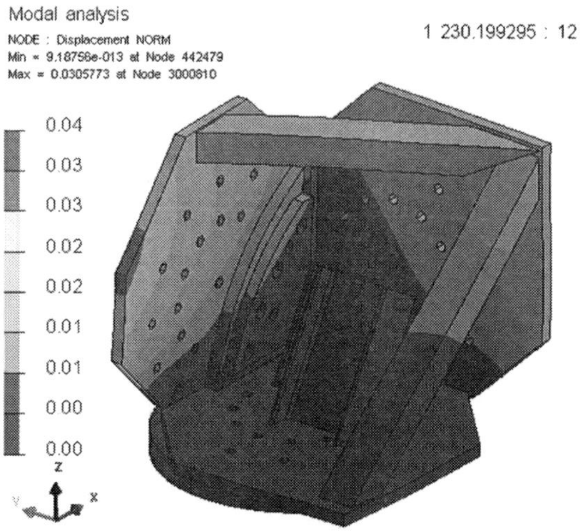

Fig. 4 Modal analysis of the optimized design

4 Durability tests results

The status of the durability test is if the integrity and functionality of the brake system is fulfilled without any failure. The prescribed vibration and shock test are on Figure 5a and 5b. The vibration testing limits are 3dB up and down the prescribed acceleration defined as the Power Spectrum Density of the acceleration. The acceleration results are in acceptable standardized limits.

There is not needed the additional acceleration sensor mounted to the brake system. The input acceleration is analyzed by the sensor mounted to the exciter's piston. This sensor is used for the results spectrum check during iteration method to control the target vibration spectrum. It is mandatory to identify the complete exciter system with the testing specimen or mechanism. It is possible to suppress natural frequencies of the hydraulic exciter and also testing specimen or mechanisms. It is recommended to use wider frequency range than prescribed spectrum. The common frequency response function (FRF) of the hydraulic exciter is used for identification. The next step is the original estimate of the input signal for the first step of iteration (the input signal response of the inverse FRF). The iteration methodology uses couple of steps to fulfil the prescribed spectrum by optimization of the inverse FRF function for the input signal of the hydraulic valve. The disadvantage of this method is the impossibility to control in real time. It is mandatory to check the result spectrum after every hour cause of the oil temperature dependency. The test breaks are allowed by the standards and are necessary in term of high temperature of the exciter system.

The best results for the hydraulic system as the shock exciter is described on Figure b. The hydraulic system behavior during the maximum peak is on the control stability limit cause of the 100% valve opening but still might be acceptable in term of fulfilling the acceleration lower limit in extremely short duration close to the 6ms.

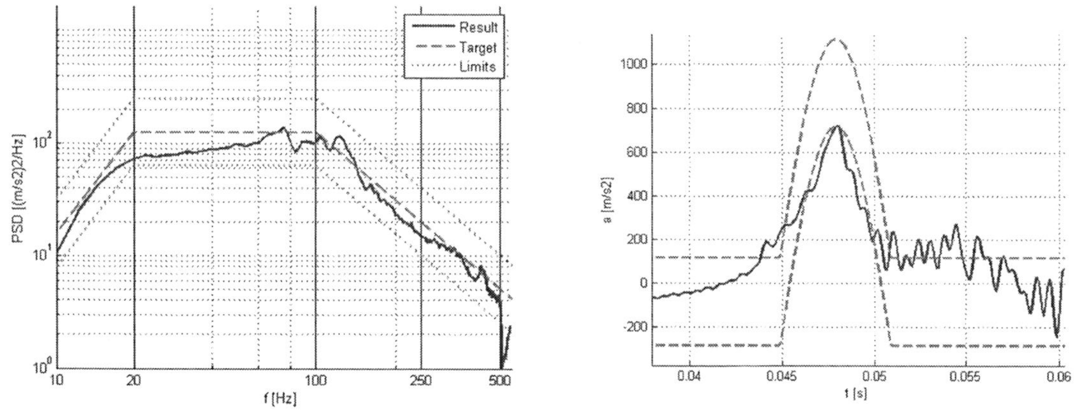

Fig. 5 a) Vibration durability test - PSD spectrum b) Shock durability test acceleration peak

4 Conclusion

The frequency analysis is mandatory for the mounting system development with the floating caliper of the brake. The Eigen modes of the mounting system have to be outside the floating caliper modes. The mass limits were fulfilled – the mounting system for all directions is after optimization 49kg. The optimized design is possible to use for vertical, longitudinal and transversal tests without disassembly the brake system and fulfils both conditions – the first Eigen mode is 230Hz (floating caliper modes till 150Hz) and the mass of the whole test component mounted to the exciter piston is 99kg (mass limit of the hydraulic unit is 100kg).

There is necessary to use frequency domain iteration to control the hydraulic exciter system. The test results for the worst acceleration level are described on Figure 5 a). The result spectrum is acceptable in prescribed limits. The shock results are at the present time on the stability limits of the used hydraulic system (Fig. 5b).

Acknowledgement

This publication was written at the Technical University of Liberec as part of the project "Innovation of products and equipment in engineering practice" with the support of the Specific University Research Grant, as provided by the Ministry of Education, Youth and Sports of the Czech Republic in the year 2016.

References

[1] Railway applications – Rolling stock equipment – Shock and vibration tests, ČSN EN 61373

[2] Environmental testing – Part 2-64: Tests – Test Fh, : Vibration, broadband random and guidance, ČSN EN 60068-2-64

[3] Hydraulic Exciter Control for Shock and Vibration Testing, Applied Mechanics and Materials (Volume 827), Chapter 1 , Page 105, ISSN: 1662-7482

Modal Analysis of Small Masonry Columns

T. Plachý [1,a], J. Hubka [1], M. Polák [1]

[1] *Czech Technical University in Prague, Faculty of Civil Engineering, Thákurova 7, CZ-166 29 Prague, Czech Republic*

[a] *plachy@fsv.cvut.cz*

Abstract: The paper presents modal analysis of four small brick columns. This non-destructive method was used for detection of damage and quality of two different mortars. The small brick columns were made from four bricks and mortar. Two of them were made using one type of mortar and the second two using another type. The multi reference modal analysis was used for investigation of the differences in dynamic behavior. Conclusion summarizes the comparison of dynamic behavior of columns and damage detection in some of them.

Keywords: brick; masonry structure; mortar; modal analysis; vibration; experiment.

1 Introduction

The modal analysis is well known method for modal parameters identification not only for large structures [1] but also for small ones. It can be also used for monitoring of dynamic behavior changes in time [2,3] or for damage identification of a specimen [4]. The paper presents multi reference modal analysis used for stiffness comparison of two types of mortar and for damage identification of the small masonry columns.

2 Tested Specimens

The four masonry columns were prepared for the purpose of this test. Each column was made from four bricks. For the connection of bricks two different types of mortars were used. The first one was composed from 70% of limestone binder (L) and 30% of metakaolin/cement binder (MK), the next part was crushed brick (CB) from the brickyard Bratronice and the last one was sand (S) from Zálezlice. This is why the first two specimens were named as LMK_SCB_1 and LMK_SCB_2. The second two specimens were made from the mortar without crushed bricks. Thus they were named as LMK_S_1 and LMK_S_2.

3 Measurement System of the Modal Analysis

We did a visual inspection of the columns before starting the modal analysis and we found some cracks in mortar on specimens LMK_S. Therefore the measurement system with two different positions of the reference response transducers was used for the better damage detection – one was placed in the left upper corner of the column and the other in the left lower corner of the column. The measurement system Brüel & Kjær consisted of two acceleration transducers 4519-003, the impact hammer 8206 and the measurement station Front-end 3560-B-120. The measurement was carried out on two perpendicular vertical sides of the column in a chosen net of points. The response transducers were placed to two reference positions and then the specimen was stroked by an impact hammer in all chosen points. The Frequency Response Functions (FRF) were evaluated for all measurement points from the impact force and response records using Fast Fourier Transform (FFT). Two sets of modal parameters (one from each response transducer) were evaluated based on the resonance peaks in FRFs.

4 Modal Parameter Identification for LMK_SCB specimens

The eight natural frequencies and mode shapes were evaluated for both specimens LMK_SCB and for both reference accelerometers. The examples of the first four natural modes of the specimen No. 2 are shown in Fig. 1 and Fig. 2. These modes look like the modes of the rigid body. Modes evaluated from both reference transducers look very similar thus modes only from the upper one are presented. From the side view of the Fig. 1 and Fig.2 there is evident that mortar of the specimens is softer than bricks because displacements in mortar are bigger than on bricks.

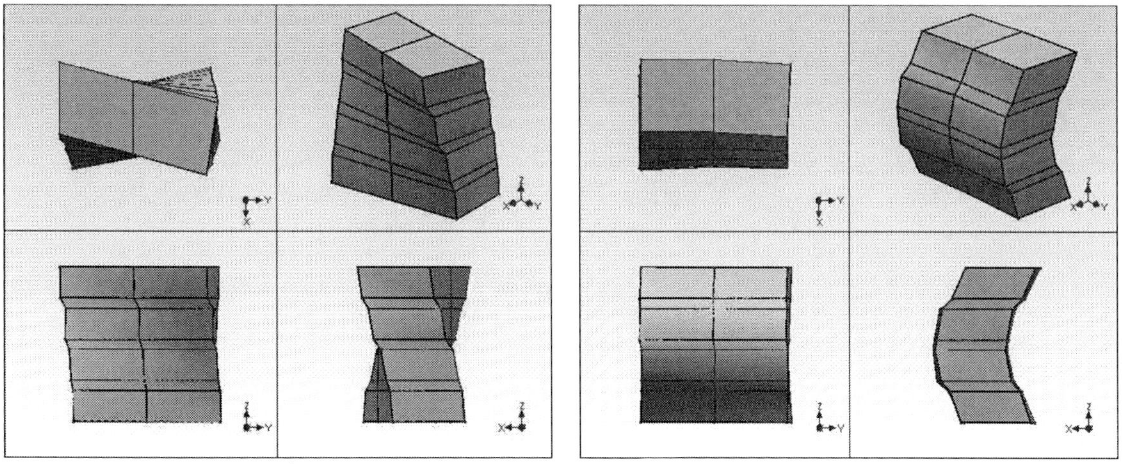

Fig. 1: The 1st and 2nd natural modes of the specimen LMK_SCB_2,f_1 = 1715 Hz and f_2 = 2298 Hz

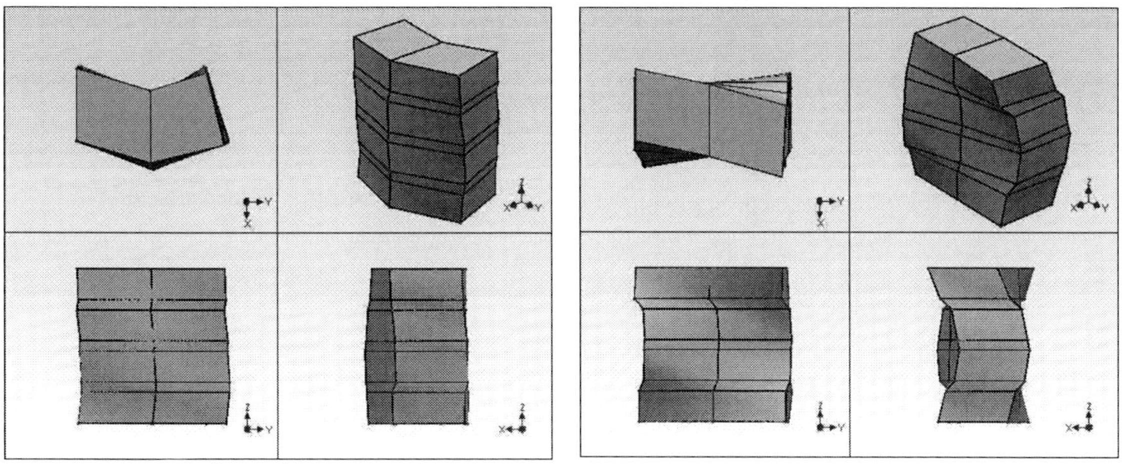

Fig. 2: The 3rd and 4th natural modes of the specimen LMK_SCB_2,f_3 = 3027 Hz and f_4 = 3247 Hz

5 Modal Parameter Identification for LMK_S specimens

The eight natural frequencies and mode shapes were evaluated for both specimens LMK_S and for both reference accelerometers. The examples of some natural modes of the specimen No. 2 are shown in Fig. 3, Fig. 4 and Fig.5. In contrary to the specimens LMK_SCB, these ones do not vibrate like the rigid body. The modes evaluated from two reference positions for the same natural frequency differs very much (see Fig.3-5). In each figure there are presented two modes for the same frequency – left one evaluated from the transducer placed in the upper left corner of the specimen and right one evaluated from the transducer placed in the lower left corner.

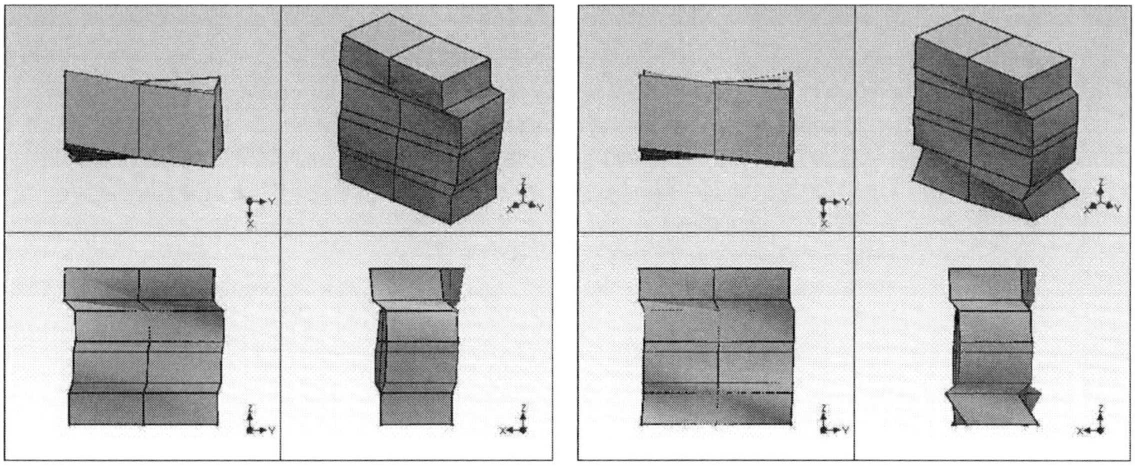

Fig. 3: The 1st natural mode of vibration of the specimen LMK_S_2, f_1 = 1153 Hz
(left – upper transducer, right – lower transducer)

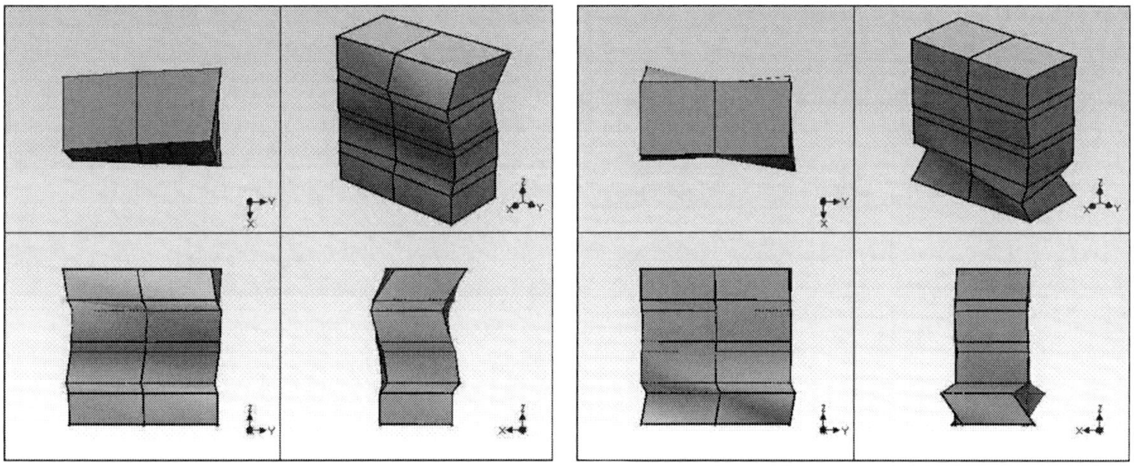

Fig. 4: The 2nd natural mode of vibration of the specimen LMK_S_2, f_2 = 1497 Hz
(left – upper transducer, right – lower transducer)

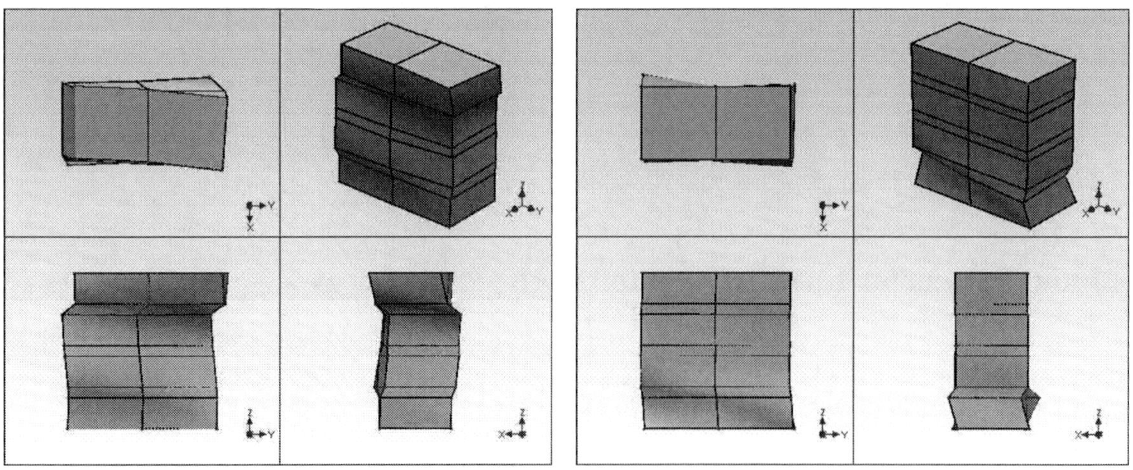

Fig. 5: The 3rd natural mode of vibration of the specimen LMK_S_2, f_3 = 1735 Hz
(left – upper transducer, right – lower transducer)

6 Damage Detection

The natural modes of the specimens LMK_SCB look like the rigid body motion without any big dissimilarity which can signalize the crack presence. There is only visible that mortar has lower stiffness than bricks.

Both specimens LMK_S have totally different dynamic behavior. The natural modes of the specimen LMK_S_2 evaluated from the two reference acceleration transducers differ very much. The upper brick vibrates especially in the modes evaluated from the upper transducer (Figs. 3-5 – left) and lower brick in the modes evaluated from the lower transducer (Figs. 3-5 – right). The excitation signals from the middle part of the column were not transferred to the upper reference transducer and were partly transferred to the lower one. Based on these results the cracks were identified in the lower and upper mortar joints.

The very similar state was observed also for the specimen LMK_S_1, for which the results evaluated from two reference transducers also differ. But in this case only upper brick vibrates independently. Thus the crack was identified in the upper mortar joint.

7 Conclusion

The modal analysis used for comparison of dynamic behavior of small masonry columns made from two types of mortar is presented. Based on the results, it is obvious that the mortar LMK_SCB is much better than LMK_S. Also the damages in the mortar joints were identified using the multi reference modal analysis. This method proved to be very useful for identifying damage and stiffness of mortar joints in these types of specimens.

Acknowledgement

This work was financially supported by Czech Technical University in Prague under the project No. SGS14/122/OHK1/2T/11 and TACR under the project No. TE01020168.

References

[1] T. Plachý, M. Polák, P. Karásek, Modal Analysis of a Footbridge Damaged by Flood, in proc.: EURODYN 2014, eds. A. Cunha, E. Caetano, P. Ribeiro, G. Müller, University of Porto, Porto, 2014, pp. 2591 - 2598

[2] J. Topič et al., Development of Mechanical Properties of Cement Paste with Different Addition of Polyvinyl Alcohol, Applied Mechanics and Materials 732 (2014) 81-54 http://doi:10.4028/www.scientific.net/AMM.732.81

[3] J. Hrůza et al., Development of Mechanical Properties of Cement Based Composites with Recycled Concrete Aggregate, Applied Mechanics and Materials 825 (2016) 11-14 http://doi:10.4028/www.scientific.net/AMM.825.11

[4] T. Plachý et al., Damage Detection and Localization on Cement Specimens, Applied Mechanics and Materials 617 (2014) 229-232 http://doi:10.4028/www.scientific.net/AMM.617.229

[5] P. Padevět, T. Otcovská, O. Zobal, Variation of material properties of cement pastes with various types of fly ash during maturation, WSEAS Transactions on Applied and Theoretical Mechanics 9 (2014) 88-96.

[6] M. Lidmila et al., Mechanical Properties of Recycled Binder / Micro-Filer Cement-Based Material, Advanced Materials Research 1054 (2014), 234-237. http://doi:10.4028/www.scientific.net/AMR.1054.234

Experimental Analysis of the Demountable System with Embedded Diaphragms

A. Polák [1,a]

[1] *Department of Building Structures, Faculty of Civil Engineering, Czech Technical University in Prague, Czech Republic*

[a] *ales.polak@fsv.cvut.cz*

Abstract: Demountable precast structures allow repeated use of reinforced concrete elements, and thus they significantly reduce the negative impact of concrete structures on the environment. The characteristic parts of the demountable column system are the "girder – column" joint and the "stiffening diaphragm – girder" joint. The dynamic characteristics of the whole structure can be significantly affected by mounting of the diaphragm on elastomeric bearings with required hardness. System design requires an entirely new structural, analytical and static solutions-based theoretical analysis and experimental verification, within which the actual conditions, external influences and effects are simulated. This article deals with the experimental analysis of the section of frame structure with flexible embedded diaphragm by elastomeric bearings of different hardness or steel screw fasteners. The experimental research was carried out within TA02010837 "Multipurpose dismantleable prefabricated reinforced concrete building system with controlled joint properties and possibility of repeated use".

Keywords: Demountable; Experimental Analysis; Diaphragm; Bearings; Dynamic Characteristics.

1 Introduction

Current dynamically developing human society has got increasing requirements on flexibility in buildings. However traditional precast systems do not correspond to this dynamics and they are characterized by limited adaptability and relatively high labor intensity when changing use. Almost all reconfigurations of the current structures due to the change in use or transfer of the building caused by production, transportation or demographic requirements are associated with full or partial demolition, as the reuse of precast elements joined by wet processes is very limited. The solution is to use structures whose rigidity is ensured by joints, which can be if necessary removed easily, without reducing the possibility of further use of precast elements.

2 Column System with Demountable Joints

Demountable joint is a characteristic part of demountable structure. These joints should allow simple and precise assembly of components, while using the structure to fulfill its structural function and subsequently enable seamless disassembly of the structure [1].

Column – girder joint (Fig. 1 and Fig. 2) used at test assembly consists of a welded special steel element which allows quick installation based on the principle of a special locking joint. The girder is then mounted on this console using similarly shaped base plate. The disadvantage of this joint is higher intensity of labor, when the special steel element is produced, but no need to use bolts. This solution enables the connection of the required number of girders on the column. From the static point of view, this joint is designed as articulated pin joint. It is designed to transfer both vertical and horizontal shear forces and torque that can be caused by depositing flor panels in the outer fields or during assembly (asymmetric loads) [2].

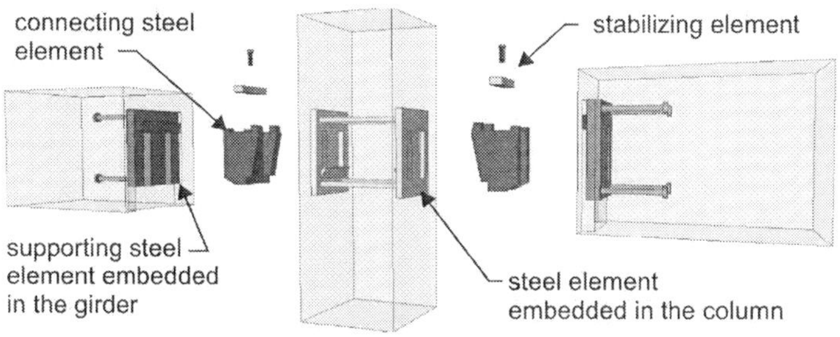

Fig. 1: Visualization of the column – girder joint.

Fig. 2: Embedded steel elements.

The connection of the diaphragm and the girder can be ensured in two basic ways. The first variant is a rigid connection using steel screw fasteners (Fig. 3). The second variant is a soft connection using elastomeric bearings of different hardness (Fig. 4) [3].

Fig. 3: Rigid connection with steel screw fasteners.

Fig. 4: Soft connection with elastomeric bearings.

3 Experimental Analysis

The aim of experimental analysis was to determine and compare the stiffness characteristic for each type of the connection between diaphragm and girders.

3.1 The Test Assembly

The test assembly of the section of frame structure consisted of two prefabricated columns (180 × 180 mm in cross-section and 2500 mm in height), two girders (180 × 240 mm in cross-section and 2860 mm in length) and embedded stiffening diaphragm (2800 mm in length, 1650 mm in height and 50 mm of thickness), which was connected with girders by elastic bearings of two different hardness (45 ShA and 85 ShA) or with steel screw fasteners. Elastic bearings are made of steel plates with welded screw fasteners, which are vulcanized in the elastomer.

The arrangement and dimensions of the test assembly are shown in Fig. 5. The materials of reinforced concrete elements are C50 / 60 and steel B500B. The steel connecting elements are made of steel S355J0.

3.2 The loading

The test assembly was loaded using a hydraulic press by forced deformations of 0.5 – 5 mm, and different frequencies in the range of 0.1 – 20 Hz. 14 linear deformation sensors (LVDT) were mounted on the test assembly. These sensors recorded total deformations of the frame (sensors 1, 2) and relative displacements between columns and girders (sensors 8, 12, 13, 14), columns and diaphragm (sensors 9, 10, 11) and girders and diaphragm (sensors 3, 4, 5, 6, 7). The test was controlled by deformation (amplitude) of hydraulic press and the force which caused the required amplitude was recorded.

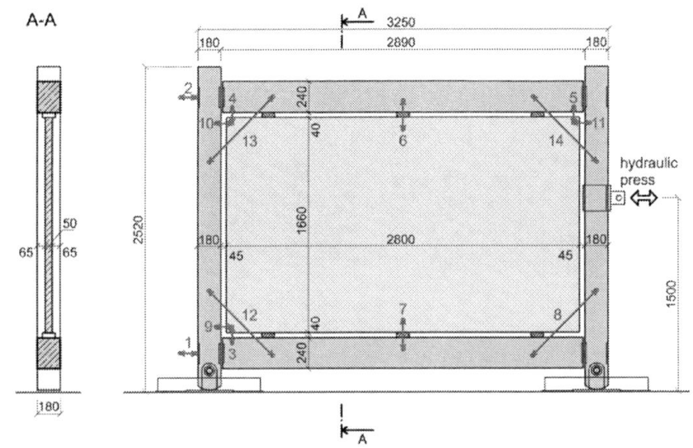

Fig. 5: The test assembly.

Fig. 6: LVDT sensors.

3.3 The Measured Values

The maximum value of the force in the hydraulic press (corresponding required amplitude and frequency) was recorded for each load case. The identified values are shown in Fig. 7.

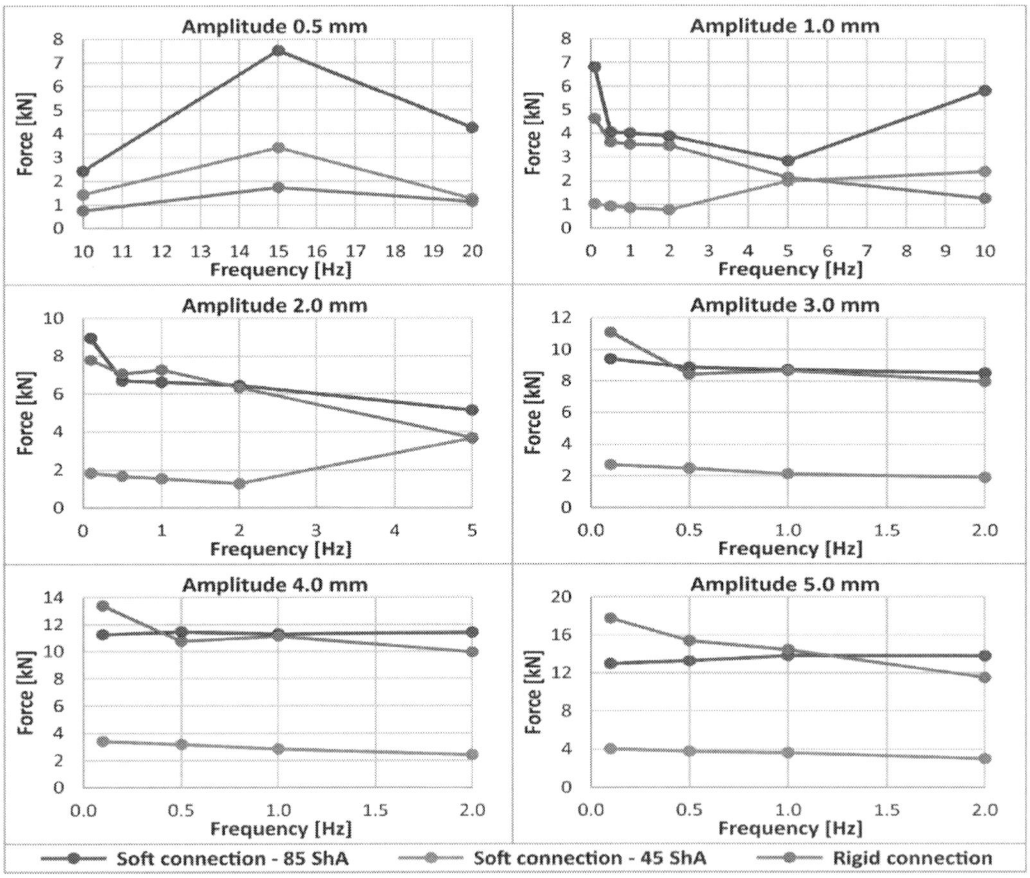

Fig. 7: Dependence of force in hydraulic press on frequency for required amplitudes

When comparing the individual types of the connection it is apparent that the values of the forces required for individual amplitudes are 3 times lower in the case of soft connection (bearings 45 ShA) than for the other two types. For small frequencies, downward strength is apparent depending on the frequency increas. This relationship is nearly linear to the value of the frequency 2.0 Hz, if the elastomeric bearings 45 ShA hardness are applied. When increasing the frequency to 5 Hz, which is a value close to the first natural frequency of this assembly, there is a significant increase in the required force. The highest values of forces are achieved with the rigid connection, wherein these forces are at low frequencies and amplitudes 4 and 5 mm in the range of 11 to 18 kN. The lowest forces were achieved with small amplitude (0.5 mm and 1.0 mm) for rigid connection and at amplitudes of 2.0 to 5.0 mm for bearings 45 ShA. The force values at different amplitudes and frequency of 5 Hz for all connections are roughly comparable.

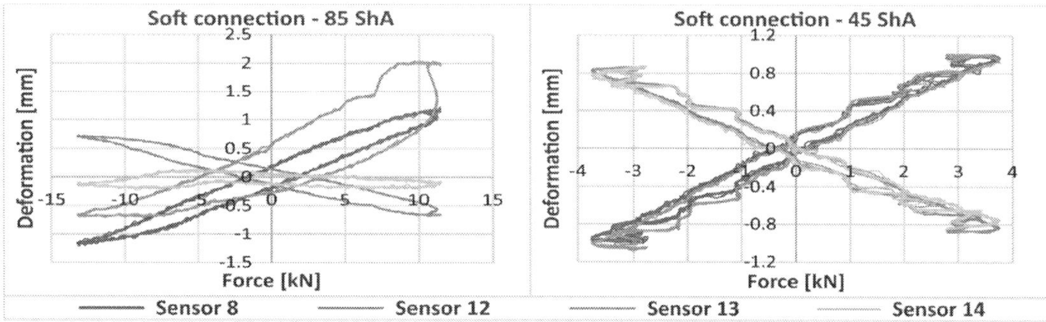

Fig. 8: The comparison of the shapes of the hysteresis curves (frequency 0.5 Hz, amplitude 5 mm)

The comparison of the shapes of the hysteresis curves for soft connection is shown on the graphs in the Fig. 8. The shapes of hysteresis loops recorded from opposite sensors are almost identical for connecting bearings with a hardness of 45 ShA. In the case of bearings with a hardness of 85 ShA, the shapes of hysteresis loops are different. This was caused by the lifting of girders due to a large stiffness of the bearing.

4 Conclusion

Selection of the hardness of the elastomeric bearing significantly affects the stiffness of the whole system. Use of bearings with high hardness can be much closer to the rigid connection of the diaphragm and girders. This can be used for damping vibrations caused by natural or technical seismicity.

Acknowledgement

The paper was written with support from TA02010837 "Multipurpose dismantleable prefabricated reinforced concrete building system with controlled joint properties and possibility of repeated use". Lead researcher of the project is prof. Ing. Jiří Witzany, DrSc. Radek Zigler, Tomáš Čejka, Jiří Karas and Pavel Kokeš contribute significantly to this project.

References

[1] J. Witzany, T. Čejka, R. Zigler, Dismantleable Joints of Load-bearing Reinforced Concrete Units of Prefabricated Concrete Building System with Controlled Stiffness, in proc.: Research and Applications in Structural Engineering, Mechanics and Computation (2013), ed. CRC Press/Balkema, University of Cape Town, Cape Town, 1503–1506.

[2] A. Polák, Experimental Verification of Demountable Precast Column System, in proc.: Applied Methods of Static and Dynamic Loads of Structures and Machines II (2016), ed. P. Padevět, Trans Tech Publications Ltd, Český Krumlov, 259–262.

[3] J. Witzany et al., Demountable Precast Reinforced Concrete Building System of Multi-Story Buildings, in proc.: Implementing Innovative Ideas in Structural Engineering and Project Management (2015), ed. Fargo: ISEC Press, Sydney, 219–224.

A Comparison of Spectators Induced Grandstand Vibrations and Fans Behavior during Two Football Matches between the Same Teams

M. Polák [1,a], T. Plachý [1], M. Verner [1]

[1] *Czech Technical University in Prague, Faculty of Civil Engineering, Thákurova 7, CZ-166 29 Prague, Czech Republic*

[a] *polak@fsv.cvut.cz*

Abstract: A synchronized motion of a considerable group of sport fans (like as some type of jumping, bouncing, swaying or hand-clapping) would cause a significant dynamic load to a grandstand structure and consequently its vibrations that could be sometimes noticeably large. The results of an experimental analysis focused on the football fans behavior and on grandstand vibrations induced by them at the AC Sparta football stadium in Prague are stated in the paper. The findings obtained from two football matches between the same teams are compared mutually and it is concluded that the fans behavior was substantially influenced by the actual position of both teams in the football league.

Keywords: football stadium; cantilever grandstand; sport fans' behavior; vibration; experiment.

1 Introduction

If a substantial group of sport spectators assembled on a grandstand moves synchronously, inconsiderable dynamic forces are frequently caused and grandstand oscillations are consequently induced which could be sometimes noticeably large. The large grandstand oscillations that are distinctly perceived or even visible may start a panic of fans, a following stampede and a crush that could result in injuries even in deaths of some fans [1]. History shows that undue vibrations of a grandstand could cause even its collapse in an exceptional situation [1].

Currently, the suitable dynamic load of grandstands is not described in any standard with acceptable precision. However, lots of new spectator's crowd load models have been developed [2, 3, 4] but adequate experiments, which could be applied for verification of the load models, have been performed less frequently [5, 6]

Two series of experiments focused on spectators induced grandstand vibrations and fans behavior during football matches were carried out at the Letná football stadium in Prague. First two experiments were performed during the 2013-14 season and then five ones were implemented during the 2014-15 season. Some results obtained from the experiments have been already published [7, 8].

The primary objective of the experiments was to investigate fans induced vibrations and fans behavior in the selected grandstand area as the experimental basis for specification and verification of new spectator's crowd load models that have been developed currently [4, 5] because of a lack of appropriate in situ experiments [7, 8].

The Letná football stadium in Prague is the home stadium of the AC Sparta Prague football club. Both series of experiments were focused on the stadium area where the most active fans of the home team (also called "Ultras") were gathered. This group of spectators was during all experiments (and still is) concentrated on the first floor of the stadium and it means the most significant potential source of the dynamic load is placed in the stadium area where the steel cantilever grandstand structure is most sensitive to vertical dynamic excitation.

The paper presents the brief summary and comparison of basic results obtained by two selected experiments concentrated on the spectators induced grandstand vibrations and fans behavior in the observed stadium area during two football matches between the same teams AC Sparta Praha and FK Teplice. The investigated matches were played in various parts of seasons and in different actual position of both teams in the football league.

Between the years 1984 and 2007, the home team AC Sparta Praha became the Czech football league champion eighteen times. It means Sparta did not win the league title only in the five seasons and there were at most one-year breaks between titles during this period. However, in three league seasons foregoing the 2012-13 one, the AC Sparta finished on the second position only.

The first investigated match was played on November 2nd 2013 and the AC Sparta team won 2:0 and because of this victory the AC Sparta became the mid-season champion and its fans were well satisfied. After the defeat, the FK Teplice team remained on the third position in the Czech first football league. At the end of the 2012-13 season, the AC Sparta won the Czech football league after three "long" years again.

The second observed match was played on March 21st 2015 in a little bit different situation. This one the AC Sparta won again but only 1:0 and the decisive goal was scored from penalty kick in the 89th minute. In time of the second match the AC Sparta was on the second position and only its victory kept real hope to become national league champion because of the distance of five points between the AC Sparta and the league leader. All mentioned circumstances affected that the home team fans celebrated this win noticeably less then after the first match. The FK Teplice team remained on the fifth position in the Czech first football league after the second match. At the end of the 2014-15 league season, the AC Sparta finished on the second position.

2 The Brief Description of the Experiment Arrangement

The grandstand dynamic response was observed at the ends of grandstand main beams where the maximal levels of induced vibrations were expected. The piezoelectric sensors Brüel&Kjær of type 4507 B005 were attached to the lower beam flanges by using magnets. The measurement system Pulse and Front-end 3050-B-040 Brüel&Kjær were used for data acquisition. The sensors were connected to the Front-end via cables. The obtained vibration records were modified by use of the Butterworth high-pass filter with cut off frequency 50 Hz.

The grandstand dynamic response was measured in three spots in vertical direction only (see Fig. 1a) during the first experiment that was performed in November 2013. The sensor positions were lightly modified for the second experiment in March 2015 (see Fig. 1b). The right spot was moved more to the middle of the observed grandstand area and one horizontally oriented accelerometer was added to the central point.

During the whole second match, the fans behavior in the observed grandstand area was recorded by a camera that was placed under main grandstand roof on the special bridge which is earmarked for journalists.

The first match was not observed by our own camera. The fans behavior only for the cases of detected large grandstand vibrations was investigated on the base of the TV video record and auditory perceptions connected with a particular behavior type. The same approach was used for the second match as well because of its verification. The compared results agreed on ninety-five percent it means their high agreement.

3 The Types of Spectator Behavior Observed by the Experiment

Eleven types of spectator behavior were recognized during observed matches that have an effect on grandstand vibrations. The designations of the basic behavior types were assumed from the reference [1]. However, other additional types were detected and distinguished in the course of the matches:

- static state - all fans sit or stand statically and watch the match only passively,
- walking and running - this type of spectator behavior occurs mostly outside the play time when occupants are coming to their seats or when they are leaving their places,
- jumping of the 1st type - the fans are jumping en masse asynchronously and independently,
- jumping of the 2nd type - the spectators are joined together in separated lines and jump synchronously on the spot,
- jumping of the 3rd type – the fans are joined together in separated lines and move horizontally by synchronous jumping from right to left and back again usually in opposite directions in adjacent lines,

- jumping of the 4st type – all spectators in the entire sector of the studied grandstand sit down at first after that they concurrently jump up on a signal of their leader and begin continuous jumping promptly,
- bouncing - the fans movement similar to jumping but the feet of persons remain in contact with the grandstand surface,
- swaying – the fans sway forward to support their shouted slogan and the entire motion is intensified moreover by arm movement,
- hand-clapping,
- goal scoring,
- Mexican wave - the sitting spectators briefly stand up and raise their arms in turn creating the appearance of a wave passing through the stadium. The fans in the studied area were standing for most of the matches therefore they created Mexican wave using a wave of arms only.

The more detailed description of the recognized types of spectator behavior is stated in the reference [8].

Tab. 1: The comparison of the maximum peak values of the vertical acceleration evaluated by the particular spectator behavior types in the middle spot.

Observed match	Hand-clapping	Goal scoring	Jumping of the 1st type	Mexican wave
First match 2013 [$m \cdot s^{-2}$]	0.32	0.96	1.11	0.49
Second match 2015 [$m \cdot s^{-2}$]	0.18	0.60	0.47	0.33

Tab. 2: The time duration of the particular types of spectator behavior which was evaluated during the second observed football match played in March 2015.

Type of spectator behavior	Minimum proportion of active fans [%]	First half [min]	First half [%]	Sec. half [min]	Sec. half [%]	Other time [min]	Other time [%]	Total time [min]	Total time [%]
Static state	80	11:53	26.6	7:26	15.4	2:08	27.6	21:27	21.3
Walking and running	100	-	-	-	-	0:59	12.7	0:59	1.0
Jumping of the 1st type	15	2:23	5.3	2:46	5.7	-	-	5:09	5.1
Jumping of the 2nd type	100	-	-	0:05	0.2	0:46	9.9	0:51	5.1
Jumping of the 3rd type	100	-	-	0:59	2.0	-	-	0:59	1.0
Jumping of the 4th type	100	1:17	2.9	-	-	-	-	1:17	1.3
Bouncing	15	5:23	12.0	5:59	12.4	1:56	25.0	13:18	13.2
Swaying	20	12:11	27.2	9:07	18.9	0:27	5.8	21:45	21.6
Hand-clapping	20	11:14	25.1	19:28	40.4	1:28	19.0	32:10	32.0
Goal scoring	85	-	-	2:13	4.6	-	-	2:13	2.2
Mexican wave	85	0:22	0.8	0:05	0.2	-	-	0:27	0.4

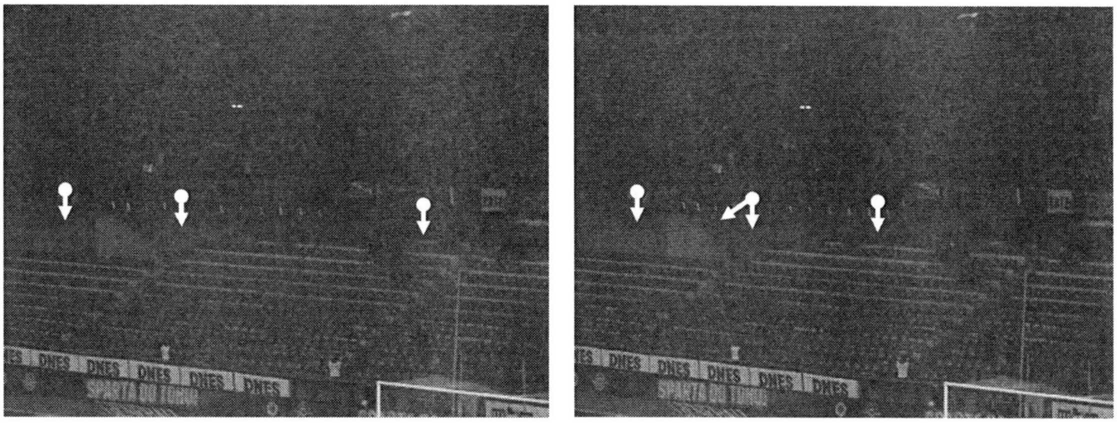

a) the sensor positions during the first match b) the sensor positions during the second match

Fig. 1: The location and orientation of the accelerometers used by the experiments performed in 2013 (on the left) and in 2015 (on the right) and the view on the observed grandstand area where the most active fans were concentrated.

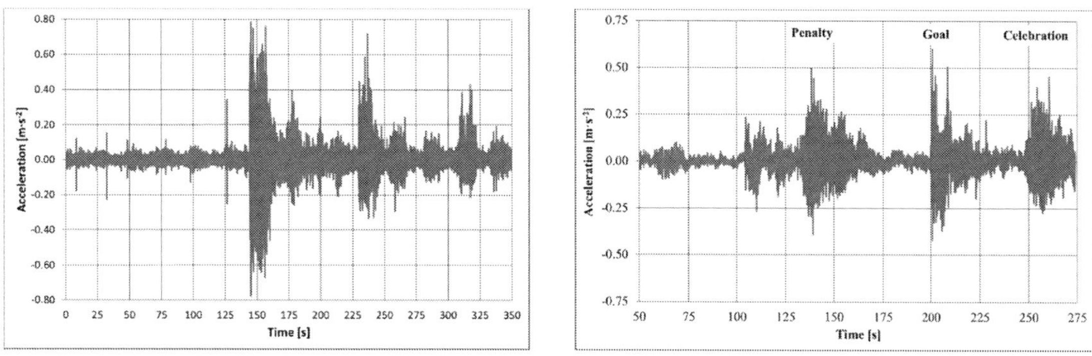

a) the first match played in 2013 b) the second match played in 2015

Fig. 2: The detail of the grandstand vertical acceleration time records from both observed matches which were measured in the middle spot in the situation when the AC Sparta scored decisive goal.

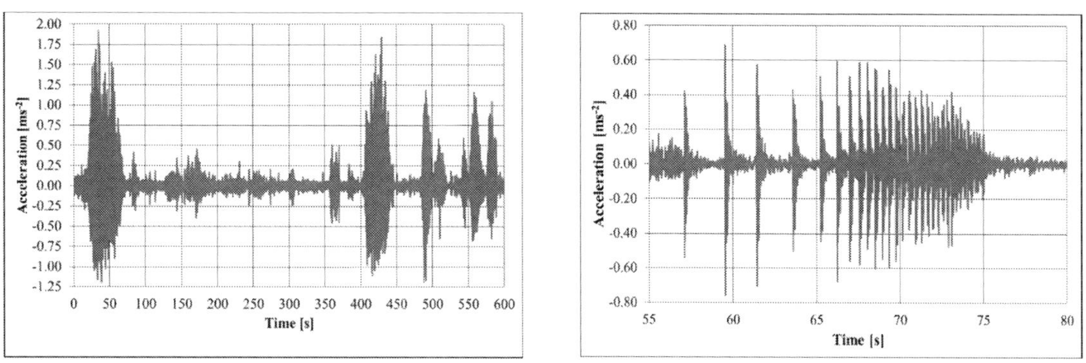

a) the first match played in 2013 b) the second match played in 2015

Fig. 3: The detail of the grandstand vertical acceleration time records from both observed matches which were measured in the middle spot in the situation when the home fans celebrated the victory after the end of the match.

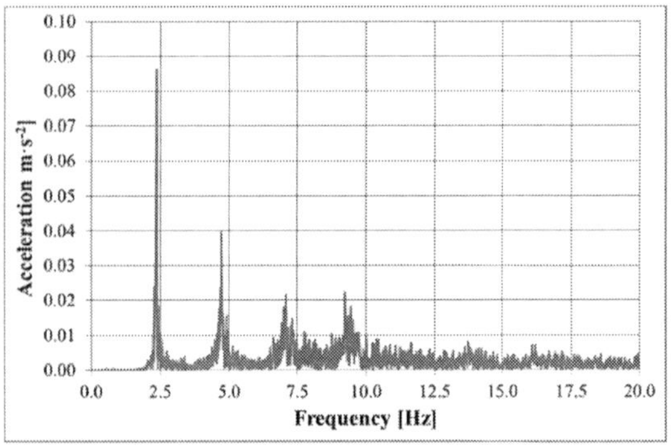

a) the first match played in 2013

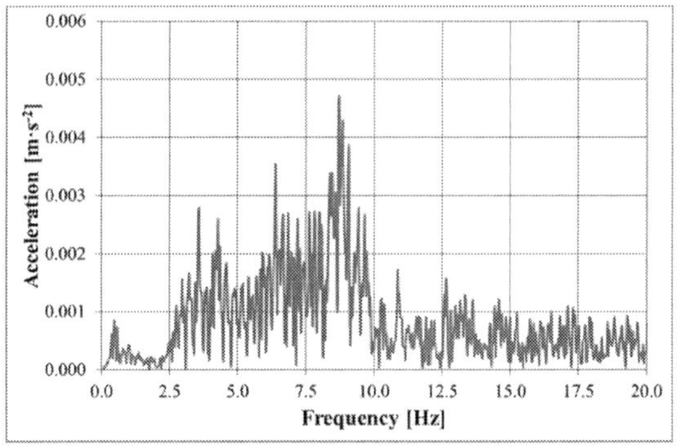

b) the second match played in 2015

Fig. 4: The frequency spectrum evaluated from the vertical acceleration shown in Fig. 3 that was measured in the middle spot in the situation when the home fans celebrated the victory after the end of the match.

4 The Brief Description of the Experiment Basic Results

The both described experiments were carried out during the whole analyzed football matches from the entrance of players to the pitch to the spectator celebration at the end. The information about a number of spectators assembled in the studied grandstand area for the first match was not available. During the second match, 910 fans were gathered in the observed grandstand sector. That was 10.7 % from the total number of spectators present at the stadium (8 502).

The comparisons of the time behavior and maximum peak values of the vertical acceleration in the middle spot for the significant spectator behavior types are shown in Fig. 2, Fig. 3 and in Tab. 1.

The two frequency spectrums are compared in Fig. 4. These spectrums were evaluated from the vertical accelerations that were recorded during home fans celebration of the victory after the end of the both matches and that are depicted in Fig. 3.

The time duration of the specific behavior types evaluated only during the second match is stated in Tab. 2 where the minimum proportion of the active fans is also mentioned.

It ensued from the Fig. 2, that the fans behavior and dynamic response of the grandstand in both matches were similar to each other in the moment when the home team scored decisive goal. The vibration from the second match could be moreover clearly divided into three phases – the award of the penalty, the goal scoring and the goal celebration.

5 Conclusion

The next four main findings resulted from the carried out experimental analysis:

Firstly, it can be stated on the base of the mutual comparison of the results evaluated from both observed matches that the intensity of the grandstand vibrations induced by fans is strongly influenced by their current mood depending on the actual league position of their supported team and on its game quality. The larger grandstand vibrations in the identical situations were clearly detected during the first match when the AC Sparta team celebrated the mid-season title (see Tab. 1, Fig. 2, Fig. 3 and Fig. 4).

Secondly, the home fans most frequently used type of cheering with hand-clapping that was practically used during a third (32.0 %) of the total play time. Nevertheless, the portion of active fans get involved in this type of support was surprisingly low (about 20 % active fans only).

Thirdly, the active spectator participation in the specific support types fluctuated significantly during the observed football matches. All of fans presented in the observed grandstand area (it means 100 % of fans) attended only on the cheering type Jumping of the 2nd, 3rd and 4th type.

Fourthly, the greatest level of the examined dynamic grandstand response was determined during the cheering types on which all of fans attended actively (Jumping of the 1st, 2nd, 3rd and 4th type).

The correctness of the above stated discoveries are going to be verified based on additional results of similar experiments that were carried out during another football matches at AC Sparta stadium (in Spring 2015) and SK Slavia Praha stadium (in Spring 2016).

Acknowledgement

This paper has been supported by the Czech Science Foundation project No. GA15-15728S.

References

[1] C.A. Jones, P. Reynolds, A. Pavic, Vibration serviceability of stadia structures subjected to dynamic crowd loads: A literature review, Journal of Sound and Vibration 330 (2011) 1531–1566, http://dx.doi.org/10.1016/j.jsv.2010.10.032.

[2] O. Rokos, J. Maca, Stochastic approach in the human-induced vibration serviceability assessment of grandstands, in proc.: Proceedings of the IX International Conference on Structural Dynamics EURODYN2014 (2014), eds. A. Cunha, E. Caetano, P. Ribeiro, G. Müller, Faculdade de Engenharia da Universidade do Porto, Porto Portugal, 2591-2598.

[3] O. Rokos, J. Maca, The response of grandstands driven by filtered Gaussian white noise processes, Advances in engineering software 72 (2014) 85 – 94, http://dx.doi.org/10.1016/j.advengsoft.2013.05.008.

[4] V. Rajic, A. Pavic, Stochastic approach to modeling of near-periodic jumping loads, Mechanical systems and signal processing 24 (2010) 3037 – 3059.

[5] P. Reynolds, A. Pavic, Vibration performance of a large cantilever grandstand during an international football match, Journal of Performance of Constructed Facilities 20 (2006) 202 – 212.

[6] A. Caprioli, M. Vanali, A. Cigada, One year of structural health monitoring of the Meazza Stadium in Milan: Analysis of the collected data, in proc.: Proceedings of the 27th Conference and Exposition on Structural Dynamics 2009 IMAC-XXVII (2009), Society for Experimental Mechanics Inc., Orlando Florida USA, 9p.

[7] M. Verner, M. Polak, T. Plachy, An Experimental Study Focused to Spectators-induced Vibrations of a Cantilever Grandstand during two Sport Matches, in proc.: Proceedings of 53rd International Conference on Experimental Stress Analysis 2015 (2015), eds. P. Padevet, P. Bittnar, Czech Technical University in Prague Faculty of Civil Engineering, Český Krumlov, 460–465.

[8] M. Verner, T. Plachy, M. Polak, An Experiment Focused on Fans Behaviour and Induced Grandstand Vibrations during a Football Match, Applied Mechanics and Materials 837 (2016) 75–78, http://dx.doi.org/10.4028/www.scientific.net/AMM.837.75.

Effect of Particle Size Distribution Waste Micronized Marble Powder on the Mechanical Properties of the Cement Composite

Z. Prošek [1,a], K. Šeps [1], J. Topič [1], P. Tesárek [1]

[1] *CTU in Prague, Faculty of Civil Engineering, Thákurova 7, 166 29 Praha 6, Czech Republic*
[a] *zdenek.prosek@fsv.cvut.cz*

Abstract: This article deals with influence of the process of micronizing marble powder on the mechanical properties of the cement composite. The tested material consisted of cement CEM I 42.5R (Radotín) and non- / micronized marble powder. The difference between individual mixtures are the use of different fineness of micronized of marble powder. The investigated mechanical parameters were dynamic modulus of elasticity, flexural strength and compressive strength for 28 days old sample.

Keywords: marble powder; mechanical properties; cement composite; recycling; micronization.

1 Introduction

Marble industry produces annually more than 10 million tons of waste [1]. Waste is dumped in nearby factories and thereby creating environmental risk and prevents the development of marble industry. The most interesting part of the waste is marble sludge (powder) for which it is almost nil usage and has a negative impact on the environment by increasing the incidence of dust, damage to flora, fauna and human resources. For these reasons, it is important to create an application this waste, which should result in improved safety and health by reducing dust in the air, lower price for storage, transport and handling of waste and the possibility of potential income from the sale of recycled materials [2].

There are several work dealing with the use of marble sludge in cement composites. It was found that adding a small amount (5 – 10 wt. %) marble sludge into cement composites will improve the mechanical properties of the material such as compressive strength and flexural strength [3, 4, 5]. This application has a positive effect on the mechanical properties of the cement composite, but is insufficient in terms of the amount of used waste. If they had used larger quantities of waste (marble sludge), there would be a rapid decrease of mechanical properties [6]. New options to eliminate or limit this effect is the use of high-speed milling from Ltd Lavaris. The resulting product is micronized marble powder. This high-speed milling process of guaranteeing separation of all grain of marble sludge and drying recycled material [7]. The high-speed milling process was successfully used for micronized of recycled concrete [8].

2 Used Materials and Samples

Mixture composed of 20 wt. % Portland cement CEM I 42.5R (Radotín) and 80 wt. % non- / micronized marble powder has been prepared for investigating the impact of micronizing. Marble powder was created by high speed milling of marble sludge from the West Bank of Jordan. Testing was performed on beams of dimensions equal to 40 × 40 × 160 mm. Six samples were tested from each mixture. Samples were demolded the day after production, and the samples were stored loosely in a laboratory environment at 22 ± 1 ° C and relative humidity 50 ± 2 %. The investigated mechanical parameters were dynamic modulus of elasticity, flexural strength and compressive strength. Mixture A contains non-micronized marble powder and serves as a reference sample. Mixtures B, C and D contain differently micronized marble powder. Amount of water was designed to maintain the same workability of fresh mixture. Consistency of the mixtures was determined using flow expansion test. The composition of individual mixtures can be seen in Table 1.

Tab. 1: Composition of the mixtures.

Set/ Material	Cement [hm. %]	Marble powder [hm.%]	Fraction of marble powder [µm]	Average grain size of marble powder [µm]	Water ratio
A	20	80	0 – 60	5.37	1.7
B	20	80	0 – 40	4.19	1.9
C	20	80	0 – 40	4.11	2.0
D	20	80	0 – 30	3.69	2.1

3 Experimental Methods and Results

Resonance method is used for non-destructive testing; its principle was explained in the works of Klapálek et al. and Topič et al. [9, 10]. Measurement of the dynamic modulus of elasticity was carried out using the measurement line from Brüel & Kjær, consisting of a measuring unit type 3560-B-X12 impact hammer type 8206 and an acceleration sensor type 4519-003 53422 assembly was connected to the control computer. The mass of all specimens was measured and the dimensions were recorded before the resonance method was applied.

The flexural and compressive strength were determined using the Heckert device, model FP100 on the 28 days old specimens. The testing was displacement controlled at a constant rate of 0.1 mm/s in the case of four-point bending and 0.3 mm/s for the compressive test. The distance between supports for four-point bending test was equal to 100 mm. The uniaxial compressive test was performed on the broken halves of the specimens with effective dimensions of 40 × 40 × 80 mm.

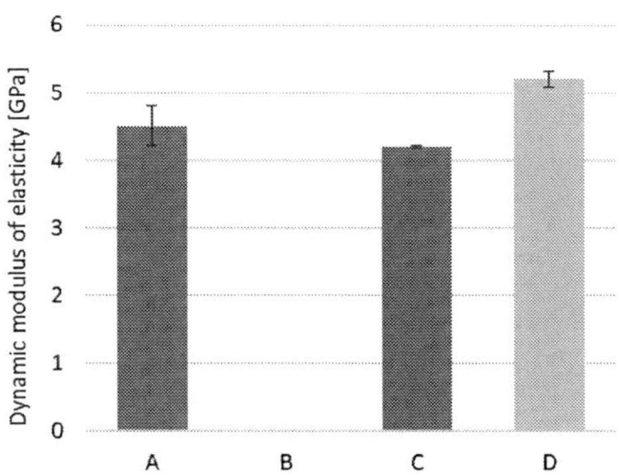

a) comparison of dynamic modulus of elasticity (with standard deviation).

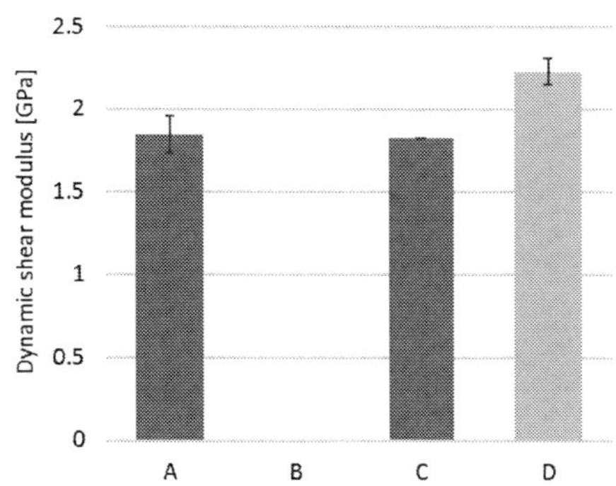

b) comparison of dynamic shear modulus (with standard deviation)

Fig. 1: Comparison of mechanical properties of the 28-day old material from/set non-destructive tests.

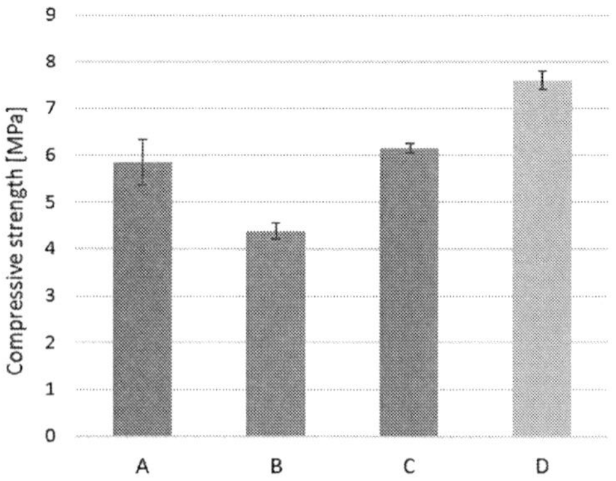

a) comparison of compressive strength (with standard deviation)

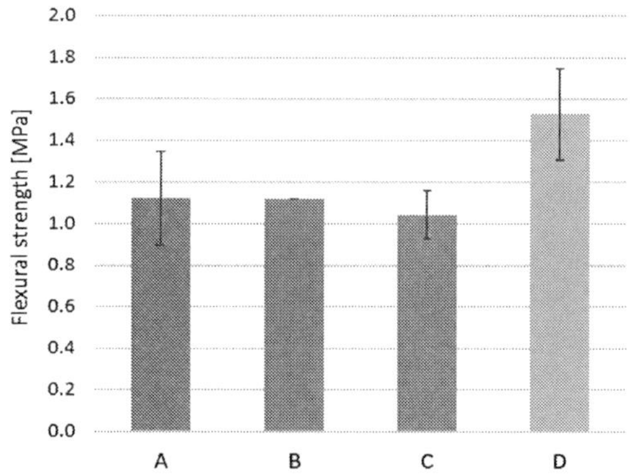

b) comparison of flexural strength (with standard deviation)

Fig. 2: Comparison of mechanical properties of the 28-day old material/set from destructive tests.

All test materials had approximately the same bulk density. The value of bulk density was 1430 ± 50 kg/m^3. The results (Fig. 1 and Fig. 2) show the effect of fineness of micronizing marble powder on the mechanical properties of the cement composite. Mixture with non-micronized marble powder (A) had the average value of compressive strength after 28 days 5.85 ± 0.45 MPa and mixture with the finely micronized marble powder (mixture D) had average value of compressive strength 7.61 ± 0.20 MPa. Samples of mixture B contained cracks and therefore, the resultant compressive strength was less than the reference sample (mixture A). Cracks in samples (B) also had an effect on the measurements using the resonance method, because it was not possible to capture the response impulses from excitation pulse "hammer" and dynamic modulus and dynamic shear modulus could not be measured. Similar trends as the compressive strength were observed in the flexural strength, dynamic modulus of elasticity and dynamic shear modulus. (Fig. 1 and Fig. 2a). Flexural strength of material (A) was equal to 1.12 ± 0.20 MPa. Set (D) had the average value of compressive strength 1.53 ± 0.20 MPa.

4 Conclusion

Using micronized marble powder was an increase in compressive strength of about 30% against the reference mixture. The results show that micronized marble powder performs the function of microfiller. The investigated materials have the potential for economic reasons, because this waste has almost zero cost, and thus it becomes very promising for the civil engineering. In the future, we will focus the possibility of involvement of micronized marble powder in the hydration process, secondly we will try to map the involvement marble powder to microstructure of cement composite and third we will focus our attention on the study of microstructure and micromechanical properties to follow the studies of e.g. [11, 12].

Acknowledgement

Financial support from the Faculty of Civil Engineering, Czech Technical University in Prague (SGS project SGS16/201/OHK1/3T/11). The authors also thank the Center for Nanotechnology in Civil Engineering at the Faculty of Civil Engineering, Czech Technical University in Prague, Faculty of Mechanical Engineering Czech Technical University in Prague and the Joint Laboratory of Polymer Nanofiber Technologies of the Institute of Physics, Academy of Science of Czech Republic, and the Faculty of Civil Engineering, Czech Technical University in Prague.

References

[1] M. A. Mashaly, A. O. Shalaby, B. N. El-Hefnawi, Characterization of The Marble Sludge of The Shaq El Thoaban Industrial Zone, Egypt and Its Compatibility for Various Recycling Applications 6 (2012) 153–161.

[2] M.S. Ibrahim, et al., Health Risk Assessment of Marble Dust at Marble Workshops, Nature and Science, 9 (2011) 144-154.

[3] A. A. Aliabdo, A. E. M. Abd Elmoaty, E. M. Auda, Re-use of waste marble dust in the production of cement and concrete, Constr. Build. Mater. 50 (2014) 28–41, http://dx.doi.org/10.1016/j.conbuildmat.2013.09.005.

[4] H. Binici et al., Durability of concrete made with granite and marble as recycle aggregates, J. Mater. Process. Technol. 208 (2008) 299–308, http://dx.doi.org/10.1016/j.jmatprotec.2007.12.120.

[5] V. M. Shelke, P. Pawde, R. Shrivastava, Effect of marble powder with and without silica fume on mechanical properties of concrete, J Mech Civ Eng 1 (2012) 40-45, http://dx.doi.org/10.9790/1684-0114045.

[6] B. Demirel, The effect of the using waste marble dust as fine sand on the mechanical properties of the concrete, Int. J. Phys. Sci. 5 (2010) 1372–1380.

[7] Z. Prošek, et al., Effect of micronized recycled materials on mechanical properties of cement mortar with crushed bricks, A R Research publication 4 (02) (2015), 120–124, ISSN 2319-8354.

[8] Z. Prošek, et al., Influence of recycled concrete composition on its elastic stiffness, Key Engineering Materials 677 (2016) 288-291, http://dx.doi.org/10.4028/www.scientific.net/KEM.677.288.

[9] P. Klapálek, L. Melzerová, Methods of non-destructive assessment of timber, Applied Mechanics and Materials 732 (2015) 369-372, http://dx.doi.org/10.4028/www.scientific.net/AMM.732.369.

[10] J. Topič et al., Effect of PVA Modification on the Properties of Cement Based Composite, Acta Polytechnica 55 (2015) 64-75, http://dx.doi.org/10.14311/AP.2015.55.0064.

[11] P. Tesárek et al., Micromechanical properties of different materials on gypsum basis, Chemicke Listy, 106 (2012) s547-s548. ISSN 0009-2770.

[12] P. Tesárek, J. Němeček, Microstructural and micromechanical study of gypsum, Chemicke Listy, 105 (2011) s852-s853.

Freeze-Thaw Durability of Fiber-Reinforced Lime-Based Mortars

M. Přinosil [1,a]

[1] *Institute of Theoretical and Applied Mechanics, Academy of Science, Prosecká 809 /76, 190 00 Praha 9, Czech Republic*

[a] *prinosil@itam.cas.cz*

Abstract: This paper presents results of experimental investigation focused on freeze-thaw durability of fiber-reinforced lime-based mortars. Evolution of structural deterioration and frost resistance were evaluated using non-destructive testing (resonance method). In this study, we consider two types of lime-based matrices (pure lime, lime-metakaolin) and two types of polyvinyl alcohol fibers in several volume fractions. The results showed that the resistance against the frost depends mainly on the matrix type and on the amount of fibers.

Keywords: Freeze-thaw durability; Lime-based mortar; Fiber composite; Fiber reinforcement.

1 Introduction

Lime mortars were widely used in the past for construction of buildings and monuments. Nowadays, lime-based materials are required for restoration of ancient structures due to compatibility with the original ones [1]. However, these materials can be often characterized by slow setting, high shrinkage, low strength and low deformation capacity [2]. Therefore, it is necessary to develop new materials, which overcome these drawbacks. For example, fiber reinforcement can improve behavior of a mortar under tensile loading, especially when the damage appears. Our intention is to systematically design the microstructure of high-performance fiber-reinforced lime-based mortar, so that imposed deformation does not localize into a single crack as in conventional mortars, but thanks to fiber reinforcement the mortar exhibits tensile strain-hardening behavior and cracking in the form of large amount of fine cracks (multiple cracking) [3]. Therefore, the mortar retains its macroscopic integrity under relatively high imposed deformations (for example due to volume changes, seismicity, etc.) and exhibits improved durability. The main objective of this work is to clarify freeze-thaw durability of developed mortars.

2 Materials and Specimens Preparation

The proposed mortars consisted of lime-based matrix and fiber reinforcement. Individual components were selected so as to meet criteria for multiple cracking and other desired mechanical properties [3, 4]. In order to ensure appropriate fiber-matrix interaction, fine grained filler represented by quartz sand with maximum particle size 0.3 mm was selected (manufactured by Sklopísek Střeleč, a.s.). The binder was represented by hydrated air lime powder Čerťák class CL 90 (made by Čertovy schody a.s.) and eventually by metakaolin Mephisto L05 (made by ČLUZ s.r.o.), which was employed especially for improvement of tensile strength of the matrix. Fiber reinforcement was represented by two types of polyvinyl alcohol fibers REC 15×12 and RSC 15×8 (both made by Kuraray Company, Ltd.) with different length (8 mm, 12 mm) and surface coating (REC fibers are coated by oil treatment leading to reduction fiber-matrix bond strength).

Ratios of individual components of the matrix were optimized with regard to the desired mechanical properties (tensile strength, fracture energy) as well as to workability of a fresh mixture [5]. Binder to aggregate mass ratio was equal to 1:3, mass ratio of lime and metakaolin to 3:1 (in lime-metakaolin mortars) and water ratio to 0.3 (calculated as ratio of weight of water and all dry ingredients).

With regard to the results of previous research [4, 6], this experiment was performed only on 4 configurations of the composition: pure lime mortar reinforced with 2.0% of RSC fibers (labeled as LS20), lime-metakaolin mortar reinforced with 2.0% of RSC fibers (LMS20) and lime-metakaolin mortar reinforced with REC fibers in volume fraction of 1.0% (LME10) and 2.0% (LME20). From each proposed mixture, a set of 6 prisms with dimensions 40×40×160 mm were prepared. Before testing, they were stored in laboratory conditions (i.e. temperature 20÷25°C and relative humidity 50÷60%) for approximately 2 years.

3 Methodology and Testing Set-Up

The investigation of frost resistance and structural deterioration of specimens was performed using adopted methodology defined in ČSN 72 2452 with non-destructive testing (resonance method) [7].

3.1 Resonance Method

This method can be used for determination of dynamic characteristics of tested materials. It is especially appropriate for homogeneous and isotropic materials, however, it can be used for heterogeneous materials like mortars and pastes as well [8]. The specimen is placed in horizontal position supported in the middle of its span. Excitation of vibrations is provided by impact hammer on one side of the specimens (in longitudinal direction), while the acceleration transducer is attached on the opposite side. The dynamic modulus of elasticity can be calculated for the specimen with a continuously distributed mass and free-free boundary conditions from the longitudinal vibration as:

$$E_{dyn} = \frac{4Lmf_l^2}{bh} \tag{1}$$

where L is the length of the specimen, m is the mass of the specimen, b and h are the cross-sectional dimensions. The first fundamental longitudinal resonant frequency f_l was evaluated by means of Bruel&Kjær device and PULSE Reflex software from signals of the applied force and the acceleration, which were transformed using Fast Fourier Transformation (FFT) to Frequency Response Function (FRF).

3.2 Investigation of Freeze-Thaw Durability

Before testing, specimens were saturated with water by immersion in bath for 24 hours. Each freeze-thaw cycle consisted of freezing at -20±3°C for 4 hours and free thawing at 20±3°C for 2 hours. All dimensions and weight of specimens were measured and dynamic modulus of elasticity E_{dyn} was evaluated using non-destructive testing before testing itself and after each 2 freeze-thaw cycles. Moreover, monitoring of visual aspect of specimens was performed. Testing was terminated when the value of E_{dyn} decreased below the 75% of initial value [7].

4 Results and Conclusions

Fig. 1 shows the evolution of dynamic modulus of tested mortars, while Fig. 2 shows the visual aspects of specimens at the end of testing.

In the case of pure lime mortar reinforced with 2.0% of RSC fibers, the dynamic modulus dropped below the limit value immediately after 2 freeze-thaw cycles. However, the specimens were compact without any visible damage (cracking, spalling of corners, etc.). After another several freeze-thaw cycles, some crack developed in specimens (Fig. 2a), but they were still compact thanks to the fiber reinforcement.

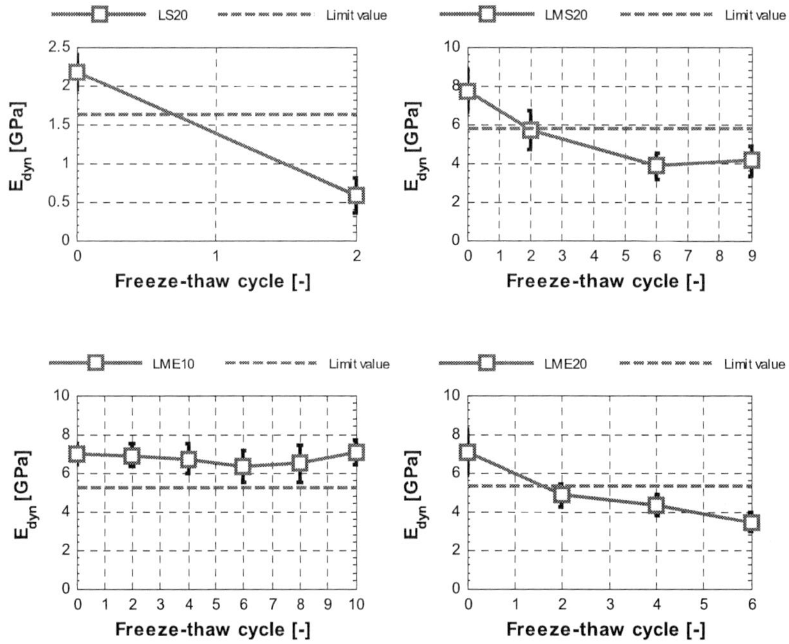

Fig. 1: Evolution of dynamic modulus of elasticity

The lime-metakaolin mortars exhibited higher initial value of the dynamic modulus than pure lime mortar, which confirms that these mortars are stiffer thanks to the addition of metakaolin. In the case of lime-metakaolin mortar reinforced with 2.0% of RSC fibers, the dynamic modulus was equal to the limit value after 2 freeze-thaw cycles and the specimens were compact without any visible damage. After 6 freeze-thaw cycles, the value dropped below the limit value, but the specimens were still compact without any visible

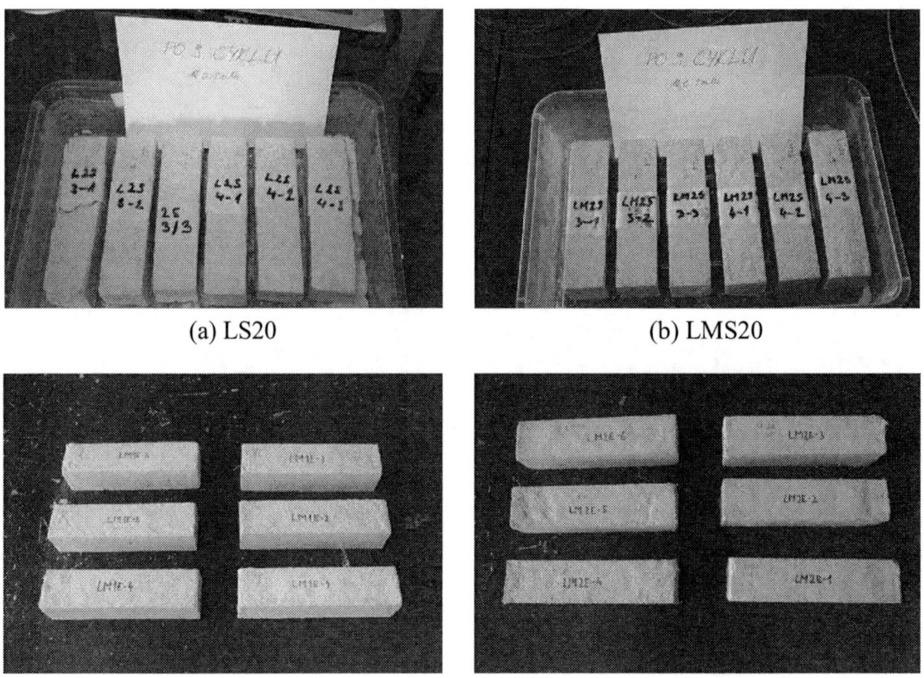

(a) LS20 (b) LMS20

(c) LME10 (d) LME20

Fig. 2: Visual aspect of tested specimens (at the end of the experiment)

damage. After 9 freeze-thaw cycles, the value of dynamic modulus was similar as previous value and the visual aspect did not change (Fig. 2b). In the case of lime-metakaolin mortar reinforced with 2.0% of REC fibers, the results were similar as the previous ones. The best results were achieved for the lime-metakaolin mortar reinforced with REC fibers in volume fraction of 1.0%. The value of dynamic modulus did not drop below the limit value during the testing in 10 cycles and the visual aspect was similar to the initial one (Fig. 2c).

5 Conclusions

The freeze-thaw durability of lime-based mortars reinforced with synthetic fibers in two volume fractions was experimentally investigated. The results revealed that the pure lime mortar suffered from temperature changes and after 2 cycles the value of dynamic modulus dropped below the limit value, however, the specimens were compact thanks to fiber reinforcement. In the case of lime-metakaolin mortar, the values of dynamic modulus of elasticity were generally higher. The results revealed that the evolution of the dynamic modulus of elasticity was similar for both types of polyvinyl alcohol fibers in the same volume fraction. Therefore, it can be concluded that the amount of fibers influenced the durability more than their type. The best results were not achieved for maximum considered volume fraction of 2.0%, but for the volume fraction of 1.0%. This conclusion is consistent with the results of mechanical tests [4, 6].

Acknowledgement

The presented research has been carried out with the financial support of the Czech Science Foundation under the project No. GB P105/12/G059 and institutional support RVO 68378297.

References

[1] J. Lanas, J.I. Alvarez, Masonry repair lime-based mortars: factors affecting the mechanical behavior, Cement and Concrete Research 33 (2003) 1867–1876, doi: 10.1016/S0008-8846(03)00210-2.

[2] A.L. Velosa, F. Rocha, and R. Veiga, Influence of chemical and mineralogical composition of metakaolin on mortar characteristics, Acta Geodynamica et Geomaterialia 6 (2009) 121–126.

[3] V. C. Li, C.K.Y. Leung, Steady-State and Multiple Cracking of Short Random Fiber Composites, Journal of Engineering Mechanics 118 (1992) 2246-2264, doi: 10.1061/(ASCE)0733-9399(1992)118:11(2246).

[4] M. Přinosil, P. Kabele, Mechanical Properties of Fiber Reinforced Lime-Based Mortars Evaluated from Four-Point Bending Test, Applied Mechanics and Materials 821 (2016) 526–531, doi: 10.4028/www.scientific.net/AMM.821.526.

[5] M. Přinosil, P. Kabele, Influence of composition on tensile and fracture properties of lime-based mortar, in proc.: Proceedings of the Conference Nano & Macro Mechanics 2013 (2013), eds. T. Plachý et al., Czech Technical University in Prague, Prague, 163–170.

[6] M. Přinosil, P. Kabele, Fracture Properties of Fiber Reinforced Lime-Based Mortar, in proc.: Proceedings of the 5th Conference Nano & Macro Mechanics (2014), eds. T. Plachý et al., Czech Technical University in Prague, Prague, 143–152.

[7] ČSN 72 2452 + Z1: Testing of frost resistance of mortar. Czech Office for Standards, Metrology and Testing, Prague, 2005.

[8] T. Plachý, P. Padevět, and M. Polák, Comparison of Two Experimental Techniques For Determination of Young's Modulus of Concrete Specimens, in proc.: Proceedings of the 5th WSEAS International Conference on Applied and Theoretical Mechanics - Recent Advances In Applied And Theoretical Mechanics (2009), eds. C. A. Bulucea et al., WSEAS Press, Athens, 68–71.

Experimental Characterization of Strain Gauges for High-Temperature Applications

J. Rosa [1,a], J. Cagáň [1], F. Rösler [1]

[1] *Výzkumný a zkušební letecký ústav, a. s. (Aerospace Research and Test Establishment), Beranových 130, 199 05 Praha – Letňany, Czech Republic*

[a] *rosa@vzlu.cz*

Abstract: The article deals with a characterization of properties and behaviour of resistive strain gauges for high temperature measurement up to 1000 °C. Vishay ZC-series strain gauges are some of a few commercionally available products, with advantage of favourable date of delivery. On the other hand, their standard high temperature material of the grid causes a high apparent strain effect for single grid gauges. That is why strain gauges with a thermal compensation were in our interest. The object of other testing was the value of gauge factor depending on temperature.

Keywords: high-temperature; strain gauge; gauge factor; apparent strain.

1 Introduction

Experimenters should have good cognizance of their equipment and used sensors. In the case of a standard and well-known measurement, one can rely on datasheets of producers and common practice. However, there are non-standard, specialized or rarely used techniques of measurement which should be verified and a primary calibration should be done. As an example, resistive strain gauge measurements can be shown. There is the common practice with foil strain gauges for low-temperature measurement, which are installed by recommended procedure, and all characteristics (as the gauge factor, coefficient of transverse sensitivity and coefficient of thermal expansion) are given by the gauge producer. Special strain gauges for high-temperature measurement have their characteristics influenced by both harsh measurement condition itself and means of sensor installation. A brief introduction was given in [1]. The strain gauge can be mounted both using various types of ceramic cement and using thermal spray alumina coats (ROKIDE® technology). Each of the connective materials has its specific coefficient of thermal expansion and an establishment of influence on the apparent strain effect should be performed. Even if a dynamic measurement is estimated only, the knowledge of the apparent strain curve is needed for correct setting of the voltage range of the measuring chain to prevent from cutting off the signal.

Classic means of thermal influence compensation, as well-known for the common foil strain gauges, don´t exist for those designated for operation in very broad ranges of temperature (up to 1000 °C). However, there are special patterns of high-temperature strain gauges which consist of two grids – in the case of their half bridge wiring they should compensate the thermal influence and be able to measure strain in good sensitivity nevertheless.

The gage factor (GF) depends on temperature and also on the means of installation probably. Although some reference data are available, it would be optimal to attest the sensitivity of the sensor for given installation and wiring. A test device based on four-point bending is usually used for professional calibrations, but its development and debugging is not easy [2, 3]. That is why our testing laboratory has not thought about similar activities. Nevertheless, an experiment was designed for obtaining of reference results. The testing was performed with using of a special thermal chamber in the form of tensile loading of a superalloy specimen equipped with tested strain gauges. The strain was measured with using of an extensometer simultaneously and voltage readings coming from strain gauge signals were compared to this reference strain.

2 Experiments – realization and results

The tested strain gauges were mounted on "dog-bone" shape specimens of Haynes 285 superalloy, which chemical composition is given in Tab. 1. The thickness of the specimens was 1 mm only, therefore strain gauge installation with using of the Rokide flame spray technology had to be performed especially carefully and a special fixture had to be made to prevent from deformation.

Tab. 1: Nominal chemical composition (in per cent) of nickel based Haynes 282 superalloy

Cr	Co	Ti	Al	Fe	Mo	Mn	Ni
20	10	2.1	1.5	max 1.5	8.5	max 0.3	as balance (cca 57)

The specimens were always equipped with K-type thermocouples for purpose of the measurement of the actual temperature near strain gauge position. Signals both from the strain gauges processed by the ESAM Traveller 2 measuring device and from the thermocouple were led to the PC equipped with the PCI-6052E DAQ card (by National Instruments) and software applications built in-house on the LabVIEW platform. These analog signals were sampled at frequency of 1 kHz and stored on the hard-disk. The off-line evaluation has provided data with the period of 1 second (mean values and standard deviations).

2.1 Apparent strain of the strain gauge with thermal compensation

The Vishay ZC-NC-G1264-120 strain gauges with two grids (overall dimensions: 2.54 x 4.55 mm) were installed with using of the PBX cement only (Fig. 1). It is supposed their using for an accurate static measurement up to 500 °C, although the tests of the apparent strain behaviour were performed in the OMEGA LUX LMF – A550 type furnace for temperature up to 750 °C. Both grids have their separate ribbon leads and their wiring is in the form of the half-bridge circuit, but the compensation grid is inactive from the point of view of mechanical loading. That is why the resulting behaviour in the mode of quarter-brigde wiring is expected, including calibration for gauge factor of 2.80.

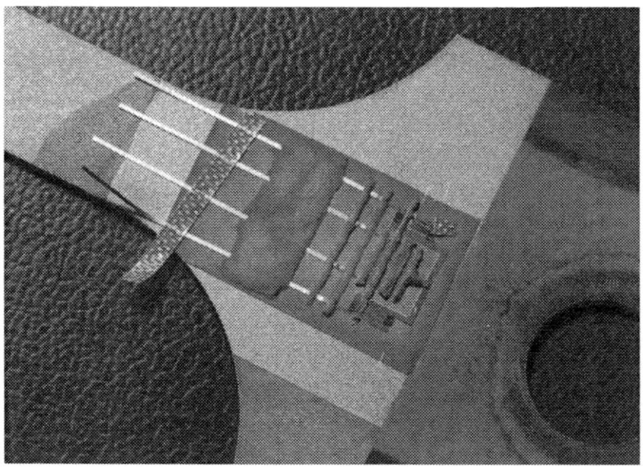

Fig. 1: Vishay ZC-NC-G1264-120 type strain gauge during instalation on the specimen.

Fig. 2: Specimen with the installed strain gauge in the OMEGA furnace.

Two strain gauges have been tested. An example of established "apparent strain vs. temperature" curves is in Fig. 3. It is evident that curves for following temperature loading cycles slightly change their positions. For given instance of a real measurement (i.e. the expected temperature range and the number of measuring runs) it should be desirable to perform a goal-directed verification. However in comparison with a typical apparent strain curve for the "single" strain gauge given in Fig. 4, it is clearly shown the suppression of the apparent strain effect and the advantage of more accurate measurement.

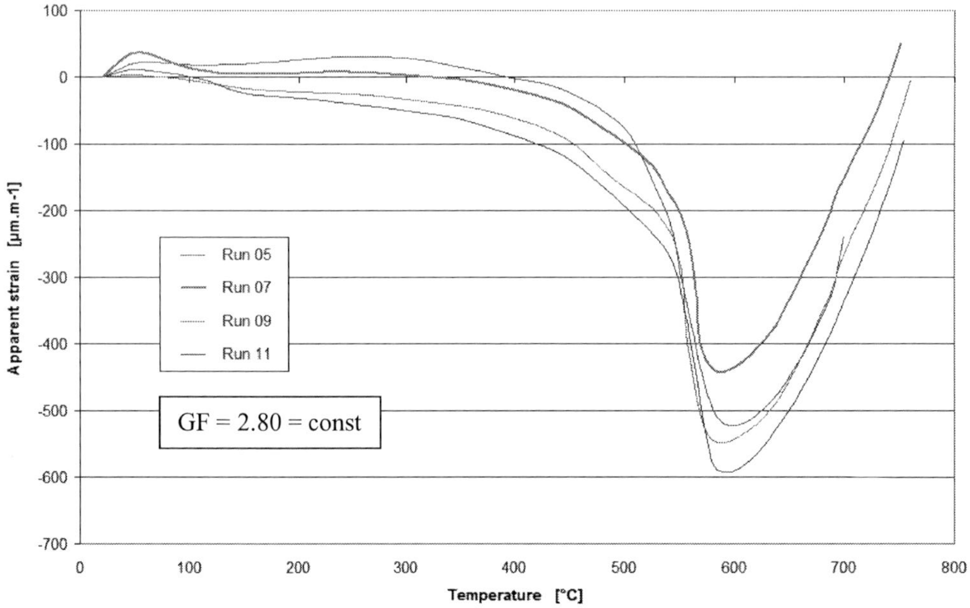

Fig. 3: Apparent strain curves for the Vishay ZC-NC-G1264-120 strain gauge, installed using of PBX cement.

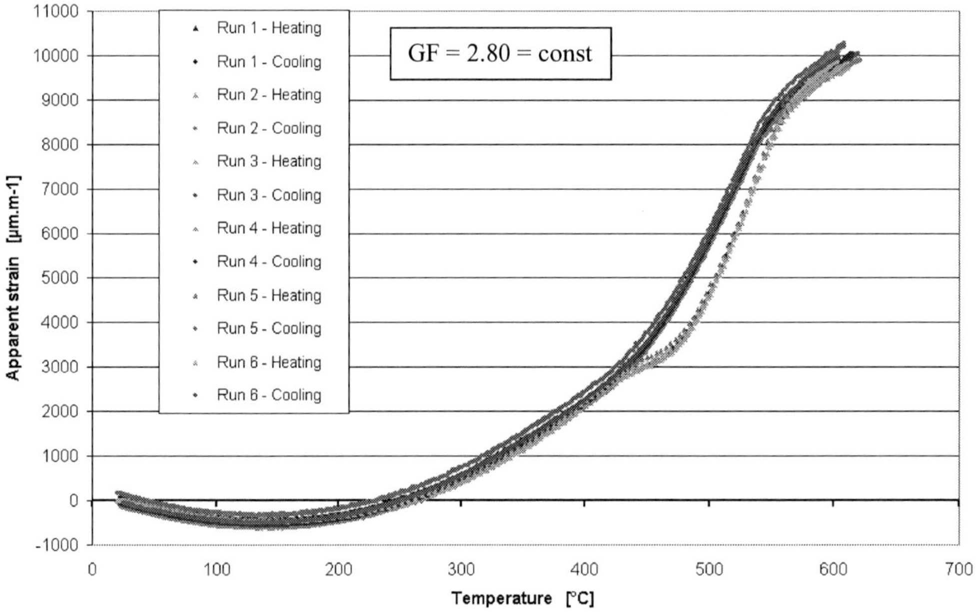

Fig. 4: Apparent strain curves for the Vishay ZC-NC-G1262-120 strain gauge, installed using of PBX cement.

2.2 Gauge factor dependence on temperature

The single grid Vishay ZC-NC-G1262-120 type strain gauges were always mounted in the total number of two pieces on opposite sides of the specimen – the installation was accomplished both with the PBX ceramic cement and with the ROKIDE flame spray coats. Three-wire configuration of the quarter bridge strain gauge circuit was always used; for wiring in the interior and near exterior of the thermal chamber a special thermal insulated wire system of AerOpak 120-3M-C(AQ)/J type by Okazaki company was used. The specimens were loaded in tensile using the INSTRON 1185 machine equipped with water cooled suspensions and the ATS 3210 thermal chamber (for heating up to 1100 °C). The Epsilon model 3548-025M extensometer is also a special device intended for high-temperature measurement. The measuring chain is the same as is described above, moreover the signals from the loading machine (i.e. tensile force, displacement of the crossbeam and strain measured by the extensometer) was adopted to our DAQ system.

In all, nine specimens in two seasons were loaded in the tensile manner within the temperature range of RT – 900 °C. The temperature in the chamber was increased by steps of 200 °C (later in the range of high temperature by steps of 100 °C for better data collection) and after temperature stabilization the tensile loading was always performed three-times up to 2100 $\mu m.m^{-1}$ indicated by the extensometer. Attempts for the temperature of 1000 °C were discarded by the reason of intensive yielding of the specimen material (nonlinear behaviour).

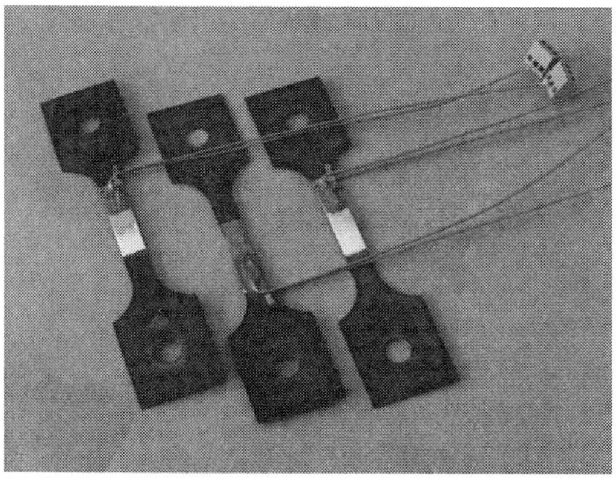

Fig. 5: Specimens with installed strain gauges and the AerOpak thermal insulated cable.

Fig. 6: Tensile loading test of the specimen – shortly after ending and opening the thermal chamber.

The basic presumption for this principle of experiment was that the extensometer as an etalon measured accurately, without slippage. That is why each tensile testing run was checked during evaluation (an example in Fig. 8) and some runs with evident irregularities were discarded.

Values of the strain gauge voltage signals were then plotted in graphs in dependence on the strain given by the extensometer and gradients of these lines were extablished. Their comparisons with gradients for the room temperature condition (i.e. for the basic gauge factor) have given corresponding changes of the gauge factor. It should be noted that the select mean of the experiment is rather difficult for correct evaluation, especially concerning to criterions for selection of linear parts of the loading curves and establishment of their gradients. Only a few specimens, strain gauges respectively, have given excellent results for each temperature conditions (Fig. 9).

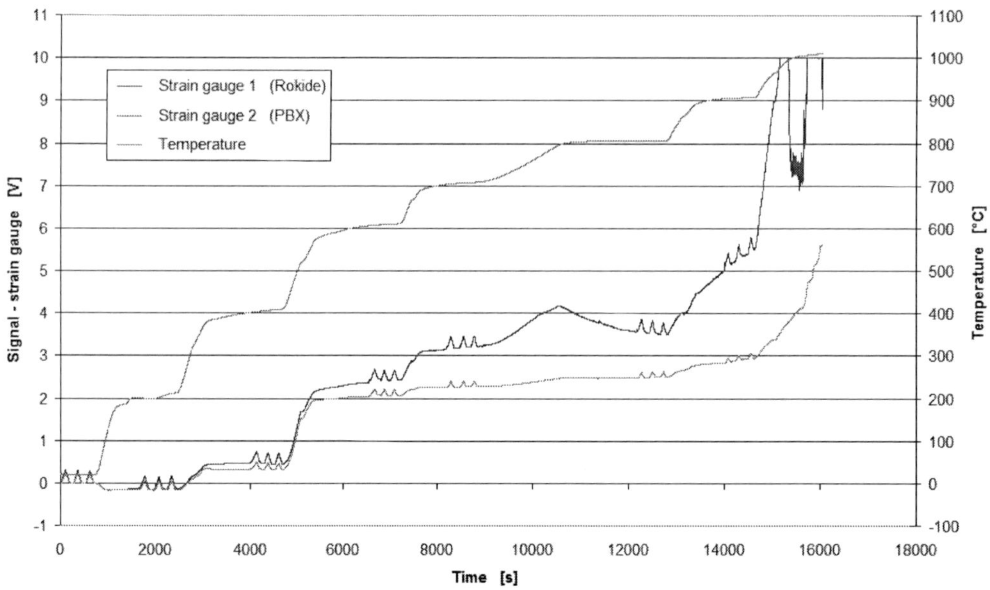

Fig. 7: Test of the specimen no. 3 – time record of the strain gauge voltage signals together with temperature. The huge effect of apparent strain in comparison with tensile loading up to 2100 μm.m^{-1} is evident.

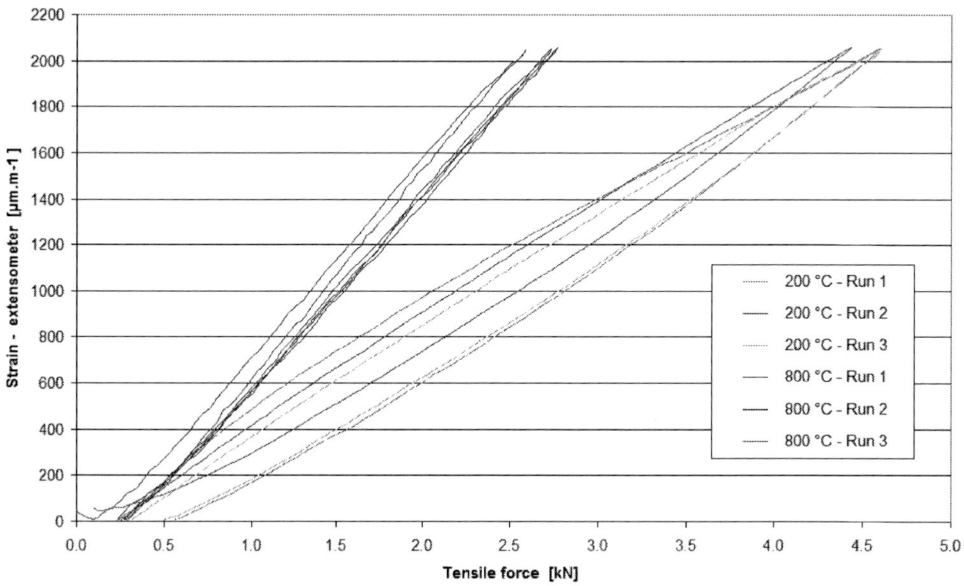

Fig. 8: Examples of tensile loading curves measured during the test of the specimen no. 3 – comparison for condition of 200 °C and 800 °C also documents the change of Haynes 282 superalloy stiffness under high-temperature condition.

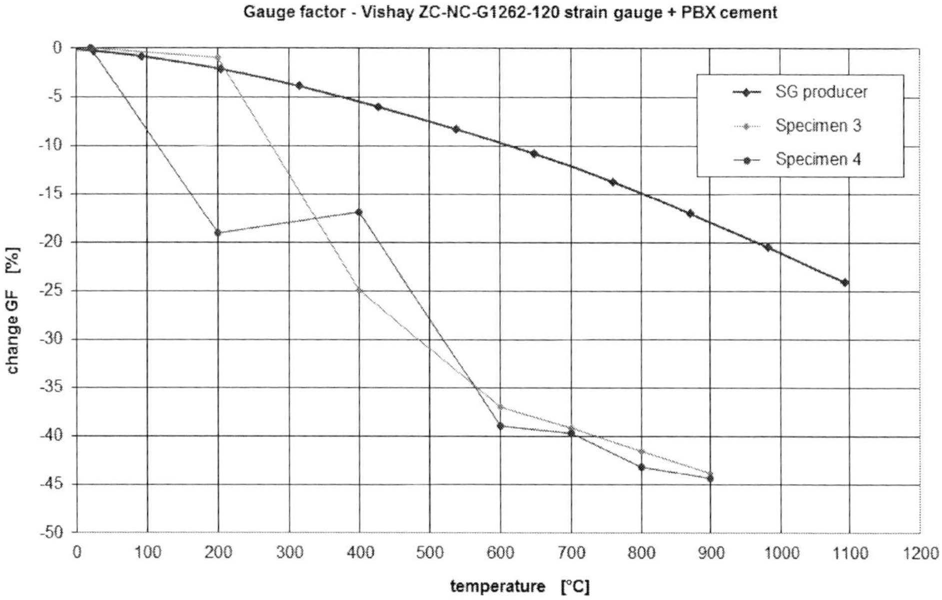

Fig. 9: Comparison of some measured results with producer´s data.

3 Conclusion

The apparent strain of the G1264 strain gauge type with the compensating grid does not exceed the value of 300 $\mu m.m^{-1}$ in all range of temperature up to 550 °C and the curves for correction of a real measurement were obtained. The temperature range above 550 °C is characterized by sharp change of apparent strain and that is why the error of correction can be higher. The results of tensile loading for the establishment of the gauge factor for G1262 tstrain gauge type were rather complicated with appreciable variability, but the most plausible results show a significant swing from gauge producer´s data. Therefore this preliminary experiment points out that more effort should be expended in this field.

Acknowledgement

This work was performed with the financial institutional support from government budget through the Ministry of Industry and Trade of the Czech Republic.

References

[1] J. Cagáň, J. Rosa, F. Rösler, Resistance Strain Gauge Technique for High-Temperature Measurement, in proc.: Experimental Stress Analysis 2014, ed. P. Polach and L. Stuna, Research and Testing Institute Plzeň, Mariánské Lázně, 13–14.

[2] M. M. Lemcoe, Development of High Temperature Strain Gages, Report NASA CR-112241, Columbus, Batelle Columbus Laboratories, 1973

[3] M. M. Lemcoe, Characterization of BCL Strain Gages for Use to 1366 K (2000 F), Report NASA CR-132739, Columbus, Battelle Columbus Laboratories, 1975

Mechanical behavior of thermoplastic rib under loading representing real structure conditions

R. Růžek [1,a], J. Šedek [1], M. Kadlec[1], P. Kucharský [1]

[1] *Aeronautical Research and Test Establishment a.s. (VZLÚ), Strength of Structure Department, Beranových 130, 199 05 Praha – Letňany, Czech Republic*

[a] *ruzek@vzlu.cz*

Abstract: Paper deals with experimental verification of mechanical behavior and strain analysis of an airframe structural part manufactured from thermoplastic composite with tailoring blanks. Investigated structural part is represented by a rib which is embedded into the door structure of the large airliner. The prevailing load mode of the rib is shear. The test rig design comply with real structure boundary conditions. Non contactless strain measurements are compared with numerical simulations. The good match of measurements with numerical simulations was achieved. The results will be applied for the new design of the rib considering maximum weight and costs with saving.

Keywords: Composite, thermoplastic, tailoring blank, contactless measurement, strain analysis.

1 Introduction

The existing composite airframe structures are almost explicitly made out of thermosets. Thermoplastic materials used for high-performance composite structures bring several advantages both in design and manufacturing. The requested performance of structural parts can be achieved using stacking of tailored blanks with combination of thermoforming process. The paper is focused on the first industrial application of composite tailored blanks in the Czech Aerospace industry. The new technology was applied to the rib used in the fuselage door structure.

2 Test specimen and methodology

The AIMS05-09-002 carbon fiber fabric with satin pattern and PPS thermoplastic resin was used for the rib manufacturing. Thickness of the rib differs depending on the layout of fabrics. Nominal thickness of the fabric is 0.31 mm. Three different lay outs were used: a) 18 layers (t = 5,58 mm) $[0/(45/0)_4/0/(45/0)_3/0]$, b) 16 layers (t = 4,96 mm) $[(0/45)_4/45/(0/45)_3/0]$ and c) 11 layers (t = 3,41 m) [0/45/0/45/0/45/0/45/0/45/0]. The rib is shown in Figure 1. The test boundary loading conditions and the test principle were derived from the real part loading transfer in the door airframe structure. The main load is a moment came up from the force applied in the door lock and the stop fitting part. The rib is connected into the internal door frame using bolts. Stiffness and deformation of the circumferential door frame was neglected. The skin influence was taking into account for the test arrangement. The scheme of the designed test jig is shown in Figure 2. The attachment allows horizontal movement of the rib during loading through a special support trolley. The rib was loaded up to failure using MTS uniaxial hydraulic facility with capacity of 1 MN. The quasi static loading was applied. Deformation was measured using contactless digital image correlation system Q-400. Test arrangement is shown in Figure 3.

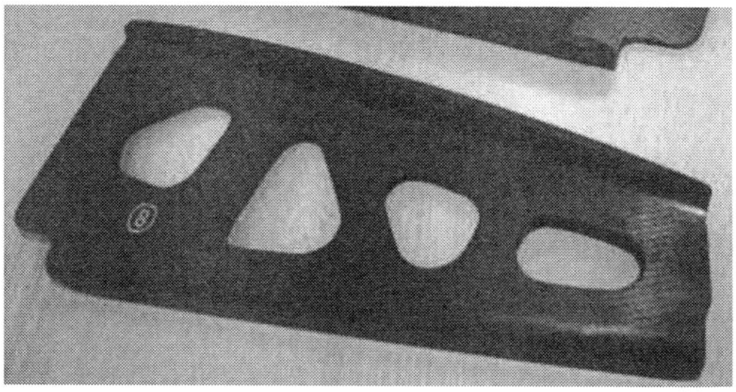

Fig. 1: Thermoplastic composite rib with tailored blanks.

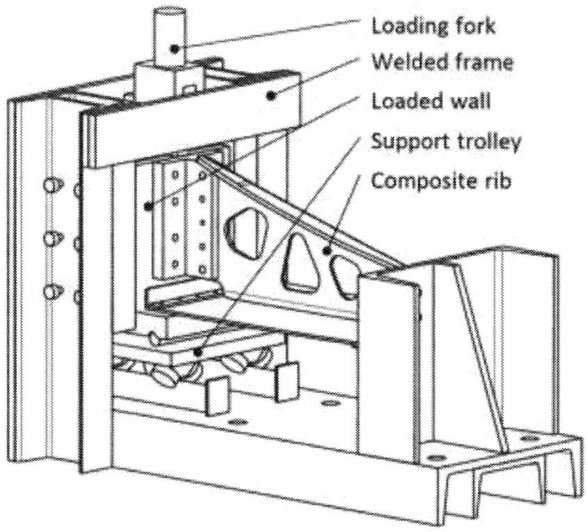

Fig. 2: Scheme of the test jig.

Fig. 3: Test arrangement with cameras of Q-400 image correlation system (left) and results image (right).

3 Results

The load-displacement dependence and relative deformations in three different axes were measured. Example of the deformation measured using Q-400 digital image correlation system in horizontal direction at 36 kN is shown in Figure 4. The relative shear deformation in the bottom section of the web along the x (horizontally) direction defined by experiment and numerical analysis is shown in Figure 5. Load vs. displacement curve is shown in Figure 6. The initial nonlinearity in the force vs. deformation curve is probably caused by assembly clearances. Then the linear loading curve up to 90 kN was measured. The second nonlinearity occurred due to stiffness changes and bearing strength when the loading force was approaching to failure.

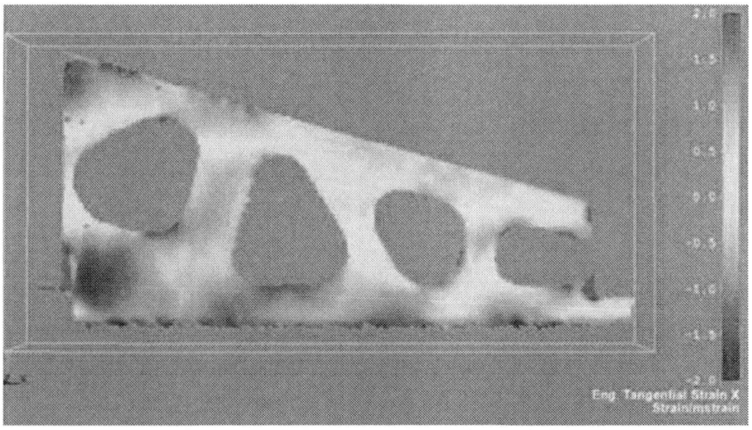

Fig. 4: Relative deformation in horizontal direction.

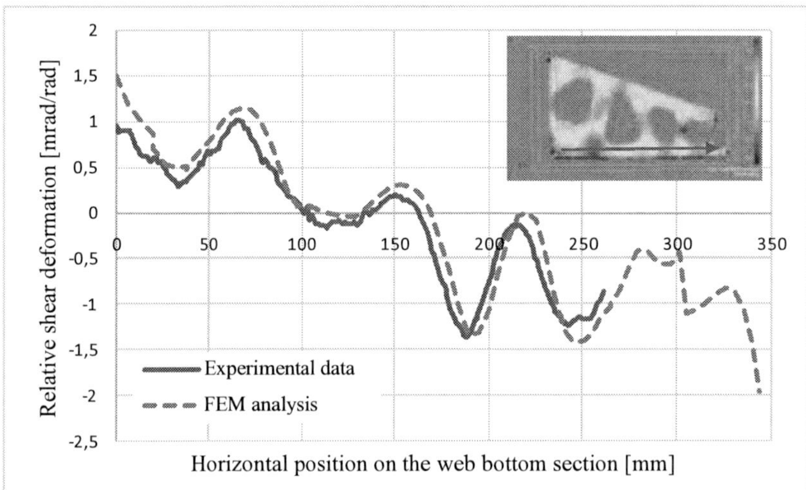

Fig. 5: Shear deformation in the web – comparison of experimental and FEM analyses.

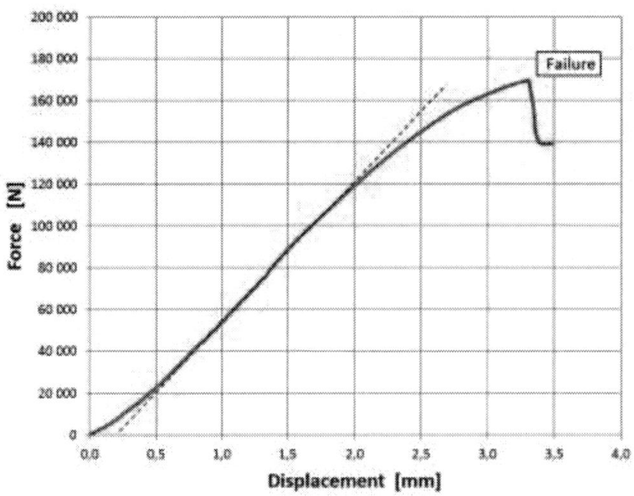

Fig. 6: Load vs. displacement curve.

The maximum force of 170 kN was achieved. The failure initiated in the end bolts fastening the rib to the surrounding structure. The failure initiated in good agreement with the prediction. Both experimentally measurements and numerical analysis confirm highest deformation in the failure area. Overview of the rib after failure is shown in Figure 7 (left) together with detailed view of damaged holes for bolts (right).

Critical areas in light of failure initiation are bolt holes located at the longer vertical part (web) of the rib. The holes placed in the upper web part are loaded in the direction out of the rib. The forces in upper part of the web invoke tensile bearing stresses in holes. The bottom holes near the skin are loaded by forces which are directed inside of the rib. These forces invoke compressive bearing stresses. Failure initiated from holes placed in the upper web part owing to lower capacity of joint in the tensile out of the rib direction.

Experimental strength of the rib was achieved multiply higher then it was required although the numerical optimization of the rib were carried out before the test. Strength optimization of the rib is the next work objective.

Fig. 7: The rib after failure.

4 Conclusion

The paper discusses experimental demonstration of mechanical behavior of the thermoplastic rib made of tailored blanks. The shear and deformation characteristics was evaluated and the good conformity with numerical predictions was achieved. The achieved strength of the rib was multiple higher then requirement. This shows evidence of the better utilization possibility of tailoring blank and the strength characteristics of the material. Therefore the next work will be focused on the strength optimization with the aim of significant weight saving.

Acknowledgement

This work was funded by the support of the Technology Agency of the Czech Republic, project no. TA03010209.

Trabecular structures of dental implants formed by 3D printing and their mechanical properties

L. Řehounek [1,a], F. Denk [1], A. Jíra [1]

[1] *Czech Technical University in Prague, Faculty of Civil Engineering, Thákurova 7, Prague*
[a] *lubos.rehounek@fsv.cvut.cz*

Abstract: This paper proposes an overview of the trabecular Ti-6Al-4V 3D-printed titanium alloy used for dental implants. Series of micro and macro mechanical tests were made to help better understand the trabecular structure and the mechanical behavior of the 3D-printed titanium alloy. The work also includes an overview of the current state of knowledge in this area as well as results obtained by experiments. Indentation tests, compression and tensile tests have been performed to obtain the values of reduced modulus and hardness.

Keywords: trabecular; 3D printing; dental implants; porous, nanoindentation.

1 Introduction

The advancement in the technology of 3D printing makes it possible to create complex geometric structures, which form the bearing scaffold for the ingrowth of bone cells into the implant [1,2]. This interconnection is beneficial for long term stability at the implantation site [3] and also helps to smoothen the material transition region, where unwanted stresses often cause large deformations, leading to implant loosening. This work is dedicated to the basic mechanical properties of the trabecular structure of the dental implant. In cooperation with ProSpon spol. s. r. o., we were able to create the structure using the M2 Cusing Concept Laser machine with a specialized input media for dental implants Rematitan Cl of the Ti-6Al-4V chemical composition.

2 3D printing technology

The first step towards creating a 3D-printed product is creating a 3D-model. This part is usually done using a computer-aided design (CAD) environment. Upon its completion, an STL model file is divided into thin cross sections [1,2] and sent to the 3D printer to be processed. Up to this point, the process is similar to the common layer-by-layer 3D-printing of plastic.

What differentiates the process of printing metals from the standard technology is using a laser beam to melt down a layer of metal powder, such as Rematitan Cl. During each cycle, the coater applies a thin layer of powder, which is processed by a laser at a pre-set melting point in a pre-determined order [4]. This process solidifies the loose powder into a 3D-layered object, such as our specimens shown on Fig.1.

a) compression test specimen

b) tensile test specimen

Fig. 1: Trabecular structure specimens for global mechanical tests.

3 Methodology

3.1 Nanoindentation

To investigate the micromechanical properties of the alloy, nanoindentation tests were made considering reduced modulus of elasticity, hardness and contact depth. The micromechanical analysis was performed using the CSM Instruments nanoindenter in the mode of directed force and repeated loading. The load program was set with consideration of eleminating surface tension and shear stifness in the atomic material structure (Tab. 1). A typical indentation curve from the cross section the trabecular structure is shown on Fig 2.

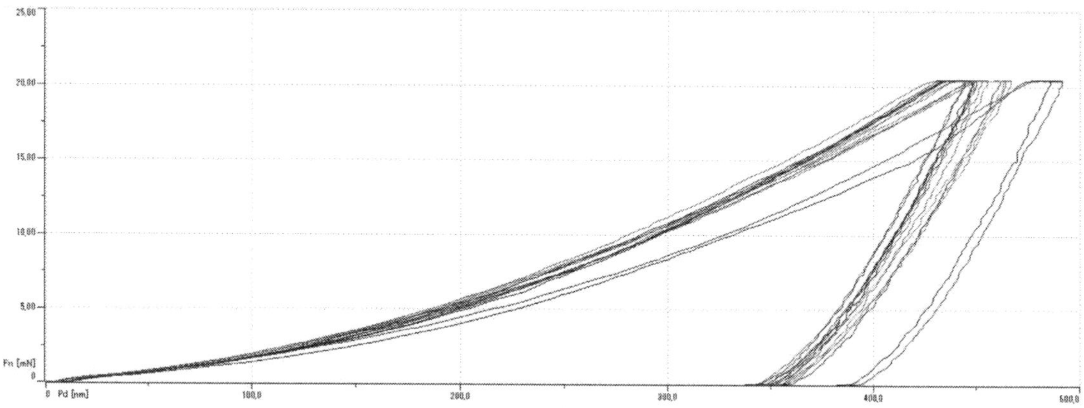

Fig. 2: A typical indentation curve of the trabecular alloy, showing the relation of the loading force F_n and the deformation of the specimen P_d.

Tab. 1: Mean values of reduced modulus E_r, hardness H_{it} and contact depth H_c in individual load cycles.

Physical property	Load cycle						
	1	2	3	4	5	6	7
H_{it} [MPa]	5190.9	5201.4	5187.8	5165.9	5190.8	5202	5190.7
Std. Dev.	251	200	281	343	379	404	395
E_r [GPa]	131	122.4	118.9	118	116.8	117.1	118.1
Std. Dev.	10	6	4	3	3	3	4
H_c [nm]	266.7	381	470.5	543.8	607.5	659.7	732
Std.dev.	7	8	13	22	25	31	45

3.2 Global mechanical tests

To investigate the global mechanical properties, we conducted compression and tensile tests. For this purpose, we used the 3D Dode-Thick [MSG] structures with dimensions of 14x14x14 mm (a cube for the compression test) and a 14x14x42 mm (a block for the tensile test). The tensile test specimen had a 14 mm trabecular middle section and end portions of homogeneous volume for ensuring a better anchor in the MTS Alliance RT-30 machine [5]. The results show an ultimate compressive strength of 31.30 MPa and an ultimate tensile strength of 32.80 MPa. Diagrams showing the stress-strain relations of the specimens obtained by tensile and compression tests are shown on Fig. 3 and Fig.4, respectively.

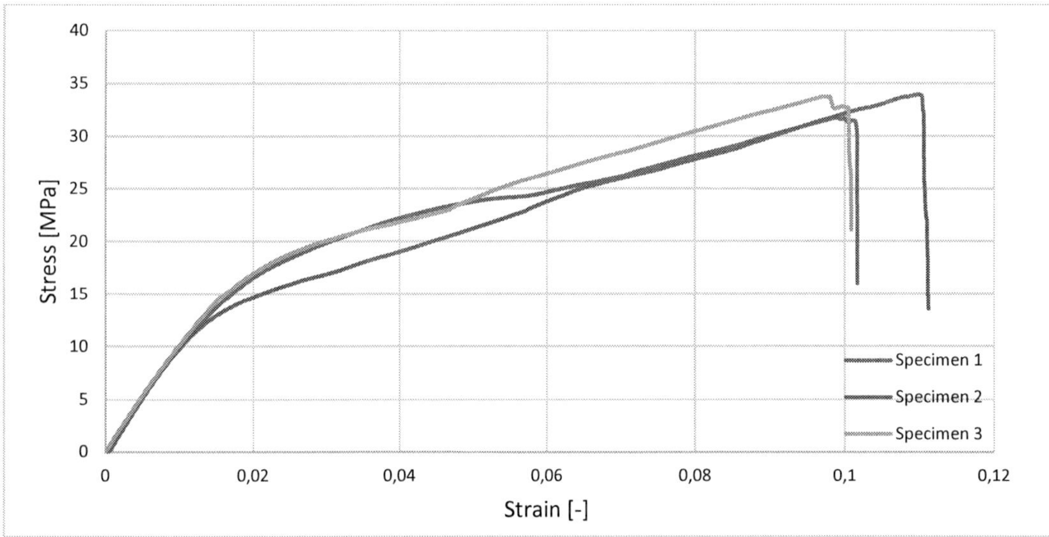

Fig. 3: Stress-strain relation diagram of the tensile test specimens

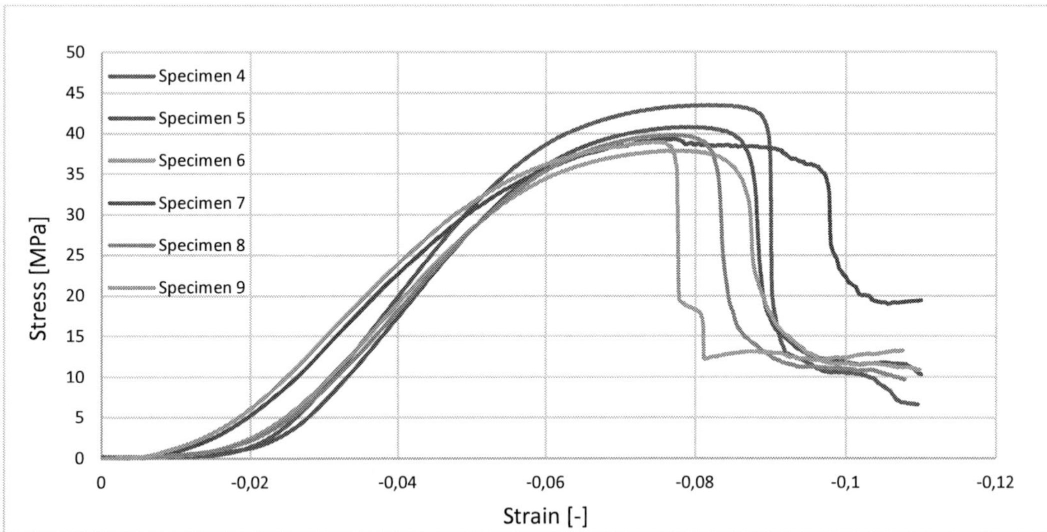

Fig. 4: Stress-strain relation diagram of the compression test specimens

From the linear part of the load curves of the diagrams shown above, we were able to calculate Young's modulus of the whole trabecular specimen (Tab.2). These values represent properties of the whole specimen, contrary to the nanoindentation experiment, where obtained values represent only the material characteristics at the micro level.

Tab.2.: Values of Young's modulus obtained by global mechanical analysis

	Young's modulus E [MPa]									
Specimen	1	2	3	4	5	6	7	8	9	Mean
Tensile test	964.66	975.91	982.20	-	-	-	-	-	-	**974.26**
Compression test	-	-	-	1043.2	1080.6	947.15	879.38	999.58	878.84	**971.46**

4 Conclusion

The micromechanical analysis has proved the properties to be dependent on the depth of the indent (size of the loading force). From a contact depth of $H_c \sim 470$ nm (corresponding load force of 30 mN), the trend of E_r and H_{it} is constant and therefore we can assume the loading force of 30 mN as a basic value. The corresponding properties of this force are reduced modulus $E_r = 118$ GPa and hardness $H_{it} = 5.187$ GPa. For comparison, the mean value of reduced modulus E_r of the stems of the commonly used implants is ~ 118 GPa and their hardness H_{it} is ~ 4.580 GPa [6]. According to the experiment and the obtained values of the micromechanical properties, we can therefore assume that the 3D printed material has very similar behaviour when compared to implants commonly used today.

The macromechanical tests have been conducted as a pilot experiment and cannot be therefore compared with other authors. However, an evaluation based on the experiment is at hand. The tensile tests have shown a bilinear stress-strain diagram with elongation at fracture ranging from 9 to 10 % in all tested specimens, which corresponds with the values given by the manufacturer. The compression tests have proven to be more diverse due to the yet unclear nature of the deformation of the trabecular structure. The elongation at fracture ranges from 7 to 10 %. The mean Young's modulus E of the trabecular specimens calculated from the linear part of the loading curves is $E_{ten} = 974.26$ MPa for the tensilte test and $E_{comp} = 971.46$ MPa.

As predicted, the analysis showed that incorporating the trabecular structure into the body of the implant has great effect on lowering the Young's modulus E of the whole implant. As opposed to regularly produced Ti-6Al-4V implants, where values of E are approximately 110 GPa, the values of E of the trabecular implant are below 1 GPa. This is a very significant reduction and the results are still to be compared with other experiments. Explaining the true mechanical behavior of the trabecular structure also requires a more detailed material model, which is the subject of further study.

Acknowledgment

The financial support by the Technology Agency of the Czech Republic (TAČR project no. TA03010886) and Faculty of Civil Engineering, Czech Technical University in Prague (SGS project No. SGS14/122/OHK1/2T/11) is gratefully acknowledged.

References

[1] F. E. Wiria et al., Printing of Titanium implant prototype, Materials & Design 31 (2010) 101–105, http://dx.doi.org/10.1016/j.matdes.2009.12.050

[2] El-Hajje, Aouni et al., Physical and mechanical characterisation of 3D-printed porous titanium for biomedical applications, J Mater Sci. 25 (2014) 2471–2480, http://dx.doi.org/ 10.1007/s10856-014-5277-2.

[3] B. Dabrowski et al., Highly porous titanium scaffolds for orthopaedic applications, J Biomed Mater Res 95B (2010) 53–61, http://dx.doi.org/ 10.1002/jbm.b.31682

[4] T. Hartmann-H'Lawatscheck, Metal Laser Melting, Laser Technik Journal 12 (2015) 41–43, http://dx.doi.org/ 10.1002/latj.201500027

[5] P. Padevět et al., Determination of the Failure on the Concrete Runway, Advanced Materials Research, 969 (2014) 109-112, http:// 10.4028/www.scientific.net/AMR.969.109

[6] A. Jíra, F. Denk, Micomechanical analysis of dental implants and their surface modification, App. mech and mat. 827 (2016) 367-370, http://10.4028/www.scientific.net/AMM.827.367

Optimization Parameters of Plaster Composites

A. Samková[1,a], P. Kulhavý [2], M. Pechočiaková[1]

[1] *Technical University of Liberec, Faculty of Textile Engineering, Czech Republic*

[2]*Technical University of Liberec, Faculty of Mechanical Engineering, Czech Republic*

[a]*Alzbeta.Samkova@tul.cz*

Abstract: This study is aimed on comparison mechanical properties changes of plaster composites reinforced by different kinds of fibrous dispersion. The trend in recent years is the substitution of conventional materials with composite materials. The reason for their using is to achieve the same (or better) properties at much lower weight. Application of these materials is constantly expanding to new sectors. Glass and basalt microfibers of a certain length were added as a reinforcing element into plasters. Another way how to reinforce plaster materials is use of polyvinyl butyral nanofibers. As the last kind of plaster reinforcement was selected leno weave of basalt. This way produced samples were tested for tensile, flexural, compressive and impact strength. Most improvements resulted in bending stresses when reinforced with glass. The results showed possible improvement in properties at reinforcing microfibers. When used basalt leno weave as reinforcement element showed the greatest improvement in the compressive strength. Using the nanofiber reinforcement led generally to a deterioration of their mechanical properties. The greatest deterioration occurred in bending stress.

Keywords: plaster composites; ThermoUM Xtra; basalt microfiber; glass microfiber; polyvinyl butyral nanofiber.

1. Introduction

A standing trend in the field of material engineering is development of materials with tailored properties allowing us their effective use. The composite materials meet this requirement and their principle is based on addition of some dispersion components into the elemental matrix. In this way we can get better material properties for particular applications [1]. The disadvantage of plaster materials is their gradual degradation, which is generated by moisture. Due to this problem physical and chemical processes could occur and they undermine the cohesive properties between the plaster and masonry, also some visual defects can be noticeable. To limit the occurrence of these defects and reduce brittleness of the plaster the fibrous reinforcement is added to the base matrix, which leads to the load transfer between the fibres and the matrix [2]. For this process the fibres length is an important factor. This so called critical length must be capable of transferring a tension caused by load transfer from the matrix through a shear stress which acts on an interface the fiber - matrix. By using the short fibres we can improve the properties in all directions of the composite. Therefore, the short glass and basalt microfibers were used in experiments [3, 2]. For the integrity of the information about all standardly used reinforcement materials also leno weave of basalt, which represents a bi-directional reinforcement, was used.

2. Materials

Variety of fiber reinforcement are nowadays used for plaster composites. The most frequently used are fibers of glass, carbon, aramid, ceramic, basalt and polypropylene. When selecting was given emphasis on the mechanical properties, the ability to withstand alkaline environment which in the plaster is, their environmental friendliness, and thermal insulating properties. Glass, basalt and carbon fiber have been chosen for the experiment after considering their individual properties. Due to the good toughness and adhesion to various surfaces for the electrostatic spinning method, the polyvinyl butyral fibers were selected. For plaster composites with micro fiber reinforcement to compute the critical fiber length:

$$l_c = \frac{R_{mV} \cdot d}{2\tau_m} \tag{1}$$

Where l_c [mm] is the critical fiber length, R_{mV} [MPa] fiber strength, d [mm] is the diameter of the fibers and τ_m [MPa] strength of the interface between the fibers and the matrix in shear.

The critical fiber length is the minimum length of fiber which is still capable of transmitting tension. This length can be achieved only with difficulty, due to the very small size. To calculate the effective length (2) of the fibers was determined by the formula [2] by W. D. Callister:

$$l = 15 \cdot l_c \tag{2}$$

These formulas are for composites with polymeric matrix. Due to the fact that in the experiment was used a different matrix, is needed for calculating the effective length adjusted so employed relationship shape factor (aspect ratio), which was calculated by the real length of the fiber reinforcement:

$$\propto = \frac{l}{d} \tag{3}$$

The final length of fiber reinforcement l_v [mm] was determined as the effective length multiplied by aspect ratio α:

$$l_v = l \cdot \alpha \tag{4}$$

The final length of the fibrous reinforcement for the glass fibers is 6.7 mm and 7.3 mm in the case of basalt fibers. The content of the fibrous reinforcement in concrete materials is usually 6%. Blending the plaster and fibers with thus high content causing problems in their bonding. For this reason the content of the fibrous reinforcement has been set to 1%, 2% and 3%. The calculation (5) of the mixing ratio [1] is based on the percentage of the components in the composite and their densities:

$$w_i = \frac{V_i \cdot \rho_i}{\sum V_i \rho_i} \tag{5}$$

Where w_i [%] is the mass fraction of the i-th component, V_i [%] the volume fraction of the i-th component and ρ_i [kg/m^3] is the density of the i-th component.

Reinforcement in the form of nanofibers was created in two ways. The first one was based on a winding a wispy smoke rising from the electrodes on the rotary drum. That led to producing of nanofibrous layer, which was inserted into test specimens. The second method based on a mixing of the nanofibrous wispy smoke directly with the plaster was used. After blend creating was necessary to obtain individual fibers and whole mixture was grinded [4]. For the last method was used leno weave of basalt 5x5 mm with a bulk density 125 g m^{-2}. Test specimens for all types of reinforcement were made according to standard DIN EN 1015-11 Methods of test for mortar for masonry.

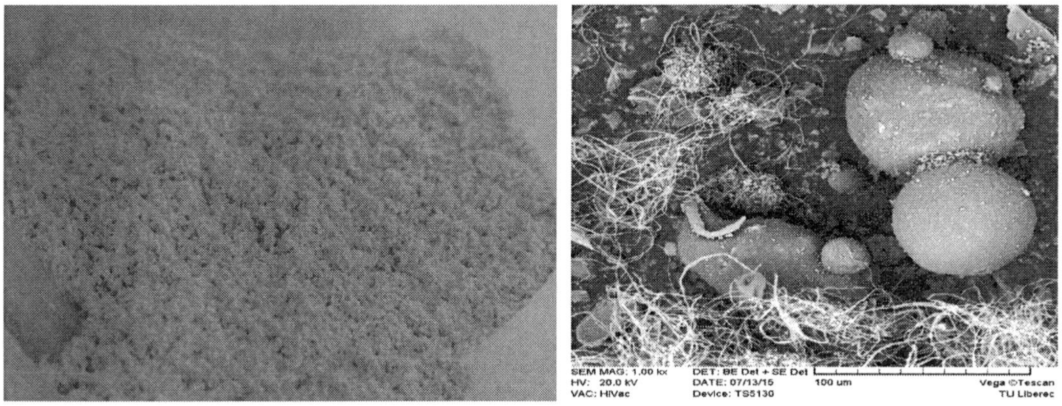

Fig. 1: Reinforcement in the form of nanofiber (photo, scanning electron microscopy)

3. Experiment

With using the standard laboratory tests for bending, compression and impact strength of the samples, it was necessary to find the changes of their mechanical properties.

Before creation of the test objects was necessary to measure the strength of the interface between the fibers and the matrix in shear. The test samples containing the bundle of fibers (glass, basalt). The individual bundles were placed into the plaster at different depths and subsequently measured on the dynamometer (Fig. 2). The results were used to calculate the aforementioned fiber critical length.

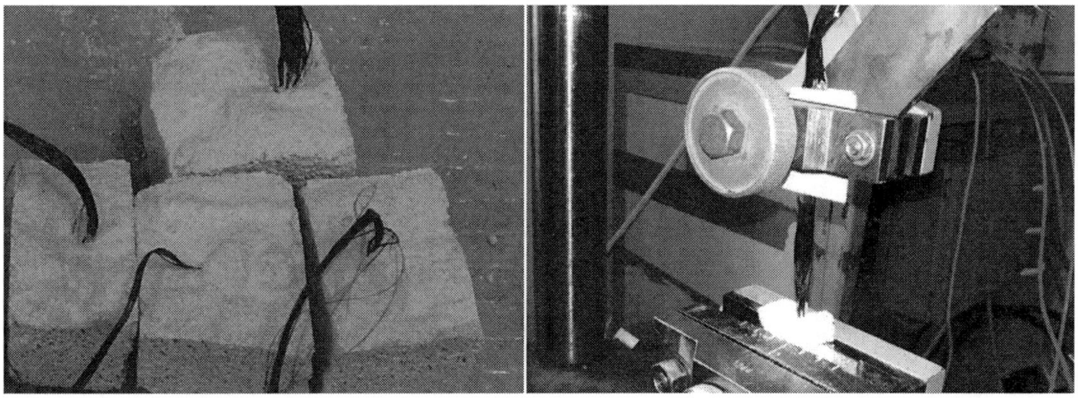

Fig. 2: Critical length of fiber and testing strength of the interface between the fibers and the matrix

The flexural strength was tested by three-point bending. The test was carried out in accordance with standard [5]. The compressive strength was tested on a dynamometer. The diameter of the compression plates was 75 mm. In both of these tests took place the burdening rate of 2 mm min^{-1} so a breach of solids occurred in the interval from 30 to 90 sec. For the method was used the standard ČSN EN 1015-11 [6]. These basic characteristics were extended to impact resistance. The energy necessary to breach test piece was measured with a Charpy hammer [7]. Based on the collected data the impact toughness was calculated (6), which represents the ability of a material to quickly absorb energy.

$$A_{\omega} = \frac{E_c}{b \cdot h} \tag{6}$$

Where A_{ω} [J/mm^2] impact toughness, E_c [J] strain energy consumed to shift the sample, b [mm] the sample width and h [mm] height of the sample.

Breach of the samples occurred with only a very small deflection. This is caused by the fact that some equilibrium processes in the material have not enough time for realization. For the method was used the standard DIN EN ISO 179-2 Plastics [7]. These basic characteristics were extended to impact resistance.

4. Result and discussion

These basic characteristics were extended to impact resistance. The values of these properties were evaluated and compared with all types of reinforcement. Results of these parameters are indicators of the lifetime of the plaster, which can be influenced by the addition of fibrous reinforcement. We found that the use of nanofibrous reinforcement in plaster composites not improve the above mentioned properties. The reason are nanofibres which are not lengthen. From figure 3 it is seen that the bending stress characteristics improving effect showed 1% content of fibrous reinforcement in both fibrous types. To increase the elastic modulus was only glass mat at 1% ratio. Other ratios exhibit a deterioration of the modulus of elasticity.

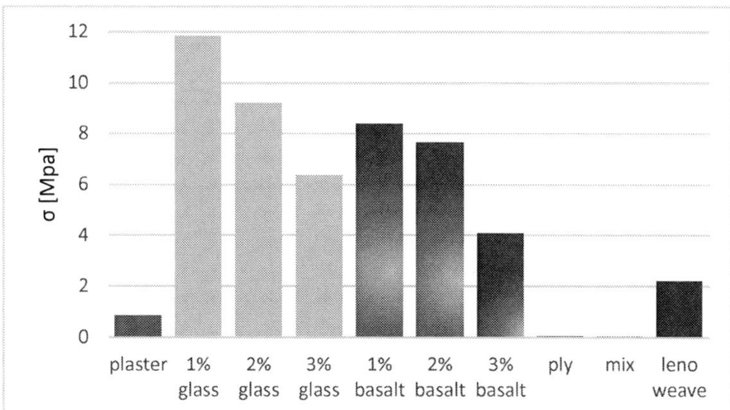

Fig. 3: Bending stress

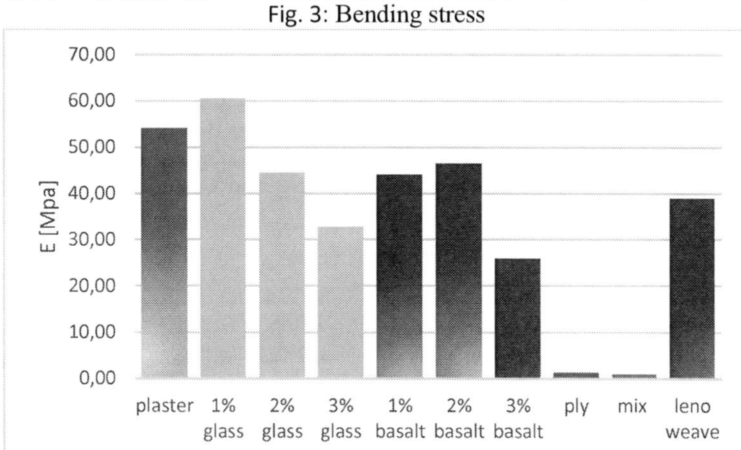

Fig. 4: Modulus of elasticity

The increasing content of fibers leads to increasing the impact resistance (Fig. 4). The greatest reinforcing effect was observed in basalt leno weave, which also prevented the destruction of the samples during the tests. Similar effect with less effect was also reflected by the sample reinforced by nanofiber ply. The best improvements in compressive strength is clearly demonstrated in samples of basalt leno weave. In Figure 6 it can be seen that the more increase the content of fibrous reinforcement in the test body, the more the compressive strength is decreasing.

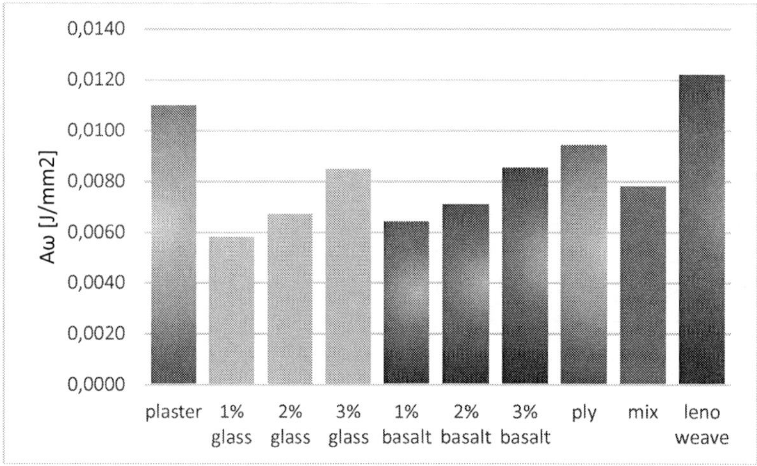

Fig. 5: Impact resistance

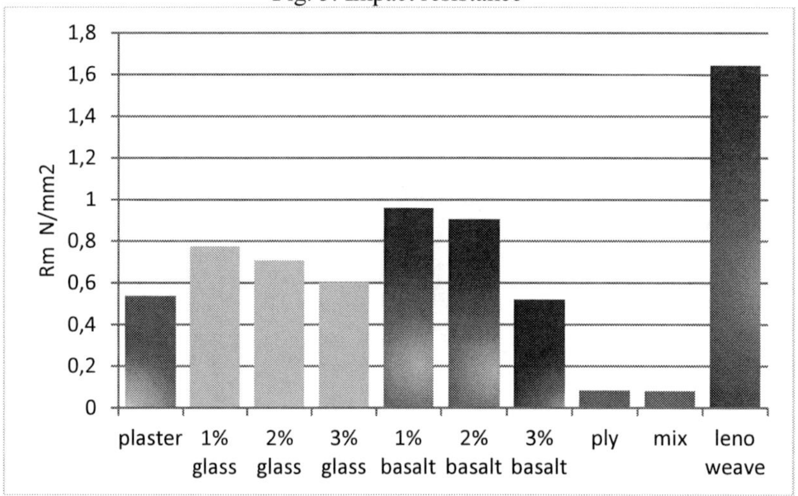

Fig. 6: Compressive strength

5. Conclusion

In this publication several sorts of fiber reinforcements and their effect on improvement of mechanical properties have been compared. For comparison the properties of the actual plaster and the reinforced plaster composites the compressive strength, bending stress, modulus of elastic and impact resistance were selected. The measuring results show that suitable fibrous reinforcement could improve all of the observed properties. Using the nanofiber reinforcement has proved as unsuitable despite the fact that the results of the impact strength is better than of basalt or glass reinforcement. To improve resistance against mechanical stress plaster is suitable basalt or glass reinforcement in the form of the leno weave or 1% by weight of the fibrous constituents.

Acknowledgments

This publication was written at the Technical University of Liberec as part of the project Options reinforcement plaster using fibrous materials, 21160 with the support of the Specific University Research Grant, as provided by the Ministry of Education, Youth and Sports of the Czech Republic in the year 2016.

References

[1] Tomková, B.: *Kompozity s textilní výztuží* [lectures on composites with textile reinforcement], TU v Liberci, 2014

[2] Calliste, W.: *Materials Science and Engineering*, John Wiley a Sons, Inc, 2007Norma DIN EN ISO 14125 *Fibre reinforced plastic composites*, Praha: Český normalizační institut, 2000

[3] Kryšková, P.: *Vliv vlákenné výztuže na mechanické vlastnosti omítkových kompozitních materiálů*, Master thesis, TU Liberec, 2014

[4] Samková, A. a spol: *Fibre reinforcement effect on plaster composite properties,* Conference Proceeding, Vol. 189, NART 2015

[5] Norma DIN EN ISO 14125 *Fibre reinforced plastic composites*, Praha: Český normalizační institut, 2000

[6] Norma ČSN EN 1015-11. *Methods of test for mortar for masonry - part 11*: Praha: Český normalizační institut, 2000

[7] Norma ČSN EN ISO 179-2 *Plastics - Determination of Charpy impact properties* - Part 2: *Instrumented impact test.*

An effect of various shielding gases on material properties of laser welded steel sheets

J.Sís [1,a], I. Černý [2,b], B. Kyselá [3,c] J. Kec [4,d] K. Kolařík [5,e], N. Ganev [6,f]

[1, 2,3,4] SVÚM a.s., Strength Department, Tovární 2053, 250 88 Čelákovice, Czech Republic
[5,6] CTU in Prague, Department of Solid State Engineering, Faculty of Nuclear Sciences and Physical Engineering, Trojanova 13, 120 00 Prague 2, Czech Republic

[a] sis@svum.cz, [b] ivo.cerny@seznam.cz, [c] kysela@svum.cz, [d] kec.jan@seznam.cz,
[e] kamil.kolarik@email.cz, [f] nikolaj.ganev@fjfi.cvut.cz

Abstract: The aim of this work was to compare mechanical properties of steel sheets made of S355J2 steel welded by diode laser with using six different shielding gases / mixtures of gases Ar, He, C12X2, He3H1, C6X1 and He50 were used as shielding gases. Residual stresses, microstructure, hardness and microhardness, notch toughness and fatigue resistance of weld joints have been tested and evaluated.

Keywords: fatigue, residual stress, laser welding, shielding gas.

1 Introduction

The welding of structural steels is one of major joining processes of many industrial areas. Using laser technologies in welding is one of many options how to achieve required results. Every technology has its own possibilities, pros and cons. The properties of laser welded joints were tested in last decades but still there are many questions. Main properties such as tensile strength, notch toughness, bending fragility etc. are more or less known and standardized. For many applications the knowledge of these properties is not enough. Fatigue properties of some types of joints made by specific conditions were tested and published [1-4]. This paper focused on testing various mechanical properties of laser welded joints where different shielding gases have been used. Except of this experimental program the residual stresses analyses were performed because of their direct connection to fatigue resistance of welded joints. The redistribution of residual stresses in weld joint [5] significantly affects the areas of fatigue cracks initiation. Shielding gas significantly affects melting and cooling rates and afterwards residual stresses and weld design. From theoretical point of view the effect of shielding gas on static and dynamic properties should be expected.

2 Experimental

In this work the effect of various shielding gases on material properties of laser welded joints prepared by high-power diode laser (HPDL) has been investigated. Power of the HPDL was set to 3,2 kW, the welding speed was 0,5 m/min and flow rate of shielding gas 15 l / min (equal rate for all types of shielding gases). The steel sheets with 4,6 mm thickness made of S355MC steel have been used as experimental material. A description of used shielding gases is in Table 1.

The XRD measurements of residual stresses were performed on PROTO iXRD COMBO diffractometer in ω-goniometer or iso-inclination mode with CrKα radiation. The line {211} of α-Fe phase was measured with the interplanar lattice spacings computed from maxima of Pearson VII functions fitted to $CrK\alpha_1$ profiles after $CrK\alpha_2$ stripping carried out by Rachinger method. The stresses were computed presuming biaxial state of residual stresses (RS) using Winholtz-Cohen method and X-ray elastic constants $\frac{1}{2} s_2 = 5.76\ TPa^{-1}$, $-s_1 = 1.25\ TPa^{-1}$.

Table 1: A description of used shielding gases

Shielding gas	Composition
Ar	argon 5.0
He	helium 4.6
C12X2	12 % carbon dioxide, 2 % oxygen, 86 % argon
He3H1	3 % helium, 1 % hydrogen, 96 % argon
C6X1	6 % carbon dioxide, 1 % oxygen, 93 % argon
He50	50 % helium, 50 % argon

The metallographic analyses and mechanical tests were performed in SVUM laboratories in compliance with international standards for weld joints. The high cycle fatigue tests were performed at high frequency testing frame Amsler HFP45 up to 10^7 cycles with loading asymmetry $R = 0$ (pure tensile). The fractographic analysis using SEM has been used for evaluation of fracture surfaces and areas of fatigue cracks initiations.

3 Results and conclusions

Analyses of residual stresses of all sheets were performed. C12X2 mixture as shielding gas shows most promising results. On the other hand, the worst results of all tested shielding gases were found in case of using He3H1 mixture.

Metallographic analysis of macro- and microstructure was performed at specimens from all sheets. In general, there are no significant differences in microstructure. Fig. 1 – 4 show typical microstructure in weld metal, weld metal-heat affected zone borderline, heat affected zone (HAZ) and base metal on sheet where argon has been used as a shielding gas. The microstructure at other sheets is very similar.

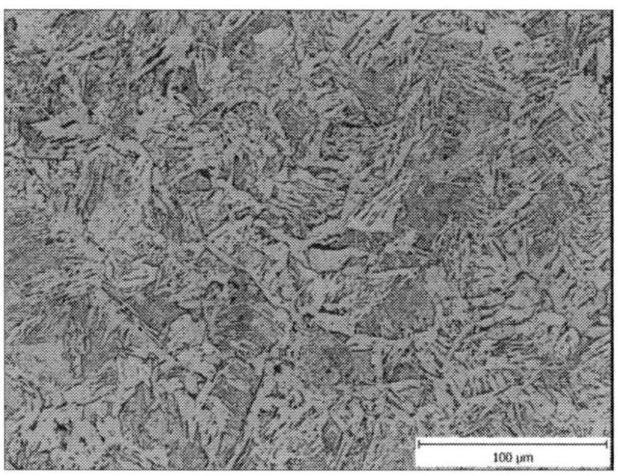

Fig. 1 The microstructure in weld metal

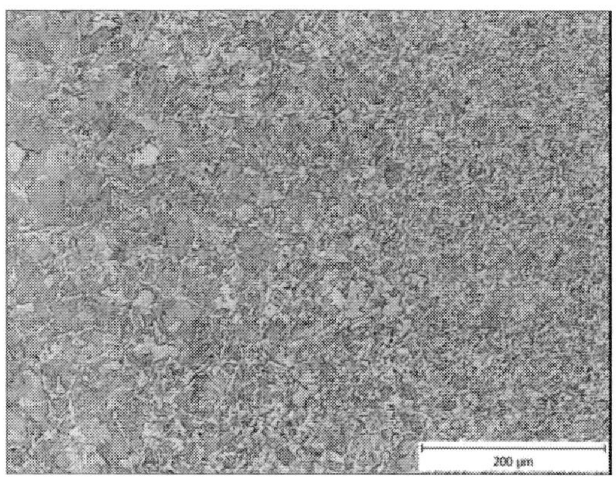

Fig. 2 The microstructure at weld metal-HAZ borderline

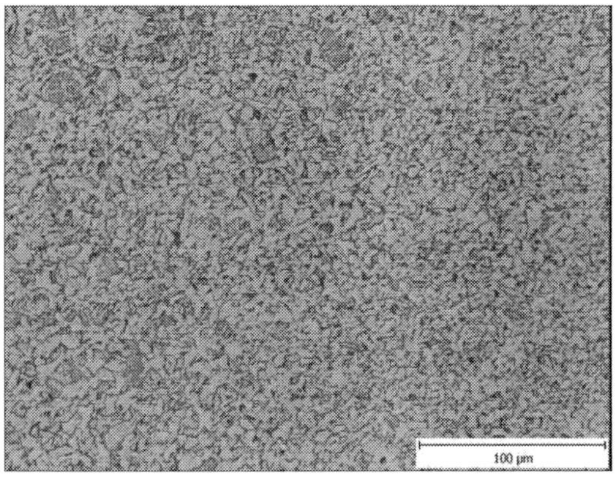

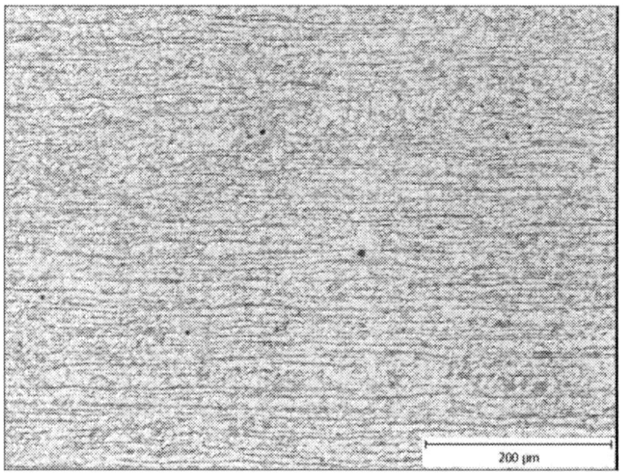

Fig. 3 The microstructure in HAZ

Fig. 4 The microstructure in base metal

The macrostructure analysis showed that all weld joints are suitable from defects presence point of view. Using of different shielding gases can affect cooling rate of welding process. This rate differences leads to different dimensions (widths) of weld metal and HAZ. The results of widths measurement are summarized in Table 2. The largest weld joint was found at sheet welded with using 0,9 mm laser beam wide and argon shielding gas. On the other hand using C12X2 mixture as shielding gas leads to smallest weld joint. In general, minimal heat influencing of material by welding is required so C12X2 is the best option from this point of view.

Table 2: Results of dimension measurements

Shielding gas	**Width of weld metal [mm]**	**Width of HAZ [mm]**	
C12X2	2,55 – 2,70	1,82 – 1,86	1,45 – 1,58
He3H1	2,47 – 3,29	1,67 – 2,06	1,75 – 2,02
C6X1	2,62 – 3,33	1,83 – 2,02	1,74 – 2,21
He50	2,50 – 3,04	1,52 – 1,76	1,87 – 2,14
He	2,87 – 3,08	1,85 – 1,91	1,85 – 1,89
Ar (0,9 mm laser beam width)	3,33 – 3,60	1,89 – 2,05	1,79 – 2,01
Ar (1,2 mm laser beam width)	2,52 – 3,17	1,74 – 2,07	1,74 – 2,09

The specimens used for metallographic analysis were also used for hardness profile measurement HV10. For S355 steel the maximal hardness 380HV is allowed. Our results showed that the effect of using different shielding gases on hardness in weld metal or HAZ is minor – see Table 3.

Table 3: Maximal values of HV10 hardness in weld joints

Shielding gas	**C12X2**	**He3H1**	**C6X1**	**He50**	**He**	**Ar (0,9 mm laser beam width)**	**Ar (1,2 mm laser beam width)**
max HV	190	196	201	197	188	187	185

At the same specimens the microhardness HV0,5 profile measurement in axis of weld metal was performed. The area of microhardness measurement is indicated at Fig. 5, the results of measurement are summarized at Fig. 6. A significant difference could be seen only in case of sheet with He50 shielding gas. The results from other sheets/shielding gases are very similar.

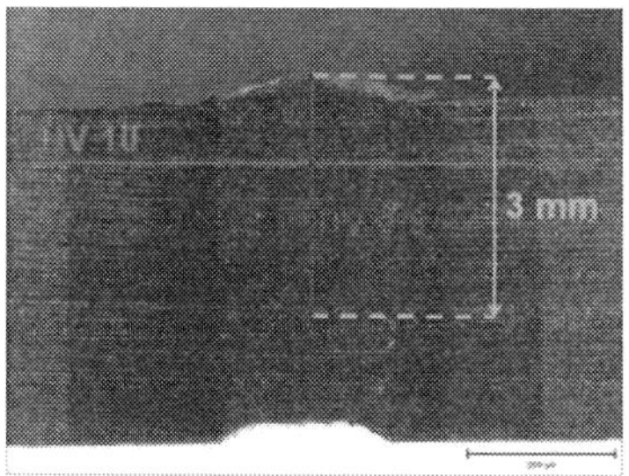

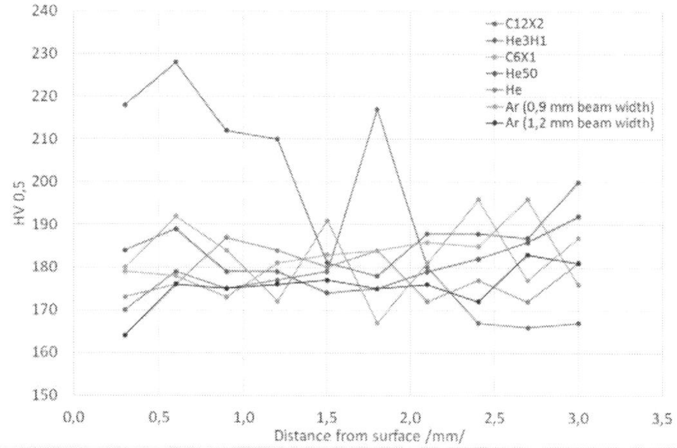

Fig. 5 and 6 Summary of microhardness HV0,5 profile measurement

The results of Charpy impact tests are summarized in Fig. 7. The results show that using argon as shielding gas leads to highest values of absorbed energy / notch toughness. The specific of argon using is also very significant when we investigate fracture surface and evaluate the ratio between areas of brittle and ductile fracture – as well seen at Fig. 7 (red columns). The comparison of different laser beam widths in using argon shielding gas shows no significant differences.

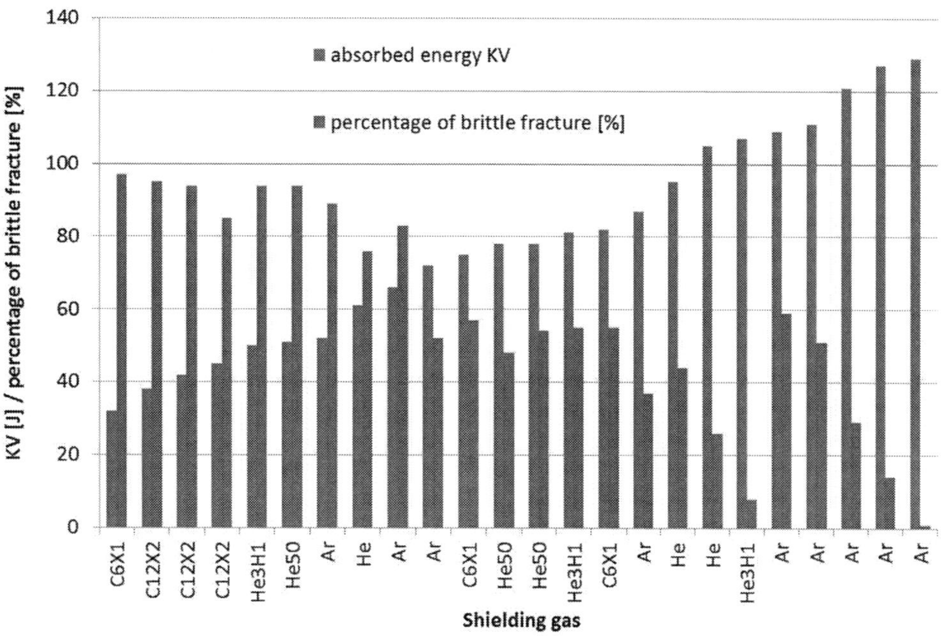

Fig. 7 An effect of shielding gas on the results of notch toughness

Two different laser beam width (0,9 and 1,2 mm) were used for welding of sheets with argon shielding gas. Specimens for fatigue tests were manufactured and tests performed. The results are summarized in Fig. 8.

It is obvious that different optics leads to different fatigue behavior and fatigue limit – σ_C = 250 MPa for 0,9 mm laser beam width and σ_C = 240 MPa for 1,2 mm laser beam width.

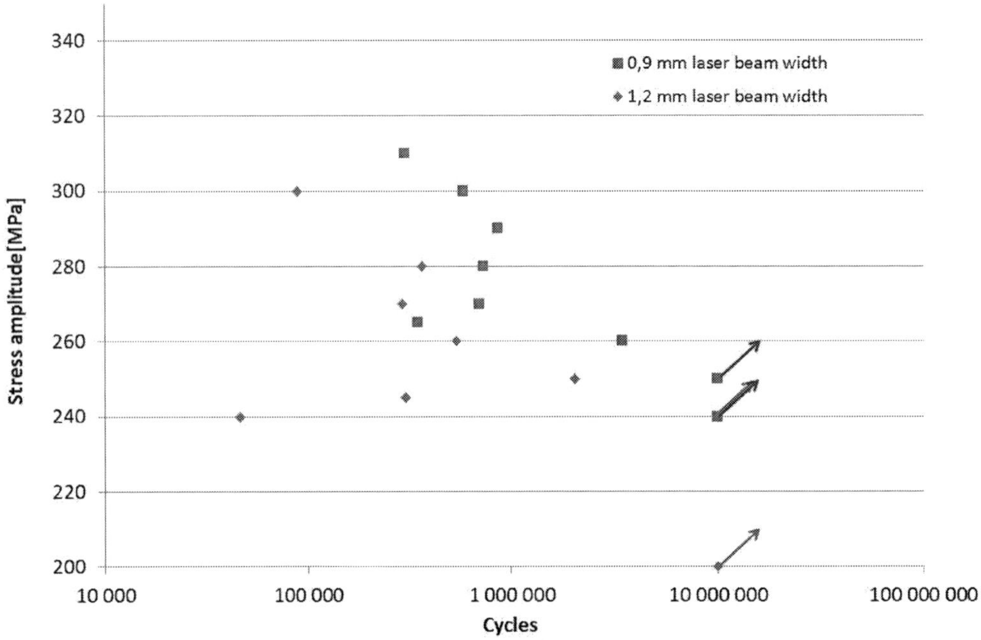

Fig. 8 The results of fatigue tests

Two specimens were prepared and tested from other sheets/shielding gases. All specimens were tested at the same stress amplitude – 255MPa – slightly above fatigue limit. The results of this comparison are summarized in Table 4. The values from sheets with argon shielding gas were statistically calculated from Wöhler curves of each sheet.

Table 4: Comparison of fatigue results for all sheets tested at the same stress amplitude

Sheielding gas	**Stress amplitude**	**Fatigue life [cycles]**	
		Specimen No. 1	**Specimen No. 1**
C12X2	255 MPa	1 123 307	10 000 000 *)
He3H1		173 512	434 404
C6X1		812 562	1 071 736
He50		906 514	1 408 651
He		496 768	682 503
Ar (0,9 mm laser beam width)		approx. 3 000 000	
Ar (1,2 mm laser beam width)		approx. 1 300 000	

) the specimen run out up to 10 000 000 cycles without failure

According to notch toughness results where argon shielded sheets show best results the fatigue results from these sheets were used as a "standard". Significantly worst fatigue resistance has been reached at specimens where He3H1 and He shielding gases were used. C6X1 and He50 sheets showed from fatigue point of view very similar results to our "standard". Using of C12X2 mixture as shielding gas lead to best fatigue resistance from all tested sheets. These results are in good agreement with residual stresses analyses where sheets welded with C12X2 shielding gas showed highest compressive residual stress in surface area. Our further work will focus on fractographic analysis of fracture surfaces after fatigue testing in combination with hardness profile in weld joint, evaluation of grinding / machining before welding on initiation of fatigue cracks and on evaluation of weld design on fatigue limit and fatigue crack initiation.

Acknowledgement

The work was carried out within the research project TA02011004 supported by Technology Agency of the Czech Republic.

References

[1] Onoro, J., Ranninger, C. "Fatigue behaviour of laser welds of high strength low alloy steels." Journal of Materials Processing Technology 68 (1997): 68-70.

[2] Farabi, N., et al. "Fatigue properties of laser welded dual-phase steel joints." Procedia Engineering 2 (2010): 835-843.

[3] Cerny, I., Sis, J. "Fatigue strength and damage mechanisms of laser welded structural carbon steel sheets." Key Engineering Materials 662 (2015): 201-204.

[4] Xu, W., et al. "Tensile and fatigue properties of fiber laser welded high strength low alloy and DP980 dual-phase steel joints." Materials and Design 43 (2013): 373-383.

[5] Kolarik, K., et al. "2D residual stress and hardness maps in steel laser weld with filler wire." Applied Mechanics and Materials 732 (2015): 9-12.

Microplasticity in Zr-2.5Nb alloy during thermal cycles

St. J. Skrzypek[1a], K. Bolanowski[2], K. Nawojowski[1], R.L.Eadie[3]

[1]*AGH-University of Science and Technology, Cracow, Poland*
[2] *Technical University in Kielce, Poland*
[1a] *corresponding: skrzypek@agh.edu.pl*
[3]*University of Alberta, Edmonton, Canada*

Abstract: Stress relaxation by micro-plastic deformation was studied in polycrystalline Zr-2.5Nb alloy under very low stress 23,74 MPa during thermal cycling in the range 70 – 370°C. The described experiment belongs to creep and hydride cracking area of material science.

A dislocation motion, residual microscopic stresses and interactions of applied stress at small scale in grains are critical in understanding how the microstructure affects the macroscopic mechanical and physical properties of materials. The back stress is attributed to movable screw and edge dislocations impeded in their movement by structural defects and by interstitial impurities.

The obtained relaxation by micro-plastic deformation following pre-stressing to applied stress was logarithmic ε = Kln ((t)+ to), with the slope i.e. velocity proportional to the time. At higher number of cycles the relation between ε and t was like asymptote i.e. dε/dt approached to constant close to zero value.

Keywords: Zr-2.5Nb; Zirconium alloy; microplasticity ; thermal cycling; strain velocity, alongation velocity.

1. Introduction

Zirconium alloys are based materials for pressure tube in nuclear reactors. They are also applied on uranium fuel shields. Both applications are connected with good mechanical, corrosion and physical properties. The last one is characterized with very low cross section for thermal neutrons.

Stress relaxation by micro-plastic deformation in polycrystalline alloys under very low stress during thermal cycling is important area of creep. The applied stress values lesser then 10% of the macro-yield level and thermal cycling condition can demonstrate real exploitation conditions particularly when delayed hydride cracking appears [1,5,7]. An industrial training and simulations of shut down of power plants can be characterize with this type of experiments. Effective plastic strains are very low in the range $\varepsilon = 10^{-3}$ and therefore some technical difficulties must be overcome in monitoring such small strain. The 10 times mechanical magnification and optical system for monitoring plastic elongation was designed and elaborated [6].

In spite of small plastic strain due to very small-applied stress it may be important in many machines and devices in technology and in exploitation. Majority of machines work under cyclic stresses where stress limit as fatigue stress is fraction usually in between 0.2-0.6 of yield point. Even so, such small stress causes gradually structural and microstructural changes, which finally causes crack nucleation in the material. The microplastic deformation is particularly important in connection between machine elements due to established technical fit where several micrometers can change type of fit and important interactions between elements.

Due to importance of microplasticity phenomenon some focus on its structural mechanisms is investigated. Among others the Bishop-Hil and uniform lower-bound models and dislocation thermally activated cross-slip are used in understanding of microplasticity structural mechanisms [2,3]. A field of tensile stress can enhance diffusion and in case of interstitial hydrogen atoms some nucleation of hydrides can appear and following their cracking gives plastic deformation due to delayed hydride cracking phenomena (DHC) [1]. Dislocation slips, twining, crystal lattice defects, diffusion and stress induced phase transformation are the primary carriers of plastic deformation for crystalline materials. A dislocation motion, residual microscopic stresses and interactions of applied stress at small scale in grains are critical in understanding how the microstructure affects the macroscopic mechanical and physical properties of materials. The back stress is attributed to movable screw and edge dislocations impeded in their movement by structural defects and by interstitial impurities. The structural relaxation mechanisms in the low-stress region are responsible for the small amount of cumulative plastic strain versus stress and time.

The micro-plastic deformation following pre-stressing of zirconium alloy Zr-2.5Nb was measured and described with mathematical functions for applied stress and thermal loading.

2. Sample preparation and experimental procedure

The samples made of Zr-2.5Nb alloy were prepared. The sample's working length was 25 mm and dimension of cross section area was 4.20x4 mm. The sample with holder (Fig.1) was placed in vertical electrical and programmable furnace. Special optical system was developed for monitoring and measuring length of the sample [6]. Temperature was measured directly on samples with NiCr-NiAl thermocouple with accuracy +-0.3 deg. Elongation of sample was measured with mechanical and optical laser system with accuracy +-0.2 μm [6].

Advanced X-ray diffraction examinations were carried out on original and used sample. After thermal cycling of sample the X-ray examinations under the same conditions were repeated in nondestructive way. The samples were under load equal to σ=23,74 MPa and continuously heated and cooled during thermal cycles within temperature range of 70-370°C. This thermal range of cycling was established due to real working temperature and to many experimental data about DHC [1,7]. Time period of a single cycle was 160 min with total cycles number equal to 100. In this paper the characterization of micorplasticity for examined alloy Zr-2.5Nb is described and considered due to meaning, mechanisms and application of empirical mathematical functions to describe obtained results.

The Figure 1 presents the schematics of sample and its holder.

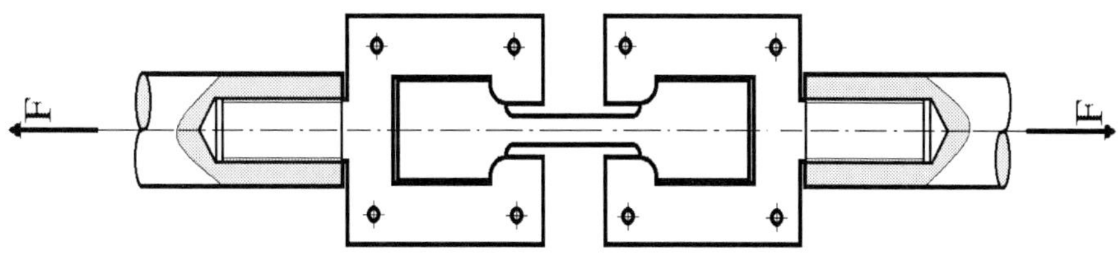

Fig. 1: The sample with holder schematics, which are inside the heat chamber of furnace.

3. Results and conclusions

Applied stress was a small fraction of yield point of this alloy i.e. 2.5%.

Two X-ray diffraction techniques were applied i.e. symmetric Bragg-Brentano (BB) and grazing incidence angle geometry. Both can characterize phase composition, state of structure and the second one can be used for residual stresses measurement [9]. Results obtained from X-ray diffraction patterns (Fig.2 and 3) show qualitative and quantitative differences between microstructure of the samples before and after the cycling experiment. The non-treated sample before experimental procedure consisted of zirconium α (hexagonal, P63/mmc, R: 03-065-3366) Sample after thermal cycling treatment consist of the same αZr as well as surface layer of zirconium dioxide ZrO_2 (Monoclinic, P21/c, R: 00-013-0307).

Illustration of results from loading and thermal cyclic experiments is shown on figure 4. As the example of strain and plastic deformation velocity of Zr-2.5Nb alloy sample under tensile stress 23,74 MPa during thermal cycling in the range of 70-370°C are presented on figure 5 - 7). Results on figures 5 and 6 can be described with logarithmic equations as:

$$\Delta L\ (t) = 9{,}6091\ln(t) + 50{,}876 \qquad (1)$$

$$\varepsilon\ (t) = 0{,}0004\ln(t) + 0{,}002 \qquad (2)$$

Where: ΔL – increment of length [μm], ε – plastic strain and t – time [min]

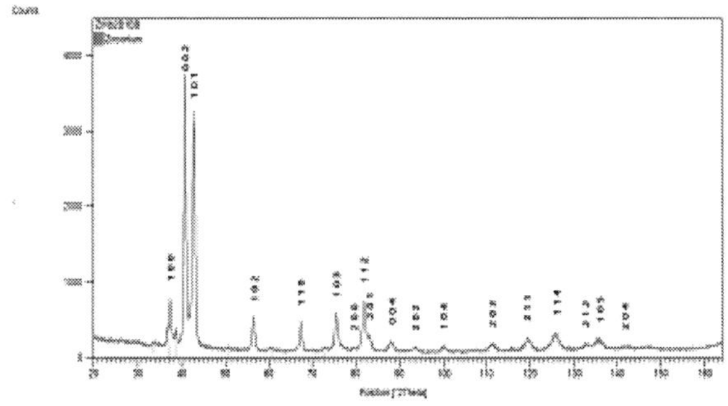

Fig.2. Diffraction pattern of Zr-2.5Nb raw sample before thermal cycling.

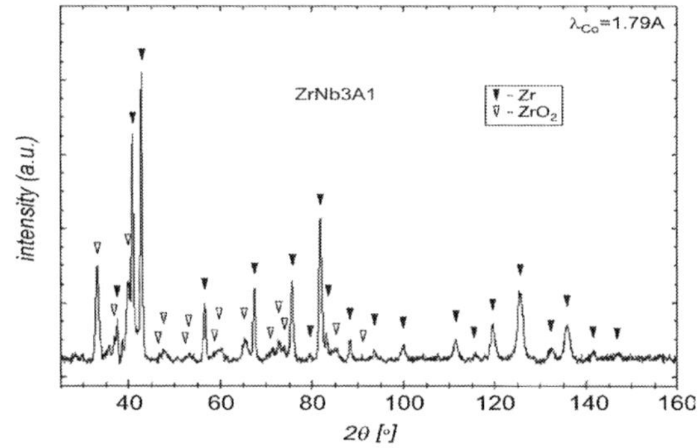

Fig.3. Diffraction pattern of Zr-2.5Nb sample after thermal cycling under 23,74 MPa tensile stress.

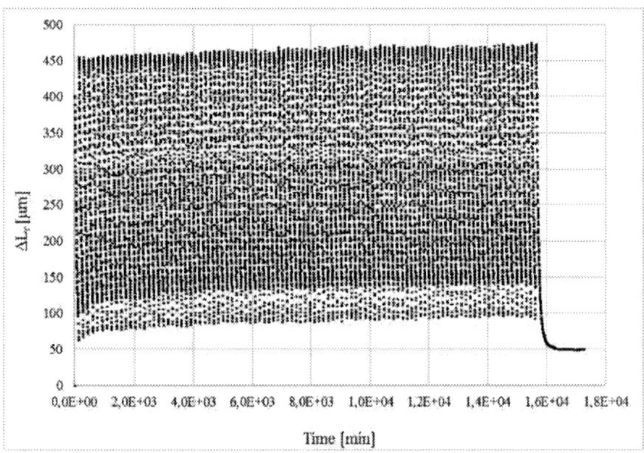

Fig. 4: Full experimantal data i.e. relation between sample's elongation and time for Zr-2.5Nb sample under tensile stress 23,74 MPa (T_{cycles}=70-370°C) .

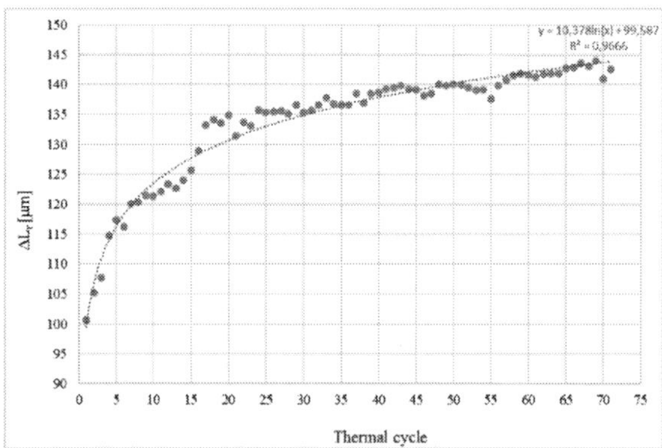

Fig. 5: The relation between elongation and time for Zr-2.5Nb sample under tensile stress 23,74 MPa at T_{cycle}=70°C with fitted log-normal distribution.

The elongation and elongation velocity of specimen at the beginning of experiment were established as $\Delta L = 1{,}01*10^{-4}$ m and $\Delta L/t = 5.36*10^{-9}$ m/s. The same data referring to the end of experimental cycling were following: $\Delta L = 1.41*10^{-4}$ m and $\Delta L/t = 1.6*10^{-10}$ m/s. These data recalculated to strain and strain rate for the beginning were: $\varepsilon=3{,}79*10^{-3}$ and $\varepsilon'=3{,}99*10^{-8}\ s^{-1}$ and after 100 cycles: $\varepsilon=5{,}46*10^{-3}$, $\varepsilon'=4{,}22*10^{-10}\ s^{-1}$.

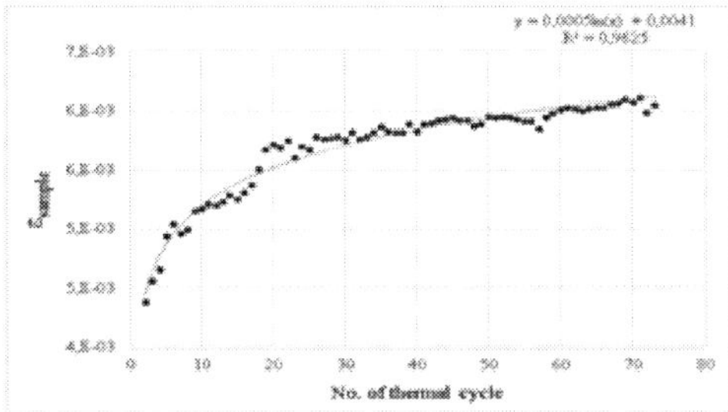

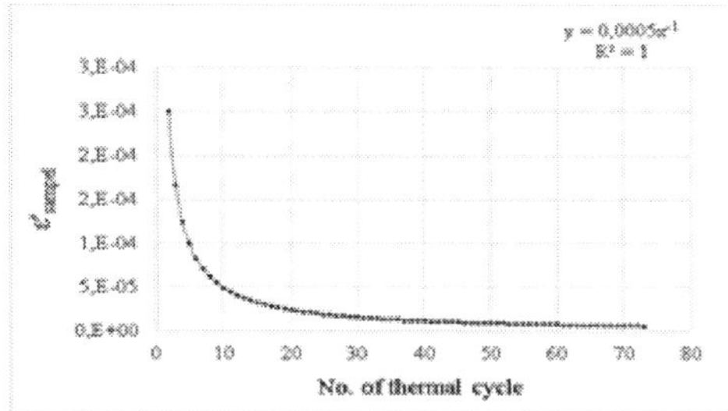

Fig. 6: The relation between strain and cycle number (time) for Zr-2.5Nb sample under tensile stress 23,74 MPa (T_{cycle}=70°C) with log-normal distribution.

Fig. 7: The relation between strain rate and cycle number (time) for Zr-2.5Nb sample under tensile stress 23,74 MPa (T_{cycle}=70°C) with fitted function.

4. Conclusion

The obtained experimental results confirmed microplastic deformation of sample made of Zr-2.5Nb alloy under small tensile stress 23,74 MPa i.e. the microplasticity occurrence. Zr-2.5Nb alloy under very low tensile stress i.e. 23,74 MPa during thermal cycling in spite plastic plastic deformation thine coating of ZrO_2 oxide was gradually growing. Increase of plastic deformation versus time (number of cycles) was a log-normal distribution. The strain and strain rate decreases approaching to very low and almost constant value versus time.

Acknowledgement

Results of research described here and this paper was supported by Faculty of Metals Eng. and Industrial Computer Science, Department of Physical and Powder Metallurgy of AGH-UST in Krakow according to project No 11.11.110.299

References

[1] G.Lin, S.J.Skrzypek, D.Li, R.L.Eadie: Monitiring Crack Advance Using Acoustic Emission and Combined Acoustic and Potential Drop in Zr-2.5%Nb, *Journal of Testing and Evaluation, JTEVA*, Vol.26, No.1, January 1998, p.15-25 (by the American Society for Testing and Materials).

[2] J.C.Bittner, B.L.Adams: Correlation of biaxial yield behavior in cooper with crystallographic texture for strains in the range $10^{-4} – 10^{-2}$, *Mat. Science and Engineering* 100 (1988) 15–22.

[3] Han-Ryong Pak, J.Chu, M.Kato, D.P.Pope: Friction stress and beck stress in cyclically deformed Ni_3Ge single crystals, ***Mat.Science and Engineering*** 100 (1988) 31-36.

[4] A.J.G. Maroto, R. Bordoni, M. Villegas, M. Blesa, A.M. Olmedo, A. Iglesias, G. Rigotti: Growth and Characterization of Ooxide Layers on Zirconium Alloys *Comision National de Energia Atomica, Buenos Aires, Argentina.*

[5] E.C.W.Perryman: Pickering pressure tube cracking expierience, *Nucl. Energy* 17 (1978), No 2, 95-105.

[6] K.Bolanowski: Private repoert, Technical Univeristy in Kielce, 2015.

[7] Delayed hydride cracking in zirconium alloys in pressure tube nuclear reactors IAEA-TECDOC-1410, Final report of a coordinated research project 1998–2002, The originating Section of this publication in the IAEA was: Nuclear Power Technology Development Section, International Atomic Energy Agency Wagramer Strasse 5 P.O. Box 100, A-1400 Vienna, Austria.

[8] L. Błaż, M. G. Akben, J. J. Jonas, E. F. Ibrahim, Mechanical Properties of Excel Alloy in the temperature range 300-1375 K, *Archives of Metallurgy*, Vol. 33, Issue 3, 1988.
[9] S.J. Skrzypek, A. Baczmański, W. Ratuszek, E. Kusior: "New approach to stress analysis based on grazing-incidence X-ray diffraction" *Journal of Applied Crystallography* 34 (2001), 427-435

Real stress arising in the backrest frame of a car seat

P. Srb [1,a], P. Kulhavý [1], A. Lufinka [1], P. Lepšík [1], V. Fliegel [1]

[1]*Technical University of Liberec, Faculty of Mechanical Engineering, Studentská 2, 461 17, Liberec 1, Czech Republic*

[a]*Pavel.Srb@tul.cz*

Abstract: At present lot of people spend still more of their life on the road. Vehicles have been becoming ever more sophisticated and the main direction of their development is placed primarily into the areas of environment, design, safety and comfort. The main current trend with regard to the environment, reducing emissions, fuel economy etc. is reduction of vehicle weight. A large amount of manufacturers and research works is concerned on a reduction of the weight of individual vehicle components primarily by application of unconventional materials. We've been finding highly sophisticated composite material which by targeted influencing of their properties can replace conventional steel materials with considerable weight reduction. As a long-time neglected part in this area we can generally mark the steel frames of car seats, especially their big mass which reduction can be a good challenge in future optimizations. In this work based on experiments with the normalized loading and numerical simulations we found how big stress and strain arise in the frame and especially their distribution. The experiment was based on loading of the back rest with a controlled torque relative to the R point of the seat. The overall deformation was monitored with using a laser and local deformation using strain gages. For a more detailed description of the emerging stresses, primarily about their distributions and directions a quasi static numerical model has been created. The found results will be applied mainly for the future construction of a composite backrest.

Keywords: car seat; real stress; Steel frame; experimental loading; simulation.

1 Introduction

Car seats are the one of very important parts of cars with a direct influence to subjective feelings, safety and ride quality. The important parameters of passenger comfort and safety are mainly properties of vibration isolation and their ability to a shock absorbing [1]. The standard seat frames are usually weldment from steel stampings and wires. The comfortable infill ensures parts from molded polyurethane foams which are being replaced with low the density foams, however, with even better properties. The properties of used foams are very important while driving a car lasting more hours, because they must sufficiently stabilize muscles especially in the area of pelvis and spine. With this requirement help us quite significantly the textile coating that doesn't have only an aesthetic role, as might appear at first, but also ensuring the rigidity of a cushion [2].

In the seats construction (Fig. 1) there are standardly some adjustable elements that can modify and set the position of the cushion, backrest, peripherals and entire position of the seat in a cabin of a car. The primary mechanisms provide longitudinally sliding guideways moving in the length of around 250 mm, controlled a rod on the lower front part of the seat. The body of the backrest is angularly adjustable via a control mechanism located on one side of the seat. For new and more luxurious types of cars all positions and angles have been controlled electronically by stepper motors typically with worm gears.

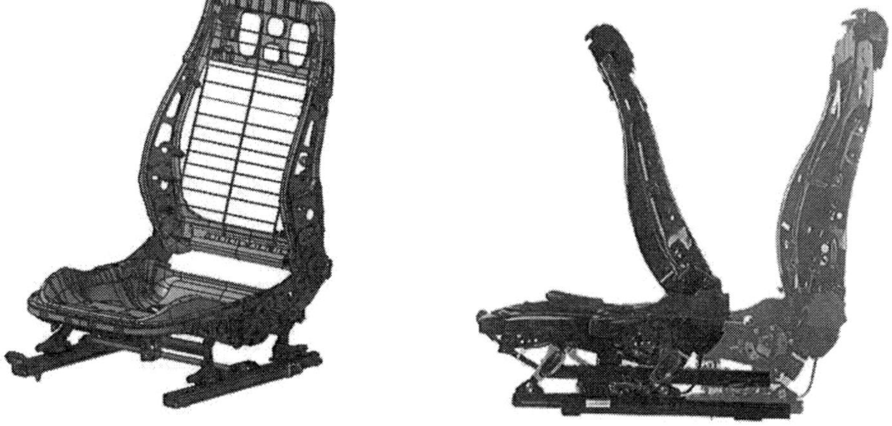

Fig. 1: The CAD model of a tested car seat in Catia and visualization of a seat shape adjustment.

2 Testing of car seats

Generally there are two group of tests. Tests that coming from the first "comfort group" like the damping, transfer function, specific pressures, breathability and thermal comfort are on safety involved indirectly, because their impact is primarily on the driver's person and his mental or physical condition. Then there is a second group of tests that cover direct mechanical testing of the seats, their parts and to the safety have an impact primarily at the time of an accident. The most commonly tested are particularly: Strength of seats, seat belts and their mounting into the floor, and static and dynamic tests of the steel frame.

2.1 Comfort - interaction with human

According to [1, 3] is driving a car action which has relatively high demands on the mental and physical capabilities of the human. A human weight and physique of each individual have a straight influence to the interaction of the body and seats. For a description of their interaction is usually used a parameter describing a distribution and size of the stress called Contact pressure, that size should not exceed the recommended health limits. According to physiological norms [4] and e.g. Vlk [5] a seat supports the driver in a relatively large contact area and transfers most of his weight. Ratios for individual parts should be approximately 64-72% on cushion, 4-16 % to backrest and the remaining approximately 15% goes as a reaction from the feet to the floor of a vehicle. It is very important, to support the lumbar spine and a minimum support in a popliteal part. Individual values were verified experimentally using X-sensor (Fig. 2) and subsequently by recalculation [6, 7] of the areas in which operate specific pressures to the resulting forces in dependencies of the cushion and backrest angle (Fig. 3).

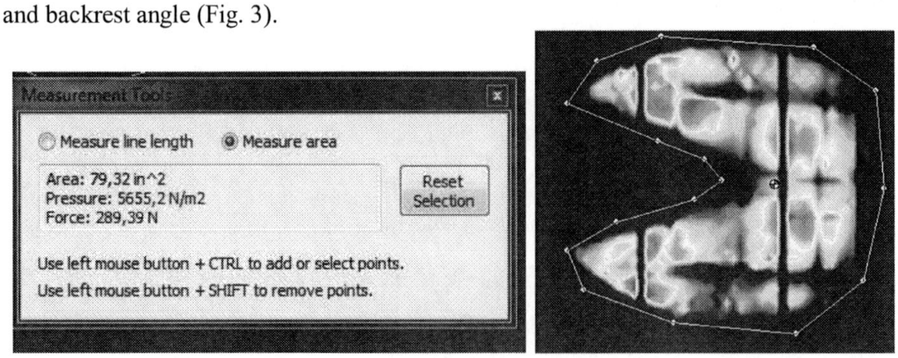

Fig. 2: Measurement of the specific pressures and subsequent determination of the forces using linear tracing of the contact region

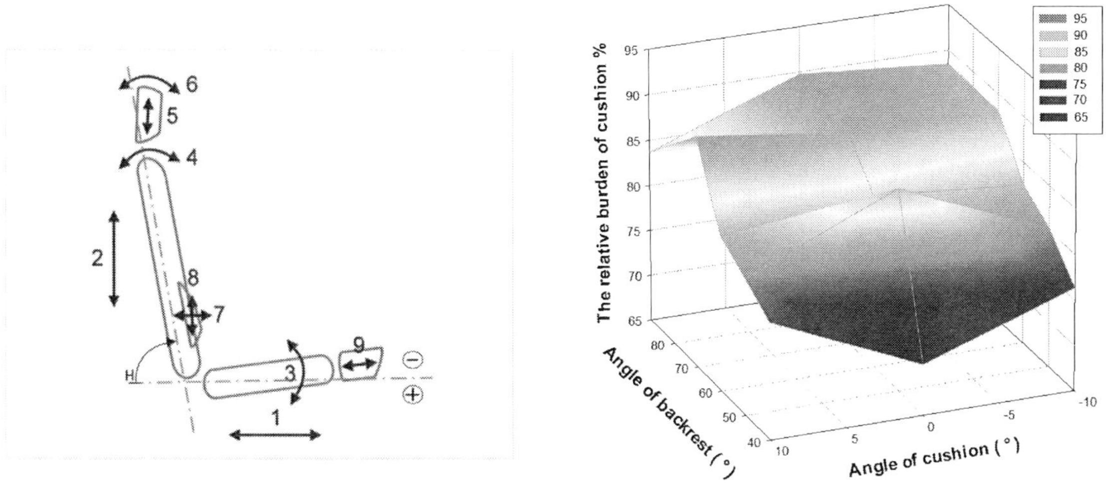

Fig. 3: The distribution of specific pressures on the seat while adjusting the angle of the seat and backrest

2.2 Mechanical tests

As already mentioned, this group includes tests that have primarily the influence to safety and in the event of an accident they should as much as possible eliminate its impacts, reduce damage to the health and save life of passengers. The most common and absolutely necessary are the tests according to [8, 9]:

Strength of seats and their fastenings to the floor. This are concentrated on a check of the angle adjustment mechanism and a frame of the backrest (Fig. 4) with torque 530 Nm relative to point H at a strain rate 150-300 mm/min. Due to the significant hysteresis of the entire structure, the loading perform several times consecutively is recommended. Then, there are static tests of the longitudinal sliding mechanism and rigidity test of the frames which are loaded 20 times the weight of the seat in the center of gravity. All this parts are also mostly checked with loads corresponding to 20 g. Further some tests on the headrest by a load-inducing the recommended torque with a maximum allowed deformation at the headrests less than 102 mm. For a dynamic test a spherical impactor with speed of 24.1 km / h is used that bump into the headrest when a deceleration on the impactor should not exceed 80 g for 3 ms.

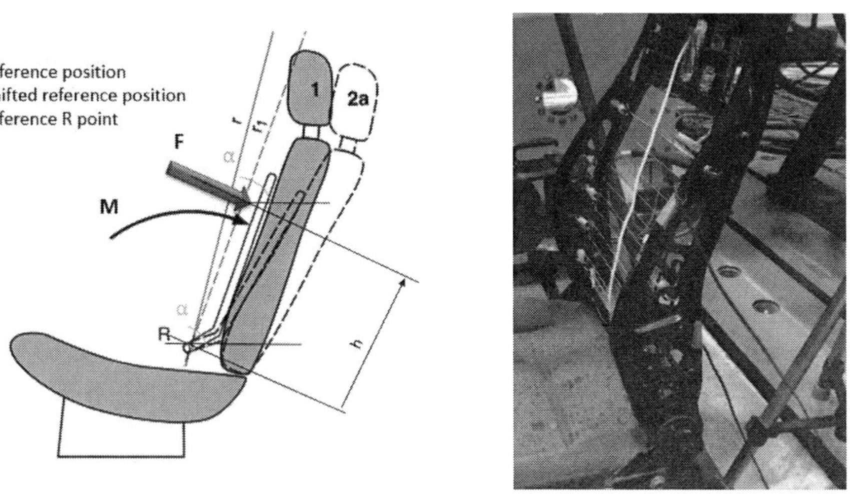

Fig. 4: Loading of the backrest (defined by a force or an equivalent torque) and position the strain gauges

3 Experiment

Considering the known dimensions of the tested seat and load from the normalized 95% male at the upper statistical percentile of weight we get, relative to the point R, a static torque about 23Nm. According to the safety standards [9], the load values ranging up to 700 N for the cushion and 400 N on the backrest, eventually a force that to the backrest cause torque 530 Nm relative to the axis of rotation.

Fig. 5: Scheme of the carried experiment

Loading on the backrest (Fig. 5) was carried out in sequence of three cycles as prescribed the standard with the feed rate of 10 mm/s. The testing force was up to 3000N (torque ~800Nm) which slightly exceeds the recommended values on that the backrests are tested. Two strain gauges were placed at the front and rear side of the frame. The front gauge (side to the steering wheel) measure elongation and the rear strain gauge measure a warping in the place of change the curvature of the frame. The entire experiment was also modeled through static contact analysis in software Ansys.

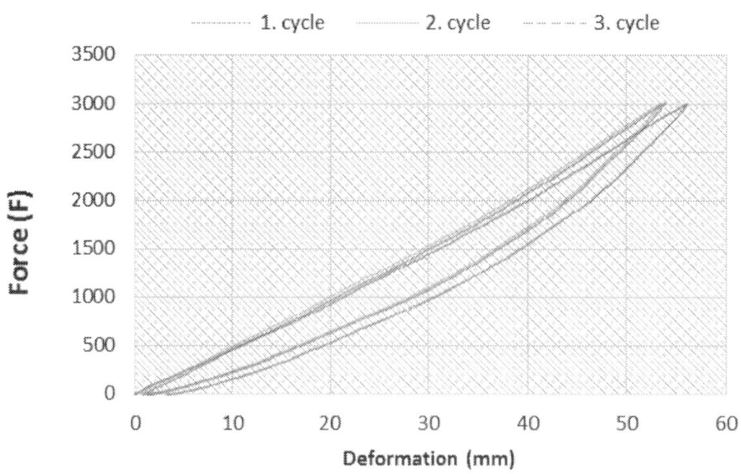

Fig. 6: Displacement in the upper part of the seat frame measured by a laser

As can be seen in the results (Fig. 6) under repeated loads was the first cycle always different and for other cycles seem the hysteresis loops to be almost identical. In the Fig. 7 there is possible to see the values of relative deformation at the front and rear side of the frame, which can help us to have at least an approximate idea about the rising deformations.

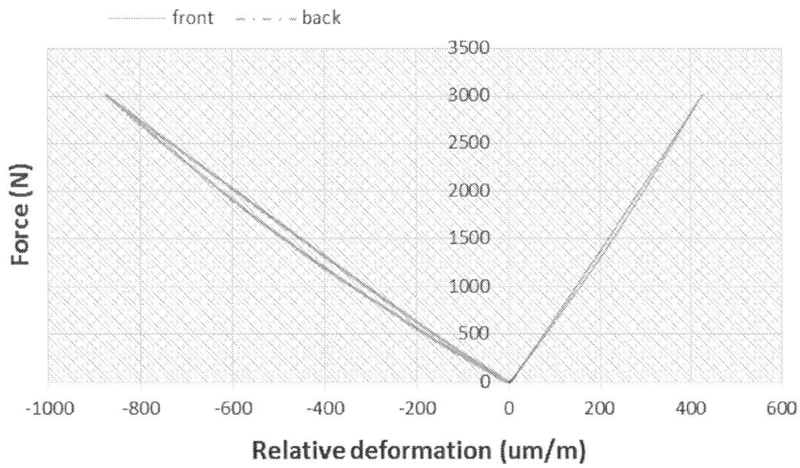

Fig. 7: The relatively deflection measured on the front and back strain gauges

4 Model

An analogical FEM model of the carried experiment carried out has been created in Ansys Software. Due to the rather complex geometry has been the CAD geometry of the backrest slightly optimized and solved as the shell element. This was especially suitable for creation the meshes and also due to time of calculation. Contacts at the connection the individual parts of the frame were formulated in multi-point constraint (MPC), which is generally suitable for "Bonded" and "No separation" behaviors. This approach is not based on the Penalty-base or Lagrange multiplier theory, it is direct and effective way for prescription contact two perfectly connected areas [10, 11]. The chosen detection method was Nodal-normal from contact with the constraint of Inside pinball radius with Couple U ROT and an automatic detection of region. The process of convergence during this model solution, has been smooth without chattering, which can sometimes appear in contact models because of used step functions. This problem is of course especially for frictional contacts.

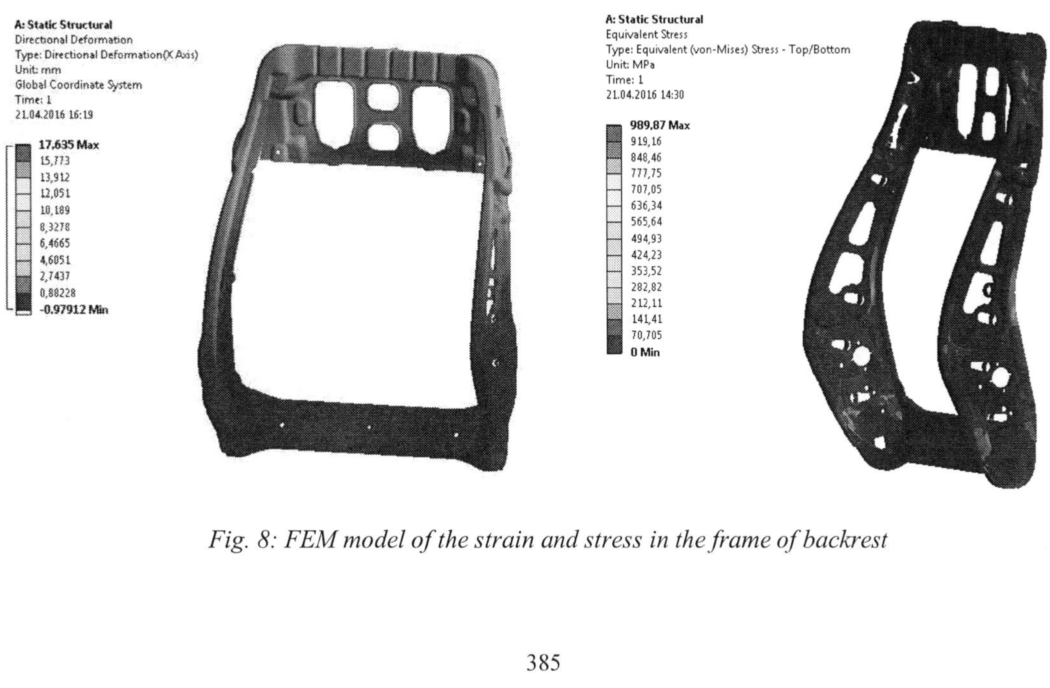

Fig. 8: FEM model of the strain and stress in the frame of backrest

5 Conclusion

The experimental results show that even with a significantly higher loading the size of the deformation on top of the frame is still just only around a half of the allowed value. Based on the FEM model results the deformations (Fig. 8), their distribution and simultaneously the individuals directions and character of stress could be clearer described. In comparison the experiment and model we could find a relatively smaller total deformation for the numerical solution. This is caused by the fact that a large portion of the deformation during experiment is caused by clearances in the backrest adjustable mechanism.

In technical literature can be found lot of works dealing with the areas of seat comfort, shock absorbing, biomechanics of human and the headrests during an impact etc. However, only several of them is devoted to a thorough description of deformations and stress in a such common part as is the frame of seater itself. The results of test conducted in this area by the car manufacturers are not easily achieved. For this reason this work has been aimed to making a sort of mapping of the fair values of stress and strain arising during the real and normalized loading. These results will be in future used primarily for comparison with proposal concepts of developed composite frames.

Acknowledgement

This publication was written at the Technical University of Liberec as part of the project "Innovation of technical systems structures with the use of composite materials" with the support of the Specific University Research Grant, as provided by the Ministry of Education, Youth and Sports of the Czech Republic in the year 2016.

References

[1] P. Kulhavy, P. Srb, M. Petru. *Numerical and experimental analysis of the real load arising in the cushion of the car seat*. In: Manufacturing technology, Vol 15 no 6, 2016

[2] P. Kulhavý, N. Kovalova, J. Vosáhlo: Numerical *Model of the Static Loading of a Stitched Seam in the Composite Cover of Car Seat*, EAN 2015, 53th, CTU in Prague, 2015

[3] DIZO, J. (2014). *Evaluation of Ride Comfort for Passengers by Means of Computer Simulation*. In: Manufacturing Technology, vol. 15 No 1, 2014

[4] Ergonomic standard *ONA 300575*

[5] *VLK, F.: Karoserie motorových vozidel. Nakladatelství Vlk, Brno, 2001,*

[6] R. Martonka, V. Fliegel. Measurement of the temperature and pressure characteristic of car seats-relationship between human *In:* EAN 2015, 53th Cesky Krumlov, CTU in Prague, 2015

[7] Neil J. Masnfeld : *Human response to vibration*, CRC Press, 2005

[8] Standard E/ECE/TRANS/505

[9] Standard EHK OSN standards to ensure the safety of passengers EHK R14, EHK R17

[10] *Practical Aspects of Finite Element Simulation*, Altair Engineering, United States, 2015

[11] Mechanical Structural Nonlinearities - *Advanced Contact*, Training Material, ANSYS, Inc. Proprietary, 2010

Plastic Behavior of the Oil Sealing Ring in a Bearing Support

F. Straka [1,a]

[1] *Doosan Škoda Power, Tylova 1/57, Plzeň 301 58, Czech Republic*
[a] *frantisek.straka@doosan.com*

Abstract: Bearing supports of steam turbines contain of many parts. One of these parts is an oil sealing ring. Oil sealing ring has function isolate the hot oil vapor inside the bearing support from mixing it with outside air. When some leakage occurs it is a sing of insufficient function of the ring. This leakage was observed on a power plant after certain operation in short time duration. The reason was plastic deformation of the sealing ring. But the main reason why it went through the plastic phase was unclear.

Keywords: temperature; elastic-plastic; deformation.

1 Introduction

The FEM analysis was done to clarify the reason and it helped to modify the design for the future operation. Numerical simulations were performed in the ANSYS 16.0 software as a multistep analysis with external temperature load and with appropriate boundary conditions. The results were compared to measured values on a disassembled oil sealing ring.

2 Analysis Setup

The 3D model was simplified for the purpose of the numerical simulation. The analysis consists of two models. The larger model – an assembly was used for definition of the temperature field and the smaller model – a part was used for the elastic-plastic static structural analysis. The larger mesh consisted of 3,035,122 mostly tetrahedral elements, the smaller model consisted of 429,048 tetrahedral elements. First approach to set up the analysis was consideration a frictional contact in casing split. It showed up a problem in this setting with highly distorted elements in the contact area. After a few unsuccessfully convergent solutions was done simplification to use upper quarter of the ring only with vertical plane of symmetry and without the problematic contact.

2.1 Temperature Field

The temperature analysis was done as the first step of the calculation. The boundary conditions were not accurate on the beginning of the simulation. Any temperature of the affected part was measured. The reliable sign of the temperature was the surface color of the oil sealing ring. The mass loss of the sealing edge helped to set up the temperature conservatively at 700 °C. The mass loss was caused by contact of sealing edge with the rotor. More thermal boundary conditions were calculated and added to the analysis. The final steady state temperature field of the assembly is on the Fig. 1. The steady state field is a simplification itself in respect of real conditions which occurred in limited time on the power plant.

The temperature analysis was done one more time with slightly different geometry. The difference was in the shape of the sealing edge (Fig. 2). The width of the sealing edge 0.3 mm was extended to the high of 2 mm. This geometry difference caused diverse temperature field (Fig. 3). The temperature difference on the end of the sealing ring (close to sealing edge) between the original and the modified geometry was around 50 °C.

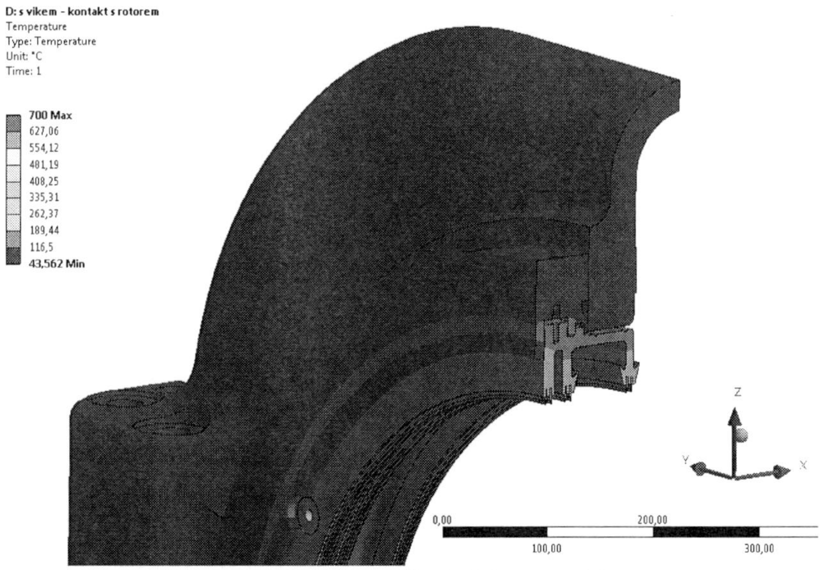

Fig. 1: Temperature field of the assembly

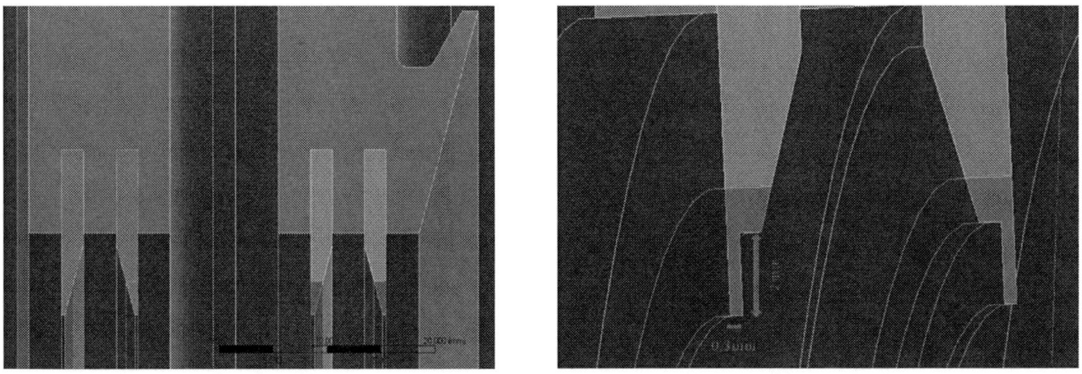

a) Original design

b) Modified design

Fig. 2: Design of the sealing edge

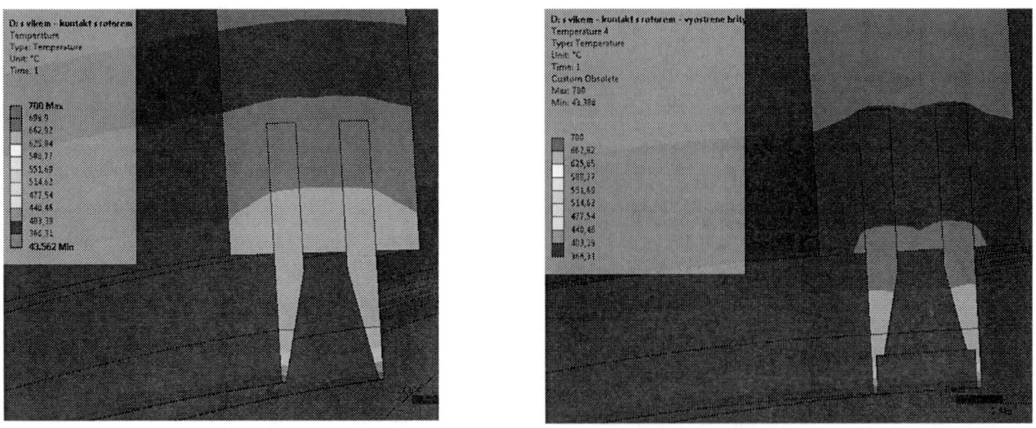

a) Temperature field – original design

b) Temperature field - modified design

Fig. 3: Temperature field

2.2 Static Structural Analysis

The temperature field was made for the assembly included the oil sealing ring. The static structural analysis was focused on the ring only. The analysis was set up from two load steps with different sets of boundary conditions. The ambient pressure on all parts was basically atmospheric and therefore it was neglected. The temperature field was imported in the first step. The next step was used for unloading of all external boundary conditions. Displacement of the oil insert and gravity remained only. The displacement was set up in respect of real conditions, especially in respect of the temperature field of the sealing ring adapter, which surrounded and supported the oil sealing ring. Due to the temperature change the support location moved. This change reflected the displacement boundary condition: A (Fig. 4). Another displacement was set in the casing split. A small area was set as frictionless with Z position locked: B (Fig. 4). In the second step (without external temperature field) total deformation and strain were observed. Static structural analysis repeated once again with different geometry described above (2.1 Temperature field) with the same set of boundary conditions.

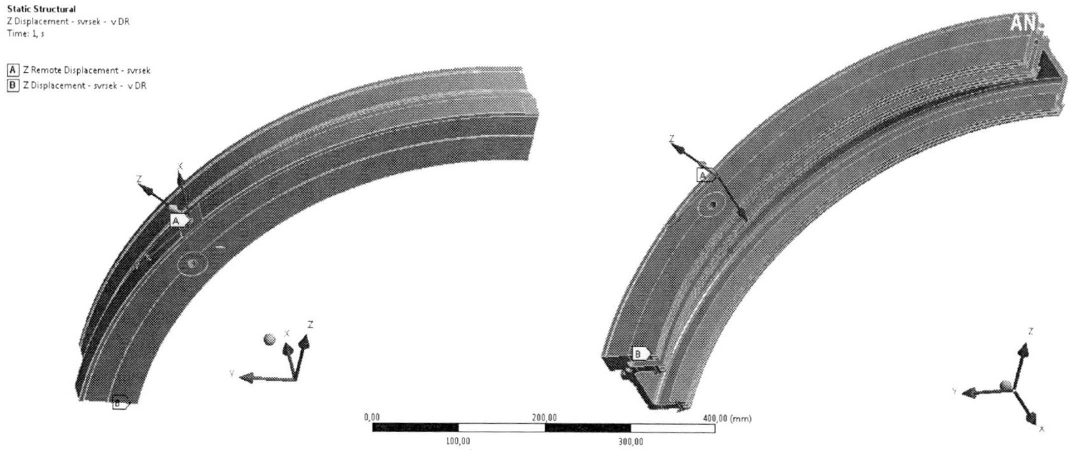

Fig. 4: Boundary condition - displacement

2.3 Material Properties

Oil sealing ring includes 1.0553 steel material properties with bilinear hardening model of plasticity. Sealing edges inside sealing rings are made from brass. Another part of the assembly is an adapter and a cover. Cover is made from material 1.0553 and the adapter has the same material properties as sealing ring.

3 Results

Ovality of the oil sealing ring was taken as a result and compared with real measured dimensions. Difference between final shape and original shape was calculated. Fig. 5 represents the shape after the elastic-plastic deformation. Ovality of the ring is calculated as vertical diameter minus horizontal diameter. The result from the analysis is 21.12 mm. The measured ovality was 16.56 mm. The result is up to 27 % higher than in reality. The result with modified geometry is 14.94 mm. It is up to 29% reduced value in shape deformation in comparison of previous FEM result.

The driven mechanism of the shape change is shown on Fig. 6. Plastic deformation is caused by the displacement and by conducted heat from the rotor during time of contact.

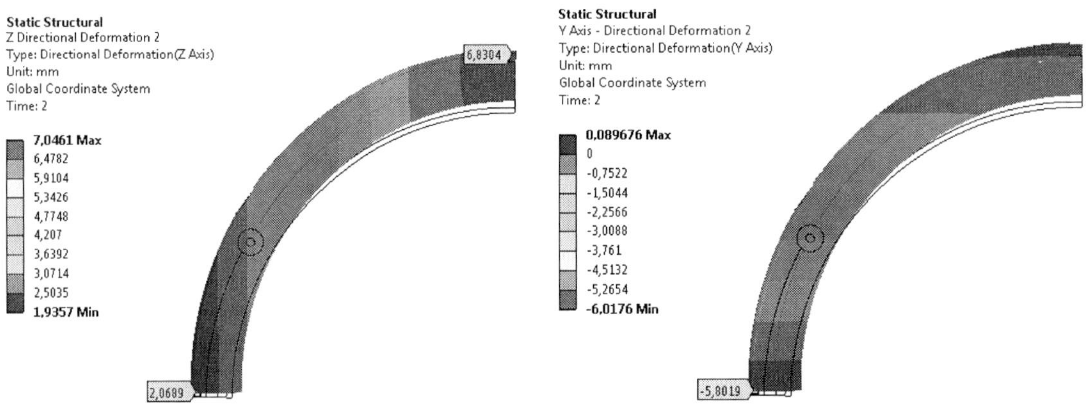

a) Vertical plastic deformation b) Radial plastic deformation

Fig. 5: Final shape after elastic-plastic deformation

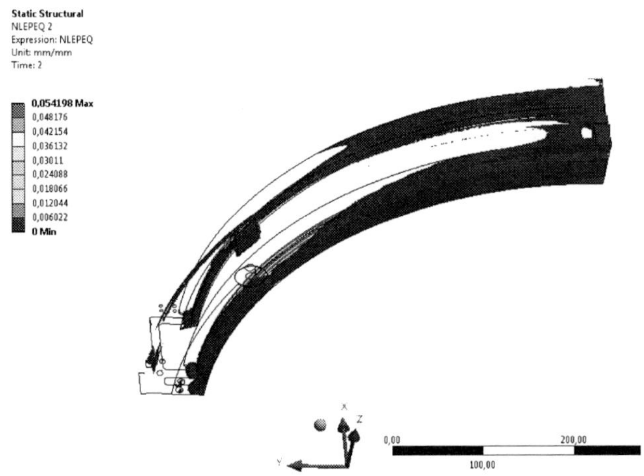

Fig. 6: Plastic area

4 Conclusion

The results from the numerical simulation were compared with the measured dimensions. The result from the static analysis has approximately 27 % higher values than measured on real oil insert. If we consider unclear and simplified boundary conditions in the analysis it is basically good result which proofed the main driven mechanism of the plastic deformation.

The result with modified geometry is up to 29 % lower than the result with original geometry. The changed design of the sealing edge has impact on absorbed heat in the sealing ring from the rotor during contact. It happens in limited time and therefore it is not fully covered in steady state thermal analysis which was done. Despite this fact these gained results show the benefit of the new design. Also the application on a power plant confirmed the improvement.

References

[1] A. Öchsner, Elasto-Plasticity of Frame Structure Elements, Southport, Springer, 2014.

[2] Retrieved from: https://support.ansys.com

Determination of spring-tensions of a vibrating bowl feeder

M. Sturm [1a], L. Pešík[2]

[1] *Faculty of Mechanical Engineering, University of Applied Sciences Zittau/Görlitz, Germany*
[2] *Faculty of Mechanical Engineering, Technical University of Liberec, Czech Republic*
[a] *m.sturm@hszg.de*

Abstract: Vibrating conveyors designed as bowl feeders are common equipment for conveying goods into production systems. These systems are designed for the supply of a certain number of goods to an individual designed interface and simultaneously and arranging the correct orientation of the goods conveyed by the same time. This type of conveyer is used various industries such as for example the automotive industry, electronical industry and medical industry. The target of this work is to find the resonance frequencies of the ready-to-operate conveyer, the resonance frequencies of the spring packages and to identify the stresses and the displacement of the spring packages under standard working conditions. In addition a comparison of the results from the test and the simulation including the interpretation of these results will be done.

Keywords: Vibrating conveyor, bowl feeder, resonance frequencies, stresses of the spring package.

1 Introduction

The bowl feeder that is used for the tests and the simulation was originally built to forward parts for the electrical industry.

Fig. 1: Sortimat bowl feeder type SMB 6 (own source).

The technical specification of the bowl feeder is as follows:

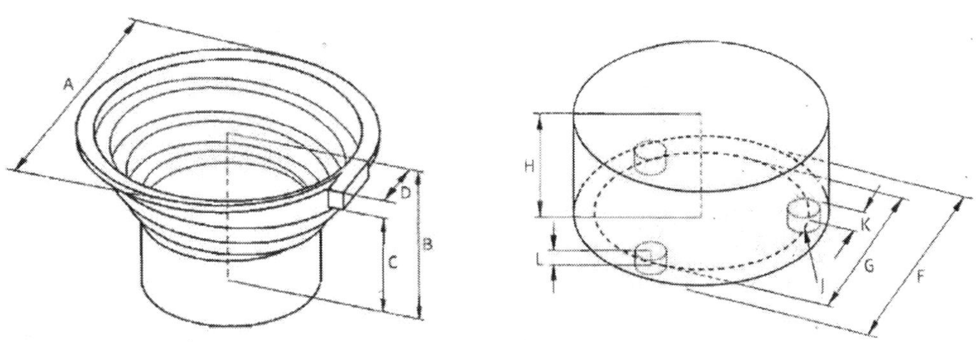

Fig. 2: Parameters Sortimat bowl feeder type SMB 6 (own source).

Tab. 1: Parameters of the bowl feeder.

A	B	C	D
700 mm	450 mm	400 mm	300 mm
F	**G**	**H**	**K**
535 mm	485 mm	190 mm	48 mm
L	**Total mass**	**Capacity**	**Moving mass**
130 mm	123 kg	12 kg	33 kg

Spring material: "Durostone", similar to the present available quality UPM S16 fiber-reinforced plastic (brand of Röchling SE & Co. KG, Germany), installed under 15° to vertical boundary line.

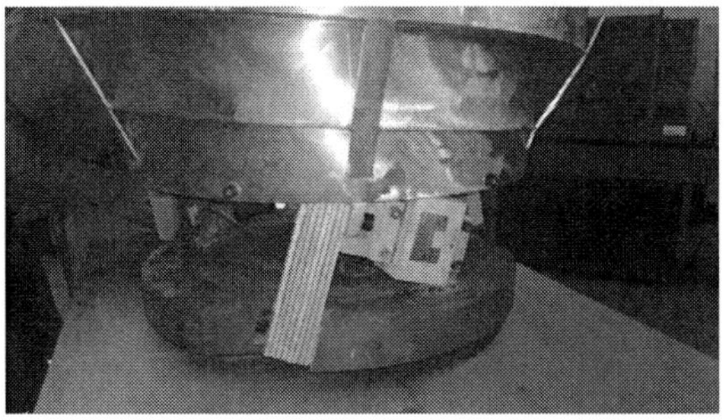

Fig. 3: Spring package with excitation device 220/230 V, 25Hz.

2 Experimental determinations of characteristic values

The first part of this elaboration deals with the experimental determination of characteristic data of the bowl feeder. The data will supply the necessary information on frequencies and amplitudes on typical operating conditions of the conveyer as well as information of the resonance frequencies of the whole system and the spring packages. Based on this information, the calculation of effective forces to the spring package can be done and using the technical values of the spring material, the tensions in the springs will be worked out.

-Excitation and measurement in radial direction

-Excitation and measurement in tangential direction

-Excitation in tangential direction and measurement in vertical direction

The primary task of a bowl feeder is the generation of micro throws and by doing this, the conveyance of goods. Therefore the last of the three mentioned excitation cases is the most significant one and will be tested. For this test the sensor is mounted vertically under the spiral path of the bowl as shown in figure 5. The excitation once is done by a rubber coated, so called Tang hammer. The excitation point is the end of the guiding channel as it has been before. The reason for this is that the major conveying direction generated by the excitation device as well is oriented tangentially. The sensor records the frequency and the acceleration of the oscillation. The analyzer and the software convert the data into acceleration over time diagram and an acceleration over frequency diagram.

2.1 Resonance frequencies of the bowl feeder

To get the necessary information on the resonance frequencies of the operational conveyer it has to be mentioned that there are different possibilities to generate and measure resonances of the conveyer system, using different fix points for the sensor as well as different exciting points and directions to bring the system into oscillation. Three exciting and measuring conditions are of interest to be analyzed, i.e. The result of the test is visualized in the following diagram.

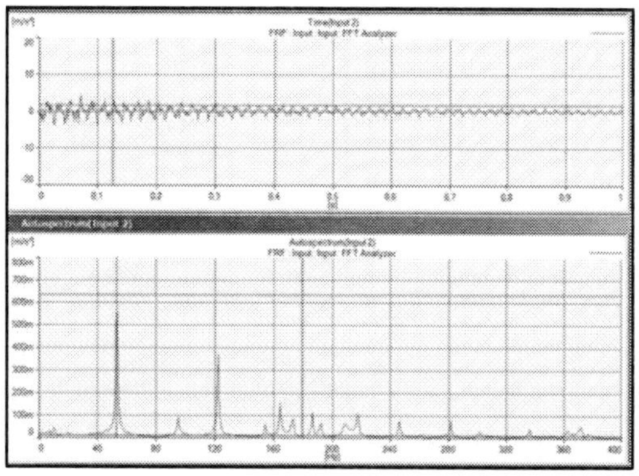

Fig. 4: Result of tangential excitation in vertical direction (own source).

The diagram shows the resonance frequencies of the conveying element (bowl) as part of the ready-to-go system with very different parameter values of the acceleration compared to the first and the second trial. It is obvious that the resonances of the system are nearly precisely matching the excitation frequency of the excitation device and multiples of it. Especially on 50Hz, which is the double of the excitation frequency, the graph has its strongest peak.

2.2 Determination of amplitudes of the operating conditions tested

For the identification of forces and tensions of the spring packages used in the bowl feeder it is necessary to do tests under working conditions. Therefore the excitation devices of the conveyer are used to get the system into vibration. Doing a pre-test it has been identified, that the maximum level to be selected is position 6 of the control dial. Trying to turn the dial furthermore up has a periodic appearing touch between spring package and exciter as result, which will influence the whole conveying characteristic of the bowl feeder negatively. The displacement of the spring package s has been measured during the vibration test by using and oscilloscope that has been turned up to the frequency of the spring package of 50 Hz. The gap between excitation device and spring package is 3 mm in non-operating condition. Turning the dial up to optimal level, the minimum gap between excited spring package and excitation device is approximately 0,3 mm. Due to the measuring possibilities and equipment a more precise measuring is not possible. This means, the maximum displacement of the spring package is 2,7 mm from the non-operating (neutral) position.

2.3 Determination of amplitudes of the operating conditions tested

To find the correct equations for the determination of the characteristic values of the spring package, the motion model of the conveying element at its mounting position to the spring packages will be described. The conveying element does a flat screw-motion as shown in the following sketch.

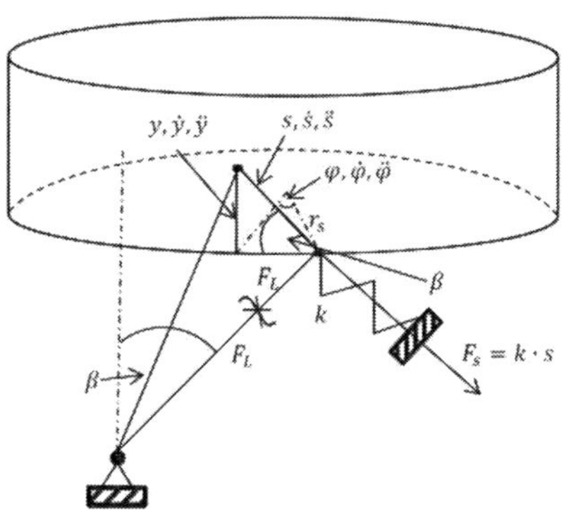

Fig. 5: Motion model of the fix point of the conveying element.

This model shows that a combined motion of rotation and vertical translation takes place after exciting the conveying element. The displacement of the point P, which describes the fixation point of the conveying element to the spring package, takes place in vertical direction by distance y and horizontal direction by the rotation angle φ. The total covered distance s finally takes place under the angle β.

$$\left(m \cdot sin^2\,\beta + \frac{J}{r_s^2} \cdot cos^2\,\beta\right) \cdot \ddot{s} + k \cdot s = 0 \qquad (1)$$

m is the mass to be accelerated, i.e. 1/3 of the total mass per spring package and decelerated by the 50 Hz frequency f of the system, ω = the pulsatance of the system, described as $2\pi f_{0,}$. J is the mass moment of inertia and m is the mass. J and m are taken from a CAD model - m = 33 kg and J = 3,9 kg m²

k is the total spring rate of the three spring packages and can be divided by 3 to get the spring rate k_1 of one spring package later.

$$k = (2 \cdot \pi \cdot f_0)^2 \cdot (m \cdot \sin^2 \beta + \frac{J}{r_s^2} \cdot \cos^2 \beta) \qquad (2)$$

Determination of the force and the moment of force:

$$F_{s1} = k_1 \cdot s_0 \qquad (3)$$

and

$$M_1 = F_{s1} \cdot l = 543\ Nm \qquad (4)$$

The bending stress finally can be determined by

$$\sigma_b = \frac{M}{W} = 339 \frac{N}{mm2} \qquad (5)$$

3 Determination of the spring tensions by simulation

The simulation will provide a confirmation respectively a variation of the displacement and the maximum bending stress in the spring package and the results will be compared and verified to the calculated results by the test.

3.1 Generation of the ANSYS model

ANSYS is a common used FEM-software, which can be used for creating the model of a complex structure to be analyzed and for the analysis process afterwards.

To get useful results it is necessary to create the model in a logical and close to reality way and to use certain material data corresponding to the real conditions existing. In case of the conveyer tested, the necessary information for modeling and analyzing the spring package are

- The dimensions of the spring package
- The force effecting the spring package
- E-module of the spring material
- Poisson-figure of the material

The dimensions of the spring package have been taken by measuring. The spring length is individually slightly modified for each bowl feeder depending on its application.

The E-module and the Poisson-figure are linked to the spring material "Durstone". This type of glass-fiber reinforced Polyester material is still available, but the manufacturer has no precise information anymore on the E-module of the approximately 5 years old material type, having been used specially for the springs. Based on the manufacturers experience, the Poisson-figure of 0,25 from the presently available "Durostone" UPM S16 and an E-module of app. 58500 MPa is selected for the simulation. The creation of the ANSYS model is done in the following steps:

- Generation of the single springs made of "Durostone"
- Generation of the distance elements and end plates made of steel
- Selection of the FEM-element type
- Generation of the FEM-screen
- Generation of the clamp situation of the spring package
- Generation of the force application
- Analysis operation

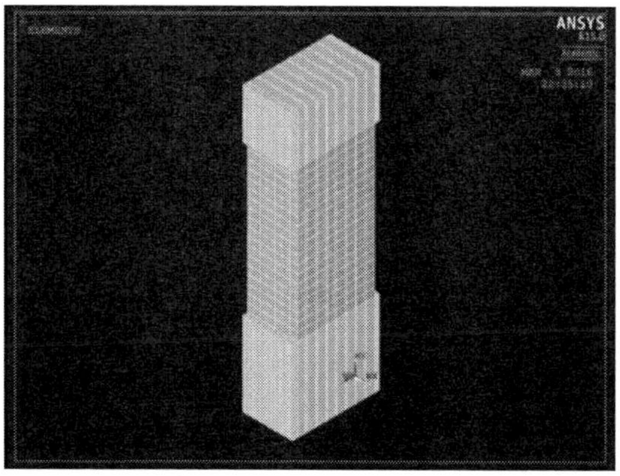

Fig. 6: Simulation of spring package with screen based on ANSYS FEM-element "solid185.

The displacement in z-direction is 2,71 mm after rounding and represents nearly the full displacement value of the simulation. This displacement represents the tangential component of the total displacement and is most interest.

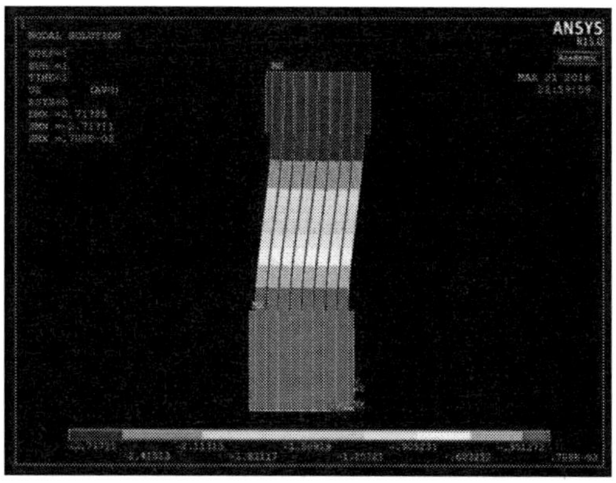

Fig. 7: Displacement in z-direction.

Looking to the existing stresses, the first stress to be analyzed is the bending stress. This is the stress that is already calculated and can be identified in the ANSYS nodal analysis as stress in y-direction, based on the coordinate system of the spring package model.

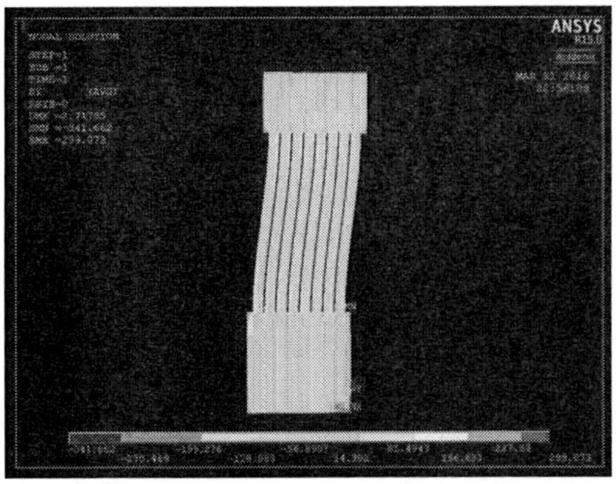

Fig. 8: Bending stress.

4 Comparison and interpretation of the results

Based on the manual calculation in chapter a bending stress of 339 N/mm² has been determined. The bending stress of the simulation supplies a value of 299 N/mm² respectively 341 N/mm² situated at the bottom clamp position of the string package. The different results seem to be mainly explained by the difference of the models used for the determination of the results. Analyzing the calculation model can be seen that the equation for the Bernoulli model. This model is different from the model used for the ANSYS simulation, particularly because the top end of the spring package in the calculation has an additional rotation degree of freedom compared to the model used in ANSYS.

5 Conclusion

This paper deals with the experimental determination and simulation of characteristic values of a bowl feeder with a focus to the displacement and stresses taking place under working conditions of the empty conveyer. It consist of an experimental determination of data of the bowl feeder such as operating frequencies and amplitudes, resonance frequencies and accelerations under different exciting situations as well as the determination of the resonance frequencies of the spring package belonging to the conveyer drive. Based on the determined maximum force, the maximum acceleration and the displacement of the spring package, bending stress of each spring package is calculated.

The determined maximum force is used to simulate displacement and stresses by the finite element software ANSYS. Differences to the previous calculations are shown. Differences between test results and simulation results are pointed out.

Finally it can be realized that the ANSYS simulation confirms the conditions measured during the test, with reference to the maximum displacement respectively amplitude of the spring package. A comparison to the simplified model of Bernoulli is done and its differences to the real installation are pointed out. The theoretical amplitude of a spring package based on the Bernoulli model is quantified

References

[1] Dresig H., Holzweißig F. (2008), Maschinendynamik 8, Auflage, Springer Verlag, Berlin.

[2] Harris C.M. (2005), Shock and Vibration Handbook, Fifth edition, McGraw-Hill, New York.

[3] Rhein-Nadel Automation GmbH: Zufuhrtechnik. [online].[Cit. 5. 2. 2013]. Available on WWW: http://rna.de/Zufuehrsysteme_Basis-geraete.mfpx?ActiveID=

[4] Nendel, K. (2008), Zweidimensionale Bewegungsformen bei Vibrationsförderern [online].. Available on WWW: http://www.vibrationsfoerdertechnik.de/ Download/ 2D-Bewegungsformen_WGTL_Tagungsband08.pdf.

Experimental and numerical analysis of impact test of carbon prepreg composite

M. Syrovatkova [1,a], M. Petru [1],

[1] *Technical University of Liberec, Studentská 2, 46117 Liberec, Czech Republic*
[a]*martina.syrovatkova@tul.cz*

Abstract: Composite structures are now increasingly used for their properties in all areas of industrial production. They gradually replace metal parts and components not only because they are lighter, but above all for their comparable and in many ways even better mechanical properties. This paper presents the experimental and numerical analysis of composite a flat samples of prepreg. The individual layers are non-crimp or interconnected plain weave. Comparison of the numerical modeling and real output of samples produced from CF prepreg, which was specially designed for absorbing the energy during impact.

Keywords: prepreg, carbon fiber, impact test, experimental analysis, numerical analysis

1 Introduction

Research and development of composite materials has wide application in many industries. The composite material is usually defined as two or more component material; the resulting properties exceed the properties of the individual parts. The continuous phase (matrix) is to bind the reinforcement to maintain its orientation and to transfer strain. Reinforcement is a carrier of strength and stiffness of the composite. Changes the composition and the orientation of the reinforcement provides a better balance between mechanical properties and weight than traditional technical materials. The advantage of fibrous composites is based on high strength of fibers in the longitudinal direction and their very small cross-section. The fibers can be used as unidirectional oriented layers without mutual interconnection (UD layer; multiaxial knitted fabric), or as interconnected system with the specified direction of orientation (fabrics, braided structure). Energy absorption is interesting and important area in terms of research and prediction of the behavior of composite materials.

2 Experimental and numerical analysis of crash test of carbon composite

Article builds on the previous experiment. Originally it was prepared four series of samples from the unidirectional carbon prepreg with epoxy matrix (nominal matrix content 38%). Table 1 contains a list of series - L, D, C, W1 including parameters). The intention was to examine the different variations of double-layer composite arrangement [3]. Impact test was conducted on a high-speed hydraulic motor. Impactor with a diameter of 20 mm was awarded the speed of 10 ms^{-1}. The samples were mounted in the axis of movement of the impactor. Impact broke the samples in the middle. The maximum force at breakdown and elongation of sample were measured; the experiments were shot with high-speed camera. Fig.1 shows the device. A new series of samples was prepared from the same unidirectional prepreg; series T and W2; parameters are listed in Table 1. Fiber layers in multilayer samples are oriented at a given angle; precise positioning enables semisolid consistency of the matrix under normal temperature of 21° C. The focus of the experiment is the examination of biaxial layer samples - the group D, C and W1. Samples designed D and C are composed of two layers of unidirectional oriented fibers, the layers mutually enclose an angle of 90°. Samples designed W1 are formed by two layers of interlaced fibers, imitate the weave fabric of plain structure. For comparison, samples were prepared: single layer L, three-layer T and the sample of aluminum sheet. The dimensions of the samples are the same. Equipment used for crash test: hydraulic motor for high speed uniaxial loading Fig.1. On the piston rod of the hydraulic motor was placed the strength tester, on the tester was mounted hemispherical impactor. The speed of the impactor for the duration of the sample penetration was constant 10 ms^{-1}. Recording was made during the test period, the position of the impactor and the maximum value of achieved strength. During the experiment was recorded with high-speed camera. The numerical model that was created in the FEM program PAM CRASH, had given the same initial and boundary conditions as an experiment. Material models for carbon composites were selected from 130 - 132 - Multi-Layered Shell Element. Initial physical properties, which were used in the model, are presented in Table 2. Sample holder was modified to remove the influence of directional orientation of the layers when clamping the sample. Impact test was conducted with the same equipment and under the same conditions. The value of energy is determined by the equation 1:

$$U = \int_{x1}^{x2} F(x)dx \tag{1}$$

Where: U - mechanical energy [J], F(x) - function of the force-deformation of the test piece [N], x - deformation of test [m].

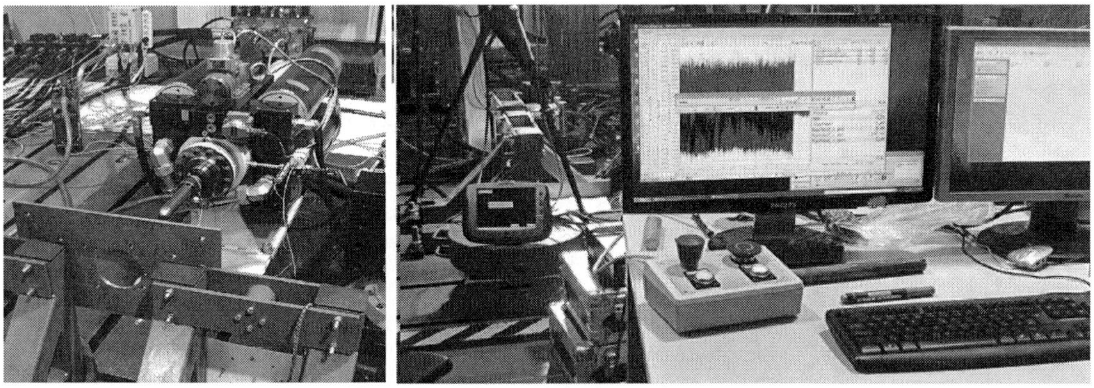

Fig. 1: Experimental device and measurement

Table 1: Parameters of experimental samples

sample	composition	thickness [mm]	areal weight [gm^{-2}]
L	1 pl;	0,16	250
D	2 pl; 0°/90°; non crimp	0,26	435
C	2 pl; ±45°; non crimp	0,30	435
W1	2 pl, plain weave	0,29	420
T	3 pl; 0°/90°/45°; non crimp	0,38	600
W2	2 pl, plain weave	0,29	425

Table 2: Physical parameters of numerical model

Material	Density [kg.m^{-3}]	Elastic modulus [GPa]		Shear modulus [GPa]		Poisson ratio [-]		Tensile strength [GPa]	Elongation [%]
		$E_{11}^{f,m}$	$E_{22}^{f,m}$	$G_{12}^{f,m}$	$G_{23}^{f,m}$	$\nu_{12}^{f,m}$	$\nu_{23}^{f,m}$		
Carbon fiber	1750±150	230	15	24	5.4	0,279	0,49	2,3±1,2	1,9±0,6
Epoxy matrix	1150±370	3.573	3.573	1,31	1,31	0,345	0,345	0,067±0,033	3,6

Numerical models were created for study of the mechanical properties of the samples of carbon composites. Models are based on analytical models, ie. Voight model, Reuss model, Chamis model and Halpin -Tsai model [2, 3]. From these simulations can be obtained elastic constants E_{22}, G_{12}, G_{23} (2-6), which can be compared with a numerical model [3]. Chamis model to calculate other elastic constants introduced into the solution to the square root of volumetric fibers $\sqrt{V^f}$; fibers are incompressibility in equations (3, 5-6), which is in accordance with the law of conservation of mass.

$$\frac{d\sigma_{11}}{d\varepsilon_{11}} = V^f \frac{d\sigma^f}{d\varepsilon^f} + V^m \frac{d\sigma^m}{d\varepsilon^m} \quad \Rightarrow \quad E_{11} = V^f E_{11}^f + V^m E^m \tag{2}$$

$$E_{22} = \frac{E^m}{1 - \sqrt{V^f}\left(1 - E^m / E_{22}^f\right)} \tag{3}$$

$$\nu_{12} = V^f \nu_{12}^f + V^m \nu^m \tag{4}$$

$$G_{12} = \frac{G^m}{1 - \sqrt{V^f}\left(1 - G^m / G_{12}^f\right)} \tag{5}$$

$$G_{23} = \frac{G^m}{1 - \sqrt{V^f}\left(1 - G^m / G_{23}^f\right)} \tag{6}$$

3 Results and conclusion

In the experiment were measured the following values: the maximum force and elongation before a crack of the sample. The results of experimental measurement were compared with FEM models that have worked with an error about of 9%. Models showed stress on samples distribution and failure of structure (Fig.2). Approximate value of the energy needed to crack the sample was calculated. Values are given in Table 3. Samples after the test are seen in Fig.3. New clamping device has helped to reduce error of measurement. All samples from the series T, W2 were broken in the middle from the impactor. Graph for describe of force versus elongation of the samples for a series of W2 is see Fig.3, where are a curves of force and the elongation of three samples for same series is similar. The measurements produced the following results as for examples: The quantity and orientation of non-crimp layers have a direct influence on the strength of the sample. Strength increases with the number of layers (D, C, T). Higher quantity of non-crimp layers reduces the flexibility of the sample and the elongation at impact (D, C, T). The highest values of strength and elongation achieved samples of plain weave (W2).

Table 3: Results of measurement

Variable/sample series	L	D	C	T	W1	W2	Al
Average of strength [kN]	0,63	3,83	4,74	4,48	2,69	4,83	1,51
Average of displacement [mm]	7,41	14,84	19,67	10,55	17,36	19,96	11,3
Average of energy [J]	-	-	-	21,30	-	38,38	-

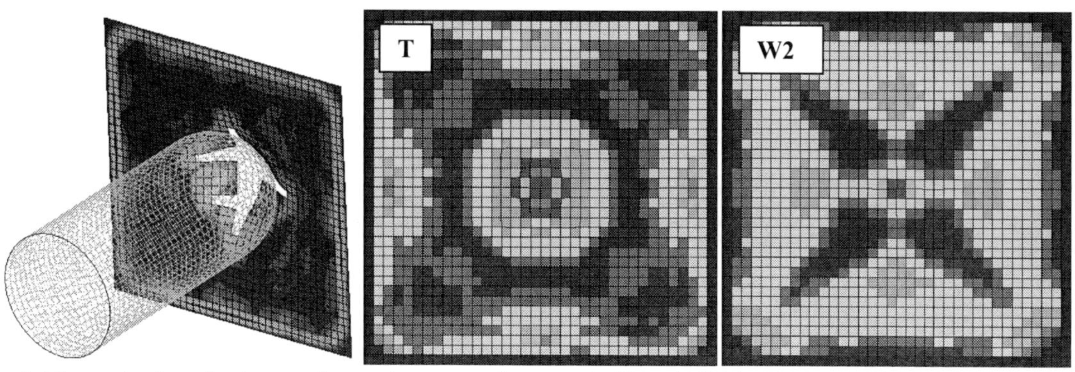

Fig.2: Numerical analysis: results major stress for sample T and W2 before a crack

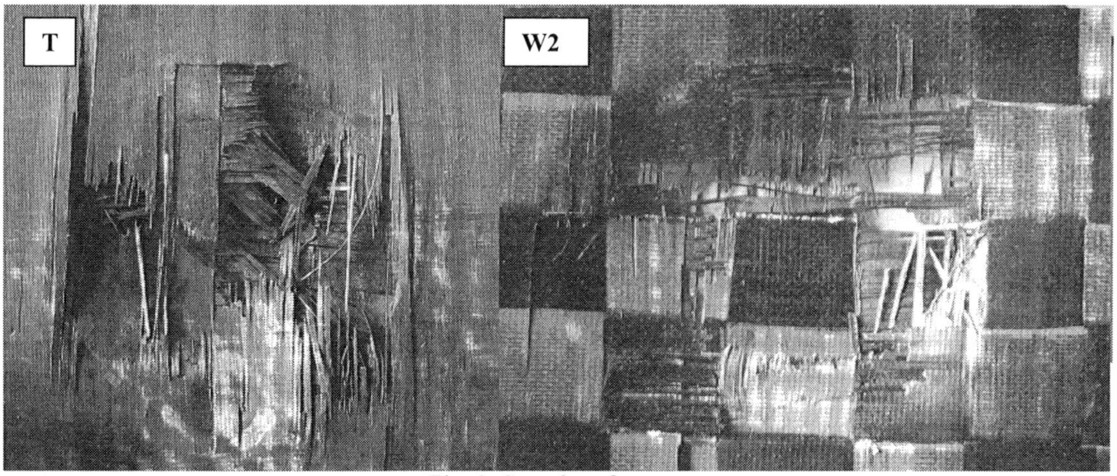

Fig. 3: Samples after test

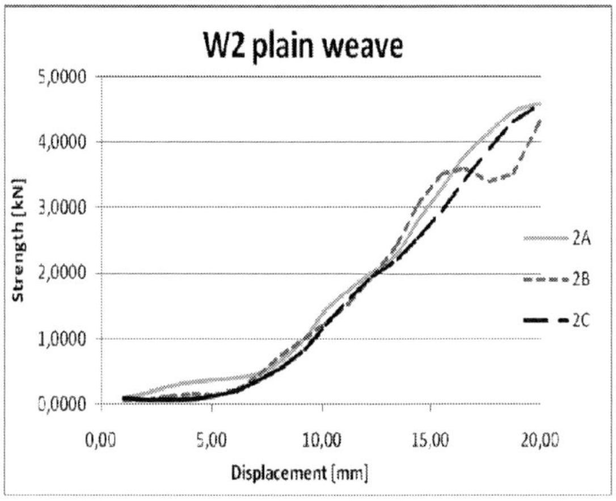

Fig.4: Graph of results of strength

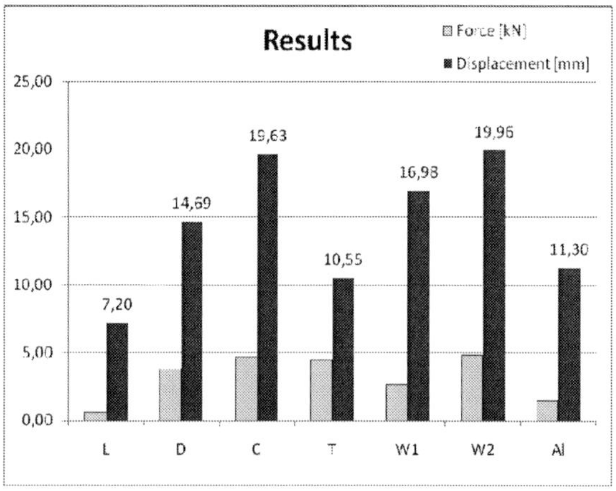

Fig.5: Graph of results: force and displacement

Acknowledgement

The results of this project LO1201 were obtained with co-funding from the Ministry of Education, Youth and Sports as part of targeted support from the "Národní program udržitelnosti I" program.

References

[1] M. Petru, M. Syrovatkova, T. Martinec, P. Lepsik, Analysis of changes in the surface quality of a UD prepregs composite due to mechanical loading, Material Science Forum, 818 (1), p.109-112, 2015.

[2] M. Petru, J. Broncek, P. Lepsik, O. Novak, Experimental and numerical analysis of crack propagation in light composite materials under dynamic fracturing, Komunikacie,16 (3A), pp. 82-89, 2014.

[3] M. Petru, M. Syrovatkova, M. Novotna, Crash test of carbon composites, Applied Mechanics and Materials, 821, p.385-391, 2015.

[4] S.S.Morye, P.J.Hine, R.A.Ducket, D.J.Carr, I.M.Ward, Modelling of the energy absorption by polymer composites upon ballistics impact, Composite Science and Technology 60 (2000) 2631-2642, received 5 April 2000, accepted 30 May 2000

Optimization of composite airframe rib using tailoring blank technology

J. Šedek [1,a]

[1] *VZLÚ – Aerospace Research and Test Establishment, Czech Republic*
[a] *sedek@vzlu.cz*

Abstract: The contribution deals with work concerning the design and optimization of composite rib intended for use in airliner door construction. New progressive tailoring blank technology of manufacturing thermoplastic composites requires specific demands on rib design. Replacing the metal machined rib by the composite one brings advantages of lower weight and effectively utilized material with minimal waste. The optimization process involved cut-out design and lamina placement. Numerical tools were utilized using finite element software ABAQUS and Hashin and LaRC failure theories were incorporated. Based on the optimization, the real part was manufactured and the static strength was evaluated. The experimental deformation and strain field measured by non-contact optical measuring system (DIC) was in good accordance with numerical analysis. Finally, the test of static strength was performed and strength according to strength theories was evaluated.

Keywords: FEM; optimization; rib; thermoplastic composite; tailored blank.

1 Introduction

Continuous fibre-reinforced thermoplastics belong to the group of high-performance composites and are increasingly utilized in design of aerospace structures. Due to their properties, thermoplastic composites are able to meet specific aerospace requirements. Thermoplastics can be manufactured by thermoforming, which is in some aspects similar to the process utilized for non-reinforced plastics. The thermo-forming process is normally conducted at a processing temperature that is higher than the melting temperature of the thermoplastic resin to allow the textile prepreg to deform. The composite part can vary in thickness, fibre orientation, material composition and shape. It can be achieved through the stacking arrangement of the tailored prepreg blanks. However, the thermoforming process of a part with variable thickness is not elementary and takes advanced demands on moulds construction and process timing, especially during heating before co-consolidation. In the work package in the project "Research and development of modern technology processes for new application of high-tech reinforced thermoplastic" supported by Technology Agency of the Czech Republic, the composite demonstrator representing the rib of the aircraft door structure was designed and manufactured.

2 Composite rib design and optimization

The rib is located opposite the console of the door lock and transfers the load resulting from the cabin pressure acting on the door to the fuselage construction. The rib is reinforced at the lower flange by fuselage skin and fixed with auxiliary and edge frame at vertical margins. Basic dimensions of the rib are determined by original construction; almost triangular shape with the height of 153 mm and the length of 333 mm. The upper and lower parts contain reinforcing flange and the lower part is in construction joined with the skin. The rib is loaded in shear through the longer vertical edge, where the moment of load is applied. The moment of load results from the stop fitting connected to the longer vertical edge and loaded by 36 kN. The upper flange transfers tension stress and the lower flange transfers pressure stress.

2.1 Lamina stacking pattern

Prepregs are stacked in the rib construction in a sequence with varying weave orientation and different blanks shape, which result in variable rib thickness. The location of the thickness change is the point of eventual difficulties in manufacturing process and interlaminar issues in service. In this point the prepreg stacking can be realized in different ways.

The simplest method to ensure a joint of adjacent layers is to add the reinforcing layers of lamina directly on the base plies and thermoform them together. By this process a simple bond joint can be formed. All plies remain straight but the interlaminar stresses in the bond joint can cause peeling of reinforcing plies. The significance of the bond joint shape made by thermoforming technology was investigated in recent work [1]. Continuous change of the thickness in tapered bond joint is preferable due to its lower interlaminar stresses in contrast to the simple lap bond joint with common end. It is also preferable that the material carries the load in entire joint when tapered.

Preferable way to realize connection of continuous and ending prepreg layers is the implementation of ending ply between continuous plies. Tests of simple and spliced joints show resistance of spliced joint against interlaminar damage under static and cyclic loading. The ending plies are covered by continuous plies which act against the increase of peel stress and prevent delamination. The construction of splice joints can also be realized in different manners. The thickness can vary in steps with straight base ply or the thickness can vary from the center plane of the laminate as shown in Fig. 1. The stacking according to the straight base ply is useful when designing the mold because one part of the negative mold can be constructed as flat basis.

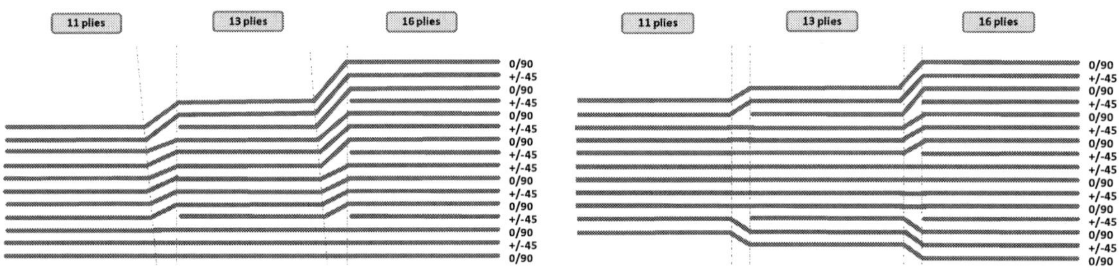

a) straight base ply

b) straight plies near the center plane

Fig. 1: Spliced joint of laminate

Four possible arrangements were specified a priori to optimization with following thicknesses: 18 plies (5,58 mm), 16 plies (4,96 mm), 13 plies (4,03 mm) and 11 plies (3,41 mm). Plies consist of carbon fabric reinforced thermoplastic prepreg with mass per unit area of 285 g/m^2 and matrix weight content of 43 %. The carbon fabric has satin weave 5HS and the matrix is polyphenylene sulphide (PPS). Stacking sequence of the demonstrator rib was realized according to the concept of straight base ply shown in Fig. 1a). The 18 plies arrangement was made in the fixture area at the higher vertical margin.

2.2 Prepreg blank tailoring

Tailoring of the prepreg blanks was conducted in order to decrease the weight and optimize load transfer through embedding cut-outs and through design of the prepregs shape. Series of finite-element (FE) analyses were carried out in order to optimize the shape and position of prepreg blanks and cut-outs. Some optimization steps are shown in Fig. 2. In basic design, the reinforcement of attaching section was applied and then, the effect of thickness variation was studied. In following steps, the cut-out effect was analysed and in pre-final steps the additive reinforcement of the margin was made where the moment of load is applied. Finally, the reinforcement was expanded up to the upper and lower flange and little correction of the thinnest area was applied. The cut-outs were designed in accordance with fibres direction in the prepreg so that the continuity of fibres was kept. It is contrast to cut-outs in metal webs, which are almost always designed circle shaped.

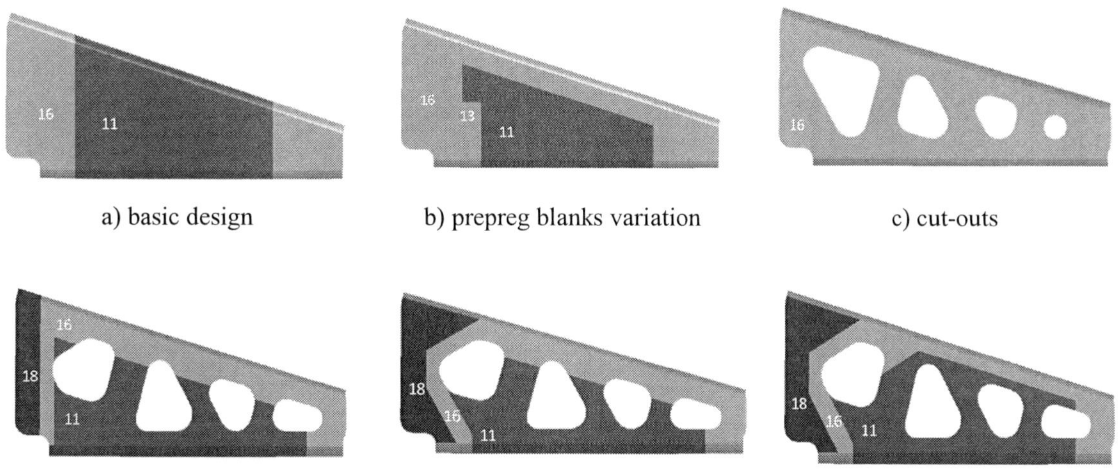

a) basic design b) prepreg blanks variation c) cut-outs

e) reinforcement of attach section f) expansion of attach section g) final design

Fig. 2: Steps of optimization design; numbers in figures indicate the number of plies.

2.3 Details of finite element model and evaluation

Finite element analyses were carried out in ABAQUS 6.13.2 software. Different lamina sections were specified according to stacking sequence. Nominal thickness of the lamina is 0,31 mm and the material constants as the elastic moduli and the strengths are specified in Tab 1. The composite was represented by shell elements of S4R a S3R type with characteristic element size of 5 mm (see Fig. 3a). Lamina stacking was prescriebed according to Fig. 1a) so only details of thickness change was not modeled properly.

Tab. 1: Material constants of the lamina t = 0,31 mm.

Material constant	E_{11} [N/mm^2]	E_{22} [N/mm^2]	E_{33} [N/mm^2]	ν_{12} [-]	ν_{13} [-]	ν_{23} [-]
Value	58000	58000	10500	0.05	0.41	0.41
Material constant	G_{12} [N/mm^2]	G_{13} [N/mm^2]	G_{23} [N/mm^2]			
Value	4100	3480	3480			
Material constant	Tensile strength [MPa]	Compressive strength [MPa]	Tensile strength [MPa]	Compressive strength [MPa]	In-plane shear strength [MPa]	
Value	750	590	750	590	105	

Throughout the optimization, the stress fields and damage variables were evaluated. The aim was to decrease stress concentration in the corners of cut-outs and to regularize the stress field across the rib. Especially principal stresses across all plies were evaluated and the damage indications according to Hashin [2] and LaRC03 [3] failure criteria were evaluated as well. After several optimization steps, the excessive stress concentrations in the corners of cut-outs were eliminated.

Critical point was indicated in the upper part near the longer vertical edge, where the moment of load is applied (see Fig. 3b). Moreover, in this point, the bold joint is located. According to the prescribed load, the bold joint transfers the load of 1.6 kN. To ensure no damage due to bearing pressure the minimal bolt diameter is 8 mm with safety factor of 2.5. So this point is weakened in addition by the hole for the bolt.

Failure criteria indicated the critical point consistently. LaRC03 indicated higher value of failure index than Hashin's criteria (0,63 versus 0,23 at the load of 36 kN). Due to the location of the critical point, it is not possible to evaluate the criteria each other experimentally. In order to do so, the critical point should be located at the joint-free area.

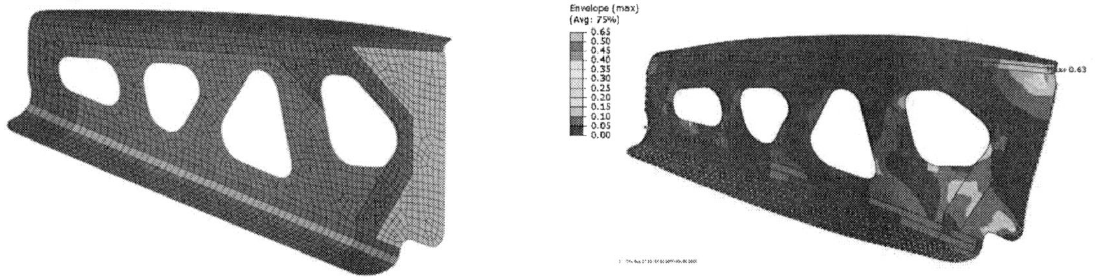

a) FE mesh – reduced shell elements

b) Index of LaRC03 failure criteria, Load 36 kN, deformation scale factor 25x

Fig. 3: FE model of the rib demonstrator.

Buckling was evaluated during optimization steps in addition. The lower flange is loaded in pressure and eventual loss of stability could occur. The load at the loss of stability considerably depends on boundary conditions which in optimizing steps do not follow the real conditions fully, however some simplification was accepted. The loss of stability of optimized alternatives was determined at the load that is 5 times greater than the specified load. Appropriate eigenshape shows buckling of the web under the cut-out near the loaded edge.

Conclusion

The rib demonstrator from high performance thermoplastic composite was designed and through FEM analysis optimized. The step-by-step optimization process was applied resulting in optimum prepreg blank tailoring and cut-outs shape design. Utilizing failure criteria the critical point was determined. The LaRC03's failure index in the critical point was higher than the Hashin's failure index.

Acknowledgement

This work was funded by the support of applied research and experimental development provided by the Technology Agency of the Czech Republic in the project ALFA TA03010209.

References

[1] J. Šedek, R. Hron, M. Kadlec, Bond Joint Analysis of Thermoplastic Composite Made from Stacked Tailored Blanks. Applied Mechanics and Materials, Vol. 827, 2015. pp.161-168, ISSN 1662-7482

[2] Z. Hashin and A. Rotem, A Fatigue Criterion for Fiber-reinforced Materials, Journal of Composite Materials, vol. 7 (1973), pp. 448–464.

[3] G.C. Davila. et al., Failure Criteria for FRP Laminates, Journal of Composite Materials, Vol. 39 (2005), pp. 323-345.

[4] J. Křena, P. Roškanin, Vývoj a zkoušky optimalizovaného žebra z vyztuženého termoplastu C/PPS, Transfer, Praha, VZLÚ, Vol. 27 (2016), pp. 24-27

The mechanical properties of fine recycled aggregate concrete after freeze – thaw cycles

M. Šefflová [1,2 a], T. Pavlů [1,2]

[1] *Czech Technical University in Prague, Faculty of Civil Engineering, Department of Building Structures, Thákurova 7, Prague, 166 29, Czech Republic*

[2] *Czech Technical University in Prague, University Centre of Energy Efficient Buildings, Třinecká 1024, Buštěhrad, 273 42, Czech Republic*

[a] *magdalena.sefflova@fsv.cvut.cz*

Abstract: This paper is focused on mechanical properties of fine recycled aggregate (FRA) concrete after freeze – thaw cycles. The use of FRA is connected with uncertainties and doubts. The most unknown aspects of FRA concrete are related to durability. This paper presents results of mechanical properties of concrete containing FRA after freeze – thaw cycles. The FRA from concrete waste originated from recycling plant in the Czech Republic was used in experiments. Four concrete mixtures were analyzed in total. The first, concrete mixture with natural sand, cement and water was prepared as a reference concrete mixture. In other mixtures, natural sand was replaced by FRA in various replacement ratios, specifically 10%, 20% and 30%. All concrete mixtures were designed with the same parameters for clear comparison. As a consequence of the quality and amount of FRA, the test results showed the influence of FRA to the freeze – thaw resistance of concrete.

Keywords: fine recycled aggregate concrete; construction and demolition waste; mechanical properties; freeze – thaw resistance; durability.

1 Introduction

The use of recycled aggregate as a partial replacement of natural aggregate in concrete is one of the possible ways how to save natural resources. The use of coarse recycled aggregate in concrete is already accepted. However, it is possible with specific restrictions [1]. The use of FRA is connected with uncertainties and doubts. It is generally agreed that the FRA influences the properties of concrete. Concrete containing FRA shows worse physical and mechanical properties [2]. The most unknown aspects of FRA concrete are related to durability [3]. This paper presents results of mechanical properties of concrete containing FRA after freeze –thaw cycles.

2 Experimental research program

2.1 Materials

Portland cement CEM I 42.5 R, natural sand, water and FRA were used in the experimental part. The FRA was obtained from recycling plant in the Czech Republic. The FRA originated from demolished concrete structures. Afterwards FRA was crushed in the laboratory jaw crusher and separated to fraction of 0 – 4 mm. There were tested geometrical and physical properties of the FRA. The physical properties were tested by pycnometric method according to ČSN EN 1097 – 6 [4] . The test results were compared with natural sand. The physical properties of FRA and natural sand are given in Tab. 1.

Tab. 1: Physical properties of recycled and natural sand

Type of aggregate	Fraction [mm]	Density [kg/m^3]	Water absorption capacity [%]
Recycled aggregate	0 - 4	2090	8,29
Natural aggregate	0 - 4	2600	2,00

2.2 Concrete mixtures proportion

A total four concrete mixtures were designed and prepared. All concrete mixtures were designed with the same w/c ratio and the same amount of cement. The first concrete mixture was reference (FRA) which did not include FRA. Natural sand was replaced by FRA in concrete mixtures FRA10, FRA20 and FRA30 in varying ratio (10 %, 20 %, 30 %). Details of designed concrete mixtures are given in Tab. 2.

Tab. 2: Concrete mix proportion, per cubic meter

Designation	REF	FRA10	FRA20	FRA30
Cement [kg]	486	486	486	486
Water [kg]	243	243	243	243
Sand [kg]	1458	1312	1166	1021
FRA [kg]	0	146	292	437

2.3 Testing

Beams of dimensions 40x40x160 mm were used for testing. The freeze – thaw resistance was tested according to ČSN 73 1322 [5]. There were carried out a total 100 freeze – thaw cycles. The freezing cycle takes 4 hours of freezing by air of minus 20°C and the thawing cycle takes 2 hours of thawing by water of 20°C.

The compressive strength, tensile strength and dynamic modulus of elasticity were measured after 50, 75 and 100 freeze – thaw cycles. The dynamic modulus of elasticity was tested nondestructive by the ultrasound method according to [6]. The tensile strength was examined on beams according to ČSN EN 12390-5 [7]. The compressive strength was examined on fragments of beams according to ČSN EN 12390-3 [8]. The eguipment, which was used for testing freeze – thaw resistence, is showed in Fig. 1.

Fig. 1: The equipment for testing freeze – thaw resistance

3 Results and discussion

3.1 The compressive strength

The graph in Fig. 2 shows development of compressive strength after 50, 75 and 100 freeze – thaw cycles.

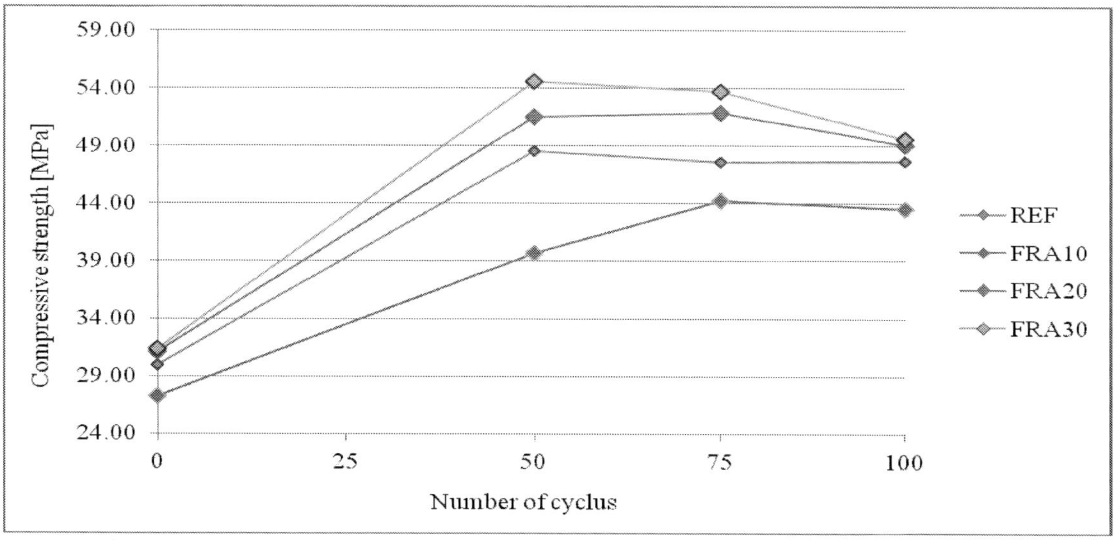

Fig. 2: Test results of compressive strength after 50, 75 and 100 freeze – thaw cycles

3.2 The tensile strength

The graph in Fig. 3 shows development of tensile strength after 50, 75 and 100 freeze – thaw cycles.

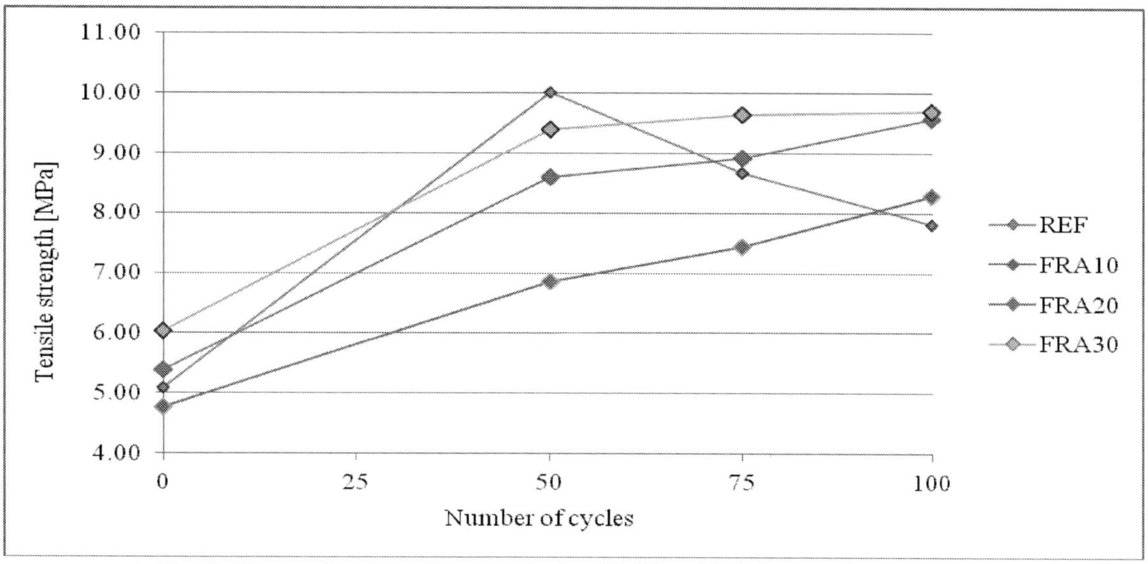

Fig. 3: Test results of tensile strength after 50, 75 and 100 freeze – thaw cycles

4 Conclusion

In this paper, experimental results of mechanical properties of FRA concrete after freeze – thaw cycles are presented and discussed. From the test results and discussions, the following conclusions are drawn:

- The use of FRA as partial replacement of natural sand influences the properties of concrete.
- The results show that the freeze-thaw resistance depends on substitution ratio of natural sand.
- The concrete samples with FRA showed good mechanical properties after 100 freeze – thaw cycles.
- It is possible tu assume good freeze – thaw resistance of FRA concrete.

Finally, it is possible to say that it was proved, good mechanical properties of concrete samples with recycled aggregate. It is possible to assume good freeze – thaw resistance of FRA concrete. Nevertheless, it is necessary to say that freeze – thaw resistance of recycled aggeragte concrete depends on many other aspects [9]. It is necessary to verify this results.

Acknowledgement

This work has been supported by TA03010501 Optimized subtile frame for energy efficient buildings and SGS15/182/OHK1/3T/11 Verification of durability and lifespan of cement composites and recycled concrete.

References

[1] ČSN EN 206. Concrete – Specification, performance, production and conformity. Prague: Czech Office for Standards, Metrology and Testing, 2014. (in Czech)

[2] L. Evangelista, et al., Mechanical behavior of concrete made with fine recycled concrete aggregates, in: Cement and Concrete Composites 29, 2007, pp. 397 – 401.

[3] Xiao, J. Z., 2005. The state-of-the-art on long-term properties of recycled aggregate concrete. Structural concrete, 21 (3), 2005, pp. 73 – 76.

[4] ČSN EN 1097 – 6. Tests for mechanical and physical properties of aggregates – Part 6: Determination of particle density and water absorption. Prague: Czech Office for Standards, Metrology and Testing, 2013. (in Czech)

[5] ČSN 73 1322. Determination of frost resistance of concrete. Prague: : Czech Office for Standards, Metrology and Testing, 1968. (in Czech)

[6] Holčapek O. et al., Destructive and nondestructive characteristics of old concrete. Advanced Materials Research 1054 (2014) 243-247.

[7] ČSN 12 390 - 5. Testing hardened concrete - Part 5: Flexural strength of test specimens. Prague: : Czech Office for Standards, Metrology and Testing, 2009. (in Czech)

[8] ČSN 12 390 - 3. Testing hardened concrete – Part 3: Compressive strength of test specimens. Prague: : Czech Office for Standards, Metrology and Testing, 2009. (in Czech)

[9] Pavlů T., Šefflová M. Experimental determination and estimation of water absorption capacity of hardened concrete with recycled aggregate: EAN 2014 - 52nd International Conference on Experimental Stress Analysis, Marianske Lazne, 2014.

Determining the causes of the solenoid coils incorrect winding

L. Ševčík [1,a], A. Lufinka [1]

[1] *Technical University of Liberec, Studentská 2, 46117 Liberec, Czech Republic*
[a] *ladislav.sevcik@tul.cz*

Abstract: Accurate placement of the wire is required during solenoid coils winding. Because many incorrectly wound coil were produced on the used winding machine, the analysis of this situation causes were to be realized. The winding machine precision measurement and simulation of the wire winding to the coil composite skeleton are described in this article.

Keywords: solenoid coil; incorrect winding; measurement; FE simulation.

1 Introduction

Accurate placement of the wire is required during solenoid coils winding. It means that the wire turns must be placed side by side, the wires must not be crossed (see Fig. 1). Because used winding machine produced a lot of incorrectly wound coil, the analysis of possible causes was realized. The homing head and coil rotation poor synchronization was the first possible cause of the incorrect winding so the machine precision measurement was first step which was carried. Because the bad winding cause was not revealed by this measurement, the winding process FEM modeling was done in the next step.

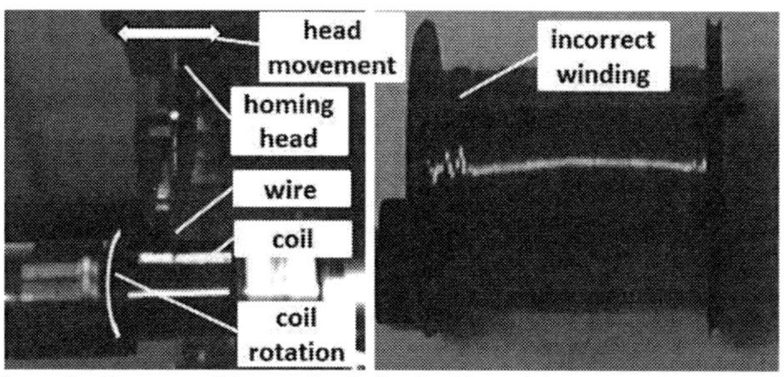

Fig. 1: The winding machine principle (left) and incorrectly wound coil (right).

2 The winding machine precision measurement

Determine of the winding machine synchronization precision was the main measurement task. The measurement arrangement is shown in the Fig.2.

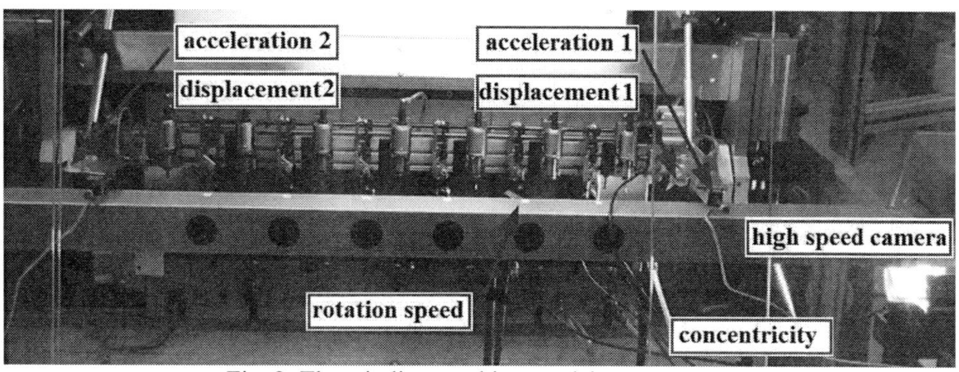

Fig. 2: The winding machine precision measurement.

One of the coils was supplemented by the aperture and rotation speed was measured by the digital laser sensor. Two very precision movement sensors were used for the homing head left and right side movement measuring and its vibrations were measured by two accelerometers as the additional information. One of the coils concentricity was measured by the contactless inductive sensor. The winding process was recorded by the high speed camera too. All sensors signals and the camera movie were synchronized by the Dewe5000 measurement unit [1].

Several cycles with different winding speeds and different head required displacement were measured and no problem with winding machine synchronization and displacement precision was found (see Tab.1).

Tab. 1: The homing head displacement error.

Measurement number	1	2	3	4
required homing head displacement	17.792	17.742	17.692	17.692
measured homing head displacement	17.750	17.693	17.690	17.690
homing head displacement error	**0.042**	**0.049**	**0.002**	**0.002**

The homing head displacement error is not greater than 0.05mm and this value is into the machine declared range. The example of the head displacement with the coil rotation synchronization is shown in the Fig.3.

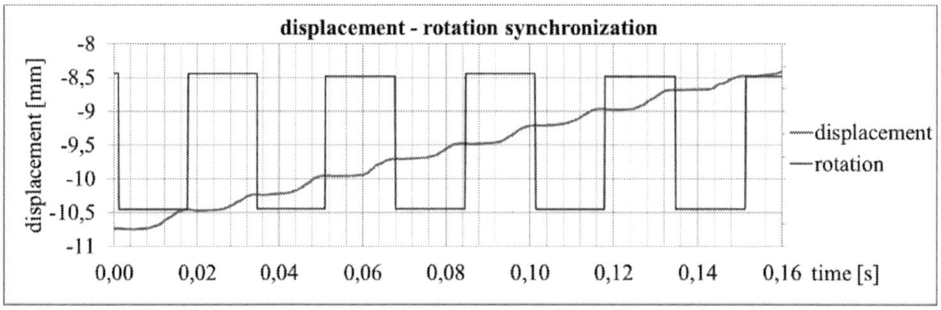

Fig. 3: The head displacement and the coil rotation synchronization.

Each coil turn is represented by one pulse from the digital sensor and regular displacement increment is evident in every step. So, the synchronization of the head displacement with the coil rotation is perfect too. It means that the winding problems are not caused by the imprecise machine function. Nevertheless some coils were incorrect wound during these measurements.

The analysis of video records showed that winding errors always start close to the coil sidewalls and it indicated a potential problem with their stiffness. The sidewall is probably pushed by the wire winding force and the wire is wound into incurred gap instead of the next layer. The homing head position and the coil winding position are out of sync, the wire is not guided perpendicular to the coil axis (see Fig.4) and incorrect winding is caused by them.

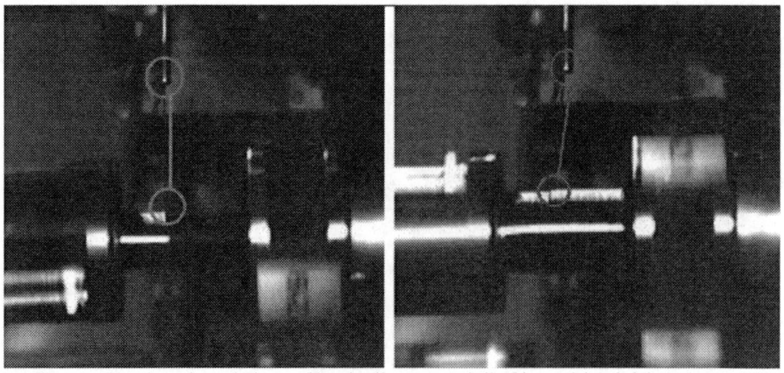

Fig. 4: The head and the coil winding good (left) and bad (right) position.

The coil sidewall dislodging could not be directly measured because no space for the sensor location is in the winding machine. Therefore the winding process was modeled by FEM for this presumption confirmation.

3 The winding process FEM modeling

The Ansys software was used for the winding process modeling [2]. Winding wire was loaded by axial force 20N. Between the wires was set frictional contact. The coefficient of friction was 0,1. Also between the wire and the composite bobbin was set frictional connections 0,1. The wire winding force is forced between the coils and the whole spool (see Fig 3), where the right part of the figure is a mesh and left part - deformation results for the wrap angle of the wire 40°. Maximum deformation is in bend of wire, but through the contact between previous coil and bobbin sidewall makes the deformation of sidewall.

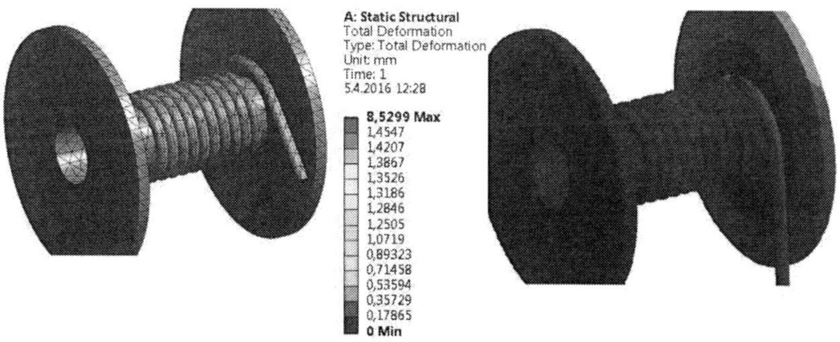

Fig. 3: The winding process FEM modeling net and deformation results.

This deformation makes the stress in wire, also in sidewall of bobbin. On the Fig. 4 is maximal reduced stress in the wire. The wire is during winding bended and tensioned. But this stress is mode bigger than in the side of bobbin. Stiffness of bobbin is too small and sidewall is easy bended. Stress in beginner of winding is given by small diameter of coil on core of bobbin and often make permanent deformation of wire. This is only static calculations; in real the speed has affects on value of stress.

Fig. 4: The winding process FEM calculation of stress HMH.

The following graph (Fig. 5) shows the results of successive static calculations. Calculations were made from the first contact with the face of the coil wire. It was also made of the wire winding 5 degrees. The calculation was deducted deformation forehead and plotted for each angle in the graph.

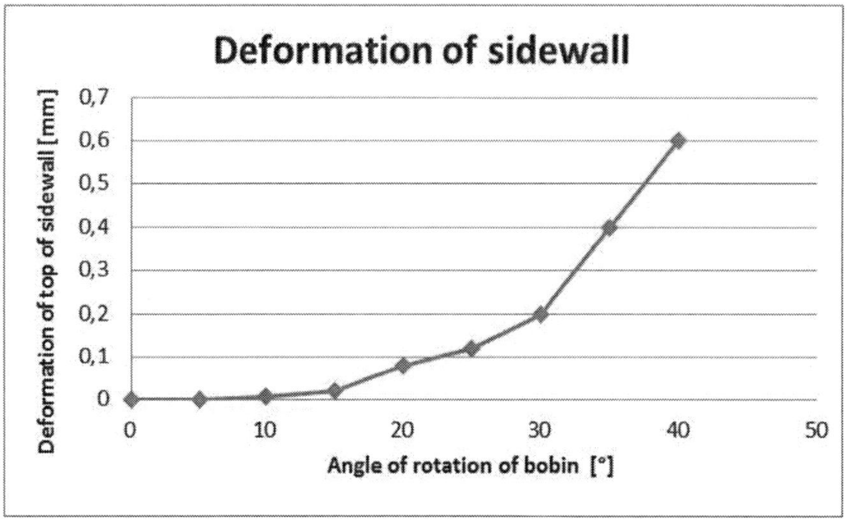

Fig. 5: Deformation of sidewall depending on contact angle.

This effect may or may not always occur. It depends on the position of the rack, on the contact points and the state of the shape geometry of the wire. In practice, the error usually occurred in the fifth layer of coils. The highest cost layer number can cause the greater the probability of errors.

4 Conclusion

The carried measurement confirmed that the winding machine works correctly, all monitored signals are into prescribed accuracy limits. Video records showed problems with the coil sidewalls, their distance increases due to the winding wire tension force. It was confirmed by FEM model, too. The coil body will have to be redesigned (sidewalls thickness increased) for the winding problem solving [3].

Acknowledgement

The results of this project LO1201 were obtained with co-funding from the Ministry of Education, Youth and Sports as part of targeted support from the "Národní program udržitelnosti I" program.

References

[1] Retrieved from: http://www.dewetron.com

[2] Ansys 15.0 reference manual

[3] The research report "Analysis of defects bobbin winding" Technical University of Liberec, 2015

Determination of Initial Shear Modulus of Woven Fabrics in Various Directions

D. Šimić Penava [1,a], Ž. Penava [2], J. Krolo [1]

[1] *Faculty of Civil Engineering, University of Zagreb, Kačićeva 26, 10000 Zagreb, Croatia*
[2] *Faculty of Textile Technology, University of Zagreb, P.b. Filipovića 28a, 10000 Zagreb, Croatia*
[a] *dianas@grad.hr*

Abstract: Shear behavior is one of the most important mechanical characteristics that contribute to performance and appearance of woven fabrics. Because of the anisotropy, shear properties of woven fabric are tested in various directions. This paper describes an experimental study of shear properties of woven fabric when shear force acts on specimens that are cut at different angles to the direction of the weft. Tests are conducted on woven fabric specimens that are fastened in two parallel clamps that are placed in tensile tester. Orthogonal plain weave cotton woven fabric was used. Based on diagram of the measured values of force and corresponding vertical displacement, shear angles and corresponding shear stresses are calculated. Initial shear modulus of woven fabrics was determined experimentally and theoretically in various directions.

Keywords: woven fabric; initial shear modulus; pure shear; anisotropy; shear angle.

1 Introduction

During shear deformation, the woven fabric yarns experience large angular variation between warp and weft yarns. Test methods shear properties of woven fabrics are listed in the existing literature [1]. To understand the mechanisms of woven fabric shear behavior, shear apparatus which measure woven fabric shear properties are described [2]. At the beginning, researches have mainly focused on the shear behavior of woven fabrics in both principal directions because it affects much woven fabric behavior, and later their attention has been paid to woven fabric shear properties in various directions. Anisotropy is the characteristics of woven fabric that affects its physical and mechanical properties [3]. Determination of shear stresses and strains in various directions involves complex mechanisms that provide information about shear properties of woven fabrics in various directions where the angles between two sets of yarns change in the intersecting points. The aim of this study is to determine the shear angle, shear forces of woven fabric using clamps that are specially designed in laboratory for measuring shear forces. Shear force is acting on woven fabric specimens that are cut at different angles to the weft direction. The influence of anisotropy to the initial shear modulus values of a woven fabric was analyzed. The degree of agreement between experimental results and calculated obtained values of the initial shear modulus was determined. The structure of the apparatus and the measurement procedures are illustrated in this paper. Woven fabric specimens that are fastened in two parallel clamps that are placed in tensile tester will be used for experimental determination of shear properties of the woven fabric.

2 Theoretical Overview

The woven fabrics are elastic orthotropic materials at very small deformation that are defined as orthotropic plates with two mutually perpendicular planes of elastic symmetry [4]. The x-axis is in the weft direction, and the y-axis is the warp direction of woven fabric. For a particular state of plane stress in element we can determine the plane in which only shear stresses act and normal stresses are zero. Such a case of stress is called a pure shear. Isolated element of woven fabric *ABCD* which is subjected to pure shear is shown in Fig. 1. During shear deformation, the woven fabric yarns experience large angular variation between warp and weft yarns.

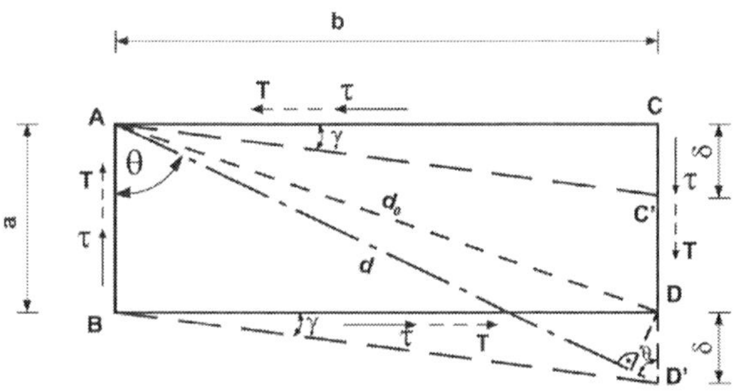

Fig. 1: Woven fabric element subjected to pure shear - rectangular shape of element.

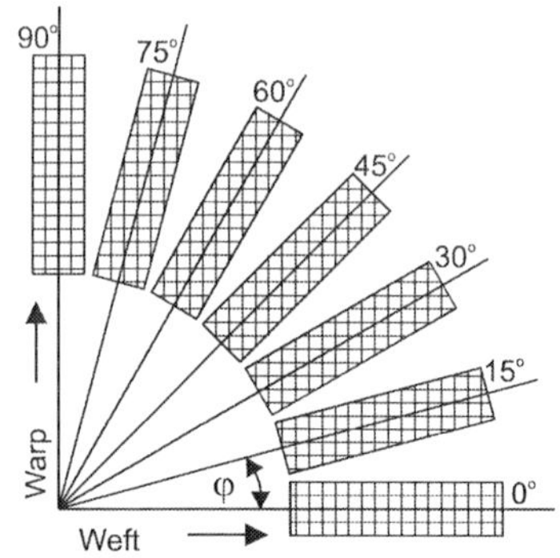

Fig. 2: Schematic view of directions (angles) of cutting specimens.

If the side AB is motionless, then under the action of shear force T which acting in the plane of the element side shear stress τ will occur and it will come to the appearance of shearing the side DC parallel with the side AB for the amount CC'=DD'=δ (mm) which is called the absolute shear (Fig. 1). Change the right angle is denoted with γ (rad) and is called the relative shear strain or shear angle. It is a measure of deformation. During the elastic deformation the shear angle γ is very small and it is equal to the absolute shear divided by the spacing between the shear plane, the Eq. (1):

$$tg\gamma \cong \gamma = \frac{\delta}{b} \tag{1}$$

Shear stress τ (N/mm^2 = MPa) of woven fabric can be directly calculated using shear forces. It is assumed that the thickness of the woven fabric t (mm) is constant during shearing. Then the shear stress is calculated using the Eq. (2)

$$\tau = \frac{T}{A} \tag{2}$$

where are: T (N) is a shear force acting on the side DC of woven fabric element as a result of stress (assuming a uniform distribution of shear stresses on the surface side), A (mm^2) is a surface side DC of woven fabric element.

2.1 Shear modulus of woven fabrics

The functional relationship between stress and strain cannot be determined theoretically, but only by experimental testing of samples made of certain materials. Mechanical properties are mainly investigated within the area of elasticity which means in terms of low load [5]. It is assumed that force- shear angle curve for woven fabric is approximate straight line before yield point. Therefore, elastic performance equation can apply here. The shear modulus is the initial, linear elastic slope of the stress-strain curve in shear. The connection between the shear angle γ and the shear stress τ is shown by Eq. (3) which represents the Hooke's law of shear and applies for elastic, homogeneous isotropic material in a linear region, i.e. where the relationship between shear stress and shear strain is linear. G (N/mm^2 = MPa) is shear modulus.

$$\tau = G \cdot \gamma \tag{3}$$

For orthotropic elastic materials, shear modulus G_{kl}, in various directions of cutting samples, i.e. in the coordinate system k, l whose axes do not coincide with the main axes x, y is obtained by using the expression for the transformation of the elastic constants, which states [5]:

$$\frac{1}{G_{kl}} = \left(\frac{1}{E_x} + \frac{1}{E_y} + \frac{2 \cdot \nu_{xy}}{E_x}\right) \cdot 4\cos^2\varphi \cdot \sin^2\varphi + \frac{1}{G_{xy}} \cdot \left(\cos^2\varphi - \sin^2\varphi\right)^2 = \frac{1}{G_\varphi} \tag{4}$$

φ is cutting angle of the samples due to the weft (Fig. 2), E_x, E_y are modulus of elasticity in two main directions (weft direction $\varphi=0°$, warp direction $\varphi=90°$); G_{xy} is shear modulus between both principal directions; v_{xy} is the Poisson's ratio. Shear modulus G_{kl} changes depending on the angle φ, and will be labeled as $G_\varphi=G_{kl}$. The numerical values of E_x, E_y, G_{xy}, v_{xy} are obtained by experimental testing of woven fabric samples in the laboratory. The theoretical treatment suggests that measurements of modulus in two directions (weft and warp direction) are insufficient to define a woven fabric's shear modulus. An investigation of the third direction is therefore necessary, and the most convenient direction is $\varphi=45°$. So, the measurements were carried out in three directions by considering samples cut along the warp, weft, and 45° directions. Shear modulus G_φ in any given direction can be predicted from the Eq. (5) when its values in the warp or weft directions and under the angle of 45° are measured:

$$\frac{1}{G_\varphi}=\left(\frac{4}{G_{45^0}}\right)\cos^2\varphi\sin^2\varphi+\frac{1}{G}\left(\cos^2\varphi-\sin^2\varphi\right)^2 \quad (5)$$

If differences in the shear modulus values between the warp and weft directions are large, we take the average value in both principal directions to calculate the shear modulus in various directions.

3 Experimental Part

The experimental study of shear properties of the woven fabric when the shear force acts on the samples that are cut at different angles 0°, 15°, 30°, 45°, 60°, 75°, 90° to the direction of the weft was conducted (Fig. 2). The values of the shear force in relation with shear angle were measured. Based on the experimentally obtained values, initial shear modulus was calculated in various directions. To carry out this study, orthogonal plain weave cotton woven fabric was available. Its fabric code is S23 and structural properties are: weft and warp yarn count is 30 tex, warp yarn density is 23 cm^{-1}, weft yarn density is 23 cm^{-1}, mass is 150 g/m^2, thickness t=0.41 mm. Before testing all samples were conditioned under standard atmospheric conditions (relative humidity 65 ± 2%, temperature of 20 ± 2 ° C). For each cutting direction, the average values obtained from five measurements will be shown in diagrams and will be used to calculate the initial shear modulus. For determination the shear properties of the fabric, clamps for shearing woven fabrics are designed, manufactured and schematically shown in Fig. 3a, and consist of left (fixed) clamp and right (movable) clamp. The force acting on the right clamp causes its vertical displacement. Left clamp is placed on the upper plate on which there is a measuring probe, and the right clamp to the lower plate on which the movable clamp is usually placed.

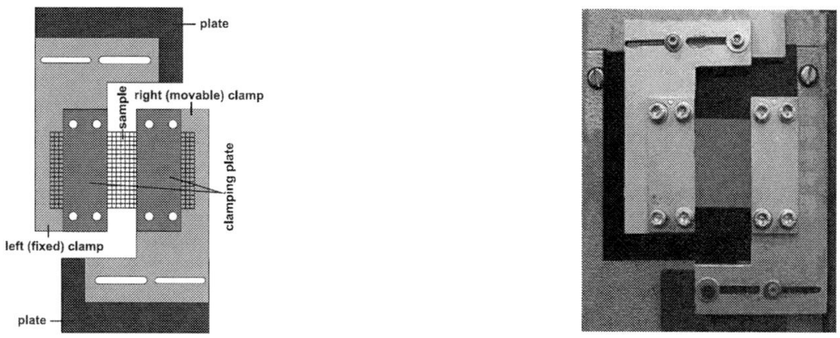

a) schematic view   b) photo

Fig. 3: Sample testing in clamps.

Distance between the left and right clamp can be adjusted in the range 0-50 mm. The maximum specimen size that can be fixed within the clamp is 75 mm. For this testing, specimens were cut with dimensions 125 x 75 mm, fastened in clamps of the instrument at a distance of b=25 mm, and subjected to a force acting in the plane of a fixed side of specimen till rupture (Figure 3b). Sample dimensions are: a=75 mm and b=25 mm.

A tensile tester Statimat M German manufacturer "Textechno" was used for testing. Two parallel clamps at a distance 25 mm are fixed on tensile tester and pulling speed of right clamp is 100 mm/min.

3.1 Overview of Testing Results

Diagrams (T- δ) of average values of test results of action of force T and the vertical displacements δ until break for samples that are cut at different angles to the direction of the weft are shown in Fig. 4.

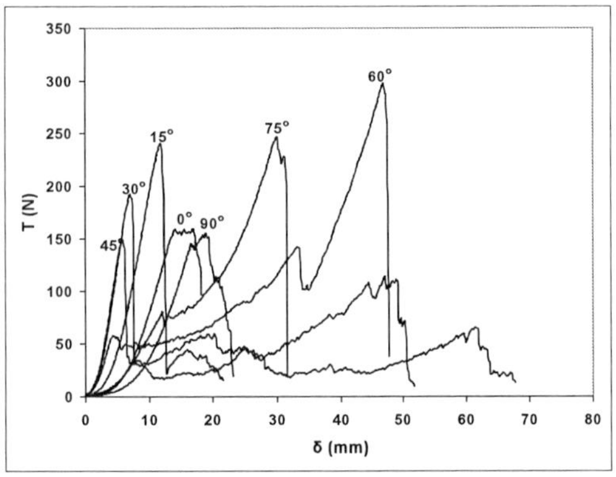

Fig. 4: Diagram T- δ (force- vertical displacement) for sample S23

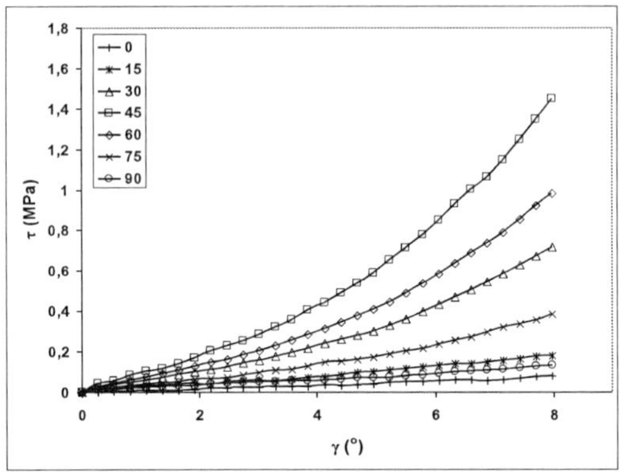

Fig. 5: Diagram τ-γ (shear stress–shear angle) for sample S23.

Based on diagrams of the measured values of force T and corresponding vertical displacement δ from Fig. 4, the average values of the shear angle γ and corresponding shear stress τ are calculated using Eq. (1) and Eq. (2). These values are shown with diagrams in Fig. 5, up to the value of the shear angle $\gamma = 8°$. During the woven fabric shearing, buckling in the woven fabric specimen occurs at a certain shear angle. According to Kawabata [6], woven fabric specimens which are subjected to the shear after exceeding the value of the shear angle $\gamma = 8°$ they tend to buckle. Shear angle $\gamma = 8°$ corresponds to vertical displacement δ=6.98 mm. The appearance of buckling causes errors in the measurement results.

In Fig. 5, at the same shear angle the highest value of shear stress τ appears in the samples which have been cut at angle of 45°. The values of shear stresses τ are very similar for complementary angles of the cutting samples. This is because balanced orthogonal plain weave fabric was considered. Shear stress τ takes on the minimum values for warp and weft directions.

3.2 Determining the initial shear modulus

Based on the experimentally obtained force-shear angles curves, the values of initial shear modulus were obtained and compared with the corresponding calculated values. Deviation in percentage between experimental and calculated values of initial shear modulus will also be calculated.

3.2.1 Experimental values of the initial shear modulus

From the presented diagrams, in Fig. 5, the values of shear force in the elastic range are used. We determined shear modulus G_φ from a particular region on force – shear angle curve that is determined by monitoring the experimental data obtained from an experimental set-up with regression control chart [7]. In this area of the curve, the relationship between shear stress and shear strain is linear. The shear modulus G_φ is defined as a slope of its shear stress- shear strain curve (τ- γ) in the elastic deformation region. The Hooke's law for shearing can be applied:

$$G_\varphi = tg\alpha = \frac{\tau}{\gamma} = \frac{T}{\gamma \cdot a \cdot t}(MPa) \qquad (6)$$

where A=a·t is the area of the sample in which shear force acting.

Using values T and γ in elastic range and using Eq. (6), the average values of initial shear modulus G_φ in relation to an arbitrary direction of cutting of the woven fabric samples are calculated. Linear regression equations are placed on the shear stress- shear strain curves in the elastic range. In Fig.6, the slope of the curve, i.e. the coefficient of line direction represents the shear modulus G_φ for sample S23 at angles 0°, 45° and 90°.

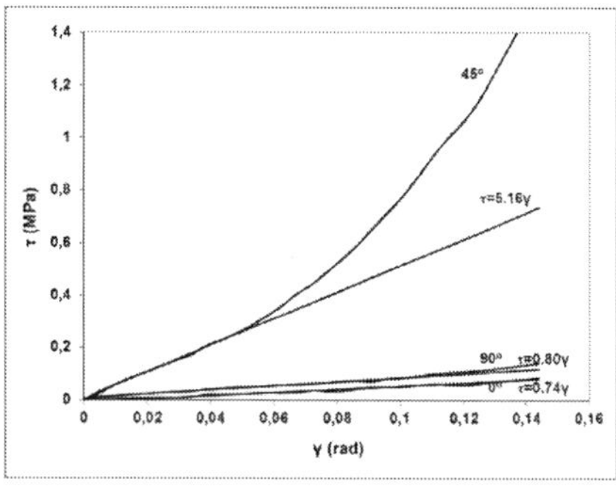

Fig. 6: Diagram of the shear stress-shear angle (τ- γ) and regression lines for woven fabric sample S23.

Tab. 1: Experimentally obtained values of shear modulus G_φ (MPa).

Sample	$G_{0°}$	$G_{15°}$	$G_{30°}$	$G_{45°}$	$G_{60°}$	$G_{75°}$	$G_{90°}$
S23	0.74	1.12	2.21	5.16	2.38	1.01	0.80

The obtained experimental values of shear modulus G_φ in dependence on the change of the cutting angle of the sample are given in Tab. 1. The diagram of experimental values G_φ for each 15° is shown in Fig. 7.

The diagram is almost symmetrical curve in relation to the angle of 45°. At that angle G_φ assumes the highest value for all woven fabric samples because initial slope of the stress-strain curve in shear is the biggest at that angle, Fig. 6. When the samples are cut in the warp direction (φ=90°) and weft direction (φ=0°) shear modulus G_φ have the lowest value because initial slope of the stress-strain curve in shear is the smallest at that angle, Fig. 6. The values G_φ in the warp and weft direction are almost equal for each woven fabric sample or it is observable that shear modulus G_φ are almost equal for complementary angles. This is because balanced orthogonal plain weave fabric was considered.

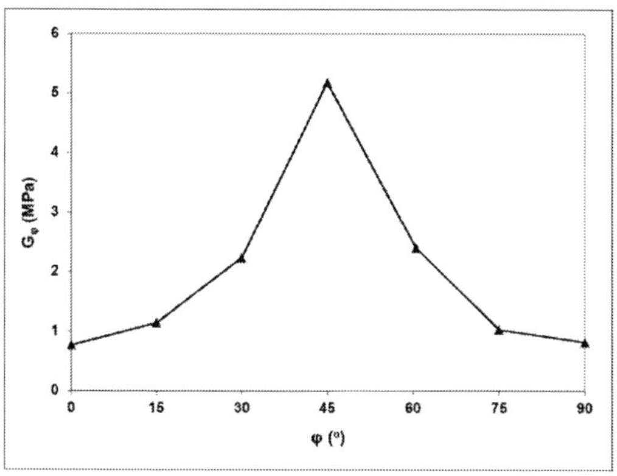

Fig. 7: Diagram of experimental values of shear modulus G_φ (MPa) for each 15° for S23.

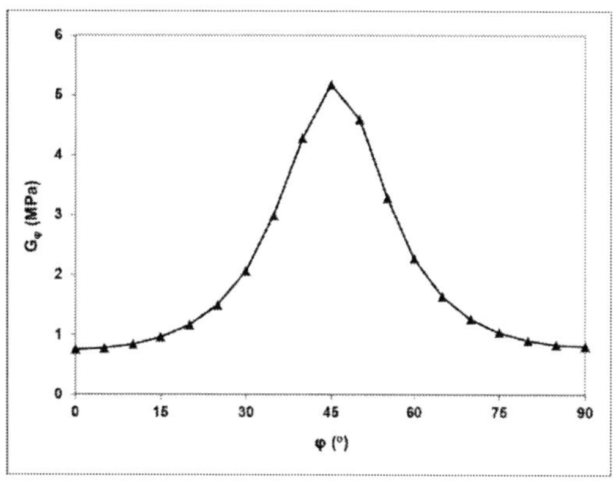

Fig. 8: Diagram of calculated values of shear modulus G_φ (MPa) for each 5° for S23.

3.2.2 Calculation of initial shear modulus in relation to an arbitrarily selected coordinate system

According to Eq. (5) and based on the experimental data $G_{0°}$ or $G_{90°}$ and $G_{45°}$ from Tab. 1, the values of initial shear modulus G_φ were calculated, depending on the change of the cutting angle of the sample. The calculated values of G_φ for each 15° are shown in Tab. 2.

Tab. 2: The theoretically calculated values of initial shear modulus G_φ (MPa).

Sample	$G_{0°}$	$G_{15°}$	$G_{30°}$	$G_{45°}$	$G_{60°}$	$G_{75°}$	$G_{90°}$
S23	0.74	0.94	2.05	5.16	2.18	1.02	0.80

The diagram of their calculated values G_φ (MPa) for each 5° is shown in Fig. 8. Tab. 3 shows deviations in percentage between experimental values G_φ from Tab. 1 and calculated values G_φ from Tab. 2. Deviations D (%) are calculated using the Eq. (7):

$$D = \frac{G_{\varphi,\exp} - G_{\varphi,calc}}{G_{\varphi,\exp}} \cdot 100(\%) \quad (7)$$

Tab. 3: Deviations D in (%) between experimental values and calculated values G_φ.

Sample	0°	15°	30°	45°	60°	75°	90°
S23	0.0	16.1	7.2	0.0	8.4	-1.0	0.0

In the warp (90°) and weft (0°) directions and under the angle of 45°, deviations in percentage between experimental and calculated values of the initial shear modulus G_φ are 0%. It follows from the Eq. (5) due to the periodicity of the sin and cos functions for these values. Negative deviation values show that the obtained calculated values G_φ are higher than the experimental values of G_φ.

4 Conclusion

Woven fabrics can be defined as orthotropic materials for which in the linear elastic range of the curve shear stress-shear angle can apply Hooke's law for anisotropic material behavior in calculating shear modulus when samples of woven fabric are cut in an arbitrary direction. The shear properties of the woven fabric are determined when the shear force acts on woven fabric specimens which are cut at different angles to the direction of the weft. Because of the anisotropy of woven fabric its shear modulus are changing in various directions. The shear modulus G_φ is almost symmetrical to the angle of 45° and the maximum value is reached exactly at that angle.

A good agreement between experimental results and the calculated obtained values of the initial shear modulus was shown, so the theoretical equations can be used with high accuracy to calculate the initial shear modulus of the woven fabric in various directions. Therefore, the measurements need to be implemented when shear force acting only on the specimens that were cut in the warp and weft direction and at an angle of 45°. The woven fabric is subjected to the shear up to $\gamma = 8°$ due to the fact that after this point woven fabric specimen tended to buckle.

References

[1] M. Nguyen, I. Herszberg, R. Paton, The shear properties of woven carbon fabric, Composite Structures, 47 (1999)1-4, 767–779.

[2] G.A.V. Leaf, Analytical Plain Weave Fabric Mechanics and the Estimation of Initial Shear Modulus, The Journal of the Textile Institute, 92 (2001) 3, 70-79.

[3] A. Alamdar-Yazdi, Shearing properties of the skewed woven fabrics, International Journal of Engineering, 18 (2005) 2, 177–185.

[4] Ž. Penava, D. Šimić Penava, Ž. Knezić, Determination of the Elastic Constants of a Plain Woven Fabrics by Tensile Test in Various Directions, Fibres and Textiles in Eastern Europe, 22(2014) 2, 57-63.

[5] R.J. Bassett, R. Postle, N. Pan, Experiment Methods for Measuring Fabric Mechanical Properties: a Review and Analysis, Textile Research Journal, 69(1999) 11, 866–875.

[6] S. Kawabata et al., Application of Objective Measurement to Clothing Manufacture, International Journal of Clothing Science and Technology, 2 (1990) 3, 18-33.

[7] B. Ozkul, D. Karaoglan, Regression control chart for determination of Young's modulus: A case study, Scientific Research and Essays 6 (2011) 30, 6393-6403.

Effect of the Shape of Geometric Discontinuity on Nature of Rayleigh Wave Back Reflection

M. Šofer [1,a], P. Ferfecki [2], J. Neugebauer [3]

[1] *VŠB - Technical University of Ostrava, Department of Applied Mechanics, Czech Republic*
[2] *VŠB - Technical University of Ostrava, IT4Innovations, Czech Republic*
[3] *PTS Josef Solnař s.r.o., Czech Republic*
[a] *michal.sofer@vsb.cz*

Abstract: The aim of our study is to show the influence of the shape of a particular type of the geometric discontinuity on the nature of Rayleigh wave back reflection. Experiments were carried out with aim of ultrasonic equipment from Olympus company, namely with Epoch 600 digital ultrasonic flaw detector and 1 MHz transducers with wedge in order to generate Rayleigh wave in used steel samples. COMSOL Multiphysics software has been used to simulate the propagation of Rayleigh wave on tested geometry as well as to estimate the losses of energy due to back reflection on tested geometric discontinuities.

Keywords: Rayleigh wave; ultrasound; geometric discontinuity; finite element method.

1 Introduction

Non-destructive testing (NDT) is nowadays widely used for various technical applications. Ultrasonic testing, as one of NDT methods, is based on the propagation of ultrasonic waves in tested object or material. From a mathematical point of view, there exist two types of waves: the bulk waves and the surface waves [1]. Bulk waves travel through the interior of the medium while surface waves propagate only at the interface between two different media. Although bulk and surface waves are fundamentally different, they are governed by the same set of partial differential wave equations [2]. The difference is, that for bulk waves, there are no boundary conditions, which need to be satisfied. Rayleigh waves, as one of the types of surface waves, are widely used in various NDT applications. Rayleigh waves are the result of interfering longitudinal and shear waves, the particle motion (Fig. 1) of the wave in the case of homogenous medium and moving from left to right is elliptical in a counter-clockwise direction along the free surface [3]. Generally, Rayleigh waves are used for detection of the surface and near-surface cracks in industrial practice. The penetration depth of surface wave is approximately equal to its wavelength and is strongly dependent on the frequency of the wave [2].

The main aim of proposed paper is to provide an experimental and a numerical study, dealing with the influence of the shape of particular type of geometric discontinuity on the nature of Rayleigh wave's back reflection.

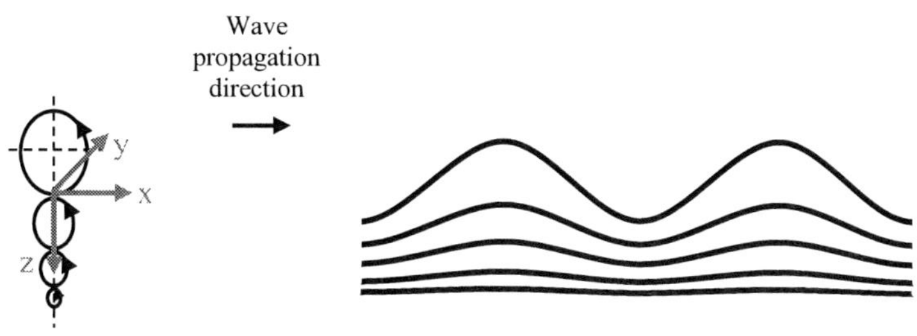

Fig. 1: Particle motion during wave propagation.

2 Mathematical Description of Rayleigh Wave Propagation Along Free Surface

The origin of mathematical relations for displacements in x and z-axis of Rayleigh surface wave propagation along the free surface is realised by an assumption of isotropic, homogeneous, linear elastic semi-space. The motion is defined as two-dimensional so that all quantities are independent on y coordinate. The displacement vector $\bar{u}$ can be with the use of Helmholtz decomposition defined as follows

$$\bar{u} = \nabla \cdot \phi + \nabla \times \bar{\psi} . \tag{1}$$

Where ϕ is a potential function, and $\bar{\psi}$ is a vector function, and this functions are defined in the following form

$$\phi = \phi(x, z) , \ \bar{\psi} = (0, -\psi(x, z), 0) . \tag{2}$$

The equation of motion expressed in terms of displacements

$$\mu \cdot \nabla^2 \cdot \bar{u} + (\lambda + \mu) \cdot \nabla \cdot (\bar{\nabla} \cdot \bar{u}) = \rho \cdot \ddot{\bar{u}} \tag{3}$$

can be separated into two wave equations, one for dilatational waves and one for shear waves

$$\nabla^2 \phi - \frac{1}{c_L^2} \cdot \ddot{\phi} = 0 , \ \nabla^2 \psi - \frac{1}{c_T^2} \cdot \ddot{\psi} = 0 . \tag{4}$$

μ and λ are Lame's constants, ρ is the mass density per unit volume, c_L is the wave speed of dilatational waves, c_T is the wave speed of shear waves, $(\cdot\cdot)$ denotes the second derivative with respect to time, and $(\bar{\ })$ denotes complex conjugation.

Let for both functions consider harmonic wave motion in form of $\phi = \Phi(z) \cdot e^{i(\omega t - kx)}$, $\psi = \Psi(z) \cdot e^{i(\omega t - kx)}$ and following boundary conditions of the components of the stress tensor on the free surface must be met $\sigma_z = \tau_{xz} = \tau_{zy} = 0$, then it is possible to derive the horizontal u and vertical v displacements

$$u = A \cdot k \cdot \left(e^{-q \cdot z} - \frac{2 \cdot q \cdot s}{s^2 + k^2} \cdot e^{-q \cdot z} \right) \cdot \sin(\omega \cdot t - k \cdot x), \tag{5}$$

$$w = A \cdot q \cdot \left(e^{-q \cdot z} - \frac{2 \cdot k^2}{s^2 + k^2} \cdot e^{-q \cdot z} \right) \cdot \cos(\omega \cdot t - k \cdot x), \tag{6}$$

$$s = \sqrt{k^2 \left(1 - \left(\frac{c_S^2}{c_T^2} \right) \right)} , \ q = \sqrt{k^2 \left(1 - \frac{1 - 2\nu}{2 - 2\nu} \left(\frac{c_S^2}{c_T^2} \right) \right)} . \tag{7}$$

A is a constant, which influences the amplitude of the wave, k is the wavenumber, c_S is the wave speed of surface waves, ω is the frequency, t is time, and ν is Poisson´s ratio.

3 Experimental Procedure

Fig. 2: Ultrasonic probes attached to acrylic wedges.

Experiments were carried out with the use of ultrasonic equipment from Olympus company, namely with Epoch 600 digital ultrasonic flaw detector and 1 MHz transducers with self-manufactured wedges in order to generate Rayleigh wave in used steel samples (Fig. 2).

Altogether, four measurements with different geometry shape in terms of the inclination angle φ, namely 60, 90, 120, and 150 degrees (Fig. 3) have been realised. All specimens were produced from the steel grade CSN 11 523 with the thickness of 15 mm, width of 65 mm, and length of 130 mm.

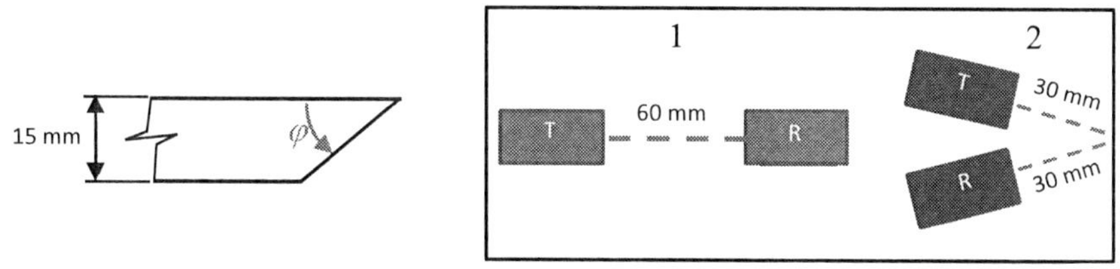

Fig. 3: Scheme of the measurement procedures (T-transducer, R-receiver).

The measurement is composed of two steps: (1) the measurement of an echo on the free surface, which is used as a reference measurement, and (2) the measurement of an echo, generated by back reflection (Fig. 3). The loss of echo related to the back reflection and compared to free surface is defined as follows:

$$L_{ech} = \left(1 - \frac{E_{re}}{E_{fs}}\right) \cdot 100\% . \tag{8}$$

L_{ech} is the loss of echo, E_{re}, E_{fs} are the echo heights (re – reflected, fs – free surface) in percentage for the stated gain, and the peaks of the oscillation in the case of numerical simulations, respectively. Needless to say, that the echo from edge back reflection had to be recalculated on the same gain value, which was used in case of measuring echo for free surface according to:

$$\Delta A = 20 \cdot \log\left(\frac{H_1}{H_2}\right), \tag{9}$$

where ΔA represents the gain in dB, and H_1, H_2 are the echo heights in percentage.

4 Results of the Numerical Simulations

Rayleigh surface wave propagation of the shape of a particular type was performed by means of a finite element code COMSOL Multiphysics. The loss of echo was obtained by solving of 2D wave propagation problem for the specified size of the inclination angle on the geometric discontinuity. The investigated specimens are made of steel with the elastic modulus of $2.0 \cdot 10^{11}$ Pa, Poisson's ratio of 0.33, and the density of 7850 $kg \cdot m^{-3}$.

The boundary condition of the prescribed displacement in the horizontal and vertical direction is defined according to Eqs. (5-6). In Fig. 4, the boundary condition is specified by the green coloured line.

In the computational model, the specimen was discretized by quadratic quadrilateral elements. The total number of degrees of freedom of the discretized specimen is approximately equal to 700 000, and its value depends on the size of inclination angle.

A generalized-alpha method for integrating of the equation of motion was applied to obtain the transient propagation of the Rayleigh surface wave.

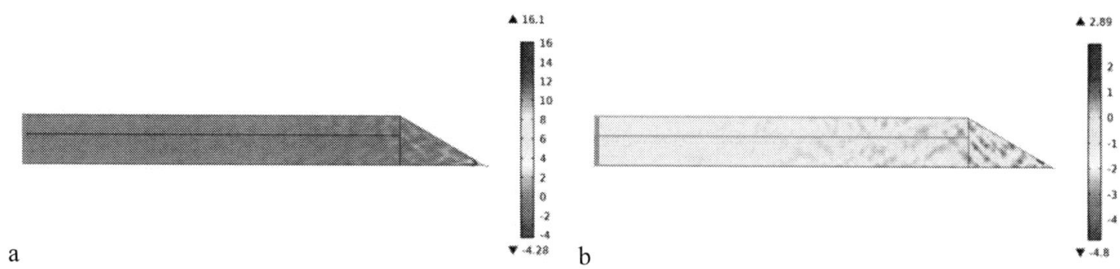

Fig. 4: Displacement in the direction-x (a) and displacement in the direction-y (b).

The distribution of the horizontal (a) and the vertical (b) displacement for the specimen with inclination angle equal to 150 degrees at time of 51.5 μs is drawn in Fig. 4. The loss of echo is determined at this time point.

5 Comparison of the Numerical and Experimental Results

Time history of the numerically computed vertical displacement component of the point on the specimen's surface in the case of the inclination angle of 60, 90 degrees and 120, 150 degrees is drawn in Figs. 5a, 5b. The peaks of the oscillation corresponding to the primary and reflected wave in the time course are indicated. The exact position of the peaks of oscillation has been determined according to the wave velocity.

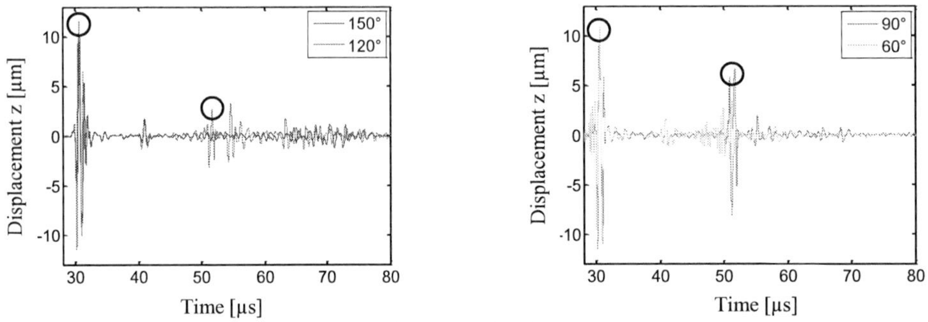

Fig. 5: History of the displacement.

The measured and the computed loss of echo of the specimens with different shape of the inclination angle are shown in Table 1. The inclination angle larger than 90 degrees will cause a significant change in the loss of echo.

Table 1: The measured and computed loss of echo.

φ [°]	60	90	120	150
$L_{ech,mea}$ [%]	35	47	81	98
$L_{ech,sim}$ [%]	34	40	73	97

6 Conclusion

Good agreement has been obtained between results of the numerical simulations and the experimental measurements in terms of analysis of Rayleigh wave back reflection on geometric discontinuities. The future work will be focused on improving the experimental and numerical procedures in order to obtain closer results between used approaches. Authors will also focus on problematics of Rayleigh wave back reflection on different configuration of surface cracks.

Acknowledgement

This work was supported by Specific Research (SP2016/145) and by The Ministry of Education, Youth and Sports from the National Programme of Sustainability II (LQ1602). The support is acknowledged.

References

[1] M. H. Sadd, Wave Motion and Vibration in Continuous Media, Kingston, University of Rhode Island, 2009.

[2] J. L. Rose, Ultrasonic Waves in Solid Media, Cambridge, Cambridge University Press, 2002.

[3] L. Rayleigh, On waves propagated along the plane surface of an elastic solid, Proceedings London Mathematical Society 17 (1885), 4–11.

Effect of a Plasticizer and Micronized Stone Powder on Mechanical Properties of a Cement Matrix

P. Tesárek [1,a], P. Holub [1], J. Havrda [1], J. Fládr [1], Z. Prošek [1,2]

[1] *Faculty of Civil Engineering, Czech Technical University in Prague, Thákurova 7, 166 29 Prague, Czech Republic*

[2] *Institute of Physic, Czech Academy of Science, Cukrovarnická 10/112, 162 00 Prague, Czech Republic*

[a] *tesarek@fsv.cvut.cz*

Abstract: The goal of this presented paper was to design a matrix based on cement. This matrix will be reinforced by different types of fibers or microfibers (e.g. on glass, polymeric or steel bases). The effect of a different composition on a flow test and macro mechanical properties as the compressive and bending strength and the dynamic Young's Modulus were tested. Two different types of additives/admixtures were tested – namely a plasticizer and a marble powder. Portland cement CEM I 42.5 R (Mokrá), natural quartz sand from Zálezlice (with fraction of 0-2.0 mm), a commercial plasticizer on a naphthalene base and a marble powder from West bank of Jordan (with fraction of 0-0.125 mm) were used for preparing of tested samples. The samples were tested after 28 days of their preparing.

Keywords: micronized stone powder; plasticizer admixtures; cement; mechanical properties; cement mortars.

1 Introduction

Glass, polymer and rock fibers have a low surface wettability with matrix (e.g. cement, lime or gypsum bases). Adhesion between the fiber surface and the matrix is problematic. Nowadays some different methods are used for modification of fiber surfaces, these ones are based on chemical or mechanical treatment [1]. Plasma treatment of fiber surfaces is newly used too [2]. Conditions of fibers and matrix adhesion depend on quality of the used matrix. Mechanical properties depend on a matrix composition – especially quality and amount of cement, a water/cement ratio and size distribution. E.g. added micronized stone powder or recycled concrete improves a workability of fresh mixtures [3, 4].

2 Materials and Specimens

Composition of designed cement matrixes (amount of components, water and the plasticizer) is shown in Table 1. Six samples for each material were prepared with dimensions of 40 × 40 × 160 mm. The first ones, which was used as reference, consist only of cement and sand (without additions), this set was denoted as (S) for sand. Two tested sets were modified by a plasticizer for reduction of the water/gypsum ratio, these sets were demoted as (P) for the plasticizer. One quarter of sand amount was replaced by a marble powder for other ones, these ones were denoted as (M) for a marble powder. Portland cement CEM I 42.5 R (Mokrá, Czech Republic), natural quartz sand from Zálezlice (Czech Republic) with fraction of 0-2.0 mm, a commercial plasticizer on a naphthalene base and a marble powder from West bank of Jordan (with fraction of 0-0.125 mm) were used for preparing of the tested samples. The marble powder was modified by grinding of waste material from a marble mine. For grinding the high speed mill from LAVARIS Ltd. (Czech Republic) was used.

Tab. 1: Composition of the tested cement matrixes.

Material [mm]	Cement CEM I 42.5R [g]	Sand (0-2 mm) [g]	Marble powder (0-0.125 mm) [g]	Plasticizer [g]	Water/cement ratio [-]
S	210	750	-	-	0.70
S-P	210	750	-	3.15	0.62
S-M	210	562	188	-	0.70
S-M-P	0.15	562	188	3.15	0.60

3 Experimental Methods and Results

The samples were tested after 28 days of their preparing. As a comparative test was used a flow test, value of the slow test was experimentally determined for 10 impulses of 105 ± 5 and 125 ± 5 mm, respectively 20 impulses of 145 ± 5 and 185 ± 5 mm (Tab. 2). A water/cement ratio (Tab. 1) was adjusted for these values. The compressive and bending strengths were determined from standard destructive tests, while the values of the Young's modulus were determined using one of resonance methods, concretely the impulse method. The compressive strength were determined on the 28 days old samples using the Heckert device, model FP100. The testing was displacement controlled at a constant rate of 0.1 mm/s in the case of three-point bending and 0.3 mm/s in compressive. The standard compressive test was performed after the bending test on the broken specimens (half of the samples) with effective dimensions of compressive area of 40 × 40 mm [5]. Measurement of dynamic modulus of elasticity was carried out on the assembly from Brüel & Kjær, basic principles of the impulse method are spelled out in e.g. Lidmila et al. and Padevěd et al. [6, 7]. Obtained experimental results, the flow test and bulk density, are summarized in Table 2. From the results it is clearly visible that the effect of the marble powder is more appreciable.

Tab. 2: Comparison of experimental results (with standard deviations).

Material	Flow test [cm]	Bulk density [kg/m^3]
S	14.6 ± 2.1	1989 ± 15
S-P	14.7 ± 3.1	1999 ±12
S-M	18.7 ± 2.5	2069 ± 25
S-M-P	18.4 ± 2.7	2123 ± 17

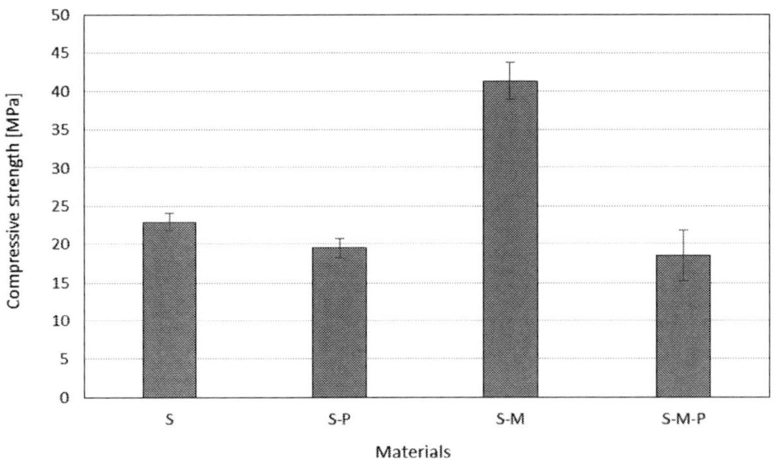

Fig. 1: Comparison of the compressive strength (with standard deviations).

The comparison of the compressive strengths and bending strengths shows Fig. 1, respective Fig. 2. The dependence between the Dynamic Young's Modulus on the bulk density shows Fig. 3. The positive effects of micronized marble powder are clearly visible from all obtained experimental results.

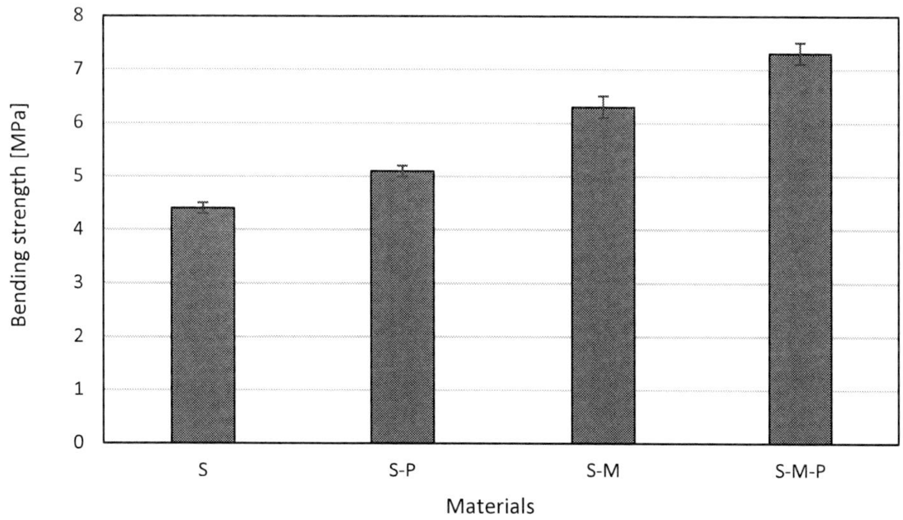

Fig. 2: Comparison of the bending strength (with standard deviations).

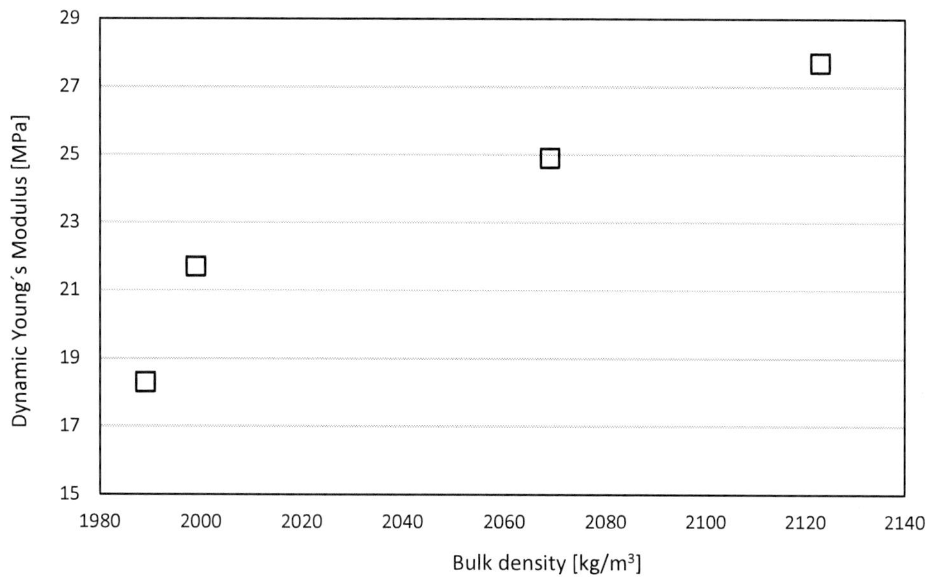

Fig. 3: Dependence of the Dynamic Young's Modulus on the bulk density.

4 Conclusion

The cement matrixes with added micronized marble powder were prepared and tested. From the view point of workability and mechanical properties, this solution was chosen instead of using plasticizers, on the other hand only one plasticizer was tested. Recycled materials are alternative for standard used ones [8, 9]. For the future work this matrix will be reinforced by different types of fibers and microfibers with various types of surface treatments [10].

Acknowledgement

This was financially supported by Czech Technical University in Prague – SGS project SGS16/201/OHK1/3T/11 and by the Czech Science Foundation research project 15-12420S. The authors also thank the Center for Nanotechnology in Civil Engineering at the Faculty of Civil Engineering CTU in Prague and the Joint Laboratory of Polymer Nanofiber Technologies of the Institute of Physics Academy of Science of the Czech Republic and the Faculty of Civil Engineering CTU in Prague.

References

[1] S. E. Elsaka, Influence of Chemical Surface Treatments on Adhesion of Fiber Posts to Composite Resin Core Materials, Dental Materials 29 (2013) 550–558, http://dx.doi.org/10.1016/j.dental.2013.03.004

[2] D. Cokeliler et al., Modification of Glass Fibers to Improve Reinforcement: A Plasma Polymerization Technique, Dental Materials 23 (2007) 335–342, doi: 10.1016/j.dental.2006.01.023

[3] K. Seps, J. Vodicka, Fibre Reinforced Concrete with Recycled Concrete and Insulation Material STERED, Special Concrete and Composites 2014, 1054 (2014) 262–266, doi: 10.4028/www.scientific.net/AMR.1054.262

[4] M. Lidmila et al., Mechanical Properties of Recycled Binder/Micro-Filler Cement-Based Material, Advanced Materials Research 2014, 1054 (2014) 234–237, doi: 10.4028/www.scientific.net/AMR.1054.234

[5] P. Padevět, P. Tesárek, P., T. Plachý, Evolution of mechanical properties of gypsum in time, International Journal of Mechanics 5 (2011) 1–9.

[6] M. Lidmila et al., Composite Material Based on Cement and PVA: Evolution of Mechanical Properties during First 28 Days, Advanced Materials Research 1054 (2014) 215–220.

[7] J. Topič et al., Effect of PVA modification on the properties of cement composites, Acta Polytechnica 55 (2015) 64–75.

[8] V. Nežerka et al., Comprehensive study on mechanical properties of lime-based pastes with additions of metakaolin and brick dust, Cement and Concrete Research 64 (2014) 17–29.

[9] J. Fládr, J., A. Kohoutková, High performance concrete experimental structure, Advanced Materials Research 1000 (2014) 269–272.

[10] A. Podstawka et al., An Experimental investigation of the impact of fiber material on fracture energy of fibre reinforced concrete, Civil-Comp Proceedings 108 (2015).

Fatigue Testing of Medial Plates for Treatment of Distal Tibia Fractures

K. Frydrýšek [1,a], G. Theisz [1], J. Demel [2], L. Pleva [2], M. Pompach [3], F. Fojtík [1]

[1] VŠB–Technical University of Ostrava, Faculty of Mechanical Engineering, Department of Applied Mechanics, 17. listopadu 15/2172, 708 33 Ostrava, Czech Republic

[2] Ostrava University Hospital, Trauma Centre, 17. listopadu 1790, 708 52 Ostrava, Czech Republic & University of Ostrava, Faculty of Medicine, Dvořákova 7, 701 03 Ostrava, Czech Republic

[3] Pardubice Regional Hospital, Trauma Centre, Kyjevská 44, 532 03 Pardubice, Czech Republic

[a] karel.frydrysek@vsb.cz

Abstract: The focus of the article is the biomechanical analysis of new locking compression medial plates. These types of plates are used in orthopedics and traumatology for internal fixation of fractures of the distal tibia. Low-cycle fatigue analyses were performed by repeated axial loading for non-fused bone with complicated comminuted fractures in the distal metaphysis. This type of fracture represents an extreme situation, where all loads between two fragments of the broken bones are carried by the plate. It is also essential to perform fatigue tests before clinical application in hospitals. From the results of the tests, the durability, loadings and fatigue crack parameters were determined and a reliability assessment was performed.

Keywords: Fatigue testing; Internal fixation; LCP; Fracture; Medial plate; Tibia; Biomechanics, Bone.

1 Introduction

In the past, open fractures and ballistic wounds involving long bone fractures were not amenable to standard fracture treatment. They usually resulted in amputation, disability or even death. Today, hospital doctors perform treatment by means of internal fixation. This is a surgical procedure that stabilizes and joins the ends of fractured bones using mechanical devices such as metal plates, pins, rods, wires or screws.

Internal fixators – which include the medial plates examined here – form an integral part of fracture treatment. Their functionality and reliability are essential not only for medical specialists, but also for patients; the plates help to shorten the duration of treatment and reduce the number of complications. It is therefore important to focus on the development of this type of implant in order to improve the quality of medical care. Among the factors affecting the success of treatment are the reliability and strength of medical implants; in the case of failure, it is necessary to repeat osteosynthesis. Unfortunately it is not possible to achieve 100% success rates. However, failure rates can be significantly reduced via the design and optimization of the implants; see [1]. It is also essential for surgeons to work effectively and to a high standard.

Medial plates are angularly stable implants; they are used primarily for extra-articular and simple intra-articular fractures of the distal tibia partially affecting the diaphysis. An example of the use of a medial plate is given in the CT images shown in Fig. 1. The general advantage of angularly stable plates is that force transfer occurs between the plate and the screws, which eliminates friction between the plate and the bone; see [2,3] and Fig. 2. Current developments are focusing on low-profile anatomically shaped plates, in which the shape and size of the plate and the orientation of the screws are adapted as much as possible to the specific anatomic location of the plate.

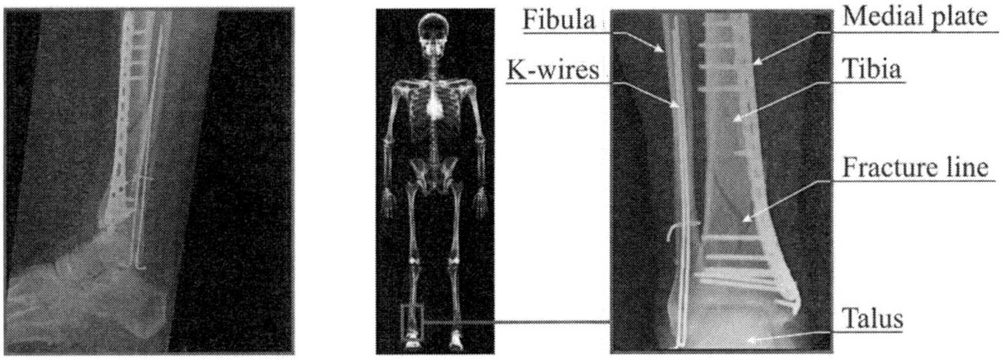

Fig. 1: X-ray images after fixation with a plate.

2 Experiments

The functionality and reliability of medial plates is essential not only for medical specialists, but also for patients; the plates help to shorten the duration of treatment and reduce the number of complications. It is therefore important to focus on the development of this type of implant in order to improve the quality of medical care. It is also essential to perform fatigue tests before clinical application.

The jig used for static and dynamic testing is shown in Fig. 2, including the medial plate fixed to the jig. The plate is fixed to a wooden model using angularly stable screws produced by the company Medin a.s., which were applied using the standard surgical procedure. Based on medical recommendations and surgical practice, the bone was substituted by a jig made of spruce wood, which is the type of wood that best approximates bone tissue. The initial tests were performed using a jig made of beech wood; however, this solution proved problematic due to the properties of the surgical tools and the hardness of the jig (it was not possible to insert the screws into the wood). The shape of the wooden model corresponds with that of a bone surface. Threaded compensation rods were screwed into the model. Thrust washers with spherical joints were screwed onto the protruding ends of the rods. These washers were inserted and fixed into casings. The casings were then clamped into the jaws of the tensile testing machine or pulsator. The jig enables rotation via spherical joints.

A total of three samples were used in order to determine the distribution of total force *Fs* and to perform the subsequent fatigue analysis. In previous experiments, one sample was tested for buckling and stability. The sample was destroyed during the test; the results of the experiment are described in [4].

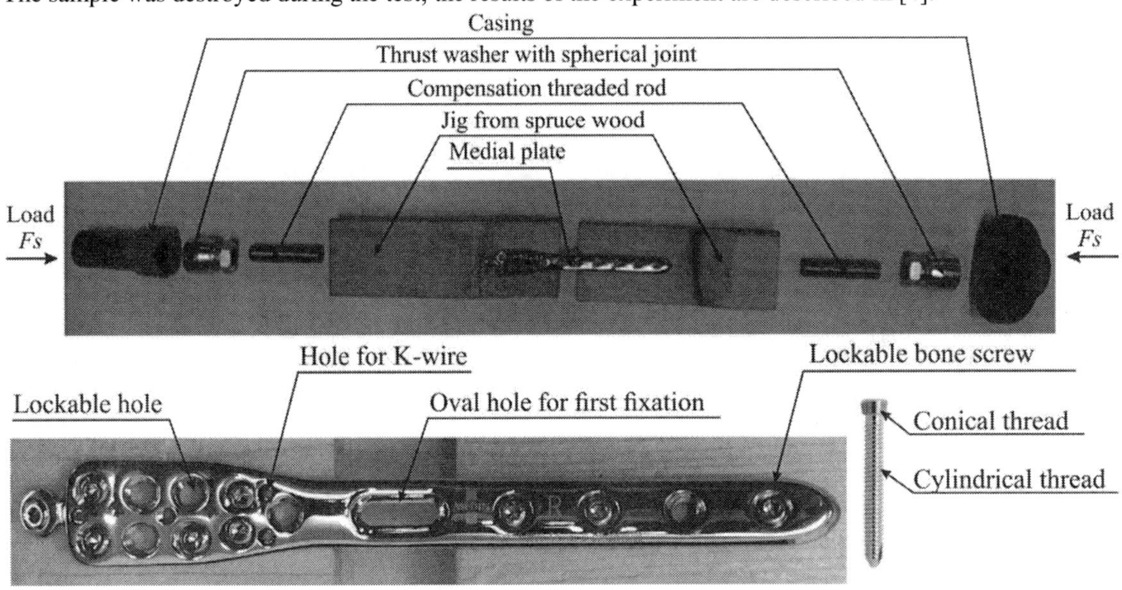

Fig. 2: Medial plate fixed to the jig, lockable screw.

2.1 LOADING OF THE PLATE USING A FORCE SENSOR

In addition to the low-cycle fatigue analysis, the distribution of the total force F_S in the bone and the plate were also analyzed for the possibilities of callus formation in bones (i.e. successful treatment).

First, static measurement was conducted using a tensometric force sensor. It was determined to what degree the total loading F_S was distributed to the bone and the plate after callus formation. The force sensor substitutes the bone at the location of a comminuted fracture; see Fig. 3. Two measurements were taken of each sample. The speed of displacement under loading 0.5 mm/min enabled sufficient control. The force was gradually increased by increments of 50 N up to a maximum value F_S = 400 N. The initial static pressure test (see [4]) indicates that the measurement was conducted in an area of elasticity. Tab. 1 gives an overview of maximum, minimum and mean values for the force transferred via the bone F_K and the force transferred via the plate F_D.

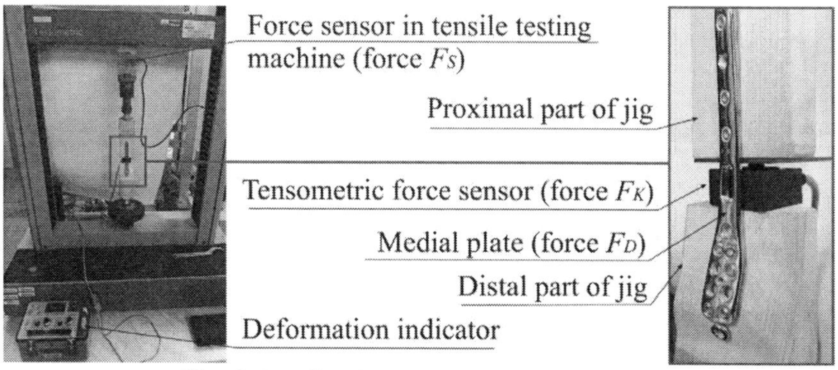

Fig. 3: Loading the plate using a force sensor.

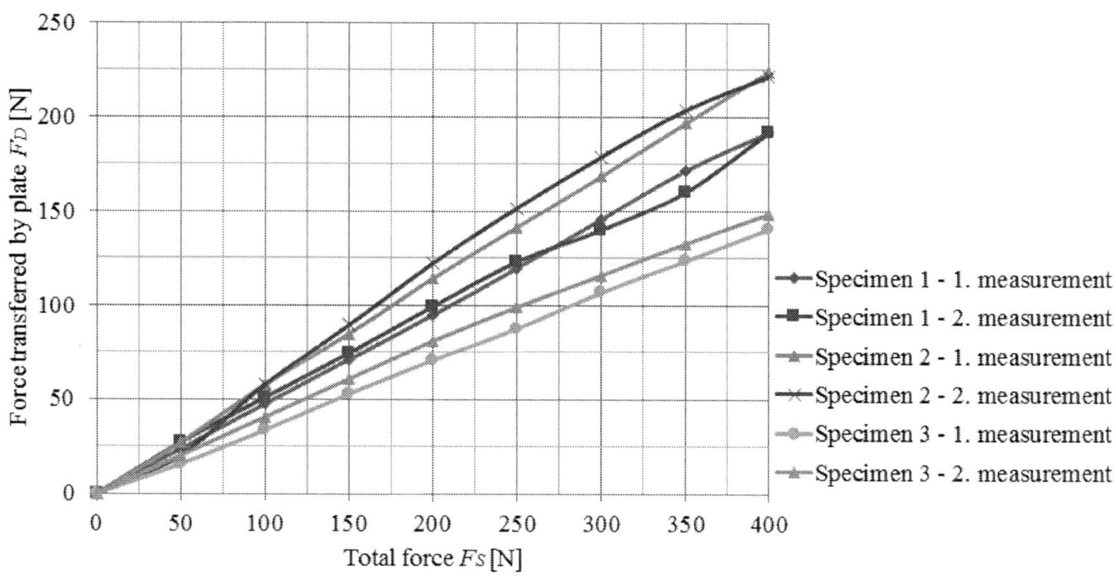

Fig. 4: Loading of the medial plate using a force sensor - Dependency of total force F_S on force in plate F_D.

For sample no. 1 the distribution of the total force between the plate and the bone is even ($F_D \approx F_K$). For sample no. 2 more of the force is transferred by the plate, and for sample no. 3 more of the force is transferred by the bone; see Fig. 4 and Tab. 1. These differences are caused primarily by geometric factors, as the individual plates were attached in slightly different ways by the doctor. This corresponds closely with reality; plates are not all applied to the bone identically, because each fracture and each osteosynthesis is different.

Total force F_S [N]	Force transferred by bone F_K [N]			Force transferred by plate F_D [N]		
	Max	Min	Mean	Max	Min	Mean
50	34.3	22.9	27.6	27.1	15.7	22.4
100	66.3	42.3	52.0	57.7	33.7	48.0
150	97.2	60.6	77.7	89.4	52.8	72.3
200	129.2	77.7	102.9	122.3	70.8	97.1
250	162.3	98.3	129.5	151.7	87.7	120.5
300	193.2	121.2	157.4	178.8	106.8	142.6
350	226.3	146.3	185.2	203.7	123.7	164.8
400	259.5	176.0	213.6	224.0	140.5	186.4

Tab. 1 – Statistical processing of measurement results.

2.2 Fatigue analysis

Fatigue tests of medial plates were conducted using an INOVA ZUZ 100 hydraulic biaxial pulsator; see Fig. 5a. The loading forces applied to the plate were determined from the diagram of the pressure test conducted on the first plate; see [4]. In the first phase, the plate was subjected to cyclical loading by repeated total pressure force $F_{S1} = 500$ N for the required number of $N_1 = 500\,000$ cycles. If the plate did not fail during the first phase, the loading force was increased to $F_{S2} = 600$ N; Fig. 5b. This reproduces the real-life situation in which a patient applies increased loading to the bone during the process of rehabilitation. The experiment does not take account of the process of bone healing that occurs during rehabilitation, when part of the loading is gradually transferred to the partially healed bone. The experiment thus reproduces an extreme situation in which the bone callus is not formed and the treatment is unsuccessful (the plate carries the entire load, i.e. ($F_D = F_S$).

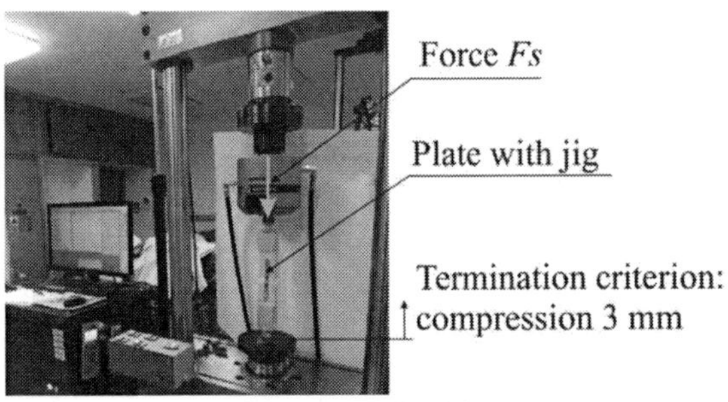

a) Inova ZUZ 100

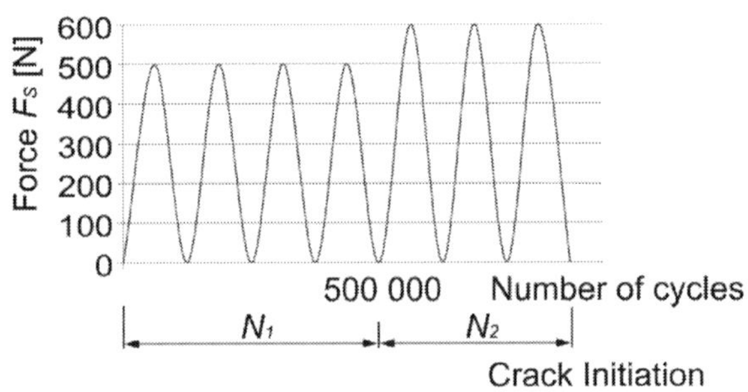

b) Course of loading

Fig. 5: Fatigue testing of the medial plate.

An example of the fatigue test results is shown in Fig. 6. Cracks appeared at the K-wire holes under the oval hole, and at the oval hole. A total of 3 medial plates for the distal tibia, with 4 holes in the proximal part, were subjected to the experiments. The results were used to determine the distribution of the total force between the bone and the plate and the fatigue resistance of the plate. The individual experiments confirm the high-risk areas, which were also determined by FEM analysis; see [4].

Sample no. 1, which was attached to the jig with only minimal clearance, achieved the best results. Failure did not occur until the second phase, after the force was increased to 600 N. Samples 2 and 3 simulated situations in which the plate does not rest entirely flush with the bone; this geometric difference has a pronounced effect on the fatigue lifespan of the plates. Sample no. 3, which was attached to the jig slightly off the axis, showed greater displacement from the beginning of the experiment; it lasted for approx. 2.6x fewer cycles ($N_1 = 161\,159$ cycles) than sample no. 2, which lasted for $N_1 = 422\,875$ cycles.

In all three samples, cracks appeared at the K-wire holes under the oval hole. A second crack occurred only in sample no. 1 at the oval hole; see Fig. 6. The criterion for terminating the experiment was the compression of the plate by 3 mm. The use of a spruce wood model appears to be sufficient; even in sample 1, which lasted $N_1 = 500\,000$ cycles at 500 N and $N_2 = 127\,692$ cycles at 600 N, the wood showed no signs of wear or failure.

Another similar fatigue analysis or FEM analysis of internal fixators is presented e.g. in [5, 6].

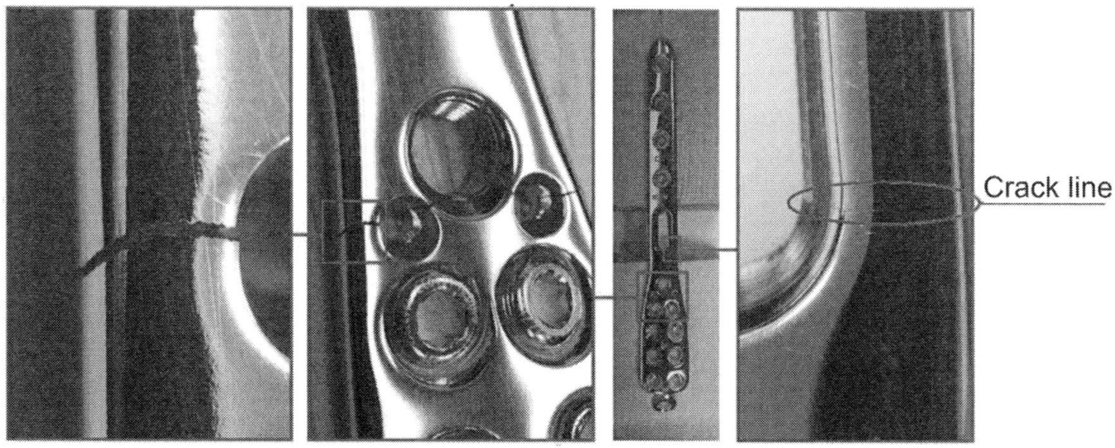

Fig. 6: Results of fatigue test (cracks) for sample no. 1 (locus minoris of the medial plate).

3 Conclusion

This article has presented a fatigue analysis of a medial plate (producer Medin a.s.) for treatment of distal tibia fractures. The experimental data confirm the FEM calculations, see [4], indicating the weakest structural point (locus minoris) of the plate. The experiment also determined the distribution of the loading force in the plate and the bone for cases of successful treatment (i.e. when bone callus is formed and part of the loading is transferred to it).

In all three samples, cracks appeared at the K-wire holes under the oval hole. A second crack occurred only in sample no. 1 at the oval hole; see Fig. 6. In view of the strength analysis (see Fig. 7 and Fig. 8) and the experimental results, alterations to the medal plate were proposed. These alterations should provide greater resistance of the plate to fatigue and loading, as they will remove the stress concentrator at the K-wire holes and enlarge the section at the oval hole.

Future research will focus on fatigue testing of plates produced using various different technologies. Instead of the existing plates – produced by forming and machining – it will be possible to produce plates by 3D printing with titanium powder (using selective laser melting technology).

The authors gratefully acknowledge the funding from the Czech projects TA03010804 and SP2016/145.

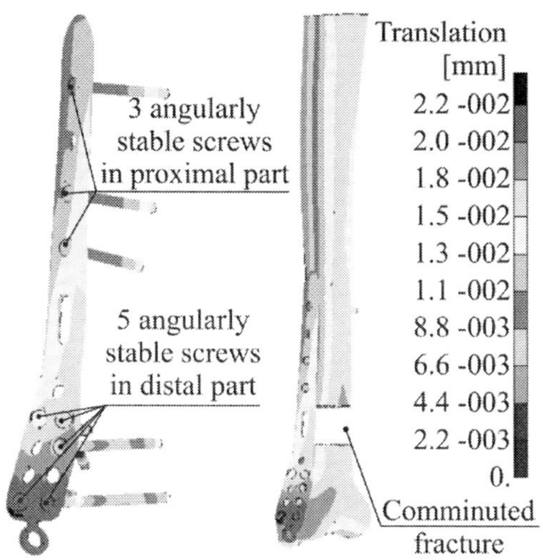

Fig. 7: Course of displacement (non-fused bone, force $F = 1$ N). [4]

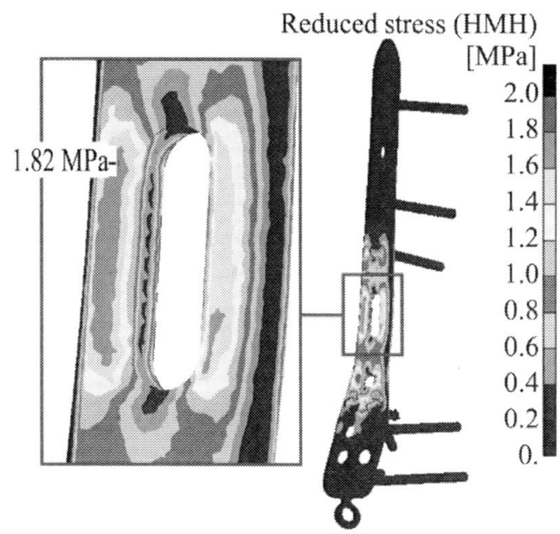

Fig. 8: Reduced stress – HMH hypothesis (non-fused bone, force $F = 1$ N). [4]

References

[1] K. Frydrýšek, L. Pleva, M. Pompach, et al. External and Internal Fixators for Treatment of Complicated Fractures in Traumatology and Orthopaedics, Engineering Mechanics, Prague, Czech Republic, ACAD SCI, 2014, p 180-183. ISBN 978-802144871-1.

[2] R. M. Greiwe, M.T. Archdeacon. Locking plate technology: current concepts, The journal of knee surgery 20 (2007), p 50-55, ISSN 15388506.

[3] J. Korner, G. Diederichs, M. Arzdorf, et al. A biomechanical evaluation of methods of distal humerus fracture fixation using locking compression plates versus conventional reconstruction plates, Journal of Orthopaedic Trauma 18 (2004), p 286-293. ISSN 08905339.

[4] G. Theisz, K. Frydrýšek, F. Fojtík, et al. Medial Plate for Treatment of Distal Tibia Fractures, in proc.: EAN 2015 – 53rd Conference on Experimental Stress Analysis (2015), ed. P. Padevet, P. Bittnar, CTU in Prague, Český Krumlov, Czech Republic, p 431-437. ISBN 978-800105735-6.

[5] M. Vilímek, R. Sedláček. Methodology for Dynamic Testing of Plates for Proximal Tibia Osteosynthesis, in proc.: Bulletin of Applied Mechanics (2011), 7 (25), p 12-14. ISSN 18011217.

[6] K. Frydrýšek, G. Theisz, L. Bialy, et al. Finite element modeling of T-plate for treatment of distal radius, in proc.: 1st European-Middle Asian Conference on Computer Modelling, (2016), 426, p 1-10. ISBN 978-331927642-7.

Cement Paste with Recycled Concrete Powder: Long Term Development of the Strength

J. Topič [1,a], Z. Prošek [1], J. Trejbal [1]

[1] *CTU in Prague, Faculty of Civil Engineering, Thákurova 7, 166 29 Praha 6, Czech Republic*
[a] *jaroslav.topic@fsv.cvut.cz*

Abstract: Presented paper deals with using of a recycled concrete powder (RCP) modified by a high speed as microfiller and partial binder replacement in cement pastes. This approach allows to exposed unhydrated cement grains to further hydration. Such use of recycled concrete powder could lead to cost reduction of a concrete mixture. The flexural and compressive strength on samples made of Portland cement containing 0, 33, and 67 wt. % of a recycled concrete powder were determined. Those mechanical properties were determined in several times during 14 months. Results show that 33 wt. % cement replacement seems like limited value, when mechanical properties in long term scale are comparable with a reference sample. Samples with 67 wt. % of RCP had significantly lower strength compared with reference samples.

Keywords: Cement; recycled concrete; cement replacement; mechanical properties, strength.

1 Introduction

Waste production and its accumulation is a worldwide environmental issue. Development of new methods and procedures for effective reuse of waste materials helps to reduce the amount of landfill sites and to protect environment and non-renewable resources [1]. Concrete is the most widely used building material. For this reason, it is necessary to solve the problem of concrete recycling. During recycling of old concrete occurs a problem with processing very fine fraction having a grain size <1 mm (powder) [2]. So far has not been found suitable application for this fine concrete waste.

Using recycled concrete powder (RCP) as a raw material for Portland clinker production [3] or as alternative heat treated binder [4] belong among several possibilities of a recycled concrete powder. Taking into account the heat consumption and the production of CO_2, which contributes to environmental degradation, the previous methods are no more environmental friendly and sustainable. On the other hand non-renewable natural resources are preserved. The possibility of using recycled concrete powder as filler for asphalt mixtures appears as environmental friendly and cheaper. The possibility to use the powder for production of geopolymeric binder [5] is also very interesting. Recent studies suggest that it is also possible to use recycled concrete powder as a partial replacement of cement [6, 7].

This paper is focused on using the recycled concrete powder RCP as a microfiller and cement replacement. In this case the old railway sleepers are recycled. In addition the fine fraction of recycled concrete is treated by high speed milling. This approach allows exposed unhydrated cement grains to further hydration [8,9]. Such use of the recycled concrete powder could lead to cost reduction of a concrete mixture.

2 Materials and Samples

The samples were made of Portland cement CEM I 42.5 R produced in Radotín (CZ) and a fine recycled concrete powder (RCP) prepared from old railway sleepers. Sleepers were crushed to particle-size fraction 0/63 mm. The final fine-grained product was obtained by grinding of separated material with a particle size 0/16 mm by using a high speed mill made by LAVARIS Ltd. Particle size distribution of RCP used for production of the tested samples was analyzed by laser device Fritsch Anlysette 22 MicroTec plus and it is presented in Tab. 1. The specific surface area of RCP measured by the Blaine method was 412 m²/kg. Particle size distribution of RCP is similar to a particle size distribution of Portland cement CEM I 42.5 R.

Tab. 1: Particle size distribution of RCP.

Grain size [μm]	0	0.5	1	2	4	8	16	32	64	128	160
Cumulative amount [wt. %]	0	1.2	3.5	8.8	19.2	36.5	59.5	82.0	96.2	99.9	100

Three sets of samples with different amount of RCP were prepared and tested (Tab. 2). Each set contained 24 prismatic samples having dimensions of 40 × 40 × 160 mm. To reach similar workability of the mixtures, the water/mixture ratio varied for each set. Consistency of the mixtures was determined by flow expansion test. The samples were cured for the next 28 days in water at the temperature of 21 ± 2 °C. After 28 days the samples were stored in laboratory conditions at the temperature of 21 ± 2 °C and the relative humidity of 50 ± 5 %.

Tab. 2: Composition of the mixtures.

Mixture	Cement (CEM I 42.5 R) [g]	RCP [g]	Water/mixture mass ratio	Flow expansion test [mm]
A (REF., 0 wt. % RCP)	1000	-	0.35	130
B (33 wt. % RCP)	670	330	0.38	130
C (67 wt. % RCP)	330	670	0.42	130

3 Measurement methods

The flexural and compressive strength were determined on the 28, 100, 287 and 409 days old samples using the Heckert device, model FP100. Each test set contained 6 samples. The testing was displacement controlled at a constant rate of 0.1 mm/s in the case of three-point bending and 0.3 mm/s during the compression test. The distance between supports during the three-point bending test was equal to 100 mm. The uniaxial compressive test was performed on the broken halves of the specimens with effective dimensions of 40 × 40 × ~80 mm.

4 Results and Discussion

The flexural strength results are presented in Fig. 1. It is obvious that higher amount of RCP in mixture after 28 days has a negative influence on the flexural strength. However, after 100 days the flexural strength of the samples B with 33 wt. % of RCP was almost 2 times higher compared with the reference sample. The samples C with 67 wt. % of RCP have the comparable flexural strength as the reference samples. After 287 days the flexural strength of the reference sample was almost same as in the previous time. In case of samples B the flexural strength was lower than in previous case and in case of samples C the decrease was almost 50 %. After 409 days the flexural strength of samples A and B decreases about 90 and 60 %. Flexural strength of samples C increase about 60 % and it was little higher than after 100 days. Based on results it can be assumed that samples B are comparable or have a little higher flexural strength in time than reference samples. Broad ranges of the flexural strength between the same samples in different times could be caused by little differences in mixing and molding because each set of the samples was mixed separately.

The compression testing results are presented in Fig. 2. The negative influence of the cement replacement on compressive strength after 28 days is significant only in case of sample C, the compressive strength was lower about 65 %. Samples with 33 wt. % of RCP have comparable compressive strength as the reference samples. Difference in the compressive strength between the samples A and B is obvious after 100 days. The strength of samples A reaches to 120 MPa and of samples B only to 78 MPa. The compressive strength of samples C is significantly lower. During time after 100 days to 409 days the compressive strength of samples

A and B declined about 45 and 25 %. The compressive strength of samples C stabilized around 30 MPa during whole time after 100 days. According to the results of compressive strength, it is obvious that cement paste with recycled concrete powder has lower strength at 100 days. But in construction practice, the values of strength at 28 days are supreme and in this case the compressive strength is comparable with the reference samples and the samples B. Moreover the decrease of compressive strength in case of the reference samples is more significant than in case of the samples B. Because of this fact the value of the compressive strength after 409 days differs only about 10 %. The compressive strength of the samples C caused by a high content of RCP, is too low for practical and economical use in construction.

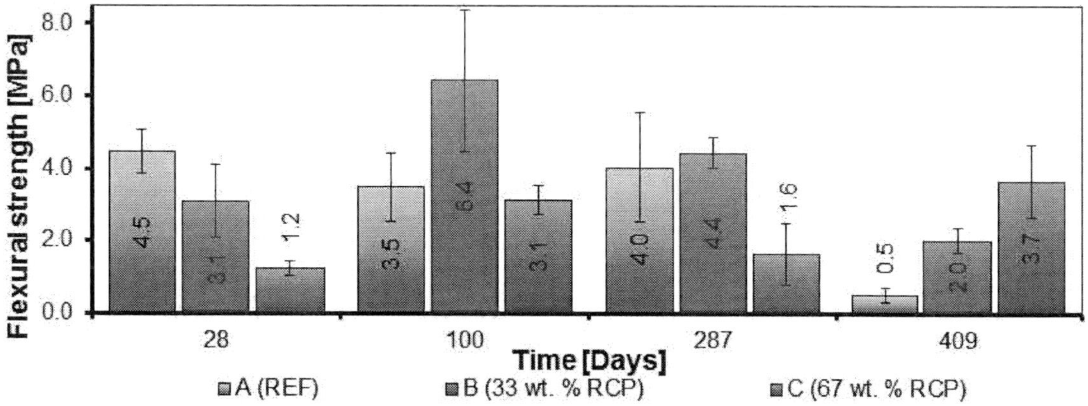

Fig. 1: Flexural strength development of the cement pastes with different amount of a recycled concrete powder.

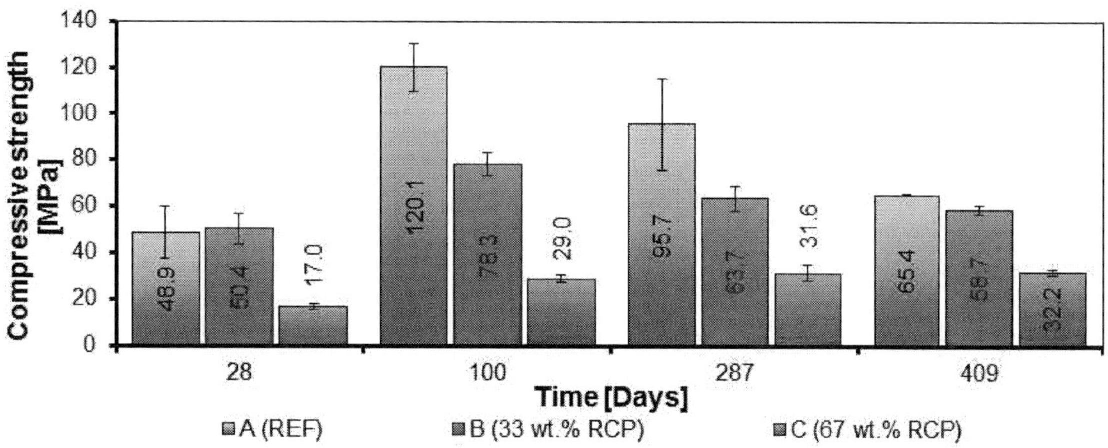

Fig. 2: Compressive strength development of cement pastes with different amount of a recycled concrete powder.

5 Conclusions

This work is focused on the development of the flexural and compressive strength of the cement paste with different amount of a recycled concrete powder (RCP) modified by high speed milling. Based on the results it can be concluded that:

- high speed milling process enables reuse of the unhydrated cement grains and some hydraulic properties and microfiller properties was achieved,
- the flexural strength of the samples with 33 wt. % of RCP was higher after 100, 287 and 409 days compared with the reference samples,

- broad ranges of the flexural strength between the same samples at different times could be caused by small differences in mixing and molding because each set of the samples was mixed separately,
- the samples with 33 wt. % of RCP have the comparable compressive strength with the reference sample,
- during time after 100 days to 409 days the compressive strength of samples A and B declines about 45 and 25 %,
- the results suggest that RCP could be used as a partial cement replacement (if added below 33 wt. %) with minimal decrease in strength, while the cost reduction is still rather significant.
- compressive strength of samples C caused by a high content of RCP, is too low for practical and economical use in construction.

Acknowledgement

Financial support from Czech Technical University in Prague – SGS project No. SGS16/122/OHK1/3T/11 and TAČR project No. TA04031256. The authors also thank the Center for Nanotechnology in Civil Engineering at the Faculty of Civil Engineering and the Faculty of Civil Engineering, CTU in Prague.

References

[1] Eurostat. Waste statistics in Europe. http://epp.eurostat.ec.europa.eu/ (last accessed in 8.2015).

[2] R. V. Silva et al., Properties and composition of recycled aggregates from construction and demolition waste suitable for concrete production. Construction and Building Materials 65 (2014), 201–217, http://dx.doi.org/10.1016/j.conbuildmat.2014.04.117.

[3] J. Shoon et al., Waste fibrecement: An interesting alternative raw material for a sustainable Portland clinker production. Construction and Building Materials 36 (2012), 391–403, http://dx.doi.org/10.1016/j.conbuildmat.2012.04.095.

[4] Z. Shui et al., Rehydration reactivity of recycled mortar from concrete waste experienced to thermal treatment. Construction and Building Materials 22 (2008), 1723–1729, http://dx.doi.org/ 10.1016/j.conbuildmat.2007.05.012.

[5] Ahmari et al., Production of geopolymeric binder from blended waste concrete powder and fly ash. Construction and Building Materials 35 (2012), 718–729, http://dx.doi.org/10.1016/j.conbuildmat.2012.04.044.

[6] Y. J. Kim et al., Utilization of waste concrete powder as a substitution material for cement. Construction and Building Materials 30 (2012), 500–504, http://dx.doi.org/10.1016/j.conbuildmat.2011.11.042.

[7] K. Šeps, I. Broukalová, Properties of cement based composite with fine ground recycled concrete, Advanced Materials Research 1000 (2014), 110–113, http://dx.doi.org/ 10.4028/www.scientific.net/AMR.1000.110.

[8] J. Topič et al., Replacement of Cement with Finely Ground Recycled Concrete: Influence on Mechanical Properties, Applied Mechanics and Materials 825 (2016), 69–72, Applied Mechanics and Materials 825 (2016), 69–72, http://dx.doi.org/10.4028/www.scientific.net/AMM.825.69.

[9] M. Lidmila et al., Mechanical Properties of Recycled Binder / Micro-Filer Cement-Based Material, Advanced Materials Research Vol. 1054 (2014), 234-237, http://dx.doi.org/10.4028/www.scientific.net/AMR.1054.234.

Analysis of Acoustic Emission Signals Generated by Cracks during Three-point Bending Test

L. Topolář [1,a], P. Rovnaník [1], L. Pazdera [1]

[1] *Brno University of Technology, Faculty of Civil Engineering, Veveří 331/95, Brno 602 00, Czech Republic*
[a] *topolar.l@fce.vutbr.cz*

Abstract: The method of acoustic emission is an experimental tool well suited for monitoring fracture processes in a material. The typical parameters of acoustic emission signals were identified from the acoustic emission records for different alkali mixtures to further describe the under the stress behaviour and failure development. Alkali-activated slag is a material which has great potential for use in building industry. An understanding of micro structure-performance relationships during stress situation is the key to true understanding of material behaviour.

Keywords: acoustic emission method; three-point bending test; energy of acoustic emission signal, the crack initiation.

1 Introduction

The Acoustic Emission Method (AEM) is an experimental tool well suited for monitoring fracture processes. The AEM can monitor changes in materials behaviour over a long time and without moving one of its components i.e. sensors. This makes the technique quite unique along with the ability to detect crack propagation occurring not only on the surface but also deep inside the material. The AEM is considered to be a "passive" non-destructive technique because usually identifies defects which they develop during the test [1].

Fracture in a material takes place with the release of stored strain energy which is consumed by nucleating new external surfaces (cracks) and emitting elastic waves which are defined as acoustic emission (AE) waves. The elastic waves propagate inside a material and are detected by an AE sensor [2]. Except for contactless sensors, AE sensors are directly attached to the surface. AE signals (Fig. 1) are detected as dynamic motions at the surface of a material are converted into electric signals [3]. Then signals are amplified and filtered. The final goal of monitoring AE phenomena is to provide beneficial information to prevent fatal fracture by correlating detected AE signals with growing fracture process or deterioration [4]. AEM is often used to detect a failure at a very early stage of damage long before a structure completely fails.

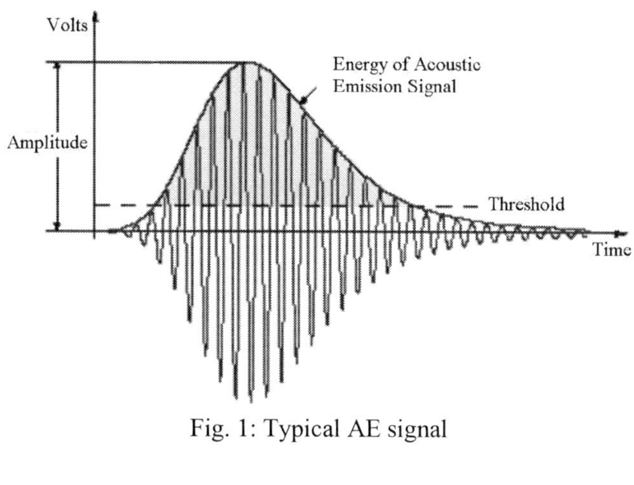

Fig. 1: Typical AE signal

Acoustic emission activity is attributed to the rapid release of energy in a material, the energy content of the acoustic emission signal can be related to this energy release. The true energy is directly proportional to the area under the acoustic emission waveform [5]:

$$\text{AE energy}_i = \int_{t0}^{t1} V_i(t)^2 dt \qquad (1)$$

i = the recorded voltage transient V(t) of a channel
t_0 = the starting time of the voltage transient record
t_1 = the ending time of the voltage transient record

Natural graphite is a versatile material with a unique structure and many different properties. Therefore, it is not only an essential component in applications such as pencils and refractory products but it is also a high-performance raw material for the plastics industry. Being a natural crystalline graphite grade, "COND" still exceeds the conventional graphite grades in terms of electrical conductivity about 10 times. Nevertheless, the COND grades possess other typical quality characteristics of natural graphite-like lubrication, thermal conductivity, and chemical stability. An enormous advantage of the COND grades are that both thermal conductivity and especially high electrical conductivity are combined in this product.

2 Experimental setup

Alkali-activated slug slurry was cast into prismatic moulds (40×40×160 mm) to set and after 20 h the samples were demoulded and immersed in water for another 27 days before testing. Prior to testing of fracture properties, the specimens were pulled out from water immersion and stored in the air at laboratory conditions (20 ± 1 °C, RH 45 ± 5 %). The three-point bending fracture tests were conducted on beam specimens with central edge notch. The initial notch was made before testing with a diamond blade saw. The depth of the notches was approximately 30 % of the depth of the specimen. The span length was equal to 120 mm. The specimens were tested at the age of 28 days. Fracture tests were carried out using a LabTest 6.1000 electromechanical testing machine within the range of 0–100 kN at a laboratory of the AdMaS Centre (Advanced Materials, Structures and Technologies), Faculty of Civil Engineering, Brno University of Technology.

Tab. 1: Composition of alkali-activated slag mixtures.

Component	**REF**	**COND 10**
Slag [g]	450	450
Water glass [g]	90	90
Quartz sand [g]	1350	1350
Water [ml]	180	135
COND 8 96 [g]	-	45
0.5% Triton X-100 [ml]	-	120
1.0% Lukosan S [ml]	-	20

During the experiment, an acoustic emission activity was recorded. The guard sensor eliminated mechanical and electrical noise. Four acoustic emission sensors were attached to the surface of the specimen with beeswax (Fig. 2). Acoustic emission signals were recorded by DAKEL XEDO measuring equipment with four IDK-09 acoustic emission sensors with a 35 dB preamplifier.

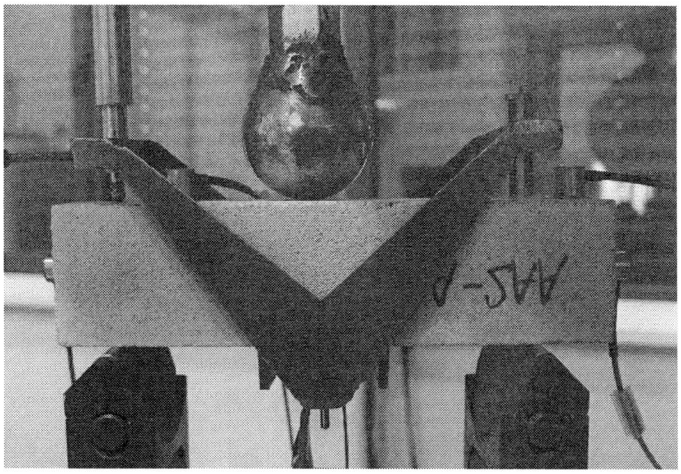

Fig. 2: Photography of experiment

3 Results

To assess the nature of the cracks, we used AE energy. For example on the Fig. 3 is relative cumulative energy AE signal and the relative load versus time. The AE energy of the signal is directly proportional to the area under the acoustic emission waveform. Fig. 3 shows that energy of AE signals starting to evolve before maximum load. That means the crack initiation and thus, release energy began at about 75% of a maximum of graphite powder COND 10. Fig. 4 shows that maximum load. The release of AE energy changed after add energy of AE signals starting to evolve while maximum load.

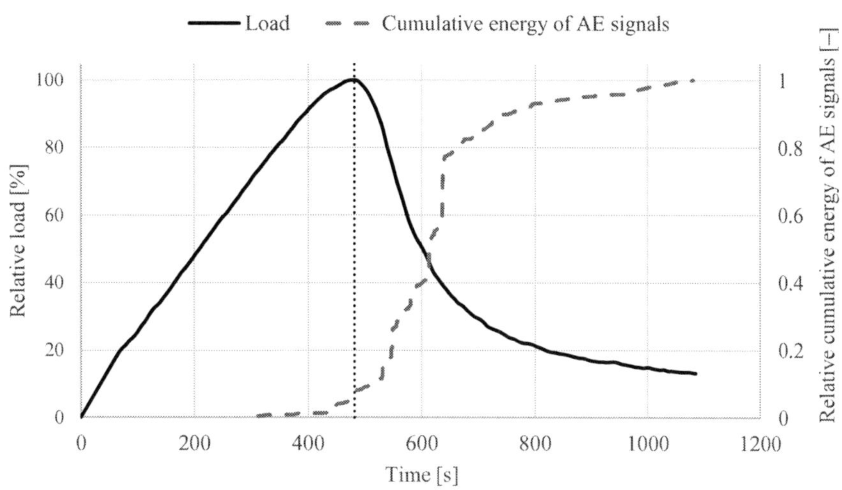

Fig. 3: Dependence of relative cumulative energy of AE signal and the relative load on time. (REF sample)

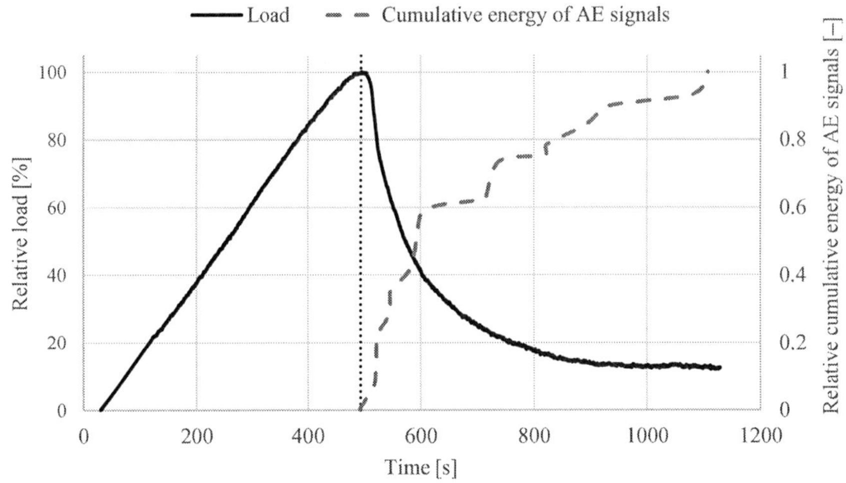

Fig. 4: Dependence of relative cumulative energy of AE signal and the relative load on time. (COND 10 sample)

4 Conclusion

The fracture properties of alkali-activated slag composites have been investigated. The acoustic emission method appears to be a suitable tool for determining the behaviour of alkali activated slag mortar specimens during stable crack propagation from an initial crack/notch. Stable crack propagation is followed by a number of AE signals until fracture takes place. The addition of graphite powder has a significant influence on the evolution of energy of acoustic emission signals. Application of acoustic emission method during the three-point bending test of alkali-activated slag mortar can help to obtain their better properties. The properties of the emerging micro cracks can ultimately be linked to the overall fracture behaviour of the materials.

Acknowledgement

This outcome has been achieved with the financial support of the Czech Science Foundation, project No. 16-00567S, and the Ministry of Education, Youth and Sports of the Czech Republic under the "National Sustainability Programme I", project No. LO1408 (AdMaS UP).

References

[1] Ch. U. Grosse, M. Ohtsu, Acoustic Emission Testing, Springer-Verlag, Berlin, 2008, DOI 10.1007/978-3-540-69972-9.

[2] M. Iwanami, T. Kamada, S. Nagataki, Application of Acoustic Emission Technique for Crack Monitoring in RC beams, JCA Proc, Cement and Concrete (1997) 51, p. 192–197.

[3] C.U. Grosse and M. Ohtsu. Acoustic Emission Testing: Basics for Research – Applications in Civil Engineering; with contributions by numerous experts. Springer, ISBN: 978-3-540-69895-1, Heidelberg, 2008.

[4] I. Havlíková., et al., Modified Cement-based Mortars: Crack Initiation and Volume Changes. Materiali in tehnologije, 2015, vol. 49, no. 4, p. 557-561.ISSN: 1580-2949.

[5] R. V. Sagar, An experimental Study on Acoustic Emission Energy and Fracture Energy of Concrete, India, 2009, National Seminar & Exhibition on Non-Destructive Evaluation, pp. 225-228

Stress and Deformation Analysis of Portable Bridge

F. Trebuňa, F. Šimčák, J. Bocko[a], M. Pástor, P. Lengvarský

Technical University of Košice, Faculty of Mechanical Engineering, Department of Applied Mechanics and Mechanical Engineering, Letná 9, 042 00 Košice, Slovakia

[a]jozef.bocko@tuke.sk

Abstract: The paper is devoted to description of strength analysis of bridge structure. Two models of bridges were assessed – original bridge with width 3040 mm and modified structure with functional width 3260 mm. The structure of the bridge is portable, positioned on truck. The bridge can be folded and positioned for passing of belt carriages and wheeled cars. The stresses and deformations in both structures were determined for basic as well as combined loadings. The bridge structures fulfill all demands from the point of view of deformations and strength.

Keywords: supporting bridge structure, pass of vehicles, finite element method, stress analysis.

1 Introduction

Portable steel bridge structure is used for preparation of ways for vehicles during floods, damages of old bridges or badly crossing ditches. They are used especially in situations where it is not economical to build new compact bridge and they are used mainly by military forces (Fig. 1).

Fig. 1: Portable bridge.

The authors analyzed stress state of original bridge structure with the width 3040 mm and modified supporting structure with width 3260 mm. The aim was to check original structure and to modify it in order to allow its folding and passing belt and wheeled carriages [1-5].

2 Analysis of the bridge

The analysis was accomplished for loading due to self-loading, transverse loading due to weight of vehicle, with the predefined parameters of axle bases. The analysis comprises also quasi-static loading due to acceleration or braking. The number of carriage crossings through the bridge were considered, too. The bridge will be loaded by vehicles with the following parameters:

- maximal allowed velocity of belt carriage – 20 km/h,
- maximal allowed velocity of wheeled vehicle – 40 km/h,
- acceleration of belt carriage 0-32 km/h – 10 s.

The contact of belt carriage with bridge is shown in Fig. 2, total mass is 50 000 kg, length of undercarriage 6 900 mm, width without boundaries 3 400 mm, distance between belts 2 790 mm and belt width 580 mm.

The whole mass of whole equipment is 29 500 kg; the mass without bulldozer equipment is 28 200 kg. In Fig. 3 are given positions and dimensions of contact forces of vehicles with the bridge.

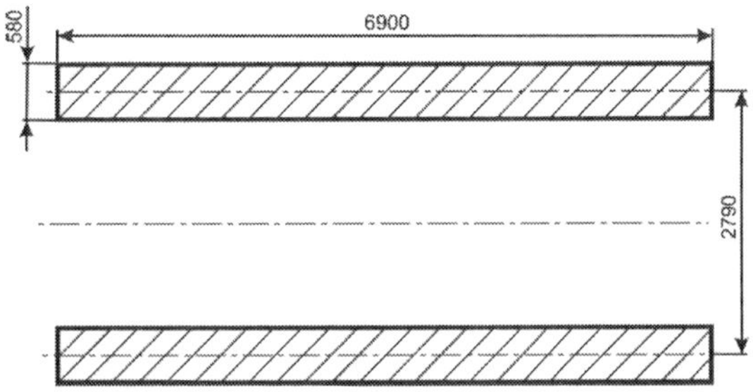

Fig. 2: Geometry and dimensions of contact surfaces of belt carriage with the bridge.

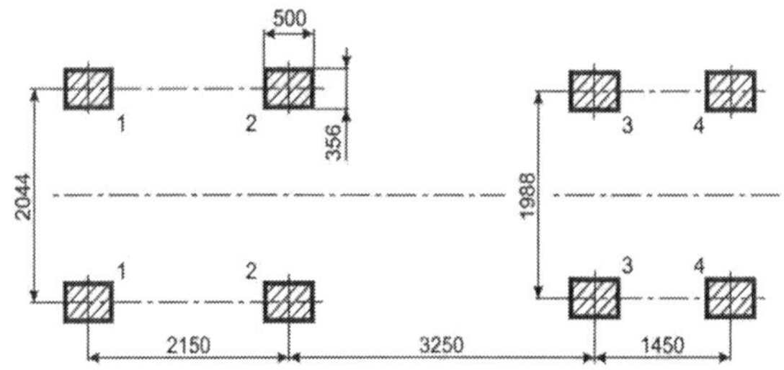

Fig. 3: Contact surfaces of wheeled car with the bridge.

In order to compute stress and deformation fields in original and new proposed bridge structures there was necessary to mesh the model of supporting structure. The computation was realized for both vertical as well as horizontal loading and the forces corresponds to loading by tank or truck. From the analysis of internal force quantities results that the highest loading due to bending moment arises in case if the tank or truck are positioned in the middle of the bridge, but the horizontal loading lead to higher stresses if the tank and truck are positioned on the boundary of bridge.

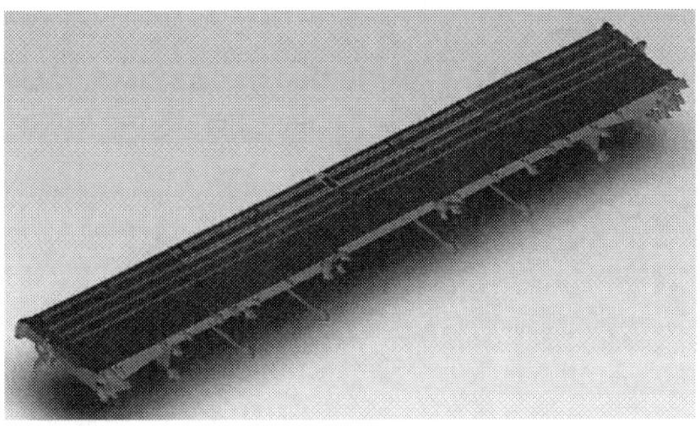

Fig. 4: Boundary conditions for vertical and horizontal loading for position of tank in the middle of the bridge.

Fig. 5: Boundary conditions for vertical and horizontal loading for position of tank on the boundary of the bridge.

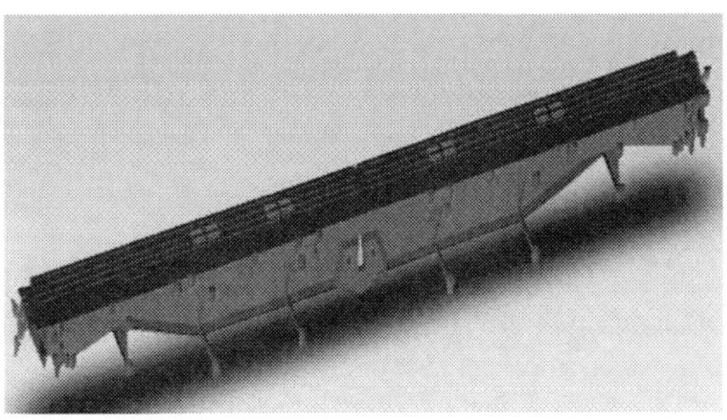

Fig. 6: Boundary conditions for vertical and horizontal loading for position of truck in the middle of the bridge.

Fig. 7: Boundary conditions for vertical and horizontal loading for position of truck on the boundary of the bridge.

2.1 Stress fields due to combined loading by self-weight of the bridge, loading by tank in horizontal and vertical direction in the middle of the bridge

The original structure of the bridge was loaded by self-weight of bridge, weight of tank including dynamic loadings in horizontal and vertical directions. The boundary conditions for loading of bridge structure are obvious from Fig. 4. In Fig. 8 and Fig. 9 are given fields of equivalent stresses and displacements of individual points of structure, respectively, due to loading by weight of tank.

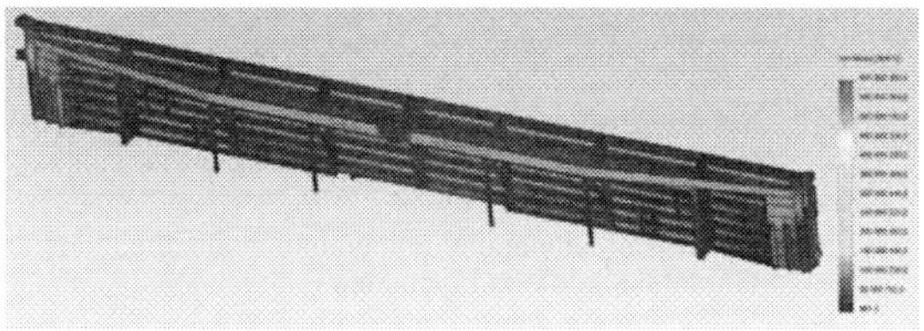

Fig. 8: Field of equivalent stresses for combined loading.

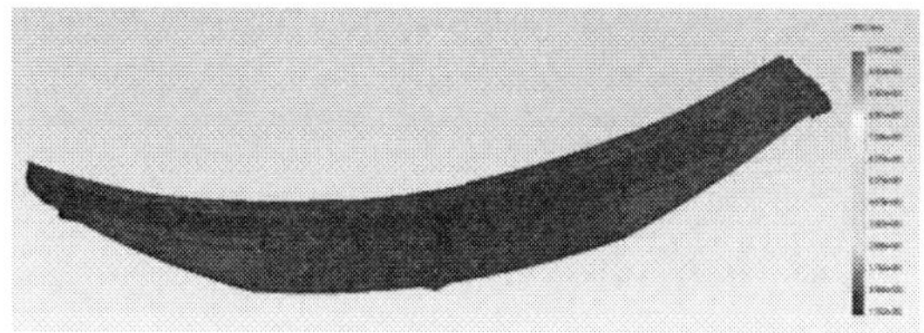

Fig. 9: Displacements of points of structure for combined loading.

New supporting structure of the bridge was in this case loaded by self-weight of bridge, weight of tank including dynamic loadings in horizontal and vertical directions. The boundary conditions for loading of bridge structure are given in Fig. 4. In Fig. 10 and Fig. 11 are given fields of equivalent stresses and displacements of individual points of structure, respectively, due to loading by weight of tank.

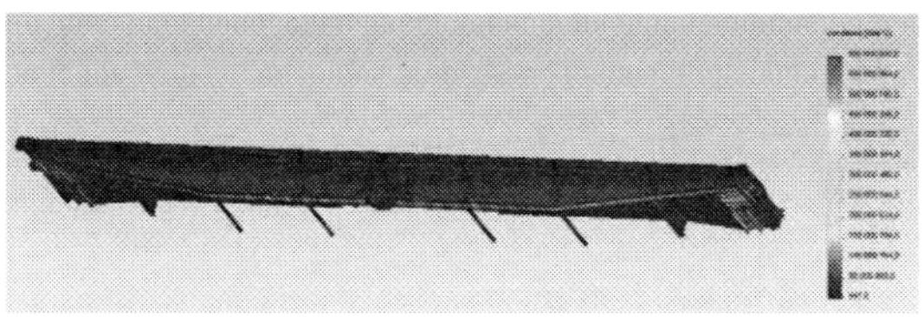

Fig. 10: Field of equivalent stresses for combined loading.

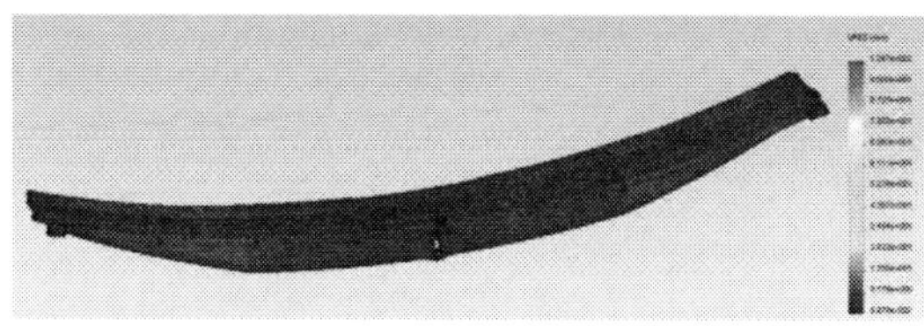

Fig. 11: Displacements of points of structure for combined loading.

2.2 Stress fields due to combined loading by self-weight of the bridge, loading by tank in horizontal and vertical direction on the boundary of the bridge

The original structure of the bridge was loaded by self-weight of bridge, weight of tank including dynamic loadings in horizontal and vertical directions. The boundary conditions for loading of bridge structure are given in Fig. 5. In Fig. 12 and Fig. 13 are given fields of equivalent stresses and displacements of individual points of structure, respectively, due to loading by weight of tank.

Fig. 12: Field of equivalent stresses for combined loading.

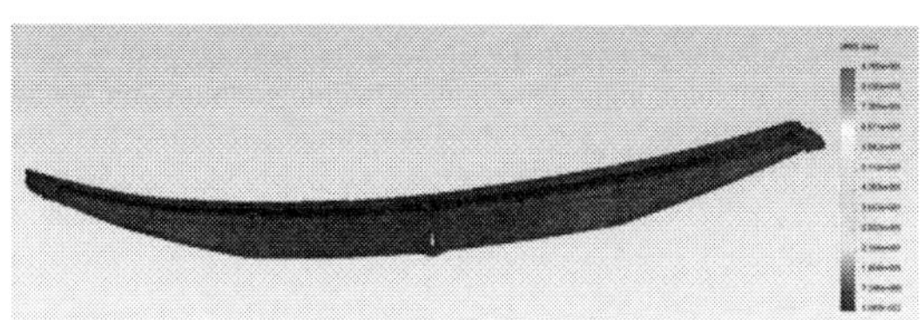

Fig. 13: Displacements of points of structure for combined loading.

New supporting structure of the bridge was in this case loaded by self-weight of bridge, weight of tank including dynamic loadings in horizontal and vertical directions. The boundary conditions for loading of bridge structure are given in Fig. 5. In Fig. 14 and Fig. 15 are given fields of equivalent stresses and displacements of individual points of structure, respectively, due to loading by weight of tank.

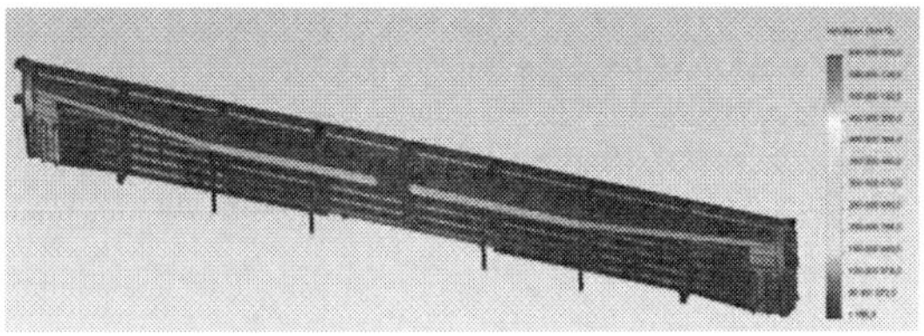

Fig. 14: Field of equivalent stresses for combined loading.

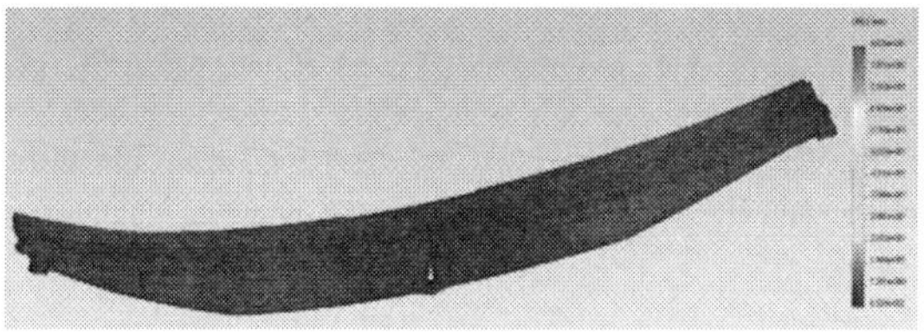

Fig. 15: Displacements of points of structure for combined loading.

2.3 Stress fields due to combined loading by self-weight of the bridge, loading by truck in horizontal and vertical direction in the middle of the bridge

The original structure of the bridge was loaded by self-weight of bridge, weight of truck including dynamic loadings in horizontal and vertical directions. The boundary conditions for loading of bridge structure are given in Fig. 6. In Fig. 16 and Fig. 17 are given fields of equivalent stresses and displacements of individual points of structure, respectively, due to loading by weight of truck.

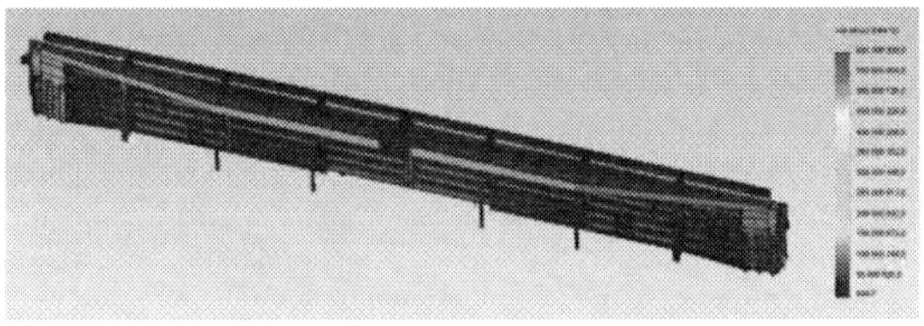

Fig. 16: Field of equivalent stresses for combined loading.

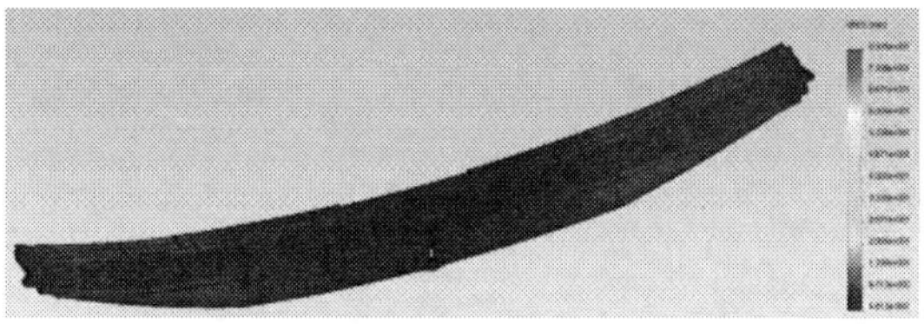

Fig. 17: Displacements of points of structure for combined loading.

New supporting structure of the bridge was in this case loaded by self-weight of bridge, weight of truck including dynamic loadings in horizontal and vertical directions. The boundary conditions for loading of bridge structure are given in Fig. 6. In Fig. 18 and Fig. 19 are given fields of equivalent stresses and displacements of individual points of structure, respectively, due to loading by weight of truck.

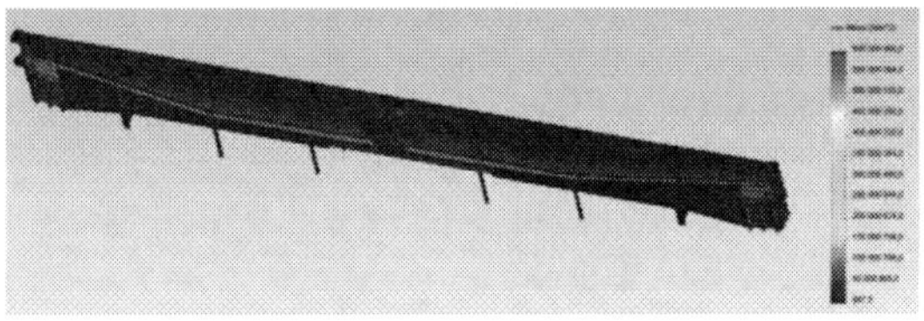

Fig. 18: Field of equivalent stresses for combined loading.

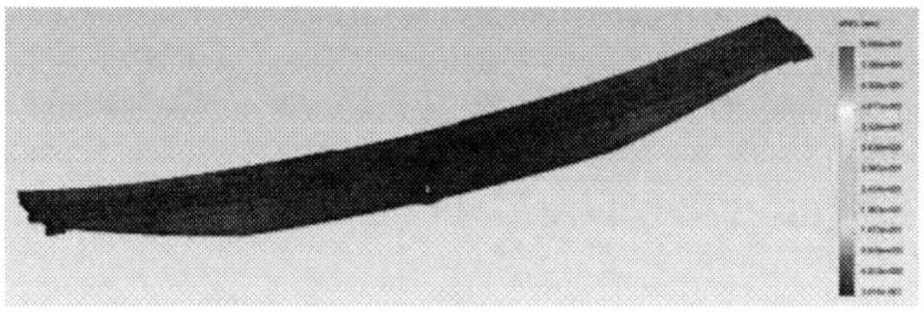

Fig. 19: Displacements of points of structure for combined loading.

2.4 Stress fields due to combined loading by self-weight of the bridge, loading by truck in horizontal and vertical direction on the boundary of the bridge

The original structure of the bridge was loaded by self-weight of bridge, weight of tank including dynamic loadings in horizontal and vertical directions. The boundary conditions for loading of bridge structure are given in Fig. 7. In Fig. 20 and Fig. 21 are given fields of equivalent stresses and displacements of individual points of structure, respectively, due to loading by weight of tank.

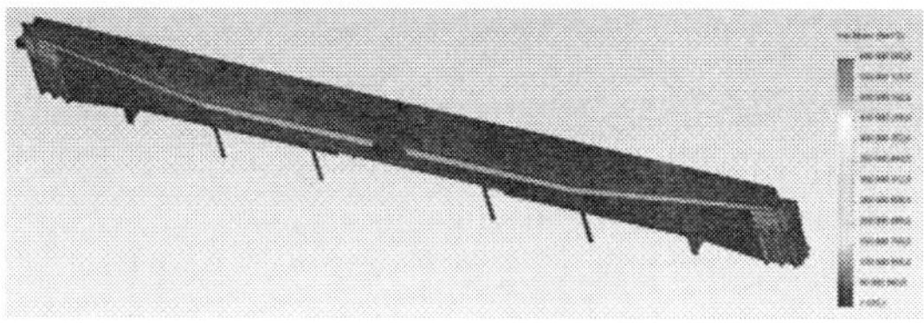

Fig. 20: Field of equivalent stresses for combined loading.

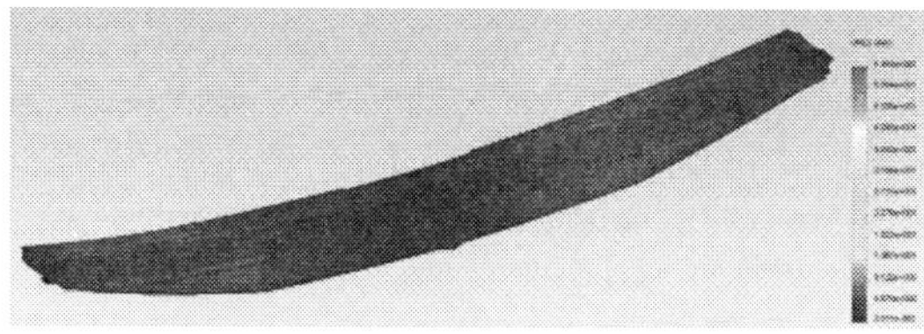

Fig. 21: Displacements of points of structure for combined loading.

New supporting structure of the bridge was in this case loaded by self-weight of bridge, weight of truck including dynamic loadings in horizontal and vertical directions. The boundary conditions for loading of bridge structure are given in Fig. 7. In Fig. 22 and Fig. 23 are given fields of equivalent stresses and displacements of individual points of structure, respectively, due to loading by weight of truck.

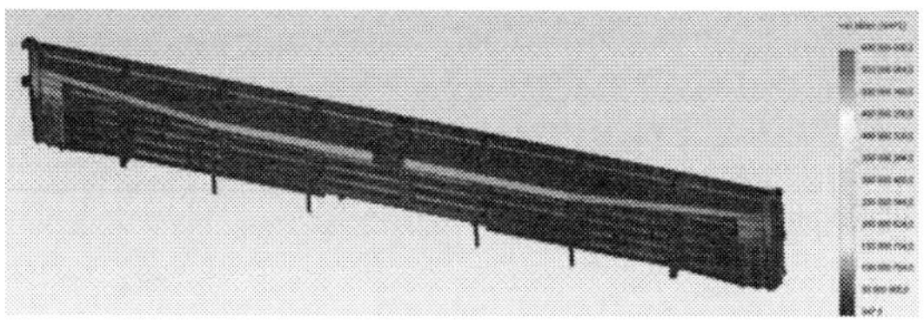

Fig. 22: Field of equivalent stresses for combined loading.

Fig. 23: Displacements of points of structure for combined loading.

3 Results and discussion

In sections 2.1 and 2.2 during analysis of combined loading due to self-weight as well as horizontal and vertical loading was found out (Figs. 10 and 14) that the stresses exceed the strength of material.

Accordingly, the authors by individual steps proposed modification of supporting structure by adding eight vertical reinforcements and triangular reinforcements between new vertical reinforcement and original flanges. These modifications led to the removing of dangerous stress levels not only in webs of main beams, but also on bottom flanges. The modifications can be seen in Fig. 24. In Fig. 25 is the field of equivalent stresses after realization of modifications. The figures clearly documents decreasing of stress levels to the allowed values in webs as well as in flanges of main beams. The vertical reinforcements and triangular elements are made of sheet of thickness 4 mm, material S700.

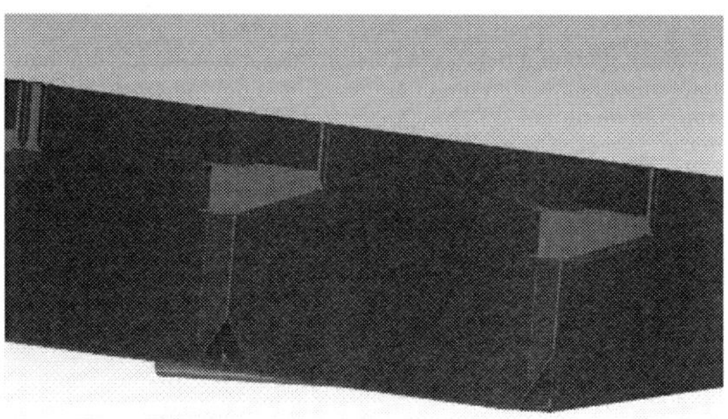

Fig. 24: Model of the bridge after modification of main beams.

Fig. 25: Fields of equivalent stresses after modification of bridge.

4 Conclusion

The analysis had shown that the highest bending stresses and deformations are in case when the tank or truck are positioned in the middle of bridge. On the other side, the horizontal loading are in case if the vehicles are positioned at the boundary of bridge. The computations of original and modified bridges were not accomplished into details, but they conclude that the main parts of left as well as right side of the bridge fulfill all demands for such structures. The stresses do not exceed yield points of materials. However, the computations do not cover all details of the bridge nor the fatigue analysis.

Acknowledgement

This research was supported by grant from the Slovak Grant Agency VEGA no. 1/0731/16 - Development of Modern Numerical and Experimental Methods of Mechanical System Analysis.

References

[1] F.Trebuňa et al., Analýza napäťových pomerov v nosných mostových konštrukciách a návrh potrebných konštrukčných zmien v kritických miestach, Košice, TUKE SjF KAMaSI, 2014.

[2] I. Fürbacher et al., Lexikon technických materiálů, Bratislava, Verlag Dashöfer, 2005.

[3] F. Trebuňa, F. Šimčák, Pružnosť, pevnosť a plastickosť v strojárstve, TU SjF Košice, Emilena, 2004.

[4] F. Trebuňa, F. Šimčák, Odolnosť prvkov mechanických sústav. TU SjF Košice, Emilena, 2004.

[5] L. Kwasniewski et al., Finite element analysis of vehicle–bridge, Finite Elements in Analysis and Design 42 (2006) 950-959.

Performance of Plasma Treated Polymer Macro-Fibers

J. Trejbal [1,a], J. Fládr [1], L. Kopecký [1]

[1] *Faculty of Civil Engineering, Czech Technical University in Prague, Thákurova 7, 166 29 Prague, Czech Republic*

Abstract: Concrix – commercial bi-components polyolefin macro-fibers (ø 0.5 mm, length 50 mm) made by BURG Group in Switzerland were plasma treated to attain the better adhesion with cement matrix. Fibers were exposed to oxygen treatment in inductively coupled plasma system by different times (from 5 to 480 seconds.). The influence of realized fiber surface treatments was observed by physical and mechanical methods: i) horizontal direct optical method enabling a measurement of fibers wettability, ii) fibers tensile strength tests, and iii) single-fiber pull-out tests from cement matrix. It turned out that fibers treated by plasma during 30 seconds reached the best ratio between their wettability and mechanical performance. While their contact angles decreased by ca. 50 %, tensile strength remained constant and cohesion with cement matrix increased by ca. 20 % in comparison with reference fibers.

Keywords: plasma; polymer fibers; reinforcement; wettability.

1 Introduction

Polymer macro-fibers used as reinforcement in cement composites are frequently applied in civil engineering. Macro-fibers as an auxiliary reinforcement can reduce the early age cracking tendency of concrete surfaces and can bestow the toughness behavior of concrete materials [1]. The principle of the fiber effect is the distribution of shrinkage into several small cracks in the case of materials based on shrinking binders, and to prevent the single crack openings after linear elastic response in the case of loaded samples [2]. Macro-fiber reinforcement also provides a compactness of materials after the crossing the material loading capacity [3]. Generally, fiber reinforced concrete is commonly used for construction of tunnels (shotcrete), fire resistance constructions, constructions exposed to dynamic loads, flat and thin construction, etc. [4, 5].

Besides a steel, polymers are an alternative material for production of fibers. Polymer macro-fibers even exhibit some material parameters better in comparison with steel fibers (tensile strength, density, corrosion resistance) [6]. In order to improve mechanical performance of composite materials containing fiber reinforcement, strong interaction between these two phases – matrix and fibers must be ensured [3]. In view of poor fiber surface wettability by water, required adhesion between their surfaces and the matrix of composite materials (containing water) is not guaranteed. Due to this reason, some different mechanical, chemical and physical modifications of fiber surfaces have been studied [7, 8]. The surface treatments must represent an affective, economic and simultaneously nondestructive method.

The treatment by low-pressure oxygen coupled plasma treatment satisfies all the requirements mentioned above and becomes popular as progressive physical method. Plasma can be defined as ionized gas (composed of electrons, ions, and neutral species) which is electrically neutral form macroscopic perspective. The principle for plasma surface modification relays on surface atoms replacement by oxygen atoms and formation of polar groups. The polar or functional chemical groups enhances the reactivity with the cement or other matrix (containing water) [1, 9].

2 Materials and Methods

2.1 Fibers and cement matrix

Polyolefin macro fibers brand named Concrix ES having diameter equal to 0.5 mm a length 50 mm were used at this study [10]. Fibers made from polyolefin composed from two parts – high strength and high modulus core guaranteeing fiber mechanical performance and shell protecting the core and ensuring good mixture processing and binding to concrete (Fig. 1). Modulus of elasticity is up to 10 GPa, density is equal to 0.91 g/cm^3.

Cement matrix samples having dimensions equal to 20×25×40 mm were prepared. The samples were composed from cement paste (CEM I 42.5 R, water to cement ratio was equal to 0.4). During the samples preparation, single fiber was submerged into each cement sample parallel to longitudinal sample axis (Fig. 2). The fiber anchor length corresponded to sample height (i.e. 40 mm). The cement samples cured and hardened 28 days in standard condition without other treatment.

Fig. 1: Scheme of fiber Concrix ES.

Fig. 2: Cement samples prepared for single-fiber pull out tests.

2.2 Experimental methods

To improve the fiber wettability, oxygen treatment in inductively coupled plasma system was done (power 100 W, O_2 flow 50 sccm). Plasma treatment process differed only in exposition time (from 5 to 480 seconds). Plasma treatment effect was observed by direct static horizontal contact angle method enabling the measurements on fibers upright submerged into demineralized water. Treated fiber tensile strength was tested using the loading frame Web Tiv Ravestein FP100 (displacement controlled, loading speed was equal to 2mm/min). The purpose of this experiment was to verify the treatment impact on fiber mechanical performance. Based on these measurements, the most effective treatment process was defined and applied for following usage.

Reference fibers (Concrix_R) and treated fibers exposed to oxygen plasma by 30 seconds (Concrix_30) were used for single-fiber pull out tests. Principle of pull out tests consisted of pulling out of fiber free end from matrix held by static part of loading frame, carrying out by the same loading frame as is mentioned above. Displacement of pulled-out fiber free-end and loading force were monitored. Six samples of each fiber type were tested.

3 Results

Contact angle size measured on reference fibers submerged into demineralized water was equal to 54.0±6.9°, while angles measured on plasma treated fibers oscillated about ca. 26°. It was clearly shown that wettability of treated fibers does not depend on treatment exposition time, while the difference between reference and plasma treated fiber wettability was significant (Fig. 3). Fibers tensile strength measured as a force causing fiber tensile breaking was roughly constant in all cases. All fibers were broken by force equal to 95-100 N without any dependence on plasma treatment time (Fig. 4).

The average force (6 samples from each fiber) obtained during pull-out tests of reference and plasma treated fibers by 30 seconds was equal to 78.4±8.2 and 96.3±10.1 N, respectively. In the case of reference samples, all fibers were pulled out from matrix, while in the case of plasma treated fibers, only two fibers showed cohesion break in an interphase area. The remaining four fibers were broken (Fig. 5).

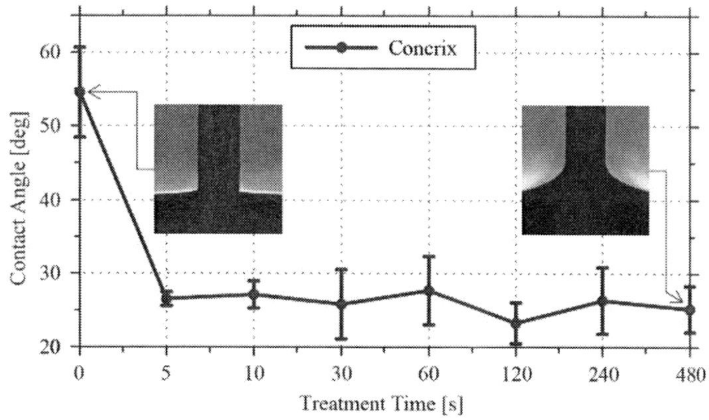

Fig. 3: Contact angle sizes depending on plasma treatment time obtained during contact angles measurements.

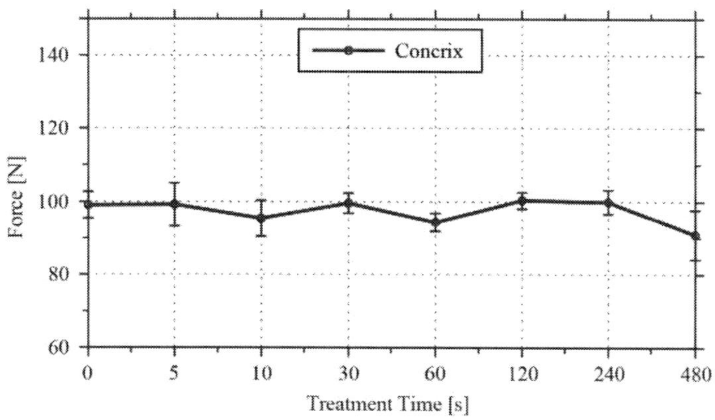

Fig. 4: Force needed to break the fibers depending on treatment time.

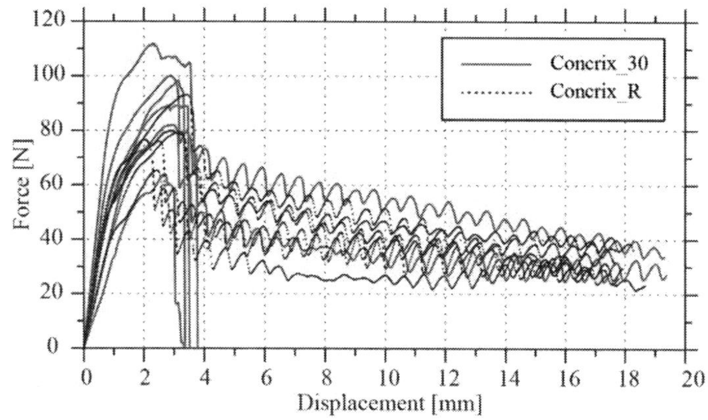

Fig. 5: Loading force depending on fiber free-end displacement obtained during pull out testing.

4 Conclusion

Polyolefin macro-fibers Concrix using as reinforcement in cement composites were plasma treated to obtain the better adhesion between their surfaces and cement matrix (CEM I 42.5 R, w/c equal to 0.4). For this purpose, oxygen treatment in inductively coupled plasma system was done. Treatment process differed in exposition time (from 5 to 480 seconds.). As a proper indicator of realized wettability treatment, horizontal direct optical method enabling contact angle measurements was used. It was found that treated fibers reached the better wettability approximately by 50 %. Tensile strength of reference and treated fibers is more or less constant. From pull-out single-fiber test from cement matrix specimens (25×20×40 mm) was shown that force needed for reference fibers pull-out was equal to 78.4±8.2 N, while in the case plasma treated fibers by 30 seconds was equal to 96.3±10.1 N.

Acknowledgement

This was financially supported by Czech Technical University in Prague – SGS project SGS16/201/OHK1/3T/11 and by the Czech Science Foundation research project 15-12420S. The authors also thank the Center for Nanotechnology in Civil Engineering at the Faculty of Civil Engineering CTU in Prague and the Joint Laboratory of Polymer Nanofiber Technologies of the Institute of Physics Academy of Science of the Czech Republic and the Faculty of Civil Engineering CTU in Prague.

References

[1] B. Felekoglu, K. Tosun, B. Baradan, A comparative study on the flexural performance of plasma treated polypropylene fiber reinforced cementitious composites, Journal of Materials Processing Technology 209 (2009) 5133–5144, doi:10.1016/j.jmatprotec.2009.02.015.

[2] P.K. Mallick, Fiber-Reinforced Composites: Materials, Manufacturing, and Design, Third Edition. CRC Press: Florida, 2007.

[3] J. Trejbal et al., Wettability enhancement of polymeric and glass micro fiber reinforcement by plasma treatment, in proc.: Nanocon 7th international conference on nanomaterials – research and application (2015), Tanger, Brno, Czech Republic, 315–321.

[4] M. Luňáček, P. Suchánek, R. Bader, Aplication of Biocomponent Syntetic Macro Fibres Concrix in Tunneling, Tunnel 8 (2012) 54–56.

[5] P.C. Tatnall, Significance of tests and properties of concrete and concrete-making materials, Georgia USA, ASTM International, 2006.

[6] Bartoš et al., Posibilities of modification PET microfibers surface layer to improve their cohesion with cement matrix, in proc.: NANS: 3rd Conference on nanomaterials a nanotechnology in civil engineering (2014), ed. P. Tesárek et al., Praha, Czech Republic, 8–11.

[7] S.E. Elsaka, Influence of chemical surface treatments on adhesion of fiber posts to composite resin core materials, Dental materials 29 (2013) 550–558, http://dx.doi.org/10.1016/j.dental.2013.03.004.

[8] M.L. Troëdec et al., Influence of chemical treatments on adhesion properties of hemp fibres, Journal of colloid and interface science 356 (2011) 303–310, 10.1016/j.jcis.2010.12.066.

[9] M.T. Kim, Study on an oxygen plasma treatment of a basalt fiber and its effect on the interlaminar fracture property of basalt/epoxy woven composites, Composites: Part B 42 (2011) 499–504, doi:10.1016/j.compositesb.2010.12.001.

[10] Retrieved from: http://www.bruggcontec.com/Portals/3/PDF/Concrix%20neu/Instruction%20sheet%20Concrix-EN.pdf

The Stress Detection and the Fatigue Lifetime of Stainless Steel AISI 316L during Three-Point Bending Cyclic Loading

M. Uhríčik [1, a], M. Oravcová [1], P. Palček [1], M. Sapieta [2], M. Chalupová [1]

[1] *University of Žilina, Faculty of Mechanical Engineering, Department of Materials Engineering, Slovakia*
[1] *University of Žilina, Faculty of Mechanical Engineering, Department of Applied Mechanics, Slovakia*
[a] *milan.uhricik@fstroj.uniza.sk*

Abstract: The article will describe investigation of the deformation of stainless steel during three-point bending cyclic loading with using thermovision. The analysis will prove different temperature response to external loading and dependence of elastic or plastic deformation development on material´s state. The input data which are necessary for this analysis we will can get from temperature field of specimen surface. Process of elastic and plastic deformation is in dependence on radiation emitted by the object. For obtain thermal fields we will use thermal camera FLIR SC7000 with the highest sensitivity. The contribution also presents results of fatigue resistance of austenitic stainless steel AISI 316L, which is used as a biomaterial, obtained at low frequency cyclic loading in the high cycle fatigue region by three- point bending test. The fracture surface of the testing sample was examined using scanning electron microscopy (SEM).

Keywords: stainless steel; stress; three-point bending; fracture surface; thermal camera.

1 Introduction

Stainless steel is the name given to a family of corrosion and heat resistant steels containing a minimum of 10.5% chromium. Just as there is a range of structural and engineering carbon steels meeting different requirements of strength, weldability and toughness, so there is a wide range of stainless steels with progressively higher levels of corrosion resistance and strength. This variety of grades results from the controlled addition of alloying elements, each offering specific attributes in respect of strength and ability to resist different environments. To achieve the optimum economic benefit from using stainless steel, it is important to select a grade of steel which is adequate for the application without being unnecessarily highly alloyed and costly. Stainless steels can be classified into the five basic groups: austenitic stainless steels, duplex stainless steels, ferritic stainless steels, martensitic stainless steels and precipitation hardened steels [1-3].

Austenitic stainless steel AISI 316L has been used over decades as a metallic biomaterial as fracture repair devices and it has also a usage for making joint replacements. Their properties exceeds in good corrosion resistance, high biotolerance and suitable mechanical properties such as relatively high strength. Their corrosion resistance depends on Cr content and on the formation of a thin passive surface oxide layer. Austenitic stainless steel forms a single fcc austenite phase from its annealing temperature (~1050 °C) to room temperature and achieves its desirable strength and fatigue resistance through strain hardening and solid solution hardening mechanisms. These joint replacements are required to have a good combination of strength and ductility because they can undergo static or cyclic loading. Implant loading conditions are harsh because they occur in an aggressive body environment and that good corrosion-fatigue resistance is required too. Fracture can occur as a result of single overload or cyclic loading at stresses below the ultimate tensile strength and even below its yield strength. Fatigue failure is a process resulting first from change of mechanical properties, then fatigue crack initiation occurs and progressive growth of crack (propagation of fatigue crack) leading to final fracture. The failure can occur either in low number of cycles ($\leq 10^4$ cycles; low cycle fatigue) or over millions of cycles (high cycle fatigue). The initiation of fatigue crack by cyclic loading is a result of dislocation interactions which lead to formation of micro voids and also dislocation run-outs which create surface irregularities such as slip bands. These irregularities such as slip bands, inclusions, surface imperfections, grain boundaries and so on act as stress concentrators promoting local crack initiation [4-6].

Stress and fatigue testing are common test methods in mechanical engineering and materials science, but provide limited information on complex structures. Thermal stress mapping provides thousands of stress measurements simultaneously, even on geometrically complex components. The deformation of structural materials is followed by thermal effects. We recognize thermoelastic or thermoplastic stress analysis, depending on whether the load creates elastic or plastic strains. Thermoelastic stress analysis describes the relation between stress changes and temperature changes of a body in specimens [7]. Considering a three-point bend test, the stress is essentially zero at the neutral axis N-N. Stresses in the y axis in the positive direction represent tensile deformation, which increases, whereas stresses in the negative direction represent compressive deformation, it decreases. The thermoplastic effect quantifies the heat generated by plastic deformation. In the elastic part it is possible under adiabatic conditions to determine the value of the first stress invariant on the material surface by measuring changes of the surface temperature. The thermoelastic effect refers to the thermodynamic relationship between the change of stress in a component under elastic loading and the corresponding change of temperature. It is simply proportional to the change in the sum of the principal stresses, if adiabatic conditions prevail.

Thermoelastic and thermoplastic effect are summarized in a 3-dimensional heat equation together with the effect of heat conduction [7, 8]:

$$\rho C_{\varepsilon} \frac{\partial T}{\partial t} = k \left(\frac{\partial^2 T}{\partial x^2} + \frac{\partial^2 T}{\partial y^2} + \frac{\partial^2 T}{\partial z^2} \right) + T_0 \sum \frac{\partial \sigma_{ij}}{\partial T} \dot{\varepsilon}_{ij}^{e} + \alpha_p \sigma_{ij} \dot{\varepsilon}_{ij}^{p} \qquad (1)$$

where

ρ	is the material density,
C_ε	is the specific heat capacity at constant deformation,
T	is the absolute temperature,
t	is the time,
k	is the thermal conductivity,
x, y, z	are spatial coordinates,
T_0	is the initial temperature,
σ_{ij}	is the stress tensor,
$\dot{\varepsilon}_{ij}^{e}$	is the rate of change of elastic deformation,
α_p	is the ratio of plastic deformation, which is converted to total heat of plastic deformation,
$\dot{\varepsilon}_{ij}^{p}$	is the irreversible part of deformation tensor.

Thermoelastic effect is known as the conversion between mechanical forms of energy and heat. This transformation occurs when stress changes within a material element alter its volume. Density of energy generated in an object is transformed into local temperature changes. If specific heat of metal is high this phenomena is insignificant in terms of temperature change. Roughly 1 MPa change in stress state causes a temperature change of 1mK in steel [9, 10].

2 Experiment

For our analysis we used a commercially available type of austenitic stainless steel, especially AISI 316L. Its microstructure is shown in Figure 1 and its chemical composition is listed in the Tab. 1

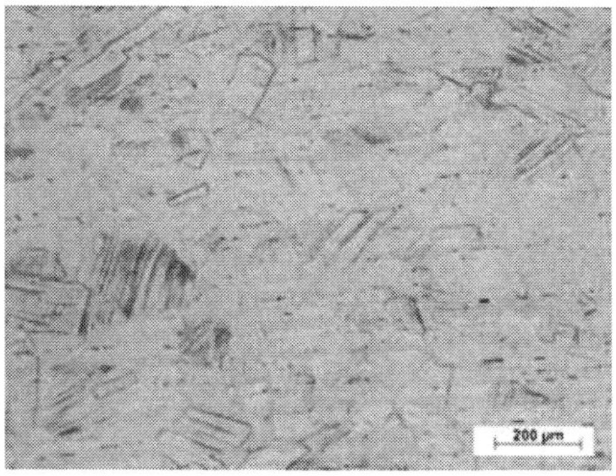

Fig. 1: Microstructure of AISI 316L

Tab. 1: Chemical composition of AISI 316L.

Elements	C	Cr	Ni	Mo	Mn	Si	S	P
[%]	0.037	17.41	9.91	2.38	1.34	0.51	0.03	0.07

Stress and fatigue testing are common test methods in mechanical engineering and materials science, but provide limited information on complex structures. For three-point bending cyclic test, we used Vibrophores Amsler 150 HFP 5100 (Fig. 2a), which is using the resonance principle with constant or variable amplitude and mean load. The testing specimens were cut from rectangular bar to dimension of 10x10x55 mm and a V-notch was placed in the middle of the testing bars. First all the testing bars were preloaded with -6 kN and then cyclic loading was applied on the testing bars. After the test, the numbers of cycles were recorded as a dependence of frequency change from number of cycles for all the testing specimens.

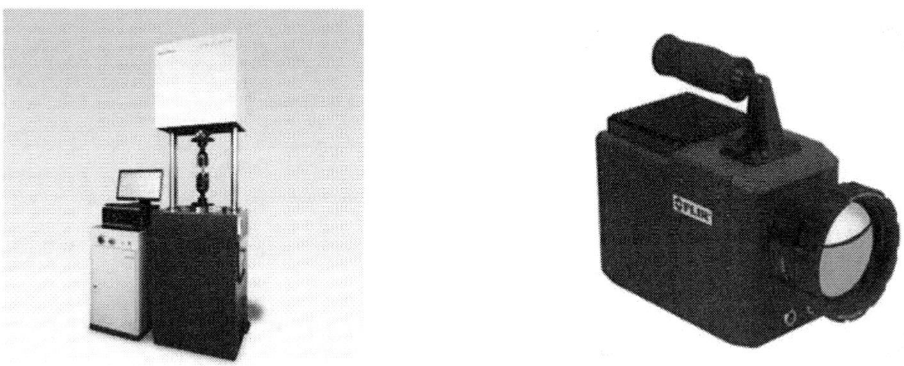

a) vibrophores Amsler 150 HFP 5100   b) thermal camera FLIR SC7500

Fig. 2: Experimental equipment

Thermal stress mapping provides thousands of stress measurements simultaneously, even on geometrically complex components. The deformation of structural materials is followed by thermal effects. For obtain thermal fields we used thermal camera FLIR SC7500 (Fig. 2b). It is very flexible camera, with high sensitivity, accuracy, spatial resolution and speed. The results of monitoring fatigue test by thermal camera are shown in Figure 3a and Figure 3b.

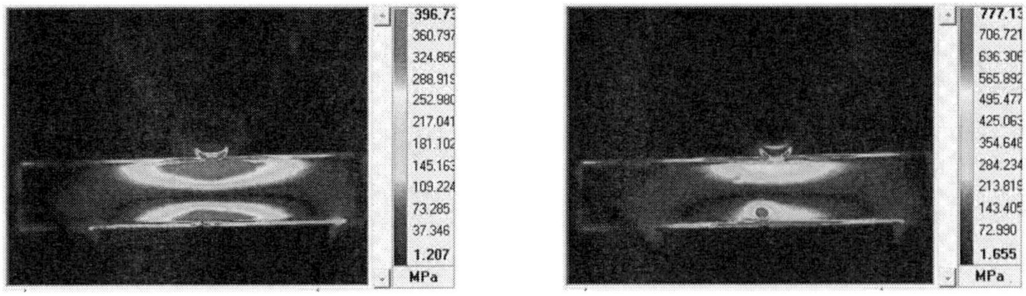

a) at the beginning of the test b) before the end of the test and after cracking

Fig. 3: Results from thermal camera for AISI 2136L during three-point bending test

3 Conclusion

The fracture's surface of the testing sample was observed by SEM. In Figure 4 is seen the macrophoto of the fracture area. After fatigue test, several initiation sites were observed under the notch, as shown in macroscopic view (Fig. 4). In Figure 5 is shown a detail of one initiation site. Also a line of cyclic deformation strengthening was observed on the fracture surface close to the final fracture. The fracture has mix fracture modes predominantly created by ductile separation with dimple morphology and also brittle intercrystalline facets, with the size of app. 50 x 90 μm, occur in the fracture and it can be seen in Figure 6 (with a detail of deformation strengthening line before final fracture in Figure 7). The intercrystalline facet was examined also on the opposite site of the sample's fracture as it can be seen in Figure 8 and Figure 9 where one site consists the salient and the opposite one consists the hollow.

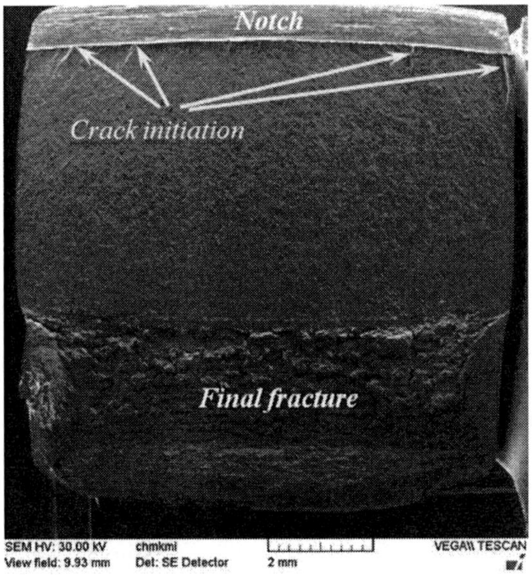

Fig. 4: Fracture surface of AISI 316L testing bar with several initiation sites, three-point bending fatigue

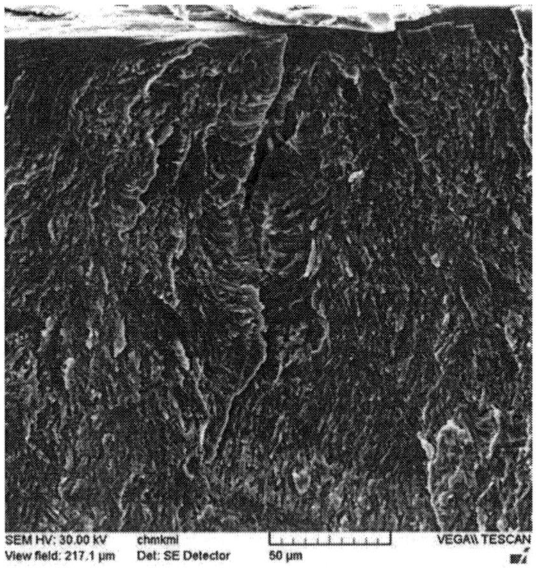

Fig. 5: Detail of one initiation site, AISI 316L, three-point bending fatigue

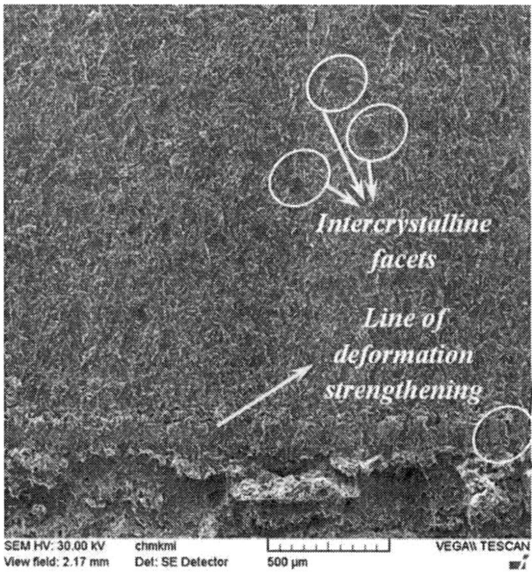

Fig. 6: Fracture surface of AISI 316L, intercrystalline fracture facets, deformation strengthening

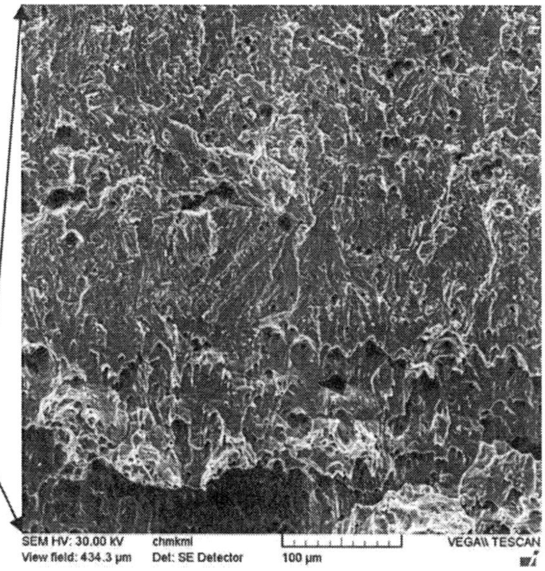

Fig. 7: Detail of deformation strengthening line before final fracture, AISI 316L

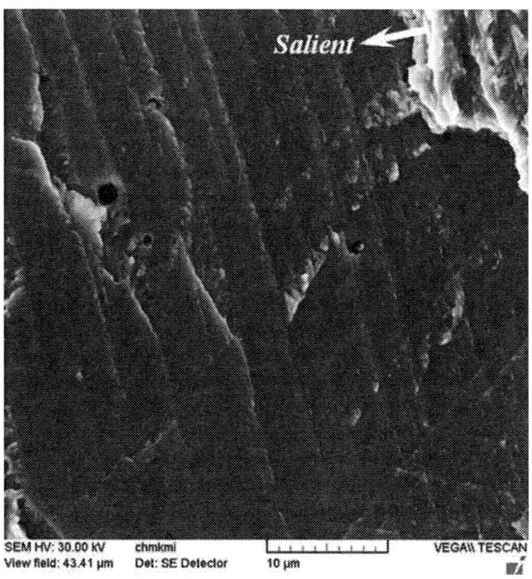

Fig. 8: Intercrystalline facet, right side of the sample

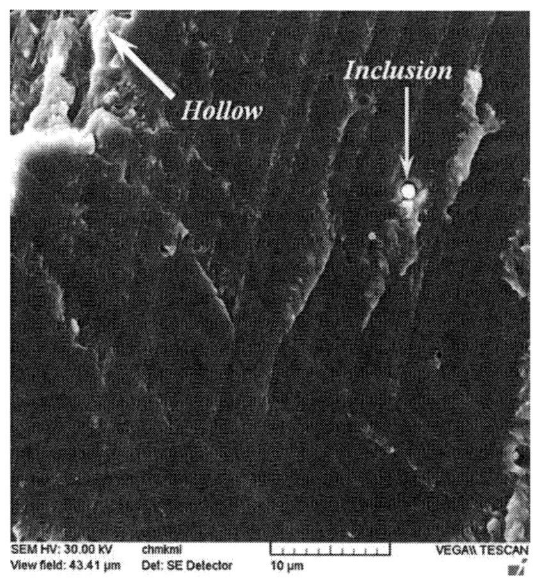

Fig. 9: Intercrystalline facet, left side of the sample

A crack was initiated on the testing bar by three-point fatigue bending test. The fracture has originated from several sites under the notch after 412 x 10^3 cycles. The failure in the fracture surface was realized by ductile dimple fracture micromechanism. The dimple dimensions were in a range of 2 - 5 microns. In the fatigue area of the fracture surface the fracture was propagated by striation mechanism. Striation spacing in this area had grown in the range of 2 - 3 µm. The striations had soft corrugated surface. The numerous striations clearly state the fact that the testing sample has failed by fatigue. Inclusions of sulphate type were observed in the structure, mainly in the dimples. On the fracture's surface annealing twins were observed and also various intercrystalline facets with slip deformation.

By thermal camera, it can give good follow of fatigue crack initiation and its progress in the monitoring and evaluating of the fatigue lifetime of the material.

Acknowledgement

This work has been supported by Scientific Grant Agency of Ministry of Education of Slovak Republic and Slovak Academy of Science, No.1/0683/15 and No.1/0533/15.

References

[1] K. V. Sudhakar, Metallurgical investigation of a failure in 316L stainless steel orthopaedic implant, Engineering Failure Analysis, Vol. 12, pp. 249-256, 2005.

[2] M. E. Stevenson, M. E. Barkey, R. C. Bradt, Fatigue failures of austenitic stainless steel orthopedic fixation devices, ASM International, Vol. 2, No. 3, pp. 57-64, 2002.

[3] F. Nový, V. Zatkalíková, O. Bokůvka, K. Miková, Gigacycle fatigue endurance of marine grade stainless steels with corrosion pits, Periodica Polytechna: transportation engineering, Vol.41, No.2, pp.99-103, 2013.

[4] V. Zatkalíková, L. Markovičová , J. Belan, T. Liptáková, Variability of local corrosion attack morphology of AISI 316Ti stainless steel in aggressive chloride environment, Manufacturing Technology, Vol.14, No.3, pp.493-497, 2014.

[5] A. S. Lima, A. M. Nascimento, H. F. G. Abreu, P. Limaneto, Sensitization evaluation of the stainless steel AISI 304L, 316L, 321 and 347, Journal of Materials Science, volume 40, 2005.

[6] L. Markovičová, L. Hurtalová, V. Zatkalíková, T. Garbacz, Evaluation of composite structures by light microscopy and image analysis, , Manufacturing Technology, Vol.14, No.3, pp.351-355, 2014.

[7] Z. Stankovičová, V. Dekýš, P. Novák, Z. Pelagić, Thermoelastic stress analysis – verification of experiment by using numerical simulation, Experimentální a výpočtové metody v inženýrství : II. Ročník conference pro mladé vědecké pracovníky, pp. 21, 2015

[8] R. Plum, et al. Extended thermoelastic stress analysis applied to carbon steel and CFRP. [online]. 1999. Avalaible online at: <http://ebookbrowsee.net/we5a3-pdf-d75918406>.

[9] P. Brémond. New developments in thermoelastic stress analysis by infrared thermography. [online]. 2007. Avalaible online at: <http://www.ndt.net/article/panndt2007/papers/138.pdf>.

[10] Z. Stankovičová, V. Dekýš, P. Novák, Thermal stress analysis of the specimen, Technológ, Vol. 7, No. 3, pp. 88-91, 2015.

Experimental Measurement and Numerical Simulations of Heat Transfer

M. Urbánek [1,a], M. Fedorko [1], D. Jánská [1]

[1] *COMTES FHT, Czech Republic*

[a] *miroslav.urbanek@comtesfht.cz*

Abstract: Industrial furnaces belong to the most common types of equipment used for heating or controlled cooling in all sectors of industry. Typically, heating takes place in a closed furnace chamber where the chamber temperature is regulated by controllers in response to information from thermocouples installed at defined points in the furnace interior. If the temperature of the charge needs to be determined, one can rely on either experimental measurement or numerical simulations. The optimal approach is to combine both. In the first case, thermocouples are attached to the charge at defined points or, sometimes, other measuring devices are provided. Temperatures at the defined measuring locations are then logged by means of a data logger or card. By contrast, numerical simulations provide information on the temperature distribution throughout the charge. The absolute values of temperature depend on the boundary conditions used, most notably on heat transfer coefficients.

Authors of this article decided to combine both approaches to tackle a simply formulated problem where a thick-walled tube was placed into a laboratory furnace and the temperatures of the chamber and charge were measured at defined points during heating. The temperature-time histories obtained were used as input into numerical simulations of the temperature field within the charge. The purpose of the measurement was to determine the temperature dependence of the heat transfer coefficient and emissivity by experiment. The calibration of heat transfer coefficient values for the simulation was based on the experimental data obtained for the defined locations. The input data for the simulations also included the dimensions of the furnace, positions of heating coils, the temperature sequence, and others. The temperature dependence of thermophysical properties of materials was determined using the JMatPro software based on chemical compositions.

The above-described experimental procedure can be used for calibrating numerical calculations of real-world heat-treatment processes.

Keywords: radiation, thermal process, finite element method

1 Introduction

In this investigation, experimental measurement involving heat transfer and subsequent evaluation using numerical simulations were carried out under the project no. LO1412 Development of the West-Bohemian Centre of Materials and Metallurgy. The objective was to determine heat transfer coefficients for a thick-walled tube using a reverse problem solved by numerical simulation.

2 Experimental measurement

The measurement was carried out in a KM30-13 electrical laboratory furnace which offers a working area of 300×300×300 mm. The furnace is fitted with 12 heating coils which provide the maximum furnace temperature of 1300 °C. A schematic description of the experiment is given in Fig. 1. The diameter and wall thickness of the tube placed in the furnace were 168 mm and 25 mm, respectively. Since the length of the tube of 320 mm was close to the furnace interior depth, which was 330 mm, the simulation could be approached as a plane problem in FEM analysis. The ends of the tube were secured against rotation. Before the tube was placed into the furnace, 6 thermocouples had been attached at defined points. Four thermocouples were on the inner surface of the tube (Fig. 1). Another two thermocouples were used for measuring the instantaneous temperature of the furnace chamber which was affected by the door being opened for placing the tube inside. The values were recorded using a logging card and an in-house script written in the Labview program. The thermocouples were welded onto the surface of the tube. The inner and outer surfaces of the tube were smooth from the tube rolling process.

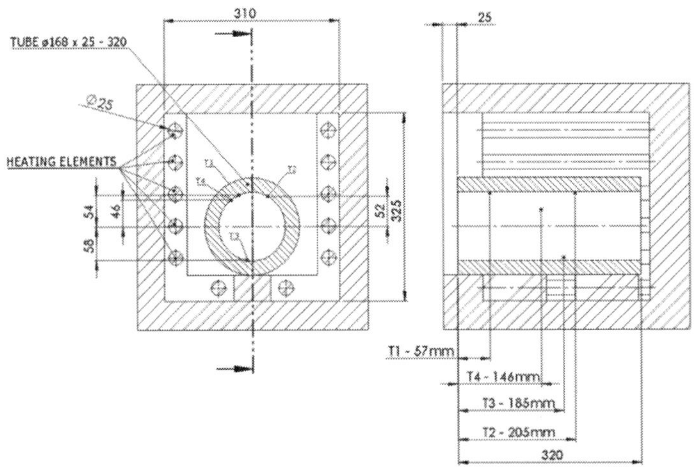

Fig. 1: Sectional view of the furnace

In the experiment, the tube was manually placed into a heated furnace. The end of the heating sequence was defined as the moment the inner diameter of the tube reached the specified temperature of 900 °C. Throughout the heating sequence, temperature was measured by means of thermocouples attached by welding. Subsequently, the tube was left to cool in the furnace. At some point during the experiment, the thermocouple no. 2 failed for unknown reasons. Figure 2 shows temperatures measured on the controller of the furnace, and the calculated input power of the furnace. The input power values were calculated from the furnace's energy consumption measured on its input terminals. The amount of energy was measured using an electric power meter. The calculation of the input power was motivated by the need for finding the energy balance of the furnace-charge thermodynamic system.

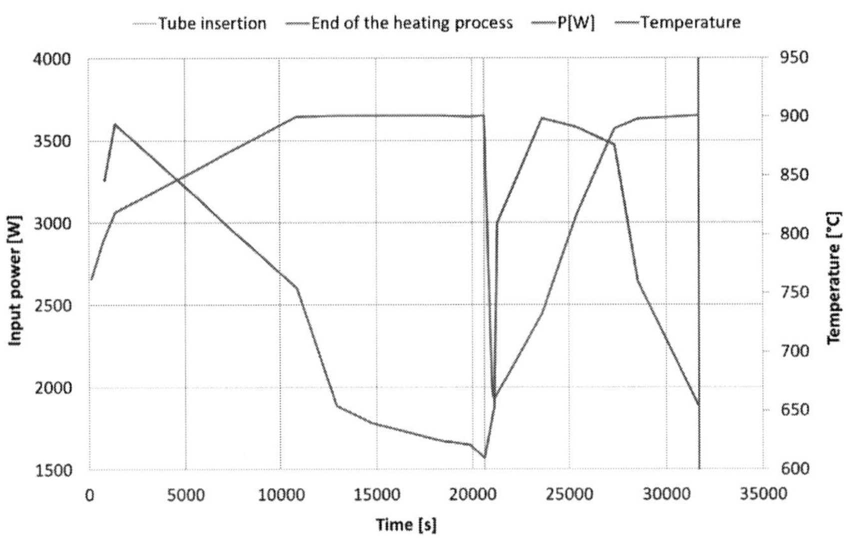

Fig. 2: Temperatures on the controller of the furnace, and the calculated input power

Temperature readings obtained from thermocouples attached at pre-defined points are plotted in Figure 3. After the charge was placed into the furnace, the air temperature dropped all the way to 228 °C. Due to the heat accumulated in the furnace, the furnace atmosphere immediately reached 320 °C and then continued to rise to the required temperature of 900 °C.

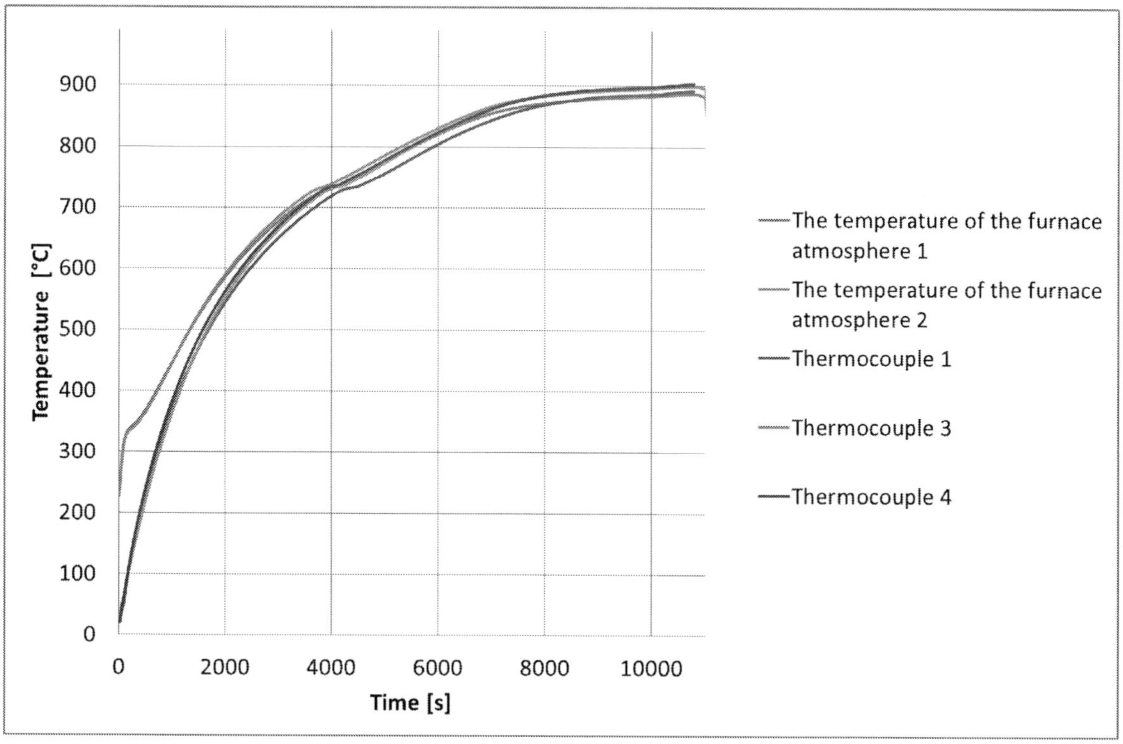

Fig. 3: Experimental results

3 Evaluation by means of FEM software

3.1 Material properties

The purpose of the laboratory furnace experiment was to measure temperatures at the defined points. Based on their values, heat transfer coefficient α [$Wm^{-2}K^{-1}$] was determined by numerical simulation. The parameters of the simulation matched the conditions of the experiment. The chemical composition of the tube material was measured with a Bruker Q4 TASMAN optical emission spectrometer.

Tab. 1: Chemical composition of ČSN 11 353.1 steel

C	Si	Mn	P	S	Cr	Co	Ni	Cu	W
0.158	0.244	0.571	0.012	0.002	0.038	0.002	0.020	0.016	<0.005

Czech Standard class 11 steels are plain structural steels with prescribed mechanical properties (yield strength R_e = 235 MPa, ultimate strength R_m = 340–440 MPa, and elongation A = 25 %). Using the JMatPro program and the measured composition as input data, temperature-dependent thermal conductivity and specific heat values were calculated. Their temperature dependence became part of the material model for numerical simulations.

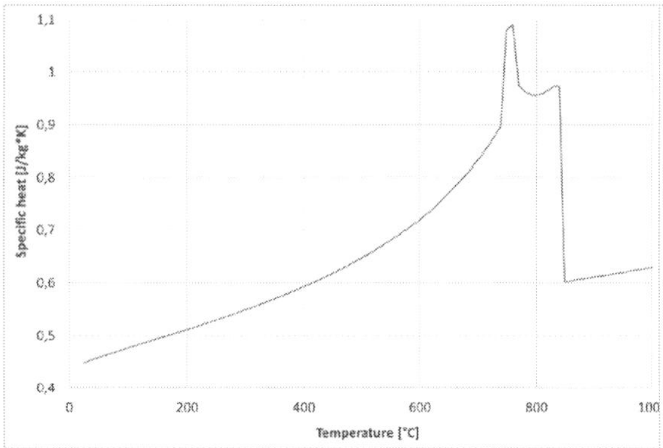

Fig. 4: Specific heat [J/kg K]

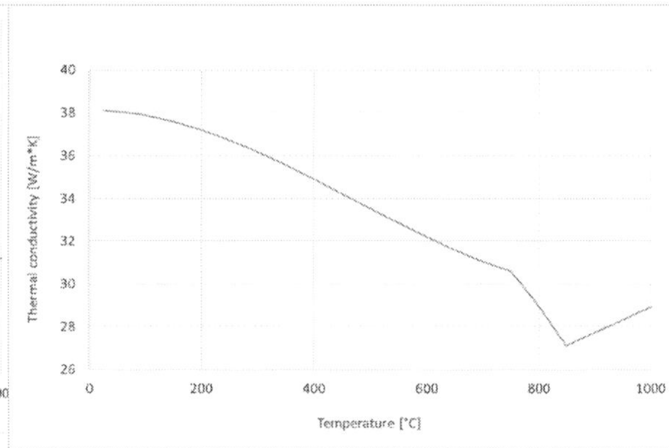

Fig. 5: Thermal conductivity [W/m K]

3.2 FEM simulation

Owing to the favourable geometry of the furnace and the tube, the simulation was first carried out as a plane problem and subsequently as a 3D problem. Using numerical modelling, the temperature field within the thick-walled tube was characterized, and absolute values of temperature were calibrated against the data measured with thermocouples at the defined points.

Generally, the crucial parameters in thermal problems include not only material properties but also heat transfer coefficients α, the subject of the present work. One can approach thermal problems in various ways: considering the heat transfer coefficient α as a parameter that accounts for radiation or by combining natural convection and radiation. The authors had already dealt with the effects of radiation on the heating of a charge, considering the relative positions of the charge and heating elements. In numerical simulations, the relative position is expressed in terms of the view factor which is based on Lambert's law and accounts for the distance and relative angle between two surfaces of an element (Fig 6.).

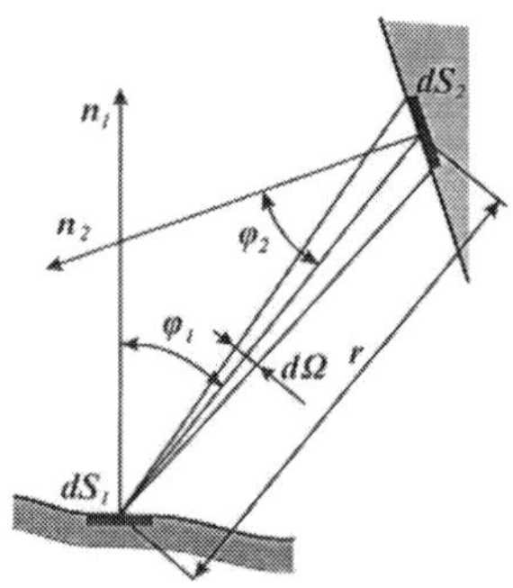

$$d^2Q_{1\to2} = A_2 d^2Q_1 = \frac{1}{\pi} A_2 E_1 \frac{cos\varphi_1 cos\varphi_2}{r^2} dS_1 S_1, \text{ where } d^2Q_{1\to2} = A_2 d^2Q_1 = \frac{1}{\pi} A_2 E_1 \frac{cos\varphi_1 cos\varphi_2}{r^2} dS_1 S_1, \text{ where } E_1 = \varepsilon_1 C_0 \left(\frac{T_1}{100}\right)^4$$

$$d^2Q_{2\to1} = A_1 d^2Q_2 = \frac{1}{\pi} A_1 E_2 \frac{cos\varphi_1 cos\varphi_2}{r^2} dS_1 S_1, \text{ where } d^2Q_{2\to1} = A_1 d^2Q_2 = \frac{1}{\pi} A_1 E_2 \frac{cos\varphi_1 cos\varphi_2}{r^2} dS_1 S_1, \quad \text{where } E_2 = \varepsilon_2 C_0 \left(\frac{T_2}{100}\right)^4$$

(1)

Fig. 6: Heat exchange between surfaces in a general relative position

3.2.1 Plane problem

For the first simulation run to determine the heat transfer coefficients, a simple model with fewer than 5000 elements was constructed. The idea was to use a "small-size" problem to refine the heat transfer coefficient values because a model with fewer elements responds very quickly to changes made to the coefficient. The model was based on certain simplifying assumptions:

- the plane problem approach was used,
- the tube model was placed in the centre of the furnace model, which was not the case in reality,
- the pads on which the tube was resting were omitted from the model,
- identical temperature profiles were used for the furnace and the heating elements.

There were two initial temperatures in the model: 25 °C for the tube, and 228 °C for other components of the model (the furnace and the heating elements). The latter is based on the previous measurement and is the initial temperature of the furnace atmosphere. Boundary conditions were set in terms of heat transfer coefficients. Two spaces were defined for the simulation. In the first one, the heat transfer between the outer surface of the tube, the furnace, and the heating elements was described. The second one described the heat transfer within the tube. In the model, these two spaces were dealt with separately. However, both used an identical temperature profile of the furnace lining (Fig. 3 – The temperature of the furnace atmosphere 1). In both spaces, heat transfer occurred by free convection and by radiation.

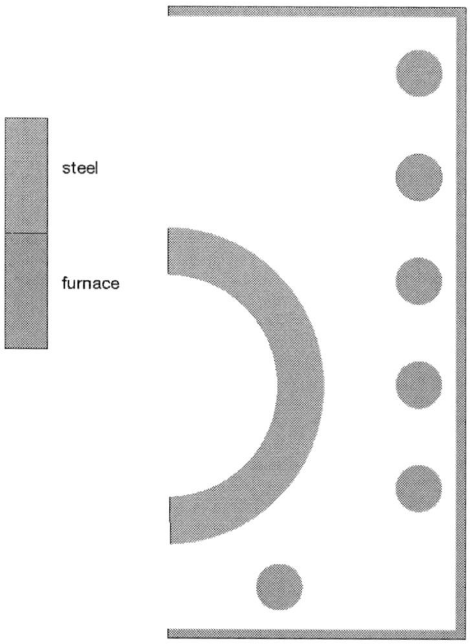

Fig. 7: Material

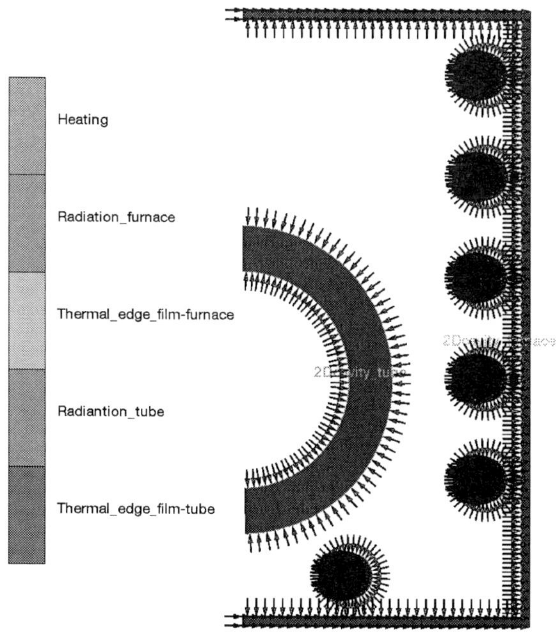

Fig. 8: Boundary conditions

Table 2 lists material properties used for characterizing parts of the furnace (Fig. 7). Table 3 gives element sizes for separate parts of the model. Using the calculated view factor (Fig. 6.) for both spaces, and the refined heat transfer coefficients for free convection given in Table 4, the temperature field at the end of the heating sequence was obtained (Fig. 9). The concrete boundary conditions are shown in Figure 8. Figure 11, which is more informative, illustrates the agreement between experimental data and simulations. The heat transfer coefficients were determined by trial and error in order to achieve the desired agreement between the measurement and the simulation.

Tab. 2: Material properties

Material properties	Units	Tube	Furnace
Specific heat	J/kg*K	graph (Fig. 4)	400
Density	kg/m^3	7850	5000
Thermal conductivity	W/m*K	graph (Fig. 5)	1
Emissivity	-----	0.97	0.97

Tab. 3: Element sizes

Geometry	Tube	Heating element	Furnace
Size of elements	variable size, outer surface: 0.13 mm inner surface: 4 mm	constant size: 1.5 mm	constant size: 5 mm

Tab. 4: Heat transfer coefficients determined

Space	Units	Furnace	Inside the tube
Heat transfer coefficient	W/m^2K	180	5
Ambient temperature	°C	defined by the furnace atmosphere temperature in fig. 3	
Radiation	J/kg	separate space	separate space

3.2.2 3D problem

Once the heat transfer coefficients were determined using the plane problem, they were employed in a 3D model. The model of the furnace was constructed on the basis of its drawings using the SolidWorks software. The 3D calculation reflected the actual position of the tube, the positions of thermocouples on the inner diameter of the tube, and those of the pads used for setting the tube's position inside the furnace. All settings of the 3D model were identical to those employed in the plane problem. In addition, the contact between the pads and the tube was defined, using a heat transfer coefficient of 100 W/m^2K.

3.3 Evaluation of results and discussion

The set-ups for the plane and three-dimensional simulations were identical. Thanks to the thorough mapping of the experiment, the results are identical as well. Heat transfer coefficients were found using 2D simulation, and then applied to the 3D problem. The out-of-centre position of the tube in the furnace had no effect on the experimental results.

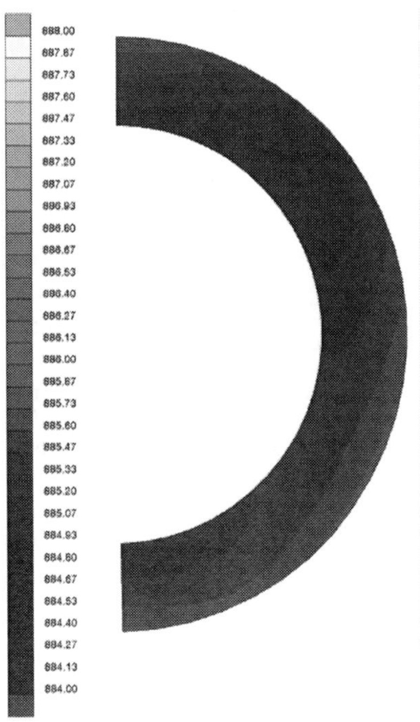

Fig. 9: Distribution of temperature – plane problem

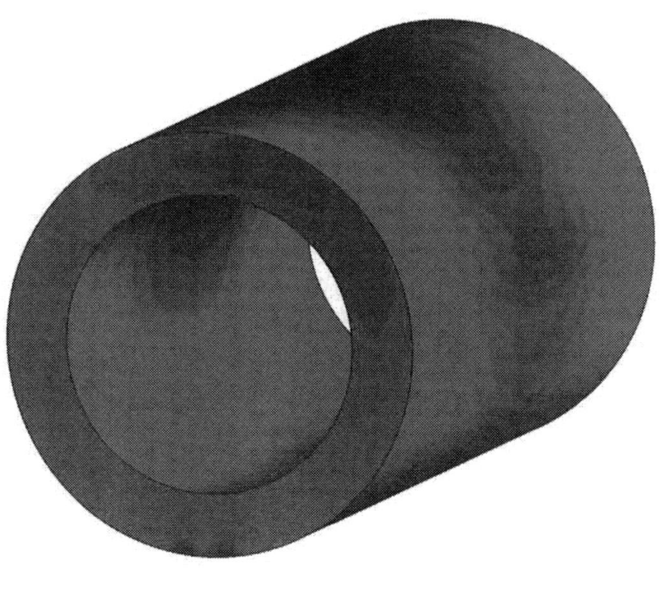

Fig. 10: Distribution of temperature – 3D problem

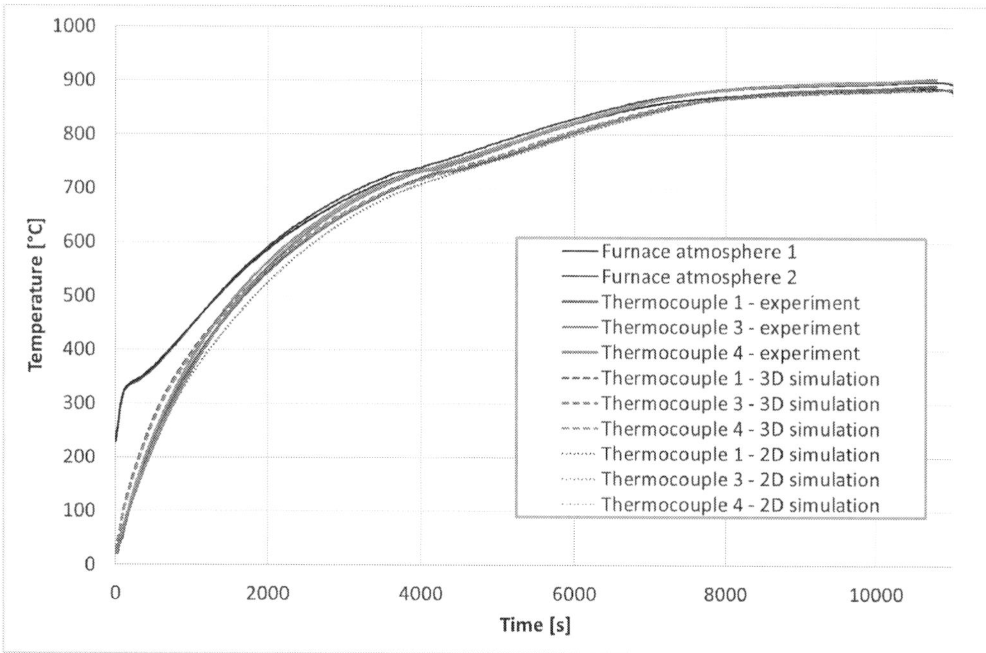

Fig. 11: Comparison of results

In the course of the experiment, a phase transformation at 727 °C was noted, which is highlighted in the graph inset (Fig. 12.). During the near-thermal arrest period, energy is consumed in austenite formation, as indicated by the Fe-Fe_3C equilibrium diagram (Fig. 12.).

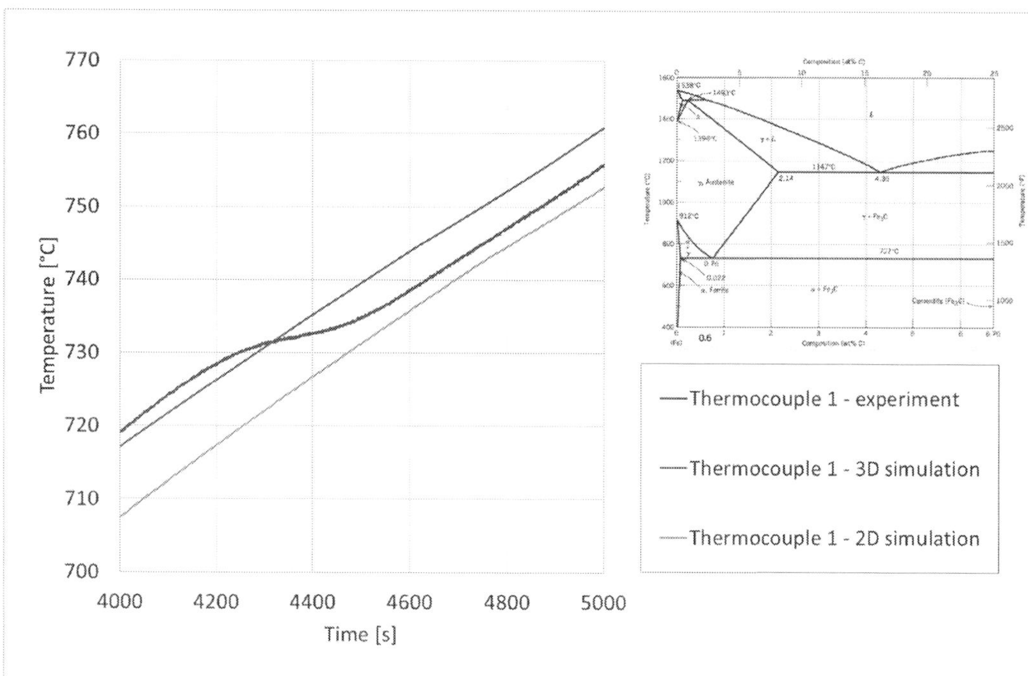

Fig. 12: Plots of experimental T1 readings and simulation results for 2D (one half) and 3D

4 Conclusion

By means of the above-described method, one can experimentally determine heat transfer coefficients in this particular case. Generally, heat transfer coefficients depend on more aspects that described here. There is no single value which could be applied in all cases. However, literature offers recommended values for specific conditions. The coefficients outlined here give a hint of the behaviour of a thick-walled tube during heating. Previously, COMTES FHT completed a range of experiments focused on determining the heat transfer coefficient α under various conditions.

Acknowledgement

This study was conducted under the project Development of West-Bohemian Centre of Materials and Metallurgy No.: LO1412, which is funded by the Ministry of Education of the Czech Republic.

References

[1] J. Džugan, M. Zemko, Input Data Influence on FEM Simulation of Steam Turbine Blades Materials Hot Forming. In: Materials Science Forum, 773 - 774(2). [online]. CH, 2013. p. 10 / 79-88. ISSN 1662-9752

[2] I. Poláková, M. Duchek, L. Maleček, FEM simulation of open die forging of a plate from material NIMONIC 80A in DEFORM 3D. In: Archives of Materials Science and Engineering, 66(1). [online]. OCSCO World Press, ul. Zygmunta Starego 6/30, 44-100 Gliwice, 2014. p. 6/31-36. ISSN 1897-2764

[3] F. Tikal, F. Jandos, M. Duchek, NUMERICAL SIMULATIONS OF EFFECTS OF FURNACE LINING RADIATION ON 34CrNiMo6 INGOTS Metal 2011: 20th Anniversary International Conference On Metallurgy and Materials, 2011, Tanger; ASM Int, Mat Informat Soc; CSNMT; VSB-TU

[4] D. Jánská, M. Fedorko, L. Maleček, M. URBÁNEK, Analýza tepelného zpracování výkovku v průběžné peci pomocí numerické simulace. In: Kovárenství, 54/2015. Brno, 2015. p. 4/13-16. ISSN 1213-9289.

[5] Marc® 2012, Volume A: Theory and User Information

Analysis of the vertical tensions on masonry structures with different combinations of load

Ludek Vejvara

Department of Mechanics, University of West Bohemia, Univerzitní 22, 306 14 Plzen, Czechia
vejvaral@kme.zcu.cz, vejvara@vejvara.cz

Abstract: The work shows the stress of the masonry structures with different combinations of load.
Keywords: Masonry, permanent and variable loads, forces, eccentricity, eurocodes

1 Introduction

Masonry structures of buildings over the past 30 years, was significantly transformed. The structures from the compact and small-sized masonry elements and ordinary mortar gradually moving on to more sophisticated products of larger dimensions. To use the bricks today is important factor still a static load rating. The total bearing capacity of masonry is not just strenght theirs masonry elements, but their combination with mortar and other technical parameters such as size of load, eccentricity of forces and rigidity. The standards (Eurocodes) reported the conditions, procedures and limitations for the design of structures, but the actual numerical solution we must make.

2 Experiment - analysis vertical tension

Finding the size and changes of the vertical stress in different locations in masonry structure of the building in terms of different load combinations is the subject of our work and that contribution. The work examines the size of the resultant force from the vertical load on individual floors or selected reference buildings. Determines the amount of eccentricity forces in the cross section and the size of the resulting tensions in different parts of cross section of the masonry structure. To determine the size of the individual values was compiled simple program involving various possible combinations of load. A program can change the thickness of masonry structures, materials and conceptions of horizontal structures and the size of the variable load on the floor.

Calculations were performed to compare the sizes of forces for the individual load combinations for ultimate limit states and serviceability limit states. Equations were used for characteristic load combination and design combinations. For the three equations for a combination of permanent and variable loads that can be use according to the standard EN 1990 - Eurocode. These are the calculations of variants of the design load values for ultimate limit state (ULS) of brickwork in compression. For minimum operating load values and the values for the long-term effects. It is a combination of quasi-permanent It is a calculation for serviceability limit state (SLS). For the loads combination by original old Czechoslovak standards CSN 730035 1986 is used to compare with Eurocodes. Load acting on the masonry is defined in the following combinations in the table no.1. Parctional load factors (coefficients) have value for the permanent load of 1.35 (($\gamma_{g,1}$) and 1.5 for variable load ($\gamma_{q,i}$). Factor $\psi_{0,i}$ is used to reduce other variable loads coefficients $\psi_{1,i}$ and $\psi_{2,i}$ for reducing variable load operating design situation.

These procedures (load combinations) always comes to comparing for one variable seven values of forces and eccentricities and four tension values (for the concentric pressure, the eccentrical and extreme pressure section)

	Combination of actions	equation	coefficient ψ
1	characteristic values	$E_d = \Sigma (G_{k,j} + Q_{k,1} + \psi_{0,\iota} * Q_{k,,\iota})$	$\psi_{0,\iota} = 0,7$
2	design values for ULS		
2a	equation 6.10 of EN 1990	$E_d = \Sigma (\gamma_{g,j} * G_{k,j} + \gamma_{q,1} * Q_{k,1} + {}_{q,i} * \psi_{0,\iota} * Q_{k,,\iota})$	$\psi_{0,\iota} = 0,7$
2b	equation 6.10a of EN 1990	$= \Sigma (\gamma_{g,j} * G_{k,j} + \gamma_{q,1} * \psi_{0,1} * Q_{k,1} + {}_{q,i} * \psi_{0,\iota} * Q_{k,,\iota})$	$\psi_{0,\iota} = 0,7$
2c	equation 6.10b of EN 1990	$= \Sigma (\xi * \gamma_{g,j} * G_{k,j} + \gamma_{q,1} * Q_{k,1} + {}_{q,i} * \psi_{0,\iota} * Q_{k,,\iota})$	$\psi_{0,\iota} = 0,7$, $\xi = 0,85$
3	values for SLS		
3a	quasipermanent	$E_d = \Sigma (G_{k,j} + \psi_{2,\iota} * Q_{k,1} + \psi_{2,\iota} * Q_{k,,\iota})$	$\psi_{2,\iota} = 0,3$
3b	frequent	$E_d = \Sigma (G_{k,j} + \psi_{1,1} * Q_{k,1} + \psi_{2,\iota} * Q_{k,,\iota})$	$\psi_{1,1} = 0,5$, $\psi_{2,\iota} = 0,3$
4	design for ULS according to CSN 730035	$E_d = \Sigma (\gamma_{g,j} * G_{k,j} + \gamma_{q,1} * Q_{k,1} + {}_{q,i} * \psi_{,\iota} * Q_{k,,\iota})$	

Fig 1– Combinations of actions for ULS and SLS

As a reference sample is considered a six-storey building with two wings. This building is a commonly-built brick building with the usual maximum allowable height of buildings. Calculations can e performed for lower and other objects.

.	Combination	Vertical Force (kN)	eccentr. of force (mm)	porovnání hodnot ku a	ku b	Compressive tension centric (kPa)	average (kPa)	minimum (kPa)	Maximum (kPa)
	2st floor								
1	characteristic	208,73	27,15	1,00	0,73	556,62	**557,46**	314,80	798,44
2a	design eq. 6.10	284,92	27,40	1,36	1,00	759,78	**771,05**	426,65	1092,90
2b	design eq. 610a	275,53	26,63	1,32	0,97	734,76	**745,34**	421,65	1047,86
2c	desigm eq.6.10b	248,14	27,95	1,19	0,87	661,70	**671,71**	365,83	957,56
3a	quasipermanent	194,14	25,43	0,93	0,68	517,70	**524,82**	307,02	728,38
3bf	frequent	198,31	25,95	0,95	0,70	528,82	**536,24**	309,24	748,40
4	old standard ČSN	225,26	28,61	1,08	0,79	600,70	**610,01**	325,74	875,66
	1st floor								
1	characteristic	227,47	24,92	1,00	0,73	606,58	**607,50**	364,77	848,40
2a	design eq. 6.10	310,21	25,17	1,36	1,00	827,23	**838,48**	494,10	1160,35
2b	design eq. 610a	300,83	24,39	1,32	0,97	802,21	**812,78**	489,10	1115,32
2c	desigm eq.6.10b	273,43	25,36	1,20	0,88	729,15	**739,14**	433,28	1025,01
3a	quasipermanent	212,87	23,20	0,94	0,69	567,66	**574,77**	356,98	789,46
3b	frequent	217,04	23,71	0,95	0,70	578,78	**586,20**	578,75	798,36
4	old standard ČSN	246,06	26,19	1,08	0,79	656,15	**665,45**	656,15	656,15

Fig. 2 - Vertical tension for a wall thickness of 375mm on six-storey building, the lowest two floors

Structural scheme and forces

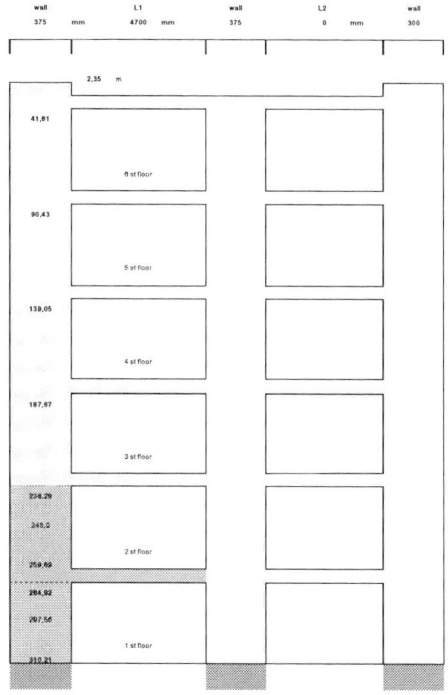

Fig.3 – structural scheme and forces in masonry – six storey building,wall 375 mm

Forces and tension in the masonry

six-storey building, wall 375 mm, panel ceilings

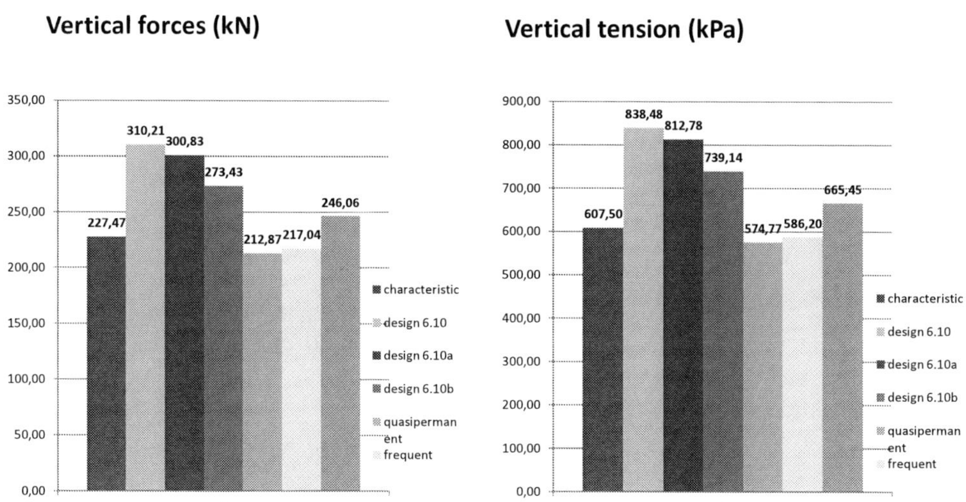

Fig. 4 - Vertical forces and tension (MPa) for a wall thickness of 375 mm on six-storey building made. The walls are of vertically perforated bricks, ceilings are monted or are from panels

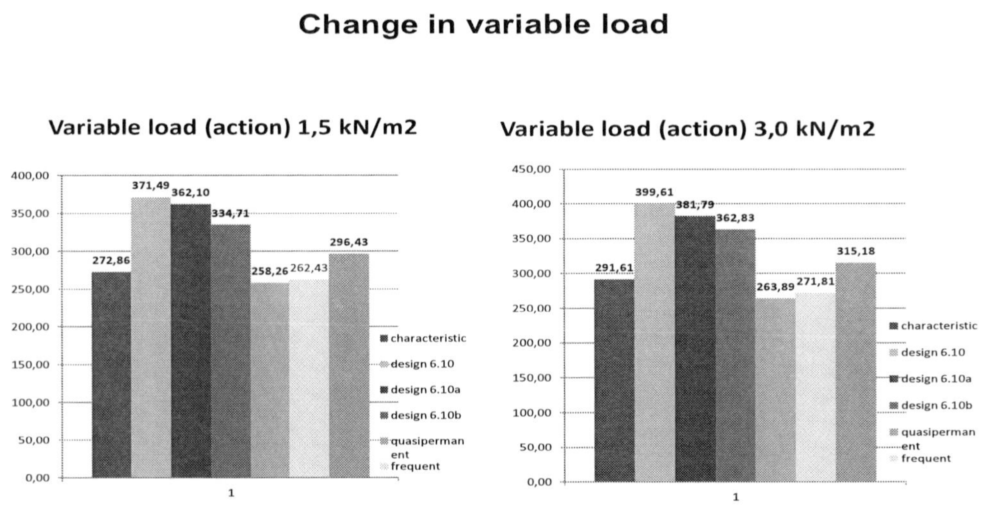

Fig. 5 - Sample results when you change the variable load (forces in kN) - six-storey

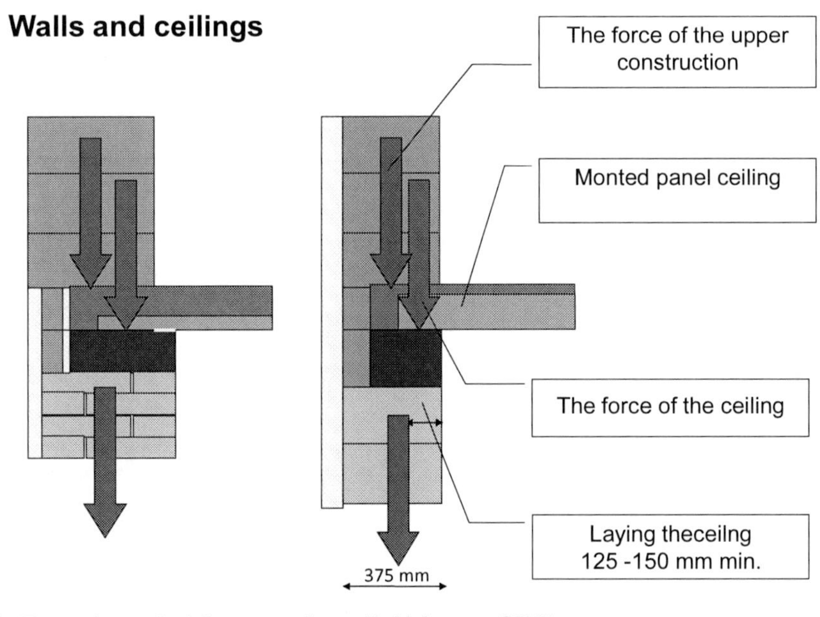

Fig. 6 - Example vertical forces on the wall thickness of 375 mm

The result is a practical verification of the size of the forces in the cross-sections of masonry buildings, including the eccentricity of forces and inferred vertical stress. Results can be calculated for different material variants and variants masonry ceiling structure. Calculations were examined in concrete ceilings of hollow panels and weighing from 2.5 kN / m2 in solid monolithic slab cross sections of 200 mm (5 kN / m2). Variable load on the floor has been considered in category A for flats (1.5 kN / m2), offices - category B (2.5 kN / m2) and areas with tables category C (3.0 kN / m2) according to the standard CSN EN 1991-1-1. Calculated values of forces and tensions give the normative variants of load combinations according to EN 1990. There are obvious differences in the results in a typical example, the limit and operational loads

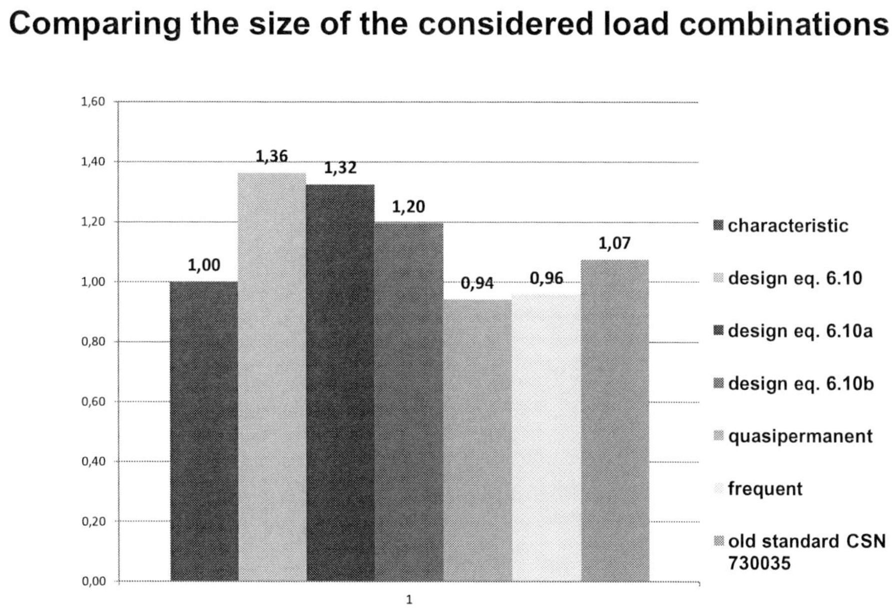

Fig. 7 - Comparing the size of the combinations – six-storey building, wall 450 mm

3 Summary

Comparison of the effects of the load combination was made for a six-storey building with a wingspan of 5 meters ceilings. It is a usual building - reference building

Limit state - carrying capacity

Results are from the calculations in differences 3-5% in the assessment of the values according to the combination equation 6.10a or 6.10 specified in CSN EN 1990. Second procedure gives the commonly used conservative results. Process according to 6.10 b is by 10 -12% lower. Procedures according to equations 6.10 is to 27 % higher than the initial proposal by Czech standards. Yet it has been implemented under this standard, many buildings for decades without problems Differences make values of load factors. Process according to equation 6.10a with a reduction of permanent loads is correct and legislative solutions.

Changes to variable loads on the floors evoke the vertical forces increase by about 7.5%.

Limit state - usability

It is also necessary to verify the effects of the load for the quasi-permanent and frequent load combination. These combinations are two variants for the minimum operating load. There should not be exceeded eccentricity of the vertical force equal to a sixth of the cross-section to avoid tense of mortars and eventual opening of the bed joints by tension.

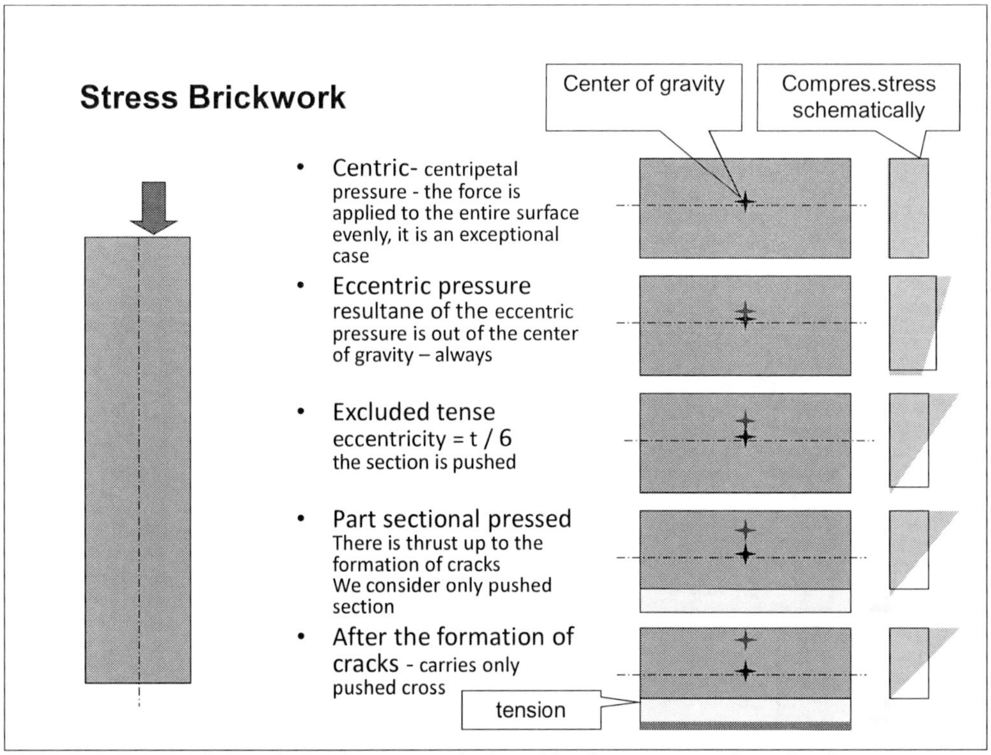

Fig. 8 - Compressive and tensile stresses induced in the masonry from load

That analysis is also used for reconstruction projects and extensions of objects in an increase in the floor can be adjusted to záakldě accurate information load factor for permanent load from the existing building

Acknowledgement

his work was supported by national financial resources of Ministry of Education Youth and Sport of Czech Republic for specific university research (SGS-2016-038)

References

Standards: CSN EN 1990, ČSN EN 1991-1-1,1-3, 1-4, CSN EN 1996-1-1

Analysis of ceiling of panel systems

Ludek Vejvara, Michal Novák

Department of Mechanics, University of West Bohemia, Univerzitní 22, 306 14 Plzen, Czechia
vejvaral@kme.zcu.cz, vejvara@vejvara.cz, novakmichal@email.cz

Abstract: The work shows the size of forces and deformations of the panels concrete structures for different combinations of load.

Keywords: Panels, concrete, permanent and variable loads, forces, eurocodes

1 Introduction

Panels systems were used in the Czech Republic for housing construction more than 30 years. Since the fifties of the last century to the years nineties was created approximately eighty separate and derived systems. An important part of these systems have always been the ceiling structure. They have been designed according to the regulations, standards and practices applicable in each period. Evaluation of the slab panels, still used in construction, according to current European standards, is the subject of the contribution. To explain - the current standards have different requirements for reliability and define a higher design load.

In West Bohemia since the end of the fifties was used three basic panel systems designated PS61, PS69 and T06B. Each system has its own ceiling structure. The paper presents and evaluates panels standard ceilings system PS69.

2 Analysis of ceiling panel system design PS692E 1989

PS 69 panel system is the most widely used system in western Bohemia since 1971.

Description of the process: at the time of construction

Loads

Assessment for the construction of the headliner status at the time used the philosophy of limit states. Load calculation was made according to CSN 730035, partial factor for permanent actions had a value of 1.1 (panel) and 1.3 (other designs). Variable load was considered normative value of 1,5 kN /m2, coefficient 1.4 Basic load combination was drawn up according to czech standard CSN 730035.

Material

Concrete was Class III (corresponds to the present C16 / 20), compressive strength of 11.5 Mpa. Reinforcement was a steel, type 10338 (290 MPa) and 10425 (375 MPa) Reinforced concrete proposal was made by the then valid CSN 731201

Description of the procedure today

Loads

Design and assessment for the construction of the ceiling panel today also utilizes the philosophy of limit states. Load calculation is performed according to EN 1991-1-1, partial factor for permanent loads have a value of 1.35. Variable load is considered normative value of 1.5 kN / m2, coefficient 5.1 Basic load combinations have been prepared in accordance with EN 1990 for permanent and temporary situations. Loads combinations:wherw for 3 equations 6.10 and char. combination according to EN 1990.

Material

Concrete meets today's C16 / 20, compressive strength 20 Mpa. Reinforcement was a steel, type 10338 (290 psi) and 10425 (375 MPa). Values are taken from ISO 13822 and related CSN 730,038th

Calculations

Assessment cross-sections is according to EN 1992-1-1 (EC 2) for ultimate bearing capacity (ULS). Is calculated deformation panel. We evaluate cracks in concrete and strength

Partitions

Given the weight partitions useful for panel construction

porousconcrete	0,65 kN/m^2
with plaster or tiles	0,99 kN/m^2
gypsum plaster board	0,25 kN/m^2

Combinations of loads

	1	¨2	3	4	5	6	7	8	9	10	11	12	13	14
Loads Types	char.	char. with partit.	perm. char.	design 6.10 1,35	design 6.10a 0,700	design 6.10b 0,850	part 6.10	part 6.10a	quasi-perm. 0,30	part	freque. 0,50	load old CSN	capac. Mu CSN	Mu EC2
	(kNm)													
Permanent	16,25	16,25	16,25	21,94	21,94	18,65	21,94	21,94	16,25	16,25	16,25			
Variable	5,53	5,53	0,00	8,29	5,80	8,29	8,29	5,80	1,66	1,66	2,76			
Partition		3,17					4,28	4,28	0,00	3,17				
Summary	21,78	24,95	16,25	30,23	27,75	26,94	34,51	32,02	17,91	21,08	19,02	26,10	40,27	40,63
comparison values														
	1,00		0,75	1,39	1,27	1,24			0,82		0,87	1,20	1,85	1,87
Reserve %					1,00									**46,4**
with part.		1,00					1,38	1,28		0,84		1,05	1,61	1,63
Reserve %								1,00						**26,87**

Fig. 1- Sizes moments from various combinations of load and ultimate bearing capacity (kNm)

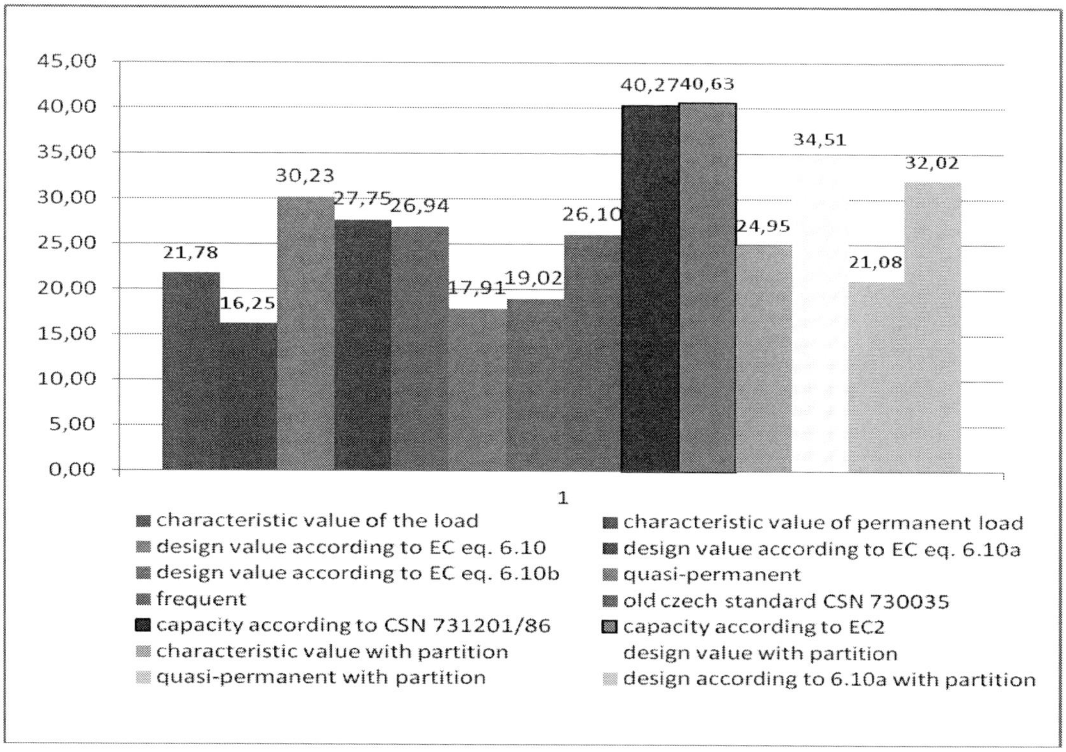

Fig.2- Sizes moments from various combinations of load and ultimate bearing capacity in the graph. (kNm)- without partitions and also with different partitions.

Static ratings results of panels ceiling structure from load increase compared to the original proposal shown in the table Reserve in load bearing capacity is given in the table.

Combination – bending moment	
characteristic load combination	1
proposal at the time of construction - old standard CSN 730035	1,20
assessment by equation 6.10 EN 1990	1,39
assessment by equation 6.10a EN 1990	1,27
assessment by equation 6.10b EN 1990	1,24
quasi-permanent combination	0,82
frequent combination	0,8
bearing capacity by original standard CSN 731201	1,85
bearing capacity by EC 2	1,87

Fig 3 - Table comparing the values of moments

Finding reserve bearing capacity is important for furtheradjust and repairs. This has implications for the assessment of modifications in flats such as loading new walls, new cores for bathroom and toilet, flooring and plaster. Reserve in load bearing capacity depends on the combination loads

Deformations

Brick partition is considered from porous concrete thickness 75 mm with plaster and tiles. This manual is satisfactory from the point of view of the ceiling deflection. Thicker and heavier partition can not use.

From the calculation for the quasi-permanent load combination (servicelive 50 years)

Panel only partition	10,59 mm
Panel with partition 75 mm	13,70 mm
Panel with partition 100 mm	14,15 mm
Limit deflection L / 250	14,02 mm

Calculated deflection for partition 75 mm is lower than the allowable deflection

3 Research results

Evaluation panels system PS69 is very important for further use objects from this system. System already used in Western Bohemia from 25 to 45 years. A further objects will be used in the future.

Ceiling panel system PS 69 statically conforms to the assessment, according to Eorocodes. The panel can aggravate extra partitions.. It can be with a brick partition wall made of porous concrete with a maximum thickness of 75 mm. Panel has a higher capacity (26,87%), but it decisions deflection. Porous concrete partition about 100 mm thickness is at the limits of applicability, it is by calculating already over the limit usability. We can use the aforementioned brick partitions with a thickness of 75 mm or partitions made of gypsum plasterboard with lower weight

Standard ceiling panel of system PS69 was tested according to Eurocodes. Other panels of system Ps69 and panels of other systems (T06B, PS61) should tested when working on projects also always.

Acknowledgement

This work was supported by national financial resources of Ministry of Education Youth and Sport of Czech Republic for specific university research (SGS-2016-038)

References

(1) Standards: CSN EN 1990, CSN EN 1991-1-1,1-3, 1-4, CSN 730038, ISO CSN 13822
(2) Czech old standards: CSN 730035, 731201/86
(2) Smejkal: Zelezobetonové konstrukce I – priklady, Zapadoceska univerzita v Plzni, 2015, ISBN 978-80261-04595-7

An Analysis of the Fans' Behavior during Two Ice Hockey Games and Its Influence on the Level of Grandstand Vibrations

M. Verner [1,a], M. Polák [1], T. Plachý [1]

[1] *Czech Technical University in Prague, Faculty of Civil Engineering, Thákurova 7, CZ-166 29 Prague, Czech Republic*
[a] *martin.verner@fsv.cvut.cz*

Abstract: Experiments focused on the football spectators behavior and grandstand vibrations were carried out at the AC Sparta football stadium in Prague in the autumn 2013 and repeatedly in the spring 2015 [1, 2]. Two experiments described in this paper continued on above mentioned ones and dealt with the same theme but on the ice hockey games. These experiments were realized at the ice hockey stadium in Pilsen and were focused again on the stadium area where the most active fans of the home team were concentrated. The grandstand in this area is constructed as the reinforced concrete frame load bearing structure. The observed grandstand vibrations were substantially smaller than at the football stadium. It was caused partially by the different grandstand structural systems and also by the dissimilar behavior of football and ice hockey spectators. Other subtypes of fans behavior were distinguished beside the types identified on the football stadium.

Keywords: ice hockey stadium; grandstand; concrete frame structure; sport fans' behavior; vibration; experiment.

1 Introduction

A considerable group of sport spectators moving synchronously could induce substantial dynamic forces. Consequently, grandstand vibrations are caused and they could be occasionally clearly perceptible or even discernible to the naked eye. There are known cases in the history when the undue vibration of a grandstand caused a panic of fans or even its breakdown in a rare situation [1].

At the present time, the satisfactory dynamic load of grandstands is not described in any standard with acceptable precision. However, new spectator's crowd dynamic load models have been currently developed that are based on stochastic approaches [2, 3, 4]. There is a lack of suitable in situ experiments that could be used as a basis for its improvement or verification [1, 5, 6], and moreover, these experiments are focused only on football matches [5, 6] or pop music concerts [6].

The basic objective of the two experiments described in this paper was to study ice hockey spectators in a selected grandstand sector, its behavior and simultaneously grandstand vibrations induced by them as the experimental basis for specification, improvement and verification of new spectator's crowd load models [2, 3]. These experiments followed on the two series of similar experiments focused on spectators induced grandstand vibrations and fans behavior during some football matches that were performed at the Letná football stadium in Prague in the 2013-14 and 2014-15 seasons [7, 8].

In the paper there is introduced the brief summary and comparison of basic results evaluated by two experiments focused on the fans behavior and its influence on the level of induced grandstand vibrations during two ice hockey games. The experiments were realized at the ice hockey stadium in Pilsen that is the home stadium of the HC Škoda Pilsen ice hockey club (see Fig. 1) during two ice hockey games against White Tigers Liberec on September 18th 2015, which HC Škoda lost 2:4, and against HC Comet Brno on October 2nd 2015, that HC Škoda won 4:3.

The grandstand load bearing structure in the observed area, where the most active fans of home team were gathered (see Fig. 2), is the reinforced concrete frame structure consisting of horizontal beams, sloping beams and columns mainly (see Fig. 2 and Fig. 3). This structure is naturally less sensitive to dynamic excitation than the steel cantilever grandstand observed in the course of the experiments performed on the Letná football stadium in Prague [7, 8].

2 The Brief Description of the Ice Hockey Stadium in Pilsen

The Pilsen ice hockey stadium was substantially reconstructed for the last time at the end of the sixties of the 20th century [9].

The basic load bearing members of the stadium roof are cables which are anchored to the perimeter reinforced concrete columns. The reinforced concrete elements that form the surface of the roof above the grandstands are supported by the cables. The roof section above the ice rink is transparent. The sloping main reinforced concrete beams are anchored on the columns too (see Fig. 3). They help to transfer horizontal load from the roof to the foundation. The T-shaped reinforced concrete beams which create the walking area of the grandstand are supported on the sloping main girders (see Fig. 3).

The capacity of the stadium decreased in the last years from 12 000 to the present 8236. The decrease is caused by changing the standing places to the sitting places. Nowadays, the capacity of standing places (2227 places) is approximately one quarter of the total stadium capacity.

Fig. 1: The exterior view of the ice-hockey stadium in Pilsen.

Fig. 2: The view on the observed grandstand sector, where the most active home fans are usually gathered.

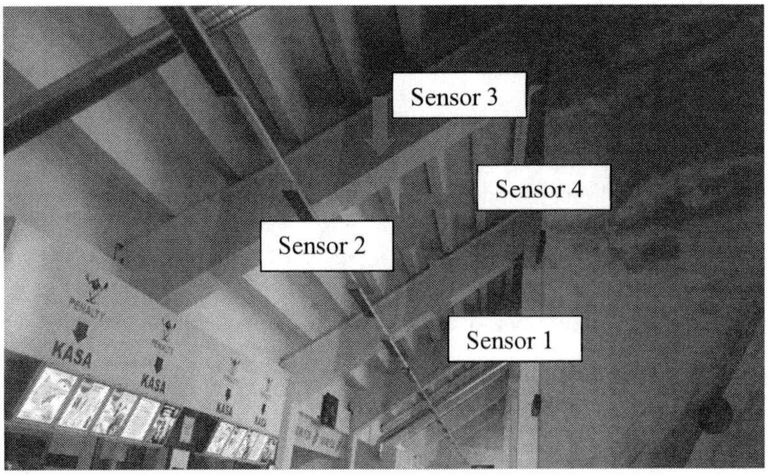

Fig. 3: The bottom view on the observed grandstand sector and the location and orientation of the accelerometers used by the carried out experiments.

Based on the experiences from the previous experiments [7, 8] it was decided to monitor the stadium sector, where the most active supporters of the home team are gathered (see Fig. 2). This section that is in Czech language also called "Kotel" is located in the part of the stadium, where the fans have to stand the whole match because of no installed seats there,

3 The Brief Description of the Experiment Arrangement

The grandstand vibrations were observed on the bottom surface of the grandstand in three points (see Fig. 3). The four piezoelectric sensors Brüel&Kjær of type 4507 B005 were used for the both experiments.

Two acceleration transducers (sensor 1 and sensor 3 in Fig. 3) were placed in the middle of the span of two sloping main grandstand girders and they recorded acceleration in the vertical direction. The remaining two transducers (sensor 2 and sensor 4 in Fig. 3) were placed in the middle of the T-shaped reinforced concrete beam, the first one measured the vertical accelerations and the second one horizontal transversal accelerations. There was chosen a T-shaped beam on which the highest concentration of the core of fans was supposed. The measurement system Pulse and Front-end 3050-B-040 Brüel&Kjær were used for data acquisition.

The results from accelerometers gave us the idea about the influence of types of spectator behavior on the levels and features of the grandstand structure vibrations. It was also important to determine the cause of the observed vibrations during the experiment. For this reason the studied area of the grandstand was monitored by camera placed on the opposite sector of the stadium.

The experiments were realized during two ice hockey games of the Ice Hockey Extra League and concretely during the second and the fourth league rounds. The first observed match was played on September 18th 2015 between the teams HC Škoda Pilsen and HC White Tigers Liberec. The home team lost 4:2 when it was losing the whole match. The goals of the home team were scored by Indrák (30 min) and Hossa (59 min) and the goals of the second team were scored two times by Radivojevič (12 and 60 min) and by Jelínek (33 and 44 min). The first observed game was broadcast live on television.

The second observed match was played on October 2nd 2015 between the teams HC Škoda Pilsen and HC Kometa Brno. The home team won 4:3 even if it was losing 1:3 during the second third. The goals of the home team were scored by Indrák (15 min), Sklenička (36 min) and two times Lev (39 and 54 min). The goals of the second team were scored by Čermák (3 min), Hruška (24 min) and by Zohorna (24 min).

In both observed games the team of Pilsen was fighting for the leading position in the League.

4 The Types of Spectator Behavior Observed by the Experiment

During two studied hockey games the very similar types of spectator behavior were observed as the types, which were monitored during the football matches [7, 8]. It means these ones - static state (all fans stand statically and watch the game passively only), walking and running, jumping, bouncing, swaying, hand clapping, goal scoring and the Mexican wave. The more detailed description of the recognized types of football spectator behavior is stated in the reference [8].

However, same differences could be found. During the football matches that are described in references [7, 8] the spectators used four types of jumping, by three of them, the spectators are joined together in separated lines and jump synchronously. The more significant difference is the limited application of these types of jumping by the ice hockey fans. In the two observed ice hockey games actually jumping as a type of cheering was not applied at all (see Tab. 1). The other noticeable difference is that the ice hockey spectators used more often the cheering type with hand clapping which could be moreover divided to three subtypes.

The first subtype of hand-clapping (the basic hand-clapping) means the spectators stay only on the spot and the applause is realized at chest level. This subtype appears most frequently in a situation when the fans praise players for a nice action.

During the second subtype of hand-clapping the frequency of the applause is controlled by a drum. The clapping frequency could be slow or could accelerate or the applause could combine different rhythms. The spectators lightly bend of their knees in the controlling rhythm in this subtype.

The third subtype of hand-clapping means the spectators stay only on the spot and their hands applaud above their heads. This subtype is impressive to look at. Fans like to use it for accentuation of the readiness to support their team.

5 The Brief Description of the Experiment Basic Results

The both described experiments were carried out during the overall game time. The behavior of spectators during breaks was not monitored.

The comparisons of the time behavior and the respective frequency spectrums of the vertical acceleration in the middle spot (the sensor 2) for the significant game situations (the opposing team scored goal, the home team scored goal, the cheering with the hand clapping of the 1st subtype, the home fans celebrate the victory) are shown in Fig. 4 to Fig. 9.

The different types of fans behavior were observed in detail in the course of the both games based on the camera records of the studied grandstand area. The evaluated time duration of the specific types of the spectator behavior and their percentage on the overall game time are stated in the Tab. 1.

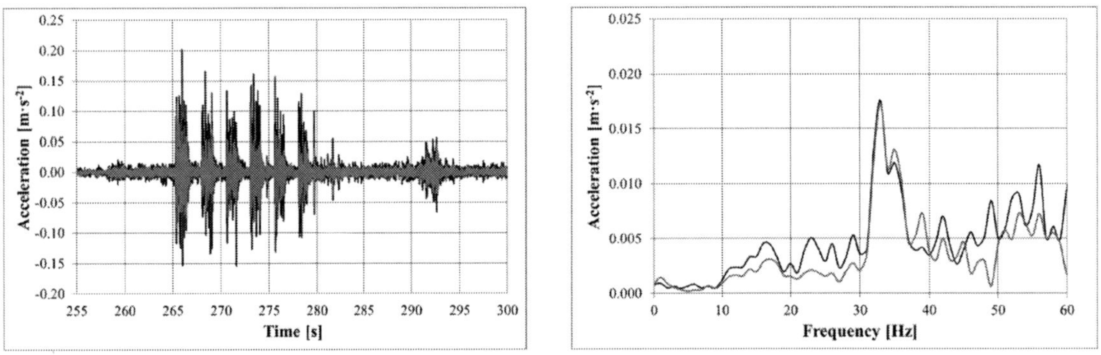

a) the acceleration time records

b) the respective frequency spectra

Fig. 4: The grandstand vertical and horizontal vibrations which were measured in the middle spot (sensor 2 and sensor 4) in the situation when the opposing team scored fourth goal in the first game.

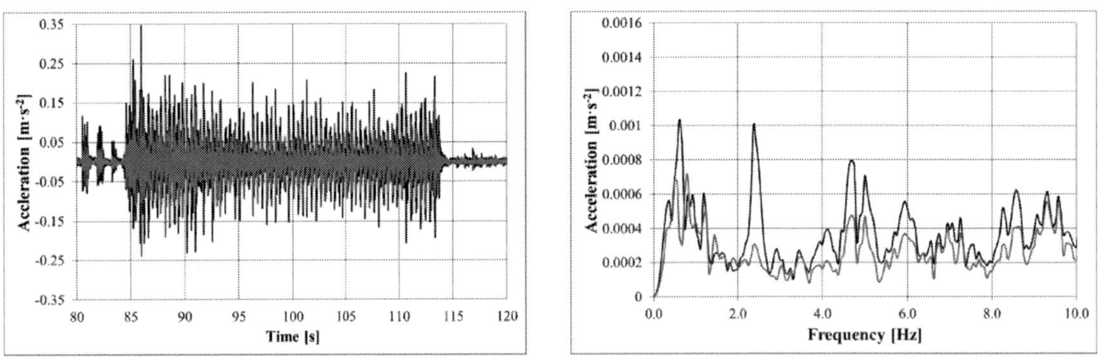

a) the acceleration time records

b) the respective frequency spectra

Fig. 5: The grandstand vertical and horizontal vibrations which were measured in the middle spot (sensor 2 and sensor 4) in the situation when the home team scored the second goal in the first game.

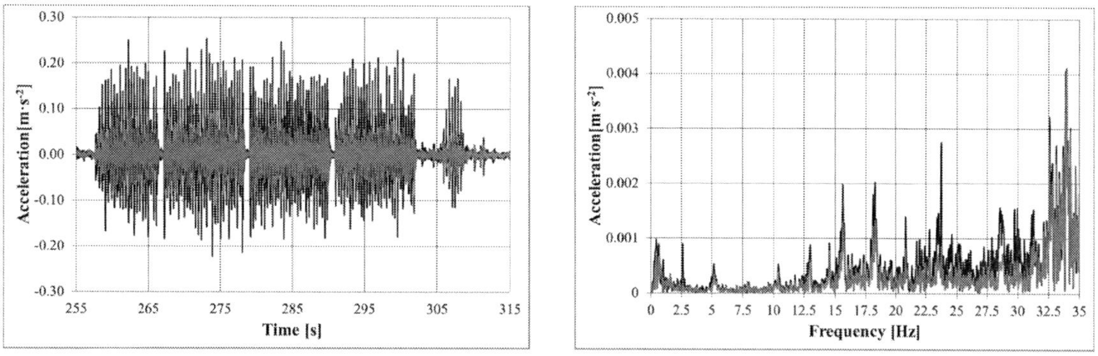

a) the acceleration time records

b) the respective frequency spectra

Fig. 6: The grandstand vertical and horizontal vibrations which were measured in the middle spot (sensor 2 and sensor 4) in the situation when the home team scored winning goal in the second game.

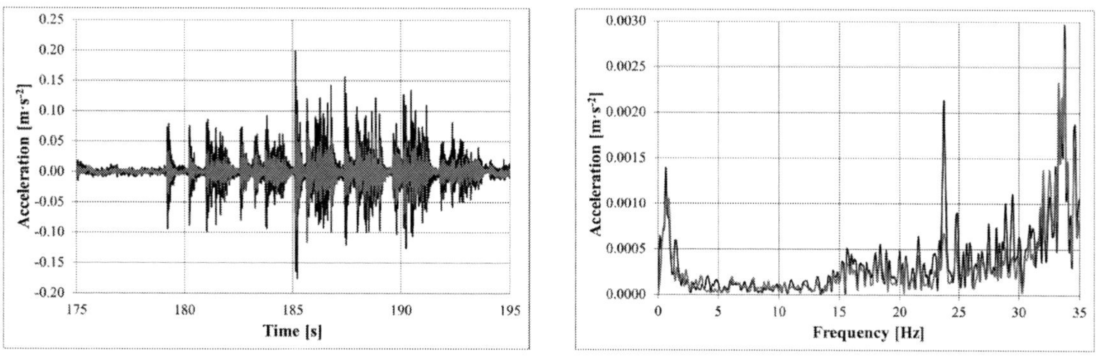

a) the acceleration time records

b) the respective frequency spectra

Fig. 7: The grandstand vertical and horizontal vibrations which were measured in the middle spot (sensor 2 and sensor 4) in the moment when the home fans gathered in the analyzed grandstand sector used the cheering with the hand clapping of the 1st subtype (the basic hand-clapping) during the second game.

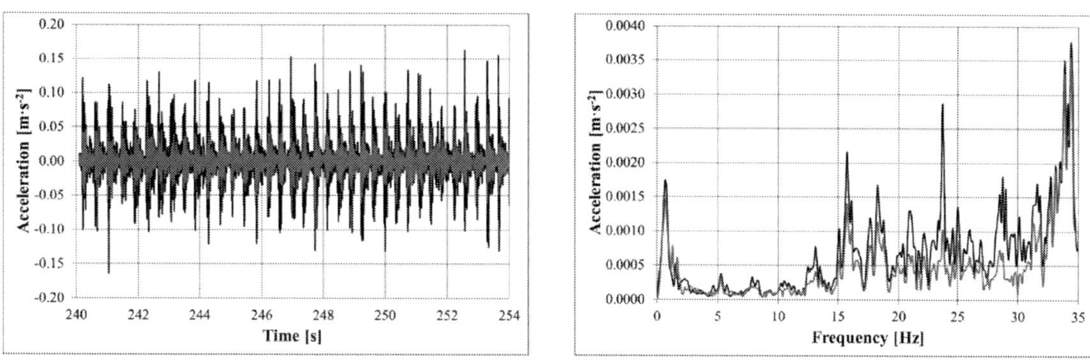

a) the acceleration time records

b) the respective frequency spectra

Fig. 8: The grandstand vertical and horizontal vibrations which were measured in the middle spot (sensor 2 and sensor 4) in the moment when the home fans gathered in the analyzed grandstand sector used the cheering with the hand clapping of the 2nd type (the gradually accelerated hand-clapping directed by the drum) during the second game.

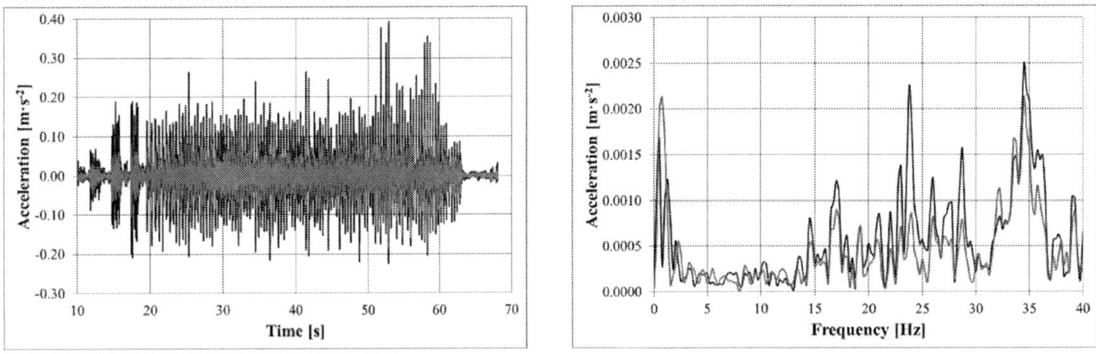

a) the acceleration time records b) the respective frequency spectra

Fig. 9: The grandstand vertical and horizontal vibrations which were measured in the middle spot (sensor 2 and sensor 4) in the situation when the home fans celebrated the win at the end of the second game.

Tab. 1: The time duration of the particular types of spectator behavior evaluated during the two studied ice hockey games.

The type of spectator behavior	The time duration for the first game [min]	The time duration for the first game [%]	The time duration for the second game [min]	The time duration for the second game [%]
Static state	27:13	23.6	21:03	14.9
Walking and running	6:20	5.5	0:51	0.6
Jumping – all types	0:00	0.0	0:00	0.0
Bouncing	1:14	1.1	1:05	0.8
Swaying	18:24	15.9	22:36	16.0
Hand clapping – all types	59:23	51.3	90:09	64.0
Goal scoring	3:00	2.6	4:49	3.4
Mexican wave	0:00	0.0	0:16	0.2

6 Conclusion

The results of the fans behavior analysis realized during two ice hockey games show that the behavior of ice hockey spectators is dissimilar in comparison with football ones. For example, other subtypes of fans behavior were distinguished beside the types, which were recognized at the football stadium [8], and it was observed that the ice hockey fans more often apply the behavior type hand-clapping that was used more than half the playing time (see Tab. 1) in both observed games.

The observed grandstand vibrations were substantially smaller than in the football stadium [7, 8]. It was caused partially by the different grandstand structural systems and also by the dissimilar behavior of football and ice hockey spectators.

It was observed that the life broadcast of a game on a television reduces noticeably the number of spectators present at the stadium. The examined grandstand vibrations were consequently smaller too during the broadcast game.

Acknowledgement

This paper has been supported by the Czech Science Foundation project No. GA15-15728S.

References

[1] C.A. Jones, P. Reynolds, A. Pavic, Vibration serviceability of stadia structures subjected to dynamic crowd loads: A literature review, Journal of Sound and Vibration 330 (2011) 1531–1566, http://dx.doi.org/10.1016/j.jsv.2010.10.032.

[2] O. Rokos, J. Maca, Stochastic approach in the human-induced vibration serviceability assessment of grandstands, in proc.: Proceedings of the IX International Conference on Structural Dynamics EURODYN2014 (2014), eds. A. Cunha, E. Caetano, P. Ribeiro, G. Müller, Faculdade de Engenharia da Universidade do Porto, Porto Portugal, 2591-2598.

[3] O. Rokos, J. Maca, The response of grandstands driven by filtered Gaussian white noise processes, Advances in engineering software 72 (2014) 85 – 94, http://dx.doi.org/10.1016/j.advengsoft.2013.05.008.

[4] V. Rajic, A. Pavic, Stochastic approach to modeling of near-periodic jumping loads, Mechanical systems and signal processing 24 (2010) 3037 – 3059.

[5] P. Reynolds, A. Pavic, Vibration performance of a large cantilever grandstand during an international football match, Journal of Performance of Constructed Facilities 20 (2006) 202 – 212.

[6] A. Caprioli, M. Vanali, A. Cigada, One year of structural health monitoring of the Meazza Stadium in Milan: Analysis of the collected data, in proc.: Proceedings of the 27th Conference and Exposition on Structural Dynamics 2009 IMAC-XXVII (2009), Society for Experimental Mechanics Inc., Orlando Florida USA, 9p.

[7] M. Verner, M. Polak, T. Plachy, An Experimental Study Focused to Spectators-induced Vibrations of a Cantilever Grandstand during two Sport Matches, in proc.: Proceedings of 53rd International Conference on Experimental Stress Analysis 2015 (2015), eds. P. Padevet, P. Bittnar, Czech Technical University in Prague Faculty of Civil Engineering, Český Krumlov, 460–465.

[8] M. Verner, T. Plachy, M. Polak, An Experiment Focused on Fans Behaviour and Induced Grandstand Vibrations during a Football Match, Applied Mechanics and Materials 837 (2016) 75–78, http://dx.doi.org/10.4028/www.scientific.net/AMM.837.75.

[9] Retrieved from: http://www.hcskoda.cz

Component Innovations Using Composite Principles

K. Vitek [1,a]

[1] *Faculty of Mechanical Engineering, Czech Technical University in Prague, Czech Republic*
[a] *karel.vitek@fs.cvut.cz*

Abstract: The paper presents the methodology of wound composite structures that are alternative to conventional components. The use of principle is based on the definition of partial functions of compact elements included. For these features are subsequently created elementary components which can be made by other technologies and of materials other than conventional compact construction. Wider properties that have a composite structure, it is advisable to check size and optimize for these composite structures could be used effectively.

Keywords: screw connection; coiled screws; screw connections securing.

1 Introduction

An important structural element is a screw connection. There is known a number of standard types of screw and nuts, where a helical thread is formed, mostly by machining or rolling, directly on the screw shank and the nut body. Such screw shank threads form notches, which significantly reduce the screw load carrying capacity. The innovative design is based on the original production technology of bolted connection form: i) the coiled screw patent of No.304377 [1] and ii) the coiled nut patent of No.304376 [2]. These new structures, see Fig. 1, have a screw thread on the screw shank or in the nut housing coiled of a rod having suitable diameter and a suitable material, which need not be identical with the material of the screw shank or the nut housing. The screw shank, or analogously the nut housing, can have different cross-sectional shapes than the rod forming the helical thread, so it is possible to design screw connections for various purposes with varying shapes or materials. The screw thread can be coiled either freely or with a pre-stress, both on the screw shank and in the nut housing and can be situated either in the slots or tightly connected - by gluing or welding. The rod cross-section, forming the thread, can be of various suitable hollow thin-walled profiles, closed or open. The hollow thread can be used for a direct lubrication or a tempering with a suitable medium. Functional samples of several types of coiled screws and nuts were made. The coiled screws and nuts provide the bolted connections having new usable advantageous properties that do not have the classic bolted connections. These include a significant material savings, because the screw shank and the nut housing can be of thin-walled profiles made of various metallic, non-metallic and composite materials. It offers the possibilities: i) to optimize properties in a wide range of the coiled screws and nuts; ii) to analyze the behavior of them under load; iii) to identify the interaction of the thread structures with the stem screw or nut housing when manufactured with: a) the freely coiled thread; b) the pressed-on thread; c) or the point-welded thread. It is necessary to define the stress state of the connection elements and to assess the knowledge bases necessary for their design and for using their properties in practice.

The usual means for the screw connection securing are standardized. These are the different types of washers (flexible, fan-shaped, toothed, with a nose, with a tab), then the castle nut cotter pins or self-locking nuts containing a friction plastic or a toothed element. The locking washers must be applied under the nut or screw heads just when assembling the bolted connections, which is not advantageous both from the mounting aspect and the securing function, because even a slight screw connection releasing could cause the locking washers stop working. As a perspective system to be used for the screw connections securing, the patent of No.302768 [3] can be applied, Fig.2, which enables the securing system application already before the assembling, or the securing washers can be also pressed on the screw connection after its installation and tightening. There were functional samples of locking washers made. Their locking pins can be designed still longer to secure the bolted connection in a higher stroke for the various tightening, which is suitable for securing them.

At the same time, it is also possible using the simpler shapes of heads of screws and nuts, as for controlling the bolted connection can serve a tool tucked into the gaps between the screw head or the nut and its locking washer.

Flexible couplings of patent No. 3052683 [4] are used to elastic connection of shafts. The inner and outer spring rings are formed by bending the helically wound coil spring after forming a circle in the torus-shaped, and both ends forming a helical spring which forms a ring, are rigidly connected, see Fig. 3. These couplings realized elastic and flexible connection and they can compensate for the large deviations connected shafts. For worm gear designs patent of No. 3052497 [5], the flexible power transmission from the drive shaft of the worm implemented on a worm wheel which is peripherally permanently attached gear teeth formed by a flexible helically coiled spring ring. The spring ring is formed by bending the helically wound coil spring after forming a circle in the torus-shaped, and both ends forming a helical spring which forms a ring, are rigidly connected. Using the principle of composite designs with flexible transmissions gear teeth can be realized escalate worm transmission, see Fig. 4.

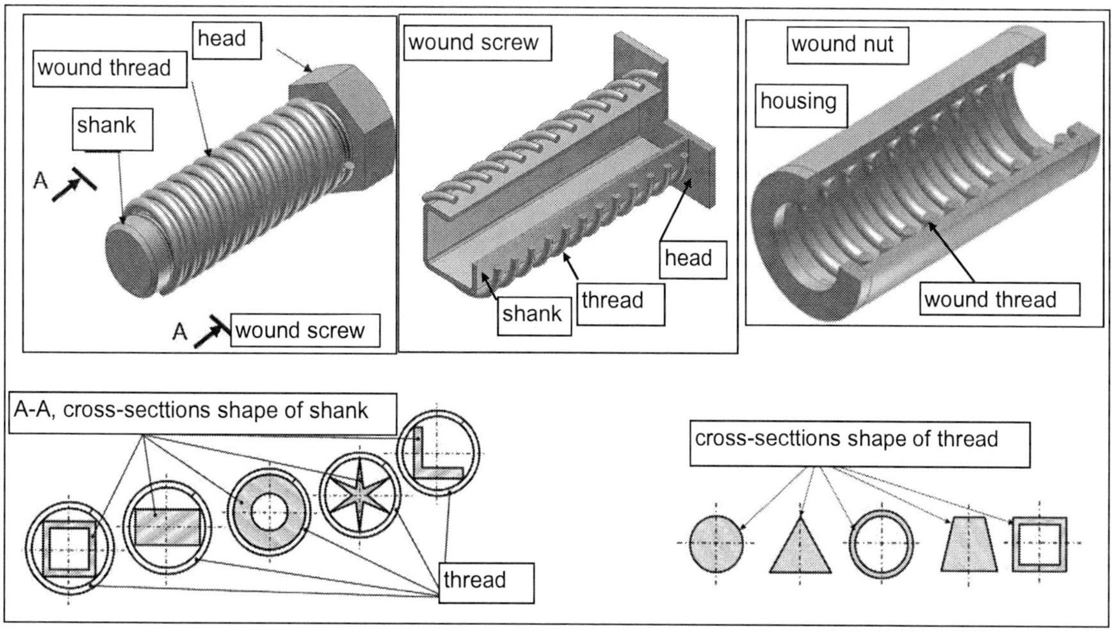

Fig. 1: Principle of wound screw and wount nut.structures .

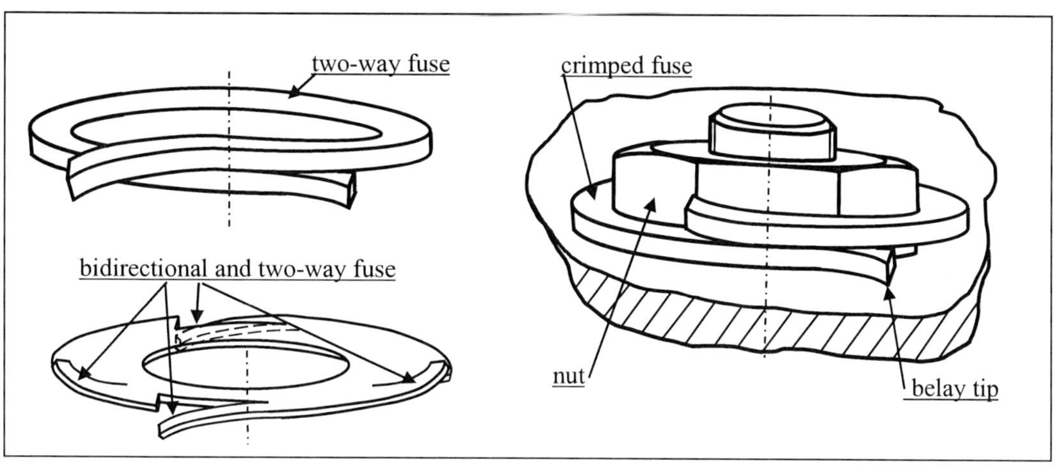

Fig. 2: Principle of assurance system of fixing bolts.

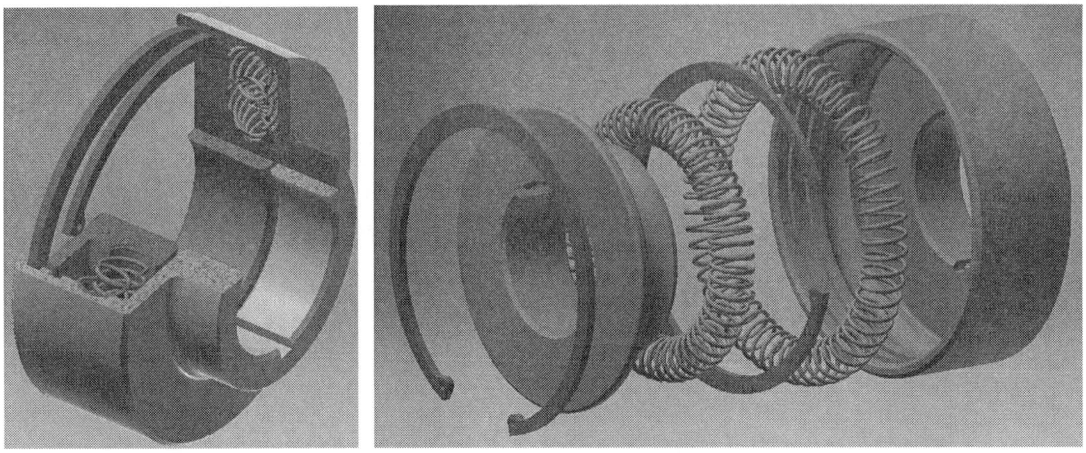

Fig. 3: Flexible coupling.

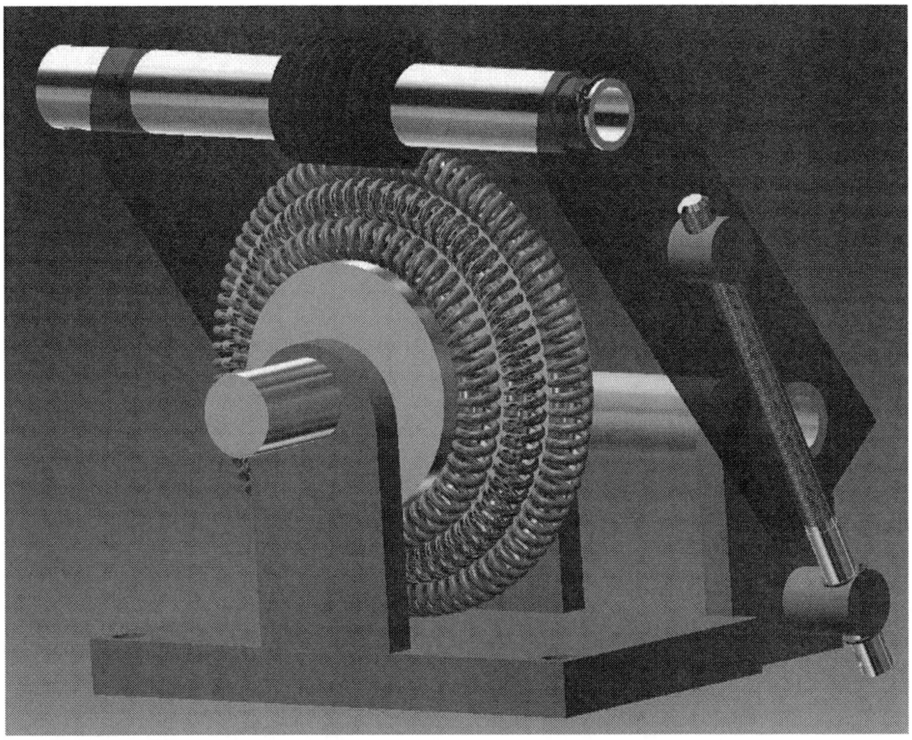

Fig. 4: The coiled worm gear with the gear shift

2 Conclusion

It is useful to analyze the assembly and structure of the locking washer properties to have necessary basic data for the design and the possible use of their more versatile functions in practice.

The inventing also requires continuous monitoring of functions of their components. The Fig. 5 illustrates an implementation of the wound thread mounted on the shaft of the screw, followed by further shown mounting the wound thread within the nut bore and is here also documented the verification of the connection of the bolt with the nut. Similarly, Fig. 6 shows an experimental model of worm gearbox, on which were verified by the basic principles of the patented design.

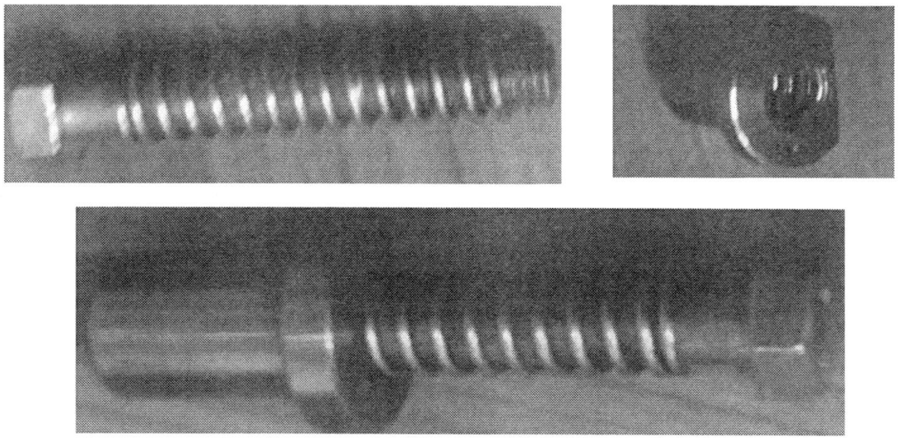

Fig. 5: Experimental model of the wound screw connection.

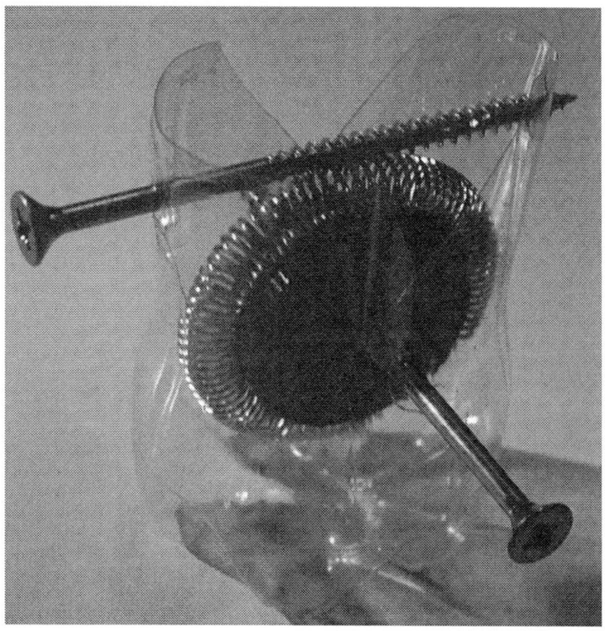

Fig. 6: The experimental model of worm gear - worm thread engages in the coiled toothing of the worm gear.

References

[1] Vítek, K.: Wound screw, Invention -patent n: 304377, Czech Patent and Trademark Office, 2014.

[2] Vítek, K.: Wound nut, Invention -patent n: 304376, Czech Patent and Trademark Office, 2014.

[3] Vítek, K.: Assurance system of fixing bolts,–patent n.: 302768, Czech Patent and Trademark Office , 2011.

[4] Vítek, K.: Flexible coupling, Invention -patent n: 305268, Czech Patent and Trademark Office, 2015.

[5] Vítek, K.: Worm gearbox, Invention -patent n: 305497, Czech Patent and Trademark Office, 2015.

Comparison of two approaches for the tensile test of single roving in polymer matrix

T. Vlach [1,2,a], L. Laiblova [1,2], M. Zenisek [1], P. Kokes [1], P. Hajek [1,2]

[1] *Faculty of Civil Engineering, CTU in Prague, Thakurova 7, 166 29 Prague 6, Czech Republic*
[2] *University Centre for Energy Efficient Buildings, CTU in Prague, Trinecka 1024, 273 43 Bustehrad, Czech Republic*
[a] *tomas.vlach@fsv.cvut.cz*

Abstract: High performance concrete is becoming more and more popular material. For extremely subtle structures are used as reinforcement durable composite materials, such as technical textiles. Elements made of this combination are called textile reinforced concrete. Modeling and designing from this material is very sensitive for boundary conditions. This study is focused on the two different methods of approaches for the determination of tensile strength and elasticity modulus of composite reinforcement using two different types of fixing to the testing machine.

Keywords: roving; polymer matrix; tensile test; modulus of elasticity.

1 Introduction

Rovings from different materials in polymer epoxy matrix are used for technical textiles formation. This advanced material is often applied as reinforcement for fine grained high performance concretes. For this particular application are used hard, durable and chemically resistant fiber materials like alkali resistant glass, basalt or carbon. This resulted in a new group of silica materials called textile reinforced concrete (TRC). TRC is used for new construction and is also possible to use this material for structures strengthening [1]. Determination of the basic mechanical parameters of each component is very important for elements designing and numerical modeling. Some types of testing are not clearly prescribed in standards. For this article about basic tensile parameters was chosen and adapted the procedure according to ACI 440.3R-04 [2] about the test methods for fibre-reinforced polymers (FRP) for reinforcing or strengthening concrete structures. This adapting is because this standard describes the same material, but with lower cross sectional area. On both ends of specimens are installed steel sleeves using epoxy resin. Because specimens preparation is very demanding and fragile during preparation and installation and specimens are susceptible to damage so the test procedure modification has been found. The method is described in [3] and has been developed for the tensile testing of single rovings without polymer matrix. The same testing procedure was applied for the tensile testing of single rovings with polymer epoxy matrix. Measured and calculated mechanical properties were the tensile strength and static elastic modulus using method [4].

2 Experimental part

2.1 Materials used for experiment

One type of roving Cem-FILL 5325 used for both approaches is made from the infinitely long alkali-resistant glass fibers. Each of them contained five specimens of composite reinforcements. From the most important properties, the roving used for experiment had a length weight (titer) of 2400 g/km (= 2400 tex), tensile strength of 1700 MPa and a modulus of elasticity of 72 GPa. Specific gravity of fibers material is 2.68 g/cm^3. From the technical data sheet it is possible to calculate theoretical maximal tensile force before breaking of all fibers in one moment and it is 1.522 kN [5]. As epoxy resin was used for experiment two part solvent free epoxy Sikafloor 156. The ratio of Part A and Part B was used 1:3 to assemble the epoxy resin according to technical data sheet. From the most important properties, the epoxy resin used for experiment had a tensile strength of 15 MPa, a modulus of elasticity of 2.0 GPa and specific gravity 1.10 g/cm^3 [6].

2.2 Tensile test

The only experiment for this work was the direct tensile test of composite reinforcement, single alkali resistant roving in matrix of epoxy resin. During test all required values for determination of tensile strength and modulus of elasticity were measured. It means the time, force, crosshead displacement and elongation of specimen using small external potentiometer [4]. For this work was chosen and adapted the procedure according to ACI 440.3R-04 [2] about the test methods for fibre-reinforced polymers (FRP) for reinforcing or strengthening concrete structures. So both ends of specimens were fitted with steel sleeves using epoxy resin because of fixing to the testing machine. The length of the specimen between steel sleeves was 300 mm and the length of each steel sleeve with diameter 20 mm was 150 mm. Specimens preparation is very demanding and fragile during preparation and installation and specimens are very susceptible to damage the test procedure modification was found. Another method for specimen preparation [3] has been originally developed for the tensile testing of single rovings without polymer matrix. Both ends of specimens are fixed to the small epoxy prims 8 x 8 x 80 mm without steel sleeves. For the epoxy prisms casting was used the silicon mold. The distance between prisms was also 300 mm.

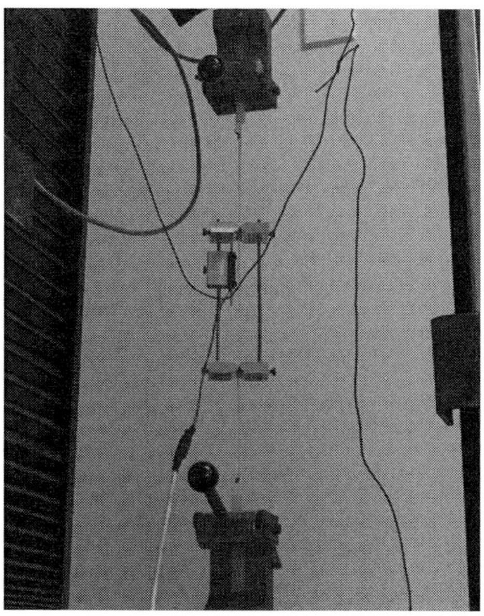

Fig. 1: General view of the specimen in sleeves made from epoxy prisms before testing.

Fig. 2: General view of the specimen in the massive steel sleeves with epoxy inside just after testing.

2.3 Results and calculations

Testing of specimens was performed in LabTest 100 testing machine. This testing machine was additionally provided with data acquisition system using external card because of using small external potentiometer for elongation measurement. The speed of loading was 2.0 mm/min. It corresponds approximately to the increase of stress 2.0 MPa/s. Specimens were loaded until the failure. The views on small potentiometer are presented in Fig. 1 for the case of epoxy prisms 8 x 8 x 80 mm on the end of specimen and in Fig. 2 for the case of steel sleeve with epoxy resin inside. The course of the test was monitored using data acquisition system that monitored the relationship of the acted force to the time of the test and deformation course. Graphical outputs were made from the courses of the tests. Three typical curves from five were chosen for each group of specimens. Force displacement curves are resented in Fig. 3 and stress-strain diagram from calculated values is presented in Fig. 4. Curves are very similar for each group of specimens.

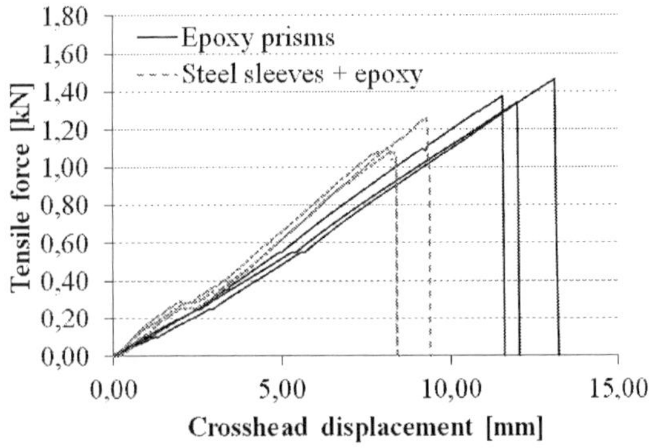

Fig. 3: Force-displacement curves using data from testing machine, both types of specimens.

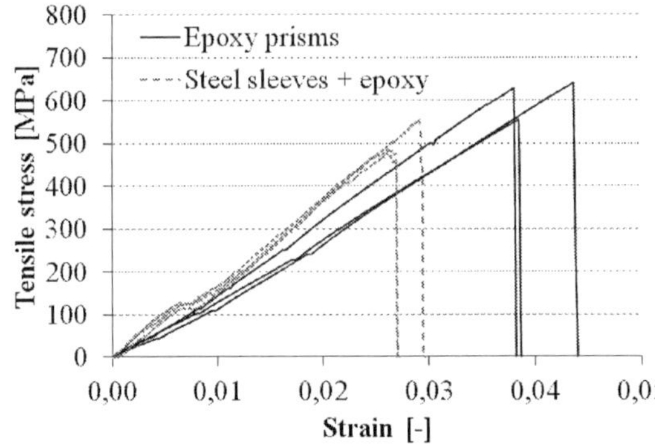

Fig. 4: Stress-strain curves using data from testing machine, both types of specimens.

Damage of specimens was similar for both types epoxy prisms and steel sleeves. They were broken always correctly approximately in the middle. Epoxy prisms were not visibly damaged during testing procedure and worked correctly. There was no visible pulling out of composite material from epoxy prisms. Specimens after testing procedure are presented in Fig. 5 and Fig. 6.

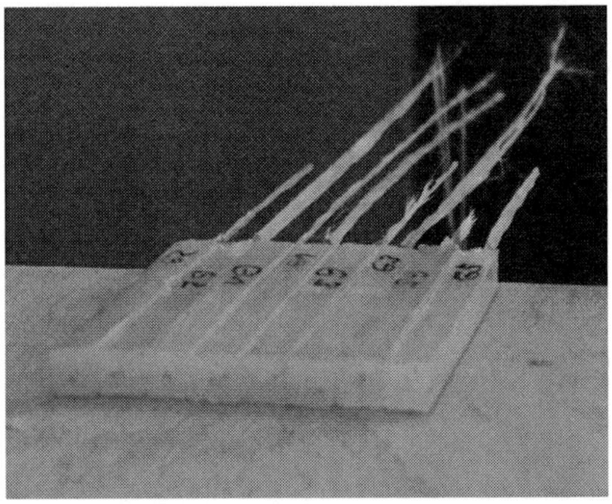

Fig. 5: View of the broken specimens in sleeves made from epoxy prisms 8 x 8 x 80 mm.

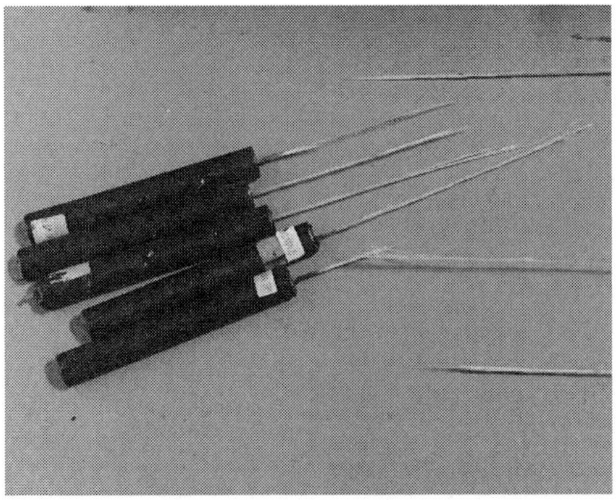

Fig. 6: View of the broken specimens in the massive steel sleeves 20/2.5 mm with epoxy inside.

Basis tensile parameters of used composite reinforcement were calculated from all measured values. All calculated results are presented in Table 1. Specimens with epoxy prisms have indication *E* and specimens with steel sleeves and epoxy inside have indication *S*. Symbol *m* in the Table 1 means the weight of specimens with length 100 mm and m_e means the calculated weight of pure epoxy, because it is known the weight of pure roving from defined length weight 2400 tex [5]. Symbol *A* and A_e mean the calculated cross sectional area of composite and pure epoxy using specific gravity of materials [5,6]. F_{max} is maximum measured tensile force just before breaking of specimen and f_t is calculated tensile strength. Elongation ε was measured using small external potentiometer monitoring the distance change between two edges of sensor with base of 130 mm. Elongation was monitored in the range of forces ΔF. Based on these results could be calculated the modulus of elasticity *E*.

Tab. 1: Results of measurement and calculations for each specimen, maximum tensile force and strength, static elastic modulus.

spec. [mm]	E 1	E 2	E 3	E 4	E 5	aver.	S 1	S 2	S 3	S 4	S 5	aver.
m [g]	0.385	0.386	0.407	0.417	0.440	**0.407**	0.398	0.405	0.405	0.428	0.420	**0.411**
m_e [-]	0.145	0.146	0.167	0.177	0.200	**0.167**	0.158	0.165	0.165	0.188	0.180	**0.171**
A [mm^2]	2.191	2.195	2.286	2.327	2.417	**2.283**	2.248	2.277	2.277	2.371	2.339	**2.302**
A_e [mm^2]	1.296	1.300	1.390	1.431	1.522	**1.388**	1.352	1.382	1.382	1.475	1.443	**1.407**
F_{max} [N]	1374	1205	1462	1283	1342	**1333**	1087	1262	1104	1132	1280	**1173**
f_t [MPa]	627	549	640	551	555	**584**	484	554	485	478	547	**509**
ε [-]	$4.234 \cdot 10^{-3}$	$4.821 \cdot 10^{-3}$	$4.066 \cdot 10^{-3}$	$4.486 \cdot 10^{-3}$	$4.678 \cdot 10^{-3}$	**$4.457 \cdot 10^{-3}$**	$3.255 \cdot 10^{-3}$	$2.842 \cdot 10^{-3}$	$2.913 \cdot 10^{-3}$	$3.003 \cdot 10^{-3}$	$2.739 \cdot 10^{-3}$	**$2.951 \cdot 10^{-3}$**
ΔF [N]	299.9	307.2	300.0	300.0	300.0	**301.4**	200.4	199.4	199.8	200.0	198.1	**199.5**
E [GPa]	32.3	29.0	32.3	28.7	26.5	**29.8**	27.4	30.8	30.1	28.1	30.9	**29.5**

3 Conclusion

Mechanical properties in this experiment are for both approaches very similar. Fig. 3 and Fig. 4 show that specimens with epoxy prisms have higher values of elongation. This is due to lower rigidity of the small epoxy prisms compared to the steel sleeves. Specimens with epoxy prisms also achieved higher maximum force just before damage. This may be due to bad sample with steel sleeves handling. They are heavy and fragile and they can be easily damaged before testing. Damage of specimens may not be visible. Prisms are lightweight and handling is much easier. Results and higher tensile strength indicates that use of the epoxy prisms gives more accurate results compared to the steel sleeves. Results of this work will be used as boundary conditions for numerical modeling.

Acknowledgement

This work has been supported by the Ministry of Education, Youth and Sports within National Sustainability Programme I, project No. LO1605. Further support was given by Student Grant Competition SGS16/131/OHK1/2T/11 – Experimental Facade Panels from UHPC as a Base for LED Display. The authors would like to acknowledge with gratitude all the financial help that was provided in order to support this research.

References

[1] O. Holcapek, F. Vogel, Bond Properties of Concrete Beams Strengthened by AR-Glass Textile and Basalt Textile Reinforced Concrete, Applied Mechanics and Materials 825 (2016) 7–10, http://dx.doi.org/10.4028/www.scientific.net/AMM.825.7.

[2] ACI 440.3R-03, Guide Test Methods for Fiber-Reinforced Polymers (FRPs) for Reinforcing or Strengthening of Concrete Structures, ACI Committee 440, 2012.

[3] R. Rypl et al., A novel tensile test device for effective testing of high-modulus multi-filament yarns, Journal of Industrial Textiles (2014), http://dx.doi.org/10.1177/1528083714521069.

[4] T. Bittner et al., Experimental Investigation of Mechanical Properties of Textile Glass Reinforcement, Applied Mechanics and Materials 732 (2015) 45–48, http://dx.doi.org/10.4028/www.scientific.net/AMM.732.45.

[5] Cem-FIL® 5325, Product Data Sheet from: http://www.ocvreinforcements.com/uploadedFiles/Reinforcement/CemFil/Doc/10010692-D_CemFil_5325_ww_12_2010_Rev7_EN_A4.pdf.

[6] Sikafloor® - 156, Product Data Sheet from: http://cze.sika.com/dms/getdocument.get/24bcfb62-e97c-3809-bd0c-84260131775e/Sikafloor-156.pdf.

Compressive Strength and Dynamic Modulus of Elasticity of the Concrete from the Dam Orlík

O. Zobal [1,2,a], T. Plachý [1], P. Padevět [1]

[1] *CTU in Prague, Faculty of Civil Engineering, Department of Mechanics, Thákurova 7, 166 29 Praha 6, Czech Republic*

[2] *Czech Technical University in Prague, UCEEB, Třinecká 1024, 273 43 Buštěhrad, Czech Republic*

[a] *ondrej.zobal@fsv.cvut.cz*

Abstract: This paper describes the analysis of the concrete properties of the body of the dam Orlik - specifically its compressive strength and dynamic modulus of elasticity. Two non-destructive methods have been used for measurement and evaluation of the dynamic modulus of elasticity - the resonant and ultrasonic methods. The concrete of the body of the dam has high values of compressive strength and dynamic modulus of elasticity after more than 50 years in operation.

Keywords: Concrete; Compressive strength; Dynamic modulus of elasticity.

1 Introduction

The Orlik Dam belongs among the most important civil engineering works in the Czech Republic. This construction was built between the years 1956 - 1961 and about 1 million m^3 of concrete [1] was consumed on the body of the dam. The dam has several functions - water regulation, electricity supply, water transport and recreation [2]. The body of the dam is straight gravitational concrete dam 81.5 m high and 450 m long [3]. Monitoring the behavior of the whole work and knowledge of the properties of the used concrete is needed with regard to the importance of this strategic construction. The mechanical and material properties of the concrete of the dam have been analyzed and this material is of high quality [4]. This paper presents results of measuring the compressive strength and dynamic modulus of elasticity.

2 Compressive Strength

The compressive strength was measured on cubes with an edge 200 mm. Cubes were cut out with a diamond saw from the cylinders with diameter 300 mm. The photo of the test cube No. 3 after the test of compressive strength is in Fig. 1 and course of loading is shown in Fig. 2.

Fig. 1:Tested cube No. 3

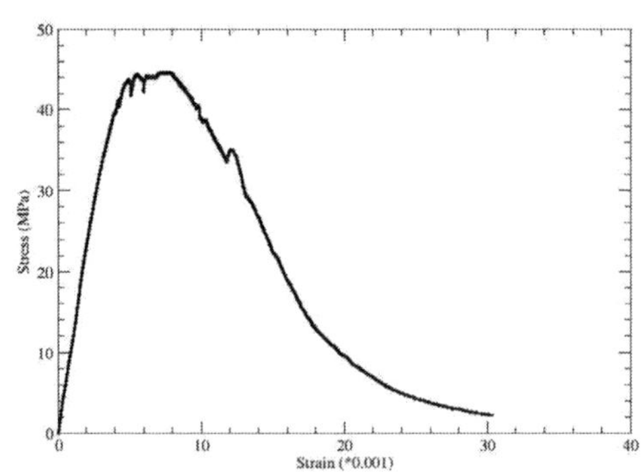

Fig. 2: Dam Orlík, Cube No. 3

Evaluation of the characteristic compressive strength was carried out according to standards EN 13791 and EN 1990 [5, 6]. Values are provided in Tab. 1.The concrete used for the body of the dam was designed as the concrete of strength class B80 (currently C8/10). The resulting strength of the concrete corresponds to a standard much higher strength class [7].

Tab. 1: The measured and characteristic compressive strength of the concrete of the body of the dam Orlik.

Designation of cube [-]	Measured f_c [MPa]	f_{ck}(ČSN EN 13791)	f_{ck} (ČSN EN 1990)
1	52	n 4	m_x 47.5
2	51	k 7	s_x 4.8
3	45		V_x 0.10096
4	42	**40.5 MPa**	**38.7 MPa**
$f_{cm}(s_x)$	**47.5 (4,8)**		

3 Dynamic Modulus of Elasticity

The dynamic modulus of elasticity was evaluated from two non-destructive methods - resonant and ultrasonic. The specimens with the different lengths were prepared for the experiment from the borehole with a diameter of 80 mm.

3.1 Resonant Method

The first used method was the resonant method. The cylindrical specimen was lying on a very soft pad (Fig. 3 – on the left) to not influence the fundamental longitudinal mode of the specimen vibration. The acceleration transducer Brüel&Kjær of Type 4519-003 was placed at the center of one of the end faces of the specimen (Fig. 3 – on the left). The end face of the concrete specimen opposite to the face, where the transducer was located, was struck by the impact hammer Brüel&Kjær of Type 8206 (Fig. 3 – on the left). Both signals, the excitation force and the acceleration, were recorded and transformed using the Fast Fourier Transform (FFT) to the frequency domain, and the Frequency Response Function (FRF) was evaluated from these signals using the vibration control station Brüel&Kjær Front-end 3560-B-120 and program PULSE 14.0. The test was repeated five times for each specimen and resultant readings were averaged. From an averaged FRF, the fundamental longitudinal resonant frequency was determined for each specimen.

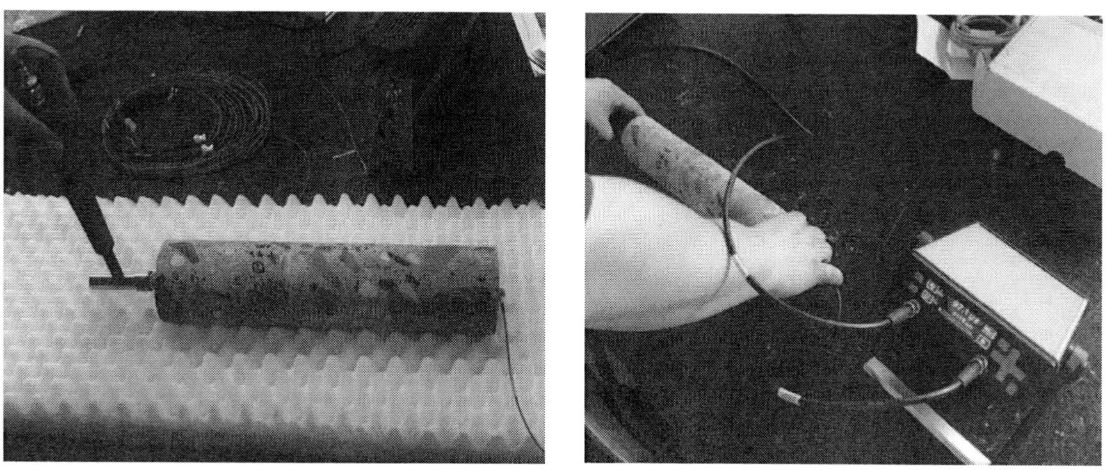

a) Resonant method

b) Ultrasound method

Fig. 3: Measurement of dynamic modulus of elasticity

Based on the equation for longitudinal vibration of the beam with continuously distributed mass with free-free boundary condition, the dynamic Young's modulus E_{dl} can be determined using the relation

$$E_{dl} = \frac{1.2732 l m f_l^2}{r^2 K} \quad , \tag{1}$$

where l is the length of the specimen [m], m is the mass of the specimen [kg], f_l is the fundamental longitudinal resonant frequency of the specimen [Hz], r is the radius of the specimen [m] and K is the correction factor for the fundamental longitudinal mode to account for the finite diameter-to-length ratio and Poisson's Ratio:

$$K = 1 - \frac{\pi^2 \mu^2 r^2}{2 l^2} \quad , \tag{2}$$

where μ is the Poisson´s ratio, r is the radius of the specimen [m] and l is the length of the specimen [m].

3.2 Ultrasound Method

The second nondestructive method used for dynamic modulus determination was the ultrasound method. The ultrasound analyzer with two probes of the frequency 55 kHz was used. One probe was placed at the center of one of the end faces of the concrete specimen (Fig. 3 – on the right). The second probe was placed at the end face of the concrete specimen opposite to the face, where the first probe was located. One probe (transmitter) sends ultrasound through the specimen and the second probe (response transducer) detects the time of signal traveling through the specimen. The dynamic Young's modulus E_{du} can be determined using the equation

$$E_{du} = \frac{m l}{\pi r^2 t^2 k_3^2} \quad , \tag{3}$$

where m is the mass of the specimen [kg], l is the length of the specimen [m], r is the radius of the specimen [m], t is the time of signal traveling through the specimen [s] and k_3 is the correction factor for 3D environment, which should be used for short specimens:

$$k_3 = \sqrt{\frac{1 - \mu}{(1 + \mu)(1 - 2\mu)}} \quad , \tag{4}$$

where μ is the Poisson´s ratio.

4 Conclusion

The compressive strength and the dynamic modulus of elasticity were verified on the concrete of the body of the dam Orlík old more than 50 years. The characteristic compressive strength of concrete has value 38.7 (40.5) MPa. The dynamic modulus of elasticity has an average value by resonance method 24.8 and by ultrasonic method 32.9. This difference is caused due to different resonant frequency for the resonant method and much higher frequency of the ultrasound signal [8]. This concrete demonstrates very good mechanical properties.

Tab. 2: The values of dynamic modulus of elasticity obtained by resonance (E_{dR}) and ultrasound (E_{dU}) methods.

Designation of	Mass	Length	Diameter	E_{dR}	E_{dU}
the specimen	[kg]	[m]	[m]	[GPa]	[GPa]
J-1-1-I	0.646	0.0657	0.0737	25.0	34.2
J-1-1-II	0.631	0.0667	0.0737	21.2	31.1
J-1-1-III	0.684	0.0694	0.0738	21.0	32.2
J-1-1-IV	0.704	0.0714	0.0738	23.7	33.0
J-1-1-V	0.676	0.0712	0.0732	24.6	32.8
J-1-1-VI	0.725	0.0716	0.0738	25.8	34.5
J-1-2-I	1.359	0.1413	0.0732	25.9	33.4
J-1-2-II	1.565	0.1601	0.0738	25.9	31.5
J-1-2-III	1.528	0.1587	0.0737	25.2	31.4
J-1-4-I	3.161	0.3215	0.0738	26.2	34.0
J-1-4-II	3.090	0.3203	0.0735	23.5	31.1
J-1-4-III	3.150	0.3205	0.0737	30.1	35.1

Acknowledgement

This work has been supported by the Ministry of Education, Youth and Sports within National Sustainability Programme I, project No. LO1605 and by the Faculty of Civil Engineering, Czech Technical University in Prague (SGS project No. SGS16/201/OHK1/3T/11).

References

[1] J. Keil: Výstavba vodního díla Orlík – sborník statí, Národní podnik vodní stavby, 1966 (in Czech)

[2] vd-tbd a.s.: Manipulační řád VD Orlík, june 2009 (in Czech)

[3] Povodí Vltavy s. p., information materials, pp. 9-10 (in Czech)

[4] O. Zobal et al.: Analysis of Concrete from the Body of the Orlík Dam after Fifty Years. In: Beton – technologie, konstrukce, sanace, 02/2014, pp. 19-25. ISSN 1213-3116 (in Czech)

[5] ČSN EN 13791: Assessment of in-situ compressive strength in structures and precast concrete components, Prague, 2007 (in Czech)

[6] ČSN EN 1990:Basis of structural design, Prague, 2004 (in Czech)

[7] ČSN EN 206-1: Concrete – Specification, performance, production and konformity, Prague, 2014 (in Czech)

[8] G. Martinček: Nedeštrukívné dynamické metódy skúšania stavebných materiálov, (in Slovak), Bratislava, SAV, 1962